SOCIETÀ ITALIANA DI FISICA

RENDICONTI

DELLA

SCUOLA INTERNAZIONALE DI FISICA
«ENRICO FERMI»

CXX Corso

a cura di T.W. Hänsch e M. Inguscio
Direttori del Corso
VARENNA SUL LAGO DI COMO
VILLA MONASTERO
23 Giugno - 3 Luglio 1992

Frontiere nella spettroscopia laser

1994

**SOCIETÀ ITALIANA DI FISICA
BOLOGNA-ITALY**

ITALIAN PHYSICAL SOCIETY

PROCEEDINGS

OF THE

INTERNATIONAL SCHOOL OF PHYSICS «ENRICO FERMI»

COURSE CXX

edited by T.W. HÄNSCH and M. INGUSCIO

Directors of the Course

VARENNA ON LAKE COMO

VILLA MONASTERO

23 June - 3 July 1992

Frontiers in Laser Spectroscopy

1994

NORTH-HOLLAND

AMSTERDAM - OXFORD - NEW YORK - TOKYO

PUBLISHED BY
North-Holland
Elsevier Science Publishers B.V.
P.O. Box 211
1000 AE Amsterdam
The Netherlands

SOLE DISTRIBUTORS FOR THE USA AND CANADA:
Elsevier Science Publishing Company, Inc.
655 Avenue of the Americas
New York, N.Y. 10010
U.S.A.

Technical Editor
P. PAPALI

Library of Congress Cataloging-in-Publication Data

International School of Physics "Enrico Fermi" 1992 : Varenna, Italy)
Frontiers in laser spectroscopy : Varenna on Lake Como, Villa Monastero, 23 June-3 July 1992 / edited by T.W. Hänsch and M. Inguscio.
p. cm. -- (Proceedings of the International School of Physics "Enrico Fermi" ; course 120)
At head of title: Italian Physical Society.
Title on added t.p.: Frontiere nella spettroscopia laser.
ISBN 0-444-81944-4
1. Laser spectroscopy--Congresses. I. Hänsch, T. W. (Theo W.), 1941- . II. Inguscio, M. III. Società italiana di fisica. IV. Title. V. Title: Frontiere nella spettroscopia laser. VI. Series: International School of Physics "Enrico Fermo". Proceedings of the International School of Physics "Enrico Fermi" ; course 120.
QC454.L3I577 1992
535.8'4--dc20 94-18500
CIP

Printed in Italy

INDICE

M. Inguscio – High-resolution and high-sensitivity spectroscopy using semiconductor diode lasers.

T. Oka – Application of laser spectroscopy to fundamental molecular species: H_3^+ and solid H_2.

K. M. Evenson – A history of laser frequency measurements (1967-1983): the final measurement of the speed of light and the redefinition of the meter.

ULTRAHIGH RESOLUTION

V. P. CHEBOTAYEV – High-resolution laser spectroscopy.

J. L. HALL – Frequency-stabilized lasers—a driving force for new spectroscopies.

FUNDAMENTAL EXPERIMENTS

C. E. WIEMAN, S. GILBERT, CH. NOECKER, P. MASTERSON, C. TANNER, CH. WOOD, DONGHYUN CHO and M. STEPHENS – Measurement of parity nonconservation in atoms.

ION TRAPS

QUANTUM OPTICS

S. Haroche – Cavity quantum electrodynamics.

E. GIACOBINO – Squeezed states of light.

CHAOS

D. KLEPPNER – Quantum chaos and laser spectroscopy.

Preface.

It has long been the tradition of the International School of Physics «Enrico Fermi» to cover important advances in spectroscopy. Several courses on this subject have been held at critical moments, when advances in the state of the art had reached a «critical mass» to become the foundation for major future advances.

An important early example is the course on «Topics on Radiofrequency Spectroscopy», which has been organized in 1960 by A. GOZZINI, including the new field of masers and maser spectroscopy. It was followed by a veritable explosion of discoveries, leading to lasers and the new field of quantum electronics. Just 3 years later, C. H. TOWNES chaired a course on «Quantum Electronics and Coherent Light», which became an important basis for a deeper understanding of lasers and coherent light-matter interactions.

The early lasers could operate only at a few rather fixed frequencies. Major progress in laser spectroscopy became possible only with the advent of broadly tunable laser sources, most notably dye lasers, in the early 1970s. These tools led to the development of powerful new spectroscopic techniques based on nonlinear light-matter interactions. In 1975, N. BLOEMBERGEN organized a memorable course on «Nonlinear Spectroscopy», which presented a basis for almost 2 decades of further advances.

Since then, tunable coherent sources have been developed in many parts of the spectrum, from the vacuum ultraviolet to the far infrared. The control of matter by electromagnetic fields, including laser cooling and trapping, has evolved into a highly developed art. Many unexpected and often subtle new phenomena have been discovered, which are leading to a much deeper understanding of light-matter interactions. In 1992, the time appeared ripe for a major review of such progress in a new course on «Frontiers in Laser Spectroscopy».

To summarize a few of the highlights, the field of general laser spectroscopy is being revolutionized by often dramatic advances in the technology of tunable sources, including the important trend towards miniaturization and towards reliable solid-state devices. The range of laser spectroscopy is expanding to ever more interdisciplinary applications, ranging from surface science to astrophysics.

Laser spectroscopy has become a powerful tool for fundamental physics research. It makes possible new tests of basic physics laws, such as parity violation, which can rival and complement experiments with giant particle accelerators.

Dramatic progress has been achieved in ultrahigh-resolution spectroscopy, opening new frontiers and creating new dreams. Precision spectroscopy of atomic hydrogen provides stringent new tests of quantum electrodynamics. Absolute frequency measurements of hydrogen resonances have yielded a new Rydberg constant which has improved in accuracy by 4 orders of magnitude since 1975.

Laser cooling and trapping of atoms and ions provides an impressive example for a new field which did not exist in 1975. Ion traps have become useful tools for metrology and fundamental studies in quantum optics. Laser cooling is now providing the lowest temperatures ever observed in a laboratory. Progress in this area has become so rapid that entirely new phenomena have been discovered in just the one year since a special course was held on this topic (E. ARIMONDO, W. PHILLIPS and F. STRUMIA, 1991).

Atom interferometers, as recently demonstrated in several successful experiments, are inspiring another new area of active research.

The field of quantum optics has seen very significant advances in the state of the art. It has become possible to experiment with just a few quanta of the electromagnetic field, and fascinating new ideas, such as quantum-nondemolition measurements of photon numbers, have emerged. Nonclassical squeezed light is becoming an accessible tool for spectroscopic experiments.

Laser spectroscopy is even beginning to shed light on the still elusive relationship between quantum mechanics and classical deterministic chaos.

In all, 16 lectures, 8 seminar speakers and 58 students from 18 countries participated in this school. We all have been inspired by the breathtaking beauty of one of the finest location in Europe, and by the rich scientific heritage of the Enrico Fermi School of Physics. And we owe special gratitude to E. MAZZI and her colleagues R. BRIGATTI and M. L. ROSSI for a perfect organization, with many warm and personal touches, which helped to make this course truly enjoyable and unforgettable.

Almost exactly one year after this school we were saddened by the news that Dr. M. SIGEL, an active participant and important contributor to this volume, had died in a tragic accident on July 4, 1993.

During the school many of us could benefit for the last time of the human and scientific interaction with V. CHEBOTAYEV, who died on September 2, 1992. He had the time to send us the manuscript of his last, brilliant lecture which is included in this volume that we dedicate to him.

T. W. HÄNSCH and M. INGUSCIO

SOCIETÀ ITALIANA DI FISICA

SCUOLA INTERNAZIONALE DI FISICA «E. FERMI»

CXX CORSO - VARENNA SUL LAGO DI COMO - VILLA MONASTERO - 23 Giugno-3 Luglio 1992

Perspectives on Laser Spectroscopy.

A. L. SCHAWLOW

Stanford University, Department of Physics - Stanford, CA 94305

Let us review the kinds of things that can be done with laser spectroscopy. We begin with the familiar properties of the light produced by lasers. In comparison with all other sources, the light from lasers is powerful, monochromatic, directional and coherent in space and time. It may be either continuous or pulsed, and more or less tunable.

Limited tunability was one of the difficulties that confronted us in the early 1960s, and it is still with us although perhaps in a different form. At that time we did not have any really tunable lasers. We had ruby, which you could tune over a few ångström by changing the temperature. Already Isaac ABELLA [1] at Columbia University had obtained a two-photon transition in cesium by using a ruby laser that happened to be able to reach the right wavelength. Around that time William TIFFANY used temperature tuning of a ruby laser to map out part of the spectrum of the bromine molecule in an experiment to try to get isotopically selective laser photochemistry [2].

The other problem is that, if the laser is tunable, you want it to be monochromatic. We still have that problem, and it is taken to extremes as we try to make the lasers very highly monochromatic. As will be discussed in this lecture, people have gone very far in that direction.

1. – Doppler broadening.

If you just have a monochromatic laser, you can already do some things. Lasers turned out to be very good sources for Raman spectroscopy. All you need is a fixed-wavelength, fairly monochromatic laser. You do not have to bother about tuning it. But, once you have a tunable laser, you can change the population of chosen atomic or molecular states. One of the things you can do with that is to eliminate Doppler broadening, at least to first order, by the method of saturation that was pioneered by HÄNSCH and BORDE [3, 4] in 1970, by polarization spectroscopy that was introduced by HÄNSCH [5], by intermodulated fluorescence of Sorem and myself [6], or by polarization intermodulated excitation (polinex) [7]. There ore other methods, too. All of them select atoms or molecules that are not moving either toward or away from the observer, and so do not have any first-order Doppler shift or broadening.

When I was a graduate student, between 1945 and 1949, I worked on optical hyperfine structure of atoms. I would really have liked to do nuclear physics, but we had no accelerator at the University of Toronto. To reduce Doppler broadening from the thermal motions of the atoms, we constructed a coarse atomic beam and bombarded the atoms with electrons to excite them. I often wished then that I could reach out and grab the atoms, and make them stand still. Nowadays, it is possible to do that, but the methods mentioned just pick out those atoms or molecules which are standing still or at most moving transversely to the direction of observation. For instance, in the saturation method shown in fig. 1, a saturating beam and a probe beam pass through a gas sample in opposite directions. The saturating beam is chopped so that it is alternately off and on. When it is on, it bleaches a path through the absorbing gas so that the probe beam can reach the detector more easily. Thus the probe beam is modulated when the two beams interact with the same atoms. These atoms can only be those which do not have a component of velocity along the direction of the light beams, or they would see the two beams as Doppler shifted to different frequencies and could not be simultaneously in resonance with both of them.

This method was enormously successful and in 1970 HÄNSCH and SHAHIN were able to get the improvement in the spectrum of hydrogen shown in fig. 2 [6]. Hydrogen, being the lightest of the stable atoms, had the largest thermal Doppler shifts, so that at room temperature even with a perfect spectrograph we would obtain the very wide profile shown for the red H_α line. We

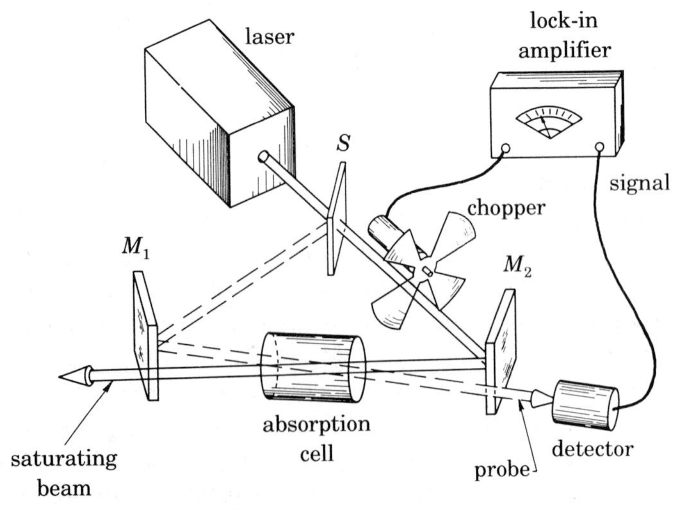

Fig. 1. – Saturation spectrometer.

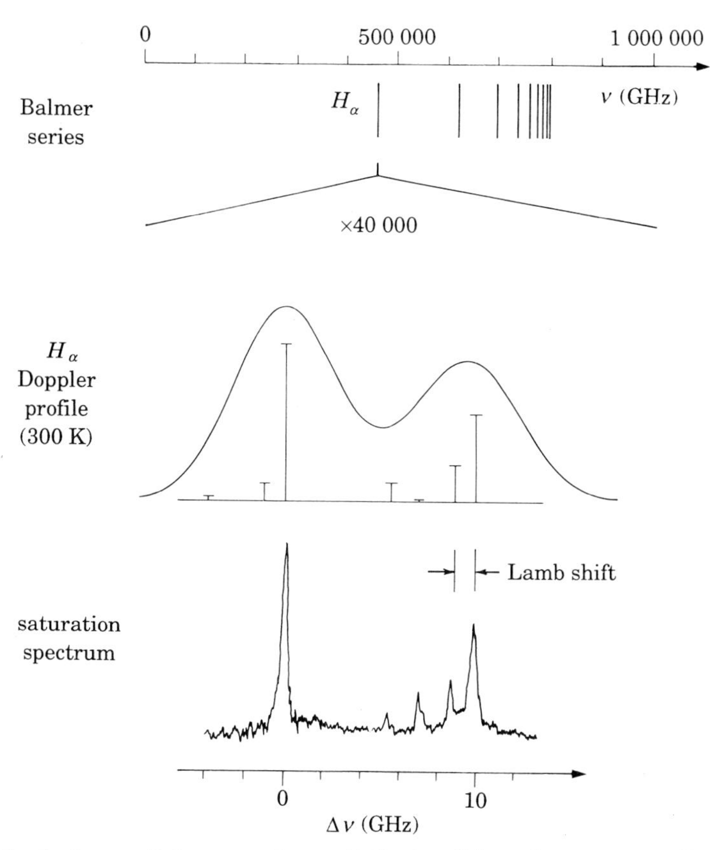

Fig. 2. – The hydrogen Balmer spectrum. At the top it is as from a conventional spectrograph. Below it, the red H_α line is expanded by a factor of 40 000 as it would be seen by an ideal conventional spectrograph, with the calculated positions of the components marked. At the bottom is an early saturation spectrum of the H_α line by HÄNSCH and SHAHIN.

knew from radiofrequency spectroscopy and theory that all the structure shown was hidden under it, but we could not resolve it. Using the saturation method, however, in the very first experiment with a pulsed dye laser HÄNSCH and SHAHIN were able to resolve the Lamb shift optically for the first time. The corresponding shift had been resolved by Gerhard HERZBERG in ionized helium, where it is eight times larger, but not in hydrogen. This led to a long series of experiments to measure the Rydberg constant more precisely.

Another way to get rid of Doppler broadening and do other interesting things is the method of two-photon spectroscopy which was proposed originally in 1970 by VASILENKO, SHISHAEV and CHEBOTAYEV [8]. Two-photon excitation was known before then as has been mentioned, but CHEBOTAYEV and his associates showed us how to eliminate Doppler broadening. This is the only method that is being used now for studies of hydrogen, because two-photon transitions can connect states that are both long lived. Of course, for two-photon excitation both the upper and the lower state have to have the same parity and usually they are forbidden to make a direct one-photon transition, and so the upper state may have a long lifetime. The method that you are probably all familiar with is that you have two beams from the same laser coming from opposite directions. As the atom moves with some speed v along the direction of the beams, one beam seems to be shifted down to $\nu(1 - v/c)$ and the other one up to $\nu(1 + v/c)$. If the atom absorbs one photon from each of these beams, the total energy absorbed by the atoms is $2h\nu$, independent of v, and so you get a Doppler-free line. Instead of just picking out a few atoms that happen to be standing still, the two-photon Doppler-free line involves all of the atoms, and it is a very sensitive method. Once we were told about it, it was very easy to do.

2. – Simplifying spectra by laser labeling.

Another thing you can do by changing the population of states is to simplify complex spectra by methods such as population labeling which we introduced [7] and by polarization labeling [9]. In these, you change the population of atoms or molecules in some particular chosen state, so that you can distinguish just those transitions involving the selected state. You can either put them up into a chosen level, and then you will be able to get absorption or emission from that level. Alternatively, you can pump them out of a lower level and depopulate that resulting in a reduced absorption from that level.

In polarization labeling, polarized light is used to pump out atoms from the chosen state with a particular orientation, leaving the remaining atoms with a complementary orientation. Then polarized light at any wavelength corresponding to transitions from the labeled level will be depolarized and can pass

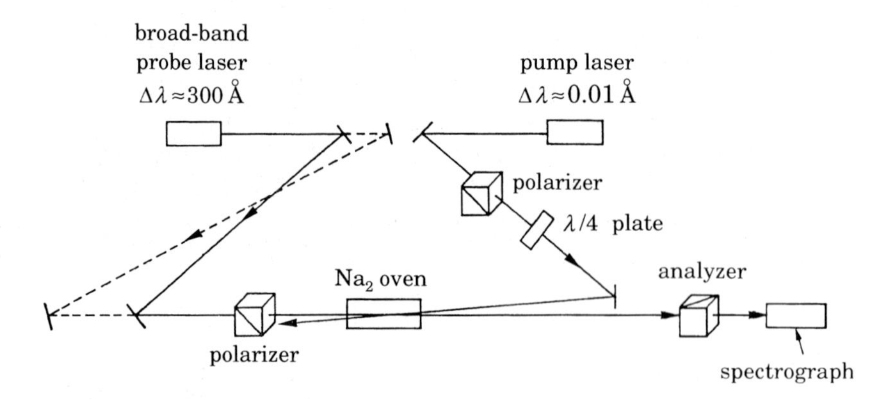

Fig. 3. – Schematic diagram of the apparatus for simplifying spectra by polarization labeling.

through the analyzer which was set to block the incoming light at all other wavelengths (fig. 3). As is well known to atomic physicists, a diatomic molecule is a molecule with one atom too many, and polyatomic molecules are even worse with their very many vibrational and rotational states. However, when the polarization labeling method is applied to a diatomic molecule such as Na_2, the observed spectrum is greatly simplified [9]. Only wavelengths corresponding to transitions from the oriented levels can pass through the crossed polarizers. Thus for each vibrational level, only a doublet is observed as in fig. 4, corresponding to transitions in which the rotational quantum number, J, increases or decreases by one unit. Then you can get the vibrational quantum number just by counting down from the end: The doublets correspond to $v = 0, 1, 2, 3, 4$ and so on. If you shift the pump laser's wavelength slightly, you get a different set. In the labeled spectrum some doublets are missing because of the Frank-Condon factors.

Thus it becomes possible to use laser labeling to analyze spectra systematically. Many higher electronic states of the molecule are accessed by using two-step labeling. Students in our group were able to find and identify

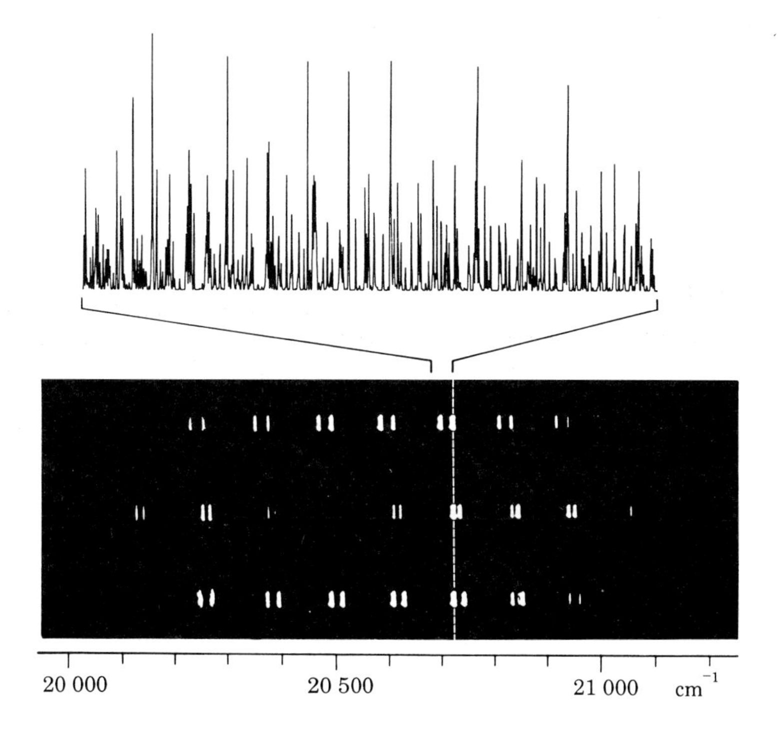

Fig. 4. – A portion of the sodium molecular spectrum. At the top it is as it would be revealed by a high-resolution conventional spectrograph. Below is a photograph of the same spectral region by polarization labeling, with several slightly different labeling wavelengths.

about two dozen levels in Na_2. Other people have gone further and used other methods but analyzing spectra is still tedious work.

3. – Observing small numbers of atoms.

One other thing you can do with the change in state population is to observe small numbers of atoms. You can excite atomic or molecular beams. I mentioned in my graduate thesis work that we had to use a rather coarse atomic beam of silver. We excited the ultraviolet resonance lines by electron bombardment and observed their hyperfine structures through a Fabry-Perot etalon and a photographic spectrograph [10]. Even with the heaviest electron bombardment we could manage, exposures of about four hours were required. With typical plate spacings of about 7 cm, the Fabry-Perot interferometer was very sensitive to the pressure of the air between the plates. Indeed, a change in atmospheric pressure of about 3 mbar was enough to blur completely the fine spacings we were trying to observe. Such small changes are only found between midnight and 4 a.m., but if a storm came during that time the exposure was ruined. An obvious solution would have been to put the interferometer in a sealed box with quartz windows. However, we were very poor in those days. I was at the University of Toronto just after World War II and the system of government funding had not been set up in Canada, so we had to make do with what old stuff we could find lying around. It was a very hard way to do experiments.

Now, of course, if you had that same atomic beam, you could use a laser to excite the atom and the experiment is much easier. You can observe very small numbers of atoms by resonance fluorescence or by ionizing atomic excited states. You can see even as little as a single trapped ion as DEHMELT and TOSCHEK and their associates have done [11]. They have shown a photograph of a single barium ion. You cannot have any less barium than one atom or ion. So that is another thing you can do.

You can also use laser light to ionize atoms or molecules from excited states and so detect their presence very sensitively by the ionization. With two or more steps of excitation using several different lasers, atoms can be raised to Rydberg states not far below the ionization threshold. From there they can be ionized completely by an electric field. Atoms in Rydberg states are very sensitive detectors for radiation, because these atoms are very large and have large cross-sections for scattering radiation or absorbing. It is thus possible to do spectroscopy on single atoms in Rydberg states.

4. – Laser cooling.

Either the saturation method or the two-photon method gets rid of first-order Doppler broadening, but there is a second-order broadening proportional to

v^2/c^2 which is about a part in 10^{11} or so. With the increase in precision that became a major problem. Still with our old dream of trying to make these atoms stand still, HÄNSCH and I thought of the method of Doppler cooling [12]. We did not do anything about it because we were interested in hydrogen particularly, and there was not and still is not a really suitable laser for cooling hydrogen. Steven CHU rediscovered the method in the early 80's and was very successful in cooling sodium and producing what he calls optical molasses [13]. The principle of this method is shown in fig. 5. Thc laser is tuned slightly below the resonance frequency of the atom, within the Doppler wing. If the atom stands still, it does not know that these beams are there. However, if it moves in some direction, it sees the oncoming light shifted up into resonance, then scatters those photons and loses momentum. It loses about one centimeter per second every time it scatters photons. A sodium atom can do that about 10^8 times per second. On the other hand, a light beam that is following the atom is shifted further down out of resonance and so it does not interact with the atom. Thus, whichever way the atom moves, it sees an oncoming beam that slows it down. Consequently, once an atom has been cooled in this way, it behaves as if it were in a very viscous medium. That is why CHU called it optical molasses.

The way he did the experiment was clever. HÄNSCH and I had calculated sodium and realized that, if we were going to start with room temperature sodium, we would have a box about a meter on each side, which would have to be

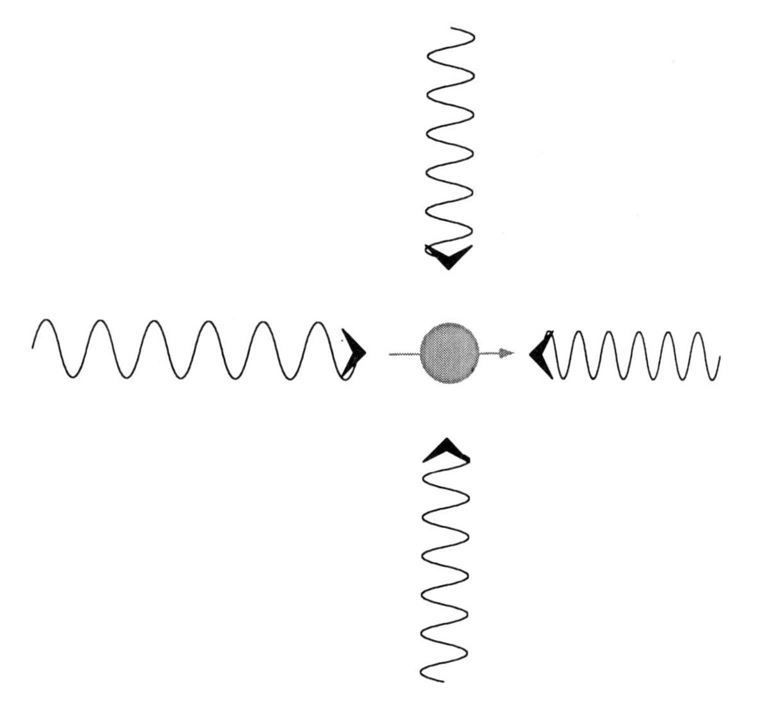

Fig. 5. – Principle of laser cooling of free atoms.

filled with laser light and that was too difficult. CHU took a sodium target and hit it with a pulse from an argon laser, evaporated a little sodium, let the fast atoms go on through and then used laser beams to make the rest slower and bring them almost to rest. Then beams from the six principal directions—up, down, left, right, front and back—would hold them there. So you would have this viscous optical molasses. Molasses is a very viscous sweet syrup, that is made from some kind of grain and is noted for its high viscosity and that is why he called the laser-damped atom cloud optical molasses. Whichever way the atoms would try to move, they were impeded by oncoming beams. There was more to it than that, as he had to shift the frequency as the cooling proceeded to stay in resonance as the atoms slowed down. He was able to reach atom temperatures of a small fraction of a kelvin.

Sometimes Nature is kind. PHILLIPS and his associates at the National Institute of Science and Technology laboratory in Gaithersburg found out that they could get even below what was calculated for this Doppler cooling [14]. The Doppler cooling would get you down to perhaps a millidegree or so absolute, but you could go much lower by using cooling beams with the proper polarizations. As explained by CHU [15] and by COHEN-TANNOUDJI and DALIBARD [16], kinetic energy was then given up to the internal states of the atom, resulting in even lower temperatures. Just recently CHU and his associates have introduced the concept of Raman cooling which really gets down to almost zero velocity and zero temperature [17]. The other methods were limited by the lifetime width of the excited states where the Raman method is not.

5. – Coherent-state superpositions.

There is a big field of coherent-state superpositions which includes hyperfine quantum beats which were first seen by HAROCHE [18]. A pulsed laser is used, with a pulse short enough to excite the atom into a superposition of several hyperfine levels. As the atoms decay by spontaneous emission, beats between these frequencies of the various hyperfine lines are observed in the fluorescence. Or you can think of it as you are setting the atoms synchronously into rotation at the frequency corresponding to the spacing of the hyperfine levels, and they emit more strongly in the direction of the observer once every cycle. Then the so-called photon echoes, probably better called wave echoes, have been seen optically again from the coherent-state superposition and modeled on the spin echoes of nuclear resonance. Sometimes people have said that there is nothing new in optical spectroscopy and that it is all done in the microwave or spin resonance. That is not quite so because things in the optical region are three-dimensional, and the direction matters. However, many concepts have been taken from radiofrequency resonance studies, particularly when only two levels need to be considered. These concepts include the π pulse, which exchanges the populations of two states and the $\pi/2$ pulse which puts the atoms into a coherent superposition of the two states.

Just recently the field of atomic interferometers has become practical because laser cooling can produce atoms so extremely slow that their de Broglie wavelength becomes comparable with the wavelength of ordinary light. Therefore, you can use many interferometric techniques with them, often in analogy with optical interferometry. CHU and KASEVICH have done a very beautiful experiment where they used the atom interferometry on very cold, slow atoms to measure the acceleration of gravity, already to about a part in a 100 million. They project atoms out of a trap, so that the atoms move slowly upward. Gravity causes the atoms to slow down as they rise, and this gravitational deceleration is measured through the Doppler shift of a Raman transition between hyperfine levels of the ground state.

6. – Semiconductor lasers.

We still have the problem of finding tunable lasers, especially ones that we can afford. Nowadays in addition to all the other kinds of lasers, we have these frustrating little semiconductor lasers. Figure 6 shows a simple semiconductor diode laser with some tunability. A diffraction grating is placed quite close to the diode, so the collimating lens can image the light diffracted by the grating back onto the small active region of the laser diode. You can do rough tuning by turning the grating, and fine tuning by moving the grating in and out because there are standing waves between the grating and the diode. The spacing between the diode and the grating

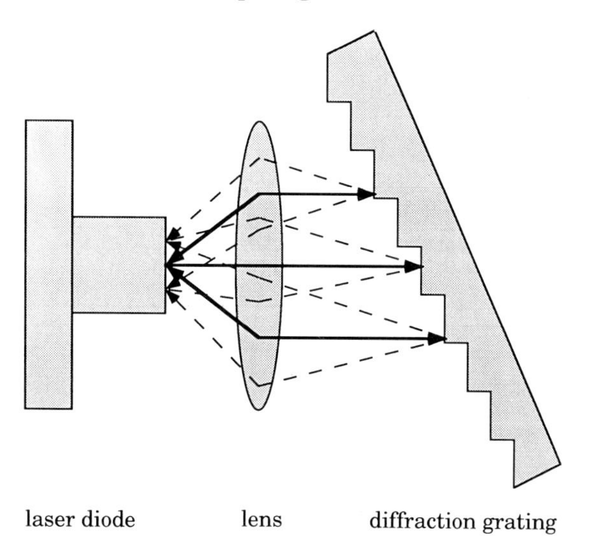

Fig. 6. – Schematic diagram of laser diode tuned by an external diffraction grating.

is made as small as possible so that the axial-mode spacing of the diode-grating combination is large enough to resolve.

One problem with the diode lasers is that they have a rather high reflectivity just from the high refractive index of the gallium arsenide or whatever the material is in the semiconductor. Therefore, you have a complicated system. With reflection standing waves here within the laser, there is the general frequency range that the laser will emit, within which the laser can be tuned to some extent by using a diffraction grating as an external mirror. There are standing waves within the laser diode, and also standing waves between the laser and the grating. It is possible to put good antireflection coatings on the laser diodes, but that is a difficult process and, if you are not careful, they can become vulnerable to moisture.

In our laboratory Guang-Yao YAN has set up a little diode laser, using a rather small spacing between the diode and the grating. He has been able to cover a range of about 30 MHz, one full wave number, in a continuous scan in the visible around 6700 ångström. I know other people have done as well or better although it is not easy even to do that well sometimes. He scanned the lithium resonance line in a hollow cathode discharge tube which was made by Hamamatsu, a commercial item. Using the saturation method, with beams coming in both directions, it was possible to resolve the hyperfine structure of the line in one continuous scan over both isotopes. Figure 7 shows the hyperfine

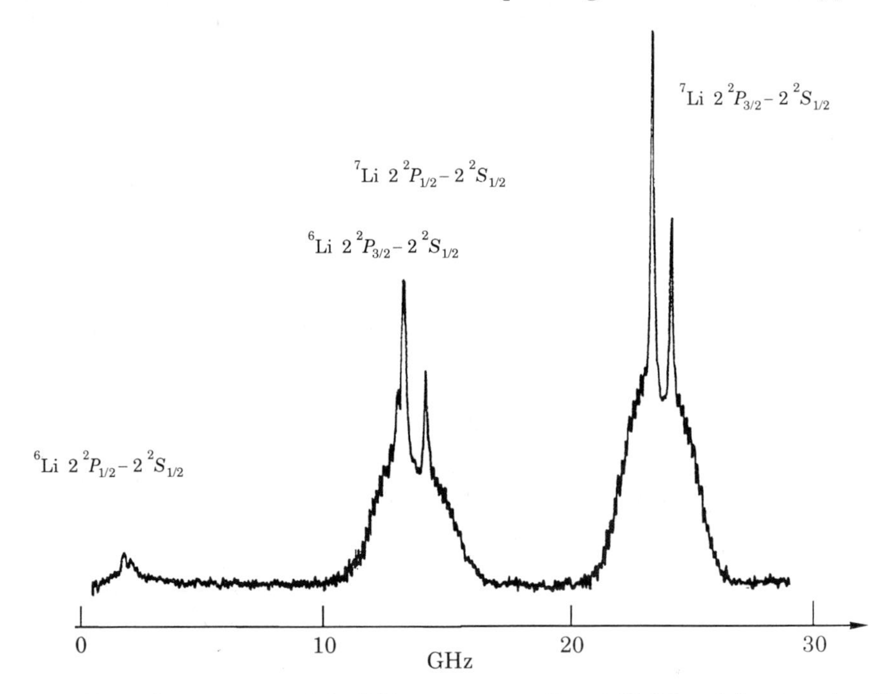

Fig. 7. – Hyperfine structure of the lithium resonance line at 6708 ångström, by saturation spectroscopy with the grating-tuned diode laser.

structures of the two lithium isotopes, lithium-6 and -7, resolved in this way. The lines are not as narrow as they should be because there is a neon buffer gas in the hollow-cathode lamp which causes pressure broadening pedestals near the base of the lines here. Semiconductor lasers will improve. It is possible to put tuning elements inside the laser and have them electronically tuned, but at the moment we have to wait.

The thing that is frustrating about semiconductor lasers is that it is like the early 1960's before the dye lasers came along. We had a lot of different lasers but not the desired wavelengths. We had many wavelengths, but we had to take what the atoms and molecules would give us. And now with the semiconductor lasers, we again have many wavelengths, but we have to take what the manufacturers will give us. They are motivated by commercial considerations and they are not going to make just every wavelength.

7. – Measurement of optical frequencies.

Laser spectroscopy has become so refined that you need to be able to make very precise measurements. In fact for the kind of thing that people do on hydrogen, wavelength measurements are no longer good enough because the dispersion of the films in the interferometers cannot be known accurately enough. Therefore, you need to measure frequencies. I always tell my students to never measure anything but a frequency. Anything else is too hard to measure and impossible to explain. Usually a wavelength is a reasonable substitute. As long as it is something you can count, and reduce the measurement to counting instead of having to try and judge the size, like specific heat, for instance, you can get accurate results. Measuring a wavelength or a frequency can be made a counting operation. But in the visible it is not so easy to tie down the spectral-line frequency to the fundamental standard of frequency. The Bureau of Standards group with EVENSON and HALL [19] have pioneered harmonics going all through the infrared region up to the visible and you can actually measure frequencies there, but it is hard work. You need a room full of lasers at least half the size of this room. HÄNSCH has a very clever method of dividing frequency differences until he gets down to something he can measure in the radiofrequency range [20].

Recently Ali JAVAN has been working on very tiny microstructures and he points out that microlithography using electron beams and also scanning tunneling microscopy can make very small structures. The one thing that has been different about even microwave work and certainly optical stuff is that we are always used to feeling that all of our circuit elements are very large compared to the wavelengths. That does not have to be the case. As he points out, you can now make something like a little $3/2$ wave antenna for visible light, which is $3/4$ of a micrometer long and perhaps 100 ångström wide and put a little gap of about 10 ångström. If the two arms are different metals, you get a rectifying contact. Particularly if one of them is a superconductor (which means cooling, of course), very high optical nonlinearities can be obtained. With that it will be possible to make mixers or dividers or parametric oscillators probably, even in the visible portion of the spectrum. And, if you can do it in the visible, you can certainly do it in any wavelength in the infrared. So this may again be a path for frequency division. But you have a factor of 100 000 or so to go from the visible down to the frequency of the cesium standard. I think that the optical standard will eventually replace the radiofrequency standard. It is my belief that it is easier to count small cycles than to divide up one big one and tell where it begins and ends exactly. So it is easier in principle to divide than to multiply.

8. – Searching for weak lines.

I have not been doing anything much with lasers lately because I have been working on a problem where I do not see any way to use a laser. I have been interested in trying to see if we can find spectral lines of some rare-earth ions in metals. Rare-earth ions behave almost like free atoms in an electrical field. You can get very sharp lines. In praseodymium fluoride we saw one line that was less than half a wave number wide in the visible. Although the transitions are highly forbidden with oscillator strengths as low as 10^{-8}, there are very many ions in a solid, about 10^{21} or more ions per cubic centimeter. Thus it is still possible to get a substantial absorption even if the light can only penetrate through a thin layer, as in a metal. In fact, in the case of the line that I mentioned in praseodymium fluoride, the $1/e$ absorption length was nine micrometers. So it occurred to us that it was worth a try to see if we could see such lines in metals. The difficulty, of course, is that metals are not transparent. You can only see through a few hundred ångström. If you calculate what absorption you might get, guessing at what the line width would be, it seems possible to observe these lines if you have a sufficiently sensitive apparatus. Twenty years or so ago, some people tried to observe absorption lines in metals, but their apparatus was only sensitive enough to detect about 1% absorption, and that was obviously not good enough. So my students Michael JONES and David SHORTT built a very sensitive conventional spectrometer. Using an optical multichannel analyzer and carefully getting rid of all noise sources, they were able to get sensitivity of a part in 10^5 or $1/1000$ of a percent [21, 22].

The reason we used a conventional spectrometer, not a laser, is we had a search problem, as we did not know quite where to look. If you look, for instance, at erbium ions, erbium 3^+ in various materials, you see that the positions of the lines vary over a range of about 100 ångström (fig. 8). There are lines around 5200 to 5300 ångström, extending over even more than 100

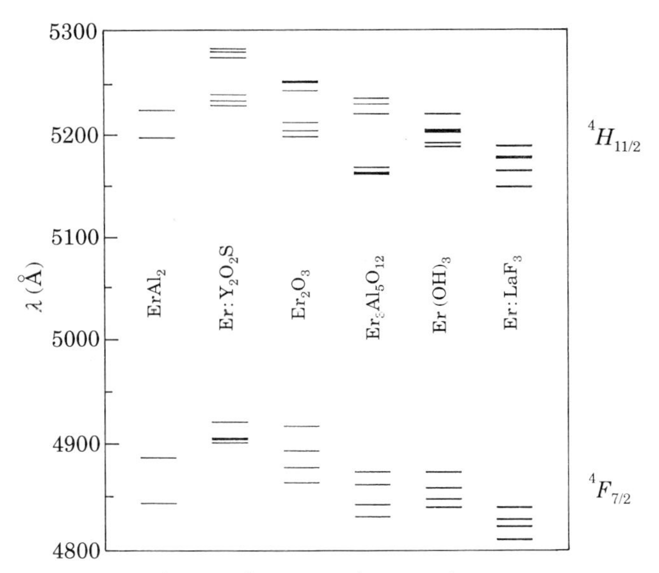

Fig. 8. – Position of erbium ${}^4F_{7/2} \leftarrow {}^4I_{15/2}$ and ${}^4H_{11/2} \leftarrow {}^4I_{15/2}$ absorption lines in several materials.

ångström. So you do not know where in this region you are going to find it. You cannot tune any low-noise laser that I know of over that range. If you could, it would take a very long time because you would want a fairly long look at each wavelength. You do not know how wide the line is going to be. Indeed, the line width is one of the things we hope to find out. It begins to look if it is going to be rather wide, so that searching might not be such a big problem. But the problem of getting a laser that is quiet enough and will have steady enough output over substantial tuning ranges is a very difficult one, and so we did not use a laser. We have seen a few lines. We have observed several transitions of neodymium ions in Nd_2CuO_4, an insulator, and also in the related superconducting metal which is obtained by replacing 7.5% or more of the neodymium ions by cerium [23]. The relative intensities of the line components change from the insulator to the metal, although the line wavelengths do not change appreciably. These intensity changes indicate the charge redistribution near the neodymium ions expected when the material becomes metallic, which alters the transition strengths of the components.

With a signal-to-noise ratio of 135 000 we have seen some both broad and narrow lines in $ErAl_2$ [24] (fig. 9). At this point we do not know which ones are really significant. They do lie in the general region where we expect to find those lines. From fig. 8 it can be seen that the erbium lines from various compounds are in that general region, but the $ErAl_2$ lines do not coincide with any of the others. Of course, we had suspected we had to worry about thin layers of oxide or fluoride on the surface, but it seems not to be any of those. This work

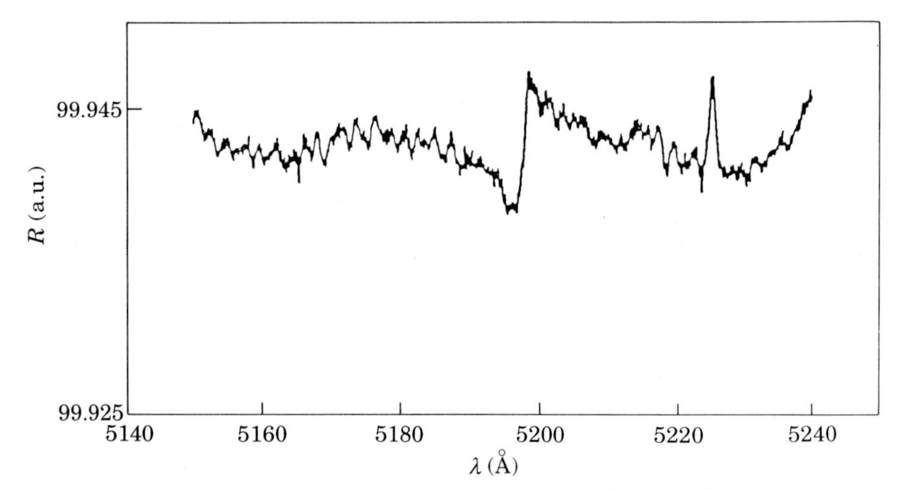

Fig. 9. – $ErAl_2$ reflection spectrum in the vicinity of the $^2H_{11/2} \leftarrow {}^4I_{15/2}$ transition. The strange line shapes are possibly due to Fano interference between the discrete and broad-band absorption.

unfortunately is going to have to discontinue shortly because I have retired, and my last student has finished his thesis. I would certainly appreciate it if somebody would show me how to search for a weak line with a laser. After all you should be able to do anything with lasers, but I do not know how to do it. However, once the lines have been located, they could be best studied by laser spectroscopy, for instance to see if they change at transition temperatures.

We will hear a lot of exciting and more modern things during this course, but this is kind of an overview of things people have done with laser spectroscopy and some of the things we cannot do with lasers.

REFERENCES

[1] I. ABELLA: *Phys. Rev. Lett.*, **9**, 453 (1962).
[2] W. B. TIFFANY, H. W. MOOS and A. L. SCHAWLOW: *Science*, **157**, 40 (1967).
[3] T. W. HÄNSCH, M. D. LEVENSON and A. L. SCHAWLOW: *Phys. Rev. Lett.*, **27**, 707 (1971).
[4] C. BORDE: *C. R. Acad. Sci.*, **271**, 371 (1970).
[5] C. WIEMAN and T. W. HÄNSCH: *Phys. Rev. Lett.*, **36**, 1170 (1976).
[6] M. S. SOREM and A. L. SCHAWLOW: *Opt. Commun.*, **5**, 148 (1972).
[7] T. W. HÄNSCH, D. R. LYONS, A. L. SCHAWLOW, A. SIEGEL, A. Y. WANG and G. Y. YAN: *Opt. Commun.*, **37**, 87 (1981).
[8] L. S. VASILENKO, V. P. CHEBOTAEV and A. V. SHISHAEV: *JETP Lett.*, **12**, 113 (1970).
[9] R. E. TEETS, R. FEINBERG, T. W. HÄNSCH and A. L. SCHAWLOW: *Phys. Rev. Lett.*, **37**, 683 (1976).

[10] M. F. CRAWFORD, A. L. SCHAWLOW, F. M. KELLY and W. M. GRAY: *Can. J. Res. A*, **28**, 558 (1950).
[11] W. NEUHAUSER, M. HOHENSTATT, P. E. TOSCHEK and H. G. DEHMELT: *Phys. Rev. A*, **22**, 1137 (1980).
[12] T. W. HÄNSCH and A. L. SCHAWLOW: *Opt. Commun.*, **13**, 68 (1975).
[13] S. CHU, L. HOLLBERG, J. E. BJORKHOLM, A. CABLE and A. ASHKIN: *Phys. Rev. Lett.*, **55**, 48 (1985).
[14] P. D. LETT, R. N. WATTS, C. I. WESTBROOK, W. D. PHILLIPS, P. L. GOULD and H. J. METCALF: *Phys. Rev. Lett.*, **161**, 169 (1988).
[15] S. CHU, D. S. WEISS and Y. SHEVY: in *Atomic Physics*, edited by S. HAROCHE, J. GAY and G. GRYNBERG (World Scientific, Singapore, 1989), p. 636.
[16] J. DALIBARD and C. COHEN-TANNOUDJI: in *Atomic Physics*, edited by S. HAROCHE, J. GAY and G. GRYNBERG (World Scientific, Singapore, 1989), p. 199.
[17] M. KASEVICH and S. CHU: *Phys. Rev. Lett.*, **69**, 1741 (1992).
[18] S. HAROCHE, J. A. PAISNER and A. L. SCHAWLOW: *Phys. Rev. Lett.*, **30**, 948 (1973).
[19] D. A. JENNINGS, C. R. POLLOCK, F. R. PETERSEN, R. E. DRULLINGER, K. M. EVENSON, J. S. WELLS, J. L. HALL and H. P. LAYER: *Opt. Lett.*, 8, 136 (1983).
[20] H. R. TELLE, D. MESCHEDE and T. W. HÄNSCH: *Opt. Lett.*, **10**, 532 (1990).
[21] D. W. SHORTT, M. L. JONES and A. L. SCHAWLOW: *Phys. Rev. B*, **42**, 132 (1990).
[22] D. W. SHORTT, M. L. JONES, A. L. SCHAWLOW, R. M. MACFARLANE and R. F. C. FARROW: *J. Opt. Soc. Am. B*, **8**, 923 (1991).
[23] M. L. JONES, D. W. SHORTT, B. W. STERLING, A. L. SCHAWLOW and R. M. MACFARLANE: *Phys. Rev.*, **206**, 611 (1992).
[24] B. W. STERLING: Ph. D. Thesis, Stanford University (1992).

LASER SPECTROSCOPY FIR $\rightarrow$ UV

Experimental Techniques for High-Resolution Laser Spectroscopy of Small Molecules and Clusters.

W. DEMTRÖDER, V. BEUTEL, H.-A. ECKEL, H.-G. KRÄMER and E. MEHDIZADEH (*)

Fachbereich Physik, Universität Kaiserslautern - D-67663 Kaiserslautern, B.R.D.

1. – Introduction.

This lecture gives a survey on experimental techniques and some applications of high-resolution stationary and time-resolved laser spectroscopy to investigations of structure and dynamics of small molecules and clusters. The following four aspects are particularly emphasized:

a) The improvement of sensitive detection techniques such as laser-induced fluorescence and resonance two-photon ionization with c.w. lasers.

b) The combination of high-resolution laser spectroscopy with cooling of molecules in collimated supersonic molecular beams and with mass-selective spectrometric detection.

c) The assignment of complex and perturbed molecular spectra by means of optical-optical double-resonance (OODR) spectroscopy.

d) The importance of time-resolved spectroscopy for measurements of radiative lifetimes, dissociative channels, intramolecular-energy redistribution and collision-induced intermolecular energy transfer.

Many researchers in various laboratories have worked on these aspects of laser spectroscopy [1-5]. The examples given below are, therefore, by far not complete but just represent a selection taken from our own work in Kaiserslautern.

(*) On leave from Department of Physics, Shahid-Bahonar University of Kerman, Kerman 76175, Iran.

2. – Sensitive detection techniques.

Laser spectroscopy in collimated molecular beams has to use sensitive detection techniques because the absorption path length (typically 1 mm) and the density of absorbing molecules (($10^5 \div 10^{10}$) molecules / cm^3 in the absorbing level $|i\rangle$) are both small and the direct measurement of the attenuation of the incident laser beam, crossing the molecular beam perpendicularly, is difficult. There are two solutions of this problem: Either the absorption path length of the laser beam is increased by multiple-reflection arrangements (fig. 1*a*), *b*)) or the absorbed laser photons are detected indirectly, for example by laser-induced fluorescence (LIF) (fig. 2*a*)), resonant two-photon ionization (fig. 2*b*)) or by transfer of the molecular excitation energy to a bolometer (opto-thermal spectroscopy (fig. 2*c*)) [6].

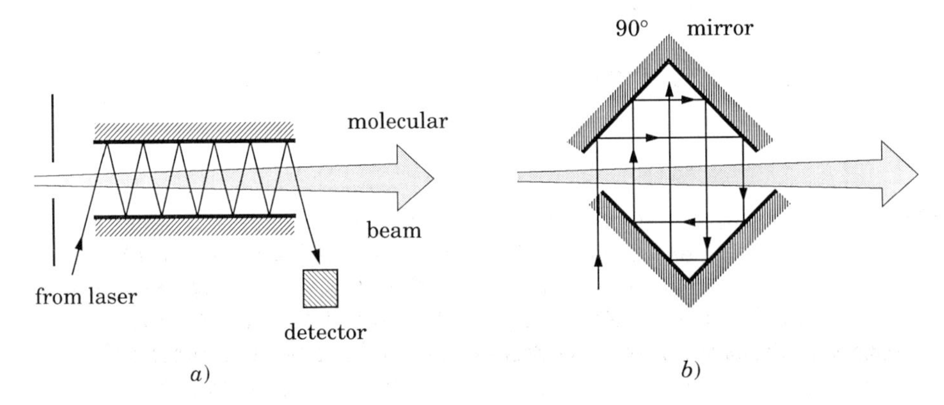

Fig. 1. – Two possible multiple-reflection arrangements for absorption spectroscopy in molecular beams.

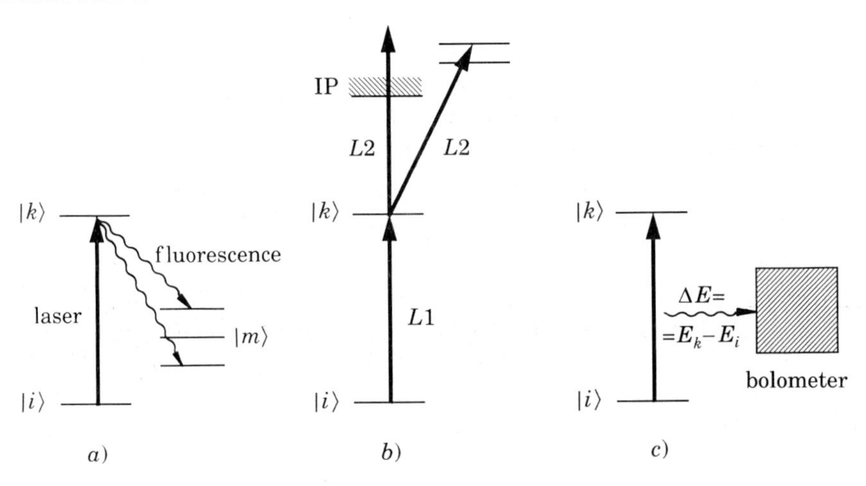

Fig. 2. – Indirect detection of absorbed photons: *a*) laser-induced fluorescence, *b*) two-step photoionization, *c*) optothermal detection.

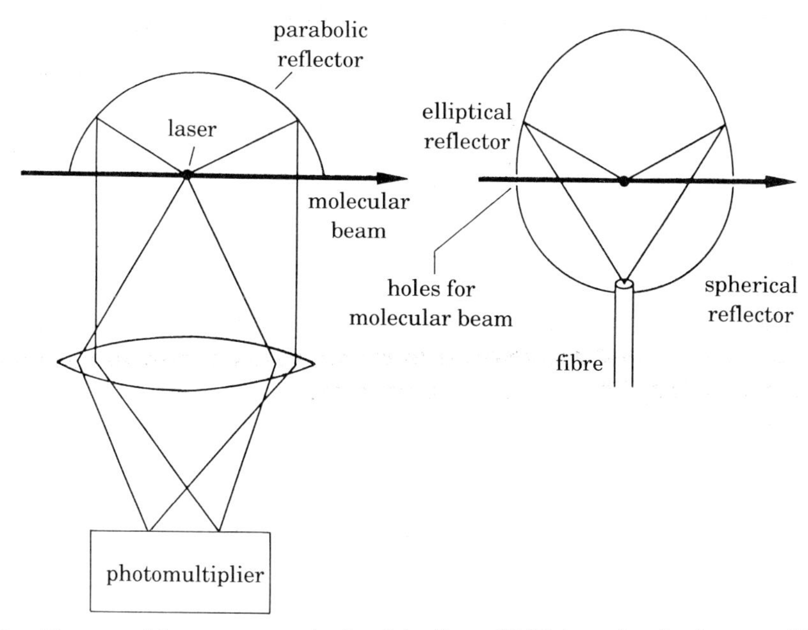

Fig. 3. – Two possible arrangements for detection of LIF in molecular beams with high photon collection efficiency.

For the LIF detection scheme the collection efficiency of the fluorescence photons emitted into all directions is essential for sensitive detection. A possible scheme with a high collection efficiency, which is used in Kaiserslautern, is shown in fig. 3. The crossing point between laser and molecular beam is located in one focus F_1 of an elliptical reflector, which images this point into the second focus F_2 where the end face of an optical fibre bundle is placed. A half-sphere reflector reflects all light emitted into the lower half-sphere back into F_1 from where it is again imaged into the fibre bundle. The other end of the fibre bundle is either placed in front of a photomultiplier or is formed into a rectangular shape adapted to the entrance slit of a spectrograph which allows the measurement of the spectrally dispersed fluorescence. With this arrangement a collection efficiency of the fluorescence photons up to 50% can be achieved.

The second detection scheme (fig. 2*b*)) is based on the population of excited molecular levels by a tunable laser $L1$ with subsequent ionization of the excited molecules by an intense laser $L2$. In case of pulsed lasers the peak power is sufficiently high to excite and ionize all molecules within the excitation volume even without tight focussing of the laser beams. Pulsed lasers, however, have two disadvantages:

1) Although all molecules can be ionized and detected during the laser pulse time Δt, they escape undetected during the time $T = 1/f$ between two successive laser pulses. For example, with a laser pulse repetition rate $f = 100\ \text{s}^{-1}$

and a typical pulse width of $\Delta t = 10^{-8}$ s the duty cycle $\Delta t/T$ is only 10^{-6} which means that in a continuous molecular beam with molecular velocities around $\overline{v} = 10^5$ cm/s and an excitation volume $q \cdot \Delta z$ with $\Delta z = 1$ cm only 10^{-3} of all molecules in the absorbing level $|i\rangle$ are really detected.

2) The spectral resolution is limited by the spectral bandwidth of the pulsed laser which is typically a few GHz. Even with single-mode pulsed lasers the Fourier-limited bandwidth $\Delta\nu = 1/\pi\Delta t \approx (40 \div 100)$ MHz cannot be surpassed.

Because of these limitations it is for many investigations essential to use single-mode c.w. lasers to obtain sub-Doppler resolution. Since the power of c.w. lasers is smaller by some orders of magnitude than the peak power of pulsed lasers, the beams of the c.w. lasers have to be focussed in order to reach sufficient intensity for the ionizing step. The difficulty for resonant two-photon ionization with c.w. lasers is the following: During the lifetime τ_K of a molecule excited into level $|k\rangle$ by the dye laser $L1$ it travels a distance $d = \overline{v} \cdot \tau_K$. In order to reach a large ionization efficiency, the molecule must be ionized by $L2$ *before* it decays by spontaneous emission into lower levels $|m\rangle$ from which it cannot be excited again by $L1$, tuned to a transition $|i\rangle \rightarrow |k\rangle$. For typical values of $\tau_K = 10^{-8}$ s, $\overline{v} = 10^5$ cm/s we obtain $d = 10$ μm. This illustrates that both laser beams must be tightly focussed and their spatial overlap has to be carefully aligned. A possible solution is shown in fig. 4, where the dye-laser beam is transported through a single-mode optical fibre. A spherical lens collimates the divergent beam which is then superimposed by a dichroic beam splitter onto the beam

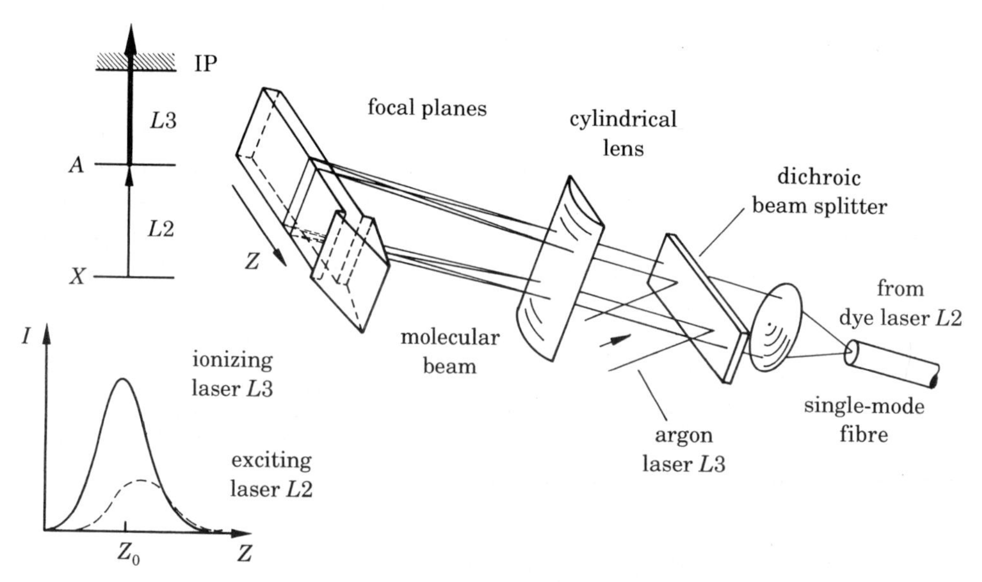

Fig. 4. – Two-step photoionization in a molecular beam using two c.w. lasers and cylindrical focussing.

from an argon laser. A cylindrical lens focusses both beams into the molecular beam where the cross-section of the focal plane ($(0.02 \times 1)\,\text{mm}^2$) is adapted to the dimensions of the collimated molecular beam. Every molecule in the beam has to pass the focal area and can be ionized by a two-step process with a high probability.

If the focal area is placed inside the ion source chamber of a mass spectrometer, the ions formed by resonant two-step ionization can be mass-selected and the ion rate $N_{\text{ion}}(m, \lambda_1)$ recorded on mass m as a function of the wavelength λ_1 of laser $L1$ yields the absorption spectrum of the neutral molecule with mass m, sensitively monitored through resonant two-photon ionization.

3. – Sub-Doppler spectroscopy of Na_3.

As an illustrative example for the application of the techniques discussed in the previous section, the sub-Doppler spectroscopy of small metal clusters [7], in particular of Na_3, is well suited. The experimental arrangement is shown in fig. 5. Sodium clusters are formed during the adiabatic expansion of a mixture of sodium vapour and argon gas (100 mbar sodium vapour, up to 10 bar argon)

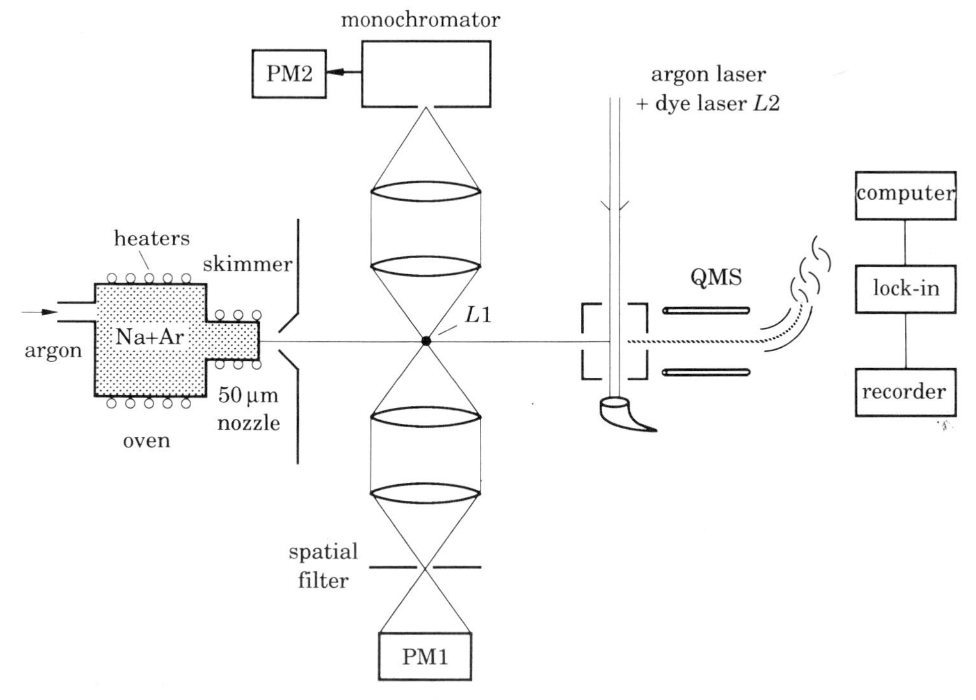

Fig. 5. – Experimental arrangement for sub-Doppler spectroscopy of metal clusters produced by adiabatic expansion of a mixture of metal vapour and argon through a small nozzle into the vacuum chamber.

through a small nozzle (50 μm diameter) into the vacuum chamber, evacuated by a 6000 l/s diffusion pump. A skimmer, about (15 ÷ 20) mm downstream from the nozzle, collimates the molecular beam and separates the oven chamber from the main vacuum chamber where the beam is crossed in a first crossing point P_1 by the beam of dye laser $L1$. The laser-induced fluorescence is observed either undispersed ($I_{Fl}(\lambda_1)$) by photomultiplier PM1 or, at a fixed laser wavelength λ_1, the dispersed fluorescence spectrum can be measured through a monochromator. In a second crossing point P_2 the molecular beam crosses the superimposed focussed laser beams of lasers $L2$ and $L3$ and the ions formed by two-step photoionization are collected and accelerated onto an open ion multiplier. The beams of $L2$ and $L3$ also pass through the ion source of a quadrupole mass spectrometer where the ions can be mass-selected.

Figure 6 shows a section of the Q-branch of the $0 \leftarrow 0$ vibronic band in the $A \leftarrow X$ system of Na_3 [8], recorded with sub-Doppler resolution in crossing point P_1. The complexity of the spectrum is due to the narrow rotational structure of Na_3 superimposed by hyperfine-structure, spin rotation splittings and pseudorotation [9]. Even at a spectral resolution of 30 MHz most of the lines are not fully resolved.

It is evident from this complex spectrum that an unambiguous analysis is not straightforward. Computer simulations showed that several sets of molecu-

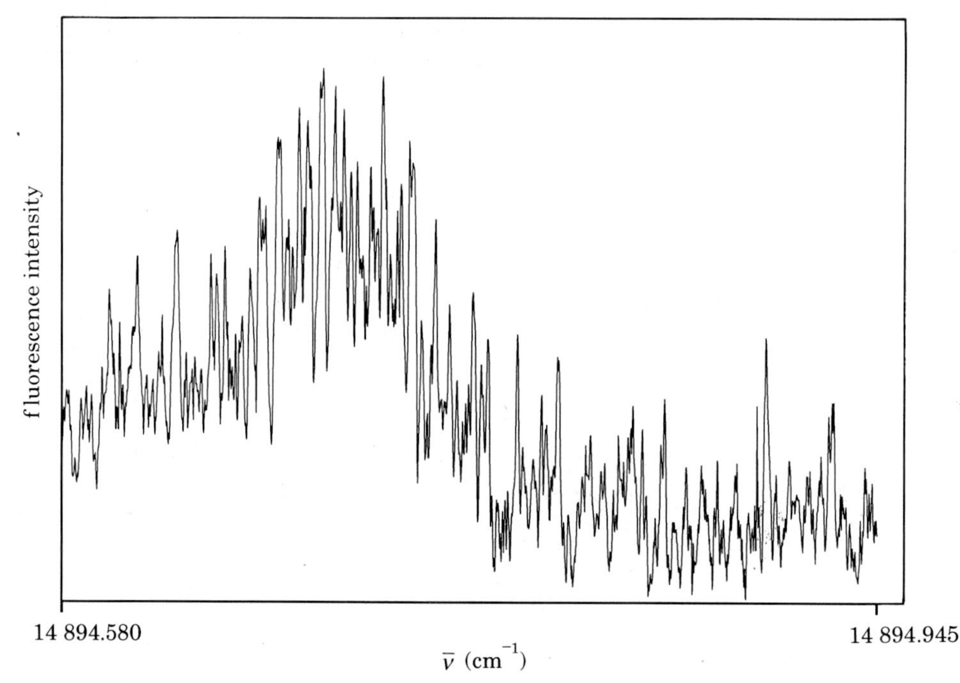

Fig. 6. – Section of the sub-Doppler spectrum of Na_3 showing part of the Q-branch in the $0 \leftarrow 0$ band of the $A \leftarrow X$ transition.

lar constants gave equally good simulated spectra. For a safe assignment, therefore, the powerful technique of optical-optical double resonance has to be applied, which works as follows (fig. 7):

Laser $L1$ is tuned to a selected transition $|i\rangle \rightarrow |k\rangle$ of the spectrum and is stabilized onto its centre wavelength. Due to saturation the population N_i of the

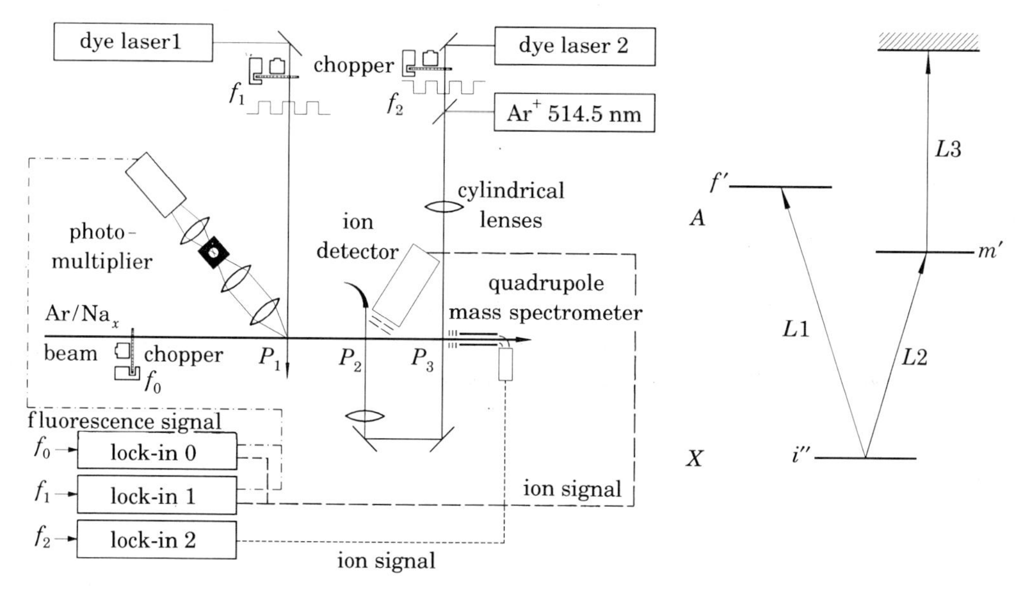

Fig. 7. – Experimental arrangement for OODR spectroscopy in molecular beams.

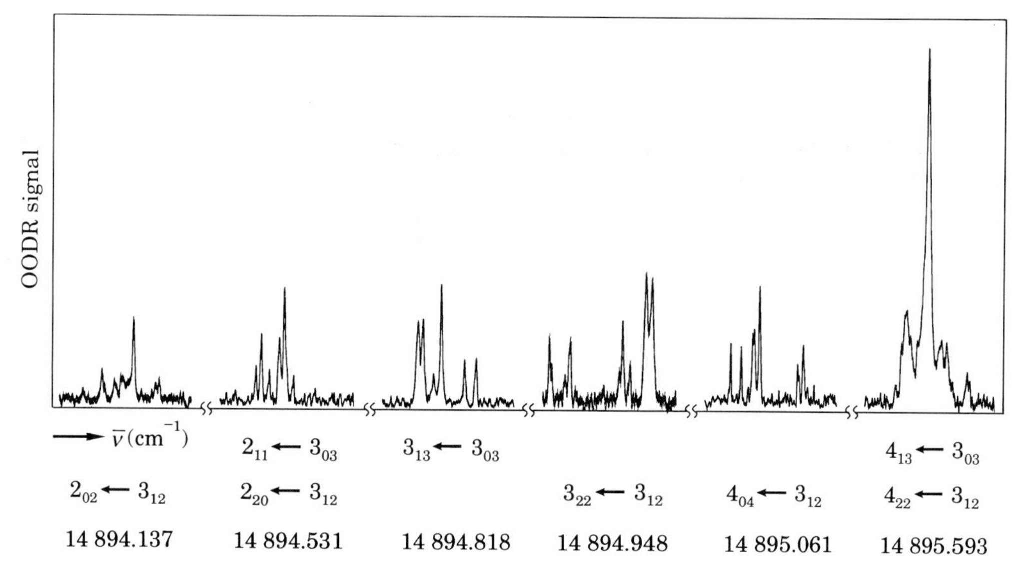

Fig. 8. – OODR spectrum of Na_3 where the pump laser $L1$ was kept on the overlapping transitions $N'_{KaKc} \leftarrow N''_{KaKc} = 4_{13} \leftarrow 3_{03}$ and $4_{22} \leftarrow 3_{12}$.

absorbing level $|i\rangle$ can be nearly completely depleted. In fact a depletion down to 10% of the thermal-equilibrium population N_{i0} has been readily achieved. If $L1$ is chopped at a frequency f_1 by a mechanical chopper, the population N_i is chopped accordingly. When the ion rate $N_{\rm ion}(\lambda_2)$ in the second crossing point is monitored through a lock-in detector as a function of the wavelength λ_2 of dye laser $L2$, only those transitions are detected which start from the labelled level $|i\rangle$ with its modulated population N_i. Such an OODR spectrum is shown in fig. 8. Each multiplet represents the hfs pattern and spin rotation splittings of a single rotational transition. The separation of the OODR signals directly reflects the energy level separations in the upper electronic state.

Many of such OODR spectra are necessary to reach a safe assignment of all transitions in the spectrum of fig. 6 and to obtain reliable values of the molecular constants, which allow the determination of the geometrical structure in lower and upper states [8].

4. – Isotope-selective spectroscopy of silver dimers.

Metal dimers play an important role in surface reactions, catalytic processes and as precursors in cluster formation. When a metal surface is bombarded with high-energy Ar^+ ions (10 keV) the sputtered particles consist of neutral metal atoms, M, dimers M_2 and larger clusters $M_n (n > 2)$. The question is now whether the dimers and clusters are emitted from the bulk or whether only atoms M and ions M^+ are directly emitted and the dimers are being formed in the vapour phase within the plume of sputtered atoms above the surface.

Measuring the internal-state distribution $N(v'', J'')$ of the sputtered M_2 dimers can answer this question [10]. If the dimers come out of the metal bulk, a thermal Boltzmann-like distribution can be expected with a temperature close to that of the bulk material. If, however, the dimers are formed by collisions within the plume of evaporated metal atoms, the higher vibrational levels should be more populated because the dimers are formed by recombination in vibrational levels close to the dissociation energy and in the expanding plume there are not sufficient stabilizing inelastic collisions which would bring the newly formed dimers into lower vibrational levels.

The experimental realization of such measurements is based on two-step photoionization with two pulsed lasers (fig. 9). The first laser $L1$ is tuned to a selected transition $|i\rangle \rightarrow |k\rangle$. With a pulsed dye laser it is not difficult to saturate this transition. This guarantees that the population N_K becomes independent of the transition probability $B_{ik}\rho_L$ and approaches the population N_i of the absorbing level $|i\rangle$. The ion signal produced by photoionization of $|k\rangle$ by the second laser $L2$ is then proportional to the ground-state population N_i of level $|i\rangle =$ $= (v_i'', J_i'')$. The ions are detected after mass selection in a time-of-flight mass spectrometer.

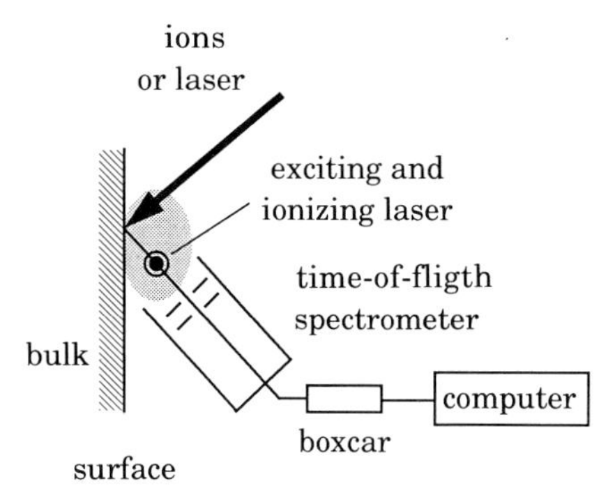

Fig. 9. – Measurement of internal-state distribution of sputtered metal clusters.

For the particular case of sputtering from silver surfaces, it turned out that no rotational analysis of silver dimers Ag_2 had been performed. In order to make correct assignments of the absorbing transitions $(v_i'', J_i'') \rightarrow (v_k', J_k')$, we had to start rotationally resolved high-resolution spectroscopy of Ag_2. This was performed in a supersonic Ar/Ag_2 beam with the apparatus shown in fig. 10. The silver metal vapour is produced in a high-temperature oven of TZM (titanium-zirconium-molybdenum alloy) that can be heated up to 2000 K. The expan-

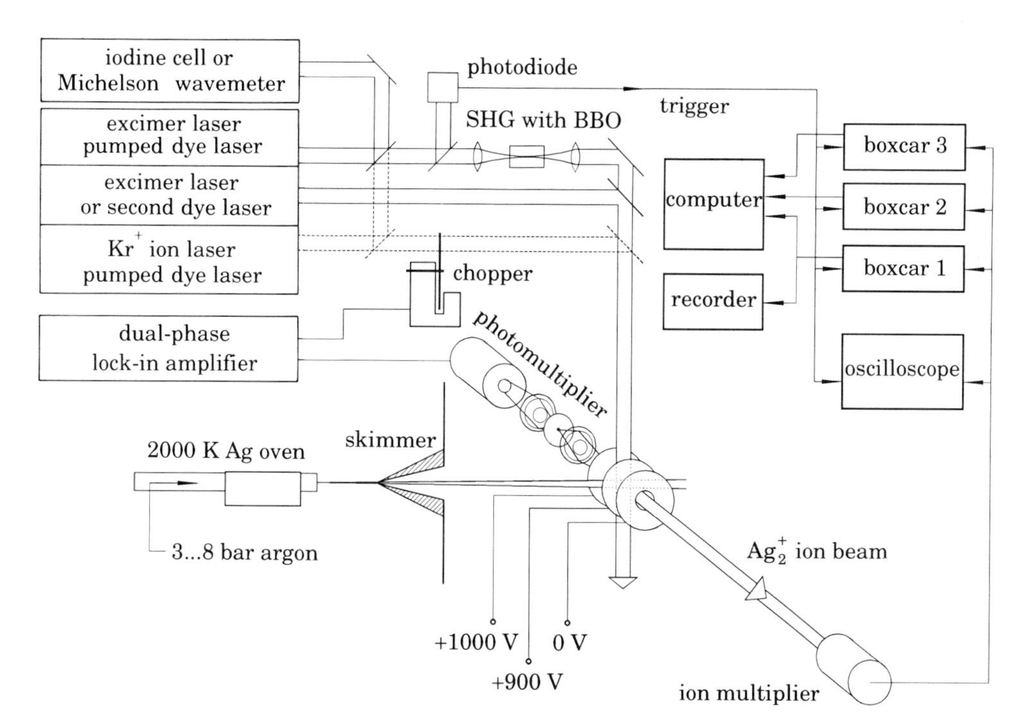

Fig. 10. – Experimental arrangement of isotope-selective spectroscopy of silver dimers Ag_2 by two-step photoionization and time-of-flight spectrometry.

sion of (2 ÷ 4) bar of argon seeded with (20 ÷ 50) mbar silver vapour, through a nozzle of 70 μm diameter, yields a supersonic beam with rotational temperatures of about 50 K and vibrational temperatures of about 200 K. The beam is collimated by a skimmer and intersected by two superimposed beams of two dye lasers 17 cm downstream of the nozzle.

The ions formed by resonant two-step photoionization are extracted by a two-stage electric field, pass through a time-of-flight spectrometer and are detected by an ion multiplier.

The ion signal recorded without mass spectrometer as a function of the wavelength λ_1 of $L1$ is illustrated in fig. 11*a*), which shows the $v' = 2 \leftarrow v'' = 0$ band of the $B \leftarrow X$ system. The spectrum consists of an overlap of three spectra of the three isotopomers $^{107}Ag^{107}Ag$, $^{107}Ag^{109}Ag$ and $^{109}Ag^{109}Ag$, as can be readily recognized from the three slightly shifted band heads.

With the time-of-flight spectrometer the three spectra can be recorded separately but simultaneously by feeding the ion multiplier output into three parallel box-car integrators with time gates adjusted to the arrival times of the three isotopomers. These isotope-selective spectra are shown in fig. 11*b*)-*d*) [11].

The detailed analysis of these and many other bands measured with this ar-

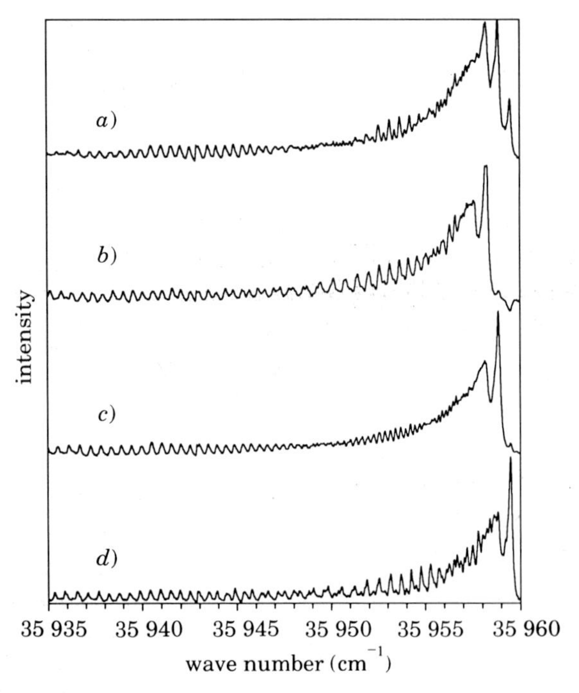

Fig. 11. – The section of the ($v' = 2 \leftarrow v'' = 0$) band of the $B \leftarrow X$ system of Ag_2. *a*) Overlap of three isotope spectra, *b*)-*d*) isotope-selective spectra recorded as ion rates $N_{ion}(m, \lambda_1)$ behind a time-of-flight spectrometer: *b*) 109-109 isotope, *c*) 107-109 isotope, *d*) 107-107 isotope.

rangement yielded the molecular constants of 5 electronic states of Ag_2 with high accuracy [12] and allowed the assignment of transitions used for the sputtering experiment. It turned out that the vibrational energy of the sputtered Ag_2 dimers was much higher (about 1 eV) than the thermal energy of the solid silver target. This proves that most of the dimers must have been formed outside the bulk in the vapour phase above the surface [13].

Of particular interest is the ionization energy of molecules. Often the onset of the ion current $N_{ion}(\lambda)$ measured by one-photon ionization as a function of the wavelength λ_1 of the ionizing laser is used for the determination of the ionization potential [14].

However, this method is not quite unambiguous, in particular when the potential curve of the ion ground state is shifted against that of the neutral ground state. In this case vertical transitions from $v''=0$ do not terminate in the lowest vibrational level $v^+=0$ of the ion state, *i.e.* the Franck-Condon factors for transitions $v''=0 \rightarrow v^+=0$ are very small and

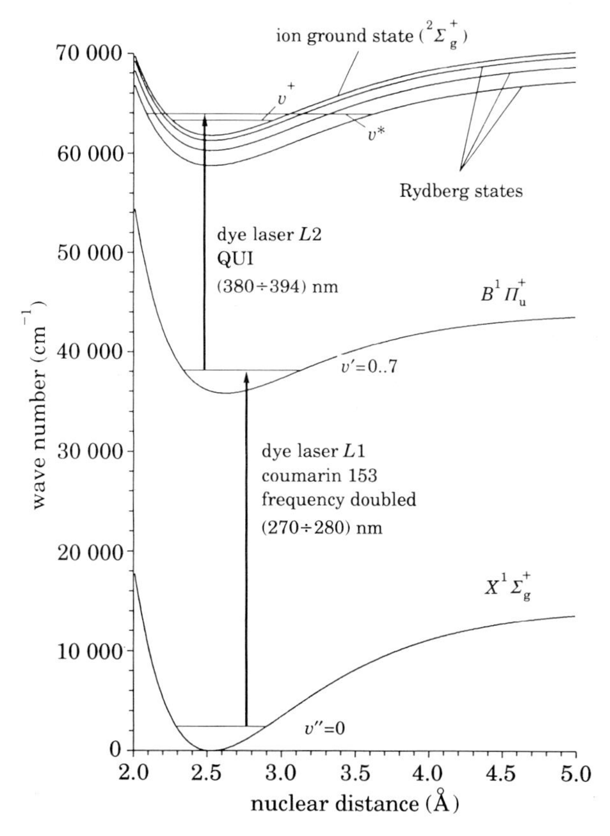

Fig. 12. – Two-step photoionization of Ag_2 molecules. The second step can either terminate in the ionization continuum or can excite onto ionizing Rydberg levels.

the onset of the ion current is difficult to be determined. Therefore, two-step photoionization with two tunable lasers gives more reliable results.

The relevant level diagram is shown in fig. 12. The first laser $L1$ populates a selected level (v, J') in the excited electronic state $B^1\Pi_u$. The second laser with a sufficiently short wavelength excites states above the ionization limit. These may be continuous states $Ag_2^+ + e^- + E_{kin}$ or bound Rydberg levels (v^*, J^*) of the neutral molecule Ag_2 which decay by autoionization [15].

The electric extraction field E, which is necessary to collect the photoions onto the ion detector, lowers the appearance potential AP against the true ionization potential IP according to

$$\text{AP} = \text{IP} - \left(\frac{e^3 E}{\pi \varepsilon_0} \right)^{1/2} . \tag{1}$$

One, therefore, has to measure the appearance potential for different electric fields E and extrapolate the measured values AP(E) towards $E \to 0$.

A more accurate method for the determination of the ionization potential is the measurement of autoionizing Rydberg series. The term value $T_n(v^*, J^*)$ of a molecular Rydberg level with principal quantum number n, vibrational level v^* and rotational quantum number J^* is given by the Rydberg formula

$$T_n(v^*, J^*) = \text{IP} - \frac{R_y}{(n - \delta)^2} + G(v^*) + F(J^*), \tag{2}$$

where δ is the quantum defect (which describes the deviation of the internal electric field experienced by the Rydberg electron from a pure Coulomb field), and $G(v^*)$, $F(J^*)$ are the vibrational and rotational term values of the Rydberg levels. If a Rydberg series $T_n(v^*, J^*)$ with $n \to \infty$ can be measured, its convergence limit

$$T^+_{\text{ion}}(v^+ = v^*, J^+ = J^*) = \lim_{n \to \infty} T_n(v^*, J^*) \tag{3}$$

yields the term value $T^+_{\text{ion}}(v^*, J^*)$ of a vibrational-rotational level $(v^+ = = v^*, J^+ = J^*)$ in the electronic ground state.

Such Rydberg series of the Ag_2 dimer are shown in fig. 13 for different values of $v^* = v^+$.

The extrapolation of the term values of the ion limits yields the true ionization potential

$$\text{IP} = \lim_{v^+, J^+ \to 0} T^+(v^+, J^+) = 61\,747 \text{ cm}^{-1} \tag{4}$$

of the Ag_2 dimer.

This extrapolation of Rydberg series allows the most accurate determination of the ionization potential, superior to all other methods [16].

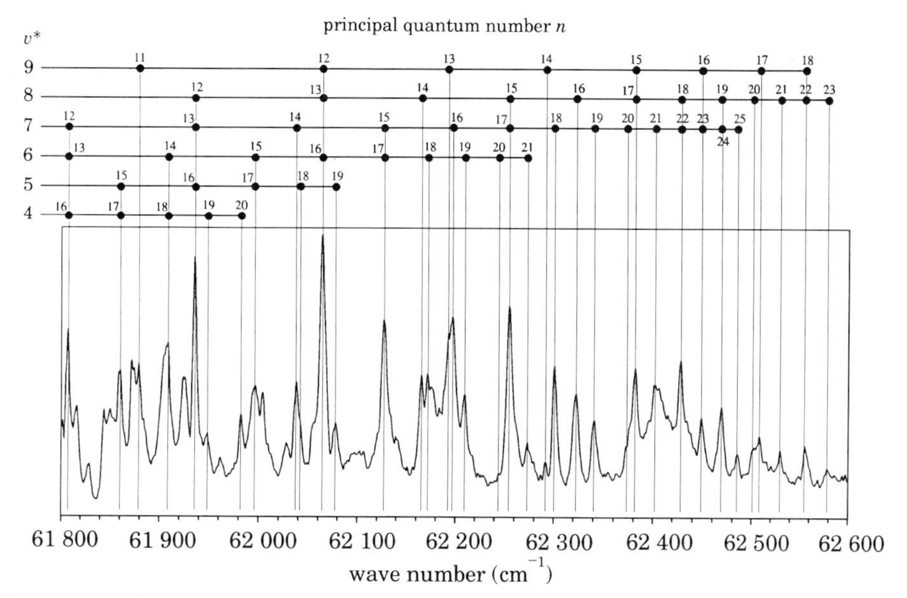

Fig. 13. – Rydberg series of the Ag_2 molecule excited by a tunable pulsed dye laser on transitions starting from the intermediate vibrational level $B\,{}^1\Pi_u(v' = 4)$.

5. – Lifetime measurements of selectively excited molecular levels.

The knowledge of transition probabilities gives very important information on the dynamics of excited molecular states, which may decay by direct radiative transitions into the ground state, by redistribution of the excitation energy among the various degrees of freedom due to intramolecular couplings between electronic and vibronic states, or by inelastic collisions with other atoms or molecules.

Lifetime measurements yield the most accurate absolute values of transition probabilities. We will discuss such measurements under different experimental conditions:

1) We assume at first that a single upper level $|k\rangle$ with excitation energy E_k has been excited by a narrow-band laser pulse. This implies that the spectral distribution $I(\lambda)$ of the laser pulse intensity does not overlap with more than one molecular absorption line (fig. 14). The decay rate of the upper-level population per second after the excitation pulse has ended is then

$$\frac{\mathrm{d}N_k}{\mathrm{d}t} = -(A_k + \sigma_k \bar{v}_r N_B) N_k \,, \tag{5}$$

which yields the exponential decay

$$N_k(t) = N_k(0) \cdot \exp[-t/\tau_{\mathrm{eff}}] \tag{6}$$

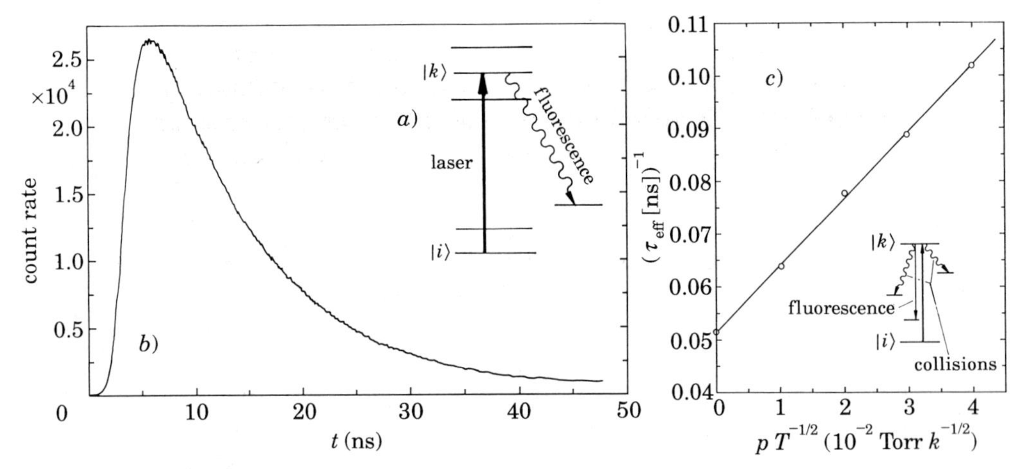

Fig. 14. – Selective excitation of a single molecular level: *a*) level scheme, *b*) exponential decay, *c*) Stern-Vollmer plot.

of the upper level with the effective lifetime τ_{eff} that depends on the radiative transition probability $A_k = 1/\tau_k^{\text{rad}}$ and the collision-induced depopulation probability $C_k = \sigma_k \bar{v}_{\text{r}} \cdot N_B$ that depends on the collision cross-section σ_k, the density N_B of collision partners B in the surrounding of the excited molecule M^* and the mean relative velocity $\bar{v}_{\text{r}}$ of the collision partners.

A Stern-Vollmer plot

$$\frac{1}{\tau_k^{\text{eff}}} = \frac{1}{\tau_k^{\text{rad}}} + \sigma_k \bar{v}_{\text{r}} N_B(t) \tag{7}$$

of the inverse measured effective lifetime against the density N_B of collision partners B gives a straight line. The slope of this line $\operatorname{tg}\alpha = \sigma_k \bar{v}_{\text{r}}$ yields the total integral cross-section of all depopulating collisions and the intersect at $N_B = 0$ gives the radiative lifetime $\tau_k^{\text{rad}} = 1/A_k$ and, therefore, also the total radiative transition probability A_k (fig. 14*c*)).

With $N_B = p_B/kT$ and $\bar{v}_{\text{r}} = (8\,kT/\pi\mu)^{1/2}$, where $\mu = m_A \cdot m_B/(m_A + m_B)$ is the reduced mass, we can write eq. (7) as a function of pressure p and temperature T in the sample cell:

$$\frac{1}{\tau_k^{\text{eff}}} = \frac{1}{\tau_k^{\text{rad}}} + \sigma_k \sqrt{\frac{8}{\pi\mu kT}}\, p_B\,. \tag{7a}$$

2) For dense molecular absorption spectra the spectral bandwith $\Delta\nu_{\text{L}}$ of the laser pulse is generally larger than the mean separation between absorption lines. This implies that several excited levels are simultaneously populated (fig. 15). Since the excitation transitions start from different lower levels, there are no definite phase relations between the wave functions of the excited levels and the simultaneous excitation of these levels is called *incoherent*. These levels de-

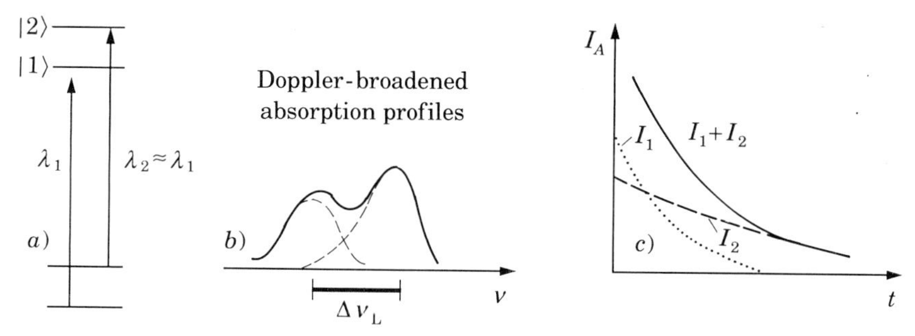

Fig. 15. – *a*) Incoherent excitation of two levels, *b*) overlap of the pump laser with two absorbing transitions, *c*) biexponential fluorescence from levels $|1\rangle$ and $|2\rangle$.

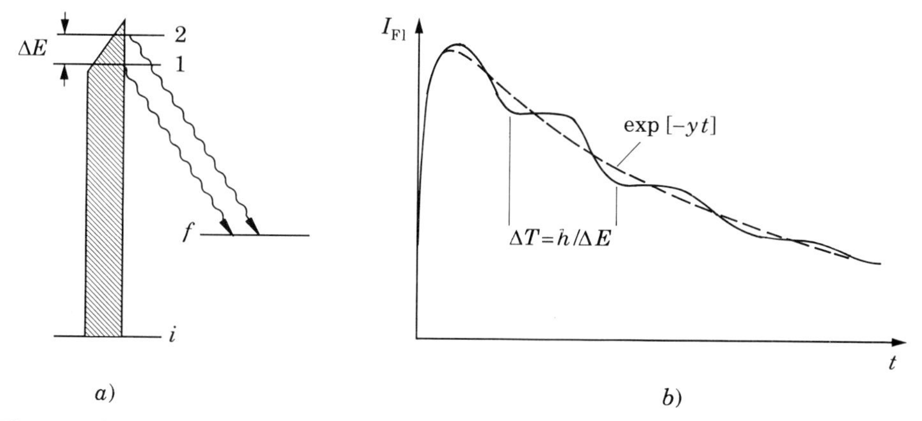

Fig. 16. – Coherent excitation of two upper levels on transitions from the same lower level: *a*) level scheme, *b*) quantum beats in the fluorescence decay.

cay independently and the total fluorescence is

$$I_{\mathrm{Fl}}(t) = \sum_k c_k I_k(t) = \sum_k c_k I_k(0) \exp[-t/\tau_k^{\mathrm{eff}}]. \tag{8}$$

In order to measure the different lifetimes, one has to fit the measured decay curve to eq. (8). This is often not very accurate, in particular if it is not known how many upper levels contribute to the fluorescence. A better way is, therefore, to separate the contributions from the different upper levels by a monochromator. The fluorescence spectrum of each upper level $|v', J'\rangle$ consists of all allowed vibrational bands $v' \to v''$ for which the Franck-Condon factors are sufficiently large. Each vibrational band consists either of two rotational lines (P and R lines with $\Delta J = J'' - J' = \pm 1$) or of only one line ($Q$ line with $\Delta J = 0$). It is, therefore, not too difficult to separate the contributions from the different upper levels $|k\rangle$.

3) If two or more upper levels are excited simultaneously on transitions starting from the *same lower level* (fig. 16), the excitation is called *coherent* be-

cause now definite phase relations exist between the wave functions $\psi_k(\boldsymbol{r}, t)$ of the upper levels. They are excited in phase at $t = 0$. However, because of their different energies E_k, the phases of their wave function

$$\psi_k(\boldsymbol{r}, t) = \phi_k(\boldsymbol{r}) \exp[-i(E_k/\hbar)t - (\gamma_k/2)t] \tag{9}$$

depend on the energy E_k and develop differently in time for the different coherently excited levels. The total fluorescence intensity observed on a transition into a lower level $|m\rangle$ is

$$I(t) = C \cdot |\phi_m|\boldsymbol{\varepsilon} \cdot \boldsymbol{\mu}|\psi(t)|^2 , \tag{10}$$

where $\boldsymbol{\varepsilon}$ is the polarization vector of the emitted fluorescence and $\boldsymbol{\mu} = e \cdot \boldsymbol{r}$ is the dipole operator. The wave function

$$\psi(t) = \sum_k c_k \phi_k(0) \exp[-(i\omega_{km} + \gamma_k/2)t] \tag{11}$$

is a coherent superposition of the simultaneously excited levels $|k\rangle$.

Inserting (9) and (11) into (10), the time-dependent fluorescence intensity on transitions of two coherently excited levels $k = 1, 2$ into a lower level $|m\rangle$ can be written as

$$I(t) = C \exp[-\gamma t](A + B \cos \omega_{21} t) \tag{12}$$

with $\gamma = (\gamma_1 + \gamma_2)/2$,

$$A = c_1^2 |\langle\phi_m|\boldsymbol{\varepsilon} \cdot \boldsymbol{\mu}|\phi_1|^2 + c_2^2 |\langle\phi_m|\boldsymbol{\varepsilon} \cdot \boldsymbol{\mu}|\phi_2\rangle|^2 ,$$

$$B = 2c_1 c_2 |\langle\phi_m|\boldsymbol{\varepsilon} \cdot \boldsymbol{\mu}|\phi_1\rangle| \; |\langle\phi_m|\boldsymbol{\varepsilon} \cdot \boldsymbol{\mu}|\phi_2\rangle| .$$

This is an exponential decay superimposed by a modulation with frequency $\omega_{21} = (E_2 - E_1)/\hbar$ which depends on the energy separation ΔE between the two coherently excited levels.

The Fourier transform of the decay curve yields the energy separations between the coherently excited levels and their level widths [17].

The experimental arrangement used for measuring lifetimes in the range 500 ps ÷ 10 μs is shown in fig. 17.

A mode-locked argon laser pumps synchronously a c.w. dye laser. If the lengths of the two laser cavities are exactly equal, the dye-laser pulses become very short (down to a few ps). For selective excitation of a single level the spectral bandwidth $\Delta\nu_L \geqslant 1/2\pi\Delta T$ should be sufficiently small. Therefore, two Fabry-Perot etalons and a birefringent filter are placed inside the dye-laser cavity in order to decrease the spectral bandwidth. This, of course, makes the pulse width ΔT larger. It is possible to achieve nearly Fourier-limited pulses with $\Delta T \approx 0.5$ ns and $\Delta\nu_L \approx 1.5$ GHz.

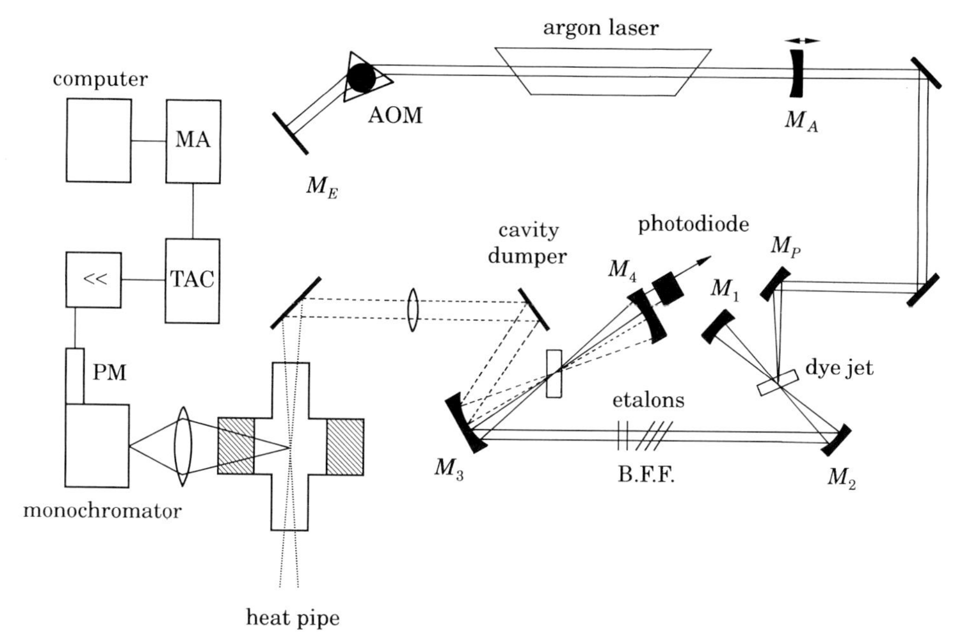

Fig. 17. – Synchronously pumped mode-locked dye laser with cavity dumper used for lifetime measurements.

For lifetimes $\tau > 12$ ns the pulse repetition rate of the mode-locked laser ($f = 80$ MHz) is too high. Therefore, an acoustic cavity dumper was used which couples only every 20th pulse out of the dye-laser resonator.

The dye-laser pulses with a repetition rate of $f = 4$ MHz were sent into a heat pipe, where the sample molecules were enclosed by argon as buffer gas. The fluorescence photons were detected after a monochromator and the time delay between fluorescence photon and subsequent laser pulse was measured with the single-photon delayed-coincidence method [2, 18].

After having presented three different situations of incoherent and coherent excitation of levels in time-resolved molecular spectroscopy, let us now discuss what can be learned from such measurements.

If lifetime measurements of molecules are performed in cells, collisions cannot be avoided and the effective lifetimes have to be measured at different pressures in order to obtain from a Stern-Vollmer plot (7) the radiative lifetime τ_k^{eff} and the total collisional deactivation cross-section σ_k. Since the mean separation of absorption lines is often smaller than the Doppler width, simultaneous incoherent excitation of several upper levels is more the rule than the exception. The fluorescence, therefore, has to be dispersed in order to separate the contributions from different upper levels $|k\rangle$.

The collision-free lifetime depends on the wave functions of upper and lower states. Often excited molecular levels are perturbed. This means a more or less strong interaction exists between different electronic states and the Born-Oppenheimer approximation, which separates the molecular wave functions into a product

$$\psi = \psi_{\mathrm{el}} \cdot \psi_{\mathrm{vib}} \cdot \psi_{\mathrm{rot}},$$

is no longer valid.

Examples of such interactions are vibronic couplings between different vibrational levels of different electronic states, spin-orbit coupling between singlet and triplet states, or couplings between bound states and continuous states resulting in predissociation.

The radiative lifetime is strongly affected by such couplings. This can be seen as follows:

Assume two Born-Oppenheimer levels $|1\rangle$ and $|2\rangle$ are coupled to each other. The total wave function of the perturbed level $|k\rangle$ is then

$$\psi_k = c_1\psi_1 + c_2\psi_2 \qquad \text{with } |c_1|^2 + |c_2|^2 = 1\,. \tag{13}$$

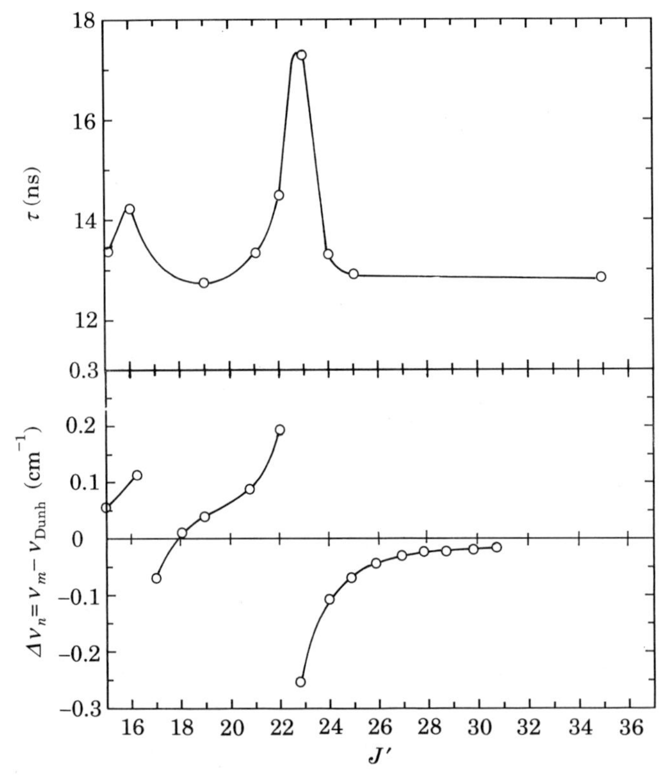

Fig. 18. – Lifetimes of rotational levels in the $A^1\Sigma_u(v' = 17)$ state of Na_2 as a function of rotational quantum number J'. The perturbations of some of the levels is caused by spin-orbit coupling with the $b^3\Pi_u$ state. The lower graph shows the corresponding energy shifts of the perturbed levels.

The transition probability A_{km} of radiative transition into a lower level $|m\rangle$ is then proportional to the square of the transition dipole moment:

$$A_{km} \propto \left| \int \psi_k \boldsymbol{r} \psi_m \, \mathrm{d}\tau \right|^2 = |\mu_{km}|^2 = c_1^2 \mu_{1m}^2 + c_2^2 \mu_{2m}^2 + 2c_1 c_2 \mu_{1m} \mu_{2m} \tag{14}$$

$$\text{with } \mu_{im} = \int \psi_i \boldsymbol{r} \psi_m \, \mathrm{d}\tau .$$

If the level $|2\rangle$ is, for example, a triplet level that is coupled by spin-orbit interaction to the singlet level $|1\rangle$, the transition matrix element μ_{2m} for a transition to a lower singlet level $|m\rangle$ is zero, *i.e.* $\mu_{2m} = 0$, and the transition probability A_{km} will decrease with increasing coupling strength, *i.e.* decreasing values of the coefficient c_1. This is illustrated by fig. 18, which shows the energy shifts of rotational levels in the $Na_2 (A\,^1\Sigma_u v' = 17)$ state that are perturbed by spin-orbit coupling with levels in the $b\,^3\Pi_u$ state.

This coupling *increases* the lifetimes, as can be seen in the upper part of fig. 18. From the change $\Delta\tau$ in lifetimes of the perturbed levels the coefficient c_1 in eq. (14) and with it the admixture of the perturbing level $|2\rangle$, *i.e.* the coupling strength, can be directly determined.

For large predissociation rates the measured broadening of the corresponding absorption lines can be used to gain information on the effective lifetime τ_k of the upper level. For lifetimes longer than 0.2 ns ($\tau_k = 1$ ns corresponds to a natural linewidth of $\Delta\nu = (2\pi\tau_k)^{-1} = 160$ MHz) Doppler-free absorption spectroscopy is necessary in order to measure changes in the natural linewidth due to perturbing effects. The measurements of lifetime changes are more reliable for quantitative determinations of transition matrix elements and coupling strengths between mutually perturbing levels. The influence of predissociation on the lifetimes of selectively excited levels (v', J') in the $D\,^1\Sigma_u$ state of the Cs_2

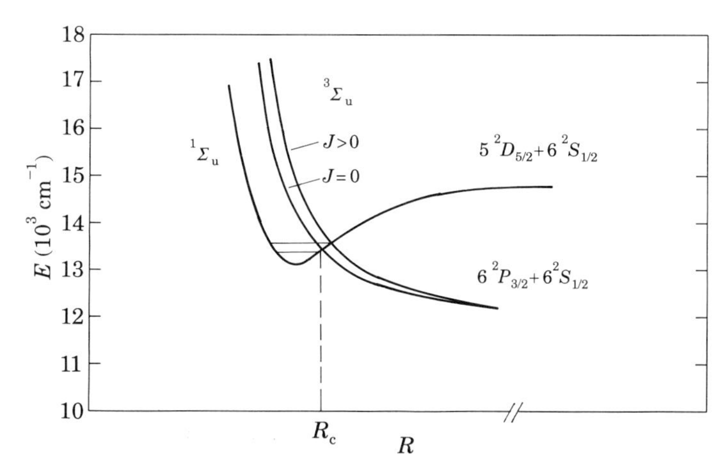

Fig. 19. – Shortening of lifetimes of levels (v', J') in the $D\,^1\Sigma_u$ state of Cs_2 by predissociation due to interaction with a repulsive $^3\Sigma_u$ state.

molecule is shown in fig. 19. The $D\,^1\Sigma_u$ state is crossed by a repulsive $^3\Sigma_u$ state and those levels with a large Franck-Condon overlap between the vibrational wave functions of the bound level (v', J') and the continuum wave function of the repulsive state show a large predissociation rate. Their lifetimes are shortened from an unperturbed value of $\tau \approx 20$ ns to values below 1 ns.

The preceding discussion has shown that lifetime measurements represent powerful means for the determination of radiative and nonradiative intramolecular-transition probabilities. They can also be used for measurements of absolute integral collision cross-sections. Together with measurements of collision-induced satellites in fluorescence spectra which are excited by c.w. lasers also absolute values of cross-sections for individual collision-induced transitions $(v', J') \rightarrow (v' + \Delta v', J' + \Delta J')$ can be obtained [2].

Such collisions play an essential role in photochemically induced chemical reactions and the study of these processes in small model molecules may help to a deeper understanding of these important processes.

REFERENCES

[1] S. Svanberg: *Atomic and Molecular Spectroscopy* (Springer, Berlin, Heidelberg, 1991).

[2] W. Demtröder: *Laser Spectroskopie,* 3rd edition (Springer, Berlin, Heidelberg, 1993).

[3] A. C. P. Alves, J. M. Brown and J. M. Hollas, Editors: *Frontiers of Laser Spectroscopy of Gases,* NATO *ASI Series,* Vol. **234** (Kluwer, Dordrecht, 1988).

[4] B. A. Gavetz and J. R. Lombardi, Editors: *Advances in Laser Spectroscopy,* I and II (Heyden, London, 1982/83); D. S. Kliger, Editor: *Ultrasensitive Laser Spectroscopy* (Academic Press, New York, N.Y., 1983).

[5] Y. Prior, A. Ben-Reuven and M. Rosenbluth: *Methods of Laser Spectroscopy* (Plenum, New York, N.Y., 1986); W. Demtröder and M. Inguscio, Editors: *Applied Laser Spectroscopy,* NATO *ASI Series B,* Vol. **241** (Plenum Press, New York, N.Y., 1990).

[6] M. Zen: *Cryogenic bolometers,* in G. Scoles, Editor: *Atomic and Molecular Beam Methods,* Vol. I (Oxford University Press, New York, N.Y., 1988), p. 254.

[7] M. L. Cohen and W. D. Knight: *The Physics of Metal Clusters, Phys. Today,* **43,** 42 (December 1990).

[8] H.-A. Eckel, J.-M. Gress, J. Biele and W. Demtröder: *J. Chem. Phys.,* **98,** 135 (1993).

[9] H.-J. Foth, J.-M. Gress, Chr. Hertzler and W. Demtröder: *Z. Phys. D,* **18,** 257 (1991).

[10] M. J. Pellin, C. E. Young, W. F. Calaway and D. M. Gruen: *Surf. Sci.,* **144,** 619 (1984).

[11] V. Beutel, M. Kuhn and W. Demtröder: *J. Mol. Spectrosc.,* **155,** 343 (1992).

[12] V. Beutel, H.-G. Krämer, G. L. Bhale, M. Kuhn, K. Weyers and W. Demtröder: *J. Chem. Phys.,* **98,** 2699 (1992).

[13] M. Wahl: Diplomthesis Kaiserslautern (1991); A. Wucher: private communication.

[14] K. I. PETERSON, P. D. DAO, R. W. FARLEY and A. W. CASTLEMAN: *J. Chem. Phys.*, **80**, 1780 (1984).
[15] V. BEUTEL, G. L. BHALE, M. KUHN and W. DEMTRÖDER: *Chem. Phys. Lett.*, **185**, 313 (1991).
[16] G. HERZBERG and CHR. JUNGEN: *J. Mol. Spectrosc.*, **41**, 425 (1972).
[17] A. BITTO and J. R. HUBER: *Opt. Commun.*, **80**, 184 (1990).
[18] B. BIENIAK, H. P. HINSKE, H. PAULUS, D. ZEVGOLIS and W. DEMTRÖDER: *Ann. Phys.*, 7 Folge, **48**, 15 (1991).

High-Resolution and High-Sensitivity Spectroscopy Using Semiconductor Diode Lasers.

M. INGUSCIO

Dipartimento di Fisica and European Laboratory for Nonlinear Spectroscopy (LENS)
Università degli Studi di Firenze - largo E. Fermi 2, I - 50125 Firenze, Italia

1. – Introduction.

As clearly discussed by Prof. SCHAWLOW in his introductory lecture [1], we spectroscopists still have the problem of finding tunable lasers and, once the laser is tunable, we want it to be more and more monochromatic. Indeed, due to the central role played by the radiation source in coherent spectroscopy, the scientific achievements strictly followed the technical development of lasers.

Semiconductor diode lasers (SDL) are certainly lasers with a wide wavelength tuning range. Although they were invented more than thirty years ago, they were not used much in atomic spectroscopy until recently [2]. In the early times, diode lasers had to be operated at low temperatures, they emitted over several cavity modes and were unreliable. Recently, semiconductor diode lasers have been rapidly improving in power, spectral purity and wavelength coverage: they can now operate c.w. at room temperature and produce tens of milliwatt on a single mode. Nowadays these lasers are showing some specific properties which make them suitable sources of radiation for very-high-resolution and/or extremely-high-sensitivity experiments. In general, an extreme simplification of sophisticated experimental schemes can be obtained. As an example let us refer to the resolution of Lamb shift of H_α transition in hydrogen. This measurement constitutes one of the cornerstones of atomic spectroscopy and the historical recording, obtained in 1972 at Stanford University by a dye-laser spectrometer, is reported both in Schawlow and Hänsch lectures [1, 3]. The same recording is reported in fig. 1, as obtained in Munich [4] nearly twenty years later by using an InGaAlP free-running laser diode.

In this lecture, we show some of the possibilities offered by semiconductor

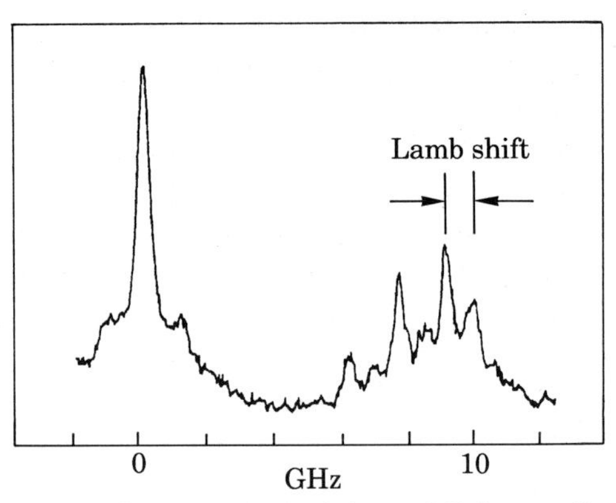

Fig. 1. – Doppler-free saturation spectrum of the red Balmer-α line of atomic hydrogen with resolved fine-structure components as recorded in Munich (ref. [4]) using a free-running InGaAlP laser diode.

diode lasers in atomic and molecular spectroscopy by describing experiments recently performed in the visible ($(650 \div 690)$ nm) and near infrared ($(750 \div 850)$ nm).

2. – Overview on diode laser characteristics and operation.

The diode laser basically works as a p-n junction biased in the forward direction. The recombination of electrons and holes takes place at the junction and leads to the emission of radiation at a frequency depending on the energy gap between the conduction and the valence bands. The reflectivity of the cleaved facets of the diode is sufficient to have dominance of stimulated emission.

In order to have continuous-wave operation of the laser, it is necessary to confine the carriers and the photons in a well-defined region in order to have enough gain. This is commonly obtained using double heterostructures, where the active region is sandwiched between two layers of semiconductors with a larger band gap. This prevents the carriers from escaping the active region, but also light is confined in the same active region because a higher band gap corresponds to a smaller index of refraction. In the first lasers, confinement was achieved only in the direction in which the current was injected (gain-guided lasers), while now also «index-guided» lasers are available, in which the active region is limited in all directions by a higher-band-gap material. It is evident that real devices have complex structures and require important technological advances [5]. More recently, also quantum-well and multiple-quantum-well lasers became available, with lower threshold currents and higher output powers compared with normal double heterostructures.

Important characteristics which are relevant for the use of SDL sources in atomic and molecular spectroscopy are spectral coverage, tunability, spectral linewidth and amplitude stability.

The emission wavelength depends on the band gap and in principle it should be possible to construct lasers at all wavelengths by choosing the proper stoichiometric abundance for the dopant element, provided that direct interband transitions can occur. For instance, one could obtain emission from 1100 to 1650 nm using InGaAsP, or cover the $(700 \div 890)$ nm and $(630 \div 690)$ nm intervals by means of GaAlAs and AlGaInP, respectively. Of course, the material must have good optical and electrical characteristics in order to minimize losses. For this reason, for instance, room temperature operation in the blu-green region was only recently demonstrated. In practice, laser diodes are commercially available only at sparsely distributed emission wavelengths and the experimentalist soon faces the problem of tuning the laser to the precise atomic or molecular absorption. At a first attempt, one can change the emitted wavelength by changing the diode temperature. Indeed, the semiconductor gap depends on the temperature and overall spans of the order of some 20 nm can easily be obtained, considering that operation at low temperatures is even more efficient and that operation down to liquid-helium temperature has been demonstrated. Tuning by temperature, however, is discontinuous, and this is a significant limitation. Changes in temperature lead to a change of the cavity length and hence of the cavity modes. Typical tuning with this effect is of the order of 0.06 nm/°C ($\sim$ 30 GHz/°C). At the same time, the gain curve shifts by about 0.25 nm/°C. Since the two temperature dependences are different, we have a continuous tuning ($\sim$ 30 GHz/°C) until the longitudinal-mode hopping caused by the gain curve shift: the general aspect is that of a staircase with sloping steps, essentially causing wavelength changes in a rather unpredictable way. From the above considerations we also learn that an extremely good thermal stability (better than 1 mK) is required in view of narrow-linewidth operation. Thermal tuning is also obtained via Joule effect by changing the current of operation. This tuning, at a rate of about 3 GHz/mA, is rather slow while a faster control, of about 300 MHz/mA, can be obtained thanks to the change in the medium refractive index caused by the change in the current density. Also in this case we deduce that an injection current stability of at least 1 μA is required to reduce effects on the linewidth.

Let us now in fact discuss the problem of spectral linewidth. An exhaustive treatment of the different factors determining the linewidth can be found in ref. [6]. However, these lasers are so small, 100 to 500 μm long, 0.1 to 2 μm thick and 2 to 20 μm wide, that the predominant broadening is caused by the quantum fluctuations of the phase as early described by SCHAWLOW and TOWNES [7], who showed the limit linewidth to be inversely proportional to the output power and to the square of the optical-mode volume. Here we report, from ref. [8], the modified formula which, by the introduction of a factor α, takes into account the dependence of the refractive index on the carrier density:

$$\Delta\nu_{\mathrm{FWHM}} = \frac{h\nu}{8\pi P_0}\left(\frac{c}{nL}\right)^2\left[aL + \ln\left(\frac{1}{R}\right)\right]\ln\left(\frac{1}{R}\right)n_{\mathrm{sp}}(1+\alpha^2), \tag{1}$$

where P_0 is the output power, L is the cavity length, R is the facet reflectivity, a gives the distributed losses in the cavity, c/n the group velocity, and n_{sp} the spontaneous-emission factor which is of the order of unity. In the case of a typical diode laser ($P_0 = 10$ mW, $L = 300$ μm, $R = 0.3$, $\alpha \sim 4 \div 5$), a linewidth of several MHz can be estimated. An increase of the cavity length or, more generally, of the cavity quality factor Q is hence needed to reduce the linewidth. As we shall see, this is at the base of the development of external optical cavities for line-narrowing purposes.

As for amplitude stability, the intrinsic stability of the structure of the semiconductor laser reduces the sources of noise. The properties of the active medium give rise to a peculiar AM and FM noise spectrum, which has been accurately investigated by HOLLBERG at NIST [9]. However, in the low-frequency range (frequencies lower than 10 MHz), typical for simple absorption spectroscopy experiments, an amplitude noise only $(10 \div 20)$ dB above the shot noise level has been measured [10], while the corresponding one for dye lasers can be even two orders of magnitude larger. At 800 nm and 5 mW power, we can compute a shot noise level of the order of 10^{-8}, hence a factor 10 dB worse yields for the SDL a relative amplitude stability of the order of $10^{-6} \div 10^{-7}$, at 1 Hz bandwidth. This impressive stability is at the base of the development of high-sensitivity absorption spectroscopy, for instance applied to molecular forbidden transitions, as we shall discuss in the last section of this contribution.

3. – Line narrowing and tuning with grating feedback.

We have seen that the linewidth of commercial SDLs is too large for applications in high-resolution spectroscopy, however it can be easily reduced by means of electronic or optical feedbacks [10].

In electronic-feedback schemes, the injection current is changed and very fast electronics is needed since the FM noise extends to high frequencies. The change in the injection current produces changes not only in the emitted frequency but also in the emitted power and this introduces an additional amplitude noise. Of course, the electronic stabilization method does not provide any control of the emission wavelength.

More popular are certainly optical-feedback methods. They are based on the simple idea of increasing the Q factor by coupling the diode to another cavity. Weak-feedback and strong-feedback regimes are both successful. The first is, for instance, ensured by the coupling with a high-Q Fabry-Perot interferome-

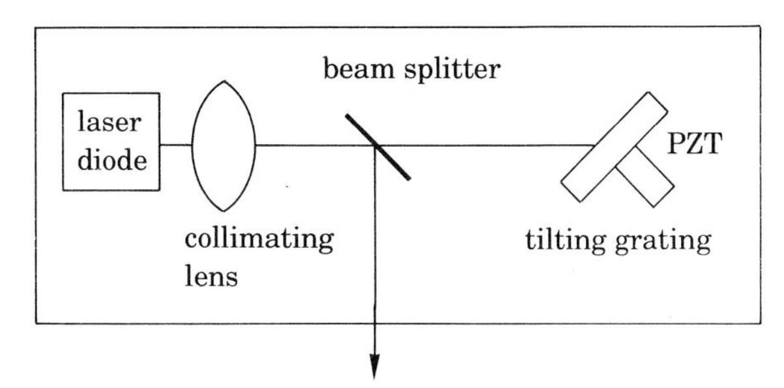

Fig. 2. – Scheme of the «pseudo-external-cavity» configuration adopted for the frequency narrowing and tuning of commercial semiconductor diode lasers.

ter, while in the second the laser is operated in an external cavity. If the feedback is selective in wavelength, it is possible consequently to control the emission. In this lecture we concentrate on the scheme using the strong feedback, which has been the most used in our laboratory after learning the art from Hollberg during his visit to LENS [11]. Our design, as the many others nowadays used all over the world, is a sort of compromise between easy construction, reliability, good spectral characteristics for high-resolution and high-sensitivity spectroscopy. Typically we can obtain a reduction by about two orders of magnitude of the emission linewidth and achieve a continuous wavelength tuning within a range of several nanometers at fixed temperature. In general one should use antireflection coating on the diode facet in order to increase the amount of feedback light coupled in. However, many of the best diode lasers available on the market are already provided with a high-reflectance coating on the back facet and a reduced-reflectance coating on the output facet. This allows one to operate the diode lasers in a pseudo-external-cavity configuration, as schematically shown in fig. 2. The first-order diffracted beam from a grating, mounted in the Littrow configuration, is fed back into the laser diode. An intracavity beam splitter ($(20 \div 30)\%$ reflectance) provides an additional output to the zeroth-order one and somewhat controls the feedback amount. Indeed a higher feedback reduces the out-coupled power, but also increases the wavelength tuning range. Further details concerning collimating lens and in general the actual construction can be found elsewhere [12], as well as the description of how the diode temperature and injection current stability stringent requirements can be fulfilled.

The simple rotation of the grating allows the coarse tuning of the extended-cavity lasers over a range of $\sim 10\,\text{nm}$. In particular, because of the finite reflectivity of the chip facets, the longitudinal modes of the solitary laser diode do not change, while the feedback from the grating allows one to select one of these modes and to drastically suppress other side modes. Fine tuning to the frequency of interest can then be achieved by slightly changing the temperature and the injection current of the diode. Continuous frequency scans of the order

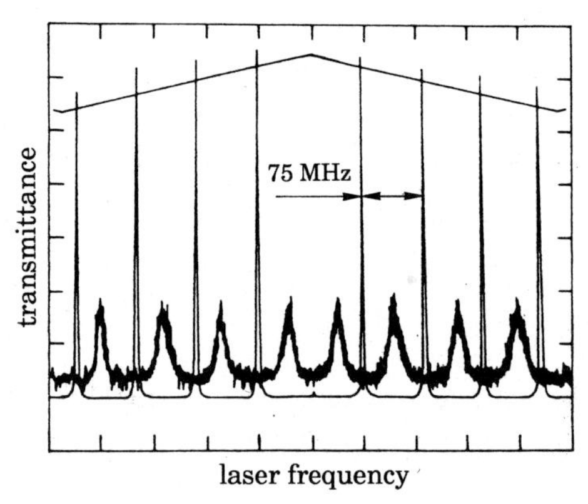

Fig. 3. – Superimposed transmission spectra from a 75 MHz free-spectral-range Fabry-Perot interferometer as the frequency of the diode laser is scanned up and down without optical feedback (broad peaks) and in the presence of optical feedback from a grating (narrow peaks). In the first case, the laser frequency was scanned up and down by sending a ramp (shown in the top of the figure) to the injection current supply. In the second case, the ramp was also sent to the PZT on the grating in order to simultaneously change the length of the external cavity.

of ~ 10 GHz are accomplished by synchronously sweeping the length of the cavity, by means of a piezoelectric transducer on the grating, and the injection current. The effect of the increased Q factor in the cavity and consequent line narrowing can be evidenced by recording the transmission peaks of a Fabry-Perot interferometer as the laser frequency is scanned. A typical situation is illustrated in fig. 3. Broad peaks are recorded by blocking the return beam from the grating (*i.e.* with the laser as is commercially available), while much narrower peaks (the figure is essentially a recording of the Fabry-Perot finesse) are obtained in the presence of the optical feedback. Figure 3 has been taken using an InGaP laser at 690 nm [13], but the results are rather general as we have experi-

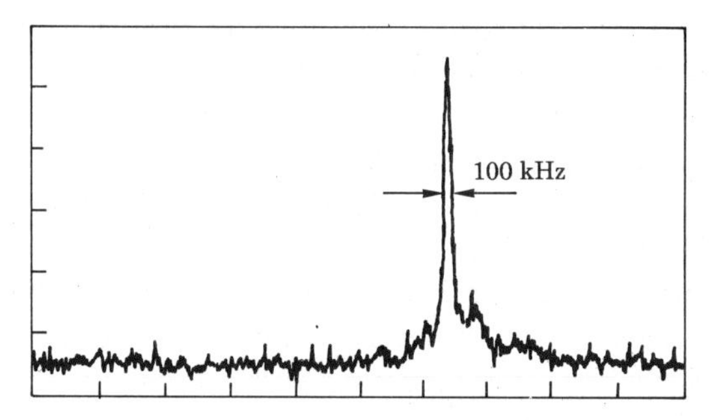

Fig. 4. – Beat note from the heterodyne between two free-running AlGaAs multiple-quantum-well lasers mounted in extended-cavity configuration. Wavelength is grating-tuned at $\lambda = 850$ nm and the two emissions are about 2.2 GHz apart. From the recording, a value of about 50 kHz can be estimated for the laser linewidth in the millisecond time scale.

enced with AlGaAs lasers operating at 780, 820 and 850 nm, both using double heterostructures and multiple quantum wells. A more accurate analysis of the spectral properties of extended-cavity semiconductor diode lasers can be performed by observing the beat note from two of them superimposed in a nonlinear detector. This is illustrated in fig. 4 for the mixing in a fast photodiode of two extended-cavity multiple-quantum-well lasers at 850 nm. In the millisecond time scale the laser linewidth is of the order of 50 kHz, corresponding to $\Delta\nu/\nu \simeq 10^{-11}$, while acoustic noise is responsible for the increase to about 400 kHz over longer time scale. Heterodyne measurements require particular care in the optical isolation between the lasers, to avoid injection locking which can be caused by even a slight feedback.

4. – High-resolution spectroscopy of atoms.

The impressive characteristics of the conductor diode laser configurations so far described have already allowed a large number of interesting applications in atomic spectroscopy. Here we shall concentrate on very few of them particularly interesting to show the potentialities of these compact laser sources.

Let us start with the investigation of forbidden transitions. This is in general an important task for atomic spectroscopy mostly related to the possibility of developing laser-based frequency markers in the visible. We chose to detect the $5s\ ^1S^0$-$5p\ ^3P_1$ intercombination transition of strontium at 689.488 nm. This wavelength is accessible to the newly available InGaP lasers and the sub-Doppler recording can provide a significant test, as the atomic line has a radiative width of only 8 kHz [13]. Strontium atoms were produced in a hollow-cathode discharge designed to allow access to the counterpropagating beams necessary for the Doppler-free recording [14]. Indeed, in spite of the weakness of the transition, a 1% absorption signal could be directly detected with a small saturation dip ($\sim 5\%$) on top, thanks to the low-amplitude noise of the diode laser. The Doppler-free contribution to the signal could be consequently enhanced using a polarization spectroscopy scheme [15]. A typical recording is reported in fig. 5. Except for the radiation source, which in that case was a more conventional dye laser, a description of the experimental details can be found in [16]. The recorded linewidth was mostly determined by collisional effects due to the discharge environment and the laser jitter contribution could be estimated to be less than 1 MHz. More recently, the same intercombination transition has been observed and investigated in a Sr atomic beam [17] for a more careful analysis of the laser linewidth.

Let us now further illustrate the potentialities of grating-tuned semiconductor diode lasers showing and discussing some applications to the spectrum of atomic oxygen. Elusive to continuous-wave excitation from the ground state, this atom can conveniently be studied by means of laser spectroscopy of transi-

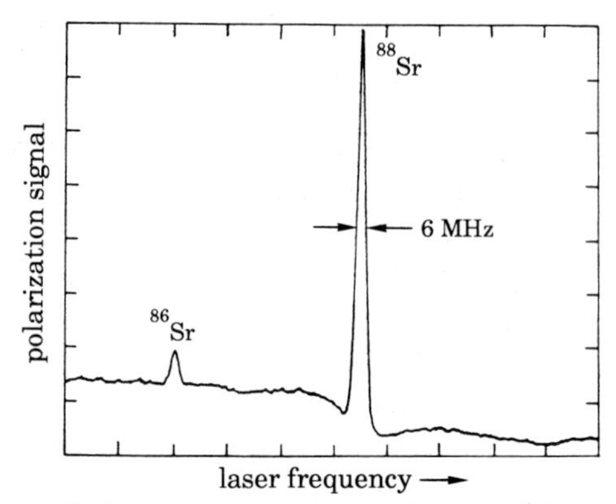

Fig. 5. – Doppler-free polarization spectroscopy of the $5s\,^1S_0$-$5p\,^3P_1$ intercombination transition of strontium at 689.488 nm as recorded with an InGaP semiconductor laser in extended-cavity configuration.

tions starting from excited levels produced in the same Ar or He discharge used for the dissociation of O_2 added in traces [18]. As evidenced in the scheme of fig. 6, transitions starting from the lowest triplet and quintet states are well suited for the investigation by means of AlGaAs lasers. The transition at 777 nm has immediately been recognized as interesting [11] because its lower level is metastable and possibly involved in a novel scheme for radiative cooling of

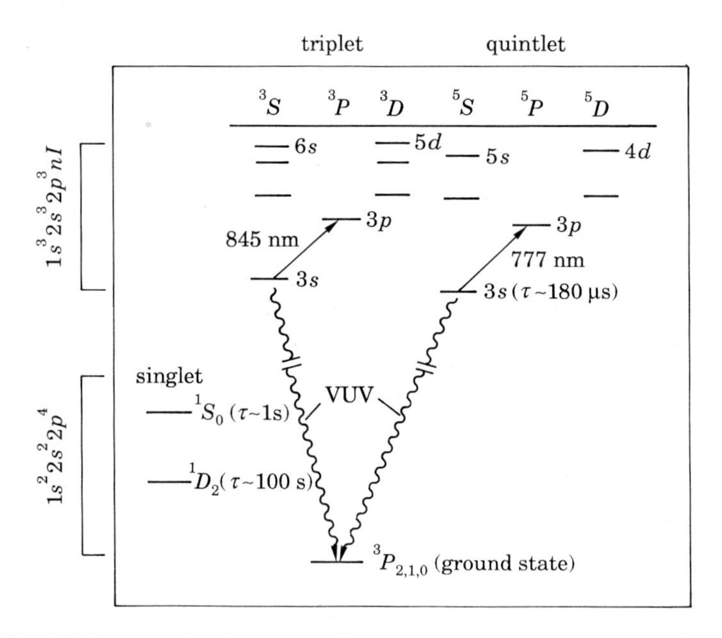

Fig. 6. – Simplified energy level scheme of atomic oxygen. Transitions studied with high resolution by means of AlGaAs diode lasers are evidenced.

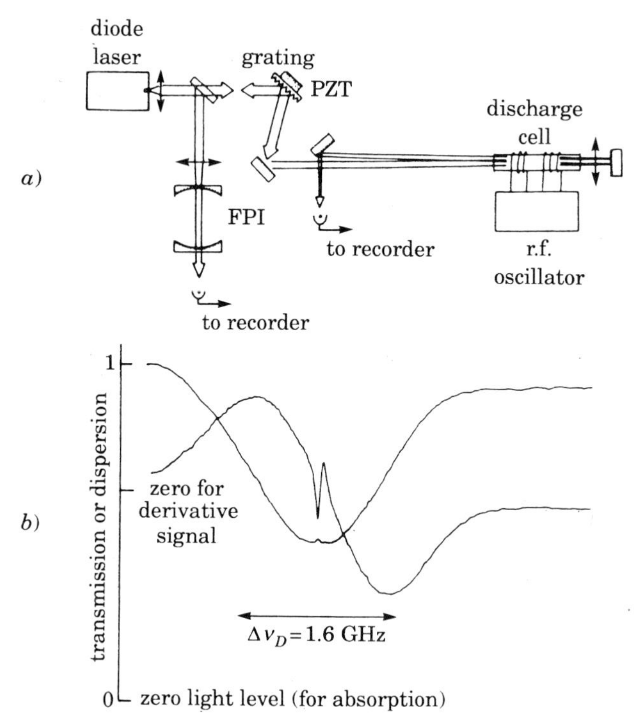

Fig. 7. – Oxygen absorption on the $^5S_2$-$^5P_3$ transition at 777 nm as detected with a grating-tuned AlGaAs semiconductor diode laser. Two counterpropagating beams are present in the cell and a saturation dip can be recorded at the line centre. This is evident in the derivative signal (*b*)) where phase-sensitive detection is applied.

oxygen. Also, the metastable level can be efficiently populated in a weak radiofrequency discharge and fig. 7*a*) reports the absorption signal as recorded on the transmitted light after passing twice through the cell: the laser beam was retroreflected by a mirror to allow saturation spectroscopy. It is worth noting the low noise level which reflects the extreme reduction in the laser amplitude fluctuations. By adding a modulation in the scanning piezo voltage, frequency modulation of the laser output can be easily achieved. This, in combination with phase-sensitive detection, allows the recording of a derivative signal and the enhancement of the saturation dip at the line centre, as displayed in fig. 7*b*). Let us now use this same atomic transition to illustrate more in general the versatility of the semiconductor laser as combined with different nonlinear spectroscopy techniques. The scheme of a general apparatus [19] is shown in fig. 8. A key role is played by the electro-optic modulator (EOM) which, in combination with the proper choice of the A and B optical devices, can act as a simple amplitude modulator, an alternate σ^+-σ^- polarizer, a π polarizer and so on. For instance, when A is not used and the EOM is placed between two parallel linear

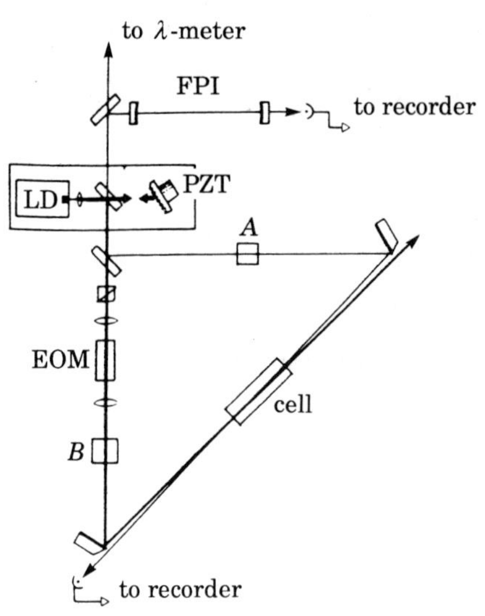

Fig. 8. – A general experimental apparatus used to combine a semiconductor diode laser with various Doppler-free spectroscopic schemes. A and B indicate different optical devices depending on the detection technique. A high-finesse Fabry-Perot interferometer (FPI) is used for the frequency scan calibration, while laser wavelength is determined by a λ-meter.

polarizers, the system is equivalent to a mechanical chopper and the scheme is essentially a Hänsch-type one. Consequently, a typical recording is shown in fig. 9*a*). As expected, only the sub-Doppler signal is recorded, while a broader pedestal is caused by the velocity-changing collisional effects [20]. The line-shape can be fitted to the sum of a Lorentzian and a Gaussian and the two contributions are separately shown in the lower part of fig. 9*b*). As is well known, the velocity-changing collision contribution to the signal can be reduced by changing the physical observable under investigation in the technique. In this respect an important example is provided by polarization spectroscopy, where the vapour birefringence, instead of the absorption, is recorded. For instance, the recording previously reported in fig. 5 for strontium showed no evidence for velocity-changing collisional pedestal, which on the contrary would have been rather important in a simple saturation spectroscopy scheme, the observed transition being originated from an atomic ground state. Coming back to the scheme of fig. 8, the pump beam can be manipulated in order to create an atomic orientation $\sum_{-J}^{+J} n_{m_j} m_j \neq 0$ or an atomic alignment $\sum_{-J}^{+J} n_{m_j} m_j^2 - (1/3)\,J(J+1) \neq 0$ with n_j and m_j, respectively, the relative population and the eigenvalue of the Zeeman sublevel J_z. Differently from the level population, observable involved in simple saturation spectroscopy, both the orientation and the alignment are significantly affected by collisions and the atoms emerging from a collision do no

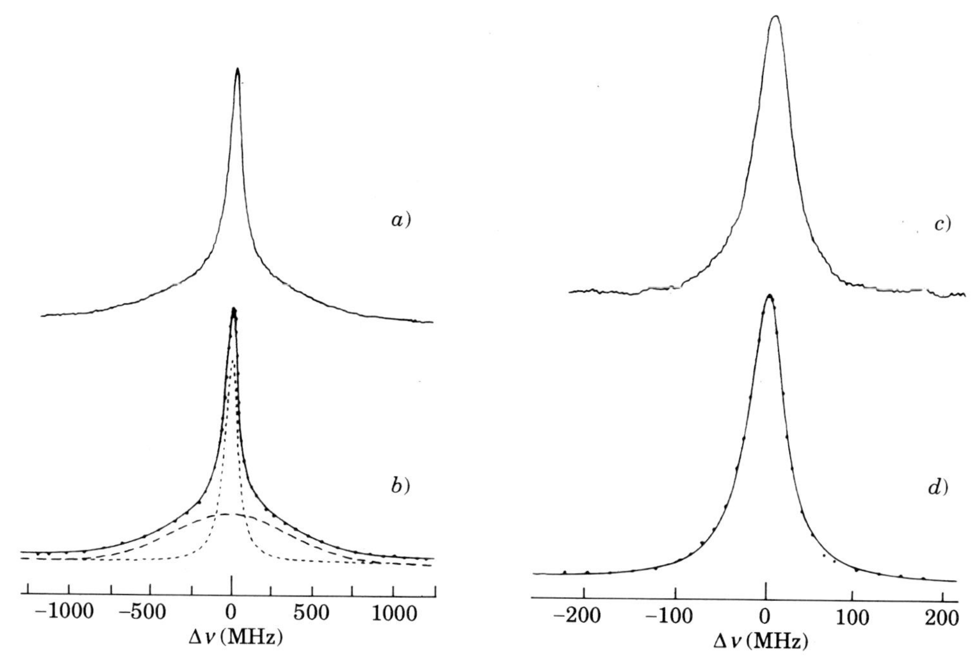

Fig. 9. – Examples of different sub-Doppler lineshapes obtained using the apparatus illustrated in fig. 8. In *a*), the saturation spectroscopy signal (Hänsch-type configuration) is the superposition of a Lorentzian and a Gaussian pedestal caused by the velocity-changing collisions, as evidenced by the fit in *b*). In *c*) the «alignment» spectroscopy configuration allows the elimination of the Gaussian contribution to the lineshape which in *d*) is fitted to a pure Lorentzian.

longer contribute to the recorded signal in the probe beam. This is clearly evident in fig. 9*c*), where the same transition of fig. 9*a*) and 7 is recorded using a configuration of «alignment» spectroscopy. The lineshape can be fitted using only a Lorentzian as shown in fig. 9*d*). The width of the Lorentzian is affected by some collisional broadening, since the experiment is performed in a cell. However, the sensitivity of the technique allowed the operation with low pressures (~ 100 mTorr of Ar buffer gas) and the extrapolation to zero pressure led to a determination of the radiative lifetime of the upper level of the transition ($^5P_3$, $\tau = 7$ ns). This lifetime is an important parameter for the real feasibility of the oxygen radiative cooling scheme proposed in [11] and analysed with a Monte Carlo simulation in [21].

The simplicity of the semiconductor lasers makes easy, and accessible to most laboratories, the operation of many of them in the same experiment. This is, for instance, useful when the experiments consist in the precise measurement of isotope shifts or atomic structures. Direct frequency measurements can indeed be obtained by heterodyning two independent lasers frequency-locked

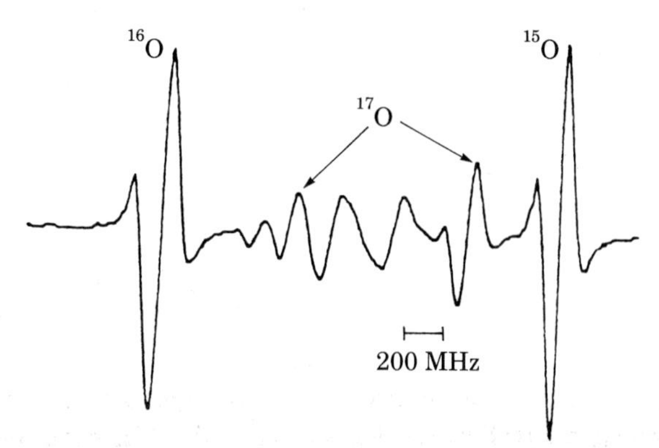

Fig. 10. – Sub-Doppler spectrum of the three stable isotopes of atomic oxygen as observed in an enriched sample ($^5S_2$-$^5P_3$ transition at 777 nm). The third-derivative signal is shown.

on two different atomic signals. For atomic oxygen this was the case for the ^{16,17,18}O isotope shift and for the hyperfine structure of ^{17}O. In fig. 10 the 777 nm transition using an enriched atomic sample [22] is recorded. A $3f$ detection scheme was used to eliminate the background slope, which instead is present in the first-derivative signal. The sample absorption was $\sim 5\%$ and the Lamb dip signal was $\sim 5\%$ of the absorption. The recording reported in the figure was obtained with a lock-in time constant of 1 ms. A servo loop (bandwidth ~ 10 Hz) was then used for frequency-locking each laser on the selected isotope signal. The scheme typically consisted of a lock-in, an integrator and a high-voltage amplifier; the output of the integrator was fed back to the high-voltage amplifier, which controlled the grating position by means of the piezoelectric transducer. The intrinsic frequency stability of such a laser system is determined by the increased Q of the optical cavity, which reduces the fast frequency fluctuations; only a relatively slow electronics is then required to correct long-term drifts. The beat note between two lasers can be observed using a spectrum analyser, as already shown in fig. 4. However, in the actual measurements with the lasers locked on the atomic third-derivative signals, a broadening of the beat note arises from the frequency modulation of the lasers and can be minimized by in-phase modulation. The accuracy of the heterodyne measurement is determined by the signal-to-noise ratio and width of the locking signal. From signals like those shown in fig. 10, isotope shifts and hyperfine structures could be measured with an accuracy of the order of 100 kHz. It is evident that heterodyne of semiconductor diode lasers can be a powerful and straightforward method for precise laser spectroscopy, in particular in combination with mixing schemes based on fast photodiodes or, even better, on GaAs Shottky diodes [9, 23, 24].

5. – High-sensitivity spectroscopy of forbidden molecular transitions.

As anticipated, the extremely good amplitude stability of semiconductor diode lasers opens the possibility of performing high-sensitivity absorption measurements directly on the light transmitted by the sample.

A first example is given by the detection and investigation of molecular oxygen at 760 nm. Transitions of O_2 in the visible-near infrared belong to the $\chi^3\Sigma_g^-$-$b^1\Sigma_g^+$ band first observed by BABCOCK and HERZBERG [27] in the atmospheric absorption from Mount Wilson. These are magnetic-dipole transitions with an absorption coefficient as low as $\alpha \sim 10^{-6}$ cm^{-1}. In the laboratory, they could be detected only by means of intracavity techniques involving dye lasers [28, 29]. The powerfulness of stable diode lasers for the detection of these forbidden transitions was early suggested to us by the recording of an absorption from the room atmosphere after a path length of only 20 cm [30]. This opened the possibility of performing more systematic investigations using an extended-cavity semiconductor diode laser spectrometer. The scheme of the spectrometer, which, as we shall discuss, is also used more in general for overtone molecular spectroscopy, is shown in fig. 11. The laser frequency is controlled by changing PZT voltage and injection current as done for atomic spectroscopy. A dither (fast modulation) is added when phase-sensitive detection is used. The radiation transmitted by the sample is detected by means of a photodiode, while a second photodiode is used to obtain a reference for the laser amplitude and to eliminate the baseline (due to the amplitude modulation) by means of differential detection. The absorption cell is 150 cm long. Typical measurements for oxygen are illustrated in fig. 12. An absorption recorded

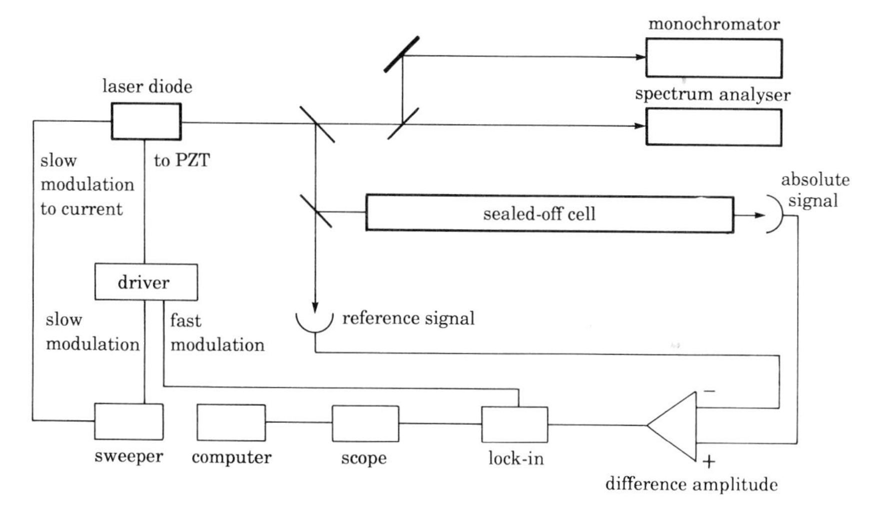

Fig. 11. – Scheme of the diode laser spectrometer used for the investigation of weak molecular absorptions.

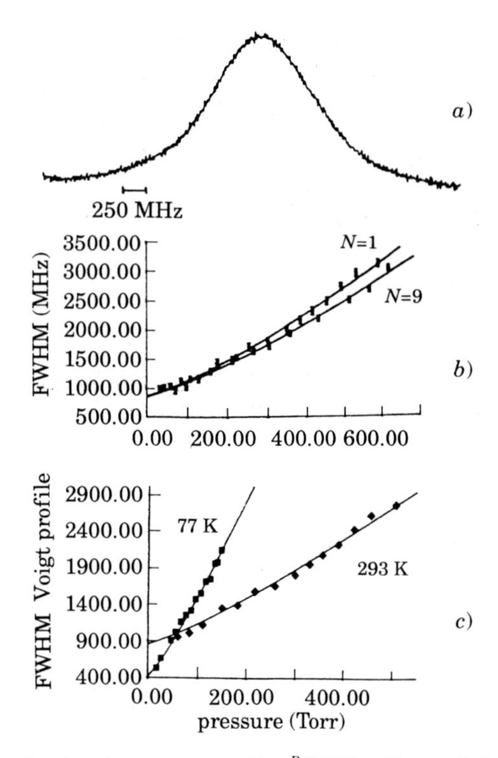

Fig. 12. – Absorption of molecular oxygen on the ${}^{P}P(K''=9)$, $v = 0\text{-}0$ component of the forbidden $\chi^3\Sigma_g^- \text{-} b^1\Sigma_g^+$ band (*a*)). In *b*) the dependence (on two different components) of the homogeneous contribution to the linewidth on the pressure is reported. In *c*) the behaviour of the collisional broadening on the same component at two different temperatures is shown.

with a gas pressure of 120 Torr is shown in fig. 12*a*). The lineshape is fitted to the convolution of a Gaussian and a Lorentzian. The latter contribution is due to collisional broadening and can be investigated at different pressures [31]. The results of different measurements in the $(0 \div 1000)$ Torr range (on two different components) are plotted in fig. 12*b*), while in fig. 12*c*) the behaviour of the collisional broadening at two different temperatures is shown. This demonstrates how important parameters for atmospheric physics can be extracted from a relatively simple spectrometer. Wider applications can certainly be found in overtone molecular spectroscopy. Fundamental vibrations of molecules produce spectra in the medium infrared. The anharmonic contribution to the potential makes possible the excitation of overtone frequencies which can occur in the visible-near infrared, where more convenient tunable laser sources and room temperature sensitive detectors are available. Specially for polyatomic molecules, where also combination vibrations are present, overtone spectra are rich and interesting to be investigated also for a better modelling of the molecular structure. However, it is also well known that the intensity of successive

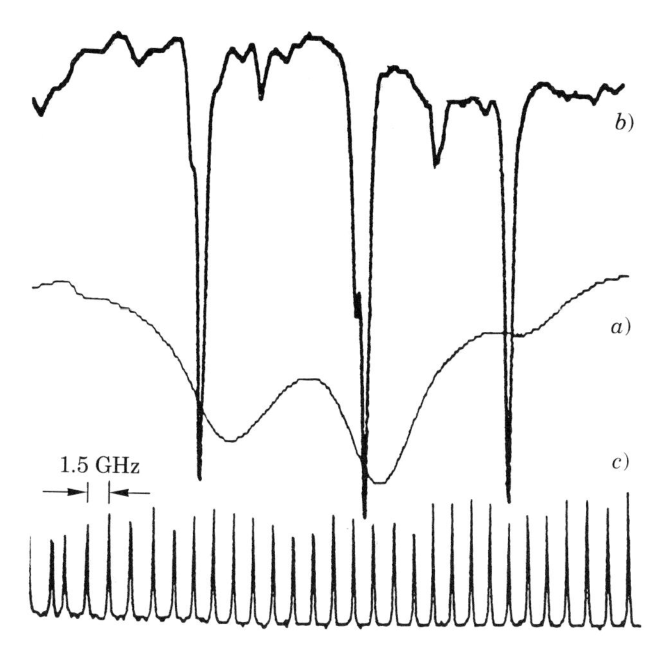

Fig. 13. – Recording of the absorption of CH_4, methane, around 886 nm by means of AlGaAs laser. Temperature and pressure are, respectively, 293 K and 100 Torr in *a*) and 77 K and 1 Torr in *b*). The total frequency scan is 42 GHz, as determined by the transmission of a reference Fabry-Perot (*c*)).

overtones and combination transitions decreases very rapidly with the increase of the number of vibrational modes and/or quanta involved. As a consequence, overtone molecular absorption in the visible is very weak ($\alpha \sim 10^{-6}$ cm^{-1} can be again a typical absorption coefficient) and constitutes an interesting challenge for the spectrometer based on our semiconductor diode lasers. A comprehensive list of the most interesting molecules and of the wavelengths at which they are accessible can be found elsewhere [32]. Let us illustrate the powerfulness of the technique using the detection of CH_4, methane, at 886 nm, where a number of combination bands are superimposed ($3\nu_1 + \nu_3$, $2\nu_1 + 2\nu_3$, $\nu_1 + 3\nu_3$, $4\nu_3$). This produces an almost continuous absorption [33] as, for instance, shown in fig. 13*a*). The spectrum is part of a systematic investigation performed at different temperatures to study the behaviour of the various components and possibly facilitate the line assignment. Indeed the room temperature recording, taken with CH_4, pressure of 100 Torr, is simplified when the temperature is lowered to 77 K, as shown in fig. 13*b*). In this case also the pressure could be reduced to 1 Torr and the linewidths are narrower because of the reduced Doppler and collisional broadenings. It is obvious that the signal-to-noise ratios observed for the detection of these overtone transitions, as well as of the oxygen forbidden transitions, can constitute a good base for the development of high-sensitivity absorption measurements. Indeed, the transitions in methane provide a system to test and compare all the existing techniques mainly based on the modulation of the laser radiation [34]. Semiconductor lasers offer the advantage of achieving an easy and fast frequency modulation of the laser output itself by acting on the injection current. For this reason the diode lasers constitute optimal candidates for detection techniques as wavelength and frequency modulation.

In the wavelength modulation technique an a.c. component is added to the injection current of the diode laser; the frequency of modulation must be smaller than the linewidth of interest, so that the absorption is probed simultaneously by a number of sidebands. If the modulation frequency is increased to exceed the linewidth, only one sideband will be absorbed at a time, giving rise to a characteristic heterodyne beat signal at the modulation frequency. If the modulation index is chosen so that only one higher sideband and one lower sideband have appreciable amplitude, we are in the case of frequency modulation spectroscopy (one-tone FM).

A variation on the theme of the FM spectroscopy with respect to the one-tone technique, previously described, is represented by the two-tone spectroscopy: in this case the laser emission is simultaneously modulated at two distinct but closely spaced angular frequencies. Also here a heterodyne signal is obtained, but it occurs at the difference frequency between the two sidebands eliminating the need of high-speed detectors and making large-linewidth detections possible.

The trick of the modulation techniques is to observe the absorbed signal with an associated noise typical of the modulation-demodulation frequency and the electronic bandwidth chosen. Because of the $1/f$ behaviour of the amplitude noise in the power spectrum of semiconductor diode laser emission, the possibility of obtaining more sensitive detections is bound to the choice of higher demodulation frequencies.

In fig. 14*a*) the direct absorbed signal of one CH_4 component without any demodulation at 100 Torr is shown (corresponding to a relative absorption of 3.6%): the associated S/N is more than 20 and the bandwidth equal to 100 Hz. In fig. 14*b*) the demodulated signal (third derivative) is shown at 1 kHz, 100 Hz of bandwidth and a pressure of 100 mTorr, corresponding to a S/N of about 10.

Finally, in fig. 14*c*) the demodulated signal with the two-tone technique at frequencies around 500 MHz and 1 Hz of bandwidth is shown (pressure of 500 mTorr) corresponding to a S/N ratio of a few thousands.

A minimum detectable absorption has been demonstrated [34] (corresponding to a S/N ratio equal to 1) for the low-wavelength modulation (1 kHz), high-wavelength modulation (100 MHz) and two-tone frequency modulation ((390 ± 5) MHz) techniques, respectively, of $4.5(1) \cdot 10^{-7}$, $9.7(3) \cdot 10^{-8}$ and $6.4(2) \cdot 10^{-8}$.

Considering the absorption limit derived by the calculated shot noise, the

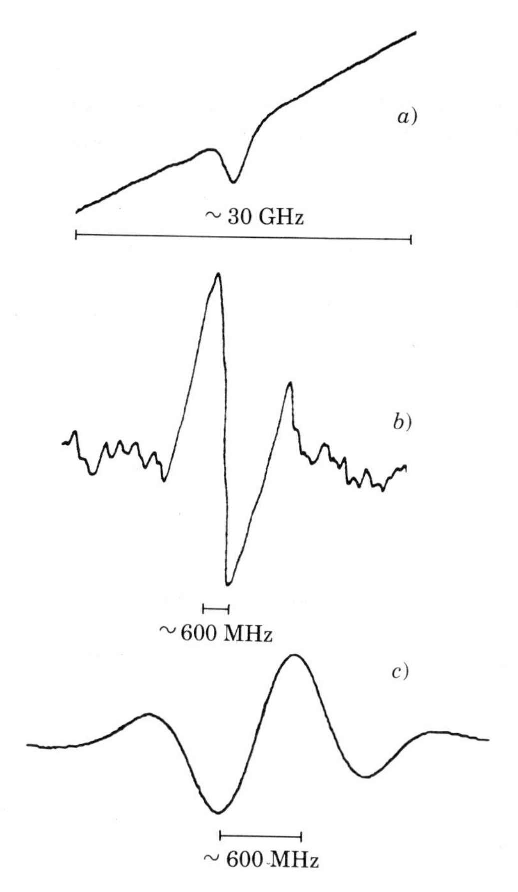

Fig. 14. – *a*) represents the direct absorbed signal of one CH_4 component without any demodulation at 100 Torr (corresponding to a relative absorption of 3.6%): the associated S/N is more than 20 and the bandwidth equal to 100 Hz. *b*) represents the demodulated signal (third derivative) at 1 kHz, 100 Hz of bandwidth and a pressure of 100 mTorr, corresponding to a S/N of about 10. *c*) represents the demodulated signal with the two-tone technique at frequencies around 500 MHz and 1 Hz of bandwidth (pressure of 500 mTorr) corresponding to a S/N ratio of a few thousands.

minimum detectable absorptions measured in the high-frequency detection techniques are about 6 dB over this value.

6. – Conclusions.

This overview should have convinced that semiconductor diode lasers are important coherent sources for atomic and molecular spectroscopy. They can potentially cover the visible and the near infrared, with gaps which can be reduced with optical feedback techniques and with the development of new materials. Coverage can be extended to blue and near ultraviolet by frequency doubling. Optical feedback techniques can drastically reduce the linewidths to the level which allows also very narrow transitions to be investigated. Available powers start to be comparable with that of dye lasers and, however, the extremely reduced amplitude fluctuations are making high-sensitivity measurements possible even with low power. Further important possibilities are opened by the possibility of fast frequency control of the laser emission. Complex experimental schemes can be simplified by the use of semiconductor diode lasers and experiments which could not even be conceived before can now be performed.

Progresses in the fascinating world of laser spectroscopy have always depended on the development of new sources of radiation and we are sure that the availability of these new laser devices will allow new important chapters to be written.

* * *

Many motivated researchers of Firenze (LENS and Department of Physics), Napoli (Department of Physics) and Pisa (Scuola Normale Superiore) have contributed to this work (M. BARSANTI, M. DE ANGELIS, P. DE NATALE, C. FORT, L. GIANFRANI, G. GIUSFREDI, F. MARIN, F. PAVONE, M. PREVEDELLI, A. SASSO) as well as many guest workers (G. DI LONARDO, K. ERNST, L. HOLLBERG, L. JULIEN, K. LEHMANN, M. ULBRICHT, L. ZINK). We owe special gratitude to F. BIRABEN, W. CHEBOTAEV, J. L. HALL, T. W. HÄNSCH, V. VELICHANSKY, C. ZIMMERMANN and S. SVANBERG, for stimulating discussions. Special thanks to F. PAVONE for a critical reading of this manuscript.

REFERENCES

[1] A. L. SCHAWLOW: this volume, p. 1.
[2] J. C. CAMPARO: *Contemp. Phys.*, **26**, 443 (1985).
[3] T. W. HÄNSCH: this volume, p. 297.
[4] Private communication (T. W. HÄNSCH and B. SCHEUMANN).
[5] C. HANKE: *High power diode lasers*, in *Solid State Lasers: Recent Developments and Applications*, edited by M. INGUSCIO and R. WALLENSTEIN (Plenum Press, New York, N.Y., 1993), p. 139.
[6] M. OHTSU and T. TAKO: *Coherence in semiconductor lasers*, in *Progress in Optics XXV*, edited by E. WOLF (Elsevier, Amsterdam, 1988), p. 191.
[7] A. L. SCHAWLOW and C. H. TOWNES: *Phys. Rev.*, **112**, 1940 (1958).
[8] D. WELFORD and A. MOORADIAN: *Appl. Phys. Lett.*, **40**, 865 (1982).
[9] See, for instance, R. FOX, G. TURK, N. MACKIE, T. ZIBROVA, S. WALTMAN, M. P. SASSI, J. MARQUARDT, A. S. ZIBROV, C. WEIMER and L. HOLLBERG: *Diode lasers and metrology*, in *Solid State Lasers: Recent Developments and Applications*, edited

by M. INGUSCIO and R. WALLENSTEIN (Plenum Press, New York, N.Y., 1993), p. 279, and references therein.
[10] C. E. WIEMAN and L. HOLLBERG: *Rev. Sci. Instrum.*, **62**, 1 (1991).
[11] G. M. TINO, L. HOLLBERG, A. SASSO, M. INGUSCIO and M. BARSANTI: *Phys. Rev. Lett.*, **64**, 2999 (1990).
[12] G. M. TINO, M. DE ANGELIS, F. MARIN and M. INGUSCIO: *Semiconductor diode lasers in atomic spectroscopy*, in *Solid State Lasers: Recent Developments and Applications*, edited by M. INGUSCIO and R. WALLENSTEIN (Plenum Press, New York, N.Y., 1993), p. 287.
[13] G. M. TINO, M. DE ANGELIS, L. GIANFRANI, M. BARSANTI and M. INGUSCIO: *Appl. Phys. B*, **55**, 397 (1992).
[14] M. INGUSCIO: *J. Phys. (Paris)*, C7, 217 (1983).
[15] C. WIEMAN and T. W. HÄNSCH: *Phys. Rev. Lett.*, **36**, 1970 (1976).
[16] M. BARSANTI, L. GIANFRANI, F. PAVONE, A. SASSO, C. SILVESTRINI and G. M. TINO: *Z. Phys. D*, **23**, 145 (1992).
[17] F. S. PAVONE, G. GIUSFREDI, A. CAPANNI, M. INGUSCIO, G. TINO and M. DE ANGELIS: *Narrow linewidth visible diode laser at 690 nm: spectroscopy of the* SrI *intecombination line*, in *Frequency Stabilized Lasers and their Applications*, edited by Y. C. CHUNG, *SPIE*, Vol. **1837**, 366 (1992).
[18] M. INGUSCIO, P. MINUTOLO, A. SASSO and G. M. TINO: *Phys. Rev. A*, **37**, 4056 (1988).
[19] M. DE ANGELIS, M. INGUSCIO, L. JULIEN, F. MARIN, A. SASSO and G. M. TINO: *Phys. Rev. A*, **44**, 5811 (1991).
[20] P. W. SMITH and T. HÄNSCH: *Phys. Rev. Lett.*, **26**, 740 (1971).
[21] G. M. TINO: *Comments At. Mol. Phys.*, **29**, 5 (1993).
[22] F. MARIN, P. DE NATALE, M. INGUSCIO, M. PREVEDELLI, L. R. ZINK and G. M. TINO: *Opt. Lett.*, **17**, 148 (1992).
[23] H. U. DANIEL, B. MAURER and M. STEINER: *Appl. Phys. B*, **30**, 189 (1983).
[24] Marco PREVEDELLI at the Max Planck Institut for Quantum Optics (Garching) has recently observed beat notes up to several hundred gigahertz from diode lasers mixed in Shottky-barrier diodes in the frame of perfecting a new method for bisecting optical frequency intervals to develop phase-coherent measurements of atomic hydrogen [25, 26].
[25] H. R. TELLE, D. MESCHEDE and T. W. HÄNSCH: *Opt. Lett.*, **15**, 532 (1990).
[26] R. WYNANDS, T. MUKAI and T. W. HÄNSCH: *Opt. Lett.*, **17**, 1749 (1992).
[27] H. D. BABCOCK and L. HERZBERG: *Astrophys. J.*, **108**, 167 (1948).
[28] W. T. HILL III, R. A. ABREU and A. L. A. SCHAWLOW: *Opt. Comm.*, **32**, 96 (1980).
[29] S. J. HARRIS: *Appl. Opt.*, **23**, 1311 (1984).
[30] M. DE ANGELIS and G. M. TINO: private communication.
[31] M. DE ANGELIS, F. MARIN, F. S. PAVONE, G. M. TINO and M. INGUSCIO: *Pure absorption spectroscopy of molecular oxygen using a* cw AlGaAs *laser*, in *Monitoring of Gaseous Pollutants by Tunable Diode Lasers*, edited by R. GRISAR, H. BÖTTNER, M. TACKE and G. RESTELLI (Kluwer Academic Publisher, Amsterdam, 1991), p. 257.
[32] K. ERNST and F. PAVONE: *Overtone molecular spectroscopy with diode lasers*, in *Solid State Lasers: Recent Developments and Applications*, edited by M. INGUSCIO and R. WALLENSTEIN (Plenum Press, New York, N.Y., 1993), p. 303.
[33] K. LEHMAN, F. PAVONE and M. INGUSCIO: unpublished.
[34] F. S. PAVONE and M. INGUSCIO: *Appl. Phys. B*, **56**, 118 (1993).

Application of Laser Spectroscopy to Fundamental Molecular Species: H_3^+ and Solid H_2.

T. OKA

Department of Chemistry, Department of Astronomy and Astrophysics and the Enrico Fermi Institute
The University of Chicago - Chicago, IL 60637

1. – Introduction.

The high sensitivity and high resolution afforded by laser spectroscopy enable us to obtain spectra that cannot be obtained by classical spectroscopic methods. We discuss two such cases here, *i.e.* the H_3^+ molecular ion and solid H_2. The high sensitivity of laser spectroscopy is exploited for the former and the high resolution for the latter. Both species are very fundamental systems. The spectrum of H_3^+ is very likely to be useful for astronomy and the spectrum of solid H_2 allows us to study condensed-phase spectroscopy from first principles.

Before going into more detail, I would like to summarize the basic principles that are used to look at molecules.

1·1. *Natural constants and orders of magnitude.* – Any quantities which appear in atomic and molecular physics can be expressed with five natural constants and nuclear parameters. The constants are: the electric charge e, Planck's constant h, the velocity of light c, the mass of the electron m and the mass of the proton M. Nuclear parameters such as their masses M_i, magnetic g factors, electric-quadrupole moments, etc. could also be expressed in terms of the fundamental constants if nuclear physics were more quantitatively understood, but at this stage we simply use them as parameters. Out of these constants, we make two dimensionless quantities, that is, the fine-structure constant

$$\alpha = \frac{e^2}{\hbar c} \sim \frac{1}{137}$$

and the Born-Oppenheimer constant

$$\kappa = \sqrt[4]{\frac{m}{M_i}} \sim \frac{1}{10}\,.$$

The smallness of these two constants allows us to treat the atomic and molecular physics by using perturbation theory. Various interactions and their orders of magnitude are listed in table I. Readers are referred to ref. [1] for their derivation. There is a simple rule of thumb to find the orders of magnitude of a specific interaction term. Express the interaction in terms of coordinate (and/or momentum) and angular momentum [2]. Assign the following orders of magnitude to each quantity and multiply:

electronic coordinate and momentum	1,
vibrational coordinate and momentum	κ,
electron spin and orbital momentum	α,
rotational angular momentum of molecules	κ^2,
nuclear-spin angular momentum	$\kappa^4 \alpha$.

Thus, for example, normalized to the electronic energy of 1, the vibrational energy is $\kappa^2 \sim 10^{-2}$, the rotational energy $\kappa^2 \sim 10^{-4}$, the spin-orbit interaction $\alpha^2 \sim 10^{-4}$, etc. This also applies to all smaller interactions. For example, the indirect nuclear spin-spin interaction proposed by Ramsey and Purcell [3] has the magnitude of $\kappa^8 \alpha^2 \sim 10^{-12}$.

Table I. – *Various interactions and their orders of magnitude* ([a]).

Orders of magnitude	Atomic interaction		Molecular interactions		Radiation
10^4	electron rest mass	α^{-2}	—		γ-ray
1	electron energy	1	—		optical
10^{-2}	—		vibration	κ^2	infrared
10^{-4}	fine structure	α^2	rotation	κ^4	microwave
10^{-6}	radiative correction	α^3	ro-vibration ([b])	κ^6	—
10^{-8}	hyperfine structure	α^4, $\alpha^2\kappa^4$	centrifugal distortion	κ^8	radiowave

(*a*) This table lists only the largest term.
(*b*) Second-order vibration-rotation interaction.

1·2. *Symmetry.* – While the consideration of orders of magnitude discussed above gives quantitative but approximate information on molecules, the symmetry argument gives qualitative but rigorous rules. In addition to the symmetry of space and time which leads in classical mechanics to conservation of momentum, angular momentum and energy [4], the quantum-mechanical symmetry operations of the time reversal T and the parity operations P play fundamental roles in microscopic physics. Thus, for example, the requirement of the time-reversal symmetry, together with Hermiticity, allows us to drop any Hamiltonian term which is odd in time operators such as terms with odd powers of angular momentum. In the electronic ground state of H_2, the parity of the level is given by $(-1)^J$, where J is the rotational angular momentum. For H_3^+ the parity of the level is given by $(-1)^k$, where k is the quantum number for the projection of the rotational angular momentum to the symmetric molecular axis [5]. Thus we see that infrared transitions due to electric-dipole moment has the rigorous selection rule of

$$\Delta k = \text{odd}\,, \tag{1}$$

and any intramolecular interaction mixes two states by the rule

$$\Delta k = \text{even}\,. \tag{2}$$

Note that these rules are rigorous if we assume that parity is conserved in electromagnetic interaction.

The most useful symmetry operations are those of nuclear permutation. Thus, according to Pauli's rule [6], a total wave function Ψ follows the rule

$$\begin{cases} P\Psi = (-1)^p\,\Psi & \text{for fermions}\,,\\ P\Psi = \Psi & \text{for bosons}\,, \end{cases} \tag{3}$$

where P is a permutation operator of identical particles and p is the parity of permutation. For H_3^+ in which the three protons are equivalent the operator (12) is odd and the operator (123) which is equal to (13) (12) is even.

The idea of the permutation group, which is also called the symmetry group (S_n for n particles), evolved from the theory of equations. The symmetry of a two-particle system such as H_2 mimics the solution of the quadratic equations

$$\chi^2 = 1\,, \qquad \chi = 1\,,\ -1\,, \tag{4}$$

which corresponds to the ortho and para symmetry, respectively. (We are considering symmetry with respect to permutation of protons only.) Likewise the symmetry of a three-particle system such as H_3^+ mimics the solution of the cubic equations

$$\chi^3 = 1, \qquad \chi = 1, \exp\left[\frac{2\pi i}{3}\right], \exp\left[\frac{-2\pi i}{3}\right]. \tag{5}$$

The first solution corresponds to ortho and the latter two to para, which are doubly degenerate.

A total wave function can be expressed as a product of the coordinate wave function and the spin wave function to a very good approximation. Thus, in order to satisfy the Pauli rule (3), the ortho-spin state of H_2 has to combine with an antisymmetric coordinate wave function (J odd) and the para-spin state with a symmetric one (J even). The stability of the ortho- and para-H_2 is ascribed to a symmetry property of the quantum-mechanical energy operator of the hydrogen molecule; it is not only invariant when the entire set of coordinates of both protons are exchanged, but also *nearly invariant* when only their Cartesian coordinates are exchanged, leaving spin coordinates unchanged [7]. This is due to the smallness of the magnetic interaction $\sim \alpha^2$ and of the nuclear magnetic moment $\sim \kappa^4 \alpha$. Likewise in H_3^+ the ortho- and para-spin species combine with rotational wave functions with $k = 3n$ and $k = 3n \pm 1$, respectively, and transitions between them are highly forbidden.

Permutation symmetry has also been used in atomic and nuclear physics, but it appears more explicitly in molecular physics because the operations are directly related to the rotational motions. For H_3^+ or any system composed of three equivalent fermions with spin $1/2$, we find that it cannot occupy the S-state, the state of zero angular momentum. The reason is that the coordinate wave function of such a state is symmetric (insensitive) to both (12) and (123) and cannot satisfy the rule (3) for spin $1/2$. Elementary-particle physicists are always dealing with three particles with spin $1/2$ (quarks) and experienced the same dilemma. While in molecular physics the *absence* of the $J = K = 0$ state is observed [8] for H_3^+, elementary-particle physicists do observe such a state. The extra dimension of color was introduced to overcome the dilemma [9].

Use of symmetry arguments for polyatomic molecules gives a beautiful perspective for molecular spectroscopy.

2. – H_3^+ in laboratory plasmas.

The most direct object of a spectroscopist is to find a novel spectrum, preferably a spectrum of simple species whose information enriches science at a fundamental level, but whose spectrum has not been seen in any spectral range. Protonated hydrogen, H_3^+, was just such a species when the high sensitivity of laser spectroscopy was applied to it [8].

2·1. *Hydrogenic ions.* – Molecular ions, *i.e.* electrically charged molecules, are rare species in the terrestrial environment because of their chemical and physical activity. The quiescent condition of the terrestrial environment, however, is an exception and most parts of the Universe are highly ionized. Free molecular ions are expected to exist in many astronomical objects where gaseous matter is highly ionized, but their temperature is sufficiently low so that they are not dissociated into atomic species.

Let us first go through an exercise on how molecular ions are produced. Consider the simplest stable molecule, H_2. From this species cations are produced either by subtracting an electron (to form H_2^+) or by adding a proton (to form H_3^+). The first process is endothermic and the energy needed for it is called ionization potential (I.P., 15.4 eV for H_2). The second process is exothermic and the energy generated from it is called proton affinity (E.A., 4.4 eV for H_2). Note that proton affinity of H_2 is about equal to the dissociation energy of H_2 (4.5 eV). The proton is bound to H_2 with the same energy as that with which two protons are bound in H_2. The H_3^+ molecular ion is a very stable system. We can make a stable anion H^- from H_2 by subtracting a proton. The H_2^- ion which is formed by adding an electron to H_2 is not stable because the electron affinity of H_2 is negative. We can regard H_2^+ and H^- also as a proton and an electron added to a hydrogen atom, respectively. The proton affinity of H is 2.6 eV and the electron affinity 0.75 eV. Altogether in hydrogen plasmas six hydrogenic species exist and their relative concentration depends critically on plasma conditions (fig. 1). The interesting hydrogenic cluster ions are outside the scope of this lecture.

It is well known that the opacity of the Sun, that is, the deviation of bulk solar spectrum from the black-body radiation, is due to H^- which exists abundantly in the solar atmosphere [10], and this applies in general to main-sequence stars. Because of the great abundance of hydrogen in the Universe other hydrogenic ions will also play major roles in astronomical processes. It was with this in mind that the spectroscopy of H_3^+ was initiated.

2·2. *The* H_3^+ *ion.* – Protonated hydrogen H_3^+ is the simplest stable polyatomic system. It is the most abundant ionic species in molecular-hydrogen plasmas and it is assumed to play a central role in the chemical evolution of molecular

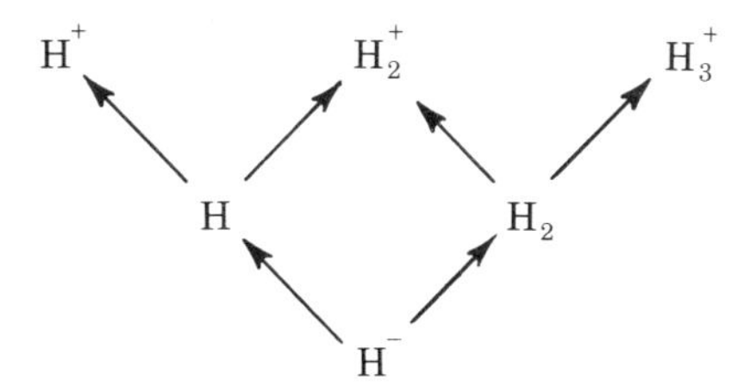

Fig. 1. – Stable hydrogenic species.

clouds which is the stage for star formation. The existence of H_3^+ has been known since early days of mass spectrometry. It was none other than J. J. THOMSON who discovered H_3^+ in his «positive rays». Readers are referred to ref. [11] and [12] for the long history of this molecular ion.

In hydrogen plasmas both in the laboratory and in space, H_3^+ is produced by the efficient ion neutral reaction

$$H_2 + H_2^+ \rightarrow H_3^+ + H\,, \tag{6}$$

which has the large Langevin rate ($\sim 10^{-9}$ cm^{3}s^{-1}) and the large exothermicity of ~ 1.7 eV (the difference of proton affinities of H_2 and H given earlier). Thus the primary ion H_2^+ produced by electron bombardment (laboratory), cosmic-ray ionization (interstellar space), etc. is quickly converted to H_3^+. Therefore, the steady-state concentration of H_2^+ is very much less than that of H_3^+. The vast amount of H_3^+ thus produced acts as the universal protonator through the very efficient proton hop reaction

$$X + H_3^+ \rightarrow HX^+ + H_2 \tag{7}$$

and initiates a chain of ion neutral reactions. Once protonated, X can react with other species of Y through

$$HX^+ + Y \rightarrow XY^+ + H \tag{8}$$

(in contrast to the radiative neutral reaction $X + Y \rightarrow XY + h\nu$, which is very inefficient). Because of this universal function of H_3^+, it is regarded as the most important species in the chemical evolution of molecular clouds [13-15].

While there had been a great many theoretical and experimental studies on H_3^+ since its discovery, the spectrum of H_3^+ was not observed until 1980 [8]. This is due to the fact that H_3^+ does not have stable electronic excited states and, hence, no optical-emission spectrum. It had to be the vibration-rotation spectrum in the infrared region [16]. It had to wait for the high sensitivity of laser spectroscopy [17]. Once the method was successfully applied to the detection of the H_3^+, it also led to an avalanche in the detection of their ionic species such as H_3O^+, NH_4^+, NH_3^+, NH_2^+, CH_2^+, CH_3^+, and many others whose spectra had not previously been known in any spectral region.

The equilibrium geometry of H_3^+ is an equilateral triangle and the three normal modes of vibration are shown in fig. 2. The ν_1 vibration is totally symmetric with respect to any permutation-inversion operators and is infrared inactive. We, therefore, use the doubly degenerate ν_2 vibration for spectroscopic observation. Instead of the ν_2 mode expressed in Cartesian displacement coordinates, we can take their linear combinations $Q_{2\pm} = [Q_{2x} \pm iQ_{2y}]/\sqrt{2}$ which have vibrational angular momentum $l = \pm 1$, pointing out of plane of the triangle (shown in fig. 2). One can consider Q_1, Q_{2+}, Q_{2-} as corresponding to the three roots 1, $\exp[2\pi i/3]$ and $\exp[-2\pi i/3]$ of eq. (5).

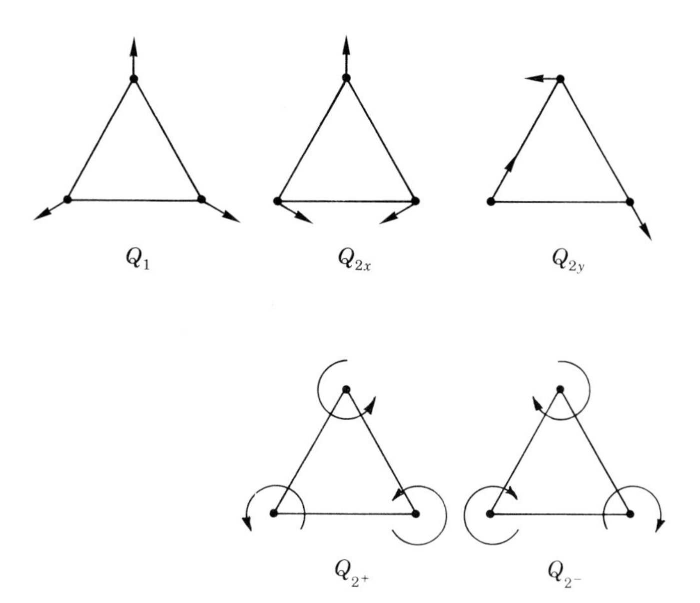

Fig. 2. – Vibrational modes of H_3^+, $\nu_1 = 3178.3\ \text{cm}^{-1}$, $\nu_2 = 2521.3\ \text{cm}^{-1}$.

The vibrational angular momentum of ν_2 is coupled very strongly to the component of the rotational angular momentum along the molecular axis through Coriolis interaction. In H_3^+, the coupling is so strong that the quantum number l of the vibrational angular momentum and the quantum number k of the projection of the rotational angular momentum are no longer good quantum numbers. However, their difference $G = |k - l|$ stays as a good quantum number. In particular, $G = 0$ (mod. 3) and $G \neq 0$ (mod. 3) represent the nearly rigorous symmetry of coordinate wave functions, which are combined with ortho ($I = 3/2$) and para ($I = 1/2$) nuclear-spin wave functions, respectively. Readers are referred to Watson's paper [18] for a detailed exposition of vibration-rotation energy levels. Note that the parity rules (1), (2) are still rigorous rules.

2·3. *Plasma spectroscopy.* – Molecular ions are produced in laboratory plasmas. The discharge tubes are typically ~ 1 m long and ~ 10 mm in diameter. The gaseous pressure is from 0.1 to 10 Torr and the current is from 0.1 to 2 A. The number density of molecular ions is limited by the number density n of electrons in plasmas which can be estimated from

$$I/S = nev\,, \tag{9}$$

where I is the current, S is the area of the cross-section of the plasma tube, e is the electric charge and v is the migration velocity [19] of electrons. For the condition of our laboratory plasmas, this formula gives an electron number density on the order of $10^{11}\ \text{cm}^{-3}$.

The plasma tube is cooled by air, water or liquid N_2 depending on ionic species and their quantum states. If we are after a H_3^+ spectrum with low rotational states, we use a liquid-N_2-cooled tube since the rotational (and translational) temperature is low. If we are after a spectrum starting from high rotational levels, we use an air-cooled or water-cooled tube. While the rotational temperature depends much on cooling, the vibrational temperature depends less on cooling but more on chemistry. Plasmas with a large amount of He usually elevate the vibrational temperature considerably. The beauty of He as buffer plasma gas is that it has an extraordinarily high ionization potential (24.6 eV) and this increases the electron temperature in plasmas, while it has a very low proton affinity (1.9 eV) and it does not deplete protons from H_3^+ into the form of HeH^+. An example of liquid-N_2-cooled plasma tube is shown in fig. 3. The tube is composed of three layers. The innermost tube of ~ 12 mm $\varnothing$ and 1 m in length contains plasmas where the electron temperature is on the order of $(20000 \div 40000)$ K. The middle tube accommodates liquid N_2 and the outer tube holds vacuum, so that the tube does not collect frost. Actually, for H_2 plasmas this type of cell is an overkill and simpler plasma tubes will suffice. This type of cell is particularly useful for many chemical plasmas such as hydrocarbon plasmas. In such discharges the intricacy of plasma chemistry is outrageous [20-22] (I call it alchemy), but this is outside the scope of this lecture.

A variety of frequency-tunable laser infrared sources are used for spectroscopy. Some of them are as follows.

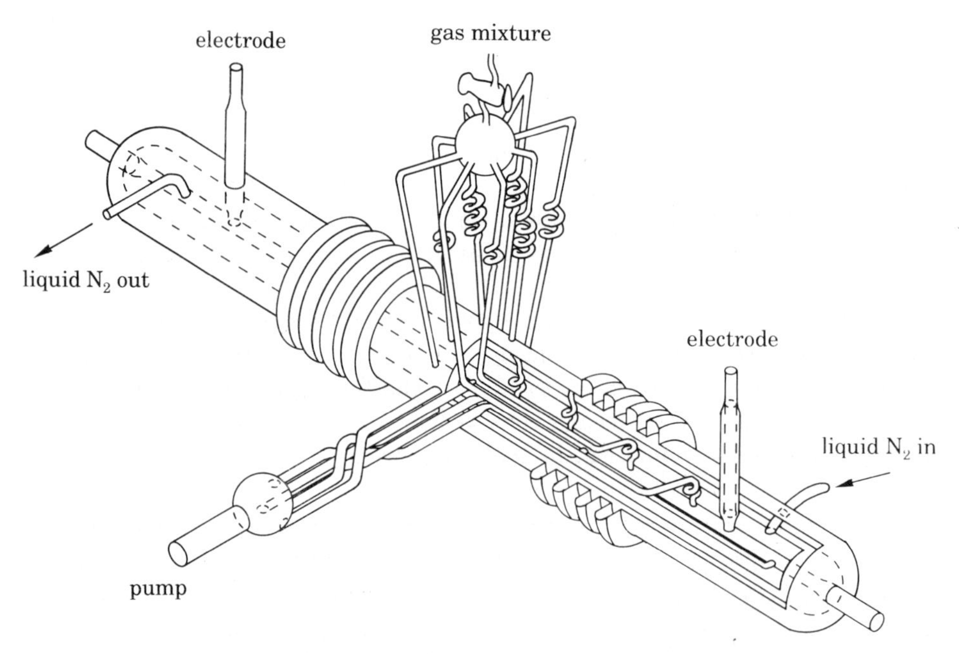

Fig. 3. – A liquid-N_2-cooled plasma tube.

a) Difference frequency system in which radiations from an argon laser and dye laser are mixed in $LiNbO_3$ or $LiIO_3$ crystals to generate the difference frequency. This system, initially developed by PINE [23], generates infrared radiation over a wide region ($(5100 \div 1800)$ cm^{-1}) continuously with reasonable power ($(3 \div 300)$ μW). The choice of this system was essential in the first discovery [8] of the H_3^+ spectrum. It is still the most powerful tool for H_3^+ spectroscopy.

b) Diode laser. PbS/CdS/PbSe/SnSe/PbTe/SnTe lasers generate tunable radiation in mid infrared within $(3500 \div 350)$ cm^{-1}. Their patchy coverage and limited tunability are often a pain in the neck, but they are easy to operate and reasonable in price and are the only source below 1800 cm^{-1} (see ref. [24] for a recent development). Recently InGaAsP communication diode lasers in the near infrared have been used for observation of the overtone band of H_3^+.

c) Other sources such as color center laser ($(4500 \div 2850)$ cm^{-1}, $(3 \div 30)$ mW), Ti:sapphire laser ($(14300 \div 9850)$ cm^{-1}, ~ 1 W), and CO_2, CO microwave sideband methods can be used also for the H_3^+ spectroscopy.

Since the number density of ions is a very small fraction of the total gas (~ 1 p.p.m.), we need high sensitivity to detect their spectrum. Usually such a high sensitivity is obtained from some molecular modulation and phase-sensitive detection. We use the Doppler effect for this purpose. Usually the Doppler broadening is the enemy of laser spectroscopists and we try to get rid of it as much as we can, but we use it here for its best purpose. (Someone's bad guy is another's good guy. Remember that many conclusions are drawn from astronomical observations that are based on this effect. Our understanding of the Universe is based on the Doppler effect.) In the velocity modulation method developed by GUDEMAN and SAYKALLY [25], plasmas are generated by using a.c. discharges with the frequency of $(3 \div 30)$ kHz. The ionic species are accelerated back and forth by the alternating field (if the ions survive chemically they travel macroscopic distances of ~ 1 cm in each phase) and thus molecular absorptions are frequency modulated due to the Doppler effect [26]. The phase-sensitive detection of the signal gives the absolute sensitivity of $\Delta I/I \sim 5 \cdot 10^{-7}$ for the time constant of 3 s. Since the absorption lines of neutral species that are much more abundant in plasmas are not modulated, the velocity modulation method gives an excellent discrimination of the ion signals as well. We have been attempting to improve the sensitivity to the shot noise limit by using the heterodyne method, but so far without success.

In order to make the most of the small fractions of molecular ions produced, we multiple pass the infrared radiation through the plasma tube. The multiple reflection has to be unidirectional so that the velocity information is not lost. The infrared beam is split into two equal parts, each of which passes four times through the plasma tube in a configuration like the ring laser. The two beams are separately detected by two matched infrared detectors whose signals are combined in opposite phases and sent to the phase-sensitive detector. The minimum detectable absorption coefficient is estimated to be $\sim 5 \cdot 10^{-10}$ cm^{-1}.

Some of you might be wondering why there is so much fuss and worry in these days when you can detect even one atom. The reason is that H_3^+ and many other molecular ions do not have electronic transitions. The spontaneous emission of H_3^+ is at least six orders of magnitude slower than that of typical electronic transitions in atoms. This is the reason why the spectrum had not been found until 1980.

2·4. *Observed spectrum.* – The lower vibrational states of H_3^+ relevant to our discussion are shown in fig. 4. The absorptions observed in the laboratory are shown by upward-pointing arrows and the emissions observed in astronomical objects are shown by downward-pointing arrows.

2·4.1. The ν_2 fundamental band. The first 15 absorption lines observed in the laboratory are shown in fig. 5 [8]. The spectrum was found after four years of single-minded preparation and search. It may sound paradoxical to nonspecialists, but high-resolution spectroscopy of a simple and light polyatomic molecule is more difficult than that of a complicated and heavy molecule. This is because for the former the spectrum extends over a wider frequency region and the convergence of the perturbation treatment is poorer. The spectral pattern in fig. 5 covers a wide wave number range and it takes a few months to cov-

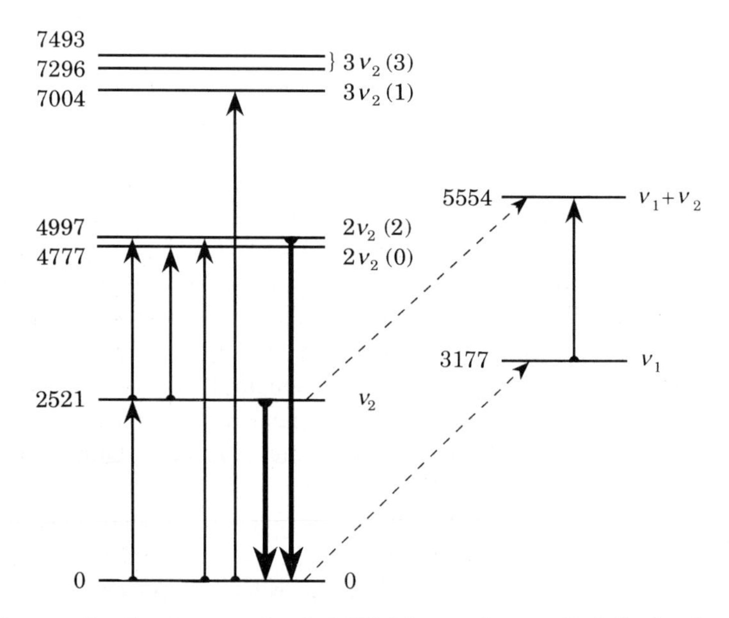

Fig. 4. – Lower vibrational-energy level of H_3^+. The number on the left of each energy level represents its energy in cm^{-1}.

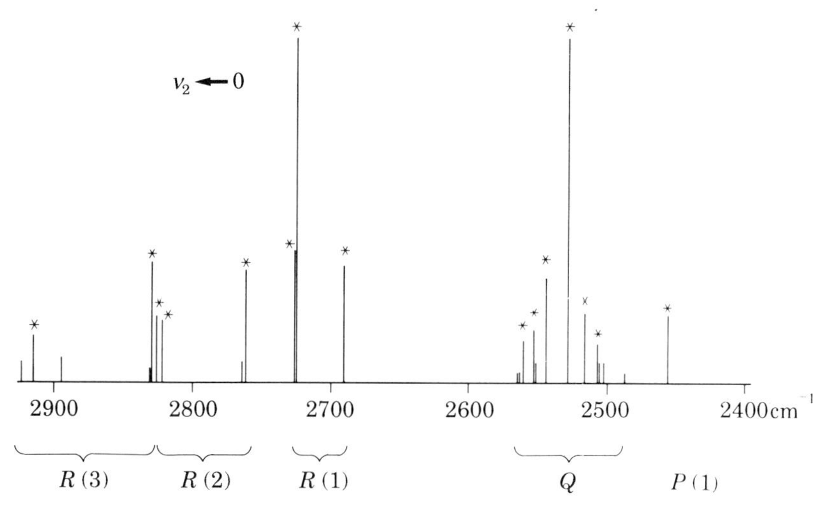

Fig. 5. – The ν_2 fundamental band of H_3^+ observed in 1980 [8].

er the whole region with high sensitivity, even if one works day and night. Unlike the usual vibration-rotation spectrum, there is no clearly discernible symmetry or regularity in the spectrum shown in fig. 5. This is due to the large interaction between vibration and rotation which totally mixes quantum states with different l and k (but same $k - l$). It was J. K. G. WATSON, the supreme theoretician in this field, who looked at this enigmatic spectrum and solved it overnight. He was helped by the *ab initio* calculations by CARNEY and PORTER, which are summarized in their classic papers [27, 28]. Their calculated value of the l-doubling constant, which was not published but had been communicated to us, played the crucial role in Watson's identification.

The most telltale feature of the spectrum in fig. 5 is the *absence* of any spectral line over an interval of more than 100 cm^{-1} at $\sim$ 2600 cm^{-1}. This indicates that the carrier of the spectrum cannot occupy the lowest rotational level with zero angular momentum ($J = K = 0$). As discussed earlier in subsect. **1·2**, this is the fingerprint of a system composed of three identical fermions with a spin of 1/2, and the feature indicated strongly that the spectrum is due to H_3^+. A similar but less extensive spectrum of D_3^+ was observed by SHY, FARLEY, LAMB and WING [29] using an entirely different, more ingenious method, and was published back to back with my paper [8].

Since the first spectrum of fig. 5 was recorded using a primitive method, the sensitivity of molecular-ion spectroscopy has increased by more than three orders of magnitude. We have continually attempted to observe more spectral lines corresponding to high rotational quantum states anticipating observation of H_3^+ in hot astronomical sources. A table recently published by KAO *et al.* [30] contains 129 lines of the fundamental band along with many more lines of other bands. We now have reached the $J = K = 15$ rotational level which is above the

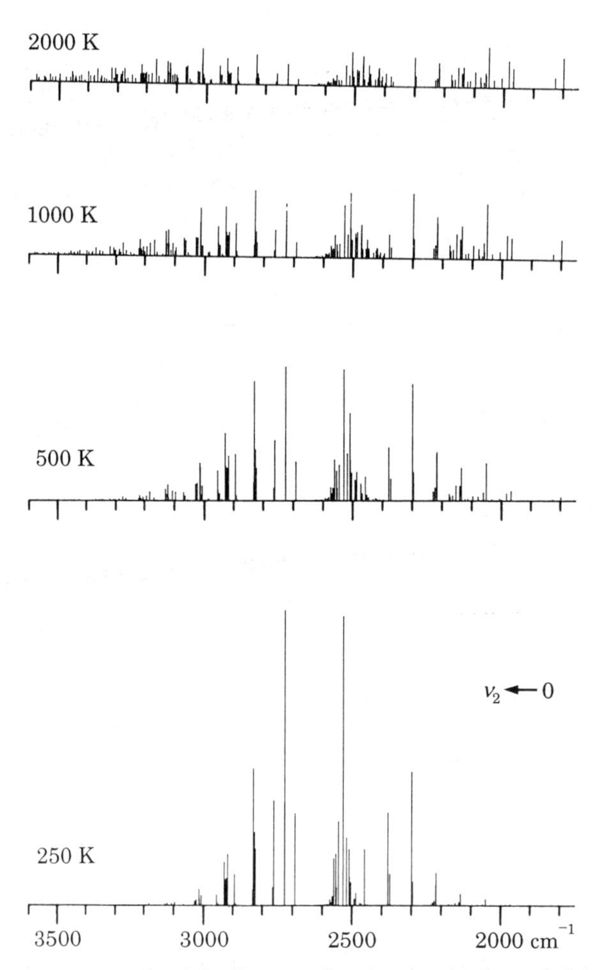

Fig. 6. – Computer-generated stick diagram for the ν_2- fundamental band at different temperatures.

lowest rotational level by 5092 cm^{-1} (0.63 eV). A computer-generated stick diagram for the H_3^+ fundamental band is shown in fig. 6 for various temperatures.

2·4.2. Hot, overtone and forbidden bands. Our second phase of H_3^+ laboratory spectroscopy was initiated in late 1987 by graduate students, Moungi BAWENDI and Brent REHFUSS, as an attempt to detect other bands of H_3^+ which are weaker than the fundamental band by a few orders of magnitude. This project was motivated purely by our curiosity about the quantum mechanics of the excited vibrational states and about the dynamic behavior of H_3^+ in plasmas. However, its results played a crucial role in the analysis of the then totally unexpected 2 μm emission spectrum of Jupiter.

BAWENDI and REHFUSS [31] observed all three hot bands $2\nu_2(2) \leftarrow \nu_2$, $2\nu_2(0) \leftarrow \nu_2$ and $\nu_1 + \nu_2 \leftarrow \nu_1$ which are allowed from the first excited states of ν_1 and ν_2. They are weaker than the fundamental band typically by a factor of ~ 50 due to the Boltzmann factors. So far 86, 24 and 39 rovibrational transitions have been identified for each respective band. The theoretical calculations of the rotation-vibration energy levels for the vibrationally excited states are not straightforward due to the strong vibration-rotation coupling and the poor convergence of the perturbation procedure as mentioned earlier. We are greatly helped by recent variational calculations by MILLER, TENNYSON and SUTCLIFFE [32]. These theorists do not use the traditional perturbational approach but solve the Schrödinger equation of the three-proton dynamics directly by using a supercomputer. Their results, based on the *ab initio* potential by MEYER, BOTSCHWINA and BURTON [33], are amazingly accurate and allow us to understand the complicated spectrum.

In usual molecules, overtone bands, that is, vibrational transitions with $\Delta v > 1$, are much weaker than the fundamental band because of the much smaller transition moments. In H_3^+, however, the overtone bands have considerable intensity because of the small mass of the proton and the relatively shallow polyatomic vibrational potential [34]. XU and GABRYS [35] observed the 2 μm $2\nu_2(2) \leftarrow 0$ first overtone band using a $LiIO_3$-based difference frequency spectrometer, and LEE, VENTRUDO and others [36] observed the 1.5 μm $3\nu_2(1) \leftarrow 0$ second overtone band using InGaAsP communication diodes. An example of the observed spectrum is shown in fig. 7 [37]. Further observation of the higher overtone band using a Ti:sapphire laser is being planned.

In principle, a transition such as $\nu_1 \leftarrow 0$ is infrared inactive (Raman active) from symmetry, but such transitions become allowed due to intricate in-

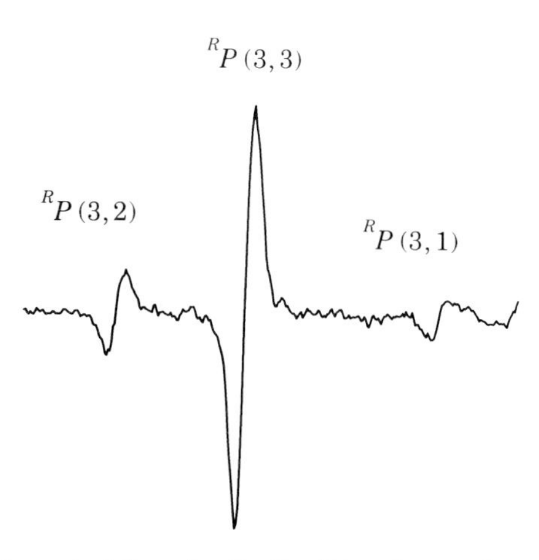

Fig. 7. – An example of spectral lines for the $3\nu_2(1) \leftarrow 0$ second overtone band of H_3^+. The three lines have frequencies 6807.714 cm^{-1}, 6807.275 cm^{-1} and 6806.633 cm^{-1}. The spectral lines were observed using an InGaAsP communication diode laser [37].

tramolecular dynamics. Such so-called forbidden transitions [38, 39] have been observed by XU, RÖSSLEIN and GABRYS [40].

3. – H_3^+ spectroscopy in space.

3·1. *Astronomy and spectroscopy.* – Since its beginning spectroscopy is closely related to astronomy. Isaac NEWTON, the theorist, laid the foundation for the principles of mechanics and dynamical astronomy, NEWTON, the experimenter, initiated spectroscopy, and NEWTON, the engineer, constructed the first reflecting telescope which is the major instrument in observational astronomy today—all within a few years [41]. When in 1814 Joseph FRAUNHOFER observed the great many absorption lines in the solar spectrum, he immediately turned his telescope to other astronomical objects [42]. He reported, «I discovered in the spectrum of this light (Venus) the same lines as those which appear in sunlight .... I have seen with certainty in the spectrum of Sirius three broad bands which to have no connection with those of sunlight». FRAUNHOFER thus initiated atomic spectroscopy and astrophysics in a single blow.

In a recent encyclopaedia article on *astrophysics* CHANDRASEKHAR notes [43]: «It is customary ... to date the birth of astrophysics with Kirchhoff's interpretation in 1859 of the Fraunhofer lines in the solar spectrum as revealing the presence of the familiar metals, such as sodium and potassium, as glowing vapors in the Sun's atmosphere .... After Kirchhoff's discovery, to speak of the composition of the stars was no longer in the realm of idle dreams; it became a problem of intense practical interest. And in the development during the twenties of the quantum theory of atomic and molecular spectra, leading astrophysicists interested in the unravelling of the spectra of the Sun and the stars ... played leading roles. Three discoveries in this context stand out: that of the identification of helium in the chromosphere of the Sun before its terrestrial identification, the identification of the negative ion of hydrogen as the source of the opacity in the solar photosphere before its isolation in its free state in the laboratory and the identification of the coronal lines of the Sun as arising from very high stages of ionization of ion».

Over many years spectroscopists were astronomers and *vice versa*: David BEWSTER, William HERSCHEL, John DRAPER, Anders ÅNGSTRÖM, William HUGGINS, Pierre JANSSEN, Norman LOCKYER, Henry DRAPER, Hermann VOGEL, Arthur SCHUSTER, Albert MICHELSON, Heinrich KAYER, Henri DESLANDRES, William PICKERING, Herman EBERT, Annie CANNON, George HALE, Henrietta LEAVITT, Alfred FOWLER, Ejnar HERTSPRUNG, Vesto SLIPHER, Walter ADAMS, Meghnad SAHA, and a great many more, the fine tradition continuing to this day by Gerhard HERZBERG and Charles TOWNES.

3·2. H_3^+ *in interstellar space.* – One of the main incentives to study H_3^+ is the pivotal role it plays in any molecular-hydrogen-dominated plasmas as discussed earlier in subsect. 2·2. Since almost all of the astronomical objects are in the state of plasmas, and since the hydrogen molecule is the dominant component in many gaseous regions, the spectrum of H_3^+ has the potential to be a powerful universal astronomical probe. The dominance of hydrogen molecules in many astronomical objects not only in our own galaxy but also in extragalactic objects is best demonstrated by the extremely strong H_2 quadrupole emission lines from the superluminous galaxy NGC 6240 shown in fig. 8. These infrared emission lines corresponding to the vibration-rotation $v = 1 \to 0$, $J + 2 \to J$ transitions carry intensities on the order of 10^8 solar luminosities [44-46]. Can you imagine—the total luminosity of a hundred million suns in a single spectral line! When HERZBERG first detected the weak quadrupole absorption spectrum in 1949, he needed high-pressure (10 atm) hydrogen gas and a path length of 5.5 km [47]. The laboratory experiment of Fink, Wiggins and Rank, first to observe the fundamental band, used comparable conditions [48]. These are weak transitions. The spontaneous lifetime of the $S(1)$ transition is 24 days. Nature does things in the most magnificent ways.

Since TOWNES and his colleagues [49] detected interstellar NH_3 in 1968, a

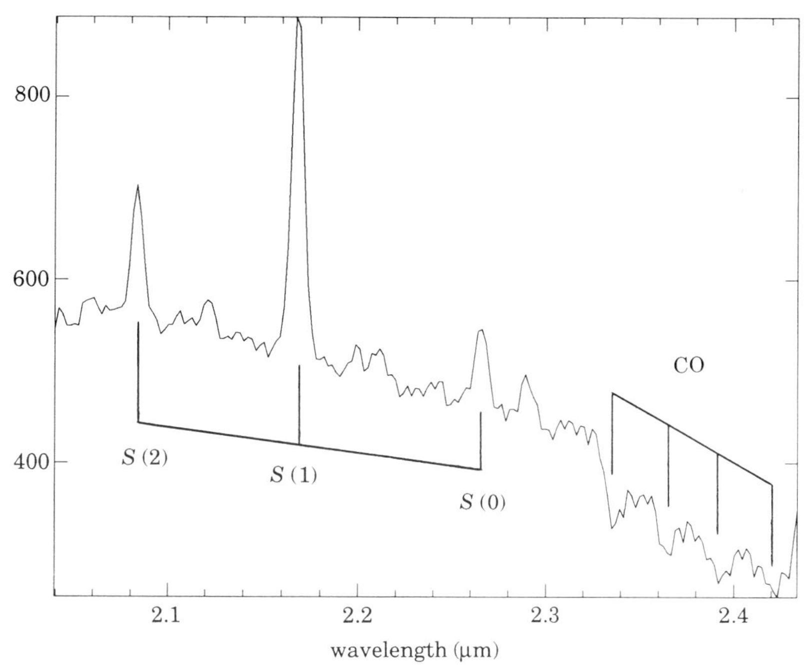

Fig. 8. – H_2 quadrupole emission spectrum from the superluminous galaxy NGC 6240. This spectrum was recorded by using the CGS4 grating spectrometer of the United Kingdom Infrared Telescope. Note that the spectrum is Doppler shifted by $Z = 0.0254$ (red shift of ~ 0.05 μm).

great many molecules have been detected and a new discipline, *molecular astrophysics*, was born. The once novel phrase «molecular cloud» is now most commonly used by astronomers as the birthplace for the stars. In attempts to explain the chemical evolution of molecular clouds, ion neutral reaction kinetics emerged as the most important mechanism because such reactions proceed under the extremely low temperature of (10 ÷ 100) K which is typical for molecular clouds. In this proposed mechanism H_3^+ plays the pivotal role of the universal protonator as discussed in subsect. 2·2, eqs. (6)-(8). While theoretical calculations [13-15] explain the relative abundances of species observed by radioastronomy reasonably well, detection of H_3^+ in interstellar space will be the most direct test of the theory. For this reason most of us agree that H_3^+ is the most important species in interstellar space yet to be detected.

Soon after my laboratory observation of the H_3^+ spectrum, I attempted its detection in the Orion molecular cloud and other astronomical objects [50]. The search continues to this day without success [51, 52]. The difficulty of detection was not unexpected; KLEMPERER, the earliest proponent of the ion neutral reaction scheme, warned me from the onset that, although the production rate of H_3^+ by cosmic-ray ionization is very high, the steady-state concentration of H_3^+ should only be on the order of 10^{-9} of the hydrogen concentration because of the extremely high reactivity of H_3^+ in reaction (7) especially with carbon monoxide.

Infrared spectrometers used for astronomical observations are not very sophisticated compared to their laboratory counterpart. «High resolution» means 10 km/s, *i.e.* a resolution of $\sim 3 \cdot 10^4$. Observation of the absorption spectrum is especially difficult with this low resolution. However, the recent improvement of the sensitivity by the use of multiple diode detector arrays is very impressive. This makes me optimistic about the fact that the discovery of H_3^+ in molecular clouds is just around the corner. Theorists predict H_3^+ column densities on the order of $\sim 10^{14}$ cm^{-2}, which is approaching to within the detectability of modern infrared spectrometers.

In the meantime, the first appearance of H_3^+ in Nature came from serendipitous discoveries in emission on an entirely unexpected astronomical object.

3·3. H_3^+ *in Jupiter.* – Beginning in September 1987, two groups of astronomers, who were primarily engaged in the studies of H_2 quadrupole emission lines in Jovian ionospheres, observed a group of strong unidentified emission lines in the 2 μm region. The spectrum observed at the McDonald Observatory by TRAFTON, LESTER and THOMPSON and published still as unidentified lines [53] is shown in fig. 9. The other group at Canada-France-Hawaii telescope observed an even more extensive spectral pattern using a Fourier-transform spectrometer [54]. The spectrum was brought to the attention of laboratory spectroscopists and theorists at the Herzberg Institute of Astrophysics. J. K. G. WATSON, who earlier analyzed the laboratory spectrum [8, 55, 56], has come

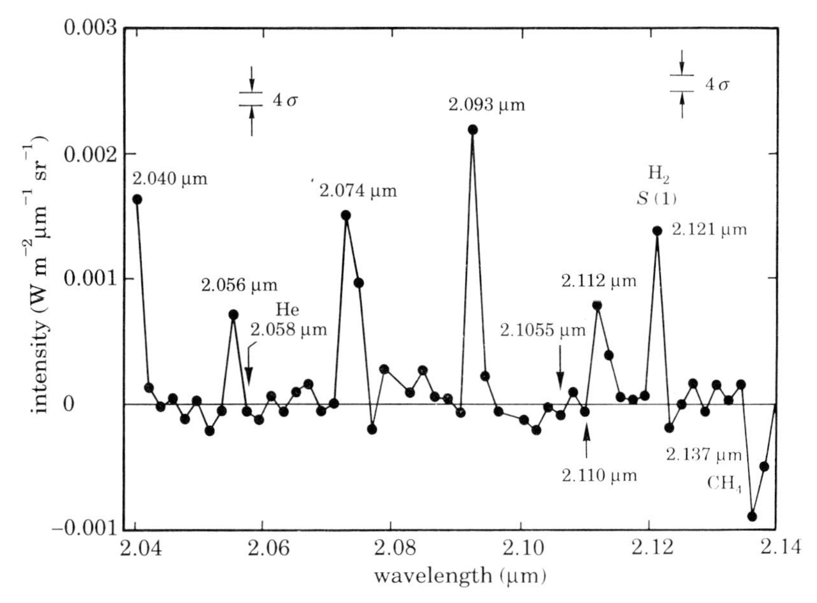

Fig. 9. – The 2 μm Jovian emission spectrum reported by TRAFTON, LESTER and THOMPSON [53]. The spectrum was recorded using the McDonald Observatory infrared grating spectrometer.

to the assignment of the spectrum as due to the $2\nu_2(2) \rightarrow 0$ overtone band of H_3^+. The laboratory spectrum of the hot band $2\nu_2(2) \leftarrow \nu_2$ observed by BAWENDI and REHFUSS [31] and a hollow-cathode emission spectrum observed by MAJEWSKI [56, 57], along with the theoretical calculations by MILLER and TENNYSON [58], supplied crucial information needed for the assignment. More details of this exciting development may be found in ref. [12].

After the 2 μm emission was attributed to H_3^+, it did not take any imagination to expect that the fundamental band at $(4 \div 3)$ μm should appear even more strongly. This emission was reported in 1990 [59-61]. As expected, the fundamental emission band is very strong and a short time of integration suffices to detect it clearly. The most remarkable thing, however, is the purity of the spectrum. Figure 10 shows the emission band in the vicinity of 2830 cm^{-1}. Note that the emission lines are almost completely free from background infrared radiation. Jupiter is still cooling and radiating much of its heat in the infrared region. At this wavelength the thermal radiation emitted by Jupiter should be stronger than the solar radiation reflected by Jupiter. This infrared background, which would otherwise be much more intense than the H_3^+ emission, is almost completely absorbed by the pressure-broadened lines of CH_4 which exists abundantly at lower altitude in the Jovian atmosphere. Only the H_3^+ spectrum emitted at much higher altitude reaches us unabsorbed. Note that this is a tiny fraction of the whole spectrum which is shown in fig. 6. Observation over a wide

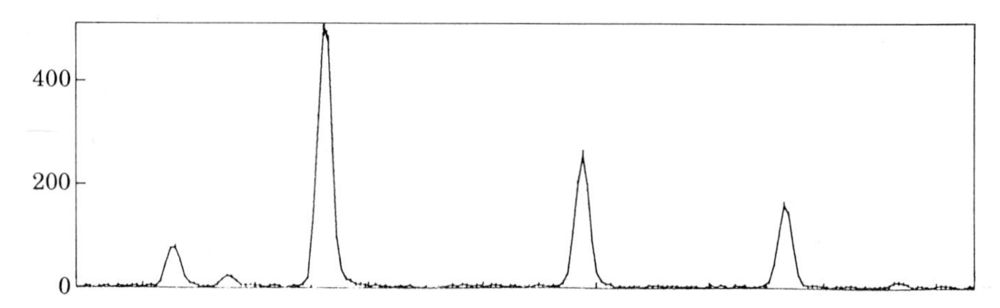

Fig. 10. – The 3.53 μm H_3^+ emission spectrum from Jupiter. The transitions are from left to right: $R(3, 2)^-$ 2832.197 cm^{-1}, $R(3, 1)^-$ 2831.340 cm^{-1}, $R(3, 3)^-$ 2829.923 cm^{-1}, $R(2, 1)^+$ 2826.113 cm^{-1}, $R(2, 2)^+$ 2823.137 cm^{-1} and $R(8, 9)$ 2821.518 cm^{-1}. The last transition is the hot band $2\nu_2(0) \rightarrow \nu_2$.

range of spectra allows us to determine the temperature of the Jovian ionosphere. The spectrum at 3.5 μm is so pure that we do not need spectrometers to observe H_3^+ emission. A narrow-band filter and an infrared camera suffice to study the morphology and temporal variation of Jovian plasmas [62, 63].

Jupiter is a huge blob of supercritical fluid composed mainly of hydrogen and helium. Much of the hydrogen is in the metallic state because of the gravitational pressure. The electrical conductivity of the inner-core hydrogen and the fast rotation of Jupiter ($\sim$ 10 h) contribute to the large magnetic moment of Jupiter ($1.6 \cdot 10^{30}$ G cm^3), which is more than four orders of magnitude higher than that of the Earth. This large magnetic moment in the path of the solar wind creates a huge magnetosphere around Jupiter which, if visible from Earth, would subtend a similar angle of sight to those of the Sun and the Moon. On a smaller scale, charged particles in the vicinity of Jupiter are trapped by this magnetic field and corotate with Jupiter with great speed. Jupiter's moon Io, with its active volcanoes, adds fuel to this gigantic rotating plasma. The electron and charged atoms and molecules move along the magnetic field and are eventually focussed at the two polar regions where intense auroras are observed. The intense H_3^+ emission is generated from these auroral regions (fig. 11). See ref. [64] for more details on the magneto-plasma activity in Jupiter.

Our recent observation [65] showed that, although the H_3^+ emission is strongest at the two polar regions, it also exists at all latitudes of Jupiter, indicating plasma chemical activity all over Jovian ionosphere. The H_3^+ emission will be a very powerful monitor to study this activity.

3·4. H_3^+ *in other objects.* – The year 1992 was a busy year for discovery of H_3^+ in other astronomical objects. In January it was claimed [66] that the two unidentified emission features in the infrared spectrum of the supernova 1987 A after 192 days of the event were due to H_3^+. This caught me by surprise because the supernova after such a short period from the explosion was the last object in which I anticipated molecular species. I then learned that the spectrum showed

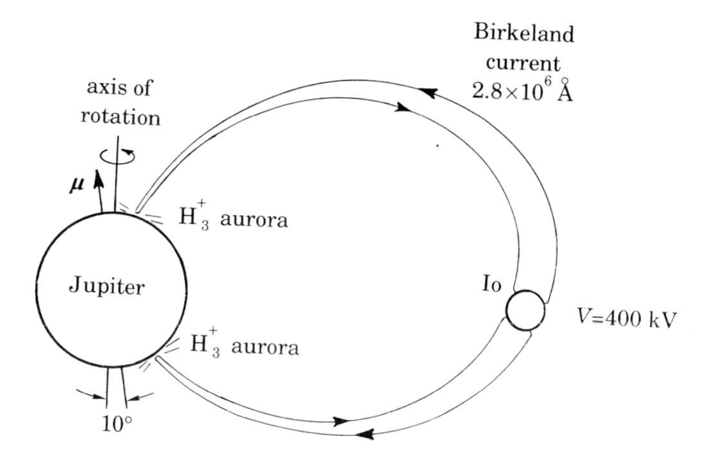

Fig. 11. – Simplified picture of the Jupiter-Io plasma power induction. The diameter of Io is magnified by a factor of 10. Other relative distances are to scale.

evidence of the existence of CO molecules in abundance. Indeed LEPP and DALGARNO [66] came up with a chemical scheme with which they showed production of a sufficient amount of H_3^+ to explain the claimed emission. This shows us that we should not disregard any gaseous astronomical objects as possible sources of the H_3^+ spectrum. As far as the spectral identification is concerned, it is nowhere nearly as certain as they are in other cases, because the fingerprint of the rotational structure is washed out by the Doppler broadening due to the rapid expansion of the gas. Nevertheless this presented us with an extremely interesting possibility which may be confirmed by observation of future events.

On April 1, 1992, the H_3^+ emission was clearly noted on Uranus [67]. Again the intensity and the purity of the spectrum greatly surprised us. The spectrum is weaker than Jupiter's by just about the amount expected from the farther distance ($\times$ 4) and the smaller diameter ($\sim 1/3$), but otherwise comparable. This is amazing in view of the much smaller magnetic moment ($\sim 1/410$) of Uranus. Obviously there is much to be learned on the plasma activity of the planetary ionosphere.

On July 19, 1992 (after the Fermi School), we detected H_3^+ emission in Saturn [68]. The three detected spectral lines are much weaker ($\sim 1/130$) than that of Jupiter and are more contaminated by the infrared background, due to the ring of Saturn, but they are clearly visible. Those and many other anticipated emission lines will be used as ground-based probe to monitor the plasma activity of Saturn.

Overall, it is fair to say that the H_3^+ spectrum provided by laser spectroscopy is beginning to be a very powerful astronomical probe to study ionized regions. In the laboratory spectroscopy, the detection of the H_3^+ spectrum has led in the last ten years to the detection of spectra of many other molecular ions such as CH_2^+, CH_3^+, $C_2H_2^+$, $C_2H_3^+$, NH_2^+, NH_3^+, NH_4^+, H_3O^+, $HCNH^+$, and many others whose spectra had previously been unknown in any spectral region. We hope the same will happen in astronomical spectroscopy, perhaps in the next 50 years.

4. – Spectroscopy of solid hydrogen.

4·1. *Beginning.* – We now switch gears and discuss our recent application of laser spectroscopy to a totally different object, solid hydrogen. It has been usually assumed that spectral lines in condensed phases are highly broadened due to homogeneous and inhomogeneous interactions and the high resolution of laser spectroscopy cannot be fully exploited unless some special techniques are employed [69, 70]. Recently we observed [71] an infrared spectral line in para-hydrogen crystal corresponding to the $J = 6 \leftarrow 0$ rotational transition at 2410.5 cm^{-1} (fig. 12). To our surprise the spectral line appeared much sharper than the Doppler-broadened and Dicke-narrowed gaseous lines of hydrogen. Since then a variety of sharp spectral lines have been observed in solid hydrogen and impurities embedded in solid para-hydrogen [72-75].

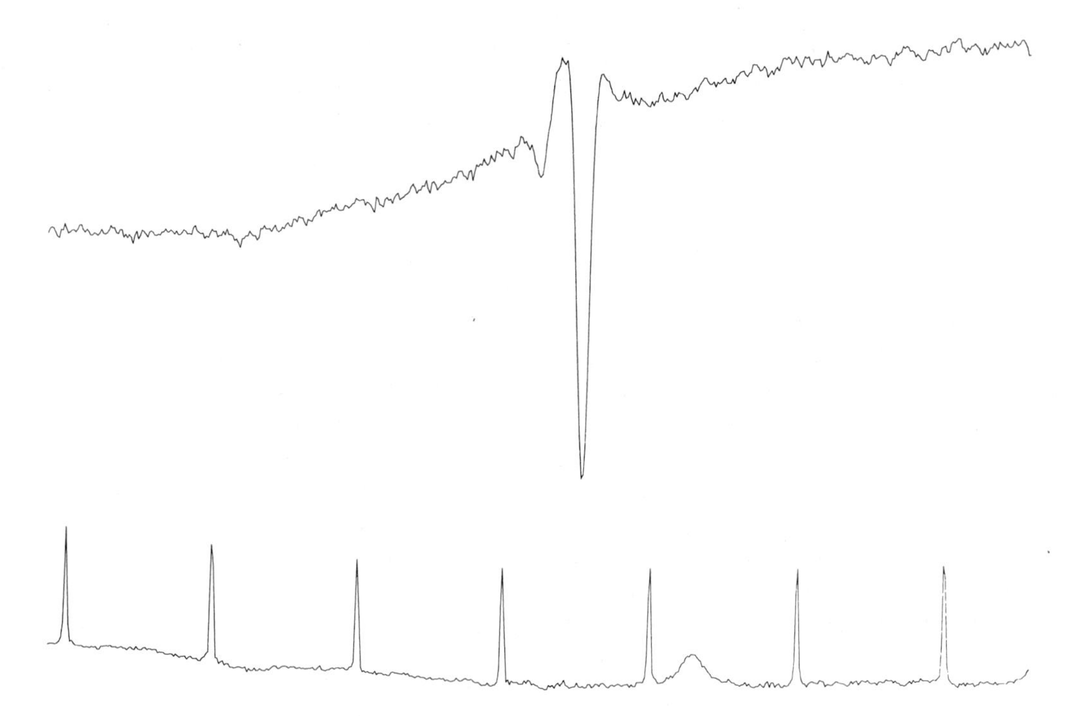

Fig. 12. – The $J = 6 \leftarrow 0$ 2^6-pole induced rotational transitional of para-H_2. The frequency is at 2410.5349 cm^{-1}. $\Delta\nu \sim 90$ MHz HWHM (ref. [71]).

Like the H_3^+ emission in Jupiter discussed in the previous section, the sharp spectral lines in solid H_2 were unexpected. With hindsight they were to be anticipated, but we are not that imaginative. We found them accidentally while preparing another experiment. We were preparing for spectroscopy of ionic species produced in solid hydrogen through electron bombardment using a van de Graaf accelerator. This experiment was conceived in our attempt to observe spectra of H_5^+ and CH_5^+, which we did not succeed, using gaseous plasmas. In going from gaseous plasmas to solid hydrogen we were prepared to make a few jumps, each of which is several orders of magnitude. The electron temperature which set the stage for chemistry drops from a few electronvolt to the helium cryogenic temperature of 4.2 K (a decrease of $\sim 10^4$). The number densities of the absorbing species increases from $\sim 10^{10}$ cm^{-3} to $\sim 10^{22}$ cm^{-3} (for hydrogen) or to $\sim 10^{19}$ cm^{-6} (for impurities). However, the observed sharpness of the spectral lines caught us by surprise. We were anticipating Fourier-transform infrared spectroscopy with the resolution $\nu/\Delta\nu$ on the order of $\sim 10^4$ and ended up laser spectroscopy with the resolution on the order of $\sim 10^8$. This was a great surprise and excitement for us because this extra resolution of four orders of magnitude allows us to study the condensed phase with unprecedented accuracy and clarity based on first principles.

4·2. *Background.* – The spectroscopic study of molecular hydrogen in condensed phase has a long history and its main stage was at the University of Toronto. Almost immediately after the Raman effect was discovered in 1928, McLennan and McLeod [76] applied the method to liquid hydrogen and observed a vibrational Raman transition $Q_1(0)$ ($v = 1 \leftarrow 0$, $J = 0 \leftarrow 0$) and two rotational Raman transitions $S(0)$ ($J = 2 \leftarrow 0$) and $S(1)$ ($J = 3 \leftarrow 1$). Their results clearly showed that H_2 has well-defined rotational quantization even in liquids. They also gave «experimental proof of the correctness of Dennison's view that hydrogen at low temperature must be regarded as a mixture of two effectively distinct sets of molecules», ortho ($J = 1$) and para ($J = 0$) hydrogen. Only two quantum states are populated and molecules are rotating freely even at $T = 0$. These results have been greatly extended using infrared and Raman spectroscopy by the Toronto group led by H. L. Welsh and J. Van Kranendonk. The great amount of fascinating results obtained from the 1950's to the '70's, which constituted their life work, are summarized in the lucid and inspiring treatise by Van Kranendonk [77]. Our work may be regarded as an extension of their results using laser spectroscopy. It should be noted that C. K. N. Patel and his colleagues have also used laser spectroscopy to systematically study overtone bands of solid hydrogen [78, 79]. Their emphasis in using the optoacoustic detection method was more on the sensitivity.

The sharpness of the spectral lines of solid hydrogen results from the purity of vibration-rotation quantum states and the slow relaxation of excited states. This was initially shown by the microwave spectroscopy of $J = 1$ pair H_2 in nearly pure para-hydrogen by HARDY and BERLINSKY [80, 81]. In this remarkable paper, using a microwave calorimetric method, they obtained sharpest lines with linewidths on the order of a few MHz. Infrared spectral lines have widths of the same order of magnitude because the Doppler broadening does not occur for solid phase. Our results should have been anticipated from Hardy and Berlinsky's experiments.

Theory was developed as experimental results accumulated. The initial work by L. PAULING [82], T. Nakamura's paper in 1955 [83] which, according to VAN KRANENDONK [77], «marks the beginning of the modern microscopic theory of the solid hydrogen», Van Kranendonk's classic papers [84, 85] on optical spectroscopy and A. B. Harris' detailed work on intermolecular and crystal interaction [86, 87] are followed by a great many theoretical papers. They are summarized in Van Kranendonk's book [77]. An excellent review on experiment and theory of solid-hydrogen crystals has also been published by SILVERA [88].

4·3. *Many-body absorption and high-ΔJ spectrum.* – One of the most fascinating aspects of solid-state spectroscopy is the many-body nature of the radiative interaction. In gaseous spectroscopy we consider the individual molecular dipole μ interacting with the radiation field E and treat the interaction $-\boldsymbol{\mu}\cdot\boldsymbol{E}$ by solving a time-dependent equation of motion. In such a treatment the intensity of the spectrum scales by $(a/\lambda)^2$ as we go from dipole (2^1-pole) to quadrupole (2^2-pole) interaction, where a is the atomic or molecular dimension and λ is the wavelength of radiation [89]. Since $a/\lambda \ll 10^{-4}$ for infrared radiation, it is clearly impossible to observe the $\Delta J = 6$ transition shown in fig. 12 which is induced by the 2^6-pole (tetrahexacontapole) moment of H_2 and should be weaker than the quadrupole spectrum by $(a/\lambda)^8 \sim 10^{-37}$. The transition is observable in solid state because the absorption mechanism is different.

The electric field due to the 2^6-pole of H_2 induces dipole moments in surrounding hydrogen molecules, and those dipole moments interact with the radiation field and absorb a photon [83]. This many-body radiative process introduces a quantitative and qualitative difference from gaseous spectroscopy. The spectral intensity due to higher-multipole interaction now scales by $(a/R)^2$, where R is the intermolecular distance in the crystal of 3.795 Å. This is a much larger number than $(a/\lambda)^2$ and makes the observation of the high-ΔJ transition possible. The symmetry argument and thus the selection rules are also different in many-body interactions. While in gaseous spectroscopy we consider only molecular symmetry, we now have to consider molecular symmetry and crystal symmetry simultaneously [90]. This leads us to M selection rules for the $J = 6 \leftarrow 0$ transition, $M = 4, 2 \leftarrow M = 0$ for a plane of polarization perpendicular to the crystal axis and $M = 3 \leftarrow 0$ for a parallel one. More discussions on this and its experimental demonstration are shown in ref. [73].

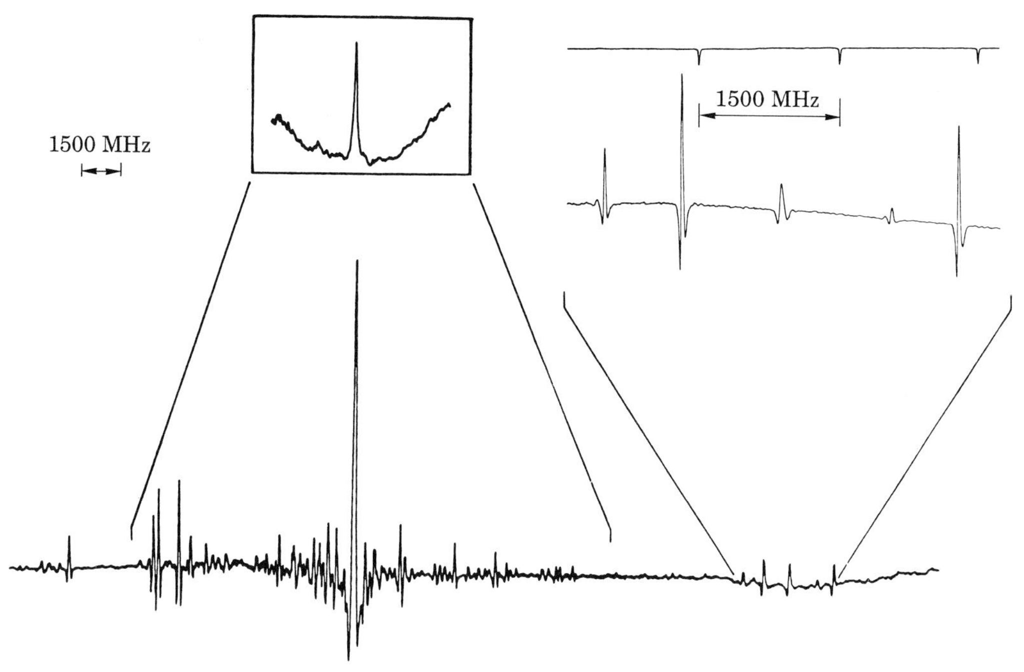

Fig. 13. – The fine structure due to intermolecular interaction of $J = 1$ H_2 molecules around the $Q_1(1)$ transition (ref. [72]).

4·4. *Fine structure due to intermolecular interaction.* – The beauty of high-resolution spectroscopy is that we can resolve all details of the spectral structure resulting from the intermolecular interaction. Figure 13 shows the $Q_1(1)$ $(v = 1 \leftarrow 0, J = 1 \leftarrow 1)$ transition which indicates an extremely rich structure due to $J = 1$ ortho-hydrogen and neighboring $J = 1$ hydrogen molecules. This is observed for ortho-hydrogen which exists as impurities in almost pure para-H_2.

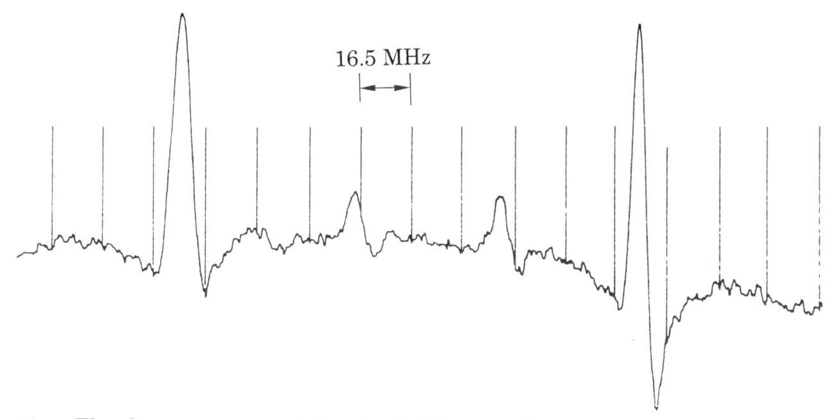

Fig. 14. – The fine structure of the D_2 $Q_1(0)$ transition.

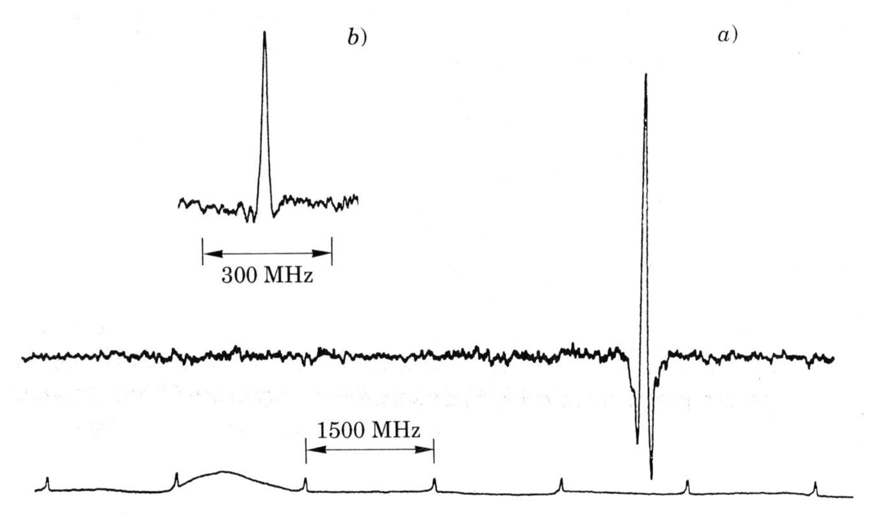

Fig. 15. – $Q_1(0)$ stimulated Raman spectrum of para-H_2 (ref. [75]).

In such a system the excitation due to infrared radiation (exciton) is localized in $J = 1$ hydrogen due to energy mismatch between ortho and para transitions. We in effect completely resolve the inhomogeneous broadening. For details see ref. [72].

Another way to localize the exciton is to use impurity molecules. Figure 14 shows the $Q_1(0)$ transition of deuterium embedded in para-H_2 crystals [91]. The structure is due to simultaneous transition of neighboring $J = 1$ H_2 which induces the spectrum. The spectral line has the sharpest width of 2 MHz half-width at half-maximum.

4.5. *Stimulated Raman spectrum.* – The exciton momentum selection rule $\Delta k = 0$ causes sharp spectral lines even in pure para-hydrogen crystals where the excitons are not localized but form exciton bands. One such example, the stimulated Raman spectrum of the $Q_1(0)$ band, is shown in fig. 15. Readers are referred to ref. [75] for more details.

REFERENCES

[1] K. SHIMODA: in *Proc. S.I.F.*, Course XVII, *Topics of Radiofrequency Spectroscopy*, edited by A. GOZZINI (Academic Press, New York, N.Y., 1960), p. 1.
[2] T. OKA: *J. Chem. Phys.*, **47**, 5410 (1967).
[3] N. F. RAMSEY and E. M. PURCELL: *Phys. Rev.*, **85**, 143 (1952).
[4] L. D. LANDAU and E. M. LIFSHITZ: *Mechanics* (Pergamon Press, New York, N.Y., 1976), § 6, 7, 9.
[5] L. D. LANDAU and E. M. LIFSHITZ: *Quantum Mechanics, Non-Relativistic Theory* (Pergamon Press, New York, N.Y., 1977).

[6] W. PAULI: *Phys. Rev.*, 58, 716 (1940).
[7] E. WIGNER: *Z. Phys. Chem. B*, **23**, 28 (1933).
[8] T. OKA: *Phys. Rev. Lett.*, **45**, 531 (1980).
[9] M. Y. HAN and Y. NAMBU: *Phys. Rev.*, **139**, B1006 (1965).
[10] S. CHANDRASEKHAR: *Astrophys. J.*, **100**, 176 (1944); **128**, 114 (1958).
[11] T. OKA: in *Molecular Ions: Spectroscopy, Structure and Chemistry,* edited by T. A. MILLER and V. E. BONDYBEY (North Holland, New York, N.Y., 1983), p. 73.
[12] T. OKA: *Rev. Mod. Phys.*, **64**, 1141 (1992).
[13] E. HERBST and W. KLEMPERER: *Astrophys. J.*, **185**, 505 (1973).
[14] W. D. WATSON: *Rev. Mod. Phys.*, **48**, 513 (1976).
[15] H. SUZUKI: in *Toward Interstellar Chemistry,* edited by N. KAIFU (University of Tokyo Press, Tokyo, 1989), p. 18, 137, 142.
[16] G. HERZBERG: *Trans. R. Soc. Can.*, **5**, 3 (1967).
[17] T. OKA: in *Laser Spectroscopy V,* edited by A. R. W. MCKELLAR, T. OKA, B. P. STOICHEFF (Springer-Verlag, Berlin, 1981), p. 320.
[18] J. K. G. WATSON: *J. Mol. Spectrosc.*, **103**, 350 (1984).
[19] A. VON ENGEL: *Ionized Gas* (Oxford University Press, Oxford, 1965).
[20] T. OKA: in *Frontiers of Laser Spectroscopy of Gases,* edited by A. C. P. ALVES, J. M. BROWN and J. M. HOLLAS, NATO ASI Series (Kluwer Academic Publishers, Amsterdam, 1988), p. 353.
[21] T. OKA: *Philos. Trans. R. Soc. London, Ser. A*, **324**, 81 (1988).
[22] M. W. CROFTON, M.-F. JAGOD, B. D. REHFUSS and T. OKA: *J. Chem. Phys.*, **91**, 5139 (1989).
[23] A. S. PINE: *J. Opt. Soc. Am.*, **64**, 1683 (1974); **66**, 97 (1976).
[24] P. CANARELLI, Z. BENKO, R. CURL and F. K. TITTLE: *J. Opt. Soc. Am. B*, **9**, 197 (1992).
[25] C. S. GUDEMAN, M. H. BEGEMANN, J. PFAFF and R. J. SAYKALLY: *Phys. Rev. Lett.*, **50**, 767 (1983).
[26] N. N. HAESE, F. S. PAN and T. OKA: *Phys. Rev. Lett.*, **50**, 1575 (1983).
[27] G. D. CARNEY and R. N. PORTER: *J. Chem. Phys.*, **60**, 4251 (1974).
[28] G. D. CARNEY and R. N. PORTER: *J. Chem. Phys.*, **65**, 3547 (1976).
[29] J. T. SHY, J. W. FARLEY, W. E. LAMB jr. and W. H. WING: *Phys. Rev. Lett.*, **43**, 535 (1980).
[30] L. KAO, T. OKA, S. MILLER and J. TENNYSON: *Astrophys. J. Suppl.*, **77**, 317 (1991).
[31] M. G. BAWENDI, B. D. REHFUSS and T. OKA: *J. Chem. Phys.*, **93**, 6250 (1990).
[32] S. MILLER, J. TENNYSON and B. T. SUTCLIFFE: *J. Mol. Spectrosc.*, **141**, 104 (1990), and references therein.
[33] W. MEYER, P. BOTSCHWINA and P. G. BURTON: *J. Chem. Phys.*, 84, 891 (1986).
[34] S. MILLER and J. TENNYSON: *J. Mol. Spectrosc.*, **128**, 530 (1988).
[35] L.-W. XU, C. M. GABRYS and T. OKA: *J. Chem. Phys.*, **93**, 6210 (1990).
[36] S. S. LEE, B. F. VENTRUDO, D. T. CASSIDY, T. OKA, S. MILLER and J. TENNYSON: *J. Mol. Spectrosc.*, **145**, 222 (1991).
[37] B. F. VENTRUDO, D. T. CASSIDY, Z.-Y. GUO, S.-W. JOO and T. OKA: unpublished.
[38] F.-S. PAN and T. OKA: *Astrophys. J.*, **305**, 518 (1986).
[39] S. MILLER and J. TENNYSON: *Astrophys. J.*, **335**, 486 (1988).
[40] L.-W. XU, M. RÖSSLEIN, C. M. GABRYS and T. OKA: *J. Mol. Spectrosc.*, **153**, 726 (1992).
[41] R. S. WESTFALL: *Never at Rest* (Cambridge University Press, Cambridge, 1980).

[42] J. FRAUNHOFER: *Denkschriften der könig̈. Akad. der Wissenschaften zur München*, V, 193 (1817), translated in *The Wave Theory of Light and Spectra*, edited by H. CREW (Arno Press, New York, N.Y., 1981).
[43] S. CHANDRASEKHAR: *Encyclopedia of Physical Science and Engineering* (Academic Press, London, 1992).
[44] R. D. JOSEPH, G. S. WRIGHT and R. WADE: *Nature (London)*, **311**, 132 (1984).
[45] G. H. RIEKE, R. M. CUTRI, J. M. BLACK, W. F. KAILEY, C. W. MCALARY, M. J. LEBOFSKY and R. ELSTON: *Astrophys. J.*, **290**, 116 (1988).
[46] D. L. DEPOY, E. E. BECKLIN and C. G. WYNN-WILLIAMS: *Astrophys. J.*, **307**, 116 (1986).
[47] G. HERZBERG: *Nature (London)*, **163**, 170 (1949).
[48] U. FINK, T. A. WIGGINS and D. H. RANK: *J. Mol. Spectrosc.*, **18**, 384 (1965).
[49] A. C. CHEUNG, D. M. RANK, C. H. TOWNES, D. D. THORNTON and W. J. WELCH: *Phys. Rev. Lett.*, **21**, 1701 (1968).
[50] T. OKA: *Philos. Trans. R. Soc. London, Ser. A*, **303**, 543 (1981).
[51] T. R. GEBALLE and T. OKA: *Astrophys. J.*, **342**, 855 (1989).
[52] J. H. BLACK, E. F. VAN DISHOECK and R. C. WOODS: *Astrophys. J.*, **358**, 459 (1990).
[53] L. TRAFTON, D. L. LESTER and K. L. THOMPSON: *Astrophys. J.*, **343**, L73 (1989).
[54] P. DROSSART, J.-P. MAILLARD, J. CALDWELL, S. J. KIM, J. K. G. WATSON, W. A. MAJEWSKI, J. TENNYSON, S. MILLER, S. K. ATREYA, J. T. CLARKE, J. H. WAITE jr. and R. WAGENER: *Nature (London)*, **340**, 539 (1989).
[55] J. K. G. WATSON, S. C. FOSTER, A. R. W. MCKELLAR, P. BERNATH, T. AMANO, F.-S. PAN, M. W. CROFTON, R. S. ALTMAN and T. OKA: *Can. J. Phys.*, **62**, 1825 (1984).
[56] W. A. MAJEWSKI, M. D. MARSHALL, A. R. W. MCKELLAR, J. W. C. JOHNS and J. K. G. WATSON: *J. Mol. Spectrosc.*, **122**, 341 (1987).
[57] W. A. MAJEWSKI and J. K. G. WATSON: unpublished.
[58] S. MILLER and J. TENNYSON: *J. Mol. Spectrosc.*, **136**, 223 (1989).
[59] T. OKA and T. R. GEBALLE: *Astrophys. J.*, **351**, L53 (1990).
[60] S. MILLER, R. D. JOSEPH and J. TENNYSON: *Astrophys. J.*, **360**, L55 (1990).
[61] J.-P. MAILLARD, P. DROSSART, J. K. G. WATSON, S. J. KIM and J. CALDWELL: *Astrophys. J.*, **363**, L37 (1990).
[62] R. BARON, R. D. JOSEPH, T. OWEN, J. TENNYSON, S. MILLER and G. E. BALLESTER: *Nature (London)*, **353**, 539 (1991).
[63] S. J. KIM, P. DROSSART, J. CALDWELL, J.-P. MAILLARD, T. HERBST and M. SHURE: *Nature (London)*, **353**, 536 (1991).
[64] A. J. DESSLER, Editor: *Physics of the Jovian Magnetosphere* (Cambridge University Press, Cambridge, 1983).
[65] T. R. GEBALLE, M.-F. JAGOD and T. OKA: unpublished.
[66] S. MILLER, J. TENNYSON, S. LEPP and A. DALGARNO: *Nature (London)*, **355**, 420 (1992).
[67] L. M. TRAFTON, T. R. GEBALLE, S. MILLER, J. TENNYSON and G. E. BALLESTER: *Astrophys. J.*, **405**, 761 (1993).
[68] T. R. GEBALLE, M.-F. JAGOD and T. OKA: *Astrophys. J.*, **408**, L109 (1993).
[69] W. M. YEN and P. M. SELZER, Editors: *Laser Spectroscopy of Solids* (Springer-Verlag, New York, N.Y., 1986).
[70] W. E. MOERNER, Editor: *Persisted Spectral Hole Burning: Science and Application* (Springer-Verlag, Berlin, 1988-1989).
[71] M. OKUMURA, M.-C. CHAN and T. OKA: *Phys. Rev. Lett.*, **62**, 32 (1989).

[72] M.-C. CHAN, M. OKUMURA, C. M. GABRYS, L.-W. XU, B. D. REHFUSS and T. OKA: *Phys. Rev. Lett.*, **66**, 2060 (1991).
[73] M.-C. CHAN, S. S. LEE, M. OKUMURA and T. OKA: *J. Chem. Phys.*, **95**, 88 (1991).
[74] M.-C. CHAN, L.-W. XU, C. M. GABRYS and T. OKA: *J. Chem. Phys.*, **95**, 9404 (1991).
[75] T. MOMOSE, D. P. WELIKY and T. OKA: *J. Mol. Spectrosc.*, **153**, 760 (1992).
[76] J. C. MCLENNAN and J. H. MCLEOD: *Nature (London)*, **123**, 160 (1929).
[77] J. VAN KRANENDONK: *Solid Hydrogen* (Plenum Press, New York, N.Y., 1987).
[78] C. K. N. PATEL and A. C. TAMM: *Rev. Mod. Phys.*, **53**, 517 (1981).
[79] Y. J. CHABAL and C. K. N. PATEL: *Rev. Mod. Phys.*, **59**, 835 (1987).
[80] W. N. HARDY and A. J. BERLINSKY: *Phys. Rev. Lett.*, **34**, 1520 (1975).
[81] W. N. HARDY, A. J. BERLINSKY and A. B. HARRIS: *Can. J. Phys.*, **55**, 1150 (1977).
[82] L. PAULING: *Phys. Rev.*, **36**, 430 (1930).
[83] T. NAKAMURA: *Prog. Theor. Phys. (Kyoto)*, **14**, 135 (1955).
[84] J. VAN KRANENDONK: *Physica*, **25**, 1080 (1959).
[85] J. VAN KRANENDONK: *Can. J. Phys.*, **38**, 240 (1960).
[86] A. B. HARRIS: *Phys. Rev.*, **82**, 3495 (1920).
[87] A. B. HARRIS, A. J. BERLINSKY and W. N. HARDY: *Can. J. Phys.*, **55**, 1180 (1977).
[88] I. F. SILVERA: *Rev. Mod. Phys.*, **52**, 393 (1980).
[89] W. HEITLER: *The Quantum Theory of Radiation* (Oxford University Press, Oxford, 1960).
[90] R. E. MILLER and J. C. DECIUS: *J. Chem. Phys.*, **59**, 4871 (1973).
[91] D. P. WELIKY, K. KERR, T. MOMOSE, T. BYERS and T. OKA: unpublished.

A History of Laser Frequency Measurements (1967-1983): The Final Measurement of the Speed of Light and the Redefinition of the Meter.

K. M. EVENSON

Time and Frequency Division, National Institute of Standards and Technology Boulder, CO 80303-3328

1. – Introduction.

At an eventful CGPM meeting on October 20th, 1983, the meter was redefined [1]:

«The metre is the length of the path travelled by light in vacuum during a time interval of 1/299 792 458 of a second».

This definition makes the unit of length depend on time; both use the same standard; and, most importantly, it fixes the value of the speed of light at exactly 299 792 458 m/s. Thus, by definition, further measurements of c after this date are not possible. With this definition, length is realized most accurately by counting fringes of a stabilized laser whose frequency has been measured.

The discovery of the laser in 1961 [2] made frequency measurements from the microwave to the visible part of the spectrum possible, because the laser possessed the necessary coherence. The final measurement of the speed of visible light which is described in this lecture was made in 1983 just before the meter was redefined. The value of c is the product of the measured frequency and vacuum wavelength of the iodine-stabilized He-Ne laser oscillating at 630 nm. This measurement of the frequency of visible light paved the way for the redefinition of the meter, because it demonstrated that, not only were direct frequency measurements in the visible possible, but the frequency measured laser also provides a source of radiation for the realization of the new meter.

Frequency measurements have been made for many years in the microwave and radio regions of the electromagnetic spectrum; they are the most accurate metrological techniques yet invented. In the visible portion of the spectrum, frequency measurements are some 10 000 times more accurate than wavelength measurements and do not have fundamental physical accuracy limitations. The reader who is interested in more details is referred to a recent review paper [3]. In the process of extending frequency measurements to the visible, the frequency of the 3.39 μm methane-stabilized He-Ne laser multiplied by its wavelength yielded a value of the speed of light with a hundredfold decrease in its uncertainty [4]. Soon afterwards, measurements of c were made in other laboratories; they were all in agreement, within the accuracy of the standard of length itself. This agreement paved the way for the new definition of length in terms of time, fixing the value of c exactly at 299 792 458 m/s.

2. – Measurement of laser frequencies.

A heterodyne method is used to measure laser frequencies; that is, a laser's radiation is beat with harmonics of a lower-frequency laser, or, at the very lowest frequency, with harmonics of a microwave oscillator. A very-high-speed harmonic-generator mixer is necessary to generate these harmonics and mix them with the other radiations to generate a radiofrequency beat note. The metal-insulator-metal (MIM) diode is the most widely used harmonic-generator mixer at laser frequencies.

If one wishes to measure a laser frequency, ν_u (calculated from a wavelength measurement), one uses a known lower-frequency laser, ν_k, approximately $1/n$ times ν_u; and by adding (or subtracting) a microwave frequency ν_m, a beat frequency ν_b (of a few hundred megahertz) is obtained from the n-th harmonic of ν_k and its microwave sideband:

$$\nu_b = \nu_u - n(\nu_k) \pm \nu_m \, .$$

All of the radiation is focused on the MIM diode, and the radiofrequency beat note is synthesized in the MIM diode itself. Because n is usually less than 10, the extension of frequency measurements from the microwave to the visible (from 0.1 to 520 THz) requires a whole series of laser frequency measurements linked together in what is now called a frequency chain. Ten lasers were used to link the microwave to the visible in the first frequency measurement of visible radiation [3].

3. – Project beginnings.

In the early 1960's, Yardley BEERS, Chief of the Radio Physics Division of the National Bureau of Standards, Boulder, Colorado, had the insight to hire or have transferred into his division several research scientists whose tasks were loosely defined: to investigate possible uses of lasers for standards. The group included Drs. J. S. WELLS, F. R. PETERSEN, D. A. JENNINGS, R. L. STROMBOTNE,

R. L. BARGER and me; each of these scientists played significant roles in the work leading to the 1983 redefinition of the meter. This redefinition reduced the number of independent units in the International System of Units (SI) from six to five. (The candela had already been defined in terms of other SI units [5]; largely brought about because of the work of Dr. D. A. Jennings with his measurements of laser power and energy.) Work in one division in NBS reduced the number of independent SI units from seven to five! The kilogram, second, ampere, kelvin and mole are the remaining independent SI units.

I began working at the Boulder laboratories in 1963 in the millimeter wave section, and had the opportunity of working with Dr. H. P. BROIDA on one of the first microwave optical double-resonance experiments [6]. These experiments made on the CN radical did not use lasers, but they did use frequency measurements for recording the microwave spectrum, and the accuracy and simplicity of directly counting frequencies became obvious to me at that time.

In the summer of 1964, Dick STROMBOTNE mentioned to me that a pulsed laser oscillating at the extremely low frequency of 890 GHz (0.34 mm) had been discovered [7], and Dick and I speculated on the possibility of directly measuring frequencies that high. In April of 1965, I wrote to my Division Chief, Y. BEERS, and the director of the Boulder Laboratories, J. M. RICHARDSON, proposing to attempt to measure this laser's frequency. The proposal was accepted, and we started work on this task on July 1st of that year.

It had been proposed that the CN radical was responsible for the laser, and we attempted to understand the mechanism responsible for the laser action [8*a*]; however, it was eventually shown to be HCN [8*b*]. In April of 1966, MULLER and FLESHER [9, 10] succeeded in obtaining c.w. laser action in a discharge of CH_3CN showing us the way to run our laser to get c.w. emission. Before we succeeded in getting our laser to oscillate, we read that Ali Javan's group at MIT had succeeded in measuring that laser's frequencies [11]; this was quite a shock, for we had not known of their efforts. In that paper, they mentioned the possibility of obtaining a greatly improved value of the speed of light from the product of the frequency and wavelength of a laser. The improvement stems from the improved accuracy in the measurement of the wavelength at these laser wavelengths. I quickly repeated MIT's measurements, confirming their results. I realized that an even more accurate value of the speed of light would come from a measurement at an even higher frequency; therefore, I decided to attempt to measure even higher frequencies.

Before I succeeded in measuring higher frequencies, a new scientist in the Boulder area convinced my new division chief, H. S. BOYNE, that I was too inexperienced to run the project, and the project was terminated. Fortunately, my section chief, D. A. JENNINGS, disagreed, and I was permitted to remain at NBS and work on the laser power and energy measurement program and a project suggested by Prof. M. MIZUSHIMA from the physics department at the University of Colorado: the measurements of the absorption of HCN laser radiation by the oxygen molecule tuned into resonance with a magnetic field. This later experiment was very successful, and we initiated a whole new field of spectroscopy: laser magnetic resonance, LMR [12]. This success convinced my division chief that I was not too inexperienced to work on the frequency measurement experiment. Dr. Ali JAVAN became a member of our division's program review panel at this time, and insisted that our frequency measurement program be restarted. It was restarted with the addition of a most competent co-worker, Dr. J. S. WELLS, who helped the project move much faster than before.

We started by attempting to measure the frequency of the 28 μm (10.7 THz) water vapor laser by generating 12 harmonics of the 337 μm HCN laser. It was not very likely that the point-contact silicon diode which Javan's group had used in their first measurement of laser frequencies would be operational at these high frequencies. Dr. Vernon DERR from NOAA advised us that the MIM diode might extend to higher frequencies than the silicon diode. We assembled a collection of different metals to test, when the news of Javan's group's successful use of this diode reached us [13]. They had used either a silver or steel base and a tungsten whisker, but soon found that nickel was a better base metal [14]. This combination is the same one we have found to be the most efficient, even after years of searching for better pairs of metals.

The earliest reference to the MIM diode which I have found is a prophetic statement in 1948 in Torrey and Whitmer's book, *Crystal Rectifiers* [15]: «It should be possible to make metal-insulator-metal rectifiers with much smaller spreading resistances than metal-semiconductor rectifiers have, consequently giving greater rectification efficiency at high frequencies» (a very prophetic statement!). We then improved the coupling to the diode [16] by using long-wire antenna theory developed for radio communication in the late 1920's. Most recently, we studied the diode extensively [17, 18], and have shown that it does mix in the visible even though it does not generate harmonics there [19].

4. – World record frequency measurements.

In the fall of 1969, we succeeded in measuring our first laser frequencies: the 28 and 78 μm water vapor laser lines [20]. In the meantime, Javan's group had made three more «world record frequency measurements» [21-23]; but, now, we were in the competition. Our main goal was to measure frequencies sufficiently high so that accurate wavelength measurements could be made and a highly accurate value of the speed of light would result from the product of the frequency and wavelength. Then, if one could measure frequencies in the visible, the meter could be defined in terms of the second; thus any stable laser whose frequency had been measured could be used as a length standard. That is, since the vacuum wavelength

TABLE I. – *World record laser frequency measurements.*

Frequency (THz)	Wavelength (μm)	Laser	Synthesis (THz)	Harmonic generator	Year	Group	Reference
0.891	337	HCN	12 × (0.074)	Si	1967	Javan	[11]
0.964	311	HCN	13 × (0.074)	Si	1967	Javan	[11]
1.540	194	DCN	22 × (0.070)	Si	1967	Javan	[21]
1.578	190	DCN	23 × (0.068)	Si	1967	Javan	[21]
2.528	119	H_2O	17 × (0.149)	Si	1967	Javan	[22]
3.557	84	D_2O	4 × (0.891) − 0.007	—	1969	Javan	[23]
3.822	78	H_2O	6 × (0.891) − 2 × × (0.804) + 3 × × (0.029)	MIM	1969	NBS	[20]
10.718	28	H_2O	12 × (0.891) + 0.029	MIM	1969	NBS	[20]
28.306	10.6	CO_2	3 × (10.7) − − 3.822 + 0.026	MIM	1970	NBS	[25]
28.360	10.6	CO_2	3 × (10.7) − − 3.822 + 0.027	MIM	1970	NBS	[25]
88.376	3.39	He-Ne	3 × (29.442) + 0.048	MIM	1971	NBS	[26]
147.916	2.0	Xe	88.376 + 30.922 + + 28.616	MIM	1974	NBS	[27]
196.780	1.5	He-Ne	147.916 + 48.862	MIM	1977	NBS	[28]
260.103	1.15	He-Ne	196.780 + 32.373 + + 30.950	MIM	1979	NBS	[29]
520.207	0.576	doubled HeNe	2 × (260.103)	$LiNbO_3$	1979	NRC-NBS	[24]

of that laser is c/ν, the meter would be realized by counting fringes (1 meter $= n$ fringes where each fringe is $\lambda/2$ long and $n = 2\nu/c$).

Our group's laser frequency measurements progressed rapidly, reaching the visible part of the electromagnetic spectrum in 1979. This world record frequency measurement is in the yellow green portion of the electromagnetic spectrum (520 THz) [24]. Table I summarizes the various frequency measurements leading to this visible frequency measurement. The synthesizing laser frequencies are shown in the synthesis column of table I. The next lower-frequency laser is used as a stepping stone to get to the higher frequency. The result is a whole chain of frequency measurements connecting lower-frequency lasers to the highest ones.

The extensions to 28 THz and 88 THz are very significant because these lasers can be locked to very narrow sub-Doppler features in CO_2 and CH_4, respectively [30, 31]. Independent measurements can then be made on the wavelength and frequency of these stabilized lasers (actually on the frequency of the molecular absorptions), and the product will yield accurate values of the speed of light. Since the lasers are locked to these supernarrow Lamb dips, they become secondary frequency standards. Significantly more accurate values of the speed of light will result because much more accurate wavelength measurements can be made at these short wavelengths. Frequency measurements are so accurate, that the accuracy in the values of c is not affected by the much smaller fractional uncertainties in the measurement of frequency compared with those of wavelengths.

5. – The speed of light.

The fortuitous coincidence of the frequency of methane (CH_4) with that of the 3.39 μm lasing line of neon meant that this laser could be locked and stabilized to the frequency of methane, thus facilitating accurate independent measurements of frequency and wavelength [32]. Once we had demonstrated that we could measure frequencies this high, we constructed our own methane-stabilized laser and proceeded to measure its frequency. This task required a year [33], and the product of its measured frequency and wavelength produced a value of the speed of light [4] with about a 100 times smaller uncertainty than that of the accepted value [34]. Although the first laser measurement of the speed of light had been completed by BAY *et al.*, three months earlier [35], it did not use a direct frequency-measuring technique. Consequently, the accuracy Bay's group achieved was about twenty times less than ours. Our value was limited mainly by uncertainties in the length standard itself (about 1.3 m/s).

Several highly accurate speed-of-light measurements were made following ours, and they are shown in fig. 1 along with the corresponding uncertainties. Most of the measurements were made in various standards laboratories throughout the world. In 1973, after the first two measurements were completed, it was obvious a laser would make a far better length standard than the krypton standard. Consequently, the Consultative Committee for the Definition of the Meter adopted a tentative value of 299 792 458 m/s to be used in a possible redefinition of the meter, so that any laser whose frequency had been measured could be used to realize the meter.

One of the striking features of these measurements is that, for the first time since the Danish astronomer ROEMER obtained the first value of c over 300 years

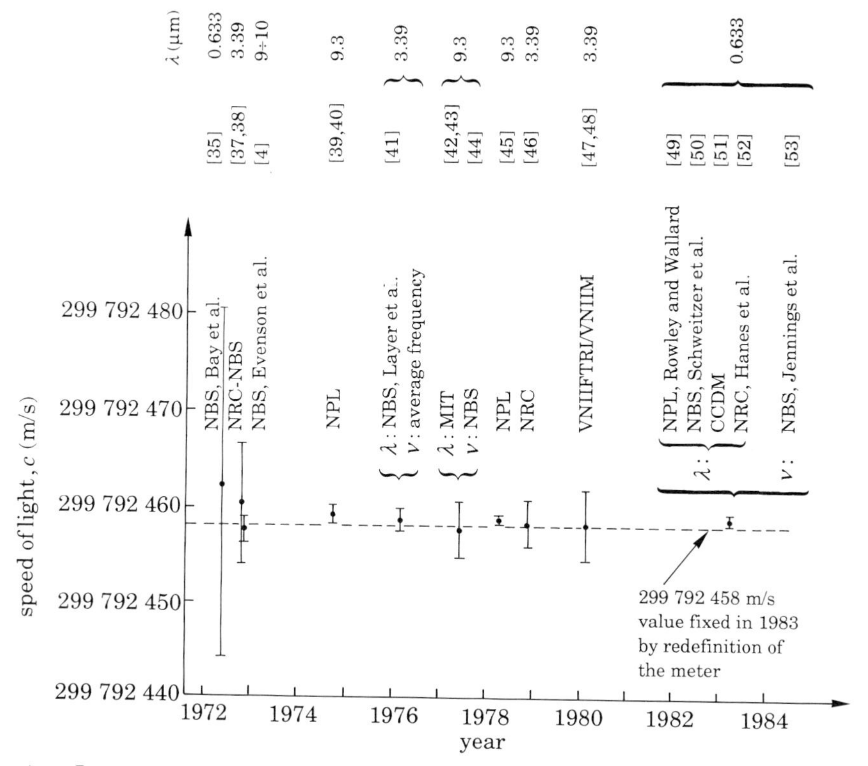

Fig. 1. – Laser speed-of-light measurements.

ago [36], the values were all in agreement within their quoted uncertainties!

The MIT measurement shown in fig. 1 depended upon a direct measurement of the 9.3 μm wavelength and, consequently, exhibits the increased uncertainty due to greater diffraction uncertainties in this wavelength measurement. The latest (1983) measurement shown is the final measurement of the speed of light and will be described later in this lecture.

6. – The fundamental constants.

The previous «best» value for the speed of light was obtained by fitting the interrelated fundamental constants to a best value for each of the constants [54]. In 1969 that value was known with an accuracy of about 100 m/s. The uncertainty in the length standard limited all of the laser measurements to an accuracy of 1.3 m/s. In 1973, the recommended value of c was used in a refit of all of the fundamental constants [55]. The most recent fit was made in 1986 [56] using this same value of c with its uncertainty equal to zero.

7. – Frequency and wavelength standards.

The accurate measurements of the various laser frequencies provide both frequency and wavelength standards throughout the laser spectral region. The vacuum wavelength is defined as exactly c/ν in the new definition of the meter to be discussed later. The frequencies of the gas lasers themselves are tunable by a few parts in a million, and consequently, even when locked to line center, provide a frequency and wavelength standard with an accuracy of a few parts in 10^8. All of the «world record» frequencies listed in table I provide such frequency and wavelength standards. In addition to these, a number of far-infrared lasers have been frequency measured to an accuracy of about one part in 10^7. These include the H_2O discharge laser and the various optically pumped lasers, such as methyl alcohol, difluoromethane, formic acid and various isotopes of these [57-61].

The really excellent standards are those in which saturated-absorption (sub-Doppler) features are used to lock the lasers. An excellent review of such lasers above 10 THz is present in Knight's paper [62]. The CO_2 laser stabilized on the low-pressure fluorescence [30] is an excellent standard because every lasing line can be stabilized. Consequently, these lines including the isotopes of CO_2 have been frequency measured [63-65]. However, the fluorescence signal does not exhibit as good a signal-to-noise ratio as does a molecule in direct absorption; thus there are more stable lines available from such molecules as SF_6 and OsO_4, and these have also been measured [66, 67]. Their disadvantage is that they occur randomly, not every lasing line exhibits an absorption, and the locked lasers are not necessarily at line center.

Methane at 88 THz is one of the most stable of the stabilized lasers, and much of the pioneering work on that laser has been performed in the laboratory of V. P. Chebotaev in Siberia, Russia. In one of their most recent measurements, the Siberian group measured the frequency of the E-line of methane to an accuracy of about one kHz (to about a part in 10^{11}) [68].

In 1976 we were transferred into the Time and Frequency Division under Jim BARNES; this transfer was most appropriate, for we were already busy making frequency measurements! We received excellent support from him and his boss, Dr. Karl KESSLER, director of the Center for Basic Standards in NBS.

8. – The measurements of iodine.

In the proposed new definition of the meter, which would be acted on at the 17th Conférence Générale des Poids et Mesures (CGPM) in 1983, several stabilized lasers whose frequencies had been measured would be referenced, and these lasers could be used to realize the new standards of length. In the proposed new single standard for frequency and length, the meter would be the distance light travels in a fraction of a second, thus the meter would be realized by counting the fringes formed in an interferometer using a laser which had been frequency measured. The most accurate wavelength measurements are made at the shortest wavelengths possible, and, for convenience sake, visible light is the best choice.

In 1982 we set up to measure such visible frequencies: the frequencies of iodine at two times the 260 THz He-Ne laser which we had previously measured less accurately when we had extended frequency measurements to the visible [24]. This time we used the hyperfine component of the $^{127}I_2$ 17-1 $P(62)$ at 520 THz (576 nm) to stabilize our dye laser [69]. We used the saturated-fluorescence-stabilized CO_2 laser as our secondary frequency references, and these were in turn measured with respect to the 88 THz methane-stabilized He-Ne laser [62]. The MIM diode was used to synthesize the sum of two harmonics of the 11.5 μm $P(52)$ CO_2 laser line plus three harmonics of the 11.5 μm $P(50)$ line equal to the frequency of the 2.3 μm line furnished by a color center laser. Lithium niobate crystals were used to double the 130 THz laser to the 260 THz He-Ne laser which was also doubled in lithium niobate to the dye laser at 520 THz. An accuracy of 1.6 parts in 10^{10} was achieved for the 520 THz line of iodine and 3.1 parts in 10^{10} for the 260 THz line of ^{20}Ne.

9. – Iodine at the red He-Ne frequency.

The He-Ne red laser at 473 THz (633 nm) had already been shown to be an excellent stabilized laser; however, its frequency had not yet been measured! We next planned its measurement using the clever scheme devised by KLEMENTYEV, MATYUGIN and CHEBOTAYEV [70]. We would resonantly sum 88, 125 and 260 THz in Ne, to synthesize 473 THz which would beat with another He-Ne laser stabilized to iodine.

By 1983, internal funding in NBS was becoming very scarce; Karl KESSLER sent $ 120 000 to us for use for the red-frequency measurement. However, our new division chief, Sam STEIN, decided that only projects funded by other agencies would continue in our division. Consequently, the money was spent evenly throughout the division, and we could not finish the measurement. We then spent three months trying to find external funding for the experiment but were unsuccessful. (The National Science Foundation does not fund NBS, and the military funds atomic-clock research but is not interested in length standards.) Thus we were shut down.

John HALL from the Quantum Physics Division of NBS, Russ PETERSEN, Don JENNINGS, I and Karl KESSLER successfully appealed to our Division Chief for three month's of funding so that the measurements might be made.

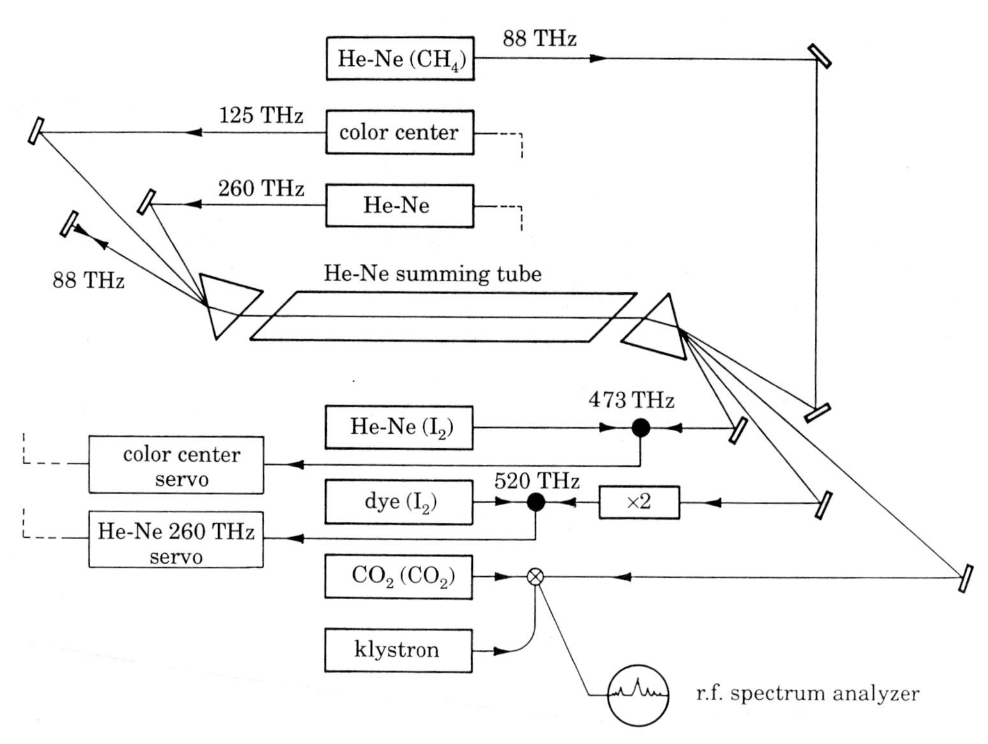

Fig. 2. – Frequency synthesis of the iodine-stabilized red He-Ne laser at 473 THz: ● Si photodiode, ⊗ MIM diode.

We assembled 8 of the world's experts: D. A. JENNINGS, C. R. POLLOCK, F. R. PETERSEN, R. E. DRULLINGER, J. S. WELLS, J. L. HALL, H. P. LAYER and me (all from NBS). These eight were all experts on laser stabilization, synthesis and metrology. We set up the experiment and synthesis steps shown in fig. 2 [53]. Each of the eight people had a most significant, but almost independent role in the experiment. For example, Howard LAYER from NBS, Washington, brought his iodine-stabilized laser with him; Cliff POLLOCK, our post-doctoral fellow, was in charge of the color center laser, Jan HALL brought his methane-stabilized laser from JILA, and similarly for the other 5 scientists. Once all of the apparatus were assembled, we were able to make all of the necessary measurements in about a week of fun and excitement. We sometimes worked from 6 in the morning to 10 at night. It was one of the most harmonious and happy scientific experiments of my career! The experiment went most smoothly, and we found the frequencies to be 473 612 340.492 and 473 612 214.789 THz for the g and i components [71]. We had been able to make the measurements before the forthcoming CGPM meeting!

10. – The final measurement of the speed of light.

The final value of the speed of light (before it was to be fixed at exactly 299 792 458 m/s by the redefinition) was obtained from the above value of the frequency, multiplied by the average value of the four best wavelengths [49-52]. The value obtained was (299 792 458.6 + 0.3) m/s (one sigma). This number is in agreement with the value to be used in the redefinition within the recognized uncertainty in the krypton lamp (1.3 m/s). Thus the way was paved for the redefinition of the meter: the value of c was reconfirmed for the last time and there were now two iodine frequencies in the visible which could be used to realize the meter. Either one of these frequencies were known to about an order-of-magnitude greater accuracy than it was possible to realize the meter using the existing standard, that from the krypton lamp.

11. – The new definition of the meter.

This was only the second redefinition of the meter since it was defined in terms of the international prototype by the first CGPM in 1889. The first redefinition was in 1960 when the meter was defined in terms of radiation of krypton-86.

This new redefinition fixes the value of the vacuum speed of light at exactly 299 792 458.000 000 0 ... m/s, without any uncertainty. The new definition eliminates an independent standard of length and reduces the number of standards by one.

A list of recommended wavelengths to be used in the realization of the meter was also given [1]; it included the 88 THz line of methane and both of the frequency-measured iodine lines described in this lecture. It also includes two more iodine lines which have been very carefully wavelength measured and are used to stabilize the argon laser. The krypton lamp itself can still be employed as can some of the other nonlaser lines used in the past [1]. The new definition is very flexible; for the greatest accuracy more accurate frequency measurements need to be made because the comparisons of wavelength can be made to a few parts in 10^{11}. However, due to a lack of funds, more measurements were not possible, and the laboratory was dismantled and most of the lasers were discarded.

12. – The rydberg.

The next and most significant spectroscopy measurements which could be made are measurements of atomic hydrogen yielding much more accurate values of the Rydberg constant which relates so many of the fundamental constants [72]. We proposed this project to E. AMBLER, the director of the Bureau of Standards, but were refused due to funding difficulties and because we had been making frequency measurements for some 15 years and it «was time to do something new». He also has been quoted as saying he did not want any more fundamental constants measured while he was director.

Direct frequency measurements of the rydberg are now being pursued by T. W. HÄNSCH in Munich, Germany. He has been active in the field for many years [73], and has made the most accurate measurements of the Rydberg constant. The uncertainty of his group's wavelength measurement has been the leading cause of the uncertainty in his recent values of the rydberg (see his paper in these proceedings).

13. – The MIM and the synthesis of far-infrared radiation.

The amazingly simple MIM diode, hand-made from a 25 μm diameter tungsten wire electrochemically etched in a 1 N NaOH solution and contacting ordinary nickel, has been at the center of our research for over 20 years. We have shown: 1) that the speed of this diode is faster than $0.7 \cdot 10^{-14}$ s, 2) it behaves as if electron tunneling through the oxide layer were responsible for its action, and 3) it is sufficiently nonlinear to be used out to 18th order. The eighteenth-order experiment (unpublished) came from the synthesis of the 26 THz $10P(50)$ line of $^{13}CO_2$ from the 16th harmonic of the 1.626 THz line of CH_2F_2 plus 9.7 GHz.

During the years we had been performing these experiments, far-infrared detectors had become much more sensitive. This increased sensitivity plus our ability to use the MIM prompted us to try to synthesize far-infrared radiation with the MIM. We were able to detect significant amounts of radiation in our liquid-helium-cooled bolometer. In fact, we generated sufficient far-infrared radiation to see the signal directly on the oscilloscope, and also to observe an absorption line of CO [74]. We had invented a whole new way of generating tunable far-infrared radiation in a region where the only other technique used microwave sidebands of FIR lasers. We have been performing tunable far-infrared spectroscopy with this new technique which covers the entire far infrared with about $1 \cdot 10^{-7}$ W of power. By comparison, the sideband technique generates more power, but has many frequency regions where there are no strong FIR laser lines. Our technique is also extremely accurate because the CO_2 lasers are stabilized and the FIR difference is accurate to about 12 kHz (far-infrared lasers are accurate to only 5 parts in 10^7, a factor of about 130 less accurate).

We made an even more amazing discovery a couple of years later: we discovered that, instead of using nickel as a base, we could use cobalt and generate microwave sidebands on the CO_2 laser difference and enjoy microwave tunability. With this third-order technique we have tens of gigahertz of tunability instead of the 100's of MHz from the high-pressure waveguide laser we used in the first 2nd-order spectrometer [75]. We are now exploiting this revolutionary technique to perform tunable high-resolution laser spectroscopy in the entire far infrared from 0.3 to 6.0 THz. The MIM diode continues to amaze us!

14. – The future.

Further work on the MIM diode will almost certainly lead to improvements in its performance. For example, we were not successful in using the MIM diode to generate harmonics of a CO_2 laser to 260 THz (1.15 μm). We were not successful in obtaining the usual negative-going whisker until we used an extremely-high-quality microscope objective. This seemed to indicate that the MIM might be operating in two modes, the usual high-speed one requiring better coupling.

The search for a new time standard is now being made in the visible with Hg^+ as one of the more likely candidates [76]. In order to adopt such a new standard, it will be necessary to directly measure frequencies in the visible (*i.e.* to directly connect frequencies in the visible with the frequency of the present cesium standard at 10 GHz). These measurements must be made to an accuracy of a few parts in 10^{14} and must be much more accurate than ours were (ours might be considered only feasibility measurements by comparison). The future of frequency measurements looks very bright and is certainly very exciting spectroscopically. Most of the standards laboratories in the world (including NIST, as NBS is now called) have major efforts in this field!

REFERENCES

[1] *Documents concerning the new definition of the metre, Metrologia*, no author given, **19**, 163 (1984).
[2] A. JAVAN, W. R. BENNETT jr. and D. R. HERRIOTT: *Phys. Rev. Lett.*, **6**, 106 (1961).
[3] D. A. JENNINGS, K. M. EVENSON and D. J. E. KNIGHT: *Proc. IEEE*, **74**, 168 (1986).
[4] K. M. EVENSON, J. S. WELLS, F. R. PETERSEN, B. L. DANIELSON, G. W. DAY, R. L. BARGER and J. L. HALL: *Phys. Rev. Lett.*, **29**, 1346 (1972).
[5] K. D. MIELENZ: *J. Res. Natl. Bur. Stand.*, **92**, 335 (1987).
[6] K. M. EVENSON, J. L. DUNN and H. P. BROIDA: *Phys. Rev.*, **136**, A1566 (1964).
[7] H. A. GEBBIE, N. W. B. STONE and F. D. FINDLAY: *Nature (London)*, **202**, 685 (1964).
[8*a*] H. P. BROIDA, K. M. EVENSON and T. T. KIKUCHI: *J. Appl. Phys.*. **36**, 3335 (1965).
[8*b*] D. R. LIDE and A. G. MAKI: *Appl. Phys. Lett.*, **11**, 62 (1967).
[9] G. T. FLESHER and W. M. MULLER: *Proc. IEEE*, **54**, 543 (1966).
[10] W. W. MULLER and G. T. FLESHER: *Appl. Phys. Lett.*, 8, 217 (1966).

[11] L. O. HOCKER, A. JAVAN, D. RAMACHANDRA RAO, L. FRENKEL and T. SULLIVAN: *Appl. Phys. Lett.*, **10**, 147 (1967).
[12] K. M. EVENSON, H. P. BROIDA, J. S. WELLS, R. J. MAHLER and M. MIZUSHIMA: *Phys. Rev. Lett.*, **21**, 1038 (1968).
[13] L. O. HOCKER, D. R. SOKOLOFF, V. DANEU, A. SZOKE and A. JAVAN: *Appl. Phys. Lett.*, **12**, 401 (1968).
[14] V. DANEU, D. SOKOLOFF, A. SANCHEZ and A. JAVAN: *Appl. Phys. Lett.*, **15**, 398 (1969).
[15] H. C. TORREY and C. A. WHITMER: *Crystal Rectifiers* (McGraw-Hill Co., New York, N.Y., 1948), p. 68.
[16] L. M. MATARRESE and K. M. EVENSON: *Appl. Phys. Lett.*, **17**, 8 (1970).
[17] E. SAKUMA and K. M. EVENSON: *IEEE J. Quantum Electron.*, QE-**10**, 559 (1974).
[18] K. M. EVENSON, M. INGUSCIO and D. A. JENNINGS: *J. Appl. Phys.*, **57**, 956 (1985).
[19] R. E. DRULLINGER, K. M. EVENSON, D. A. JENNINGS, F. R. PETERSEN, J. C. BERGQUIST, L. BURKINS and H.-U. DANIEL: *Appl. Phys. Lett.*, **42**, 137 (1983).
[20] K. M. EVENSON, J. S. WELLS, L. M. MATARRESE and L. B. ELWELL: *Appl. Phys. Lett.*, **16**, 159 (1970).
[21] L. O. HOCKER, D. RAMACHANDRA RAO and A. JAVAN: *Phys. Lett. A*, **24**, 690 (1967).
[22] L. FRENKEL, T. SULLIVAN, M. A. POLLACK and T. J. BRIDGES: *Appl. Phys. Lett.*, **11**, 344 (1967).
[23] L. O. HOCKER, J. G. SMALL and A. JAVAN: *Phys. Lett. A*, **29**, 321 (1969).
[24] K. M. BAIRD, K. M. EVENSON, G. R. HANES, D. A. JENNINGS and F. R. PETERSEN: *Opt. Lett.*, **4**, 263 (1979).
[25] K. M. EVENSON, J. S. WELLS and L. M. MATARRESE: *Appl. Phys. Lett.*, **16**, 251 (1970).
[26] K. M. EVENSON, G. W. DAY, J. S. WELLS and L. O. MULLEN: *Appl. Phys. Lett.*, **20**, 133 (1972).
[27] D. A. JENNINGS, F. R. PETERSEN and K. M. EVENSON: *Appl. Phys. Lett.*, **26**, 510 (1975).
[28] K. M. EVENSON, D. A. JENNINGS, F. R. PETERSEN and J. S. WELLS: *Laser frequency measurements: a review, limitations, extension to* 197 THz (1.5 μm), in *Laser Spectroscopy III*, edited by J. L. HALL and J. L. CARLSTEN (Springer-Verlag, Berlin, Heidelberg, New York, N.Y., 1979), p. 56.
[29] D. A. JENNINGS, F. R. PETERSEN and K. M. EVENSON: *Opt. Lett.*, **4**, 129 (1979).
[30] C. FREED and A. JAVAN: *Appl. Phys. Lett.*, **17**, 53 (1970).
[31] J. L. HALL: *IEEE J. Quantum Electron.*, QE-**4**, 638 (1968).
[32] R. L. BARGER and J. L. HALL: *Appl. Phys. Lett.*, **22**, 196 (1973).
[33] K. M. EVENSON, J. S. WELLS, F. R. PETERSEN, B. L. DANIELSON and G. W. DAY: *Appl. Phys. Lett.*, **22**, 192 (1973).
[34] K. D. FROOME: *Proc. R. Soc. London*, **247**, 109 (1958).
[35] Z. BAY, G. G. LUTHER and J. A. WHITE: *Phys. Rev. Lett.*, **29**, 189 (1972).
[36] M. ROEMER: *J. Scavans*, 233 (December 7, 1976).
[37] K. M. BAIRD, H. D. RICCIUS and K. J. SIEMSEN: *Opt. Commun.*, **6**, 91 (1972).
[38] J. D. CUPP, B. L. DANIELSON, G. W. DAY, L. B. ELWELL, K. M. EVENSON, D. G. MCDONALD, L. O. MULLEN, F. R. PETERSEN, A. S. RISLEY and J. S. WELLS: *The speed of light: progress in the measurement of the frequency of the methane stabilized* He-Ne *laser at* 3.39 μm, in *Proceedings of the Conference on Precision Electromagnetic Measurements* (*Boulder, Colo.*, 1972), p. 79.

[39] T. G. BLANEY, C. C. BRADLEY, G. J. EDWARDS, B. W. JOLLIFE, D. J. E. KNIGHT, W. R. C. ROWLEY, K. C. SHOTTON and P. T. WOODS: *Nature (London)*, **251**, 46 (1974).
[40] T. G. BLANEY, C. C. BRADLEY, G. J. EDWARDS, B. W. JOLLIFE, D. J. E. KNIGHT, W. R. C. ROWLEY, K. C. SHOTTON and P. T. WOODS: *Measurement of the speed of light:* I. *Introduction and frequency measurement of a carbon dioxide laser*, and II. *Wavelength measurement and conclusion, Proc. R. Soc. London, Ser. A*, **355**, 61, 89 (1977).
[41] H. P. LAYER, R. D. DESLATTES and W. G. SCHWEITZER jr.: *Appl. Opt.*, **15**, 734 (1976).
[42] J. P. MONCHALIN, M. J. KELLY, J. E. THOMAS, N. A. KURNIT, A. SZOKE, A. JAVAN, F. ZERNIKE and P. H. LEE: *Opt. Lett.*, **1**, 5, 140 (1977).
[43] J. P. MONCHALIN, M. J. KELLY, J. E. THOMAS, N. A. KURNIT, A. SZOKE, F. ZERNIKE, P. H. LEE and A. JAVAN: *Appl. Opt.*, **20**, 736 (1981).
[44] F. R. PETERSEN, D. G. MCDONALD, J. D. CUPP and B. L. DANIELSON: *Accurate rotational constants, frequencies and wavelengths from* $^{12}C^{16}O_2$ *lasers stabilized by saturated absorption*, in *Proceedings of the Vail Conference, Laser Spectroscopy*, edited by R. G. BREWER and A. MOORADIAN (Plenum, New York, N.Y., 1973), p. 555.
[45] P. T. WOODS, K. C. SHOTTON and W. R. C. ROWLEY: *Appl. Opt.*, **17**, 1048 (1978).
[46] K. M. BAIRD, D. S. SMITH and B. G. WHITFORD: *Opt. Commun.*, **31**, 367 (1979).
[47] V. M. TATARENKOV, V. G. IL'IN, V. I. KIPARENKO, V. K. KOROBOV and S. B. PUSHKIN: *Meas. Tech.* (USSR), **23**, 108 (1980).
[48] V. P. KAPRALOV, G. M. MALYSHEV, P. A. PAVLOV, V. E. PRIVALOV, YA. A. FOVANOV and I. SH. ETSIN: *Opt. Spectrosc.* (USSR), **50**, 34 (1981).
[49] W. R. C. ROWLEY and A. J. WALLARD: *J. Phys. E*, **6**, 647 (1973).
[50] W. G. SCHWEITZER jr., E. G. KESSLER jr., R. D. DESLATTES, H. P. LAYER and J. R. WHETSTONE: *Appl. Opt.*, **12**, 2927 (1973).
[51] Rapport: Comité Consultatif pour la Définition du Mètre, 5th Session, p. M54, Bureau International des Poids et Mesures, Sèvres (1973).
[52] G. R. HANES, K. M. BAIRD and J. DEREMIGIS: *Appl. Opt.*, **12**, 1600 (1973).
[53] D. A. JENNINGS, C. R. POLLOCK, F. R. PETERSEN, R. E. DRULLINGER, K. M. EVENSON, J. S. WELLS, J. L. HALL and H. P. LAYER: *Opt. Lett.*, **8**, 136 (1983).
[54] B. N. TAYLOR, W. H. PARKER and D. N. LANGENBERG: *Rev. Mod. Phys.*, **41**, 375 (1969).
[55] E. R. COHEN and B. N. TAYLOR: *J. Phys. Chem. Ref. Data*, **2**, 663 (1973).
[56] E. R. COHEN and B. N. TAYLOR: *J. Phys. Chem. Ref. Data*, **17**, 1795 (1988).
[57] F. R. PETERSEN, K. M. EVENSON, D. A. JENNINGS, J. S. WELLS, K. GOTO and J. J. JIMÉNEZ: *IEEE J. Quantum Electron.*, QE-**11**, 838 (1975).
[58] F. R. PETERSON, K. M. EVENSON, D. A. JENNINGS and A. SCALABRIN: *IEEE J. Quantum Electron.*, QE-**16**, 319 (1980).
[59] F. R. PETERSEN, A. SCALABRIN and K. M. EVENSON: *Int. J. Infrared Millimeter Waves*, **1**, 111 (1980).
[60] E. C. C. VASCONCELLOS, F. R. PETERSEN and K. M. EVENSON: *Int. J. Infrared Millimeter Waves*, **2**, 705 (1981).
[61] M. INGUSCIO, G. MORUZZI, K. M. EVENSON and D. A. JENNINGS: *J. Appl. Phys.*, **60**, R161 (1986).
[62] D. J. E. KNIGHT: *Metrologia*, **22**, 251 (1986).
[63] F. R. PETERSEN, E. C. BEATY and C. R. POLLOCK: *J. Mol. Spectrosc.*, **102**, 112 (1983).
[64] A. CLAIRON, B. DAHMANI and J. RUTMAN: *IEEE Trans. Instrum. Meas.*, IM-**29**, 268 (1980).

[65] L. C. BRADLEY, K. L. SOOHOO and C. FREED: *IEEE J. Quantum Electron.*, QE-**22**, 234 (1986).
[66] CH. J. BORDÉ, M. OUHAYOUN, A. VAN LERBERGHE, C. SOLOMON, S. AVRILLIER, C. D. CANTRELL and J. BORDÉ: *High resolution saturated spectroscopy in* CO_2 *lasers. Application to the* ν_3 *bands of* SF_6 *and* OsO_4, in *Laser Spectroscopy IV*, edited by H. WALTHER and K. W. ROTHE (Springer-Verlag, Berlin, 1979), p. 142.
[67] A. CLAIRON, B. DAHMANI, A. FILIMON and J. RUTMAN: *IEEE Trans. Instrum. Meas.*, IM-**34**, 265 (1985).
[68] V. F. ZAKHAR'YASH, V. M. KLEMENT'EV, M. V. NIKITIN, B. A. TIMCHENKO and V. P. CHEBOTAEV: *Ž. Tekh. Fiz.*, **53**, 2241 (1983).
[69] C. R. POLLOCK, D. A. JENNINGS, F. R. PETERSEN, J. S. WELLS, R. E. DRULLINGER, E. C. BEATY and K. M. EVENSON: *Opt. Lett.*, **8**, 133 (1983).
[70] V. M. KLEMENTYEV, YU. A. MATYUGIN and V. P. CHEBOTAYEV: *JETP Lett.*, **24**, 508 (1976).
[71] D. A. JENNINGS, R. E. DRULLINGER, K. M. EVENSON, C. R. POLLOCK and J. S. WELLS: *J. Res. Natl. Bur. Stand.*, **92**, 11 (1987).
[72] P. ZHAO, W. LICHTEN, ZHI-XING ZHOU, H. P. LAYER and J. C. BERGQUIST: *Phys. Rev. A*, **39**, 2888 (1989).
[73] T. W. HÄNSCH: *High resolution spectroscopy of hydrogen*, in *The Hydrogen Atom*, edited by G. F. BASSANI, M. INGUSCIO and T. W. HÄNSCH (Springer-Verlag, Berlin, Heidelberg, 1989), p. 93.
[74] K. M. EVENSON, D. A. JENNINGS and F. R. PETERSEN: *Appl. Phys. Lett.*, **44**, 576 (1984).
[75] K. M. EVENSON, D. A. JENNINGS and M. D. VANEK: *Tunable far infrared laser spectroscopy*, in *Frontiers of Laser Spectroscopy of Gases*, edited by A. C. P. ALVES, J. M. BROWN and J. M. HOLLAS (Kluwer Academic Publishers, Amsterdam, 1988), p. 43.
[76] J. C. BERGQUIST, F. DIEDRICH, W. M. ITANO and D. J. WINELAND: Hg^+ *single ion spectroscopy*, in *Laser Spectroscopy IX*, edited by M. S. FELD, J. E. THOMAS and A. MOORADIAN (Academic Press, Inc., Boston, Mass., 1989), p. 274.

Laser Spectroscopy in the Far-Ultraviolet Region.

B. P. STOICHEFF

Department of Physics, University of Toronto
Ontario Laser and Lightwave Research Centre
Toronto, Ontario, M5S 1A7, Canada

1. – Introduction.

The availability of tunable lasers in the visible and infrared wavelength regions has made possible significant advances in atomic and molecular spectroscopy, as demonstrated in the programme of the present discussions on *Frontiers in Laser Spectroscopy*. However, progress in far-ultraviolet spectroscopy, namely, in the vacuum ultraviolet (VUV, $(200 \div 100)$ nm) and extreme ultraviolet (XUV, $(100 \div 20)$ nm) has not kept pace. This region of the spectrum lacks lasers and especially tunable lasers, so that experimental spectroscopy at these short wavelengths continues to experience considerable difficulties. In spite of these shortcomings, some progress has been made in producing coherent VUV and XUV radiation by frequency mixing of laser radiation in nonlinear media, and radiation tunable over the wavelength range 200 to ~ 50 nm is now available for spectroscopy.

It is my purpose to review briefly the basic theoretical concepts and general experimental methods for generating tunable VUV and XUV radiation, and to discuss applications in atomic and molecular spectroscopy, using examples from my laboratory.

The observation of second-harmonic generation (SHG) in 1961 by FRANKEN *et al.* [1] was a crucial step leading to the eventual production of coherent radiation in the VUV region. This observation was quickly followed with the classic theoretical paper on second- and third-order nonlinear susceptibilities by ARMSTRONG *et al.* [2]. Third-harmonic radiation (THG) was demonstrated by MAKER *et al.* [3] in crystals, glasses and liquids, and the major problem of generating even shorter-wavelength radiation (due to the limited transparency of many of the nonlinear solids to the region above ~ 200 nm) was resolved when NEW and WARD [4] succeeded in producing THG in a number of gases. HARRIS and MILES [5, 6] then demonstrated that high conversion efficiency of THG and of sum frequency mixing could be obtained by using phase-matched metal vapors as nonlinear media, and that efficiency could be significantly improved by resonance enhancement.

A seminal contribution which now forms the basis for generating coherent VUV and XUV radiation, tunable over broad regions, was made by HODGSON *et al.* [7]. Their concept was to make use of four-wave sum mixing, $2\omega_1 + \omega_2 = \omega_0$, together with the advantage of resonance enhancement. In the experiment, they used two dye lasers, one tuned to a two-photon resonance in a nonlinear metal vapor (strontium), and the other, tunable over a broad frequency range ω_2, such that $(2\omega_1 + \omega_2)\hbar$ corresponded to the energy from the ground state to a broad autoionizing state of the metal vapor. In this way, they succeeded in generating coherent radiation tunable over a relatively broad region near 190 nm. Such radiation is now produced from 200 to ~ 50 nm by this technique, using metal vapors and the rare gases as nonlinear media. The resulting radiation is coherent, monochromatic and directional, and thus has all of the characteristics of laser radiation, except high intensity. Nevertheless, the intensities achieved to date are sufficient for many applications in absorption and fluorescence spectroscopy.

2. – Résumé of theory.

Laser-driven VUV sources are based on third-harmonic generation (THG) or 4-wave sum mixing (4-WSM) in nonlinear media. These processes are usually described by the induced macroscopic polarization of the medium which, of course, is dependent on the polarizabilities of the atomic or molecular systems when irradiated by intense laser light. It is well known that the polarization of a medium in the presence of a monochromatic field $E(r, t) = \Sigma(\omega_i)$ can be written as

$$(1) \quad \overline{P}(\omega_i) = \chi^{(1)}(\omega_i) \cdot \overline{E}(\omega_i) + \sum_{jk} \chi^{(2)}(\omega_i = \omega_j + \omega_k) \cdot \overline{E}(\omega_j) \cdot \overline{E}(\omega_k) + \\ + \sum_{jkl} \chi^{(3)}(\omega_i = \omega_j + \omega_k + \omega_l) \cdot \overline{E}(\omega_j) \cdot \overline{E}(\omega_k) \cdot \overline{E}(\omega_l) + \ldots ,$$

where $\chi^{(n)}$ are the susceptibility tensors of n-th order. The lowest-order term producing nonlinear effects is $\chi^{(2)}$. However, this tensor has nonzero components only in noncentrosymmetric systems; isotropic media such as cubic crystals, liquids and gases do not exhibit quadratic nonlinearities. For third-order processes such as THG and 4-WSM we need be concerned only with $\chi^{(3)}$, whose principal term may be written [2, 8]

$$(2) \quad \chi^{(3)}(\omega_0 = \omega_1 + \omega_2 + \omega_3) = \frac{3e^4}{4\hbar^3} \frac{r_{ga} r_{ab} r_{bc} r_{cg}}{(\Omega_{cg} - \omega_1 - \omega_2 - \omega_3)(\Omega_{bg} - \omega_1 - \omega_2)(\Omega_{ag} - \omega_1)} .$$

Here $r_{ga} = \langle g | r | a \rangle$ is the electric-dipole matrix element between the ground

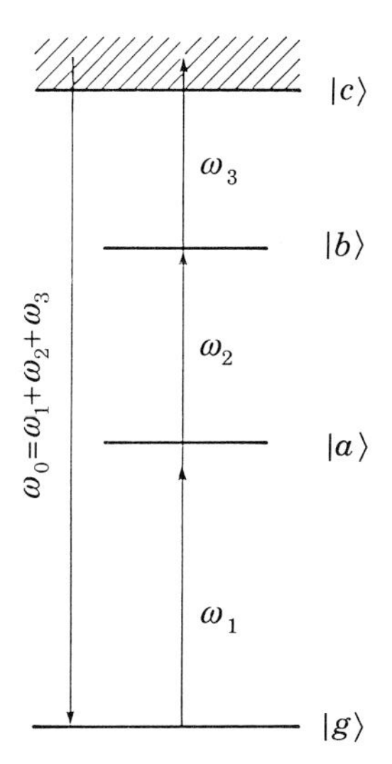

Fig. 1. – Diagram of 4-WSM process $\omega_0 = \omega_1 + \omega_2 + \omega_3$ with a 2-photon resonance $\Omega_{bg} = \omega_1 + \omega_2$.

state $|g\rangle$ and an excited state $|a\rangle$, having a lifetime Γ_a, and $\Omega_{ag} = \omega_{ag} - i\Gamma_a/2$ is the energy difference between states $|a\rangle$ and $|g\rangle$, e is the electronic charge and $\hbar = h/2\pi$, with h being Planck's constant.

A representation of a system having energy levels $|g\rangle$, $|a\rangle$, $|b\rangle$, $|c\rangle$, with applied frequencies ω_1, ω_2, ω_3, is given in fig. 1, as an aid to visualizing the possible virtual transitions and properties of eq. (2). Equation (2) shows that $\chi^{(3)}$ will be resonantly enhanced whenever the applied frequencies, ω_1, ω_2, ω_3, are such that the real part of the resonance denominator vanishes, namely when $(\Omega_{ag} - \omega_1) = 0$, or $(\Omega_{bg} - \omega_1 - \omega_2) = 0$, or $(\Omega_{cg} - \omega_1 - \omega_2 - \omega_3) = 0$, corresponding to one-, two- or three-photon resonance, respectively. If any of ω_1, ω_2, ω_3 is set equal to a resonance frequency (Ω_{ag}, etc.), $\chi^{(3)}$ will be enhanced, but the incident radiation will be strongly absorbed. If, however, $\omega_1 + \omega_2$ is equal to a 2-photon resonance (Ω_{bg}), the incident radiation at $\omega_1 + \omega_2$ is expected to be only weakly absorbed by the 2-photon transition, while the resonance enhancement of $\chi^{(3)}$ could be just as strong as for the 1-photon resonances.

For third-harmonic generation (THG), $\chi^{(3)}$ given in eq. (2) simplifies to

$$\chi^{(3)}(\omega_0 = 3\omega) = \frac{3e^4}{4\hbar^3} \frac{r_{ga} r_{ab} r_{bc} r_{cg}}{(\Omega_{cg} - 3\omega)(\Omega_{bg} - 2\omega)(\Omega_{ag} - \omega)} . \tag{3}$$

When 2ω approaches resonance, $\chi^{(3)}$ undergoes strong ($> 10^4$) enhancement. For efficient THG, collinear phase matching is necessary, that is, the refractive index $n(3\omega) = n(\omega)$ in order to yield a maximum effective interaction length. Limited tunability (without the loss of resonance enhancement) is achieved by varying the incident frequency ω over the linewidth Γ_b^{-1}.

For generating tunable radiation by 4-WSM, the process $\omega_0 = 2\omega_1 + \omega_2$ is of interest. Strong enhancement is again achieved by tuning $2\omega_1$, to a parity-allowed 2-photon resonance Ω_{bg}, and $\chi^{(3)}$ becomes

$$(4)\ \chi^{(3)}(\omega_0 = 2\omega_1 + \omega_2) = \frac{3e^4}{4\hbar^3}\frac{1}{\Omega_{bg} - 2\omega_1}\sum_{ca}\frac{r_{ga}r_{ab}r_{bc}r_{cg}}{(\Omega_{ag} - \omega_1)(\Omega_{cg} - 2\omega_1 - \omega_2)}.$$

Tunability and further enhancement is then obtained by selecting ω_2 so that $2\omega_1 + \omega_2$ corresponds to the ionization continuum or to broad autoionizing levels above the ionization limit. In such a situation, $\chi^{(3)}$ may be written as

$$(5)\quad \chi^{(3)}(2\omega_1 + \omega_2) = \frac{3e^4}{4\hbar^3}\frac{r_{ab}r_{ga}}{(\Omega_{bg} - 2\omega_1)(\Omega_{ag} - \omega_1)}C\int\frac{r_{gc'}r_{bc'}\,\mathrm{d}\omega}{\Omega_{c'g} - 2\omega_1 - \omega_2},$$

where now $r_{gc'}$ and $r_{bc'}$ are the matrix elements of the dipole moment between the ground state $|g\rangle$ and the perturbed continuum state $|c'\rangle$, and between the states $|b\rangle$ and $|c'\rangle$. The integration takes into account the contribution of the autoionizing states as well as of the ionization continuum, and C is a normalization constant. More detailed treatments of the relevant theory including phase matching, saturation effects and conversion efficiencies have been given by JAMROZ and STOICHEFF [9] and by VIDAL [10].

3. – Sources for generating tunable, coherent radiation.

Several experimental arrangements for generating tunable VUV and XUV radiation have been developed, and are reviewed briefly in [11]. Here, I will describe two systems, one begun in 1975 at the University of Toronto, that uses primarily metal vapors as nonlinear media [12], and an arrangement, developed in several laboratories, for high brightness and monochromaticity. As an example of the latter, a recent source, set up at the University of California at Berkeley, using rare gases as nonlinear media, will be described [13].

3·1. VUV *source developed at Toronto.* – A general outline of the experimental arrangement used in our laboratory, including a primary excitation source, two dye lasers for generating ω_1 and ω_2, heat pipes for producing the nonlinear metal vapor, a monochromator and detection system, is shown in fig. 2. Initially, the primary excitation source was a N_2 laser (Molectron UV-1000), which was suitable for VUV generation at $\lambda > 140$ nm. For generating shorter-wavelength radiation, a transverse-discharge excimer laser was developed [14, 15]. In the present application, radiation produced by the excimers KrF at 249 nm, XeF at 350 nm and XeCl at 309 nm is used for pumping the tunable dye lasers.

Two tunable dye lasers, using gratings (of 2400 lines/mm) at grazing incidence, are operated at frequencies ν_1 and ν_2. Both dye lasers are pumped simul-

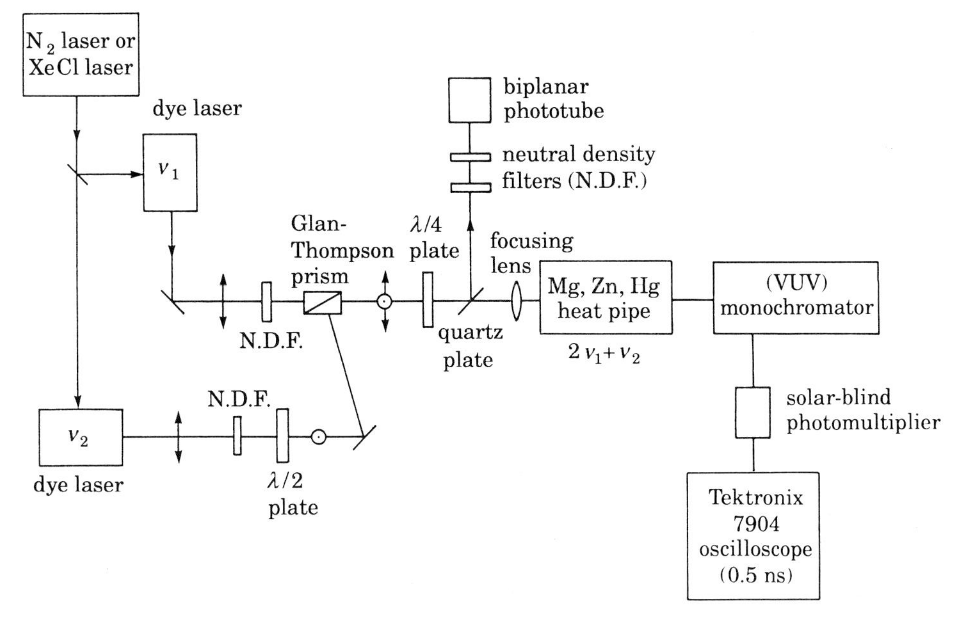

Fig. 2. – Schematic diagram of the overall experimental arrangement.

taneously, and a variety of available dye solutions provide tunable laser radiation from ~ 340 to 800 nm. Output powers of up to 10 kW and ~ 8 ns duration are obtained for the N_2-laser pumping, and up to ~ 40 kW and ~ 4 ns for the XeCl-laser pumping. Linewidths are typically $< 0.3\ cm^{-1}$ in the ultraviolet region and $\sim 0.1\ cm^{-1}$ for $\lambda > 400$ nm (although this may be reduced to $0.01\ cm^{-1}$, with a shorter cavity and a grating angle of ~ 180°, but at the expense of greatly diminished intensity). Both ν_1 and ν_2 beams are plane-polarized horizontally. The ν_2 beam polarization is rotated by 90° with a half-wave plate (or with two mirrors for $\lambda < 400$ nm), and then the two beams are spatially overlapped in a Glan-Thompson prism. This is the beam configuration when an S state is used for resonance enhancement. For enhancement using a D state, both beams are circularly polarized by a $\lambda/4$ plate, to eliminate THG without inhibiting the 4-WSM process.

For generating VUV radiation using XeCl pumping, amplifiers are added to each arm of the dye-laser systems (fig. 3). Each oscillator-amplifier system receives half of the XeCl laser power, with 15% of this allocated to the oscillator and the remainder to the amplifier. To allow for cavity build-up time of the oscillator, the pump beam to the amplifier is delayed (by traversing a longer path) for an interval of (4 ÷ 7) ns. Both oscillators emit powers of (1 ÷ 10) kW, and amplified powers are typically (50 ÷ 200) kW, depending on the dye efficiencies. In this way, any desired wavelength in the range 660 to 330 nm can be produced in each dye system, with XeCl (308 nm) pumping.

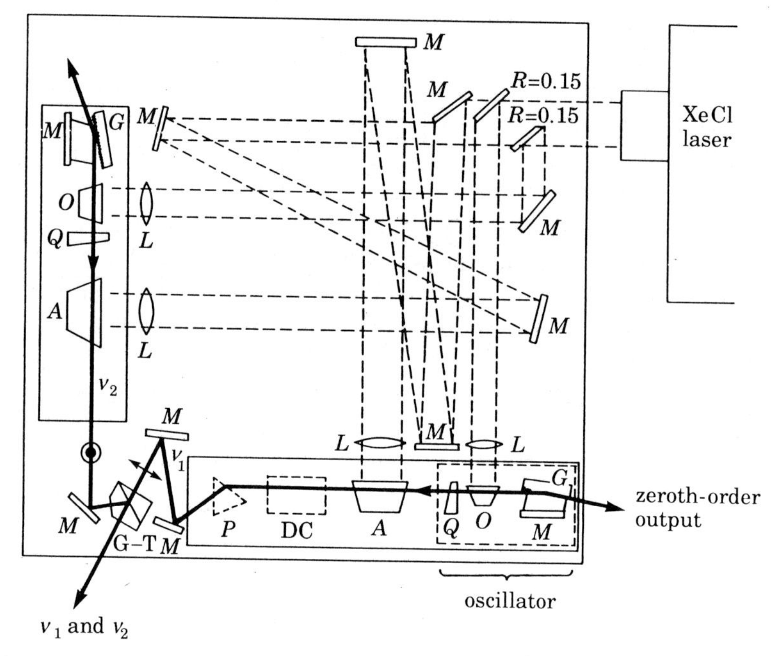

Fig. 3. – Diagram of the excimer laser and tunable dye oscillator-amplifier systems used to provide radiation at ν_1 and ν_2 for frequency mixing in Mg, Zn and Hg vapors. All mirrors labelled M are front-surface, totally reflecting; mirrors R are 15% reflecting. The oscillators each consist of a grating G, a dye cell O and a quartz wedge Q. The dye-cell amplifiers are labelled A, DC is a doubling crystal, P is a quartz prism to separate the fundamental and harmonic beams, G-T is a Glan-Thompson prism, and L designates a cylindrical lens.

Two important features, one in each dye-laser system, bear further discussion. When a fixed wavelength $\lambda_1 < 330$ nm is required for two-photon resonances, second-harmonic radiation is generated in a potassium dihydrogen phosphate (KDP) crystal having a tuning range $(320 \div 350)$ nm, or in a barium borate (BBO) crystal which extends the tuning range to ~ 200 nm. A quartz prism separates the fundamental and harmonic beams. To achieve a smooth, linear scan for accurate high-resolution spectroscopy, the scanning mechanism of an infrared spectrometer (Perkin-Elmer, model 099) has been incorporated in the ν_2 oscillator to finitely rotate the cavity mirror. A change of gears permits selection of scan rates of 1 to $30\,\text{cm}^{-1}/\text{min}$, with linearity accurate to 0.1%.

Magnesium vapor was the first choice of this laboratory as a nonlinear medium, based on its known VUV absorption spectrum beyond the ionization limit at 162 nm, and because generation could be achieved with radiation from N_2-pumped dye lasers. Several brief reports describe this work [16]. The first describes third-harmonic generation (THG) and 4-WSM with continuous tunability from 140 to 160 nm. The resulting VUV radiation was emitted in 4 ns pulses of $\sim 10^{11}$ photons/pulse (corresponding to an efficiency of $\sim 0.2\%$) and had a linewidth of $\sim 0.2\,\text{cm}^{-1}$. Tunability was extended to shorter wavelengths, to ~ 120 nm, and to longer wavelengths, as far as 174 nm. These early experiments demonstrated that 4-WSM in Mg vapor could provide an efficient source of coherent VUV radiation that is monochromatic, directional and tunable over the broad wavelength region $(120 \div 174)$ nm, a range of $\sim 25\,000\ \text{cm}^{-1}$.

To generate radiation of shorter wavelengths, the prime candidates as nonlinear media were Zn and Hg, since they have higher ionization limits than Mg (162 nm) corresponding to $\lambda \sim 132$ and 119 nm, respectively, and they also have strong and broad autoionizing resonances. Zinc vapor was next selected for study. A report by JAMROZ *et al.* [17] describes the generation of continuously tunable VUV radiation from 140 to 106 nm, a range of $\sim 23\,000$ cm^{-1}. Mercury vapor was found to be more efficient than Zn vapor in the region below 120 nm [18], and provides XUV radiation to ~ 85 nm [19].

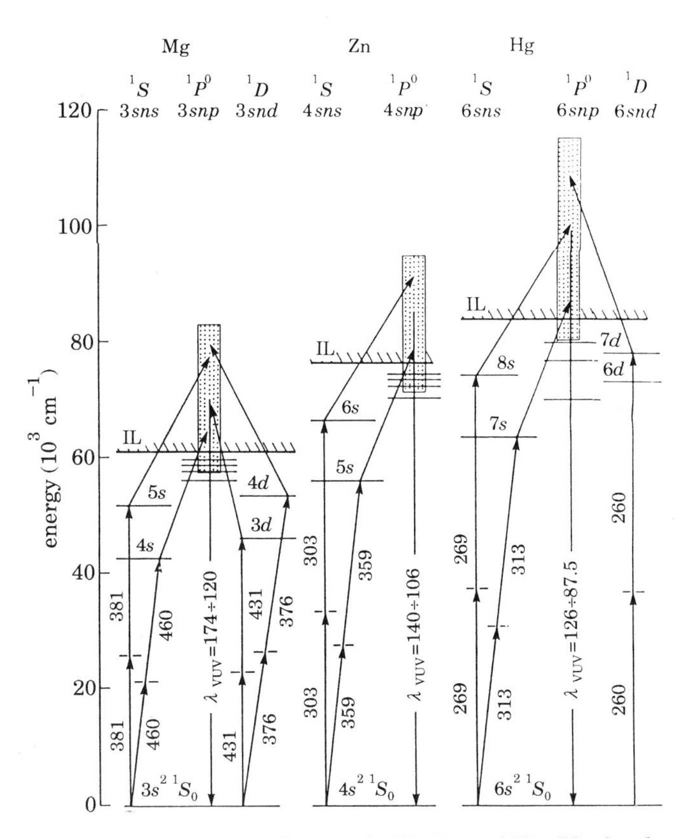

Fig. 4. – Partial energy level diagrams for atomic Mg, Zn and Hg. The levels used for two-photon resonance enhancement of 4-WSM are shown (along with corresponding wavelengths in nanometers). The regions of ionization continua and broad autoionizing levels that contribute to the tunability of these sources are indicated by the hatched areas.

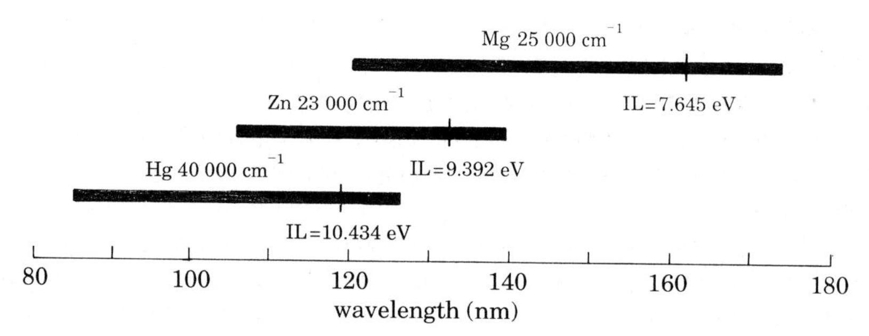

Fig. 5. – Wavelength and wave number ranges, above and below ionization limits (IL) over which tunable coherent radiation has been obtained using Mg, Zn and Hg vapors.

Thus Hg vapor has become an important source for tunable radiation from 125 to ~ 85 nm, spanning a range of ~ 40 000 cm^{-1}.

In fig. 4 is a schematic diagram of energy levels representing the 4-WSM in Mg, Zn and Hg vapors for generating tunable radiation over the wavelength region of ~ 175 to 85 nm (fig. 5). The metal vapors are prepared and contained in heat pipes of the form shown in fig. 6. The concept of heat pipes for use in spectroscopy has been discussed by VIDAL and COOPER [20]. For Mg and Zn vapors, heat pipes of a simple design were used at first, with heating of the central sections being provided by electrical heaters. Higher stability, and more uniform vapor density over prolonged periods, was obtained by replacing the electrical heaters with a vertical heat pipe surrounding the central parts of the Mg and Zn

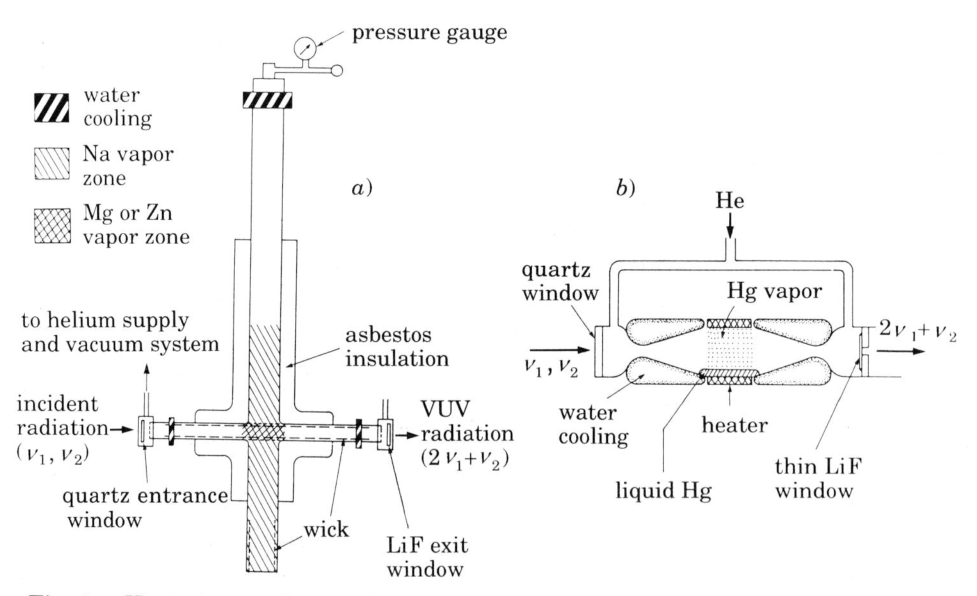

Fig. 6. – Heat pipes used to produce stable densities of *a*) Mg or Zn vapors and *b*) Hg vapor.

heat pipes, as shown in fig. 6*a*). Sodium metal serves as the working vapor to heat the Mg and Zn, with Ar gas in the Na heat pipe used to control the temperatures of the inner heat pipes. Under these conditions, stable operation can be maintained for several hundred hours before being limited by growth of crystals near the cooled ends, and typical operating lifetimes are ~ 1000 h before new wicks are required.

The heat-pipe oven for Hg vapor is a simple cell of Pyrex glass, as shown in fig. 6*b*). For 4-WSM at $\lambda = 120$ to ~ 105 nm (which is the limit of LiF transmission) LiF windows ~ 1/2 mm thick are used; for $\lambda < 105$ nm, a glass capillary array was found to be an efficient XUV window, with 50% transmission [19]. The generated VUV and XUV radiation is analyzed with a grating spectrometer (McPherson 225 or 234) and detected by a solar-blind photomultiplier (EMR 510G-08-13), fitted with a LiF window. For absolute-intensity calibration, an ionization chamber is placed between the heat pipe and the monochromator. The chamber is used with xylene or diethyl sulfide vapor, or Xe gas, depending on the wavelength of the radiation being generated.

3·2. *Sources of high brightness and monochromaticity.* – Laser-driven sources of high brightness for VUV and XUV generation have been developed by several groups [21]. All of these schemes use radiation from a narrow-band or single-mode dye laser, followed by pulse amplification with excimer, or doubled Nd:YAG laser radiation, and frequency mixing in a gas cell or pulsed free jet.

A schematic diagram of a recent system developed by CROMWELL, TRICKL, LEE and KUNG [13] is given in fig. 7. The starting point of the system is a ring dye laser (Coherent 699-29, Autoscan), pumped by an argon ion laser. The emitted radiation has a bandwidth of 1 MHz, and is continuously tunable over the visible region. This radiation is transmitted to a series of three dye-amplifier cells that are pumped by frequency-doubled beams from a Nd:YAG laser (Quantel model 592). Care is taken to ensure that the temporal profiles of the pump and dye lasers are as smooth as possible, and near Gaussian, in order to achieve the narrowest bandwidth. After amplification, the visible radiation generally has the following properties: energy > 100 mJ, bandwidth ~ 90 MHz (FWHM) and pulse duration ~ 7 ns. The visible radiation is then doubled to produce a UV beam of ~ 30 mJ energy tunable over the range (220 ÷ 310) nm and having a bandwidth of ~ 140 MHz. Finally, frequency mixing to produce VUV and XUV radiation tunable from 74 to 124 nm is carried out in a pulsed free jet of the rare gases, or N_2, or CO, and the desired radiation is separated from the UV pump beam by dispersion in a monochromator (McPherson model 225). The output achieved is $> 10^{11}$ photons/pulse with a bandwidth of 210 MHz, in a pulse of duration < 7 ns.

The use of a pulsed jet of gas as the nonlinear medium, introduced by KUNG [22] and by BOKOR *et al.* [23], has several important advantages for gener-

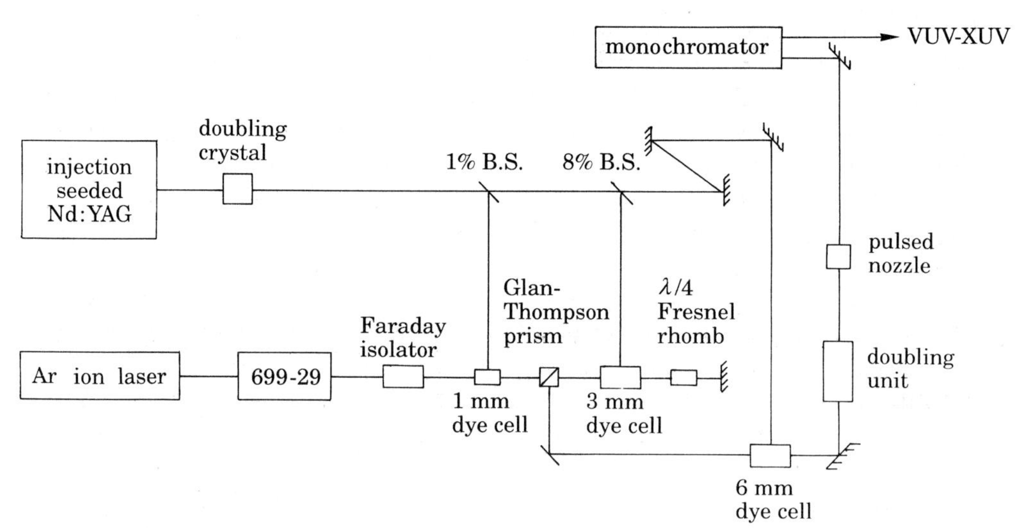

Fig. 7. – Schematic diagram of a high-brightness laser system with ultra-narrow bandwidth VUV and XUV radiation [13].

ating XUV radiation. There are no windows to contain the nonlinear medium, hence there is no material absorption. The gas density can be high at the nozzle, resulting in improved frequency-tripling efficiency, while the volume of gases is kept to a minimum, and large pumping speeds are not required to maintain a suitable vacuum in the experimental chamber. Alternative schemes for containing the nonlinear medium when working in the XUV region, that use pinholes, rotating pinholes, or capillary arrays, necessarily require more elaborate, differential pumping systems.

4. – Spectroscopic studies in the far ultraviolet.

As already noted, tunable, coherent radiation is now available over the range 50 to 200 nm. The wavelength regions for radiation generated with metal vapors are reviewed in ref. [11, 12, 19], and those produced with the rare gases in a series of papers by WALLENSTEIN and colleagues [24].

These laser-driven sources emit 10^8 to 10^{11} photons/pulse, in pulses of duration 1 to 10 ns. The minimum bandwidths obtained at 100 nm are 0.2 cm^{-1} with metal vapors [25] and 0.007 cm^{-1} with rare gases [13], resulting in resolving powers of $5 \cdot 10^5$ to $1.5 \cdot 10^7$, up to 100 times the values achieved with grating instruments. Thus these are sources of high intensity and excellent monochromaticity, with many potential applications in physics, chemistry and biology.

Amongst these possible applications are high-resolution spectroscopy and measurement of radiative lifetimes of selected levels of atoms and molecules.

Here, I would like to review several examples of spectroscopic studies carried out in my laboratory, that were extremely difficult, if not impossible to carry out with other VUV and XUV sources. These include: high-resolution studies of spectra of the rare-gas dimers Xe_2, Kr_2 and Ar_2; measurement of radiative lifetimes of selected rovibronic levels of CO, as well as of the rare gases; determination of the dissociation energy of H_2; and generation of second-harmonic radiation at Lyman-α in atomic H with reduced absorption.

4'1. *Spectra of rare-gas excimers.* – The electronic spectra of the rare-gas dimers have been a subject of interest for many years, mainly because these dimers are model systems for studying van der Waals interactions, and, when pumped by energetic electrons, they are laser media operating in the VUV. However, the spectroscopic constants for the relevant electronic states of these dimers are not yet established. Thus, in the early 80's in our laboratory, a series of investigations was begun on the spectra of Xe_2, Kr_2 and Ar_2. Two techniques were combined for this work: a pulsed supersonic jet to produce rotationally and vibrationally cold dimers, and 4-WSM to provide the necessary VUV radiation for fluorescence-excitation spectroscopy [26]. In this way, it was found possible to resolve rovibronic structures in several band systems of these dimers and their isotopes in the region of 104 to 150 nm. From these spectra, the relevant molecular constants were determined for the ground and three lowest (stable) excited states, and the corresponding potential-energy curves were derived.

A diagram illustrating a typical experimental arrangement is given in fig. 8,

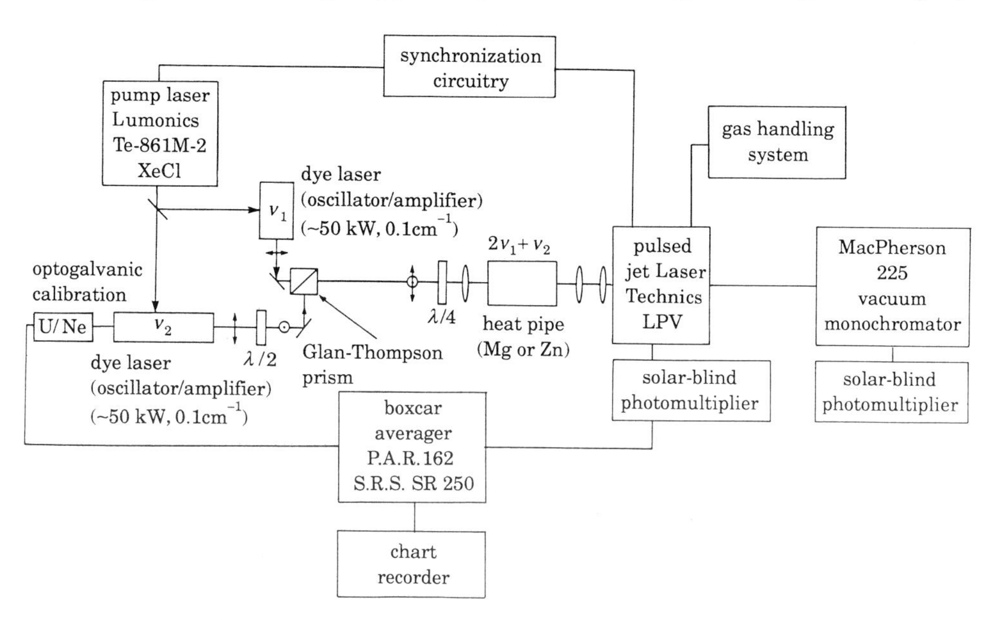

Fig. 8. – Experimental arrangement for investigating spectra of Ar_2.

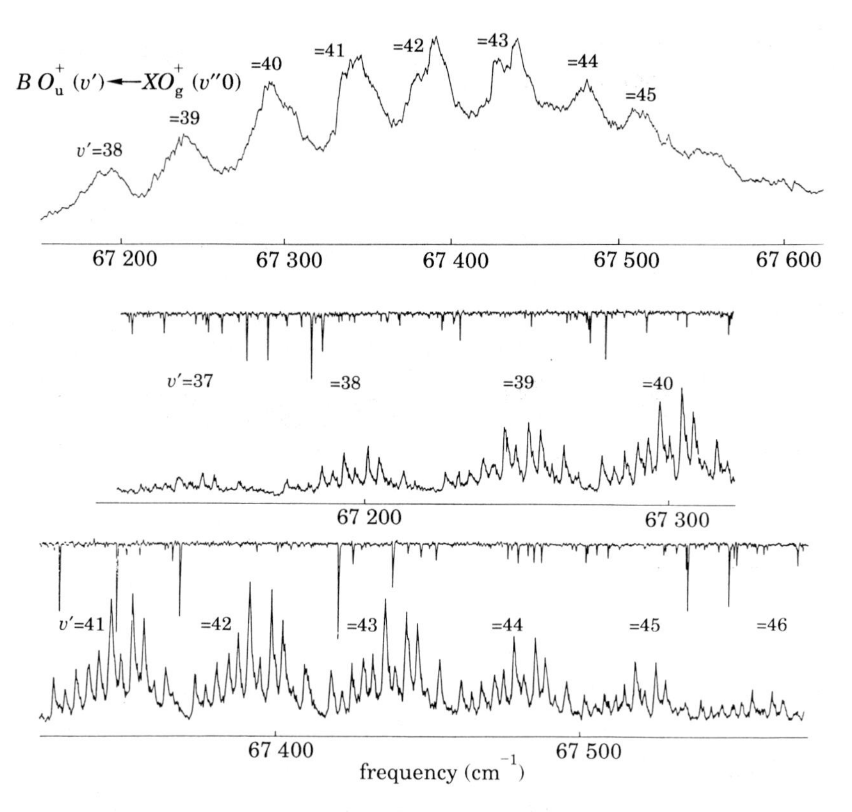

Fig. 9. – Fluorescence-excitation spectrum of the *B*-*X* band system of Xe_2, at 148.5 nm probed ~ 2 mm from the nozzle (upper) and ~ 15 mm from the nozzle (middle and lower).

indicating all of the necessary components for 4-WSM, sample preparation, synchronization of jet and laser pulses, and detection and recording of spectra, as well as wavelength calibration. It is instructive to examine spectra of cold as well as warm dimers using the same experimental methods. This was possible with the gas jet by exciting the dimers at different distances from the nozzle, and spectra of the 148.5 nm band are shown in fig. 9 for comparison. The spectrum of dimers excited as close as possible to the nozzle (~ 2 mm) consists of broad bands similar to spectra obtained in cooled-cell (~ 150 K) experiments. When excited ~ 15 mm from the nozzle, the spectrum shows 10 vibrational bands with considerable structure within each band. Each band exhibits the same pattern of narrow components, with widths corresponding to temperatures of ~ 10 K. These resolved features are vibronic bands of the many isotopes of Xe_2, and their resolution and analysis has led to the unambiguous quantum numbering of the observed bands. This numbering is essential for determining the molecular constants.

At the low temperatures obtained with the supersonic jet, only the $v'' = 0$ vibrational level is populated. Thus measured frequencies can be described by the relation (in cm^{-1})

$$(6) \qquad \nu_i = T'_e + \rho_i \omega'_e (v + 1/2) - \rho_i^2 \omega'_e x'_e (v' + 1/2)^2 - \\ - (21.12\,\rho_i)/2 + (0.65\,\rho_i^2)/4 - (0.003\,\rho_i^3)/8 .$$

Here, T'_e is the electronic energy of the upper state, ω'_e and $\omega'_e x'_e$ are the vibrational frequency and anharmonic constant, respectively, and ρ_i is the ratio $[\mu(^{129,\,132}Xe_2)/\mu_i]^{1/2}$, where μ_i is the reduced mass of a particular dimer i. The ground-state constants $\omega''_e = 21.12$ cm^{-1}, $\omega''_e x''_e = 0.65$ cm^{-1} and $\omega''_e y''_e = 0.003$ cm^{-1} were evaluated by FREEMAN *et al.* [27]. From an analysis of the measured frequencies of nine isotopes in each of the eight most intense bands of fig. 9, the vibrational numbering was determined, and the constants were evaluated. Two other band systems of Xe_2 at 130 and 150 nm were recorded and analyzed, yielding spectroscopic constants for two additional excited states.

Corresponding isotopic structure was observed in the spectra of Kr_2 and Ar_2 and served to identify the vibronic bands. While the available resolution was not sufficient to resolve rotational structure in the spectra of Xe_2, it was possible to resolve such structure in spectra of Kr_2 (fig. 10) for the first time, and to determine the internuclear separation in the ground and excited states. Also, the very much lighter mass of Ar_2 resulted in clear resolution of rotational structure even at low dispersion (fig. 11). The high-resolution spectrum of one of the bands, shown in fig. 12, led to the identification of the symmetry of the Ar_2 excited states. Here three rotational branches are clearly resolved, and interpreted as *P, Q, R* branches. From this result it was possible to establish that Hund's case (*c*) is the

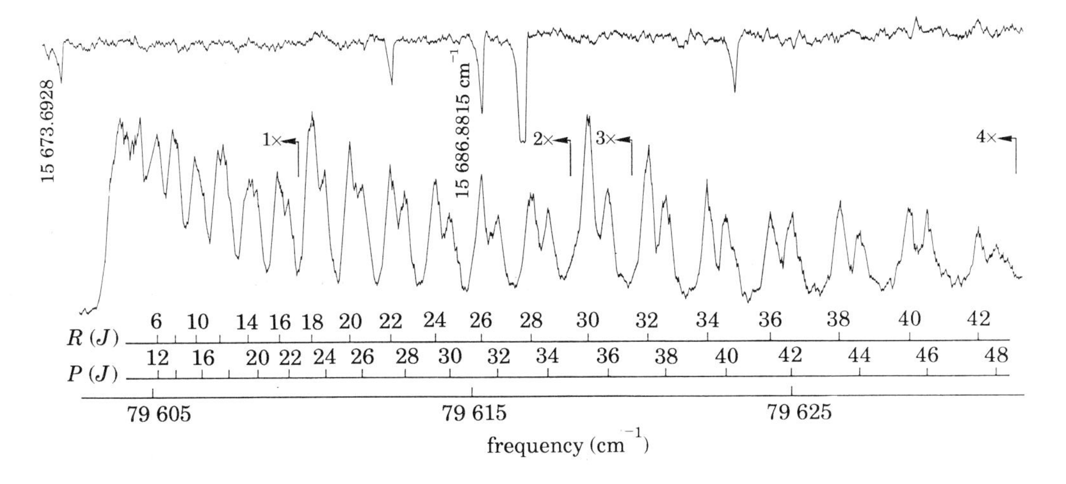

Fig. 10. – Rovibronic structure in the 34-0 band of the *B-X* system of Kr_2 at 125 nm.

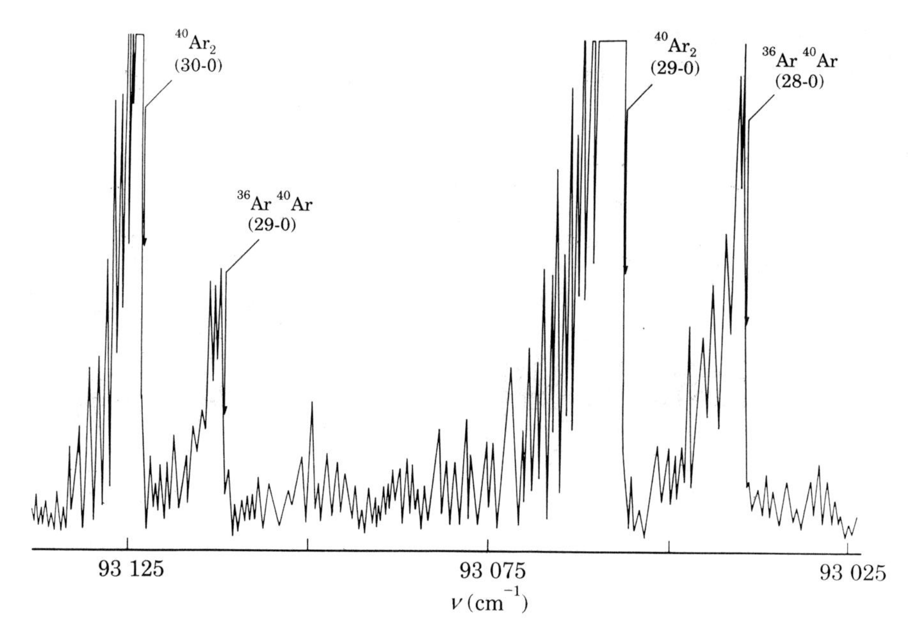

Fig. 11. – A small portion of the B-X band system of Ar_2 showing vibrational and rotational structures of two isotopic species, $^{40}Ar_2$ and $^{36,40}Ar_2$.

dominant coupling for the high vibrational levels of the A state of Ar_2, and thus its symmetry should be designated as $A1_u$ in place of $A\,^3\Sigma_u^+$ [25].

Finally, potential-energy curves for the ground state and three lowest bound states were calculated for each of the dimers [26, 28, 29]. Those for Ar_2 are shown in fig. 13. Since data for the lowest vibrational levels were available for the ground and C states, these potential curves are deemed to be accurate. However, the potential-energy curves for the A and B states are much less reli-

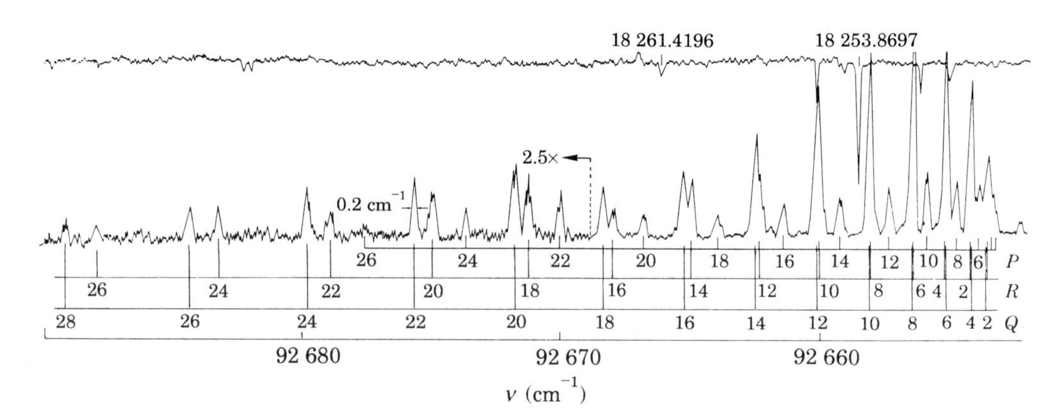

Fig. 12. – Spectrum of the $A1_u$-$X0_g$, 25-0 band of Ar_2 clearly showing P, Q and R branches.

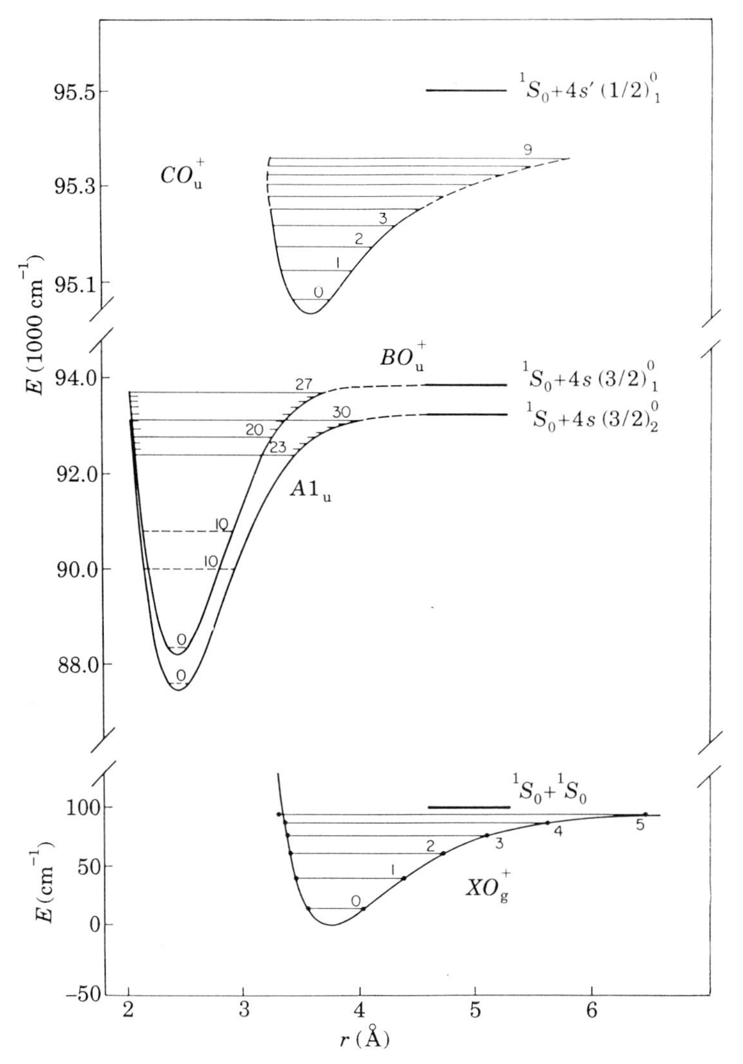

Fig. 13. – Calculated potential-energy curves for Ar_2 showing the observed vibronic levels and separated-atom states at the dissociation limits.

able, since only high vibrational levels (> 20) were accessible in the experiments, and long extrapolations to $v = 0$ were necessary.

4'2. *Radiative lifetimes of rovibronic levels in the* $A^1\Pi(v = 0, 1)$ *state of* CO. – The first investigation to be carried out with the VUV source described in subsect. 3'1 was the state-selective VUV excitation of rovibronic levels in the $A^1\Pi(v = 0)$ state of CO to determine radiative lifetimes of some known perturbed levels. Initial experiments were performed in 1976 by PROVOROV *et al.* [30], and, later, more extensive measurements by MAEDA [31]. Earlier studies of the spectra of CO had shown that the $A^1\Pi$ state is strongly perturbed by nearby triplet states. For the $v=0$ rovibronic levels, the perturbations are caused by the $e^3\Sigma^-$ $(v=1)$ and $d^3\Delta_i(v=4)$ states, while for the $v=1$ levels only the $d^3\Delta_i(v=5)$ state is involved in perturbing the lowest rovibronic levels. Thus, in the theoretical treatment of the problem, we need only consider the interaction of a singlet state $S_0(A^1\Pi)$ with a triplet state $T_0(e^3\Sigma^-$ or $d^3\Delta_i)$. Near a perturbation there is a mixing of states S_0 and T_0, resulting in the linear combination states

$$|S\rangle = \alpha|S_0\rangle + \beta|T_0\rangle, \qquad |T\rangle = -\beta|S_0\rangle + \alpha|T_0\rangle. \tag{7}$$

Here, $|S\rangle$ and $|T\rangle$ refer to the perturbed singlet and triplet states, and $|S_0\rangle$ and $|T_0\rangle$ to the unperturbed states. The mixing coefficients α and β are related by $\alpha^2+\beta^2=1$. Thus the singlet state takes on some of the characteristics of the triplet state, and *vice versa*. The respective decay rates, Γ_S and Γ_T, may be written as

$$\Gamma_S = \alpha^2\Gamma_{S_0} + \beta^2\Gamma_{T_0} \quad \text{and} \quad \Gamma_T = \alpha^2\Gamma_{T_0} + \beta^2\Gamma_{S_0}, \tag{8}$$

or, alternatively, in terms of the unperturbed radiative lifetimes τ_{S_0} and τ_{T_0} as

$$\Gamma_S = \tau_S^{-1} = \alpha^2\tau_{S_0}^{-1} + \beta^2\tau_{T_0}^{-1} \quad \text{and} \quad \Gamma_T = \tau_T^{-1} = \alpha^2\tau_{T_0}^{-1} + \beta^2\tau_{T_0}^{-1}. \tag{9}$$

It is known that $\tau_{S_0} \sim 10$ ns and $\tau_{T_0} \sim 3$ μs. FIELD [32] and, more recently, LEFLOCH *et al.* [33] have calculated values of α^2 and β^2 from analyses of observed perturbations in the high-resolution spectra of CO, and derived theoretical radiative lifetimes for the singlet and triplet levels.

For this study, VUV radiation generated in Mg vapor was focused into a stainless-steel cell to excite fluorescence in flowing CO at a pressure of 50 mTorr. Radiation emitted at right angles to the incident VUV beam was detected directly with a solar-blind photomultiplier fitted with a LiF window. For each measurement, the signal was transmitted to a boxcar-averager and gated-integrator system, used in the scanning mode with a window time of 2 ns. Initially, the fluorescence-excitation spectra of the 0-0 and 1-0 bands, $A^1\Pi(v = 0, 1)$-$X^1\Sigma(v=0)$, were recorded by tuning the VUV radiation over the wavelength ranges 154.4 to 155.5 nm and 150.9 to 151.8 nm, respectively. The exciting radiation was then tuned to each of the unblended rotational lines and the decay of fluorescence intensity with time was measured. For each decay curve, ~ 2000 pulses at 8 Hz were integrated with the boxcar-integrator system. It became evident that the total instrumental response time significantly broadened the VUV excitation pulse beyond the 2 ns duration of the dye-laser pulses, and that it was necessary to correct for the instrumental profile, in order to determine the fluorescence decay times τ.

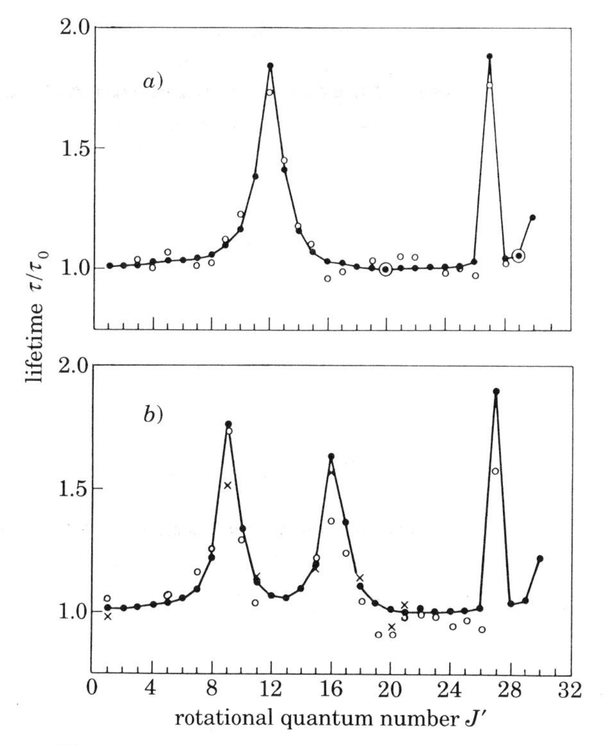

Fig. 14. – Graphs of lifetimes normalized by the unperturbed lifetime τ_0 plotted as a function of rotational quantum number J' for the $v = 0$ level of CO *a*) obtained from *Q*-branch lines (● theory, ○ experiment) and *b*) from *P*- and *R*-branch lines: ○ *P* branch, ×*R* branch, ● theory.

For the $v = 0$ level, measurements of fluorescence decay rates were made for the unblended lines of the *P*, *Q* and *R* branches, and radiative lifetimes for levels up to $J' = 29$ were determined. Values in the range 9.8 to 10.8 ns were considered to designate lifetimes of unperturbed levels, and the average of 26 such values gave an unperturbed lifetime, $\tau_0 = (10.5 \pm 0.5)$ ns at a CO pressure of 50 mTorr. Values of lifetimes normalized by the unperturbed lifetime, that is τ/τ_0, are plotted as a function of rotational quantum number J' in fig. 14. Theoretical values (τ/τ_0) are also shown for comparison [32, 33]. From fig. 14, it is clear that large perturbations occur for levels $J' = 12$ and 27 in *Q*-branch transitions, and for levels $J' = 9$, 16 and 27 in *P*- and *R*-branch transitions. The observed lifetimes are almost double the unperturbed lifetime of 10.3 ns. The overall good agreement with theoretical lifetimes for perturbed and unperturbed levels over the range of $J' = 1$ to 29 is also evident.

The intensity of the 1-0 band was found to be lower than that of the 0-0 band, and most of the *R*-branch lines were blended with *P*- or *Q*-branch lines. Measurements of fluorescence decay rates were made for all possible transitions from

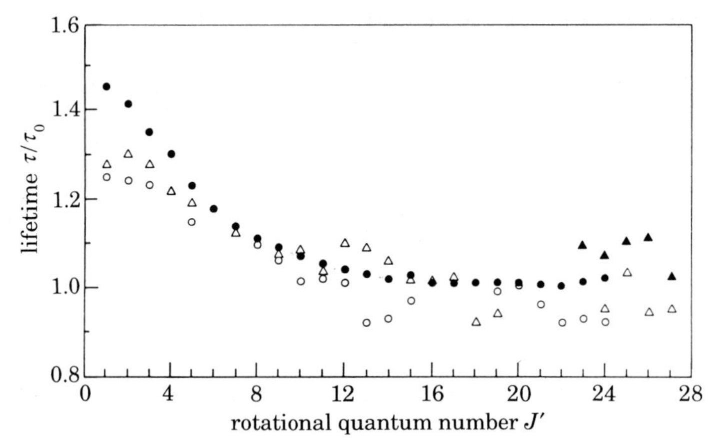

Fig. 15. – Graphs of lifetimes normalized by the unperturbed lifetime τ_0 plotted as a function of rotational quantum number J' for the $v = 1$ level obtained from P- and Q-branch lines: ○ P branch; △ Q branch; ● theory, $J' \leqslant 22$; ▲ theory, $J' > 22$.

levels up to $J' = 27$. Those values below 12.0 ns were considered to be unperturbed, giving an average $\tau_0 = (10.7 \pm 0.7)$ ns (for a CO pressure of 50 mTorr). Values of τ/τ_0 are plotted *vs.* J' in fig. 15, along with theoretical values (τ/τ_0) provided by FIELD. While these experimental values exhibit more uncertainty than obtained for the 0-0 band measurements, they are in reasonable agreement with the theoretical values. It is evident that perturbations occur only for the low values of J' in the $v = 1$ level, with τ/τ_0 decreasing from ~ 1.3 to 1.0 in the range of $J' = 0$ to 8.

The selective excitation of individual rovibronic levels for the $v = 0$ and 1 bands of the $A^1\Pi$ state of CO carried out with a tunable, pulsed VUV source has yielded radiative lifetimes which are in excellent agreement with theoretical values. Such good agreement attests to the validity of the theory for strongly and weakly perturbed levels of CO, as well as to the potential importance of the high-resolution state-selective excitation source used in this research.

4·3. *Dependence of* Ar_2 *radiative lifetimes on internuclear distance.* – In a continuation of the study of the rare-gas excimers, described in subsect. 4·1, the radiative lifetimes of vibronic levels of the $A1_u$ states of Ar_2, Kr_2 and Xe_2 were measured [34]. As already noted, because of the relative positions of the potential-energy curves for the strongly bound excimer states and the shallow ground states (fig. 13), only high vibronic levels of the $A1_u$ states are accessible for fluorescence excitation. Thus only the levels $v' = 24$ to 30 for Ar_2, $v' = 32$ to 38 for Kr_2 and $v' = 36$ to 43 for Xe_2 could be investigated. For all of these levels, the fluorescence decay curves exhibited single-exponential decays with time (fig. 16). The measured lifetimes were found to be essentially constant for these high vibronic levels, with average lifetimes of $\tau = (160 \pm 10)$ ns for Ar_2,

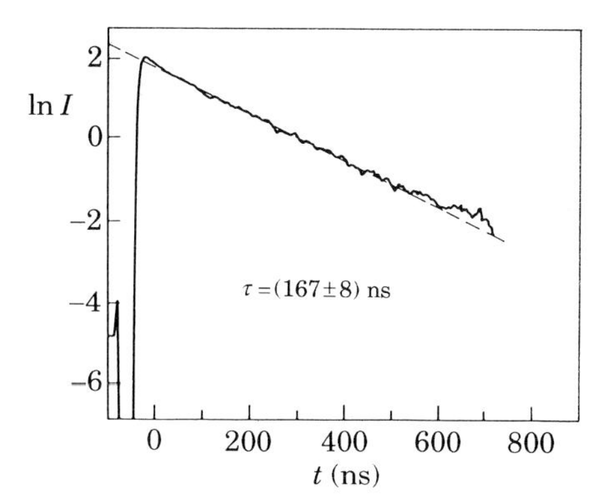

Fig. 16. – Typical curve of fluorescence intensity *vs.* time illustrating single exponential decay for $v' = 26$ of the $A1_u$ state of Ar_2.

(55 ± 5) ns for Kr_2 and (47 ± 5) ns for Xe_2. These results for the high vibronic levels differ significantly from the values for low levels, namely, 2.9 μs, 264 ns and 99 ns, respectively, measured by earlier investigators. (They used charged particles and synchrotron radiation for fluorescence excitation of these excimers formed at relatively high pressures in cell experiments. At these pressures, rapid vibrational relaxation occurs, resulting in fluorescence emission from low vibrational levels of the excited states.) Thus the differences in lifetimes imply reductions by factors of 20 for Ar_2, 5 for Kr_2 and 2 for Xe_2, in going from $v' = 0$ to $v' \sim 20 \div 40$ in the $A1_u$ states.

While differences of a factor of two in radiative lifetimes for vibrational levels of the same electronic state are not uncommon in molecular spectroscopy, a factor of 20 is unique. MADEJ *et al.* [34] explained this large difference for Ar_2 by calculating the dependence of the electronic transition moment $\mu(R)$ on internuclear distance, given in fig. 17. In the united- and separated-atom limits, transitions are forbidden; but at intermediate distances, spin-orbit coupling causes transitions to be weakly allowed. It was found that the derived values of $\mu(R)$ increase rapidly, from $\sim 1 \cdot 10^{-2}$ a.u. at 4.5 a_0 (corresponding to $v' \sim 0$) to

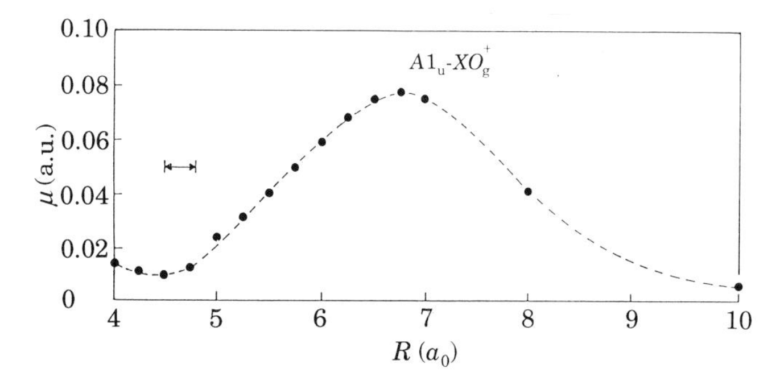

Fig. 17. – Calculated transition moment $\mu(R)$ for the A-X transition in Ar_2.

TABLE I. – *Comparison of measured and calculated radiative lifetimes of vibrational levels of the* $A1_u$ *state of* Ar_2 [34].

v'	$\tau_{calc}\,(10^{-6}\,s)$	$\tau_{expt}\,(10^{-6}\,s)$
0	8.6	3.2 ± 0.3
5	1.9	—
10	0.88	—
15	0.58	—
20	0.31	—
24	0.21	0.17 ± 0.02
25	0.20	0.16 ± 0.01
26	0.18	0.17 ± 0.01
27	0.17	0.16 ± 0.01
28	0.16	0.16 ± 0.01
29	0.17	0.16 ± 0.01
30	0.17	0.17 ± 0.01

$\sim 8 \cdot 10^{-2}$ a.u. at 6.8 a_0 (for $v' \sim 30$). These values were used to calculate spontaneous-emission probabilities, which give the reciprocals of the radiative lifetimes. A comparison of the calculated and measured lifetimes is made in table I. There is excellent agreement for high vibronic levels and a discrepancy of a factor of three for $v' = 0$, which can be corrected by an increase of $\sim 5\%$ in the equilibrium internuclear distance for the $A1_u$ state (presently only known to an accuracy of $\sim 10\%$). Similar calculations for Kr_2 and Xe_2 yielded good agreement with measured values at high vibronic quantum numbers.

4·4. *Dissociation energy of molecular hydrogen.* – The experimental determination of a precise value for the dissociation energy of molecular hydrogen in its ground electronic state has long remained a challenge for spectroscopy. This parameter is important as a test of basic molecular theory, which today includes calculations of relativistic, radiative and nonadiabatic corrections.

While experimental values were available as early as 1926, all attempts to observe the onset of the dissociation continuum, whether in emission or absorption, were plagued by overlapping molecular bands in this region. Prior to the application of laser techniques to this problem, the best values were those reported by HERZBERG [35], $D_0(H_2) = 36\,118.6\ cm^{-1}$, and by STWALLEY [36], $D_0(H_2) = (36\,118.3 \pm 0.5)\ cm^{-1}$. More recently MCCORMACK and EYLER [37] have

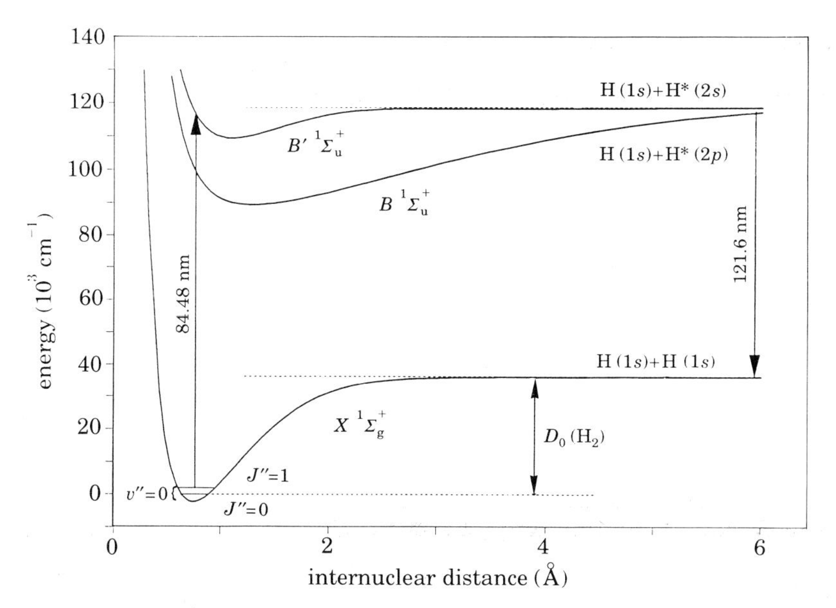

Fig. 18. – Energy level diagram showing the ground $X(^1\Sigma_g)$ and excited states $B(^1\Sigma_u)$ and $B'(^1\Sigma_u)$ involved in the photodissociation of H_2.

used high-resolution, double-resonance, laser spectroscopy and reported a value of $(36\,118.26 \pm 0.20)$ cm^{-1}.

Experiments on this problem have recently been carried out in this laboratory [38] based on fluorescence-excitation spectroscopy near 84.5 nm, with detection by emission of Lyman-α radiation at 121.6 nm (fig. 18). Spectra were obtained as the XUV radiation was tuned through the higher vibrational levels of the B and B' states and into the so-called second dissociation continuum to yield $H_2(B'^1\Sigma_u) = H(1s) + H(2s)$ at $(118\,377.06 \pm 0.04)$ cm^{-1}.

The experimental arrangement for generating coherent, tunable radiation in the region of 84.5 nm was essentially the same as that shown in fig. 8, except that here the nonlinear medium was a pulsed jet of Kr gas. Radiation at ~ 424 nm with a bandwidth of 0.2 cm^{-1} was amplified and frequency-doubled in a β-barium borate crystal to provide radiation at $2\nu_1$ which is resonant with the level $4p^5 5p[0, 1/2]_0$ of Kr. For ν_2, a single-longitudinal-mode dye laser was used to generate radiation with a bandwidth of 0.02 cm^{-1} and tunable in the region of 414 nm. The two beams were synchronized and spatially overlapped, then focused with a lens (of 25 cm focal length) just below the nozzle of a pulsed jet of Kr gas, to generate tunable XUV radiation at $2\nu_1 + \nu_2$, in pulses of ~ 4 ns duration. This radiation was incident (at right angles) on a nearby vertical beam of H_2 gas, in the region of supersonic expansion, ~ 50 mm below the nozzle of a second jet. A solar-blind photomultiplier was positioned perpendicular to the XUV and H_2 beams, to detect the 121.6 nm fluorescence radiation.

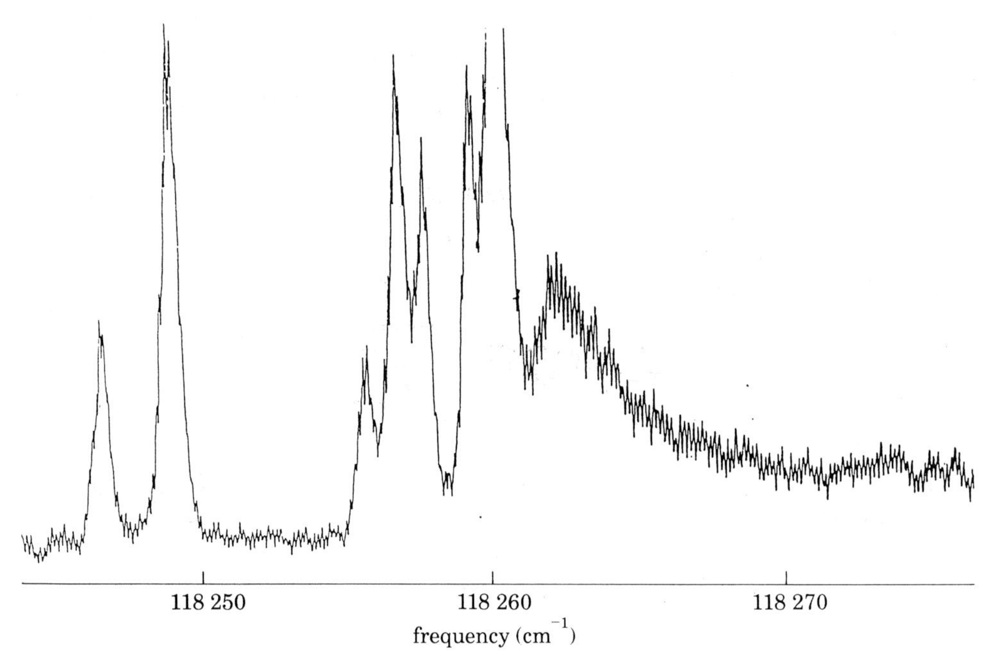

Fig. 19. – Fluorescence-excitation spectrum in the vicinity of the dissociation limit of H_2 obtained immediately after laser excitation.

In preliminary experiments, the spectra revealed intense molecular lines which completely obscured the threshold of the continuum (fig. 19), just as in the earlier experiments [35]. In order to obtain an unobstructed view of the threshold region, the fluorescence-excitation spectrum was investigated with delayed detection of fluorescence radiation. For this purpose, the metastable H(2s) atoms formed in the photodissociation process were quenched with an electric field which coupled the 2s and 2p levels, resulting in emission of Lyman-α radiation. The field was switched on ~ 200 ns after the excitation pulse, a delay sufficient to allow molecules in the B and B' states to decay by radiation, before the long-lived 2s atoms were quenched. In this way, it was possible to observe the threshold of the second dissociation limit clearly, as illustrated in fig. 20.

A value of (118 258.57 ± 0.04) cm^{-1} was obtained for the dissociation threshold measured from the $J''=1$, $v''=0$ level of the ground state, by drawing a straight line through the linear portion of the onset using least squares, as shown in fig. 20. To this value was added the rotational spacing of 118.49 cm^{-1} between the $J''=0$ and $J''=1$ levels to give the second dissociation limit of 118 377.06 cm^{-1}. On subtracting the 2s-1s separation of 82 258.95 cm^{-1}, we obtain for the dissociation energy of H_2 in its ground electronic state, $X(^1\Sigma_g)$, the value $D_0(H_2) = (36\,118.11 \pm 0.08)$ cm^{-1}. This experimental value is in agree-

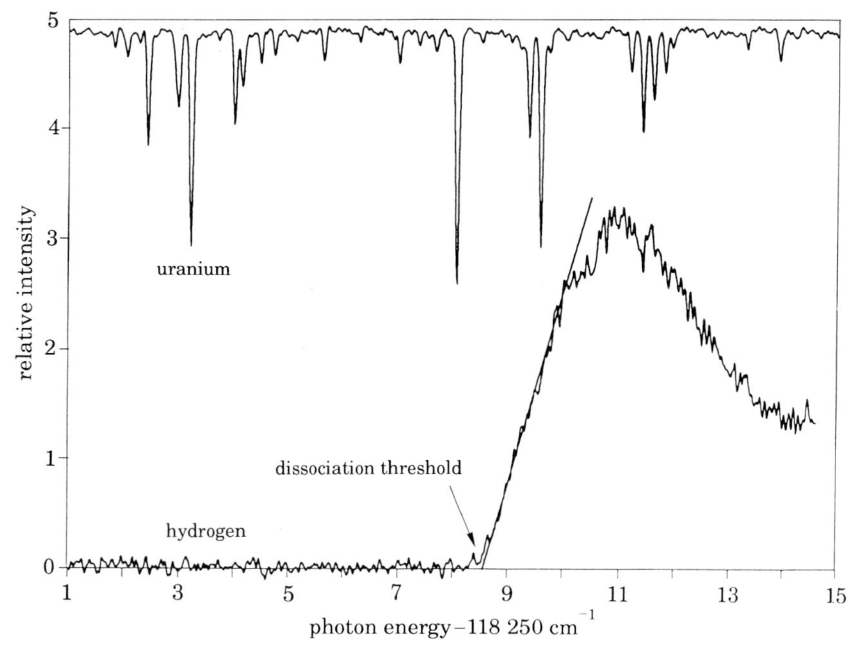

Fig. 20. – Fluorescence-excitation spectrum of delayed fluorescence in the vicinity of the dissociation limit of ortho-H_2. At the top is the optogalvanic uranium spectrum used for calibration.

ment with the latest theoretical calculation of $(36\,118.088 \pm 0.10)\ cm^{-1}$, which includes nonadiabatic, relativistic and radiative corrections [39]. A list of the experimental and most recent theoretical values of $D_0(H_2)$ is given in table II.

When the present value of $D_0(H_2)$ is combined with experimental values for the ionization potentials of H $((109\,678.764 \pm 0.01)\ cm^{-1})$ and of H_2 $((124\,417.512 \pm 0.016)\ cm^{-1})$, in eq. (10):

$$D_0(H_2^+) = D_0(H_2) + IP(H) - IP(H_2), \tag{10}$$

TABLE II. – *Theoretical and experimental values of* $D_0(H_2)$.

Method	$D_0(H_2)$ in cm^{-1}	Reference
theory	$36\,118.09 \pm 0.10$	KOLOS *et al.* [39]
experiment	$36\,118.3$	HERZBERG [35]
experiment	$36\,118.6 \pm 0.5$	STWALLEY [36]
experiment	$36\,118.26 \pm 0.20$	MCCORMACK and EYLER [37]
experiment	$36\,118.11 \pm 0.08$	present [38]

a value of $D_0(H_2^+) = (21\,379.36 \pm 0.08)\ cm^{-1}$ is obtained for the dissociation energy of the ion H_2^+. Again, this is in good agreement with the theoretical value of $21\,379.348\ cm^{-1}$ [40].

In summary, the use of pulsed XUV excitation with delayed detection has led to the solution of a fundamental and long-standing problem in molecular physics.

4·5. *Second-harmonic generation* (SHG) *at* 121.6 nm *in atomic hydrogen.* – It is well known that the second-order susceptibility, $\chi^{(2)}$, vanishes in the dipole approximation for isotropic media. Nevertheless, the experimental generation of second-harmonic radiation in the absence of external fields has been reported for a number of atomic gases, namely barium, lithium, mercury, potassium, sodium, zinc and, most recently, hydrogen. These observations have been explained by the contributions of quadrupole or magnetic-dipole moments, by intensity gradients of the incident laser beam, by electric-field-induced harmonic generation (with the field caused by ions in 3-photon ionization) and by collision effects. For some atoms there is good agreement between theoretical calculations and experimental observations; for others, the observed results remain unexplained or even appear to contradict the proposed models. The generation of SH radiation with the application of electric or magnetic fields to remove the symmetry of free atoms or nonpolar molecules has also been observed, and appears to be well understood.

Recently, we examined the problem of SHG in atoms, with the simplest system, atomic hydrogen [41]. With incident radiation at 243 nm, corresponding to the 2-photon transition $2^1S \leftarrow 1^1S$, Lyman-α radiation at 121.6 nm was generated. The choice of atomic hydrogen was motivated by the simplicity of the theory and calculations, by the forbidden quadrupole and magnetic-dipole transitions, and by the availability of an efficient nonlinear crystal, β-barium borate (having transmission just below 200 nm) which was used to generate the incident radiation at 243 nm. The transition matrix elements for the $n = 2 \leftarrow 1$ transition of H are large, thus permitting the use of low gas density with the reduction or elimination of pressure effects. In this way, it was possible to demonstrate that the dominant process for inducing SH radiation in H is a charge separation field, arising from 3-photon ionization. Moreover, it was possible to measure this field, and to compare its value with that calculated from theory.

The experimental arrangement is shown in fig. 21. A dye laser, pumped by a XeCl excimer laser, emitted tunable radiation near 486 nm, which was doubled to 243 nm. Hydrogen atoms were generated in a d.c. glow discharge of H_2 gas, and directed into the experimental chamber as a beam, through a 0.9 mm nozzle. The gas density at the nozzle was $\sim 10^{14}\ cm^{-3}$, with the background pressure of the chamber maintained at $2 \cdot 10^{-6}$ Torr. The tunable UV beam was focused midway between the (nozzle) electrode and a fine grid, spaced 1.3 mm apart, and used to apply a variable d.c. electric field.

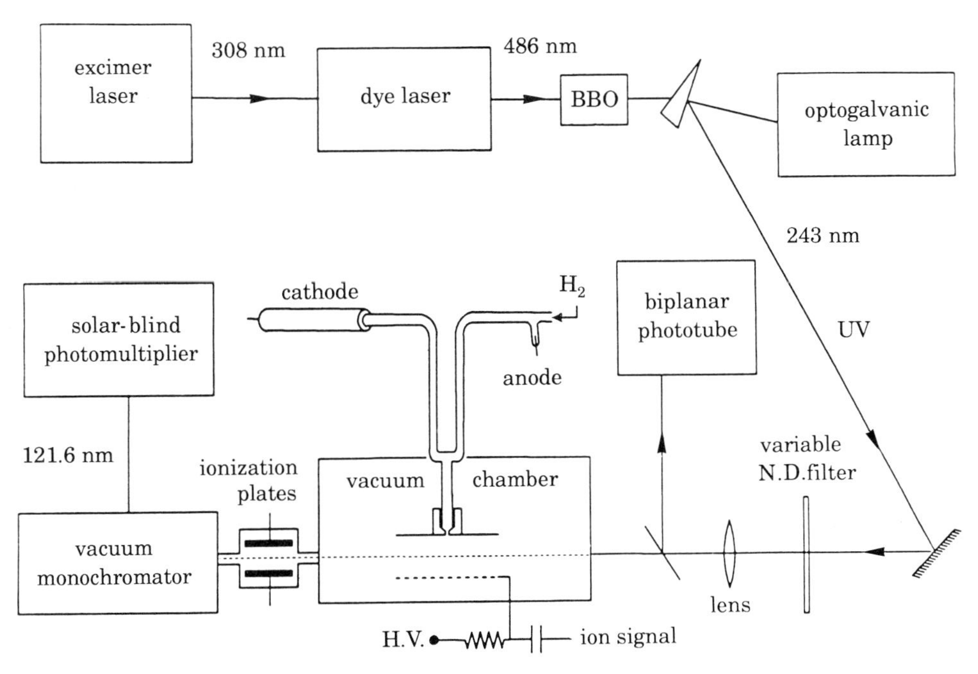

Fig. 21. – Experimental arrangement for generating SH radiation at 121.6 nm in a hydrogen atom beam, with and without an applied electric field.

Measurements of SH radiation intensity were made with applied electric fields, E_a, that ranged from 0 to $\pm$ 14 kV/cm. The results are shown in fig. 22 for incident laser power of 43 kW. The SH intensity increased quadratically with E_a up to ~ 5 kV/cm, approached saturation at ~ 7 kV/cm (with output of ~ 40 mW), and then decreased at higher fields. It is clearly shown that the SH intensity is not zero when $E_a = 0$, but ~ 850 mW. As expected for SHG, the intensity with, as well as without, an applied electric field increased quadratically with increasing incident laser power. All measurements were carried out with linearly polarized laser radiation; the SH intensity dropped by a factor > 20 when circularly polarized radiation was incident on the H beam.

An important clue to the mechanism leading to SHG was the observation of the SH intensity profile at $E_a = 0$. The far-field profile was a doubly peaked structure, with a dip in the center. Such nonuniformity is a sign of a radial electric field produced by the laser radiation, which in turn points to the creation of a charge separation field (CSF). This CSF was attributed to the motion of free electrons and protons that result from 3-photon ionization of the hydrogen atoms. Indeed the presence of ions was confirmed experimentally by the detection of an ion current only when 243 nm radiation was incident on the H beam.

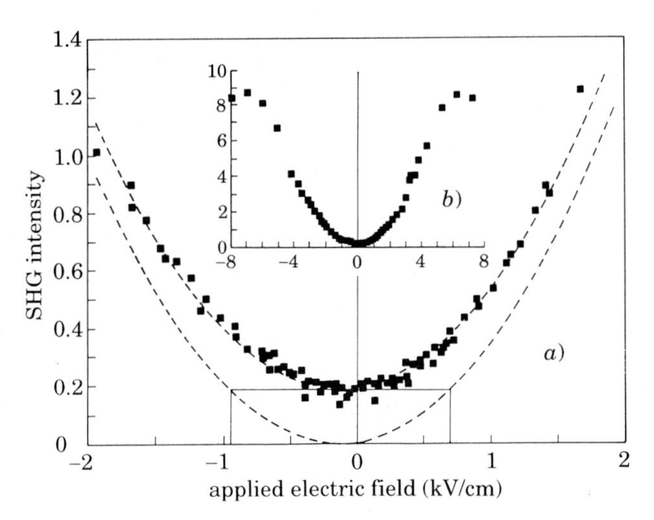

Fig. 22. – SHG at 121.6 nm plotted as a function of applied electric field *a*) for fields between − 2 and + 2 kV/cm, and *b*) for fields up to ±8 kV/cm.

Measurement of ion current exhibited a cubic dependence on laser power, and provided an estimate of the instantaneous ion density ($\sim 4.5 \cdot 10^{13}$ cm^{-3}), for use in calculating the CSF under the experimental conditions.

An estimate of the CSF was made from the experimental data of fig. 22. While the observed low-field data were fitted to a parabola, as shown in fig. 22*a*), it was also possible to fit the high-field data to a parabola with apex at zero SH intensity for $E_a = 0$. This could be thought of as a representation of SH intensity dependence on the applied d.c. field, assuming that no other fields were present. A line drawn through the experimental measurements near $E_a =$ $= 0$ and parallel to the field axis meets the latter parabola at (820 ± 70) V/cm. This value is the average CSF in the interaction region, created by the laser beam through 3-photon ionization. A value for this CSF was calculated based on the mechanism of charge separation [42], arising from the initial 1.7 eV kinetic energy of the ejected photoelectrons, which carries them away from the laser beam axis. In this calculation it was also necessary to include the motion of the protons. A calculated CSF of ~ 1 kV/cm was found, in good agreement with the value derived experimentally.

This experiment with atomic hydrogen has turned out to be of some consequence on two accounts. Firstly, it is the only known example of SHG for which the CSF has been measured experimentally, as well as confirmed by calculation. Secondly, as discussed below, it led to direct experimental observation of SHG of Lyman-α radiation with reduced absorption. This is relevant to a topic of current interest in quantum optics, namely, gain and lasing without population inversion.

4·6. SHG *of Lyman-α radiation with reduced absorption.* – Recently there has been considerable discussion concerning the possibility of obtaining stimulated emission without population inversion. In 1989 HARRIS [43] developed a theory to show that, when two upper levels of a four-level system are purely lifetime broadened and decay to the same continuum, there will be a destructive interference in the absorption profile of lower-level atoms which is not present in the stimulated-emission profile of upper-level atoms, thus resulting in laser gain without inversion. The upper levels are coupled by spontaneous decay to the same final continuum or discrete level, and the broadening can be caused by autoionization, tunneling, or radiative decay. It was also predicted [44] that such transparency could be induced by active coupling of the two upper levels with a strong electromagnetic field, and that, in nonlinear media, this process could resonantly enhance the nonlinear susceptibility and at the same time induce transparency and a zero in the contribution of the resonance transition to the refractive index [45]. Work in our laboratory demonstrated that coupling of metastable and upper levels by an applied d.c. electric field yields the same characteristics as coupling with an electromagnetic field. In addition, experimental evidence was found, using the behaviour of SHG, that coupling of the $2s$ and $2p$ levels in atomic hydrogen leads to reduced absorption at the center of the Stark-split components [46].

In an external field, SHG in atomic systems can be treated by the third-order nonlinear susceptibility, or as a second-order process where the electric field is explicitly included in the expression for $\chi^{(2)}$. The latter was selected, and calculations of the linear and nonlinear susceptibilities describing single-photon absorption and SHG were carried out using bare $1s$, $2s$ and $2p$ states as a basis set. These are shown in the energy level diagram of fig. 23. A d.c. electric field is applied to couple the metastable $2s$ with the $2p$ state which radiatively decays to the $1s$ ground state. A laser field of frequency ω_a, incident on the system, coherently drives the dipole at $\omega_c = 2\omega_a$ and generates SH radiation. The susceptibilities are given by

$$\chi^{(1)}(-\omega_c;\,\omega_c) = \frac{N}{2\varepsilon_0\hbar}\,\frac{\Delta\overline{\omega}_{21}}{\Delta\overline{\omega}_{21}\Delta\overline{\omega}_{31} - |\Omega_{32}|^2}\,\mu_{13}\mu_{31}\,, \tag{11}$$

$$\chi^{(2)}(-\omega_c;\,\omega_a,\,\omega_a) = \frac{N}{\varepsilon_0\hbar^2}\,\frac{\Omega_{32}}{\Delta\overline{\omega}_{21}\Delta\overline{\omega}_{31} - |\Omega_{32}|^2}\,\mu_{13}\sum_j \frac{\mu_{2j}\mu_{jl}}{\omega_j - \omega_a}\,. \tag{12}$$

Here, N is the atomic density, $\Delta\overline{\omega}_{21} = \omega_2 - \omega_1 - 2\omega_a$, $\Delta\overline{\omega}_{31} = \omega_3 - \omega_1 - \omega_c - - i\Gamma/2$, Ω_{32} denotes the Stark shift, μ_{ij} are matrix elements, and the ω_i represent frequencies of levels i.

From these formulae it can be shown that, if the decay rate of the $2s$ state is zero, then $\chi^{(1)}$ becomes zero at the bare $2s$ position (since $\Delta\overline{\omega}_{21} = 0$). That is, the medium becomes perfectly transparent to the SH radiation at $\omega_c = 2\omega_a$. This behaviour arises from the destructive interference between two Stark-mixed

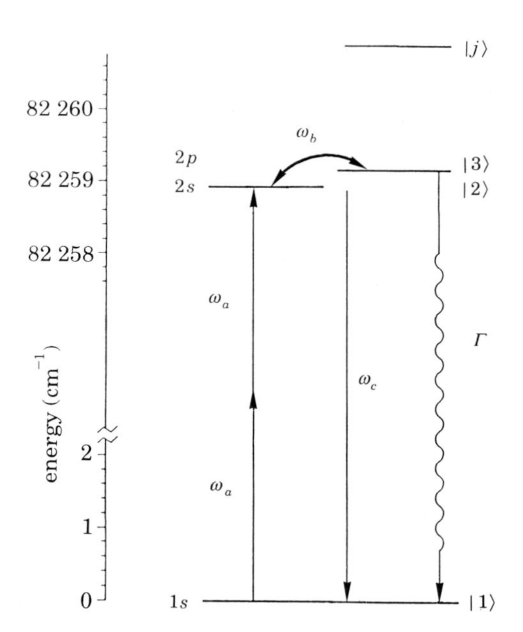

Fig. 23. – Energy levels of atomic hydrogen involved in the SHG process.

states via spontaneous-emission processes. On the other hand, $\chi^{(2)}$ shows a completely different behaviour due to the constructive interference in the SHG process. This is clearly seen in the tuning characteristics of $|\chi^{(2)}|^2$, Im $\chi^{(1)}$ and Re $\chi^{(1)}$ illustrated in fig. 24. With increasing d.c. field up to ~ 5 kV/cm, $|\chi^{(2)}|^2$ grows and the tuning curve splits into two components above fields of ~ 9 kV/cm. Meanwhile, the peak of Im $\chi^{(1)}$ decreases with increasing field, and the absorption profile splits into two peaks above ~ 5 kV/cm. Note especially that, on resonance (*i.e.* at the center of the Stark-split components) at 13 kV/cm, $|\chi^{(2)}|^2$ remains at $\sim 50\%$ of the peak value, while Im $\chi^{(1)}$ decreases to $\sim 3\%$ (limited by residual absorption due to the Doppler tail). Thus the atomic medium becomes almost totally transparent at the center while maintaining an appreciable nonlinear susceptibility. Moreover, the resonance contribution to the refractive index, Re $\chi^{(1)}$, takes a zero value at the center, thus satisfying the phase-matching condition. These results demonstrate that the d.c. field SHG in atomic hydrogen may provide ideal conditions for high conversion efficiency at 121.6 nm.

As a test of these characteristics, experiments were carried out with the same arrangement as used for SHG in hydrogen (fig. 21), but with applied d.c. electric fields in the range of $(0 \div 25)$ kV/cm. Second-harmonic radiation at 121.6 nm was detected by a solar-blind photomultiplier after being dispersed by a monochromator, and photoion current was observed simultaneously, as a monitor of the photon absorption. The measurements shown in fig. 25 confirm the basic characteristics of $|\chi^{(2)}|^2$ and $\chi^{(1)}$ predicted theoretically. In conclu-

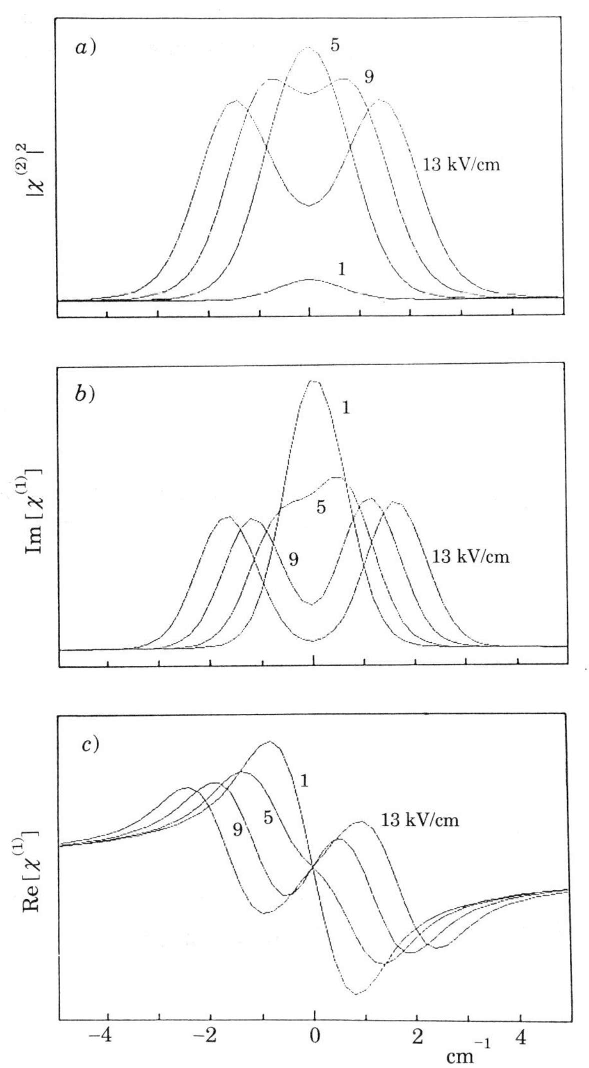

Fig. 24. – Calculated characteristics of $|\chi^{(2)}|^2$ and $\chi^{(1)}$.

sion, this investigation has provided evidence for enhanced SHG with reduced absorption at Lyman-α. A conversion efficiency of $\sim 10^{-6}$ was obtained in a 2 mm interaction length, at a hydrogen density of 10^{14} atoms/cm^3. More recent measurements at higher density and longer interaction lengths have shown that the SHG intensity scales according to the relation $I_{\mathrm{SHG}} \propto (NL)^2$. Thus, with the product $NL = 4 \cdot 10^{14}$ cm^{-2}, a peak power of 6 W was achieved, corresponding to an efficiency of almost 10^{-4} [47]. Further increase in NL should improve the efficiency by several orders of magnitude, and result in a powerful source of coherent Lyman-α radiation.

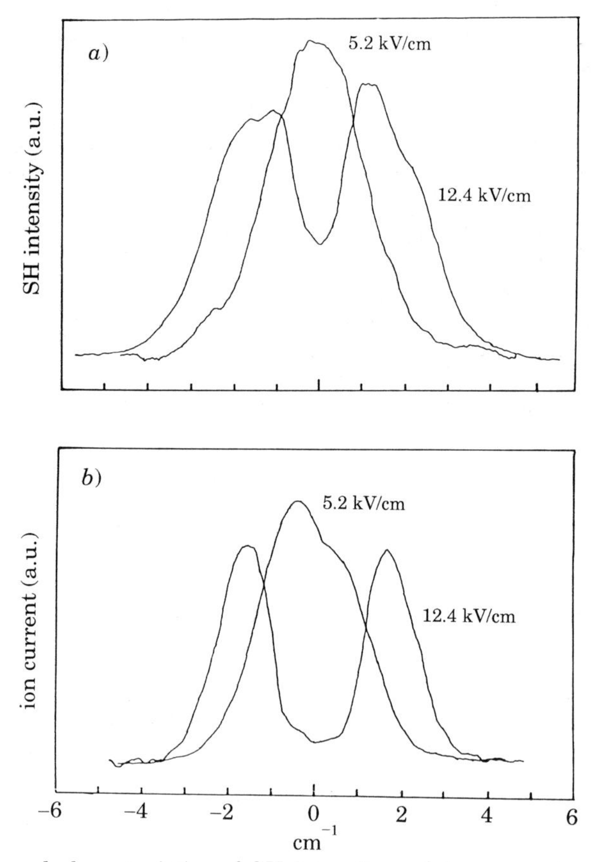

Fig. 25. – Observed characteristics of SH intensity and ion current.

* * *

The development of the VUV/XUV source at the University of Toronto and its application to the spectroscopic problems that I have discussed were carried out mainly by young students such as you. I am delighted to acknowledge their dedication and contributions.

REFERENCES

[1] P. A. FRANKEN, A. E. HILL, C. W. PETERS and G. WEINREICH: *Phys. Rev. Lett.*, **7**, 118 (1961).

[2] J. A. ARMSTRONG, N. BLOEMBERGEN, J. DUCUING and P. S. PERSHAN: *Phys. Rev.*, **127**, 1918 (1962).

[3] P. D. MAKER, R. W. TERHUNE and C. M. SAVAGE: in *Quantum Electronics III*, edited by P. GRIVET and N. BLOEMBERGEN (Columbia University Press, New York, N.Y., 1964), p. 1559.
[4] G. H. C. NEW and J. F. WARD: *Phys. Rev. Lett.*, **19**, 556 (1967).
[5] S. E. HARRIS and R. B. MILES: *Appl. Phys. Lett.*, **19**, 385 (1971).
[6] R. B. MILES and S. E. HARRIS: *IEEE J. Quantum Electron.*, QE-**9**, 470 (1973).
[7] R. T. HODGSON, P. P. SOROKIN and J. J. WYNNE: *Phys. Rev. Lett.*, **32**, 343 (1974).
[8] B. J. ORR and J. F. WARD: *Mol. Phys.*, **20**, 513 (1971).
[9] W. JAMROZ and B. P. STOICHEFF: in *Progress in Optics XX*, edited by E. WOLF (North-Holland, Amsterdam, 1983), p. 325.
[10] C. R. VIDAL: in *Tunable Lasers*, edited by L. F. MOLLENAUER and J. C. WHITE (Springer-Verlag, Heidelberg, 1987), p. 57.
[11] B. P. STOICHEFF: *Pure Appl. Chem.*, **59**, 1237 (1987).
[12] P. R. HERMAN, P. E. LAROCQUE, R. H. LIPSON, W. JAMROZ and B. P. STOICHEFF: *Can. J. Phys.*, **63**, 1581 (1985).
[13] E. CROMWELL, T. TRICKL, Y. T. LEE and A. H. KUNG: *Rev. Sci. Instrum.*, **60**, 2888 (1989).
[14] T. J. MCKEE, J. BANIC, A. JARES and B. P. STOICHEFF: *IEEE J. Quantum Electron.*, QE-**15**, 332 (1979).
[15] A. J. ANDREWS, A. J. KEARSLEY and C. E. WEBB: *Opt. Commun.*, **20**, 265 (1977).
[16] S. C. WALLACE and G. ZDASIUK: *Appl. Phys. Lett.*, **28**, 449 (1976); T. J. MCKEE, B. P. STOICHEFF and S. C. WALLACE: *Opt. Lett.*, **3**, 207 (1978).
[17] W. JAMROZ, P. E. LAROCQUE and B. P. STOICHEFF: *Opt. Lett.*, **7**, 148 (1982).
[18] P. R. HERMAN, P. E. LAROCQUE and B. P. STOICHEFF: *J. Chem. Phys.*, **89**, 4535 (1988).
[19] P. R. HERMAN and B. P. STOICHEFF: *Opt. Lett.*, **10**, 502 (1985).
[20] C. R. VIDAL and J. COOPER: *J. Appl. Phys.*, **40**, 3370 (1969); H. SCHEINGRABER and C. R. VIDAL: *Rev. Sci. Instrum.*, **52**, 1010 (1981).
[21] H. EGGER, T. SRINIVASAN, K. HOHLA, H. SCHEINGRABER, C. R. VIDAL, H. PUMMER and C. K. RHODES: *Opt. Lett.*, **5**, 282 (1980); *Appl. Phys. Lett.*, **39**, 37 (1981); R. WALLENSTEIN and H. ZACHARIAS: *Opt. Commun.*, **32**, 429 (1980); L. CABARET, C. DELSART and C. BLONDEL: *Opt. Commun.*, **61**, 116 (1987).
[22] A. H. KUNG: *Opt. Lett.*, **8**, 24 (1983).
[23] J. BOKOR, P. H. BUCKSBAUM and R. R. FREEMAN: *Opt. Lett.*, **8**, 217 (1983).
[24] R. HILBIG, G. HILBER, A. LAGO, B. WOLF and R. WALLENSTEIN: *Comments At. Mol. Phys.*, **18**, 157 (1986); R. HILBIG, A. LAGO and R. WALLENSTEIN: *Opt. Commun.*, **49**, 297 (1984); R. HILBIG and R. WALLENSTEIN: *Opt. Commun.*, **44**, 283 (1983).
[25] P. R. HERMAN, A. A. MADEJ and B. P. STOICHEFF: *Chem. Phys. Lett.*, **134**, 209 (1987).
[26] R. H. LIPSON, P. E. LA ROCQUE and B. P. STOICHEFF: *J. Chem. Phys.*, **82**, 4470 (1985).
[27] D. E. FREEMAN, K. YOSHINO and Y. TANAKA: *J. Chem. Phys.*, **61**, 4880 (1974).
[28] P. E. LAROCQUE, R. H. LIPSON, P. R. HERMAN and B. P. STOICHEFF: *J. Chem. Phys.*, 84, 6627 (1986).
[29] P. R. HERMAN, P. E. LAROCQUE and B. P. STOICHEFF: *J. Chem. Phys.*, **89**, 4535 (1988).
[30] A. C. PROVOROV, B. P. STOICHEFF and S. C. WALLACE: *J. Chem. Phys.*, **67**, 5393 (1977).
[31] M. MAEDA: *Jpn. J. Appl. Phys.*, **24**, 717 (1985).

[32] R. W. FIELD: Ph.D. Thesis, Harvard University (1972).
[33] A. C. LEFLOCH, F. LAUNAY, J. ROSTAS, R. W. FIELD, C. M. BROWN and K. YOSHINO: *J. Mol. Spectrosc.*, **121**, 337 (1987).
[34] A. A. MADEJ, P. R. HERMAN and B. P. STOICHEFF: *Phys. Rev. Lett.*, **57**, 1574 (1986); A. A. MADEJ and B. P. STOICHEFF: *Phys. Rev. A*, **38**, 3456 (1988).
[35] G. HERZBERG: *J. Mol. Spectrosc.*, **33**, 147 (1970).
[36] W. C. STWALLEY: *Chem. Phys. Lett.*, **6**, 241 (1970).
[37] E. F. MCCORMACK and E. E. EYLER: *Phys. Rev. Lett.*, **66**, 1042 (1991).
[38] A. BALAKRISHNAN, V. SMITH and B. P. STOICHEFF: *Phys. Rev. Lett.*, **68**, 2149 (1992).
[39] W. KOLOS, K. SZALEWICZ and H. J. MONKHORST: *J. Chem. Phys.*, **84**, 3278 (1986).
[40] L. WOLNIEWICZ and T. ORLIKOWSKI: *Mol. Phys.*, **74**, 103 (1991).
[41] L. MARMET, K. HAKUTA and B. P. STOICHEFF: *Opt. Lett.*, **16**, 261 (1991); *J. Opt. Soc. Am. B*, **9**, 1038 (1992).
[42] D. S. BETHUNE: *Phys. Rev. A*, **23**, 3139 (1981).
[43] S. E. HARRIS: *Phys. Rev. Lett.*, **62**, 1033 (1989).
[44] A. IMAMOĞLU and S. E. HARRIS: *Opt. Lett.*, **14**, 1344 (1989).
[45] S. E. HARRIS, J. E. FIELD and A. IMAMOĞLU: *Phys. Rev. Lett.*, **64**, 1107 (1990).
[46] K. HAKUTA, L. MARMET and B. P. STOICHEFF: *Phys. Rev. Lett.*, **66**, 596 (1991); *Phys. Rev. A*, **45**, 5152 (1992).
[47] K. HAKUTA, L. MARMET and B. P. STOICHEFF: in *Laser Spectroscopy X*, edited by M. DUCLOY, E. GIACOBINO and G. CAMY (World Scientific, Singapore, 1992), p. 301.

SPECTROSCOPY AND SURFACE EFFECTS

Surface Spectroscopy by Nonlinear Optics.

Y. R. SHEN

Department of Physics, University of California and Materials Sciences Division Lawrence Berkeley Laboratory - Berkeley, Cal. 94720

1. - Introduction.

The advances of laser spectroscopy have promoted material studies to an exciting new level. Its extremely high sensitivity and superlative spatial, spectral and temporal resolution have opened the door to many interesting possibilities. One may wonder if laser spectroscopy can also be expediently used for surface studies. Surface science is known to be a field of great importance in many disciplines ranging from physics, chemistry, biology to modern electronics. The progress of surface science and technology relies heavily on our ability to probe and characterize surfaces and interfaces [1]. Thus new surface probes are always in great demand. Driven by the need, a number of laser-based spectroscopic techniques have actually been developed as surface probes in recent years. For example, laser-induced desorption [2], photoacoustic spectroscopy [3] and photothermal deflection spectroscopy [4] have been adopted for spectroscopic studies of adsorbates and surface states. Resonance fluorescence [5] and multiphoton ionization [6] have been used to measure energy redistribution in molecules scattered or desorbed from a surface. Stimulated Raman gain spectroscopy [7] and coherent anti-Stokes Raman scattering [8] have also been found to be sensitive enough to yield the Raman spectrum of a molecular monolayer. Most of these techniques, however, suffer from either insufficient sensitivity or difficulty to discriminate against the bulk contribution to the signal. More recently, optical sum frequency generation (SFG) has been proven to be a most versatile tool for surface spectroscopy [9]. As a coherent nonlinear optical process, it possesses all the advantages of a laser spectroscopic technique. Moreover, it has an intrinsic surface specificity and is applicable to any interface accessible by light.

Optical SFG here describes a process in which two input laser beams at frequencies ω_1 and ω_2 interact and generate an output at the sum frequency $\omega = \omega_1 + \omega_2$. Being a second-order wave-mixing process, it is forbidden in a medium with inversion symmetry [10]. Yet SFG is always allowed at an interface where the inversion symmetry is necessarily broken. This makes SFG highly surface specific for interfaces between centrosymmetric media. More generally, one can expect surface and bulk to have different symmetries so that with appropriate input and output beam polarizations the bulk contribution, even if it is allowed, can still be strongly suppressed [11].

As a surface probe, SFG is sufficiently sensitive to detect a submonolayer of atoms or molecules. With tunable input beams, it provides a viable method for surface spectroscopy. Because of the intrinsic ability of SFG to discriminate against bulk contribution, the measurement of a surface spectrum is now possible even in the presence of a strong overlapping bulk spectrum. Being a second-order process, SFG is also sensitive to the average polar orientation or arrangement of atoms and molecules. Furthemore, SFG is capable of remote-sensing measurements in a hostile environment and is applicable to almost all interfaces, including the buried interfaces. With the help of ultrashort laser pulses, it permits the study of surface dynamics with a time resolution down to subpicosecond. These very unique features of SFG make the technique most unusual and powerful as a surface spectroscopic tool.

Surface SFG was first developed as a natural extension of surface optical second-harmonic generation (SHG) which is itself a viable surface probe [12]. With a tunable laser, HEINZ *et al.* first showed that the SHG can be used to obtain the spectrum of electronic transitions of an adsorbed molecular monolayer [13]. It is obvious that SFG with either ω_1 or ω_2 or both tunable can provide much more flexibility and spectroscopic information than SHG. The latter is not suitable for surface vibrational spectroscopy, because the output appears in the infrared region where the sensitivity of available photodetectors is not sufficient to allow the detection of a monolayer. SFG, on the other hand, is ideally suited for surface vibrational spectroscopy. With ω_1 in the infrared and ω_2 in the visible, the visible output can be easily detected by sensitive photomultipliers. This was first demonstrated on molecular monolayers adsorbed on solid substrates [14]. The technique since then has attracted much attention. It has been successfully applied to a variety of interfacial systems including surface states of metals [15] and surface molecular vibrations of pure liquids [16]. This lecture will provide a brief survey of the progress in the field.

The lecture is organized as follows. Section **2** briefly reviews the basic theory of surface SFG. Emphasis is on how the SFG measurements are related to surface nonlinearity which can yield information about surface or interfacial properties. Theoretical difficulties in evaluating surface response from a more microscopic point of view are discussed. Section **3** describes the experimental arrangement and numerical estimate of a typical surface SFG measurement. Section **4** surveys the various SFG experiments that have provided interesting new results. Particular attention is drawn to those from which the information deduced is not obtainable by other means.

2. – Theory.

The basic theory for surface SHG and SFG has been described in detail elsewhere [12, 17]. Here, we will give only a brief review outlining the key results. For simplicity, we consider only SFG from reflection from an interface, depicted in fig. 1, and assume that the interface is between a linear medium (1) and a nonlinear medium (2).

As shown in fig. 1, we can model the interface as a 3-layer system. Let $\boldsymbol{P}^{\mathrm{s}}$ be the surface nonlinear polarization (per unit area) induced in the interfacial layer and $\boldsymbol{P}^{\mathrm{B}}$ the bulk nonlinear polarization (per unit volume) induced in medium 2. The radiation field generated by $\boldsymbol{P}^{\mathrm{s}}(\omega)$ and $\boldsymbol{P}^{\mathrm{B}}(\omega)$ can be obtained by solving the wave equation with proper boundary conditions for the three-layer system in the limit of an infinitesimally thin interfacial layer [18]. Alternatively, $\boldsymbol{P}^{\mathrm{s}}(\omega)$ can be treated as a thin sheet of radiating dipoles [19].

Consider first the ideal case of $\boldsymbol{P}^{\mathrm{B}}(\omega) = 0$ and let

$$\boldsymbol{P}^{\mathrm{s}}(\omega = \omega_a + \omega_b) = \overleftrightarrow{\chi}_{\mathrm{s}}^{(2)} : \boldsymbol{E}_{\mathrm{L}}(\omega_1)\boldsymbol{E}_{\mathrm{L}}(\omega_2), \tag{1}$$

where $\overleftrightarrow{\chi}_{\mathrm{s}}^{(2)}$ is the surface nonlinear susceptibility and $\boldsymbol{E}_{\mathrm{L}}(\omega_1)$ and $\boldsymbol{E}_{\mathrm{L}}(\omega_2)$ are the input fields in the interfacial layer. One can then show that the SF radiation

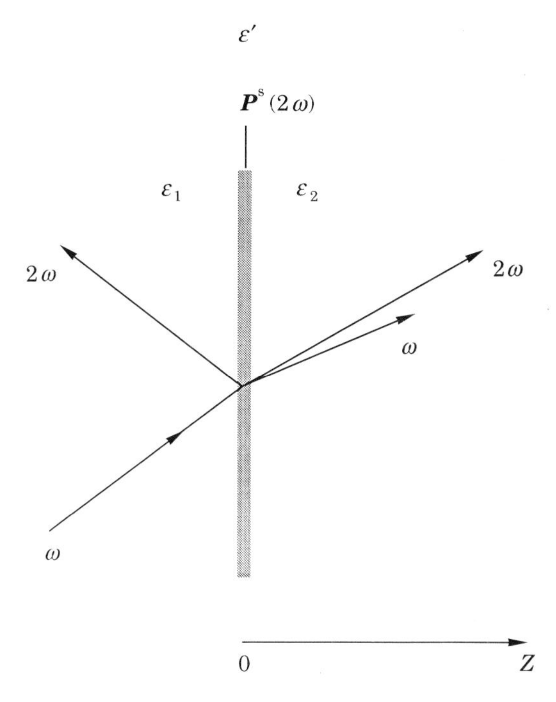

Fig. 1. – Geometry of second-harmonic generation from an interface in the reflected and transmitted directions. The polarization sheet P^{s} is imbedded in a thin layer of dielectric constant ε'.

field in medium 1 generated by $\boldsymbol{P}^{\mathrm{s}}$ has the expression

$$(2)\quad \begin{cases} E_{\mathrm{p}}(\omega) = i(2\pi\omega/c)\Bigg[L_{xx}(\omega)\,\chi^{(2)}_{\mathrm{s},\,zjk}L_{jj}(\omega_a)\,L_{kk}(\omega_b) + \\ \qquad + \dfrac{k_x(\omega)}{k_{1z}(\omega)}\,L_{zz}(\omega)\,\chi^{(2)}_{\mathrm{s},\,xjk}L_{jj}(\omega_a)\,L_{kk}(\omega_b)\Bigg]E_{\mathrm{I}j}(\omega_a)\,E_{\mathrm{I}k}(\omega_b), \\ E_{\mathrm{s}}(\omega) = i\,\dfrac{2\pi k_1^2(\omega)}{k_{1z}(\omega)\varepsilon_1(\omega)}\,L_{yy}(\omega)\,\chi^{(2)}_{\mathrm{s},\,yjk}L_{jj}(\omega_a)\,L_{kk}(\omega_b)\,E_{\mathrm{I}j}(\omega_a)\,E_{\mathrm{I}k}(\omega_b). \end{cases}$$

Here, the subscripts p and s refer to beam polarizations. $\boldsymbol{E}_{\mathrm{I}}(\omega_i)$ is the input field at ω_i in medium 1, and L_{ii} is a transmission Fresnel coefficient for the field acting as a macroscopic local-field correction:

$$(3)\quad \begin{cases} L_{xx}(\omega_i) = \dfrac{2\varepsilon_1(\omega_i)\,k_{2z}(\omega_i)}{\varepsilon_2(\omega_i)\,k_{1z}(\omega_i) + \varepsilon_1(\omega_i)\,k_{2z}(\omega_i)}, \\ L_{yy}(\omega_i) = \dfrac{2k_{1z}}{k_{1z} + k_{2z}}, \\ L_{zz}(\omega_i) = \dfrac{2\varepsilon_1 k_{1z}(\varepsilon_2/\varepsilon')}{\varepsilon_2 k_{1z} + \varepsilon_1 k_{2z}}. \end{cases}$$

If $\varepsilon_1 = \varepsilon' = \varepsilon_2$, then $L_{ii} = 1$ and eq. (2) correctly reduces to the well-known results for a radiating dipole sheet imbedded in a uniform dielectric medium. Thus the result in eq. (2) is physically transparent [19]: the radiation field can be directly obtained from that generated by an induced polarization sheet with proper Fresnel coefficients to take into account the field ratios in different media. Then extension to cases with different geometries is straightforward. Equation (2) is still valid only if $\boldsymbol{k}(\omega)$ is properly chosen and the appropriate L_{ii}'s are used. For example, in the geometry of fig. 1, the SF field generated from the interfacial layer and propagating in medium 2 is again given by eqs. (2) and (3) with $k_1(\omega)$ and $k_2(\omega)$ interchanged and $k_{iz}(\omega)$ replaced by $-k_{iz}(\omega)$.

The solution of the wave equation including $\boldsymbol{P}^{\mathrm{B}}$ in medium 2 can also be readily found [20]. It can be shown that the above result still holds if we simply replace $\boldsymbol{P}^{\mathrm{s}}(\omega)$ by an effective surface polarization $\boldsymbol{P}^{\mathrm{s}}_{\mathrm{eff}}(\omega)$ (for SF reflection into medium 1)

$$(4)\quad P^{\mathrm{s}}_{\mathrm{eff},\,j} = p^{\mathrm{s}}_j + \frac{iP^{\mathrm{B}}_j}{[k_{2z}(\omega) + k_{2z}(\omega_a) + k_{2z}(\omega_b)]\,F_j(\omega)}$$

and $\overleftrightarrow{\chi}^{(2)}$ by an effective surface nonlinear susceptibility $\overleftrightarrow{\chi}^{(2)}_{\mathrm{s,\,eff}}$

$$(5)\quad (\chi^{(2)}_{\mathrm{s,\,eff}})_{ijk} = (\chi^{(2)}_{\mathrm{s}})_{ijk} + \frac{(\chi^{(2)}_{\mathrm{B}})_{ijk}}{[k_{2z}(\omega) + k_{2z}(\omega_a) + k_{2z}(\omega_b)]\,F_i(\omega)\,F_j(\omega_a)\,F_k(\omega_b)},$$

where $F_j(\Omega) = \varepsilon_2(\Omega)/\varepsilon'(\Omega)$ for $j = z$ and $F_j(\Omega) = 1$ for $j = x, y$. The SF output intensity can now be calculated from eq. (2) and is given by

$$(6) \quad I(\omega) = c\varepsilon_1(\omega)|E_1(\omega)|^2/2\pi =$$

$$= \frac{8\pi^3\omega^2\sin^2\theta_\omega}{c^3[\varepsilon_1(\omega)\varepsilon_1(\omega_a)\varepsilon_1(\omega_b)]^{1/2}}|\boldsymbol{e}^\dagger(\omega)\cdot\chi^{(2)}_{\mathrm{s,\,eff}}:\boldsymbol{e}(\omega_a)\boldsymbol{e}(\omega_b)|^2 I_{\mathrm{I}}(\omega_a)I_{\mathrm{I}}(\omega_b).$$

Here, $\boldsymbol{e}(\Omega) \equiv \overleftrightarrow{L}\cdot\hat{e}(\Omega)$, $\hat{e}(\Omega)$ is the unit vector describing the polarization of the field at frequency Ω, θ_ω is the reflection angle of SFG from the surface normal, and $I_{\mathrm{I}}(\omega_i)$ is the input laser intensity at ω_i. If the input is in the pulse form with pulse width T and the beam overlapping cross-section at the interface is A, then the SF output, in terms of photons/pulse, becomes

$$(7) \quad S(\omega) =$$

$$= \frac{8\pi^3\omega\sin^2\theta_\omega}{c^3\hbar[\varepsilon_1(\omega)\varepsilon_1(\omega_a)\varepsilon_1(\omega_b)]^{1/2}}|\boldsymbol{e}^\dagger(\omega)\cdot\chi^{(2)}_{\mathrm{s,\,eff}}:\boldsymbol{e}(\omega_a)\boldsymbol{e}(\omega_b)|^2 I_{\mathrm{I}}(\omega_a)I_{\mathrm{I}}(\omega_b)AT.$$

The three-layer model to describe an interfacial system is clearly only an approximation and the dielectric constant ε' for the interfacial layer is certainly not well defined. The difficulty of properly describing an interface and its optical response is well known in condensed-matter physics. It is, however, clear that, on the microscopic scale, both the optical field and the dielectric constant cannot vary abruptly and the interfacial layer must have a finite thickness [17]. With z normal to the interface, we can assume that $\varepsilon(z)$ varies from ε_1 to ε_2 along z across the interface. The field component parallel to the interface is continuous (*i.e.* independent of z) across the interface, but the component along z takes the form $E_z(z) = D_z/\varepsilon(z)$, where the displacement current component D_z is continuous across the interface. We then recognize that, in the above derivation, we should actually have $P_i^{\mathrm{s}}(\omega)$ defined as

$$(8) \qquad P_i^{\mathrm{s}}(\omega) \equiv \int_{\mathrm{int}} f_i(\omega, z)P_i^{\mathrm{B}}(\omega, z)\,\mathrm{d}z/F_i(\omega),$$

where $F_i(\omega) = F_i(\omega, z) = 1$ for $i = x, y$, and $F_i(\omega) = \varepsilon_2(\omega)/\varepsilon'(\omega)$ and $f_i(\omega, z) = \varepsilon_2(\omega)/\varepsilon(\omega, z)$ for $i = z$. The integration is across the interfacial layer. With $P_i^{\mathrm{B}}(\omega, z) = (\chi_{\mathrm{B}}^{(2)}(z))_{ijk}\,E_j(\omega_a, z)\,E_k(\omega_b, z)$, we find

$$(9) \quad P_i^{\mathrm{s}}(\omega) \equiv \int_{\mathrm{int}} f_i(\omega, z)(\chi_{\mathrm{B}}^{(2)}(z))_{ijk}f_j(\omega_a, z)f_k(\omega_b, z)\,\mathrm{d}z\,\frac{E_{2j}(\omega_a)E_{2k}(\omega_b)}{F_i(\omega)}.$$

It is readily seen that $\chi^{(2)}_{\mathrm{s},ijk}$ in eq. (2) should now be replaced by

$$\chi^{(2)}_{\mathrm{s},ijk}(\omega) \equiv \int_{\mathrm{int}} f_i(\omega, z)(\chi^{(2)}_{\mathrm{B}}(z))_{ijk} f_j(\omega_a, z) f_k(\omega_b, z)\, \mathrm{d}z / F_i(\omega) F_j(\omega_a) F_k(\omega_b), \tag{10}$$

which provides a physical and more rigorous definition for the interfacial surface nonlinear susceptibility. Through F's, the unphysical quantity ε' is eliminated in the final result.

More correctly, $\boldsymbol{P}^{\mathrm{B}}(\omega, z)$ also depends on the field derivatives (*i.e.* a nonlocal response in general), and can be expanded into a multipole series

$$\boldsymbol{P}^{\mathrm{B}}(\omega) = \boldsymbol{P}^{(2)}_{\mathrm{D}}(\omega) - \nabla \cdot \overleftrightarrow{Q}^{(2)}(\omega) - \frac{c}{i2\omega} \nabla \times \boldsymbol{M}^{(2)}(\omega) + \dots, \tag{11}$$

where $\boldsymbol{P}^{(2)}_{\mathrm{D}}, \overleftrightarrow{Q}^{(2)}$ and $\boldsymbol{M}^{(2)}$ denote electric-dipole polarization, electric-quadrupole polarization and magnetization, respectively [17]. We shall neglect $\boldsymbol{M}^{(2)}$ and higher-order multipoles in the following discussion. By approximating $\boldsymbol{P}^{(2)}_{\mathrm{D}}$ and $\overleftrightarrow{Q}^{(2)}$ as

$$\begin{cases} \boldsymbol{P}^{(2)}_{\mathrm{D}}(\omega, z) = \overleftrightarrow{\chi}^{(2)}_{\mathrm{D}}(\omega, z) \colon \boldsymbol{E}(\omega_a, z) \boldsymbol{E}(\omega_b, z) + \\ \qquad + \overleftrightarrow{\chi}^{(2)}_{\mathrm{Q}a}(\omega, z) \colon \nabla \boldsymbol{E}(\omega_a, z) \boldsymbol{E}(\omega_b, z) + \overleftrightarrow{\chi}^{(2)}_{\mathrm{Q}b}(\omega, z) \colon \boldsymbol{E}(\omega_a, z) \nabla \boldsymbol{E}(\omega_b, z), \\ \overleftrightarrow{Q}^{(2)}(\omega, z) = \overleftrightarrow{\chi}^{(2)}_{\mathrm{Q}}(\omega, z) \colon \boldsymbol{E}(\omega_a, z) \boldsymbol{E}(\omega_b, z), \end{cases} \tag{12}$$

and substituting eq. (11) into eq. (8), we find

$$\begin{aligned} P^{\mathrm{s}}_i(\omega) &= \int_{\mathrm{int}} f_i(\omega, z)[\boldsymbol{P}^{(2)}_{\mathrm{D}}(\omega, z) - \nabla \cdot \boldsymbol{Q}^{(2)}(\omega, z)]_i \, \mathrm{d}z / F_i(\omega) = \\ &= \frac{1}{F_i(\omega)} \left\{ \int_{\mathrm{int}} \left[f_i(\omega, z) P^{(2)}_{\mathrm{D}i}(\omega, z) + Q^{(2)}_{iz}(\omega, z) \frac{\partial f_i(\omega, z)}{\partial z} \right] \mathrm{d}z - \right. \\ &\left. - [f_i(\omega, z) Q_{iz}(\omega, z)]^{0^+}_{0^-} \right\} = \frac{1}{F_i(\omega)} \left\{ \int_{\mathrm{int}} f_i(\omega, z)(\chi^{(2)}_{\mathrm{D}})_{ijk} f_j(\omega_a, z) f_k(\omega_b, z)\, \mathrm{d}z + \right. \\ &+ \int_{\mathrm{int}} \left[\frac{\partial f_i(\omega, z)}{\partial z} (\chi^{(2)}_{\mathrm{Q}})_{ijk} f_j(\omega_a, z) f_k(\omega_b, z) + f_i(\omega, z)(\chi^{(2)}_{\mathrm{Q}a})_{ijk} \frac{\partial f_j(\omega_a, z)}{\partial z} f_k(\omega_b, z) + \right. \\ &\left. + f_i(\omega, z)(\chi^{(2)}_{\mathrm{Q}b})_{ijk} f_j(\omega_a, z) \frac{\partial f(\omega_b, z)}{\partial z} \right] \mathrm{d}z + \\ &\left. + [f_i(\omega, 0^-)(\chi^{(2)}_{\mathrm{Q}}(\omega, 0^-))_{ijk} f_j(\omega_a, 0^-) f_k(\omega_b, 0^-) - (\chi^{(2)}_{\mathrm{Q}}(\omega, 0^+))_{ijk}] \right\} E_{2j}(\omega_a) E_{2k}(\omega_b). \end{aligned} \tag{13}$$

This leads to

$$(14)\qquad \chi^{(2)}_{\mathrm{s},\,ijk}(\omega=\omega_a+\omega_b)=\frac{1}{F_i(\omega)F_j(\omega_a)F_k(\omega_b)}\cdot$$

$$\cdot\left\{\int_{\mathrm{int}} f_i(\omega,z)(\chi^{(2)}_{\mathrm{D}})_{ijk}f_j(\omega_a,z)f_k(\omega_b,z)\,\mathrm{d}z+\int_{\mathrm{int}}\left[\frac{\partial f_i(\omega,z)}{\partial z}(\chi^{(2)}_{\mathrm{Q}})_{ijk}f_j(\omega_a,z)f_k(\omega_b,z)+\right.\right.$$

$$\left.+f_i(\omega,z)(\chi^{(2)}_{\mathrm{Q}a})_{ijk}\frac{\partial f_j(\omega_a,z)}{\partial z}f_k(\omega_b,z)+f_i(\omega,z)(\chi^{(2)}_{\mathrm{Q}b})_{ijk}f_j(\omega_a,z)\frac{\partial f_k(\omega_b,z)}{\partial z}\right]\mathrm{d}z+$$

$$\left.+[f_i(\omega,0^-)(\chi^{(2)}_{\mathrm{Q}}(0^-))_{ijk}f_j(\omega_a,0^-)f_k(\omega_b,0^-)-(\chi^{(2)}_{\mathrm{Q}}(0^+))_{ijk}]\right\}.$$

The above derivation allows us to clearly identify the physical origins of the various terms in eq. (14). The first term in the curly brackets arises from the electric-dipole contribution (or local response). Because of the broken inversion symmetry at the interface, this term is nonvanishing even if the bulk is centrosymmetric. It can be dominating if the interfacial structure is highly noncentrosymmetric. The second term is electric quadrupole in nature (or nonlocal response). If the linear dielectric constants of the two bonding media are matched, then the fields are continuous across the interface, $f=1$, and this quadrupole term should vanish. The last term, contained in the square brackets, comes from the structural disparity between the two bonding media. It is completely determined by the bulk parameters of the two media, and can in fact be called a bulk contribution.

Calculations of $\varepsilon(\Omega, z)$, $\overleftrightarrow{\chi}^{(2)}_{\mathrm{B}}$ and $\overleftrightarrow{\chi}^{(2)}_{\mathrm{s}}$ are clearly difficult, because of lack of information about the interfacial structure and the corresponding electronic properties. Formally, we can write, for the electric-dipole contribution to $\chi^{(2)}_{\mathrm{s}}$ [21],

$$(15)\qquad (\chi^{(2)}_{\mathrm{D}})_{ijk}=\int\left\{\sum_{g,n,n'}\left(-\frac{e^3}{\hbar^2}\right)\left[\frac{\langle g|r_i|n\rangle\langle n|r_j|n'\rangle\langle n'|r_k|g\rangle}{(\omega-\omega_{ng}+i\Gamma_{ng})(\omega_b-\omega_{n'g}+i\Gamma_{n'g})}+\right.\right.$$

$$\left.\left.+\frac{\langle g|r_i|n\rangle\langle n|r_k|n'\rangle\langle n'|r_j|g\rangle}{(\omega-\omega_{ng}+i\Gamma_{ng})(\omega_a-\omega_{n'g}+i\Gamma_{n'g})}+6\text{ others terms}\right]\rho^0_{gg}\right\}g(\Omega)\,\mathrm{d}\Omega\,,$$

where ρ^0_{gg} is the population in the $\langle g|$ state, ω_{ng} and Γ_{ng} are the transition frequency from $\langle g|$ to $\langle n|$ and the associated damping constant, and $g(\Omega)$ is a distribution function of the parameter, or a set of parameters, Ω, with $\int f(\Omega)\,\mathrm{d}\Omega =$ $= N(z) =$ number of molecules per unit volume at z. We have assumed that the interfacial layer is composed of a set of localized molecules. Extension to the more general case is fairly straightforward, but is not essential for the discussion here. Similar expressions are found for $\overleftrightarrow{\chi}^{(2)}_{\mathrm{Q}}$, $\overleftrightarrow{\chi}^{(2)}_{\mathrm{Q}a}$ and $\overleftrightarrow{\chi}^{(2)}_{\mathrm{Q}b}$ with one of the matrix elements in each term of eq. (15) replaced by a quadrupole matrix element $\langle l|r_\alpha r_\beta|m\rangle$. The microscopic expression for $\overleftrightarrow{\chi}_s^{(2)}$ then shows that we can study resonant transitions of an interfacial system from the resonant dispersion of $\overleftrightarrow{\chi}_s^{(2)}$. This is the basis of surface SFG spectroscopy.

It is interesting to see the relative importance of surface and bulk contributions to the reflected SFG and relative importance of $\overleftrightarrow{\chi}_D^{(2)}$ and $\overleftrightarrow{\chi}_Q^{(2)}$ within $\overleftrightarrow{\chi}_s^{(2)}$. Unfortunately, the answer depends very much on the interfacial system. The following discussion is only meant to provide a general guideline. Assuming that the bulk media are centrosymmetric, we have $\chi_B^{(2)} \sim k\chi_Q^{(2)} \sim ka\chi_D^{(2)}$, where a is the size of a relevant electronic orbit. Across the interface, we expect $(\mathrm{d}f/\mathrm{d}z)/f \sim d$ which is roughly the thickness of the interfacial layer. We then find from eq. (14) that the electric-dipole contribution to $\chi_s^{(2)}$ is larger than the electric-quadrupole contribution by a factor of $\sim d/a$. Since d is at least the length of one molecular monolayer, we should have $d/a \gtrsim 1$ and possibly much larger than 1 if $d \gg a$. In comparing the surface and bulk contributions to SF reflection, we note that from eq. (14)

$$(\chi_s^{(2)})_{ijk} \sim (\chi_D^{(2)})_{ijk}\, d \sim (\chi_Q^{(2)})_{ijk}\,(d/a), \tag{16}$$

and then, from eq. (5), we find that the ratio of the surface to the bulk term in $(\chi_{s,\,\mathrm{eff}}^{(2)})_{ijk}$ is approximately $(2d/a)F_iF_jF_k$, which is larger than 1 with $\varepsilon_2 > \varepsilon'$. Therefore, the surface contribution to SF reflection appears stronger. If F's are significantly larger than 1, then the surface contribution should dominate. In many cases, the spectra of the interface and the bulk are different; the surface and bulk contributions can then be easily distinguished from their spectral features.

Finally, we briefly discuss here the characteristic features of surface SFG spectroscopy. As a second-order process with dominating contribution from the interface, the SF output is sensitive to the polar orientation and arrangement of the interfacial layer. The surface specificity of SFG allows spectroscopic measurements of interfacial layers, even in the presence of a strong bulk absorption. The pronounced resonances observed in SFG are expected to be of both one-photon and two-photon allowed transitions (both Raman- and infrared-active transitions if resonances are in the infrared). Double resonances can be observed if both ω_a and ω_b are tunable. As a coherent nonlinear optical spectroscopic technique, the spectral profile of SFG resembles that of the well-known coherent anti-Stokes Raman spectroscopy (CARS) [22]. Unlike conventional spectroscopy, the SF output is proportional to $|\chi_{s,\,\mathrm{eff}}^{(2)}|^2$, which depends on both the real part and the imaginary part of $\chi_{s,\,\mathrm{eff}}^{(2)}$. Even close to a resonance, the nonresonant contribution to $\chi_{s,\,\mathrm{eff}}^{(2)}$ may not be negligible. We can write

$$|\overleftrightarrow{\chi}_{s,\,\mathrm{eff}}^{(2)}|^2 = \left| \overleftrightarrow{\chi}_{\mathrm{NR}}^{(2)} + \sum_q \frac{\overleftrightarrow{A}_q}{\omega_a - \omega_q + i\Gamma_q} \right|^2 \tag{17}$$

assuming that only ω_a is near resonances at ω_q. Equation (17) shows that the profile of a resonance can vary significantly depending on the relative amplitude and phase of χ_{NR} and A_q. Only when χ_{NR} is negligibly small and the resonances are far apart, will unequivocal resonant peaks appear in the SFG spectrum. More generally, eq. (17) has to be used to fit the observed spectrum in order to deduce the characteristic parameters of the resonances.

Transient surface spectroscopy is certainly also possible with SFG. With picosecond or subpicosecond pulse excitation, both longitudinal and transverse relaxations of a surface resonance can be measured. This will be discussed in more detail in sect. 4.

3. – Experimental considerations.

Equation (7) allows us to have a crude estimate on the signal strength of surface SFG and helps us decide on which types of lasers are needed for the experiment. Consider a case with $I(\omega_a) \sim I(\omega_b) \sim 10^9\ W/cm^2$, $T \sim 10$ ps, $A \sim 1\ mm^2$ (corresponding to laser pulse energy of 100 μJ/pulse and fluence of 10 mJ/cm² which is presumably below the surface damage threshold), and $|\chi^{(2)}_{s,\,eff}| \sim 10^{-15}$ e.s.u. (typical value for a molecular monolayer). Substitution of these numbers into eq. (7) yields a signal stregth of $\sim 10^5$ photons/pulse. Since the noise of a detection system can be lower than 0.1 photon/pulse, we can have a signal-to-noise ratio of 10^6. This example gives a crude idea of how sensitive the surface SFG spectroscopy can be. The laser pulses assumed in the example can be readily obtained from a high-power picosecond laser system. For infrared-visible SFG spectroscopy, tunable laser pulses in the infrared can be derived from stimulated Raman scattering, optical parametric amplification or difference frequency generation pumped by high-power picosecond lasers in the visible [14, 23]. Nanosecond laser pulses can also be used to yield a sufficient signal-to-noise ratio [14, 24].

A typical experimental setup is sketched in fig. 2. The two input laser beams must be well overlapped at the interface. The SF output is highly directional as a result of wave vector matching parallel to the interface. Spatial filtering can, therefore, be used to help discriminate background noise arising from laser scattering or fluorescence. Either an interface filter or a spectrometer is usually employed to selectively detect the SF output and further improve the signal-to-background ratio. As suggested in eq. (7), the signal strength should increase with decrease of the cross-sectional area A, but the improvement is eventually limited by laser-induced damage of the interface.

Infrared-visible SFG spectroscopy has attracted much attention of researchers because of the scarcity of infrared surface spectroscopic techniques. This requires a relatively high-power tunable infrared laser source which is most easily obtained by optical parametric amplification or difference frequency

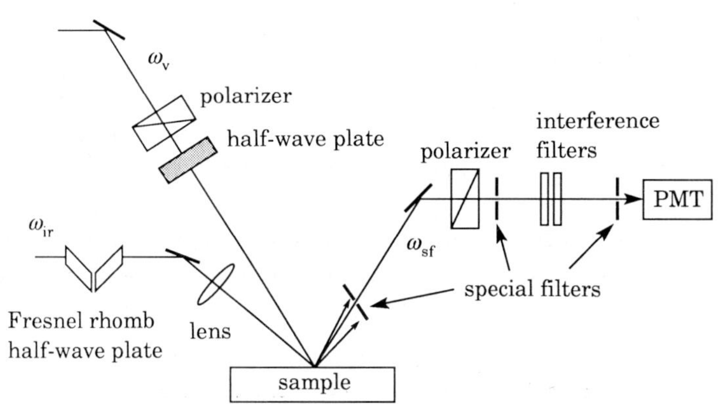

Fig. 2. – A typical experimental setup for SFG.

generation. With $LiNbO_3$, BBO, LBO and KTP as nonlinear mixing crystals, the infrared output can have a tuning range down to $\sim 4\ \mu m$ [25-27]. With $AgGaS_2$, it can be extended to $\sim 12\ \mu m$, and with $AgGaSe_2$ possibly to $\sim 18\ \mu m$ [28]. Further extension to lower infrared frequencies becomes extremely difficult. A free-electron laser is then needed to cover the rest of the infrared range.

4. – Applications.

Surface infrared-visible SFG spectroscopy was first demonstrated and used to obtain vibrational spectra of adsorbed molecular monolayers [14]. The results were very encouraging as the monolayer spectra can be readily observed with any substrate. Figures 3 and 4 give two representative examples. The former displays the SFG spectra of C-H bond stretching vibrations of three different molecular species adsorbed on fused quartz [29]. The different stretching modes are clearly resolved and the spectra are different for different species. In fig. 4, the SFG spectra of the CH_3 stretch modes of an OTS (octadecyltrichlorsilane, $CH_3(CH_3)_{17}SiCl_3$) monolayer adsorbed on silicon, aluminum and silicate are presented [30]. They confirm the possibility of applying SFG spectroscopy to monolayers on any substrate.

To illustrate the surface specificity of SFG spectroscopy, we use the results in fig. 5 as an example [31]. Here the spectra in the C-H stretch region are presented for three liquid/solid interfacial systems: hexadecane/silica, hexadecane/OTS/silica and CCl_4/OTS/silica, where OTS denotes a full monolayer of OTS adsorbed on silica. The first spectrum shows no CH peaks even though liquid hexadecane is known to have a strong infrared absorption in this spectral region arising from the large number of CH_2 stretch modes in these molecules. The second spectrum shows the CH stretch peaks of the terminal methyl (CH_3)

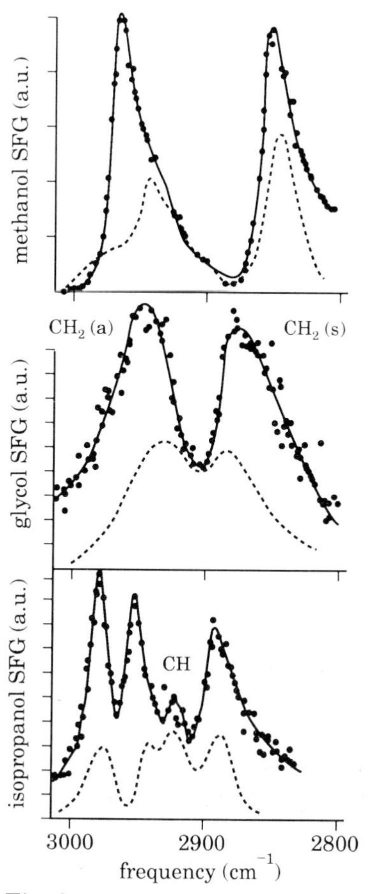

Fig. 3.

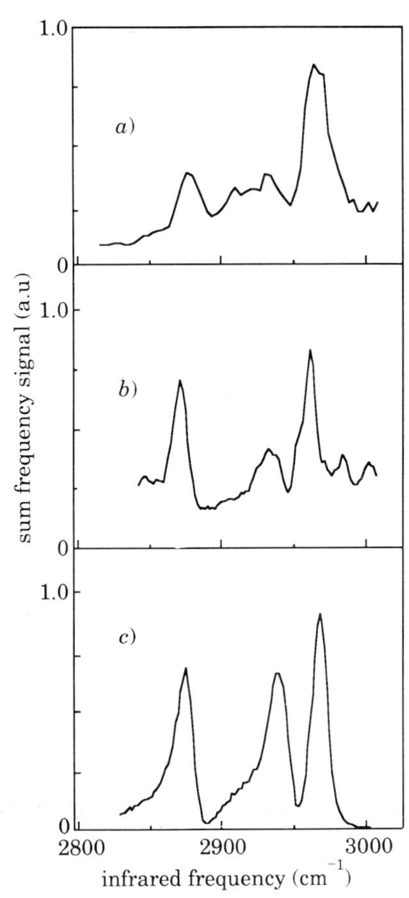

Fig. 4.

Fig. 3. – Sum frequency generation spectra as a function of infrared input frequency for three adsorbed molecular species on glass. Top frame: methanol, CH_3OH, the two peaks at 2840 and 2960 cm^{-1} correspond to the CH_3 symmetric and antisymmetric stretch vibrations, respectively. Middle frame: ethylene glycol, $C_2H_4(OH)_2$, the symmetric and antisymmetric CH_2 stretches appear at 2875 and 2935 cm^{-1}, respectively. Bottom frame: isotropyl alcohol, $(CH_3)_2CHOH$, the two peaks at 2950 and 2980 cm^{-1} are the degeneracy-lifted CH_3 asymmetric stretches. The peak at 2885 cm^{-1} CH_3 is the symmetric stretch, and the peak at 2920 cm^{-1} is the CH stretch. The bold lines are the bulk liquid Raman spectra. (After ref. [29].)

Fig. 4. – Sum frequency spectra of *a*) OTS on silicon, *b*) OTS on freshly evaporated aluminum and *c*) OTS on glass, as described in the text. (After ref. [30].)

group of OTS. The CH_2 stretch modes of the alkane chain are absent presumably because, with the chain straight and normal to the interface, they are forbidden in SFG by the macroscopic symmetry in the CH_2 group distribution. The third spectrum is nearly identical to the second one, even though hexade-

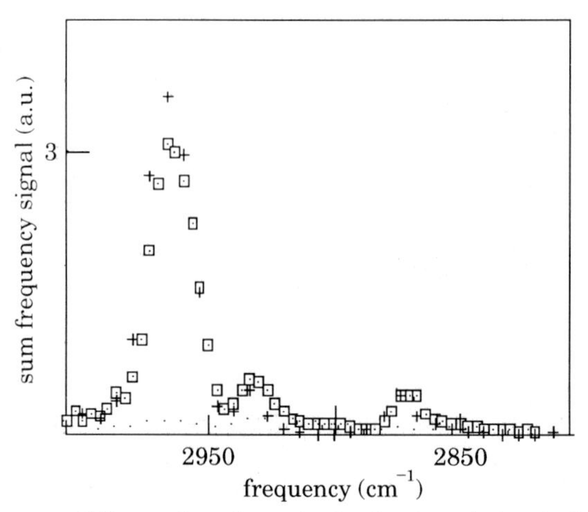

Fig. 5. – SFG spectra at different interfaces in the (p-vis, p-i.r.) polarization combination. Dots: silica-hexadecane interface. Dotted squares: silica-OTS-CCl_4 interfaces. Crosses: silica-OTS-hexadecane interface. (After ref. [31].)

cane is now replaced by CCl_4, again indicating that the spectrum definitely comes from the OTS monolayer instead of the bulk. This example, therefore, shows most convincingly that the surface SFG spectroscopy is indeed highly surface specific.

In discussing surface spectroscopy, it is natural to ask how SFG spectroscopy compares with linear optical spectroscopy. From the sensitivity point of view, the two techniques are comparable; one could be better than the other depending on the interfacial system. SFG, however, has a number of very unique advantages. Surface specificity is one of them. Capability to determine polar orientations at an interface is another. Transient SFG to probe surface dynamics also seems to provide more flexibility than infrared spectroscopy.

In the following, we shall focus our discussion on SFG experiments that have yielded interesting new information about various interfacial systems. They serve as examples to illustrate the power and potential of surface SFG as a unique and versatile spectroscopic tool.

4·1. *Buried interfaces.* – Unlike most of other surface techniques, surface SFG can be used to probe buried interfaces. The liquid/solid interface we mentioned earlier is a good example. It shows how surface specificity of SFG enables us to measure the spectrum of an interfacial layer, even in the presence of a strong bulk absorption in the same spectral region.

Solid/solid interfaces are of great importance in modern technology. Again, surface SFG spectroscopy is ideally suited to study such interfaces. This was first demonstrated by HEINZ *et al.* to probe interfacial states of CaF_2/Si(111) [32], which is a prototype semiconductor/insulator interface. The observed spectrum is depicted in fig. 6. The main peak at 2.4 eV arising from

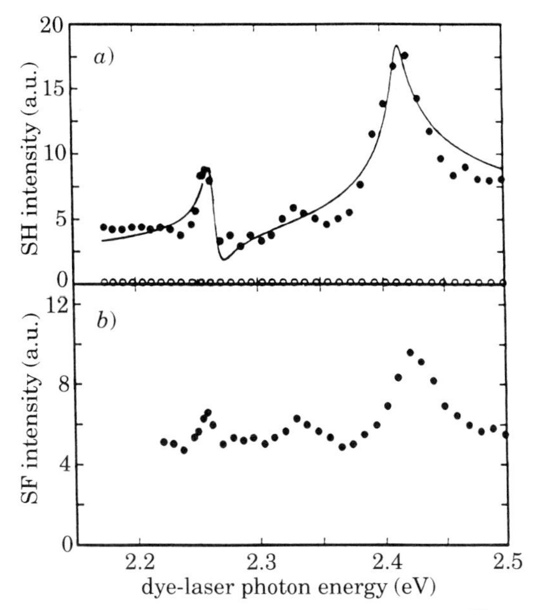

Fig. 6. – Resonant three-wave mixing signals associated with ${}^{s}\chi^{(2)}_{zzz}$ as a function of the energy of a photon from the tunable dye laser. *a*) Results for the SHG process and *b*) the case of SFG involving mixing the output of the dye laser with a photon of fixed energy (1.17 eV). The filled symbols refer to signal from the CaF_2/Si(111) sample, the open symbols to a Si(111) surface covered by the native oxide. The solid curve in *a*) is a fit to theory. (After ref. [32].)

electronic transitions between the filled and empty interface states can be identified as the band gap in the interfacial electronic band structure. This band gap energy is in marked contrast to the band gaps of the bulk Si (1.1 eV) and CaF_2 (12.1 eV). The lowest peak at 2.26 eV is suggested to be associated with the $n = 1$ surface exciton, but the middle peak at 2.32 eV is not clearly identified.

Liquid/liquid interfaces can also be studied by surface SFG. No such experiment has yet been reported.

4'2. *Conformational changes of adsorbed molecules.* – The polarization dependence of surface SFG can also yield useful information. It allows us to deduce the resonant $\chi^{(2)}_{\mathrm{s},\,ijk}$ components for a spectral peak and learn about the geometry of the corresponding vibrational mode. Consider an adsorbed molecular monolayer with its vibrational resonances probed by SFG. Each spectral peak can be assigned to a resonance involving a group of atoms within the molecule. We then have [12]

$$\chi^{(2)}_{\mathrm{s},\,ijk} = N_{\mathrm{s}} \sum_{\xi,\,\eta,\,\zeta} \langle G^{\xi\eta\zeta}_{ijk} \rangle \alpha^{(2)}_{\xi\eta\zeta}\,, \tag{18}$$

where $\alpha^{(2)}_{\xi\eta\zeta}$ is the nonlinear polarizability associated with the vibrational resonance of the specific group of atoms and $\langle G_{ijk}^{\xi\eta\zeta}\rangle$ is an average geometric transformation factor depending on the average orientation of the group. If $\alpha_{\xi\eta\zeta}^{(2)}$ is known and $\chi_{s,ijk}^{(2)}$ is measured from SFG, we can find $\langle G_{ijk}^{\xi\eta\zeta}\rangle$.

Figure 7 describes the SFG spectra of the CH stretch modes of PDA (pentadecanoic acid, $CH_3(CH_2)_{13}COOH$) molecules adsorbed on water with three different surface densities [33]. The two different input/output polarization combinations yield very different spectra, suggesting that the molecules must be oriented. An analysis of the spectra for the full monolayer (22 Å^2/molecule) using eq. (18) indicates that the terminal CH_3 group of the alkyl chain has its axis tilted at ~ 35° from the surface normal. This implies that the chain is straight and normal to the surface. As the surface density of PDA decreases, the CH_3 peaks in the spectra diminish, while peaks assigned to CH_2 stretch modes begin to emerge and increase in strength. The alkyl chains must have started to tilt and develop kinks in the form of trans-gauche defects.

In another example, shown in fig. 8, a DMOAP (n, n-dimethyl-n-octadecyl-3-amino propyltrimethoxysilychloride, $CH_3(CH_2)_{17}(Me)_2\text{-}N^+(CH_2)_3Si(OMe)_3Cl^-$) monolayer on glass is seen to have a similar SFG spectrum of CH stretch modes as a half-monolayer of PDA (47 Å^2/molecule) on water [34], because the surface

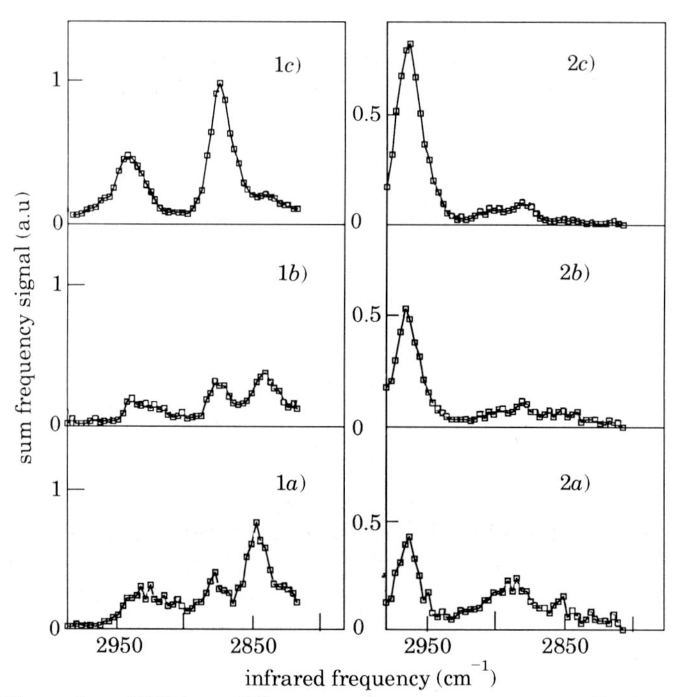

Fig. 7. – SFG spectra of PDA at different surface coverages normalized per molecules. 1a)-1c) were taken with the s-visible, p-i.r. polarization combination. 2a)-2c) were taken with p-visible, s-i.r.: a) 47 Å^2/molecule, b) 34 Å^2/molecule, c) 22 Å/molecule. (After ref. [33].)

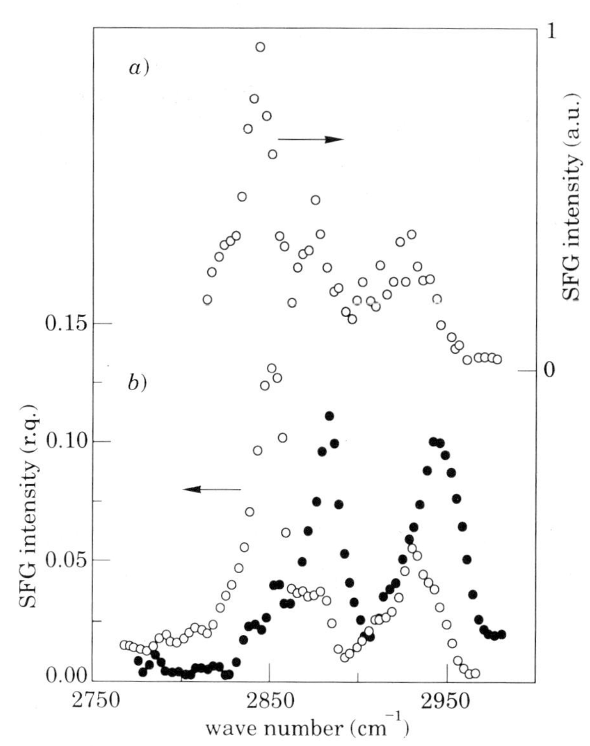

Fig. 8. – *a*) Spectrum of PDA monolayer on water surface at density of 47 Å/molecule. *b*) Comparison of the SFG spectrum of DMOAP on clean glass (open circles) and the spectrum of the same sample after deposition of 0.7 monolayer of 8 CB (filled circles). (After ref. [34].)

densities of alkyl chains in the two cases are nearly the same. With a 0.7 monolayer of 8 CB (4′*n*-octy-4-cyanobiphenyl, $CH_3(CH_2)_7(CH_4)_2CN$) coadsorbed with DMOAP on glass, however, the CH_2 peak at 2850 cm^{-1} is almost completely suppressed, and the peaks associated with CH_3 stretch modes grow. This indicates that the coadsorbed 8 CB must have effectively interacted with DMOAP molecules and removed the trans-gauche defects in the alkyl chains of DMOAP [34].

The above conformational changes of alkyl chains cannot be observed by infrared spectroscopy. In the latter case, the spectrum is overwhelmingly dominated by the CH_2 stretch modes and the trans-gauche defects along the alkyl chain are difficult to detect. With SFG, the CH_2 modes are almost completely suppressed by symmetry when the alkyl chains are straight.

4·3. *Polar orientations of molecules at interfaces.* – The nonlinear susceptibility $\chi_D^{(2)}$ of molecules is in general a complex quantity, namely, $\chi^{(2)} = |\chi^{(2)}| \exp[i\phi]$. Unlike the linear optical response, the phase ϕ changes by π

when the polar orientation of the molecules is reversed. Thus SHG and SFG measurements that can yield $\chi_D^{(2)}$ should also allow us to determine the polar orientation of molecules [35]. In the usual experiments, only the output power or energy of SHG and SFG is measured, which permits the deduction of $|\chi_D^{(2)}|$. However, if the output is made to interfere with a reference beam in an interferometer setting, then ϕ can also be deduced [36]. In the case of SFG, interference of the resonant $\chi_D^{(2)}$ with the nonresonant backward $\chi^{(2)}$ could also let us deduce ϕ for the resonant $\chi_D^{(2)}$. This then allows the determination of polar orientation of a specific group of atoms in the molecules [37].

Consider the example of 8 CB molecules adsorbed on glass. The cyano biphenyl chromophore of the molecules is known to be responsible for SHG, while the CH_3 terminal group of the alkyl chain gives the dominant contribution to the SFG spectrum in the CH stretch region. The SHG measurement shows that the chromophore is oriented with a tilt of $\sim 70°$ from the surface normal and CN is attached to the glass [38]; yet it does not provide any information about the orientation of the alkyl chain. The phase measurement of SFG at the CH stretch resonance is able to conclude that the CH_3 group of the chain is pointing away from the glass, thus implying that the chain is also pointing away from the glass [37]. We note that, in the determination of absolute polar orientations using SHG or SFG, molecules of known orientations are always needed to serve as references.

4·4. *Pure-liquid surfaces.* – There is certainly no lack of studies on pure crystalline surfaces in surface science. Surface states of metal and semiconductor surfaces, for example, have been extensively investigated. Transitions involving surface states can be probed by various types of surface spectroscopy [39]. Here, SFG spectroscopy could also be very helpful [40], as it can give excellent signal-to-background ratio because of its surface specificity, can provide better spectral resolution, and can probe state symmetries with more flexibility.

In all respects, liquid surfaces are certainly not less important than crystalline surfaces. So far, however, they have hardly been explored because of lack of experimental techniques. Let us consider, for example, the vapor/liquid interface of a pure liquid. Since molecules at the interface experience a very different environment than those in the bulk, they can be oriented differently at the interface (instead of randomly oriented as in the bulk). The physical and chemical properties of a liquid surface should depend critically on how the surface molecules are oriented. The molecular vibrational spectrum influenced by environment could be different for molecules at the interface and in the bulk. Such information cannot be obtained by conventional surface techniques and SFG spectroscopy appears to be the only probe for such a study. We discuss here the cases of water and methanol [41, 42]; both are polar liquids with hydrogen bonding. Hydrogen bonds between molecules are expected to play an im-

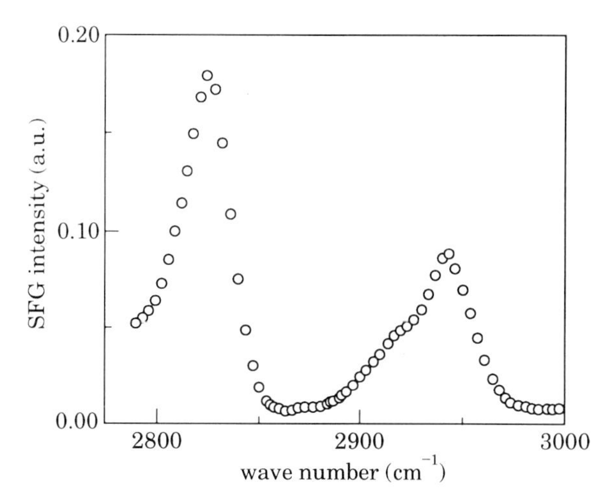

Fig. 9. – SFG spectrum obtained from the liquid/vapor interface of pure methanol. The beam polarizations are s, s and p for the sum frequency, visible (0.532 μm) and infrared beams, respectively. (After ref. [41].)

portant role in determining the polar orientation of molecules at the vapor/liquid interface.

Figure 9 gives the surface SFG spectrum of the CH_3 stretch modes of methanol at the vapor/methanol interface [41]. Only the (s, s, p) input/output polarization combination yielded a clear spectrum; this immediately suggests that the methanol molecules at the interface are oriented. Various measurements were conducted to ensure that the spectrum was indeed dominated by the surface electric-dipole contribution. It was actually found that the electric-quadrupole contribution from the bulk and from the interface is negligible. A phase measurement on the SF output was also carried out and compared with that from a monolayer of methoxy on glass. The result indicates that the CH_3 groups of methanol molecules at the interface are pointing out of the liquid as one would expect from maximization of the number of hydrogen bonds between molecules at the interface in order to lower the free energy [43]. From analysis of the spectra in fig. 9, we can conclude that the orientational distribution of the surface methanol molecules is rather broad. These results agree fairly well with the predictions of molecular-dynamics simulation by MATSUMOTO and KATAOKA [43].

Water is the most important liquid in life. For this reason, many theoretical calculations are available predicting orientation of water molecules at the vapor/water interface [44]. Maximization of the number of hydrogen bonds of molecules at the interface would suggest that the H_2O molecular planes are not much tilted from the surface plane, and essentially all H terminals are hydrogen bonded. This would predict the absence of a free OH stretch mode in the vibrational spectrum of the vapor/water interface. The observed SFG spectrum

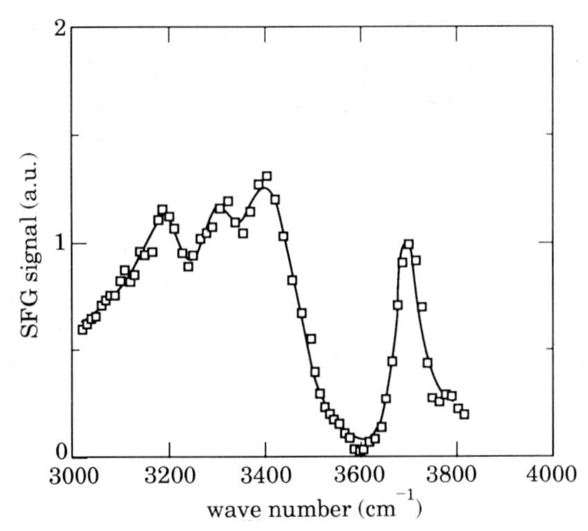

Fig. 10. – SFG spectrum of pure water liquid/vapor interface with beam polarization combination sum-frequency-s, visible-s, infrared-p polarized. The fit to theory is shown as a solid line. (After ref. [42].)

from the interface is shown in fig. 10 [42]. The sharp peak at 3690 cm^{-1} can be unequivocally assigned to the free OH stretch mode and the broad band at lower frequencies to the hydrogen-bonded OH stretch modes. Since no water molecules in the bulk can have free OH bonds, the spectrum must come from polar-oriented molecules at the interface. Energy consideration as well as the appearance of only a single free OH peak in the spectrum suggests that the molecules are oriented with one H sticking out of the liquid. An estimate of the signal strength indicates that $\sim 20\%$ of a water monolayer has this orientation. The polarization dependence of the SFG spectrum is consistent with a broad orientational distribution. An analysis of the SFG spectrum in fig. 10 using eq. (17) also finds that $\chi^{(2)}$ for the hydrogen-bonded OH modes and for the free OH modes are opposite in sign. This implies that the bonded OH bonds of the surface molecules are pointing into the liquid. Unlike the methanol case, these results for water show little agreement with predictions from the molecular-dynamics simulation by MATSUMOTO and KATAOKA [43]. The calculation indicates that the number of water molecules at the interface with OH pointing towards the vapor cannot be more than 4% of a monolayer and most of them are on the vapor side. The discrepancy between theory and experiment is presumably due to the unrealistic interaction potential between water molecules used in the calculation.

4·5. *Hydrogen on silicon and two-phonon bound states.* – The understanding of interaction of hydrogen with silicon surfaces is of great importance to modern electric technology. Chemical vapor deposition of Si from silane [45], oxidation of Si by H_2O [46] and nitridation of Si by NH_3 [47] all involve H adsorption on Si. Surface vibrational spectroscopy of H/Si is certainly a most direct method to probe the interaction of H and Si. Indeed, infrared reflection spectroscopy has been employed to study the H-Si stretch vibration [48]. In order to have a sufficient signal-to-noise ratio, however, a multiple-internal-reflection geometry must be used. In ref. [48], for example, a sample of $(0.5 \times 19 \times 38)$ mm^3 beveled at both ends was prepared, making it possible for the i.r. beam to bounce a total of 75 times inside the sample. With SFG spectroscopy, a single reflection is sufficient to yield the same vibrational spectrum with a good signal-to-noise ratio [49]. Even the $v = 1 \rightarrow v = 2$ transition of the H-Si stretch vibration can be easily observed [50].

Figure 11 depicts the SFG spectra of the fundamental ($v = 0 \rightarrow v = 1$) and hot-band ($v = 1 \rightarrow v = 2$) transitions of the H-Si stretch vibration of H/Si(111) [50]. Two infrared laser beams were used in the experiments: one pump and one probe. The pump beam pumped the $v = 0 \rightarrow v = 1$ transition to a saturation level. With $v = 0$ and $v = 1$ states both occupied, the tunable probe beam then probed via SFG the $v = 0 \rightarrow v = 1$ and $v = 1 \rightarrow v = 2$ transitions. Figure 11 shows that, as expected, the hot-band transition can only be observed after pumping. The peak for the fundamental transition diminishes upon pumping since it is proportional to the population difference between $v = 0$ and $v = 1$. On the other hand, the hot-band peak directly reflects the population in $v = 1$.

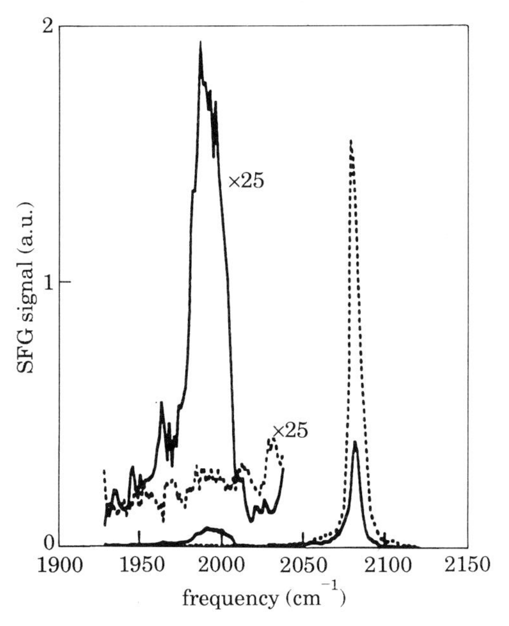

Fig. 11. – SFG spectra of the fundamental and hot-band transitions of stretch vibration of H/Si(111). The solid line is obtained 30 ps after the 7 ps pump beam at 2084 cm^{-1}, pumping the fundamental transition. The dashed line is obtained with no pump. (After ref. [50].)

It is down-shifted by $\sim 90\ cm^{-1}$ from the fundamental peak, presumably because of anharmonicity in the stretch vibration. However, this value is significantly larger than the $68\ cm^{-1}$ shift induced by anharmonicity for an isolated Si-H bond in silanes [51]. The difference can be explained if the lateral interaction of the H-Si bonds of H/Si is taken into account and a phonon picture is used [50]. The hot-band peak is identified as from the zone center transition of a one-phonon state to a two-phonon bound state. The latter is dominated by the combination of two phonons at the zone edge because of the high density of states there. As a result of phonon dispersion (due to lateral interaction of molecular vibrations), the zone edge phonon is lower by $\sim 11\ cm^{-1}$ than the zone center phonon. Without anharmonicity, the two-phonon state would be at an energy $2 \times 11\ cm^{-1}$ below the energy of two zone center phonons. The anharmonic interaction between phonons further reduces the energy of the two-phonon state by $68\ cm^{-1}$, resulting in a total down-shift of $\sim 90\ cm^{-1}$ as was observed. The anharmonic interaction tends to localize and bind phonons [52]. Here, for H-Si vibrations, anharmonicity is significantly larger than phonon dispersion, indicating that the observed two-phonon state is a well-defined two-phonon bound state. This is the first demonstration of the existence of two-phonon bound states on a surface.

4.6. *Hydrogen on diamond.* – SFG spectroscopy can also be employed to study hydrogen adsorption on diamond [53]. This is again a system of great technological importance. Diamond film growth by chemical vapor deposition requires the knowledge of H adsorption on diamond as an intermediate step [54]. To obtain the surface vibrational spectrum of the CH modes of H/diamond, infrared spectroscopy can in principle also be used, but a multiple-reflection geometry is still needed to achieve a detectable signal. This is, however, not a realistic scheme for diamond. Unlike Si, large diamond samples are prohibitively expensive. SFG spectroscopy then becomes a rather unique tool for such a study. Electron energy loss spectroscopy of H/diamond has been attempted [55], but the result is very much limited by the poor spectral resolution.

Figure 12 describes the SFG spectra of the C-H stretch vibration of H/C(111) [53]. Like in Si, a single sharp peak is observed. The polarization dependence of the spectrum indicates that hydrogen atoms sit on top of C. With a full monolayer coverage of C(111) by H, the relaxed C(111) surface plane is ideally terminated by H. From the observed H-C vibrational frequency, the CH bond length is estimated to be 1.1 Å. Anharmonicity in the CH stretch vibration has been calculated by Zhu and Louie from first principles [56]. The value obtained suggests that, as in the case of Si, a two-phonon bound state of H/C(111) should exist, which can certainly be probed by SFG spectroscopy.

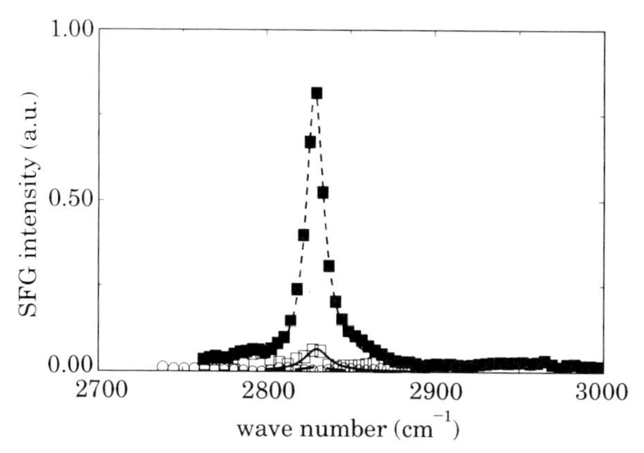

Fig. 12. – SFG spectra of stretch vibration of a well-annealed H-terminated C(111)-(1 × 1) surface, obtained with different polarization combinations: Solid squares for the s (sum frequency, output)-s (visible, input)-p (infrared, input) polarization combination, open squares for the pss combination, and open circles for the sps combination. (After ref. [53].)

4·7. *Dynamics of surface vibrations.* – SFG spectroscopy with ultrashort laser pulses is also best suited for studies of ultrafast surface dynamics. It can be used, for example, to probe desorption or surface reaction of selective adsorbed species and pumping or relaxation of surface states. We consider here measurements of longitudinal and transverse relaxations of surface molecular vibrations [49, 50, 57-61].

Longitudinal (or population) relaxation of a mode is directly connected to energy transfer in and out of the mode. Vibrational-energy transfer between a molecular adsorbate and a substrate is a fundamental process governing the dynamics of surface reactions. Ultrafast infrared spectroscopy has been developed to study such processes, but the poor signal-to-noise ratio has limited most of the investigations to powder or porous materials with high surface areas [62]. Transient SFG spectroscopy allows relatively easy measurements on molecules adsorbed at well-defined flat surfaces, including crystalline metal [57-59] and semiconductor surfaces [49, 50, 60]. A representative example is the relaxation study of the stretch vibration of H/Si(111) [49, 50].

As described earlier in subsect. 4·4, the $v = 1$ state of the H-Si vibration of H/Si(111) can be pumped by an infrared pulse; the resulting population in the $v = 1$ state can be probed by SFG via the hot-band ($v = 1 \rightarrow v = 2$) transition [50]. By measuring the SFG signal *vs.* the time delay between the pump and probe pulses, we can learn directly how the population in $v = 1$ decays. The measurement in ref. [50] shows that the decay is exponential with a relaxation time $T_1 \simeq 800$ ps at room temperature. This lifetime is long compared to what one would normally expect from chemisorbed molecules on semiconductors or metals. It is due to the fact that the H-Si stretch vibration is well decoupled from the H-Si bending modes and Si surface phonons because of frequency mismatch. Energy transfer from the vibrational mode to electronic excitations is also forbidden since the Si band gap is much larger than the vibrational frequency. The long relaxation time has been attributed to four-phonon energy transfer from the H-Si stretch vibration to the low-frequency H-Si bending modes and surface Si phonon modes [49, 50].

The above longitudinal relaxation of the H-Si stretch vibration has also been measured by SFG probing directly the recovery of the strength of the fundamental ($v = 0 \rightarrow v = 1$) transition after saturation and pumping of the same transition [49]. The results appear to be the same (see fig. 13). This scheme has the advantage that the pump and probe infrared pulses have the same frequency so that only one infrared laser is needed. However, the SFG signal in this case is related to the difference of population between $v = 0$ and $v = 1$. Only if there are no intermediate states shelving the population in the relaxation process, can the relaxation of the $v = 1$ state be unambiguously deduced from the result. This is generally not true, but happens to be the case for H/Si(111) [49]. Thus it is obvious that pump and (SFG) probe experiments with two tunable infrared lasers are more appropriate. If transitions from excited states of different modes can be probed after pumping of a selected vibrational mode, then the relaxation pathway of the vibrational excitation can be clearly identified.

Nevertheless, to avoid experimental complication, most researchers have employed a single i.r. laser in their relaxation measurements. HARRIS and coworkers have used the technique to study relaxation of the symmetric CH

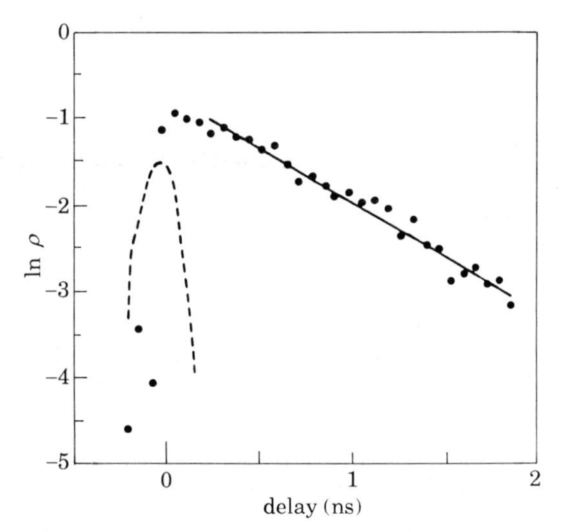

Fig. 13. – Logarithm of $\rho = 1 - r^{1/2}$, where r is the ratio of the probe SFG signal probing the $v = 0 \rightarrow v = 1$ transition with the pump on and off. The straight line is a least-squares fit giving $T_1 = 795$ ps. The dotted line is the cross-correlation of the pump with the visible pulse (temporal resolution). (After ref. [49].)

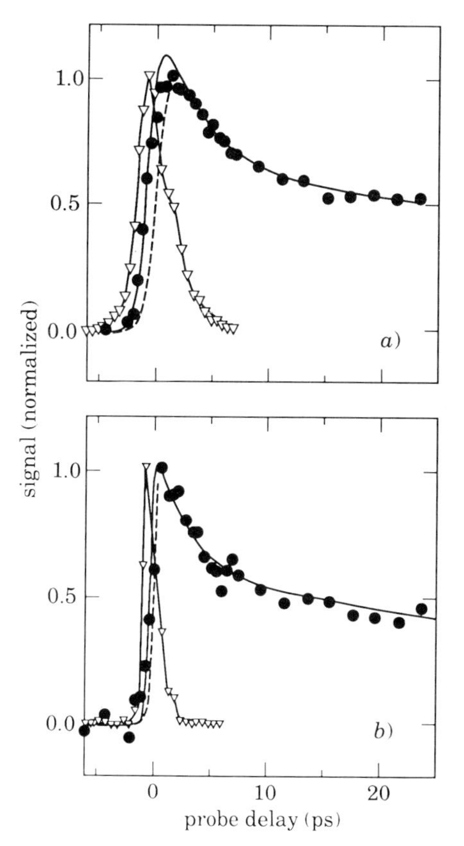

Fig. 14. – Transient vibrationally resonant SFG signals at 300 K in experiments carried out with pulse widths (FWHM) *a*) $\tau_p = 2.5$ ps or *b*) $\tau_p = 1.0$ ps and corresponding visible-infrared cross-correlations. Solid points, normalized transient SFG signals $1 - [S(\tau_d)/S_0]^{1/2}$; solid lines, fits to transient SFG signals according to a three-level model; lines with inverted triangles, visible-infrared cross-correlation. (After ref. [58].)

stretch vibration of methyl thiolate (CH_3S) on Ag(111) [58]. A biexponential decay curve was found (fig. 14). The result indicates that an intermediate state is involved in the relaxation from $v = 1$ to $v = 0$. The relatively fast relaxation (< 100 ps above 100 K) suggests that it is dominated by intramolecular relaxation rather than energy transfer to electron-hole pair excitations in the Ag metal. This is believed to be generally true for polyatomic molecules adsorbed on a surface if the molecules are sufficiently larger. For smaller molecules, such as CO, on metals, the electron-hole pair excitation mechanism becomes dominant [63], as was found by HARRIS and co-workers for the vibration of CO on Cu(100) using transient SFG [59] and by BECKERLE *et al.* for the vibration of CO on Pt(111) [64].

Transverse (or dephasing) relaxation of a surface vibration can also be studied by SFG [61]. One would first use the i.r. pulse to resonantly excite the vibrational wave (known as transverse excitation). The visible laser pulse then comes in after a time delay to mix with the vibrational wave and generate the SF output. The SF is expected to decay with time as the vibrational wave is dephased. From the decay, the dephasing relaxation time can be deduced. The above description is, however, only valid if the vibrational transition is dominated by homogeneous broadening [22]. Otherwise, the decay time gives nothing but the inverse of the inhomogeneous linewidth. It is known that, for an inhomogenously broadened transition, dephasing relaxation can only be monitored by coherent transient processes such as photon echoes [65]. To observe photon echoes from surface vibrations could be difficult because of the limited sensitivity of infrared detectors. However, with the help of SFG, it becomes readily possible [61, 66]. The scheme is as follows.

Two short i.r. pulses, separated by τ, resonantly excite an inhomogenously broadened surface vibrational mode, where τ is less than or comparable to the dephasing time T_2. At a time τ, after the second pulse, the inhomogenously dephased surface vibrations rephase and form a giant oscillating dipole (or vibrational wave) which then radiates and generates the photon echo. It is possible to detect this i.r. echo pulse by an up-conversion method via SFG in a nonlinear crystal [66]. Alternatively and more directly, one can use SFG to probe *in situ* the appearance of the rephased surface vibrational wave [61, 66], *i.e.* SFG from mixing of the visible probe pulse with the rephased vibrational wave. In a recent experiment, GUYOT-SIONNEST has successfully used the second scheme to observe photon echoes from the H-Si stretch vibration of H/Si(111) [61]. A dephasing time of 85 ps at 120 K was observed (fig. 15). This corresponds to a homogeneous linewidth of $0.12\ \text{cm}^{-1}$, which is less than the observed spectral

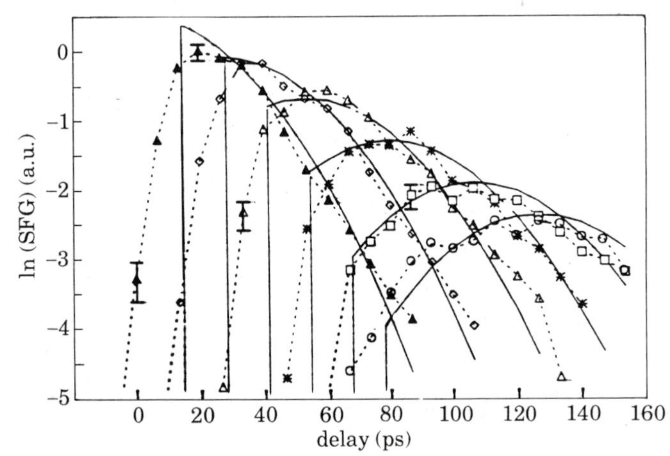

Fig. 15. – Logarithmic plot of the surface SFG echo signal from the stretch vibration of H/Si(111) as a function of visible probe. The data points connected by dashed lines are shown for several delays between the two i.r. pulses: 15 ps, ▲; 28 ps, ◇; 42 ps, △; 55ps, *; 68 ps, □; 78 ps, ○. The solid lines are theoretical fits. (After ref. [61].)

width of $0.3\,cm^{-1}$, indicating that the spectral line is inhomogenously broadened. Quadratic anharmonic coupling of the H-Si vibration with Si surface phonons at $\sim 190\,cm^{-1}$ is believed to be mainly responsible for the dephasing.

* * *

This work was supported by the Director, Office of Energy Research, Office of Basic Energy Sciences, Materials Sciences Division of the US Department of Energy under contract No. DEAC03-76SF00098.

REFERENCES

[1] See, for example, A. ZANGWILL: *Physics at Surfaces* (Cambridge University Press, Cambridge, 1988); G. SOMORJAI: *Chemistry in Two Dimensions: Surfaces* (University Press, Ithaca, N.Y., 1981).
[2] T. J. CHUANG and H. SEKI: *Phys. Rev. Lett.*, **49**, 382 (1982).
[3] F. TRAGER, H. CONFAL and T. J. CHUANG: *Phys. Rev. Lett.*, **49**, 1720 (1982).
[4] A. C. BOCCARA, D. FOURNIER and J. BADOZ: *Appl. Phys. Lett.*, **36**, 130 (1980); M. A. OLMSTEAD and N. M. AMER: *Phys. Rev. Lett.*, **52**, 1148 (1984).
[5] See, for example, H. ZACHARIAS, M. M. T. LOY and P. A. ROLLAND: *Phys. Rev. Lett.*, **49**, 1790 (1982).
[6] See, for example, J. HAEGER, Y. R. SHEN and H. WALTHER: *Phys. Rev. A*, **31**, 1962 (1984).
[7] J. P. HERITAGE and D. L. ALLARA: *Chem. Phys. Lett.*, **74**, 507 (1980).
[8] C. K. CHEN, A. R. B. DE CASTRO, Y. R. SHEN and F. DE MARTINI: *Phys. Rev. Lett.*, **43**, 946 (1979).
[9] See, for example, Y. R. SHEN: *Nature (London)*, **337**, 519 (1989).
[10] See, for example, Y. R. SHEN: *Principles of Nonlinear Optics* (J. Wiley, New York, N.Y., 1984), Chapt. 2.
[11] T. STEHLIN, M. FELLER, P. GUYOT-SIONNEST and Y. R. SHEN: *Opt. Lett.*, **13**, 389 (1988).
[12] See, for example, Y. R. SHEN: *J. Vac. Sci. Technol. B*, **3**, 1464 (1986); *Annu. Rev. Mater. Sci.*, **16**, 69 (1989); *Annu. Rev. Phys. Chem.*, **40**, 327 (1989); G. L. RICHMOND, J. M. ROBINSON and V. L. SHANNON: *Prog. Surf. Sci.*, **28**, 1 (1988); T. F. HEINZ: in *Nonlinear Surface Electromagnetic Phenomena*, edited by H. E. PONATH and G. I. STEGEMAN (North Holland, Amsterdam, 1991), p. 353.
[13] T. F. HEINZ, C. K. CHEN, D. RICARD and Y. R. SHEN: *Phys. Rev. Lett.*, **48**, 478 (1982).
[14] X.-D. ZHU, H. SUHR and Y. R. SHEN: *Phys. Rev. B*, **35**, 3047 (1987); J. H. HUNT, P. GUYOT-SIONNEST and Y. R. SHEN: *Chem. Phys. Lett.*, **133**, 189 (1987); A. L. HARRIS, C. E. D. CHIDSEY, N. J. LEVINOS and D. N. LOIACONO: *Chem. Phys. Lett.*, **141**, 350 (1987).
[15] M. Y. JIANG, G. PAJER and E. BURSTEIN: *Surf. Sci.*, **242**, 306 (1991); M. Y. JIANG, G. PAJER, E. BURSTEIN, M. YAGANEH and A. YODH: *Bull. Am. Phys. Soc.*, **37**, 652 (1992).
[16] R. SUPERFINE, J. Y. HUANG and Y. R. SHEN: *Phys. Rev. Lett.*, **66**, 1066 (1991); R. SUPERFINE, J. Y. HUANG, Q. DU and Y. R. SHEN: in *Proceedings of the X International Conference on Laser Spectroscopy, 1991,* edited by M. DUCLOY, E. GIACOBINO and G. CAMY (World Scientific, Singapore, 1992), p. 117.

[17] P. GUYOT-SIONNEST, W. CHEN and Y. R. SHEN: *Phys. Rev. B*, **33**, 8254 (1986); P. GUYOT-SIONNEST and Y. R. SHEN: *Phys. Rev. B*, **35**, 4420 (1987); *Phys. Rev. A*, **38**, 7985 (1989), and references therein.
[18] N. BLOEMBERGEN and P. S. PERSHAN: *Phys. Rev.*, **128**, 606 (1962).
[19] Y. R. SHEN: *Annu. Rev. Phys. Chem.*, **40**, 327 (1989).
[20] C. C. WANG: *Phys. Rev.*, **178**, 1475 (1969).
[21] J. A. ARMSTRONG, N. BLOEMBERGEN, J. DUCUING and P. S. PERSHAN: *Phys. Rev.*, **127**, 1918 (1962).
[22] See, for example, ref. [10], Chapt. 15.
[23] P. GUYOT-SIONNEST, P. DUMAS, Y. J. CHABAL and G. S. HIGASHI: *Phys. Rev. Lett.*, **64**, 2156 (1990).
[24] J. MIRAGLIOTTA, R. S. POLIZZOTTI, P. RABINOWITZ, S. D. CAMERON and R. B. HALL: *Chem. Phys.*, **143**, 123 (1990); *Appl. Phys. A*, **51**, 221 (1990).
[25] A. LAUBERAU, L. GREITER and W. KAISER: *Appl. Phys. Lett.*, **25**, 87 (1974); A. SEILMEIER, K. SPANNER, A. LAUBEREAU and W. KAISER: *Opt. Commun.*, **24**, 231 (1978).
[26] J. Y. HUANG, J. Y. ZHANG, Y. R. SHEN, C. CHEN and B. WU: *Appl. Phys. Lett.*, **57**, 1961 (1990); J. Y. ZHANG, J. Y. HUANG, Y. R. SHEN, C. CHEN and B. WU: *Appl. Phys. Lett.*, **58**, 213 (1991); J. Y. HUANG, Y. R. SHEN, C. CHEN and B. WU: *Appl. Phys. Lett.*, **58**, 1579 (1991).
[27] H. VANHERZEELE: *Appl. Opt.*, **29**, 2246 (1990).
[28] T. ELSAESSER, H. LOBENTANZER and A. SEILMEIER: *Opt. Commun.*, **52**, 355 (1985); T. ELSAESSER, H. LOBENTANZER, A. SEILMEIER, W. KAISER, P. KOIDL and G. BRANDT: *Appl. Phys. Lett.*, **44**, 383 (1984); X. Y. FAN and R. L. BYER: in *New Lasers for Analytical and Industrial Chemistry*, edited by A. F. BERNHARDT: *Proc. Soc. Photo-Opt. Instrum. Eng.*, **461**, 27 (1984), and references therein.
[29] J. H. HUNT, P. GUYOT-SIONNEST and Y. R. SHEN: in *Laser Spectroscopy VIII*, edited by W. PERSON and S. SVANBERG (Springer-Verlag, Berlin, 1987), p. 253.
[30] R. SUPERFINE, P. GUYOT-SIONNEST, J. H. HUNT, C. T. KAO and Y. R. SHEN: *Surf. Sci.*, **200**, L445 (1988).
[31] P. GUYOT-SIONNEST, R. SUPERFINE, J. H. HUNT and Y. R. SHEN: *Chem. Phys. Lett.*, **144**, 1 (1988).
[32] T. F. HEINZ, F. J. HIMPSEL, E. PALANGE and E. BURSTEIN: *Phys. Rev. Lett.*, **63**, 644 (1989).
[33] P. GUYOT-SIONNEST, J. H. HUNT and Y. R. SHEN: *Phys. Rev. Lett.*, **59**, 1597 (1987).
[34] J. Y. HUANG, R. SUPERFINE and Y. R. SHEN: *Phys. Rev. (Rapid Commun.) A*, **42**, 3660 (1990).
[35] K. KEMNIZ, K. BHATTACHARYYA, J. M. HICKS, G. R. PINTO, K. B. EISENTHAL and T. F. HEINZ: *Chem. Phys. Lett.*, **131**, 285 (1986).
[36] R. K. CHANG, J. DUCUING and N. BLOEMBERGEN: *Phys. Rev. Lett.*, **15**, 6 (1965); J. J. WYNNE and N. BLOEMBERGEN: *Phys. Rev.*, **188**, 1211 (1969); H. W. K. TOM, T. F. HEINZ and Y. R. SHEN: *Phys. Rev. Lett.*, **51**, 1983 (1983).
[37] R. SUPERFINE, J. Y. HUANG, and Y. R. SHEN: *Opt. Lett.*, **15**, 1276 (1990).
[38] C. S. MULLIN, P. GUYOT-SIONNEST and Y. R. SHEN: *Phys. Rev. (Rapid Commun.) A*, **39**, 3745 (1989).
[39] See, for example, A. GOLDMAN, V. DOSE and G. BORSTEL: *Phys. Rev. B*, **32**, 1971 (1985).
[40] M. Y. JIANG, G. PAJER and E. BURSTEIN: *Surf. Sci.*, **242**, 306 (1991).
[41] R. SUPERFINE, J. Y. HUANG and Y. R. SHEN: *Phys. Rev. Lett.*, **66**, 1066 (1991).
[42] R. SUPERFINE, J. Y. HUANG, Q. DU and Y. R. SHEN: in *Laser Spectroscopy X*, edited by M. DUCLOY, E. GIACOBINO and G. CAMY (World Scientific, Singapore, 1992), p. 117.

[43] M. MATSUMOTO and Y. KATAOKA: *J. Chem. Phys.*, **90**, 2390 (1989).
[44] C. Y. LEE and H. L. SCOTT: *J. Chem. Phys.*, **73**, 4591 (1980); M. MATSUMOTO and Y. KATAOKA: *J. Chem. Phys.*, **88**, 3233 (1988); R. M. TOWNSEND and S. A. RICE: *J. Chem. Phys.*, **94**, 2207 (1991); M. A. WILSON, A. POHOVILLE and L. R. PRATT: *J. Phys. Chem.*, **91**, 4873 (1987).
[45] F. BAZSO and P. AVOURIS: *Appl. Phys. Lett.*, **53**, 1095 (1988).
[46] H. IBACH, H. WAGNER and D. BRUCHMANN: *Solid State Commun.*, **42**, 457 (1982).
[47] L. KUBLER, E. K. HLIL, D. BOLMONT and G. GEWINNER: *Surf. Sci.*, **183**, 503 (1987); R. J. HAMERS, P. AVOURIS and F. BOZSO: *Phys. Rev. Lett.*, **59**, 2071 (1987).
[48] P. DUMAS, Y. J. CHABAL and G. S. HIGASHI: *Phys. Rev. Lett.*, **65**, 1124 (1990).
[49] P. GUYOT-SIONNEST, P. DUMAS, Y. J. CHABAL and G. S. HIGASHI: *Phys. Rev. Lett.*, **64**, 2156 (1990).
[50] P. GUYOT-SIONNEST: *Phys. Rev. Lett.*, **67**, 2323 (1991).
[51] L. HALONEN and M. S. CHILD: *Mol. Phys.*, **46**, 239 (1982).
[52] F. BOGANI: *J. Phys. C*, **11**, 1283, 1297 (1978); J. C. KIMBALL, C. Y. FONG and Y. R. SHEN: *Phys. Rev. B*, **23**, 4946 (1981); T. HOLSTEIN, R. OBACH and S. ALEXANDER: *Phys. Rev. B*, **26**, 4721 (1982); F. BOGANI, R. GIUA and V. SCHETTINO: *Chem. Phys.*, **88**, 375 (1984).
[53] R. P. CHIN, J. Y. HUANG, Y. R. SHEN, T. J. CHUANG, H. SEKI and M. BUCK: *Phys. Rev. (Rapid Commun.) B*, **45**, 1522 (1992).
[54] R. C. DE VRIES: *Annu. Rev. Mater. Sci.*, **17**, 161 (1987), and references therein; S. MATSUMOTO, M. HINO and T. KOBAYASHI: *Appl. Phys. Lett.*, **51**, 737 (1988); Y. CONG, R. W. COLLINS, F. G. EPPS and H. WINDISCHMANN: *Appl. Phys. Lett.*, **58**, 819 (1991).
[55] B. J. WACLAWSKI, D. J. PIERCE, N. SWANSON and R. J. CELOTTA: *J. Vac. Sci. Technol.*, **21**, 368 (1982).
[56] X. ZHU and S. LOUIE: *Phys. Rev. B*, **45**, 3940 (1992).
[57] A. L. HARRIS and N. J. LEVINOS: *J. Chem. Phys.*, **90**, 3878 (1989).
[58] A. L. HARRIS, L. ROTHBERG, L. H. DUBOIS, N. J. LEVINOS and L. DAHR: *Phys. Rev. Lett.*, **64**, 2086 (1990); A. L. HARRIS, L. ROTHBERG, L. DAHR, N. J. LEVINOS and L. H. DUBOIS: *J. Chem. Phys.*, **94**, 2438 (1991).
[59] A. L. HARRIS, N. J. LEVINOS, L. ROTHBERG, L. H. DUBOIS, L. DAHR, S. F. SHANE and M. MORIN: *J. Electron Spectrosc. Relat. Phenom.*, **54/55**, 5 (1990); M. MORIN, N. J. LEVINOS and A. L. HARRIS: *J. Chem. Phys.*, **96**, 3950 (1992).
[60] M. MORIN, P. JAKOB, N. J. LEVINOS, Y. J. CHABAL and A. L. HARRIS: *J. Chem. Phys.*, **96**, 6203 (1992).
[61] P. GUYOT-SIONNEST: *Phys. Rev. Lett.*, **66**, 1489 (1991).
[62] See, for example, E. J. HEILWEIL, M. P. CASASSA, R. R. CAVANAGH and J. C. STEPHENSON: *Annu. Rev. Phys. Chem.*, **40**, 143 (1989).
[63] G. P. BRIVIO and T. B. GRIMLEY: *J. Phys. C*, **10**, 2352 (1977); *Surf. Sci.*, **89**, 226 (1979); B. N. J. PERSSON: *J. Phys. C*, **11**, 4251 (1978); B. N. J. PERSSON and M. PERSSON: *Solid State Commun.*, **36**, 175 (1980).
[64] J. D. BECKERLE, M. P. CASASSA, R. R. CAVANAGH, E. J. HEILWEIL and J. C. STEPHENSON: *Phys. Rev. Lett.*, **64**, 2090 (1990); J. D. BECKERLE, M. P. CASASSA, E. J. HEILWEIL, R. R. CAVANAGH and J. C. STEPHENSON: *J. Electron Spectrosc. Relat. Phenom.*, **54/55**, 17 (1990).
[65] N. A. KURNIT, I. D. ABELLA and S. R. HARTMANN: *Phys. Rev. Lett.*, **13**, 567 (1964).
[66] X.-D. ZHU and Y. R. SHEN: *Appl. Phys. B*, **50**, 535 (1990).

Gas Manipulation by Light.

L. Moi

Dipartimento di Fisica dell'Università degli Studi di Siena
via Banchi di Sotto 55, 53100 Siena, Italia

Introduction.

It is known since the beginning of this century that light can affect the motion of atoms and gases. The first experimental evidence goes back to the thirties when O. R. Frisch was able to deflect a sodium atomic beam shined by a powerful resonant lamp [1]. This was a manifestation of the resonance radiation pressure (RRP) that, as is well known, is not the only way the light acts on matter. During the last few years a lot of work has been done on this subject and light has been used to control the diffusion of gases and the velocity of atoms [2, 3]. In the present lecture the analysis will be limited to the light-induced diffusion of a gas that is kept in thermal contact with the environment. Therefore, all the cooling processes will be neglected.

Let us consider the situation sketched in fig. 1: a cylindrical cell, filled with a low-density gas, is illuminated over all its section by a laser beam, that can be resonant with the gas. The actual realization is a capillary cell with an internal diameter comparable to that of the laser beam. This experimental set-up optimizes the spatial laser-gas coupling and permits an effective analysis of the gas diffusion induced by the following effects: *a*) the resonant-radiation pressure, due to the momentum transfer from the photons to the atoms; *b*) the light-induced drift (LID), due to the momentum transfer between atoms. LID is, in fact, effective in the presence of a velocity-selective excitation and of a transport property dependence on the internal state of the atoms. In the case of alkali atoms, LID can be observed by filling the cell with a second gas, not resonant with the laser radiation and at a much higher pressure. The studies of these effects give information on the transport properties of the gas, on the atom-atom and gas-wall interactions and they permit to realize and to maintain nonequilibrium regimes in the gas. Moreover, many applications, spanning from the isotope separation to astrophysical problems, can be experienced.

The lecture gives a general survey of these subjects and a more detailed

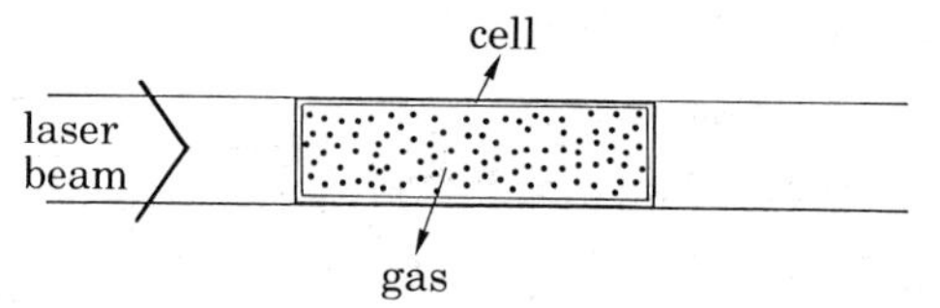

Fig. 1. – Sketch of the experimental set-up.

analysis of the experimental results obtained with the alkali vapours. The resonance radiation pressure in a gas will be discussed first, followed by the description of the general features of the light-induced drift effect. Some applications of LID will be discussed with particular attention to the production of vapour jets, to the gas drift induced by white light and to the isotope separation. A last issue is represented by a short introduction of new effects concerning the atom-surface interaction in silane-coated cells.

1. – Resonance radiation pressure in a gas.

When a two-level atom is irradiated with a resonant laser beam, the force

$$F_z = \frac{h}{\lambda\tau}\,\frac{I_{\rm L}/I_{\rm S}}{1+2I_{\rm L}/I_{\rm S}}\,, \tag{1}$$

where z is the propagation direction of the laser beam, is exerted on it. τ is the natural lifetime of the excited state, λ the laser wavelength, $I_{\rm L}$ the laser power density and $I_{\rm S} \simeq \pi hc/\lambda^3\tau$ the saturation power density. This force pushes the atom along the propagation direction of the light beam and, under favourable experimental conditions, can appreciably modify the gas diffusion [4, 5]. While it is relatively easy to put in evidence the RRP in an atomic beam, the same is not true in a gas, where the collision rate and hence the thermalization rate are so fast, if particular solutions are not adopted, to cancel any RRP effect. In an atomic beam, once the Doppler shift is compensated, the resulting acceleration is constant, so the atoms experience a velocity change directly proportional to the time T they are on resonance with the laser. In a vapour confined in a capillary, the atoms are accelerated as soon as they leave the surface but they thermalize again on it.

When the effect is eventually observed, it is necessary to shift the theoretical description from that of a ballistic motion of the atoms to that of the diffusion of the whole gas. This is made by introducing appropriate diffusion equations. If the whole gas is considered, the force on it results proportional to the fraction

χ of atoms on resonance with the laser, and, if a monochromatic radiation is used,

$$\chi = \frac{\Delta\nu_{\text{eff}}}{\pi\Delta\nu_{\text{D}}} = \frac{\Delta\nu_{\text{n}}(1 + I_{\text{L}}/I_{\text{S}})^{1/2}}{\pi\Delta\nu_{\text{D}}}, \tag{2}$$

where $\Delta\nu_{\text{eff}}$ is the effective linewidth and $\Delta\nu_{\text{n}}$ is the natural one. χ is in general very small and, in case of sodium atoms, it is smaller than 10^{-2} when $I_{\text{L}} = I_{\text{S}}$, due to the hyperfine structure of the ground-state level. Therefore, all efforts directed to increase F_z have to be focused to the maximization of χ. Under the saturation condition and $\chi = 1$, each sodium atom is submitted to a force F_z given by

$$F_z = \frac{2h}{3\lambda\tau}, \tag{3}$$

where the factor 2/3 results by considering the sodium level degeneracies. In order to have χ close to unity and multiply the net force acting on the gas by one or two orders of magnitude, two different approaches can be followed, either by power-broadening the absorption line or by increasing the laser bandwidth $\Delta\nu_{\text{L}}$. The first solution imposes very high laser power densities, not easy to obtain with a c.w. laser. In this case in fact c.w. laser powers of the order of 30 W are needed. Moreover, these high power densities could trigger secondary processes which might obscure the RRP effect. The second solution, instead, requires weaker laser sources but with peculiar spectral characteristics.

In the more general case, the diffusion equation will have the coefficients depending on the spatial and temporal co-ordinates, but under optically thin condition, the vapour drift is described by a differential equation of the following form:

$$\frac{\partial n}{\partial t} = D\frac{\partial^2 n}{\partial z^2} - v\frac{\partial n}{\partial z}, \tag{4}$$

where D, the diffusion coefficient, and v, the drift velocity, are assumed independent of z. The diffusion coefficient is given by

$$D = \frac{kT}{mf}, \tag{5}$$

where f is the collisional thermalization rate. The drift velocity is determined by the balance between the friction force and the RRP force

$$mfv = \frac{kT}{D}v = F. \tag{6}$$

The solution of eq. (4) depends on the boundary conditions and has been solved for a finite cell with the metal reservoir at one end. The reservoir imposes the boundary condition of constant density for $z = L$ (L is the cell length), while at

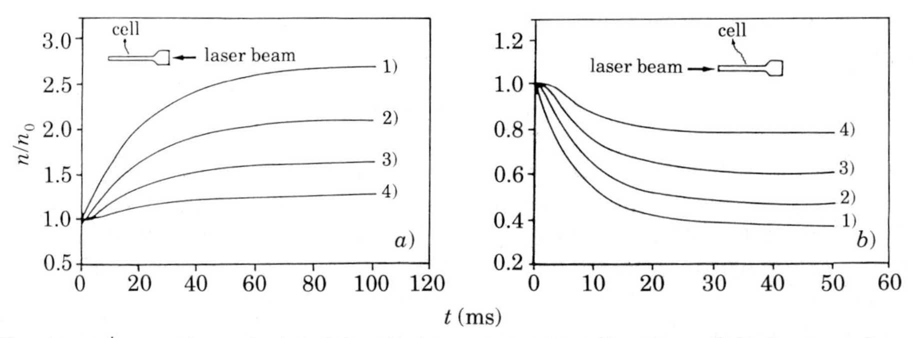

Fig. 2. – n/n_0 *vs.* time calculated for the two propagation directions of the laser as shown in the insets, by assuming $v = 20$ m/s and $D = 8$ m^2/s. The curves correspond to different positions along the capillary: 1) $z_0 = 10$ cm, 2) $z_0 = 20$ cm, 3) $z_0 = 30$ cm, 4) $z_0 = 40$ cm [5].

the other end the flux vanishes,

$$n(z = L, t) = n_{\text{res}} = \text{const}, \tag{7}$$

$$\Phi = vn - D\frac{\partial n}{\partial z} = 0 \qquad \text{for } z = 0. \tag{8}$$

The asymmetry determined by the reservoir position is reflected in the solution of the diffusion equation, which depends on the propagation direction of the laser beam. In fig. 2 the ratio n/n_0, where n_0 is the unperturbed vapour density, is reported as a function of time. When the laser enters from the reservoir side, it pushes the atoms inside the capillary by increasing the vapour density in it (fig. 2*a*)). When, on the contrary, it comes from the opposite direction, it pushes out the atoms, by lowering the vapour density in the capillary (fig. 2*b*)).

The major experimental problems to be solved in order to observe the RRP effect in a gas are essentially two: *a*) the usually very poor laser-vapour coupling, as already discussed; *b*) the large atom-wall friction. They have been solved by using a laser with a special cavity configuration («lamp-laser» [6]) and by adopting suitable coatings [7].

The light source has to be powerful and, at the same time, able to instantaneously excite all the atoms regardless of their speed. The first requirement can be easily fulfilled by a laser, while the second one is naturally achieved by a spectral lamp. As a lamp has not enough spectral-power density, the solution has to be found by modifying a laser source. The «lamp» or «white» laser is a possible solution. The idea consists in reducing the longitudinal-mode frequency splitting to a value which is smaller than the homogeneous linewidth of the absorbing atoms. When this condition is satisfied, the laser radiation is «seen» by the atoms as «white», *i.e.* as a powerful lamp. The frequency separation between the longitudinal modes $\Delta\nu_{\text{M}}$ is inversely proportional to the laser

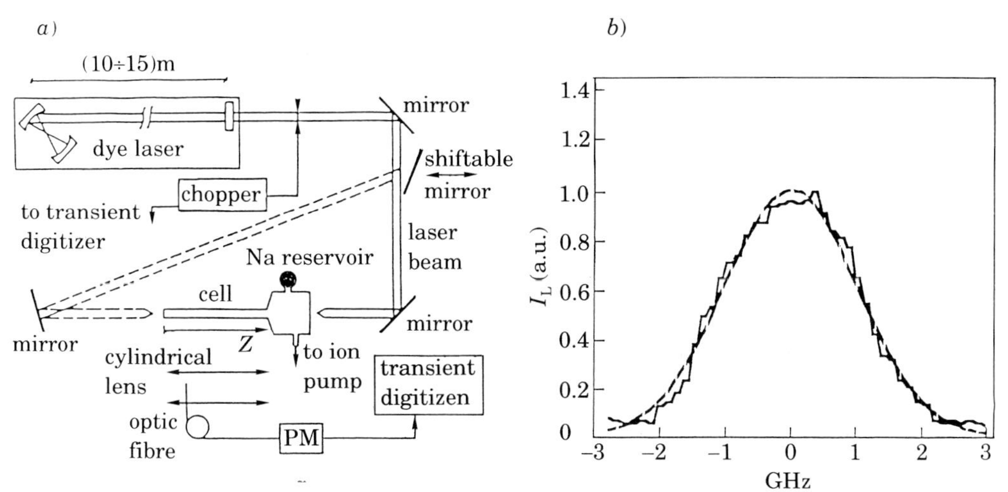

Fig. 3. – *a*) Sketch of the experimental apparatus used to study the resonance radiation pressure (PM = photomultiplier), *b*) laser linewidth of the «lamp-laser» as resolved by a Fabry-Perot interferometer. The broken curve represents the best fit by assuming a Maxwellian profile.

cavity length L_c. When L_c is large enough to have $\Delta\nu_M$ equal to or smaller than the homogeneous linewidth $\Delta\nu_H$ of the absorption line and the total laser bandwidth $\Delta\nu_L$ is comparable with $\Delta\nu_D$, then the maximum spectral coupling is realized. Therefore, the «lamp-laser» is a broad-band dye laser with a long-cavity configuration, where, in the case of the $\Delta\nu_H = 10$ MHz of sodium atoms, long means up to more than 15 m. This schematization is valid until the two-level configuration can be applied to the atom, if not, the mode structure of the laser can induce new effects like, for example, the coherent population trapping [8]. The lamp-laser is not the only approach to the problem, other solutions can be adopted [9, 10].

The «lamp-laser» is obtained by modifying a commercial standing-wave dye laser. It is important to remark that, when L_c increases, the total bandwidth decreases because the number of longitudinal modes remains roughly the same. By adding an etalon to the cavity a $\Delta\nu_L = (3 \div 4)$ GHz and a total laser power $W_L = 200$ mW have been obtained for $L_c = 15$ m. In fig. 3*b*) the laser bandwidth, resolved by a Fabry-Perot interferometer, is reported. The curve represents the envelope of few hundreds of longitudinal modes that are not resolved by the interferometer. The broken curve gives the best fit of the experimental curve obtained by assuming a Maxwellian profile. This laser has also been used in an atomic-beam cooling experiment, as proposed by Moi [11], and a deceleration of the atomic beam has been observed [12].

The sodium vapour is confined in a cell made from a 50-cm-long pyrex capillary (the internal diameter is 0.2 cm) welded to a larger cylindrical body (the diameter is 2 cm) to which the sodium reservoir is linked. The cell is permanently connected to an ion pump in order to keep the background gas pressure below 10^{-8} Torr. This precaution is necessary to eliminate LID contributions. The cell, which has optically polished windows at both ends, has been coated by using an ether dimethylpolysiloxane solution. The coating has the property of reducing, to negligible values, the adsorption energy of sodium atoms at the surface [7]. It lowers the friction of the vapour at the cell walls and avoids the formation of a sodium layer on the surface. The vapour drift is detected through the induced fluorescence signal variations. An optical fibre is placed sideways, in front of the cell image produced by a cylindrical lens placed as shown in fig. 3*a*), where a sketch of the experimental apparatus is reported. The fibre can be moved along z to monitor the vapour drift as a function of the position z. The fluorescence signal is detected by a photomultiplier and processed by a transient digitizer. A chopper, driven by a function generator, controls the laser beam admission to the cell.

In fig. 4 two fluorescence signals obtained for the two laser propagation directions are shown. This figure clearly shows the drift of the sodium vapour in the capillary cell by radiation pressure. In fig. 4*a*) the vapour density decreases to less than 30% of the initial value, and in fig. 4*b*) the vapour density increases by more than a factor two. The fluorescence signals have the behaviour predicted by calculations, as shown in fig. 2. From the best fit of the experimental curves the following values have been obtained for the diffusion coefficient and the drift velocity: $D = (8.4 \pm 1.6)\ \mathrm{m^2/s}$, $v = (39 \pm 6)\ \mathrm{m/s}$ (fig. 4*a*)); $D = (4.3 \pm \pm 1.4)\ \mathrm{m^2/s}$, $v = (22 \pm 7)\ \mathrm{m/s}$ (fig. 4*b*)). By substituting these values in eq. (6), the result is $F \simeq 0.4\ F_{\max}$, which is in agreement with the force calculated from the laser intensity and by assuming $\chi = 1$. The D and v values, instead, turn out to be larger than estimated and this discrepancy might indicate that the collisions with the cell walls are not completely thermalizing.

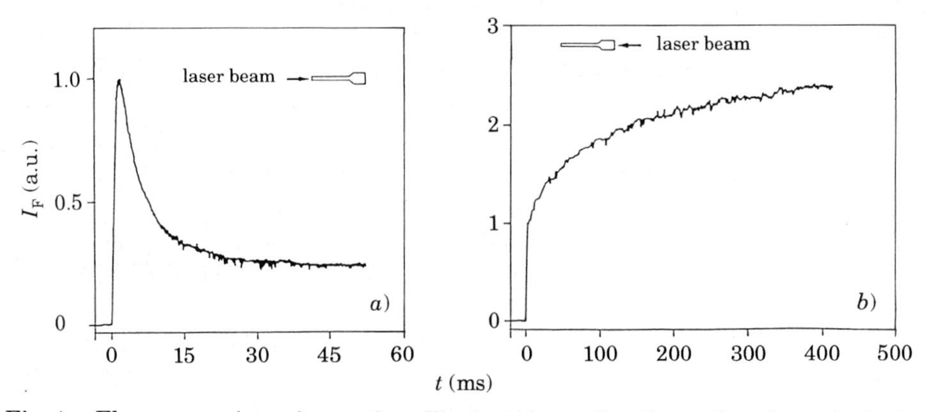

Fig. 4. – Fluorescence intensity *vs*. time. The laser beam direction is that shown in the insets, $I_L = 120$ mW, $L_c = 15$ m. *a*) $z_0 = 24$ cm, $D = (8.4 \pm 1.6)\ \mathrm{m^2/s}$, $v = (39 \pm 6)\ \mathrm{m/s}$; *b*) $z_0 = 35$ cm, $D = (4.3 \pm 1.4)\ \mathrm{m^2/s}$, $v = (22 \pm 7)\ \mathrm{m/s}$ [5].

2. – Light-induced drift.

The simple apparatus, consisting of a laser resonant with a gas confined in a capillary cell, has been already considered. This apparatus allowed us to observe and to analyse the resonance radiation pressure. We want to verify now if there are modifications of the gas dynamics due to the simultaneous presence in the cell of a buffer gas (for example, a noble gas). The experiments show that there are changes of the vapour diffusion dynamics and that they are huge and spectacular. The RRP effect is, in fact, rapidly washed out by the increased collision rate and overwhelmed by a new effect, known as light-induced drift (LID). LID is produced by the combined action of a laser velocity-selective excitation and the collisions between atoms of different species, which perturb the Maxwellian distribution of the atomic velocities. An intuitive explanation of LID is the following: the Maxwellian velocity distribution, calculated along the laser beam propagation direction, is symmetric with respect to $v = 0$. This implies that, across any section of the cell, the flux of atoms going in one direction is perfectly balanced by that going in the opposite direction. This symmetry is conserved, if the atoms are not perturbed, when a laser excites the vapour. In fact, in this case and neglecting the RRP effect, the laser excitation produces two fluxes, one of excited atoms and the other of ground-state atoms. These two fluxes move in opposite directions but with equal and opposite velocities. In a cell containing a buffer gas, the atoms are perturbed by the collisions, and, if the diffusion coefficient of the atoms in the excited state is different from that of the ground-state atoms, as is usually the case, the two fluxes will have different velocities. The consequence is an unbalance between the two fluxes, which results in a macroscopic movement of the gas in one or in the other direction, depending on the velocity class selected by the laser. The vapour diffusion becomes very sensitive to the laser detuning and the effect of a macroscopic diffusion of the vapour to either sides of the cell is spectacular and easily visible by naked eyes. Figure 5 gives a schematic representation of the different features of the resonance radiation pressure and of the LID effect.

In LID the photons only label a specific velocity class and the collisions with the buffer gas transform the random atomic motion into ordered motion, *i.e.* the drift. Net transfer of photon momentum is not involved, in fact equal but opposite momenta are exchanged between the absorbing atoms and the buffer gas. LID has been predicted in 1979 by GEL'MUKHANOV and SHALAGIN [13] and it has been experimentally demonstrated in the same year by ANTSYGIN *et al.* [14]. Since LID is based on transfer of atomic momenta rather than of photon momenta, the LID pressure can be orders of magnitude larger than the radiation pressure [5]. LID can be in either direction relative to the laser beam propagation. This last is well described by the drift velocity, which depends on the relative difference between the diffusion coefficients of the ground and excited states, respectively, the laser intensity, the fraction of excited atoms and, if not

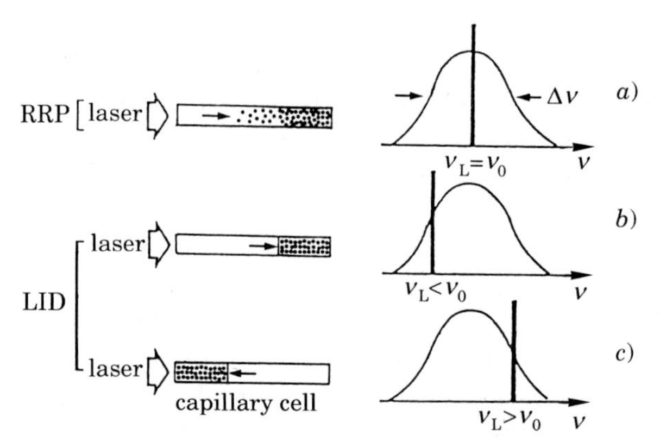

Fig. 5. – Schematic representation of unidimensional RRP and LID effects. The curves to the right represent the Doppler-broadened absorption line profiles. The bold lines represent the laser excitation.

eliminated, by the friction of the atoms at the wall surface. This last problem has been almost completely solved by coating the cell with paraffin [15] or silane compounds [7], which have a very small adsorption energy [16]. Using a one-dimensional random-walk argument and assuming complete thermalization after a single collision, the drift velocity v, for a two-level atom, is given by

$$v = -\nu_L \frac{n_e}{n} \frac{\sigma_e - \sigma_g}{\sigma_g}, \tag{9}$$

where n_e and n are the number of atoms in the excited state and in the ground state, respectively, and σ_e and σ_g the relative collision cross-sections. The ratio n_e/n can be increased by using broad-band lasers simulating «white» radiation [6] or by using frequency-modulated dye lasers [9]. The ratio including the collision cross-sections can be modified by more than a factor ten by choosing the proper buffer gas. Drift velocities of the order of 30 m/s can be achieved when the «lamp-laser» and a heavy buffer gas are used [17].

Equation (9) gives an intuitive picture of the effect, but it gives too high values when applied to alkali atoms. A more accurate theoretical approach has been made, where the multilevel structure of the alkalis has been satisfactorily simplified to a four-level scheme [18]. The LID features have been deeply studied both theoretically and experimentally and the results and the references can be found, for example, in ref. [3]. Here in the following the most recent applications of LID will be discussed.

2·1. *Light-induced vapour jets.* – In a capillary cell the LID is affected by the adsorption of the atoms at the cell walls, which leads both to a very lengthy establishment of a steady-state vapour condition inside the capillary and to a rather small drift velocity. A detailed analysis of these surface effects has been made both theoretically by NIENHUIS [19] and experimentally by GOZZINI *et al.* [20], who were able to measure the sodium-pyrex adsorption energy. The experimental value is $E_a = (0.71 \pm 0.02)$ eV. The problem of the atom-wall interaction has been solved by adopting paraffin or silane coatings. Unfortunately such layers, which cannot be heated up to high temperatures (the maximum is about 500 K for dimethyl-polysiloxane), can be used only for elements that have a comparatively low temperature of evaporation and low chemical reactivity. As it is rather difficult to find new layers that are able to resist both high temperatures and strong chemical aggression, any further experimental development of LID was restrained. This limitation is very detrimental, considering that the existing c.w. laser sources can excite a large part of the alkali metals, alkaline-earth metal elements and rare-earth elements, and that new applications of LID seem possible. An attempt to study LID far from the cell walls has been performed by GOZZINI *et al.* [21], who adopted a spherical cell and observed sodium vapour compression at its centre induced by six counterpropagating laser beams.

A different approach has been proposed by ATUTOV *et al.* [22]. In this case a large cell, kept at room temperature and filled with a buffer gas, is used. The cell has in its middle part an active atom source that, when heated up, produces a spatially well-delimited vapour cloud. When a properly detuned laser beam crosses this cloud, a vapour jet expanding in «the free space» represented by the large buffer gas volume is produced by LID. This is made possible by the fact that the induced drift velocity can be much higher than the thermal diffusion one [15, 17]. Therefore, the time the vapour spends diffusing out of the light beam is long enough to allow the vapour to drift along the laser beam, thus creating the vapour jet. The experiment outline is sketched in fig. 6. The cell is a glass tube of 6.5 cm in diameter and 33 cm long with a Na vapour source in the middle. The vapour source consists of a small glass tube (0.3 cm in diameter,

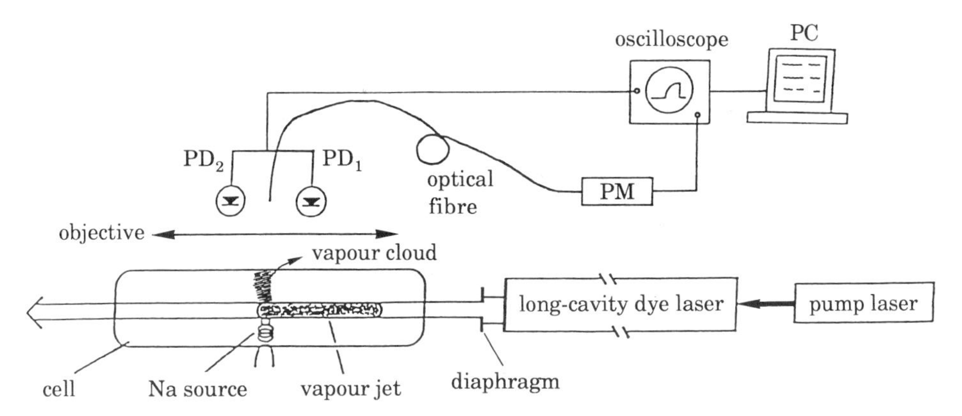

Fig. 6. – Sketch of the experimental apparatus used in the vapour jet experiment. PC = personal computer, PM = photomultiplier, $PD_{1,2}$ = photodiodes.

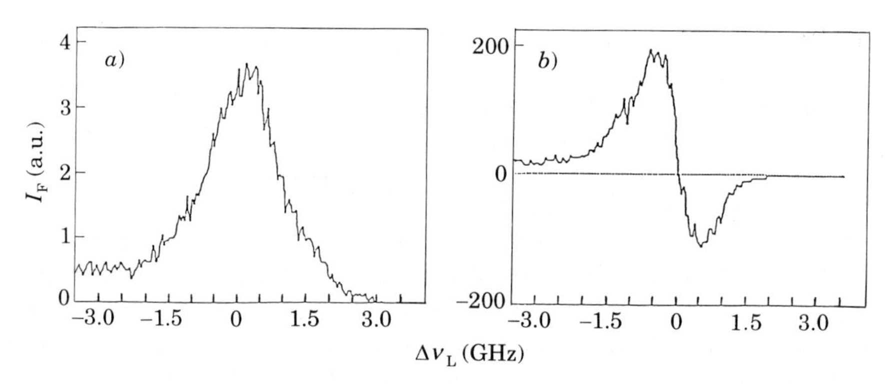

Fig. 7. – *a*) Fluorescence signal and *b*) DDS signal as a function of the laser detuning $\Delta\nu_L$ [22].

1 cm in length), containing metallic Na, placed in a vertical position, and heated up independently of the cell, which is kept at room temperature. The cell is filled with 44 Torr of krypton as a buffer gas. The used laser has a long cavity configuration (about 5 m), a total bandwidth of the order of $(1 \div 2)$ GHz and a maximum power equal to about 270 mW. This corresponds to a mean power density of about $(10 \div 20)\,\text{mW/cm}^2$ for each longitudinal mode. The laser beam, tuned to the D_2 line, is sent through the cell along its axis and slightly above the vapour source. Its diameter is about 0.5 cm and can be reduced by a diaphragm. An image of the vapour jet is made on the plane of a differential detector system (DDS), consisting of two photodiodes placed symmetrically with respect to the vapour source position $z = 0$. When the laser is on, the vapour cloud around the source is easily observable by naked eyes when illuminated by the resonant laser light. An optical fibre that is connected to a photomultiplier and positioned in the middle of the DDS gives a signal proportional to the fluorescence in front of the vapour source. An unbalance between the two photodiodes, that is proportional to the induced asymmetry of the vapour cloud, is obtained as a signature of the vapour jet. In fig. 7, two curves, as a function of the laser detuning, are shown. Figure 7*a*) represents the sodium fluorescence in front of the vapour source. This curve is symmetrical because the vapour density is kept fixed at that point. Figure 7*b*) shows the DDS signal taken at a distance $2z_0$ between photodiodes equal to 3.4 cm. This signal has a dispersive behaviour and demonstrates the existence of the vapour jet going from one side to the other of the vapour cloud. The effect is clearly visible by naked eyes.

These features can be described by solving the diffusion equation

$$\frac{\partial n}{\partial t} + \operatorname{div}\left[\boldsymbol{v}\,n + \nabla(Dn)\right] + \frac{n}{\tau} = 0\,, \tag{10}$$

where τ is the decay time of the vapour density n due to chemical losses. The detailed solution of this equation, which is valid if the velocity distributions remain approximately Maxwellian and if light pressure is ignored, is in general very complicated. A crude but useful model can be, anyway, proposed. It can be shown that the drift velocity is given by

$$v = v_0 \frac{d_b^2}{d_d^2} \left\{ 1 - \exp\left[-\left(\frac{d_d}{d_b} \right)^2 \right] \right\}, \tag{11}$$

where v_0 is the drift velocity on the beam axis, d_b is the full diameter of the laser beam and d_d is the diameter of the laser beam after the diaphragm. By assuming D and v independent of the cylindrical co-ordinates and $1/\tau = 0$, it results

$$n = n_0 \exp[-z/L_1], \qquad z > 0, \tag{12}$$

$$n = n_0 \exp[-z/L_2], \qquad z < 0, \tag{13}$$

with

$$(L_{2,1})^{-1} = \sqrt{\frac{v}{2D} + \frac{\gamma}{D}} \pm \frac{v}{2D} \tag{14}$$

and

$$\gamma^{-1} = \frac{d_b^2}{16\,D}. \tag{15}$$

The experimental differential signal, normalized to the signal from the photomultiplier, can be derived from these calculations once it is expressed by

$$I = \frac{I(z_0) - I(-z_0)}{I_0} = \exp[-z_0/L_2] - \exp[-z_0/L_1]. \tag{16}$$

In fig. 8 I as a function of the laser power and of the diaphragm diameter is shown. The agreement between the experimental data and the theoretical curves is satisfactory. The derived drift velocity results comparable to that obtained in the capillary cells.

2·2. *White-light-induced drift.* – It has been affirmed that LID is effective only if a velocity-selective excitation is present. A new development of the field consists in the utilization for LID experiments of noncoherent and non-monochromatic sources. This is made in order to study new excitation schemes and drift mechanisms, as proposed by ATUTOV and SHALAGIN [23]. One of this is, for example, the fluorescent-light-induced drift (FLID), in which the velocity-selective excitation is provided by a pressure-shifted fluorescence line [24]. The white-light-induced drift (WLID), where the initially flat excitation spectrum is modified by the absorption of an optically thick sample, is another one. They suggested that WLID might be responsible for the observed anomalies in the atmospheres of the so-called «chemically peculiar stars» as well as of, for example, Titan's atmosphere. This last is in fact enriched in deuterium by a fac-

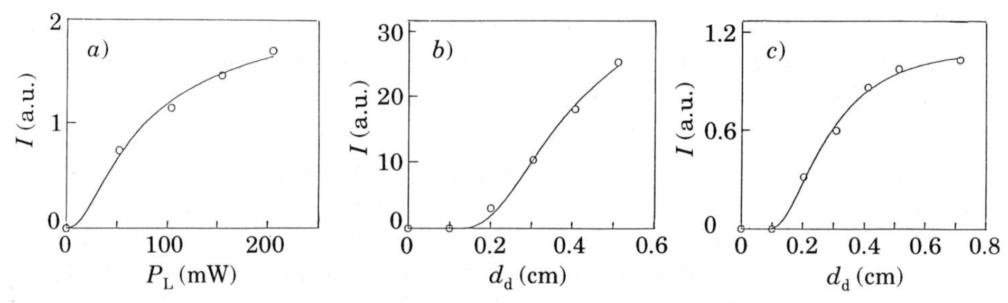

Fig. 8. – Plots of the DDS signal I as a function a) of the laser power P_L, b) and c) of the laser diameter d_d. The open points represent the experimental measurements and the solid curve is the best fit obtained from eq. (16). a) $d_d = 0.5$ cm; b) $P_L = 116$ mW, $v = 6.4$ m/s; c) $P_L = 212$ mW, $v = 14.2$ m/s.

tor > 3 relative to Jupiter and Saturn. Theories exist that claim to explain all that in terms of radiation pressure, chemical reactions and other, but the results are not satisfactory. By assuming the spectral distribution of the light emitted by the core of the star to be white, ATUTOV and SHALAGIN argued that in an optically thick vapour, where two species are embedded in a buffer gas, LID can occur when the two species have overlapping absorption lines. As the gas is optically thick, the spectral distribution of the light depends on the distance from the source (see fig. 9). The fact that the Fraunhofer lines overlap is the root cause of the LID effect: there is a slight reduction of the excitation of species A in the high-frequency Doppler wing and similarly for species B in the low-frequency Doppler wing. If for both species the collision cross-section increases upon excitation, the two species will drift in opposite directions. It has been theoretically demonstrated by POPOV *et al.* [25] that WLID may be effective in particular mixtures of two atomic species, for example, of the two Rb isotopes, whose absorption lines partially overlap. WLID can also be observed in one-component gas systems [26], when atoms with Λ level configuration are considered. In this last case the excitation asymmetry is enhanced by hyperfine optical pumping. WLID has been experimentally demonstrated by GOZZINI *et*

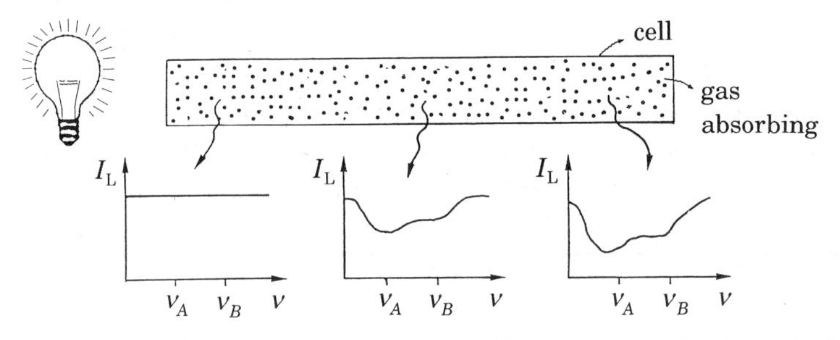

Fig. 9. – Schematic representation of the modification induced to the initially flat excitation spectrum by the optically thick vapour. I_L is the spectral-lamp intensity; ν_A and ν_B are the resonance frequencies of the two atomic species.

al. [27] on sodium vapour immersed in a noble gas. In the case of the single species with Λ energy level scheme, WLID has a different origin. Let us consider sodium that, because of its hyperfine ground-state structure, can be assumed as a typical example. Due to the different degeneracies of the two ground-state hyperfine levels, two dips of different intensities are created in the excitation spectrum. These two peaks grow up, as the vapour becomes more and more optically thick, with different rates and their difference is amplified by the optical pumping. This induces a positive feedback that modifies the excitation spectrum in such a way that a velocity-selective excitation is induced for the atoms in the two hyperfine levels. The consequence is that two unbalanced fluxes of atoms moving in opposite directions are generated and a drift of the vapour can be observed in the laser propagation direction.

The observation of both FLID and WLID is difficult when conventional lamps, which have a very low spectral-power density I, are used. In the experiment of ref. [27] this problem has been avoided by using the «lamp» laser. In this case, because the absorption linewidth is collisionally broadened, a 5 m cavity laser is long enough to simulate the «lamp-light» condition. The other requirement is a laser bandwidth so large to be assimilated to a white source. To obtain it, the intracavity etalons have been eliminated and only the birefringent filter has been maintained in its place. With these simple modifications a 25 GHz total bandwidth, which is about ten times larger than the Doppler-broadened absorption line, has been measured with the help of a Fabry-Perot interferometer. The typical spectral laser power density is $1\ \mathrm{W/cm^2\,GHz}$, that has to be compared with the $2 \cdot 10^{-3}\ \mathrm{W/cm^2\,GHz}$ of a 1 kW Xe lamp. Moreover, in the practice it turns out to be much lower, *i.e.* of the order of $10\ \mu\mathrm{W/GHz}$ [18]. In fig. 10*a*) two signals from the Fabry-Perot interferometer are reported. Curve 1) represents the unperturbed laser spectrum, while curve 2) shows the laser spectrum modified by the sodium vapour absorption. This second curve has been obtained by inserting a sodium cell between the laser and

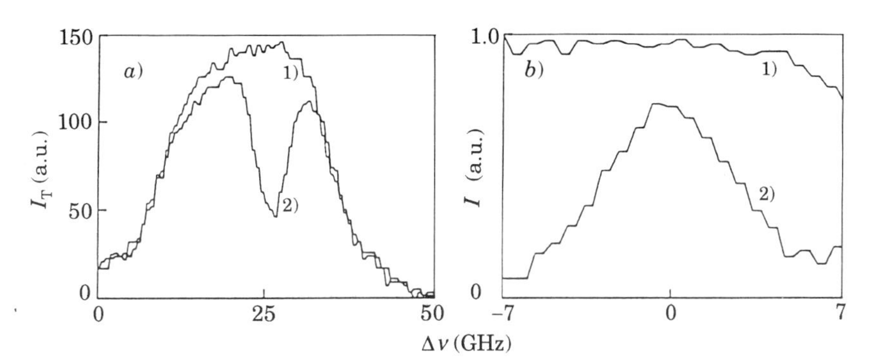

Fig. 10. – *a*) Laser lineshape resolved by a Fabry-Perot interferometer: 1) unperturbed, 2) modified by the absorption of the reference cell. *b*) Particular of the absorption region: 1) normalized spectral laser profile, 2) absorption sodium line profile.

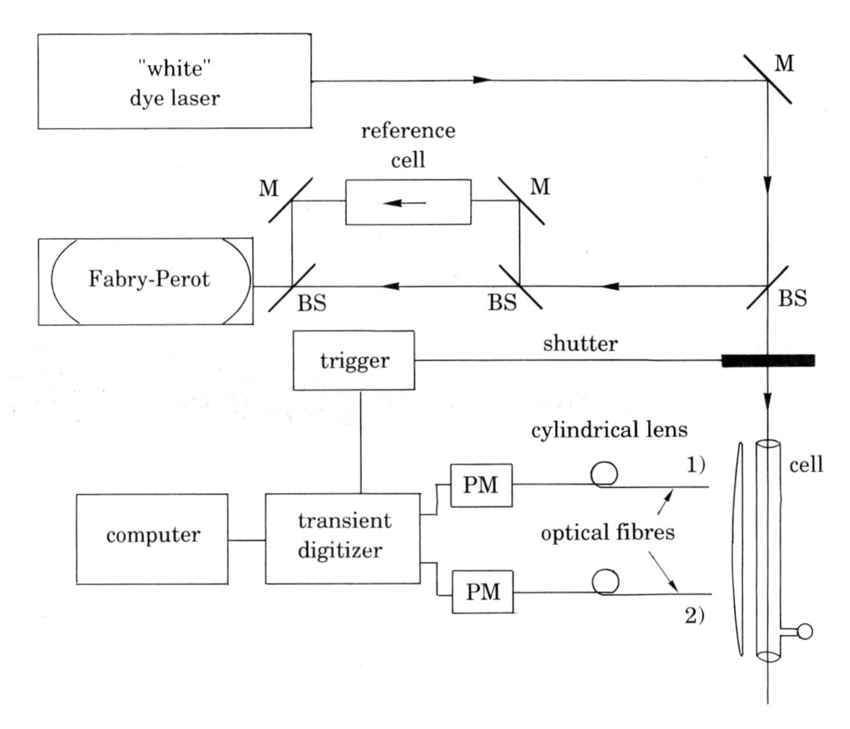

Fig. 11. – Sketch of the apparatus used in the WLID experiment. PM = photomultiplier, M = mirror, BS = beamsplitter.

the interferometer. In fig. 10*b*) a detailed view of the absorption region is reported. Here curve 1) represents the laser profile normalized to its maximum, while curve 2) is obtained by the relation $I = I_T/I_L$, where I_L is the direct laser intensity and I_T the absorbed one. From these measurements the «white»-light assumption results totally justified.

The experimental apparatus is sketched in fig. 11. The laser beam is sent to a capillary cell filled with sodium and a few Torr of neon or kripton. The cell is coated and heated up in the temperature range (140 ÷ 250) °C. The fluorescence is collected at right angles and focused on two optical fibres, one placed at the cell entrance, the other one in front of its middle part. The laser is switched on/off by a mechanical chopper and the fluorescence temporal evolution is monitored and stored by a transient digitizer connected to a personal computer. In this case, contrary to the procedure followed in the other LID experiments, the laser cannot be detuned in order to see the drift effect. Therefore, the only way to check if possible fluorescence variations are due to WLID is to change the vapour density, *i.e.* to increase the temperature of the cell. By increasing the optical thickness of the vapour, in fact, the laser spectrum is modified and at a given temperature WLID will start pushing the atoms. In fig. 12 two experimental curves are shown in the case of the Na/Ne mixture. The two signals

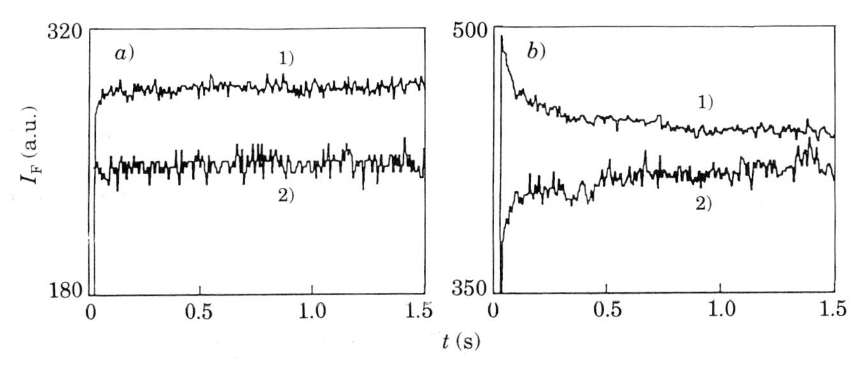

Fig. 12. – Sodium fluorescence evolution from the cell entrance (curve 1)) and from the cell end (curve 2)). *a)* $T = 140$ °C, *b)* $T = 250$ °C.

have been taken at two different temperatures corresponding to optically thin and optically thick vapour condition, respectively. In fig. 12*a*) the vapour has a low density and no fluorescence variation is observed, as the laser spectrum is not appreciably modified. When the temperature increases, the absorption of the vapour modifies the radiation spectrum and WLID becomes effective. The decreasing fluorescence at the cell entrance (fig. 12*b*), trace 1)) demonstrates the drift of the vapour toward the cell end where, on the contrary, an increasing fluorescence is observed (trace 2)). An estimate based on the time spent to reach the stationary state leads to a drift velocity of the order of few cm/s, that is about one hundred times slower than in the case of the optically thin regime [17]. This result is consistent with the evaluation reported by ARKHIPKIN *et al.* [28], by remembering that in our case $I_L = 1$ W/cm^2 GHz and the buffer gas is neon which has a diffusion factor about ten times slower than kripton. In fig. 13 fluorescence signals obtained from a cell containing Na and kripton as a

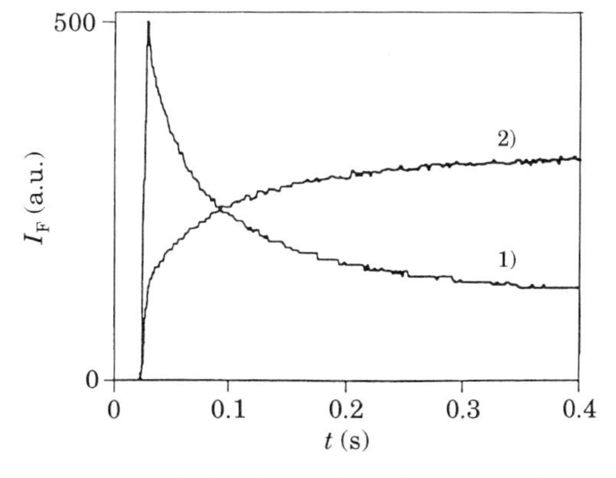

Fig. 13. – Sodium fluorescence evolution from the cell entrance (curve 1)) and from the cell end (curve 2)). $T = 140$ °C.

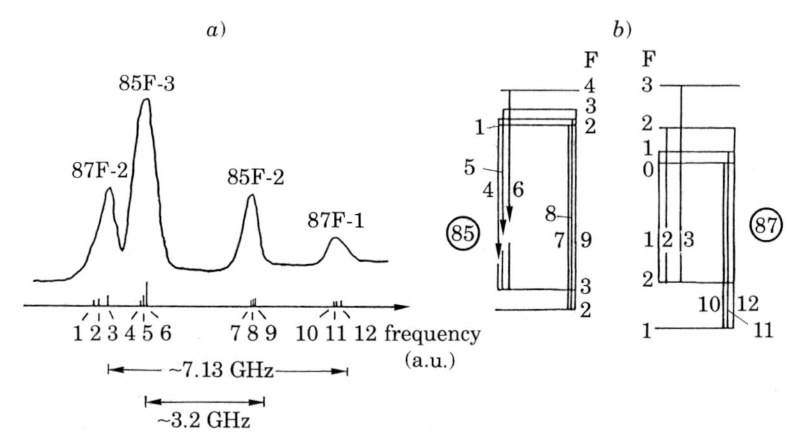

Fig. 14. – *a*) Spectrum of the D_2 rubidium line; numbers on the frequency axis correspond to the transitions as reported in *b*) energy level scheme for the two rubidium isotopes.

buffer gas are reported. In this case the fluorescence decay is much faster and the intensity variations much larger than before, as expected.

2·3. LID-*induced isotope separation.* – The effective control of the gas diffusion by LID suggests its application to the isotope separation. Among the alkali atoms, rubidium is a good candidate. In fact it has two stable isotopes, ^{85}Rb and ^{87}Rb with a relative concentration equal to 72% and 28%, respectively. Moreover, they have a very favourable absorption spectrum which shows, in the case of D_2 excitation, four well-separated peaks, where each one of them is due to the absorption of one isotope, as shown in fig. 14.

A theoretical evaluation of the drift velocity as a function of the laser detuning can be made by schematizing the Rb atoms to three-level atoms. Two of them correspond to the hyperfine structure of the ground state and the last one to the excited level. This approach has been proposed by HAVERKORT *et al.* [29], who discussed the case of sodium first. For sodium atoms, the scheme uses four levels because the fine structure of the excited atoms cannot be neglected. By solving the rate equations of the density matrix $\rho_i(v)$ and by remembering that the drift velocity v is given by

$$v = \sum_i \int \rho_i(v)\, v\, \mathrm{d}v\,, \tag{17}$$

v can be calculated as a function of the laser detuning. In fig. 15, v is reported for both isotopes and for a laser power density equal to 10 W/cm^2 and a buffer gas pressure of 5 Torr of kripton. From this figure it results that, for some spe-

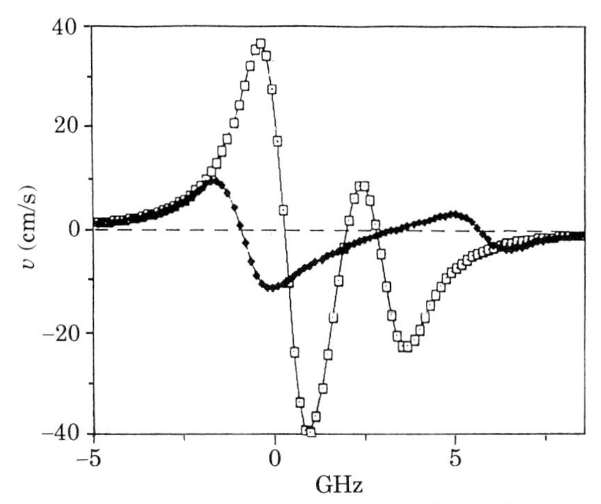

Fig. 15. – Drift velocity of the two Rb isotopes calculated as a function of the laser detuning: □ ^{85}Rb, ● ^{87}Rb. The laser power density is 10 W/cm^2 and the buffer gas pressure 5 Torr of Kr.

cific laser detunings, the two isotopes drift in opposite directions, by making the isotope separation even more suitable. We have directly checked these theoretical curves by looking at the fluorescence variations induced by a diode laser tuned across the D_2 resonance line. The laser beam is sent in a silane-coated capillary cell. The fluorescence is collected at right angle by an optical fibre coupled to a photomultiplier. The cell is heated up enough to have an optically thin vapour. When LID is effective, the whole vapour is pulled or pushed along the capillary and fluorescence variations are consequently detected. Obviously in this way only the whole vapour drift can be followed and not the drift of the single isotope. Systematic measurements of v have been performed and the obtained v values follow well the theoretical curves of fig. 15. This fact can be nicely verified by sweeping the laser frequency once very slow and once fast, and by making the difference between the fluorescence signals obtained in the two cases. In fact, they differ for the LID contribution that makes the vapour drift during the slow sweeping. In fig. 16*b*) the obtained result is reported: the fluorescence shows increasing and decreasing peaks corresponding to vapour compression or dilution in the place of the observation point. This signal is compared with the absorption spectrum (fig. 16*c*)) and with the theoretical curve showing the sum of the drift velocities of the two isotopes (fig. 16*a*)). The agreement between the curves *a*) and *b*) is very good confirming, in first approximation, the validity of the theoretical approach [30]. Similar observations have been already done by Streater *et al.* [31], who used the «optical machine gun» method. These results seem to confirm the possibility of obtaining, even with a relative facility, the isotope separation. But new problems make it difficult, as first observed by Hamel *et al.* [32], who were not able to achieve an effective

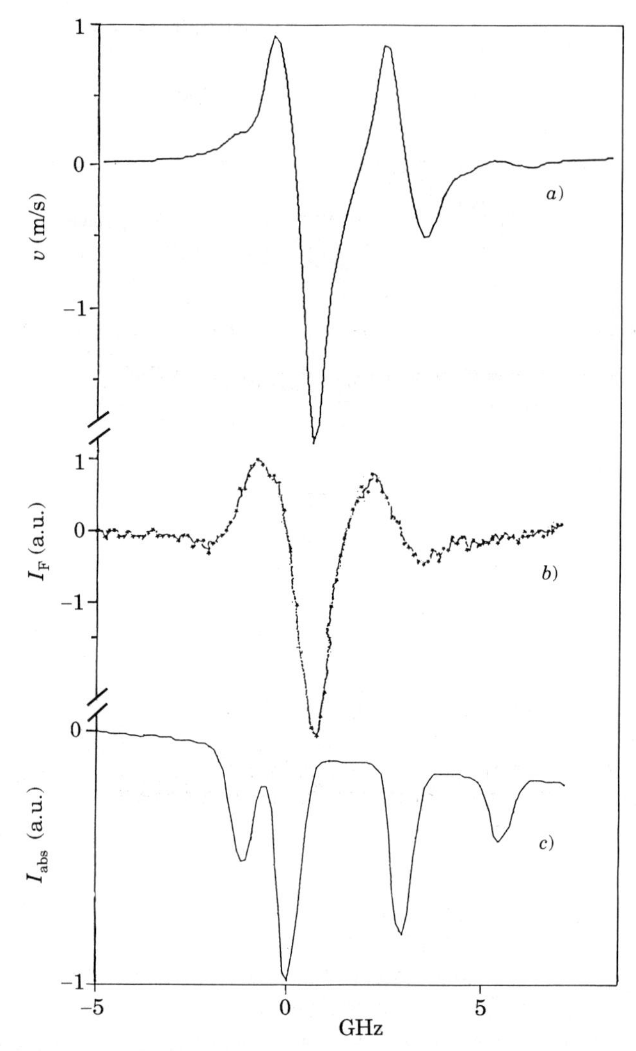

Fig. 16. – *a*) Calculated drift velocity of a natural isotope mixture of Rb as a function of the laser detuning, *b*) difference between two fluorescence signals obtained with slow and fast sweep of the laser frequency, *c*) absorption spectrum obtained from a reference cell.

isotope separation in an optically thick vapour. The reason of this negative result has to be ascribed to the desorption of rubidium of natural isotopic composition from the cell walls, that are normally covered by an alkali monolayer. We have repeated the experiment by adopting the silane coating described before and by working under optically thin conditions.

The Rb vapour is confined in a cell consisting of two cylinders (radius about 2 cm and length about 3 cm) connected by a capillary 5 cm long, as shown in fig. 17. The internal diameter of the capillary is about 2 mm. The cell is filled

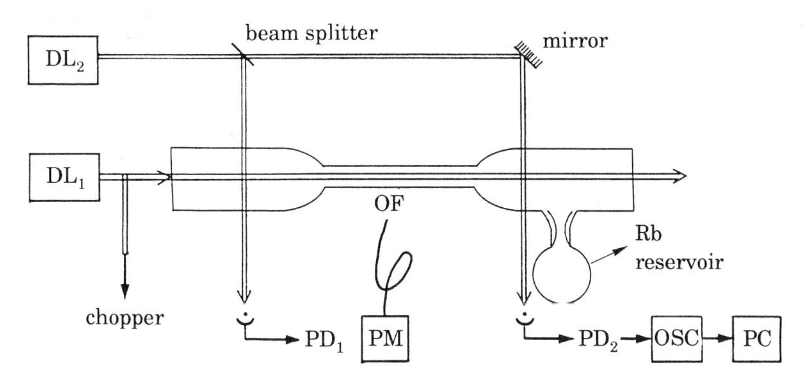

Fig. 17. – Sketch of the cell used in the isotope separation experiment: PM = photomultiplier, $DL_{1,2}$ = diode lasers, $PD_{1,2}$ = photodiodes, PC = personal computer, OSC = oscilloscope.

with few Torr of a buffer gas and the metal reservoir is welded to one of the two cylinders. The capillary should then work as an optical tap, by allowing the migration of one isotope from one side of the cell to the other and, *vice versa*, for the other isotope. Two diode lasers are used, one to induce the vapour diffusion and the second to monitor the isotopic composition. DL_1 and DL_2 have both a maximum power of about 30 mW and may be tuned to the D_2 line. DL_1 is tuned to optimize the LID effect, while DL_2, strongly attenuated, is swept across the D_2 absorption spectrum to monitor the relative concentrations of the two isotopes. Two photodiodes, PD_1 and PD_2, are used either to have a differential signal or, separately, to have absolute absorption from each side arm of the cell. A photomultiplier connected to an optical fibre monitors the fluorescence arising from the capillary. Huge density variations have been obtained in the cell part without the metal reservoir, but no evidence of isotope enrichment has been observed, even after laser expositions of about one hour. Improvements have been tried by using the two lasers collinear and detuned so as to reduce hypefine pumping, but still without appreciable modification of the relative isotope concentrations in the cell bulbs. A possible explanation of this failure can be again found in the participation of the cell walls to the process. This hypothesis is confirmed by the new observations made, which reveal an important role played by the coating [33].

2·4. *Light-induced atom desorption.* – During the attempts made to obtain effective isotope enrichment in one section of the cell, some new and very interesting effects have been observed. The most striking one is that reported in fig. 18, where the absorption of the diode laser, crossing one of the two cell bulbs and tuned to one Rb line, is reported as a function of time [33]. As is evident from the figure, the light level changes both when the room light (*i.e.* the neon lamps hanging from the ceiling) is switched on or off and when a flash-

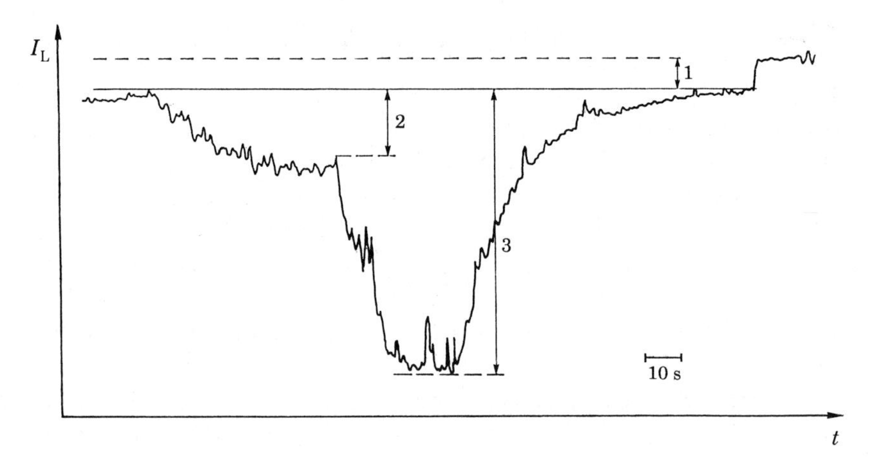

Fig. 18. – Transmission of a coated cell at room temperature. The laser frequency is kept on resonance with the Rb atoms. 1 corresponds to the cell transmission in the dark; 2: transmission when the ceiling light is on; 3: flash-lamp + ceiling light on.

lamp illuminates the cell. When both lights are off, the cell transmission goes back to the starting level. These changes of the cell transparency correspond to Rb density variations. From fig. 18 it results that the Rb density increases by about one order of magnitude. The increasing time constant depends on the light intensity, while the decay time is determined by the cell geometry and the gas pressure. The signal in discussion has been obtained with the cell at room temperature. Similar modifications are obtained when, instead of the room light, the laser tuned out of resonance illuminates the cell. In this case it has been checked that the density variations depend linearly on the laser intensity and on the cell surface irradiated.

This effect can be explained in terms of a light-induced atom desorption (LIAD) from the coated cell surface. Similar observations have been made by GOZZINI *et al.* [34] with sodium atoms by adopting a coating similar to ours. They observe a strong sodium fluorescence at room temperature when a resonant laser hits the cell. A possible explanation of the desorption effect is given in ref. [34], where the formation of anions and cations soluted in the coating are supposed to generate the observed density increase. Due to the very weak power densities applied and to the nonresonant character of the phenomenon, the heating of the surface can be excluded. Another explanation can be found in the possible excitation of the Rb adatom, as observed in the photoemission of sodium atoms from sapphire surfaces [35]. In this last case, anyway, the effect is very small and triggered by much higher photon flux.

Such new features open a new field of investigation, *i.e.* the detailed analysis of the atom-surface interactions that has been studied very little with the laser spectroscopy techniques.

3. – Conclusions.

The possibility of gas manipulation by laser light has been clearly demonstrated. We presented the general features together with the experimental and theoretical results of both the resonance radiation pressure and of the light-induced drift effects. We reported the last results on the production of vapour jets expanding in noble gas and we showed the possibility of getting LID by «white»-light excitation. This last result opens a new possible explanation to the unsolved problem of the «chemically peculiar stars». Finally we showed the difficulties found in the achievement of isotope separation by LID, because of the surface effects. The new effects obtained when a coated surface is irradiated by nonresonant light open new interesting perspectives to the field. This last point in fact links the RRP and LID effects to the new field of the gas-surface interactions, that deserves a deeper and more extended study.

* * *

The author is grateful to M. BADALASSI and P. MANNUCI for their technical assistance.

REFERENCES

[1] O. R. FRISCH: *Z. Phys.*, **86**, 42 (1933).
[2] See the issue on laser cooling and trapping of atoms, *J. Opt. Soc. Am. B*, **6**, 2020 (1989).
[3] See the *Proceedings of the International Workshop on «Light Induced Kinetic Effects on Atoms, Ions and Molecules»*, edited by L. MOI, S. GOZZINI, C. GABBANINI, E. ARIMONDO and F. STRUMIA (ETS, Pisa, 1991).
[4] J. H. XU and L. MOI: *Opt. Commun.*, **67**, 282 (1988).
[5] S. GOZZINI, G. PAFFUTI, D. ZUPPINI, C. GABBANINI, L. MOI and G. NIENHUIS: *Phys. Rev. A*, **43**, 5005 (1991).
[6] J. LIANG, L. MOI and C. FABRE: *Opt. Commun.*, **52**, 131 (1984).
[7] J. H. XU, M. ALLEGRINI, S. GOZZINI, E. MARIOTTI and L. MOI: *Opt. Commun.*, **63**, 43 (1987).
[8] G. ALZETTA, A. GOZZINI, L. MOI and G. ORRIOLS: *Nuovo Cimento B*, **36**, 5 (1976).
[9] M. C. DELIGNIE, H. I. BLOEMINK, A. H. DE BOER and E. R. ELIEL: *Phys. Rev. A*, **42**, 1560 (1990).
[10] I. C. M. LITTLER, S. BALLE and K. BERGMANN: *J. Opt. Soc. Am. B*, 8, 1412 (1991).
[11] L. MOI: *Opt. Commun.*, **50**, 349 (1984).
[12] S. GOZZINI, E. MARIOTTI, C. GABBANINI, A. LUCCHESINI, C. MARINELLI and L. MOI: *Appl. Phys. B*, **54**, 428 (1992).
[13] F. KH. GEL'MUKHANOV and A. M. SHALAGIN: *JEPT Lett.*, **29**, 711 (1979).
[14] V. D. ANTSYGIN, S. N. ATUTOV, F. KH. GEL'MUKHANOV, G. G. TELEGIN and A. M. SHALAGIN: *JEPT Lett.*, **30**, 243 (1979).

[15] S. N. ATUTOV, ST. LESJAK, S. P. PODJACHEV and A. M. SHALAGIN: *Opt. Commun.*, **60**, 41 (1986).
[16] M. A. BOUCHIAT and J. BROSSEL: *Phys. Rev.*, **147**, 41 (1966).
[17] S. GOZZINI, J. H. XU, C. GABBANINI, G. PAFFUTI and L. MOI: *Phys. Rev. A*, **40**, 6349 (1989).
[18] J. E. M. HAVERKORT and J. P. WOERDMAN: *Ann. Phys. (Leipzig)*, **47**, 519 (1990).
[19] G. NIENHUIS: *Opt. Commun.*, **62**, 81 (1987).
[20] S. GOZZINI, G. NIENHUIS, E. MARIOTTI, G. PAFFUTI, C. GABBANINI and L. MOI: *Opt. Commun.*, 88, 341 (1992).
[21] S. GOZZINI, D. ZUPPINI, C. GABBANINI and L. MOI: *Europhys. Lett.*, **11**, 207 (1990).
[22] S. N. ATUTOV, S. GOZZINI, C. GABBANINI, A. LUCCHESINI, C. MARINELLI, E. MARIOTTI and L. MOI: *Phys. Rev. A*, **46**, R3601 (1992).
[23] S. N. ATUTOV and A. M. SHALAGIN: *Sov. Astron. Lett.*, **14**, 284 (1988).
[24] F. WITTGREFE, J. L. C. VAN SAARLOS, S. N. ATUTOV and E. R. ELIEL: *J. Phys. B*, **24**, 145 (1991).
[25] A. K. POPOV, A. M. SHALAGIN, A. D. STREATER and J. P. WOERDMAN: *Phys. Rev. A*, **40**, 867 (1989).
[26] A. D. STREATER: *Phys. Rev. A*, **41**, 554 (1990).
[27] S. GOZZINI, C. MARINELLI, E. MARIOTTI, C. GABBANINI, A. LUCCHESINI and L. MOI: *Europhys. Lett.*, **17**, 309 (1992).
[28] V. G. ARKHIPKIN, D. G. KORSUKOV and A. K. POPOV: preprint N640F, USSR Academy of Sciences, Siberian Branch, L. V. Kirensky Institute of Physics (1990).
[29] J. E. M. HAVERKORT, H. G. WERIJ and J. P. WOERDMAN: *Phys. Rev. A*, **38**, 4054 (1988).
[30] M. MEUCCI, E. MARIOTTI, C. MARINELLI, P. BICCHI and L. MOI: *Europhys. Lett.*, **25**, 639 (1994).
[31] A. D. STREATER, J. MOOIBROEK and J. P. WOERDMAN: *Opt. Commun.*, **64**, 137 (1987).
[32] W. A. HAMEL, A. D. STREATER and J. P. WOERDMAN: *Opt. Commun.*, **63**, 32 (1987).
[33] E. MARIOTTI, M. MEUCCI, P. BICCHI, C. MARINELLI and L. MOI: to be published.
[34] A. GOZZINI, F. MANGO, J. H. XU, G. ALZETTA, F. MACCARRONE and R. A. BERNHEIM: *Nuovo Cimento D*, **15**, 709 (1993).
[35] A. M. BONCH-BRUEVICH, T. A. VARTANYAN, A. V. GORLANOV, YU. N. MAKSIMOV, S. G. PRZHIBEL'SKII and V. V. KHROMOV: *Sov. Phys. JEPT*, **70**, 604 (1990).

ULTRAHIGH RESOLUTION

High-Resolution Laser Spectroscopy.

V. P. CHEBOTAYEV †

Institute of Laser Physics - Novosibirsk-90, Russia

1. – New possibilities of superhigh-resolution spectroscopy.

The main application of the superhigh-resolution laser spectroscopy technique is connected with the carrying-out of fundamental experiments in physics and the creation of optical frequency standards. Recently we have considered some other applications of these methods. They were applied to the interaction of excited particles with a surface, the generation of the time-stable light pulses with a duration of $10^{-14 \div 15}$ s, the development of atomic-optical and atomic interferometers. The elaboration of these applications gave new possibilities for superhigh-resolution spectroscopy. We shall dwell on this in the first half of the lecture.

1·1. *Narrowing of multiphoton absorption resonances.* – Here we shall describe the simple methods, in the technical sense, that allow an increase of the time of the coherent interaction of particles with the field at (fixed) certain dimensions of the field and, consequently, to obtain the narrowing of the resonances. The particle diffusion in the field due to collisions without phase perturbation permits to increase the time of interaction. The second one is due to the use of trapped particles.

In this and the following subsections we shall consider two possibilities for increasing the duration of the interaction of particles with a field and for eliminating the influence of transit effects. Obviously, only elastic scattering without phase mismatch can be used. For the two-level particles the method is well known in the microwave range (Dicke narrowing). It permits to eliminate the Doppler broadening when the free-path length of the particle is comparable with the wavelength. In the optical range there are many examples in which the cross-section of elastic collisions without phase mismatch is larger than the

† Deceased on September 2, 1992.

cross-section of inelastic collisions. As a result, narrowing of a Doppler contour has been observed. Unfortunately, the application of this effect for obtaining narrow lines in the optical range proved to be impossible. The narrowing of a Doppler contour requires very high gas pressures at which the free-path length of a particle is of the order of or less than a wavelength. At atmospheric pressures the contribution of inelastic processes remains considerable; it is, therefore, hardly possible to obtain very narrow lines (by 10^4 to 10^5 times narrower than the Doppler width). A maximum narrowing of a Doppler contour by about 10 times has been obtained in H_2 [1].

Quite a different situation can occur when Doppler-free two-photon resonances (TPR) are used. The elimination of the Doppler effect allows the use of a low gas pressure at which the free-path length of the particles is much greater than the wavelength and, as has been said above, the homogeneous width is small. Since the process of two-photon absorption is independent of the particle velocity, an elastic collision without phase mismatch will not result in resonance broadening. Since with these collisions the residence time of a particle in the field increases (it is determined by the time of molecule diffusion in a light beam), one can observe a narrowing of the resonance as compared with the case where the width is determined by transit effects. Such a possibility was investigated in [2]. The two-photon resonance width is $\gamma \sim \tau_d^{-1} + \Gamma_{21}$, Γ_{21} is the homogeneous width of the $1 \to 2$ transition allowing a radiation decay of levels 1 and 2 and inelastic collisions, τ_d is the time of particle diffusion in a light beam. Since τ_d and Γ_{21} are proportional to the gas pressure P, then $\gamma \sim \alpha/P + \beta P$. The dependence $\gamma(P)$ is given in fig. 1. The minimum width is given by $\gamma_{min} \sim \sqrt{\alpha\beta} = \sqrt{\Gamma_{21}/\tau_d} = \text{const} \cdot P$.

Another possibility of TPR collision narrowing considered in [3] is connected with the use of cells, in which the relaxation of excited particles in collisions with the walls is very small. Elastically reflected from the walls over the dipole moment of a phase break, the particles cross a beam many times. The time of coherent interaction of the particles with a field is increased and the broadening

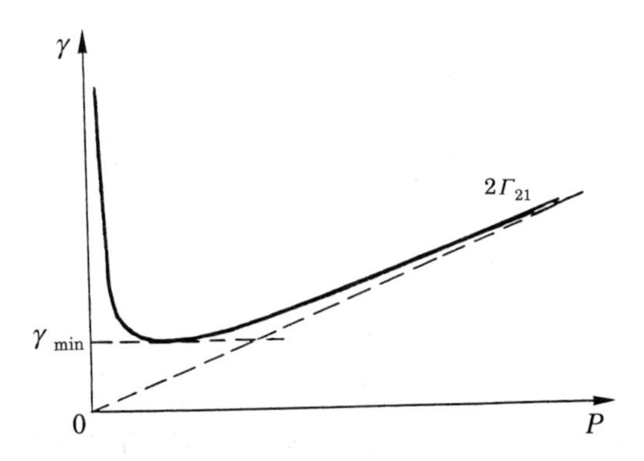

Fig. 1. – Qualitative dependence of the two-photon resonance width γ on pressure P.

connected with the transit effect is removed. The width of the TPR is just determined by the inelastic-scattering channel. However, there is a physical difference from the known effect of one-photon absorption narrowing in a storage cell which is the base of the operation of a hydrogen maser [4]. In the microwave band the narrowing of a Doppler contour may be observed at cell dimensions less than a wavelength. In the optical range this condition is practically impossible to be fulfilled. The use of TPR permits to adopt a cell with dimensions much greater than the wavelength. The narrowing of TPR using the rotational-vibrational transitions of simple molecules (H_2, N_2) has been considered in [3].

Analogous effects can be used in the narrowing of the nonlinear resonances of stimulated Raman scattering, free of the Doppler broadening [5]. Figure 2 shows the scheme of the transitions which are responsible for the resonances. Usually the resolution of SRS and CARS methods is very strongly influenced by the Doppler effect.

At least there are two mechanisms of Doppler-free interactions that contribute to the resonance excitation of the Raman oscillations ρ_{eg} of the medium. The first is double SRS (fig. 2*a*)). The difference frequency Doppler shift $(\pm)(k_1 - k_2)v$ acquired in the first SRS atomic interaction with copropagating waves at ω_1 and ω_2 is completely compensated by the shift $(\pm)(k_1 - k_2)v$ in the second SRS interaction with mirror-reflected light waves. The latter interaction is synchronous with the first one. The other mechanism (fig. 2*b*)) involves

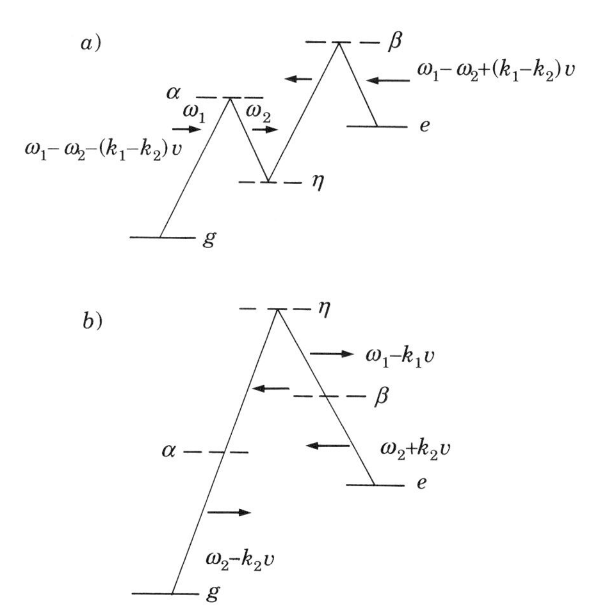

Fig. 2. – Possible transitions for obtaining Doppler-free Raman resonances: *a*) double stimulated Raman scattering, *b*) two-photon absorption and two-photon emission of counterpropagating photons.

two Doppler-free processes: stimulated two-photon absorption of the counter-propagating waves at frequency ω_1, and, simultaneously, stimulated emission of two oppositely directed photons at frequency ω_2. The absence of the Doppler effect for this excitation of ρ_{eg} is due to its absence in each elementary step.

1·2. *Application of the optical methods to microwave spectroscopy.* – The methods of nonlinear laser spectroscopy can be used in microwave spectroscopy to increase the resolution and to improve the quality of the frequency standards. We shall demonstrate here two possibilities connected with the use of two-photon resonances and resonances of saturated absorption. Figure 3 shows the scheme of the two-frequency beam spectrometer. Atoms in the ground states, atoms of cesium for example, go along the waveguide where they interact resonantly with the two counterpropagating waves with frequencies ω_1 and ω_2, respectively. If the length of the waveguide is much greater than the wavelength of the transition, the absorption line for each wave has a Doppler broadening and a shift of about $(1 \div 10)$ kHz (*). The homogeneous line width is determined by the time of interaction and can be considerably less. Each of the waves interacts with atoms whose velocities are $v_1 = \Omega_1/k$ and $v_2 = -\Omega_2/k$, where Ω_1 and Ω_2 are the detunings of the frequencies. The saturated-absorption

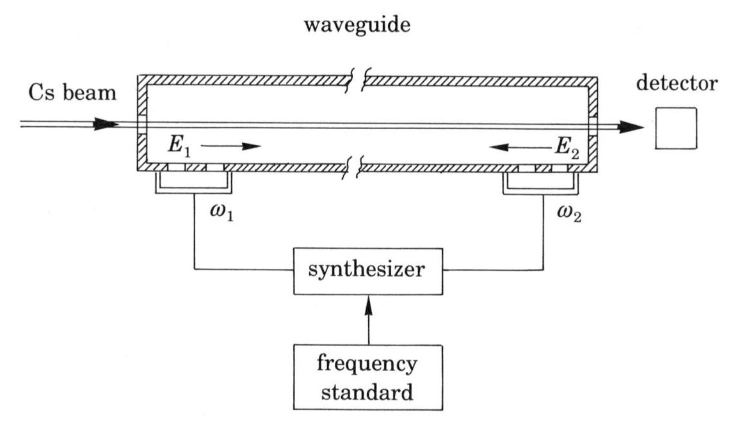

Fig. 3. – Scheme of the microwave two-frequency beam spectrometer.

(*) The method of separated oscillatory fields suggested by RAMSEY [6] permits to eliminate this Doppler effect and to obtain the resonant structures corresponding to the time of flight of a particle between the fields. But the resonances in the separated fields are interfering by nature, and their maximum position depends directly on the phase difference between the fields. The realization of Zacharias' proposal of atomic fountains decreases the influence of this phase difference, but it gives immediately the dependence of resonance maximum position on the direction of the gravitational field. Such a dependence can be used for a very careful adjustment of the gravitational axis.

resonance appears when both waves interact with the same atoms $v_1 = v_2 = \Omega_1/k = -\Omega_2/k$. Owing to the asymmetric distribution of the velocity projections, the saturated-absorption resonance has the largest amplitudes with $v_1 = v_2 = v_0$, where v_0 is the thermal velocity of the particle. Correspondingly, the frequencies of the contrary waves ω_1 and ω_2 are different, and the half-sum will be equal to the frequency of the transition [7]. The method permits to eliminate the second-order effect. As was noted, the effective selection takes place due to the effects of saturation of the longitudinal velocities and, consequently, the second-order Doppler effect is under control [8].

If a particle beam has a small divergence and $\Delta/k \approx u$ (where u is the average thermal velocity $\Delta\omega_1\omega_2$), then the selection of atoms over the velocity projections means selection over the absolute velocities. The expression of the absorption resonance is given as [8]

$$\alpha/\alpha_0 \sim 1 - \frac{G}{2} - \frac{G}{2}\,\frac{\Gamma^2}{\{\Omega + [(\Delta/2kv_0)^2 + \theta_0^2](\omega_{12}v_0^2/2c^2)\}^2 + \Gamma^2},$$

where Γ is the line half-width, $\Omega = (\omega_1 + \omega_2)/2 - \omega_{12}$, α_0 is the absorption, G is the saturation parameter, ω_{12} is the transition frequency, ω_1 and ω_2 are the generation frequencies. The resonance centre is shifted by the value

$$\Omega = -\theta_0^2\,\frac{v_0^2}{2c^2}\,\omega_0 - \left(\frac{\Delta}{2}\right)^2 \frac{1}{2\omega_{12}}.$$

Under the transit time condition $\Gamma \simeq \Delta(kL)$ (L is the length of the waveguide), $\theta_0 \ll 1$ is the beam divergence, the shift caused by the second-order Doppler effect depends on the value Δ only.

In the case in which the two-photon resonance is used, the counterpropagat-

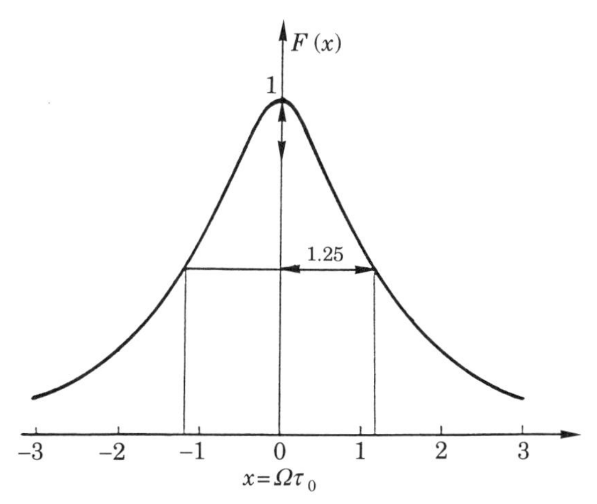

Fig. 4. – The shape $F(x)$ of two-photon absorption resonance in the longitudinal beam, $\tau_0 = L/v_0$.

ing waves must have the same frequencies and be equal to half of the transition frequency $\omega = \omega_1 = \omega_2 = \omega_{12}/2$. The cavity whose frequency is tuned to the frequency of the field can be used instead of the waveguide. The shape of two-photon absorption was considered in [9]. It is shown in fig. 4. The influence of the second-order Doppler effect in the microwave region is negligible. Its influence at the position of the resonance maximum must be necessarily accounted for. When the transit width is larger than the quadratic shift for a particle with average thermal velocity, the shift of the resonance maximum is $\delta\omega = -3(v_0/c)^2\,\omega/2$. For the diffusing beam the influence of slow particles affects the shape of the second derivative, the frequency of which is determined by the homogeneous width of the two-photon transition. Particles with velocity $v = \Gamma_{21}L$ make the main contribution here. Therefore, the shift of the maximum of the second derivative will be $(\Gamma\tau_0)^2$, *i.e.* less than the maximum resonance shift.

1·3. *Method of separated optical fields and atomic interferometry.* – The method of separated fields was developed in the fifties for obtaining narrow resonances in the microwave region [6] and for the investigation of relaxation processes using the quantum echo [10]. Note that both directions (which have appeared, approximately simultaneously, in the fifties) were developed independently for a long time. The main problem of the first method was to obtain narrow resonances, and that of the second direction was the investigation of relaxation processes in an inhomogeneously broadened medium. The significant difference was in the manner of the realization of experiments. In the first case, the systems with homogeneously broadened line were the object under investigation. In the second one, the use of media with inhomogeneously broadened line was principally important. Evidently, the difference in investigation purposes and their character explains the independent development of both methods. In the seventies, the method of separated optical fields has started to be developed for two- and three-level atoms, and two-photon absorption [11-13]. The unification of both ideas in the optical region turned out to be fruitful to develop both the superhigh-resolution spectroscopy and the transient coherent methods.

We shall briefly show the possible way of elimination of the influence of the Doppler effect in the optical range. A particle dipole moment after interacting with the first travelling-wave field in a drift region accumulates the Doppler phase $\Delta\varphi = kv_z t$, where t is the drift time. As a result of the *nonlinear* interaction with the second field, the phase of the *nonlinear* part of the dipole moment of the particle has a phase jump. If the second field is a travelling wave, then the phase jump $\Delta\varphi$ will be equal to the double of the phase difference of the fields at the point z_2, $\Delta\varphi = 2(\varphi_2 - \varphi_1) = 2k(z_2 - z_1) + 2\Omega T$; here z_2 and z_1 are the point coordinates, where the particles interact with the fields, T is the particle transit time. One may easily see that at a distance $X = L$ from the first field the

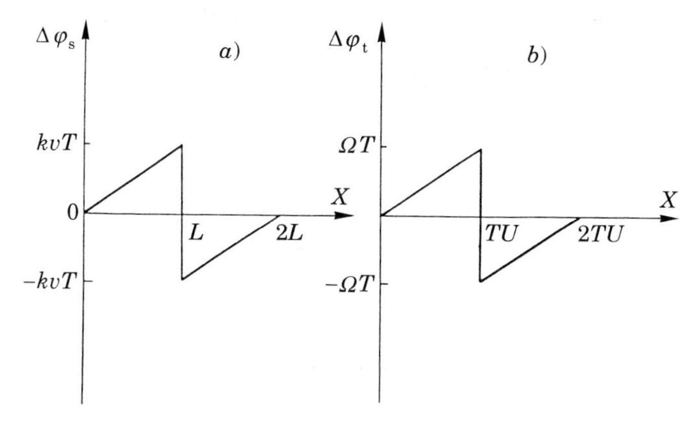

Fig. 5. – The dependence of the phase difference between of the dipole moment of a particle and the field at a distance: *a*) spatial part, $\Delta\varphi_s$; *b*) time part, $\Delta\varphi_t$.

phases of the particle dipole moment and of the field will coincide independently of the particle velocity. Thus the phase jump $\Delta\varphi$ compensates the stored Doppler shift and the polarization created by the first field is transferred at a distance $X = L$. The process may be considered as the development of a spatial quantum echo. If the second field is formed by two colliding waves (standing wave), then the phase jump will be equal to $\Delta\varphi = \pm 2kz_2$. At a distance $X = L$ the phase of the particle dipole moment will coincide with a wave phase spreading in opposition to the first wave. The absence of a component dependent on the detuning frequency ($2\Omega T$) in the phase jump is important to obtain a signal frequency dependence in separated optical fields.

It proved to be convenient to connect the phase jump with the amplitude modulation of the dipole moment of the particle which appears to be due to the nonlinear interaction with the second field [11, 14]. If the particle beam divergence is very small, the phase of the field during the interaction can be considered as constant. It is determined by the point of entrance of the particle into the field (fig. 5). If the first field is a standing wave with transverse dimension a, we have a spatial harmonic in the distribution of the dipoles along the axis of the field immediately after the interaction, $D_1(z) = N(z)\,W\cos(kz_1)$, where $N(z)$ is a smooth function describing the distribution of particles in the ground state, W is the probability of dipole excitation for the particle after the interaction with the first field. The spatial harmonic of the dipole distribution is conserved at a small distance from the first field. At a distance $L = \lambda/\theta$ (θ is the beam divergence) from the first field the space harmonic of polarization disappears. After the interaction with the second field, the initial dipole moment has an additional spatial modulation: $D_2(z) = D_1(z)[1 + A\cos^2(kz_2)]$, where the coefficient A depends on the amplitude of the second field and on the time interaction of the particle with the field. Now we shall consider the spatial distribution of dipoles at a distance X from the second field. Taking into account the ratio (fig. 5) $Z_2 = Z_1 + (Z_3 - Z_1)L/(L+X)$, we shall define the amplitude $D_3(z)$ of the dipole moment at the point Z_3:

$$D_3 = D\cos(kz_1)[1 + A\cos^2[k(z_1 + (z_3 - z_1)L/(1+X))] =$$

$$= D\cos(kz_1)[1 + A\{1 + \cos[2k(z_1 + (z_3 - z_1)L/(L+X))]\}/2].$$

Due to the finite divergence of the beam, the particles arrive at the point Z_3 with different values of Z_1. Averaging over the coordinate Z_1 will give a result if the amplitude $D_3(z_3)$ will not contain oscillating terms as $\cos(kz_1)$. This is possible only when $X = L$. This condition corresponds to the conditions for the observation of the space echo. Apparently the observed approach can be applied to different physical processes in the presence of amplitude modulation in space and time and gives an obvious explanation of the echo mechanism. A clear demonstration of the described phenomenon can be done with the help of the incoherent light source and two amplitude gratings. The spatial frequency of the second grating is two times as large as that of the first one (fig. 6). By illuminating the first grating with a collimated light beam, one can observe its shadow at a long distance (we have not taken into account the diffraction of light). The shadow from the first grating disappears if we use a wide light source. The condition for the observation of the shadow is analogous to the condition for the observation of the space harmonic: the distance to the screen X must be of about

$$X < d/\theta, \tag{1}$$

where d is the grating period and θ is the divergence of the beam. At $X > d/\theta$ the shadow (space intensity distribution) cannot be observed. If the second grating is placed between the screen and the first grating, then the shadow is restored at a distance equal to the distance between the gratings. The explanation for this is analogous to that given above. Diffraction of light must be taken into consideration at the decrease of the grating period. The condition for the observation of the diffraction effect is analogous to condition (1) in the physical sense: $X\lambda/d > d$. The spread of the shadow at a distance X due to diffraction divergence at the angle $\theta = \lambda/d$ must be comparable to d. First let us consider the collimated light beam. The light at different wavelengths is diffracted at different angles. For first order of diffraction the angle θ_1 is equal to $\theta_1 = \lambda/d$. At a distance $X < D/\theta_1$ (D is the transverse size of the beam) the diffracted beams overlap. Therefore, they produce an interference pattern with period $s = \lambda/\theta_1 = d$. This means that the diffraction does not destroy the space harmonic of the intensity distribution. This conclusion is valid for a nonmonochromatic source. The role of the second grating remains the same: it helps to eliminate the influence of the divergence of the initial beam and to restore the spatial harmonic at $X = L$. The model in which the obtained spatial structure of the in-

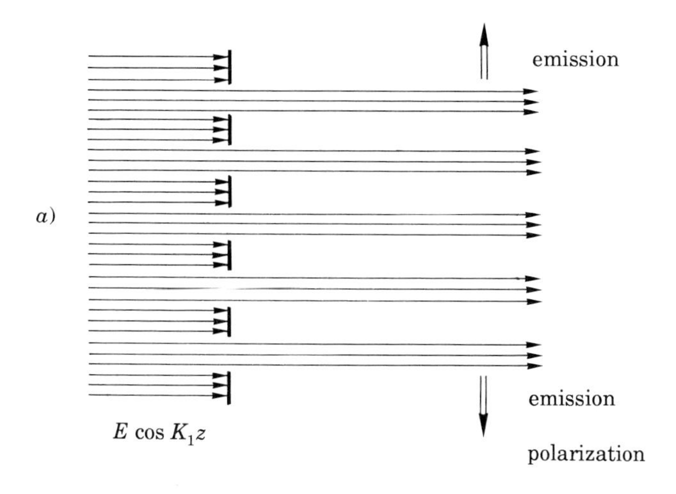

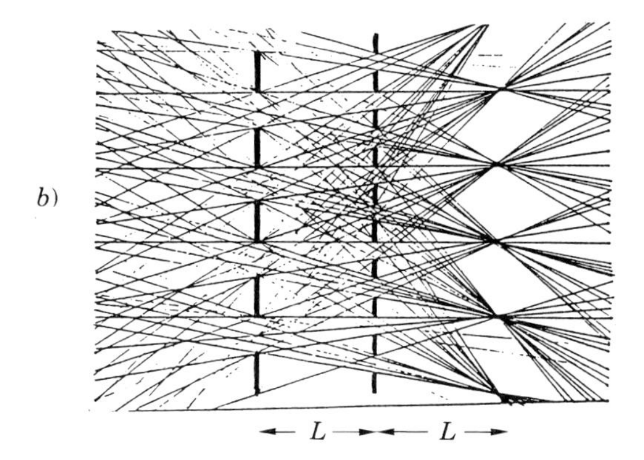

Fig. 6. – Space distribution of particles after interaction with amplitude gratings: *a*) parallel beam of particles, *b*) wide beam of particles.

tensity distribution does not depend on the wavelength (interference in the white light) is quite adequate to our echo model.

We directed our attention to the above model of the method of separated optical fields because the optical diffraction and the diffraction of particles are similar in many aspects. At high resolution the diffraction of particles due to the wave features of matter should be taken into account [15]. Using the above optical consideration, we can arrive at the conclusion, without any exact calculation, that the atom diffraction does not change the pattern of resonances in separated fields. No wonder that such a calculation gives a result which coincides with the result of the observation of the particle motion in the classical (geometric) trajectories [16, 17]. A theoretical consideration in connection with the experimental observation of the resonances in separated fields assuming the wave features of particles was done in [17-19].

The recoil effect, the second-order Doppler effect and so on must also be taken into consideration at the increase of the resolution (distance between fields). Scattering of particles in the drift region, caused by different physical reasons, must be taken into account. For example, the deviation of the particle at the angle 10^{-4} rad at a distance of 10 cm leads to shifting the particle along the Z axis at a distance of 10 μm. This corresponds to a variation of the phase difference between the field and the dipole $\simeq 60$ rad ($\lambda = 1$ μm).

Let us consider the recoil effect whose influence was investigated theoretically in [16]. Then the recoil effect should be taken into account in the case in which the recoil frequency shift $\varDelta = \hbar k^2/2M$ is compared with T^{-1} that is usually considerably less than the transit width $1/\tau$ if $a \ll L$. This means that the influence of the recoil effect during the interaction is not important. The change of the particle velocity after the interaction with the first field is also not essential because the corresponding phase change is compensated by the phase jump due to the nonlinear interaction with the second field. After the interaction with the second field an additional change of the particle velocity takes place (fig. 7). During the drift between the second and the third field the particles suffer an additional Doppler shift equal to $2\,\Delta v_z kT$, where Δv_z is the change of the particle velocity due to the absorption or the emission of a photon. This shift may be compensated by varying the frequency of the field by a value equal to $2\Omega T = 2k\,\Delta v_z T$. Hence $\Omega = \mp k\,\Delta v_z$ is the resonance shift due to the recoil effect. If a particle in the drift region is affected by some forces, the particle deflection produces an extra phase shift. It may be compensated by the frequency detuning.

The peculiarities of the interaction of particles with separated optical fields

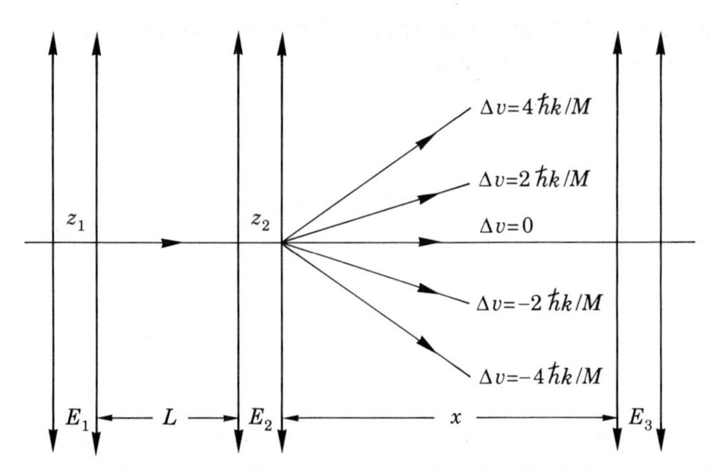

Fig. 7. – Interaction of particles with three separated optical fields in consideration of the recoil effect.

are considered in detail in the review [20], where experimental results are also obtained. Here we shall dwell briefly on an important side of the method, which is connected with atomic interferometry.

Two principal different types of atomic interferometers can be considered. The scheme of the first one (atomic-optical interferometer) corresponds to the observation of resonance phenomena in separated optical fields. That type of interferometer is conveniently called atomic-optical interferometer (AOI). The AOI has two arms (fig. 8*a*)). In one of them the phase memory is transferred by the particles. The phase of the coherent radiation at a distance $2L$ from the first field is hardly connected with the phase of the optical field. As the particle velocity is much less than the velocity of light, such an interferometer as a large delay line [14]. The coherent emission in separated optical fields was observed in [21] using the optical heterodine technique. Note that usually the shape of the absorption line is detected. The coherent part of the absorption depends directly on the phase difference between field and dipole. The result is proportional to the product of the amplitudes of field and dipole moment. Therefore, correctly

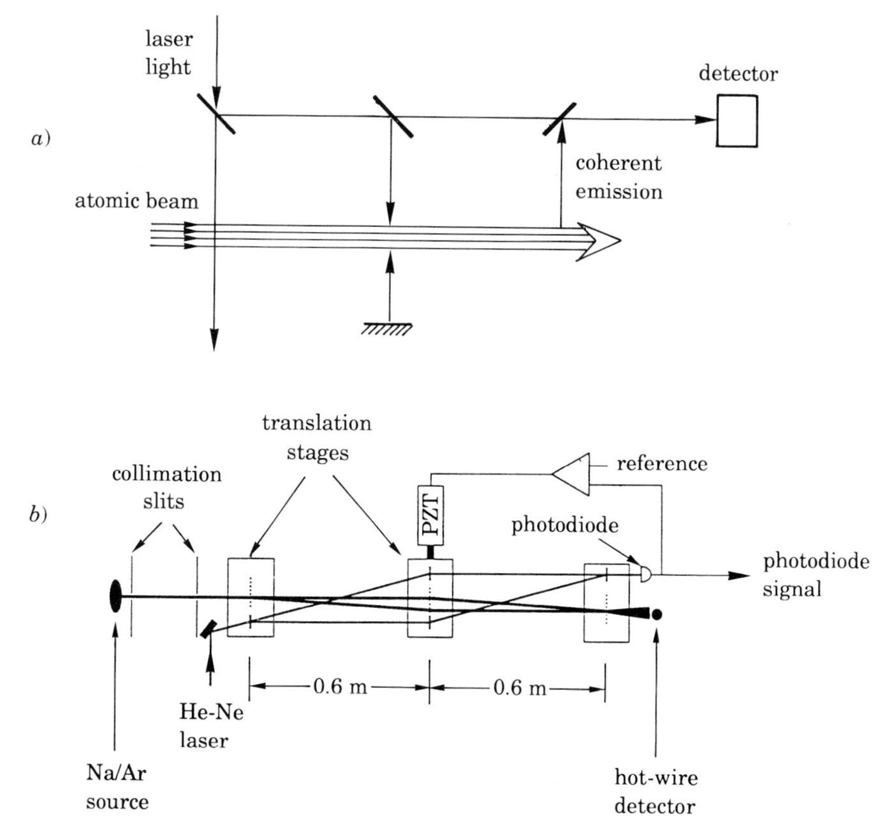

Fig. 8. – Scheme of atomic-optical interferometer (*a*)) and atomic interferometer (*b*)).

the system should be considered as a phasemeter. At high resolution the diffraction of the particles should be taken into account. As was mentioned before, the quantum-mechanical scattering calculation gives the same result as that of the classical analysis.

The second type of atomic interferometer is absolutely based on the wave features of matter. The main problem for the creation of such an interferometer is connected with obtaining coherent particle beams. The principles of different atomic interferometers using the standing optical waves as a coherent splitter of particles were considered in [15] (fig. 8*b*)). The basic idea of an atomic interferometer is close to that of a neutron interferometer. In [22] the use of a mechanical selector and a mechanical diffraction lattice plate with cuts and a period of 0.2 μm for the creation of an atomic interferometer is shown. Recently the successful operation of such an interferometer with separated coherent atomic beams was demonstrated in [23].

1·4. *Time Fourier high-resolution spectroscopy.* – The method of phase synchronization of lasers is widely used in superhigh-resolution spectroscopy and for the creation of optical clocks and optical frequency standards. It permits to transfer the frequency characteristics from a very stable laser to another one. The phase synchronization of lasers with different frequencies gives the possibility of increasing the set of synchronized frequencies. It permits to decrease the time width of the pulse. A possible scheme for obtaining an attosecond pulse is shown in fig. 9. Our interest in the generation of very short pulses is connected with its application for developing a new method of superhigh-resolution spectroscopy.

All the known methods are based on the resonant interaction between particles and optical fields, hence a highly monochromatic laser radiation is essential in these methods. The observation of resonances is carried out for a time which is much longer than the oscillation period of the quantum transitions under investigation.

A new spectroscopic method based on using pairs of pulses of radiation with a line width which is broader than the measured quantum transition frequency interval, but with a stable interpulse time, has been considered in [24]. The duration of the interaction between a particle and a single pulse is shorter than the oscillation period of the quantum transition. Hence, the result of the atomic interaction with two short pulses is defined by the phase of the free oscillations of a dipole moment at the time of arrival of the second pulse. In other words, the particle makes the transition from one energy level into the other one synchronously with the atomic oscillations. The synchronized quantum transitions are very accurately determined in time, and may be applied to direct precision measurements of time and frequency.

The phenomenon of synchronization of quantum transitions is not critically dependent on the nature of the pulse perturbation. We shall explain it by the

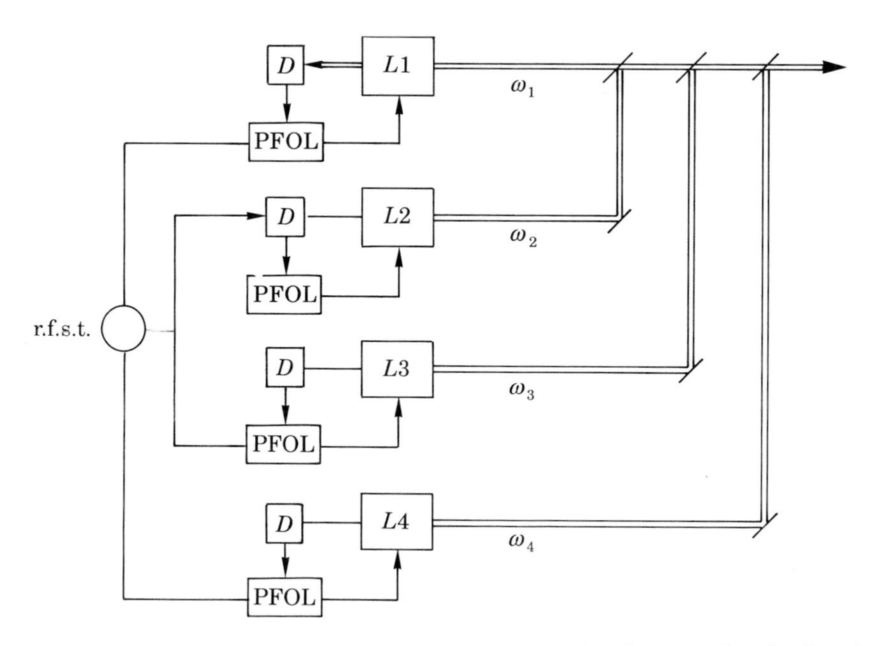

Fig. 9. – Scheme of generation of attosecond pulses using the phase synchronization of different lasers. The pulse repetition rate of each is locked to the frequency r.f. standard(s), D is the detector, PFOL the electronic system of phase-frequency offset lock.

simple example of an atom interacting with two d.c. pulses of an electric field. We write the latter in the form

$$E(t) = Eg(t) + E'\, g(t - T),$$

where $g(t)$ is the pulse shape, and T is the interpulse time. The field-induced dipole moment of a 2-level atom may be written in the form

$$d(t) = a_{21}\, d_{21}^* \exp[-i\omega_{21} t] + \text{c.c.}, \tag{2}$$

where a_{21}, d_{21} and ω_{21} are the probability amplitude, the dipole moment and the frequency of the transition $|1\rangle \rightarrow |2\rangle$, respectively (the atom is considered to be in the state $|1\rangle$ when $t = -\infty$). Both $d(t)$ and d_{21} are assumed to be projections in the field direction. The probability a_{21} is equal to

$$a_{21} = (i/\hbar)\, d_{21} \int_{-\infty}^{t} \mathrm{d}t'\, E(t') \exp[i\omega_{21} t']. \tag{3}$$

If the duration τ of the atomic interaction with a single pulse is much shorter than the period of atomic oscillations $2\pi/\omega_{21}$, then the function $g(t)$ in (2) and (3) may be replaced by $\tau\delta(t)$, where $\delta(t)$ is Dirac's delta-function. This means that we probe the atomic-dipole moment at times $t = 0$ and $t = T$. For $t > T$ we find from (2), (3) the dipole moment as the superposition of dipole moments excited at $t = 0$ and $t = T$:

$$d(t) = id_{21}^{*}\,\tau\Omega_{21}\exp[-i\omega_{21}t][1 + (E'/E)\exp[i\omega_{21}T]] + \text{c.c.}, \tag{4}$$

where $\Omega_{21} = Ed_{21}/\hbar$, and the probability of the transition $|1\rangle \rightarrow |2\rangle$ is

$$|a_{21}|^2 = \tau|\Omega_{21}|^2[1 + (E'/E)^2 + 2(E'/E)\cos(\omega_{21}T)]. \tag{5}$$

As we see from (4) and (5), both $d(t)$ and $|a_{21}|^2$ have maxima at times T that are integer multiples of the period $2\pi/\omega_{21}$. Thus, when an atom interacts with two short pulses, the quantum transition $|1\rangle \rightarrow |2\rangle$ may be synchronized with its natural oscillations. The effect is also shown to be very distinctive with equal pulse amplitudes ($E' = E$).

Instead of d.c. pulses of electric or magnetic fields, a.c. pulses may be used as well. The case of a.c. pulses is of interest for optical transitions. It is clear that a carrier frequency of an a.c. pulse may take any value satisfying the inequality $\omega \gg \tau^{-1}$, ω_{21}. There are only two pulse parameters of principal importance, *i.e.* the duration of a perturbation pulse, which should be shorter than an atomic-oscillation period, and the interpulse time, which should be stable and a multiple of the period.

At present, advanced methods for the generation of ultrashort laser pulses provide the possibility of obtaining light pulses of $(10 \div 100)$ fs duration. This allows one to carry out experiments for the synchronization of IR and FIR quantum transitions and direct measurements of time with an accuracy of the order of $10^{12} \div 10^{13}$. As the light frequency $\omega \gg \omega_{21}$, the excitation of the quantum transition is realized by the two-photon Raman process. The consideration with an analysis of the Raman interaction between an atom and a single light pulse of any intensity was done in [25].

In the method under consideration the light source has a very large line width, and the value of its carrier frequency is unimportant. Only the stability of the pulse delay times is significant. There are at least two possibilities to realize the spectrometer. In the first case, an optical delay line may be used to form the terminating pulse. The delay time $T = L/c$ (where L is the delay line length) may be changed continuously with a high accuracy. Unfortunately, the

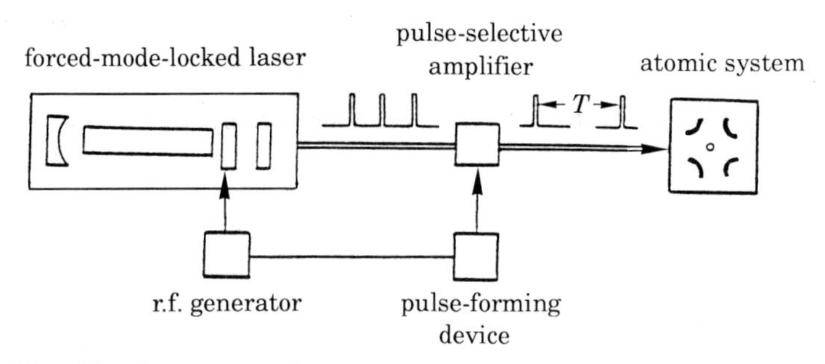

Fig. 10. – Time Fourier superhigh-resolution laser spectrometer.

absolute accuracy of the delay time measurement is limited here by the length measurement accuracy. If the transition frequency is known, then, according to the new definition of the metre, the delay time T and the length L may be directly measured. The second possibility is given by the laser spectrometer scheme shown in fig. 10. This spectrometer allows one to measure the absolute values of both the delay time T and the transition frequency ω_{12}. The spectrometer is based on the use of ultrashort pulses generated by the force-mode-locked laser. The pulse repetition frequency is determined by the frequency Ω of the r.f. generator, which controls the intracavity amplitude modulator. The optical pulse amplifier locked to the r.f. generator allows one to form time-separated pulses. Their delay time will be a multiple of the laser interpulse time, *i.e.* $T = n\omega_{21}^{-1}$. Time-tuning of T can be realized by tuning the frequency Ω. Such a system may serve as a standard for time and frequency simultaneously. Since the frequency ω_{12} is stable, it is possible to stabilize the delay time T and consequently the r.f. generator frequency Ω. The method of stabilization of the interpulse time has already been considered.

2. – Second-order Doppler-free spectroscopy.

When the homogeneous width of a line is comparable with the line shift caused by the second-order Doppler effect (SODE), the influence of the SODE on the shape of the saturated-absorption line is visible. Here, the influence of the SODE upon resonance under the transit time conditions is essential. The resonance frequency of a particle and, hence, its interaction with a field depend not only on the projection of the velocity v_z but also on the projection of v_r. There exists a mechanism of inhomogeneous broadening of the Lamb dip. If the transit effects are not important (the so-called transit parameter $\beta = \Gamma\tau_0 \gg 1$, where Γ is the homogeneous line half-width, $\tau_0 = a/v_0$ is the transit time through the beam for a particle with average thermal velocity v_0, a is the beam radius), the contribution of particles to the resonance is proportional to the density of the particles with a given velocity. Therefore, substituting Ω for $\Omega + \Delta v_r^2/v_0^2$, where $\Delta = (1/2)\,\omega v_0^2/c^2$ is the second-order Doppler shift for the particle with thermal velocity, with the formula describing the saturation resonance

$$\alpha_n = \alpha\left[1 - G\left(1 + \frac{\Gamma^2}{\Gamma^2 + \Omega^2}\right)\right], \qquad G \ll 1,$$

and averaging over the velocity distribution $W(v_r) = 2(v_r/v_0)\exp[-v_r^2/v_0^2]$, we get a simple expression for the line shape

$$\text{(6)} \qquad \alpha_n = \alpha_0(1 - G(1 + 0.25))\int_0^\infty [1 + (\delta + \overline{\omega} v^2/v_0^2)^2]^{-1}\, W(v_r)\,\mathrm{d}v_r\,,$$

where $\delta = \Omega/\Gamma$, $\overline{\omega} = \Delta/\Gamma$. The integral in (6) can be written in the form

$$\int_0^\infty [1 + (\delta + x)^2]^{-1} \exp[-x/\overline{\omega}]\, dx \,.$$

At $\overline{\omega} \gg 1$, the resonance is described by the simple formula

$$\alpha_n = \alpha_0 (1 - G(1 + \Gamma/4\Delta)(\pi/2 - \operatorname{arctg}(\Omega/\Gamma))\, R)\,, \tag{7}$$

$$R = 1 \text{ at } \Omega > 0\,, \qquad R = \exp[\Omega/\Delta] \text{ at } \Omega < 0\,.$$

The maximum of the resonance is shifted towards the red region. Its frequency is $\Omega_m = -(\Delta\Gamma/\pi)^{1/2}$. The resonance half-width is equal to $\Delta \ln 2$. A strongly expressed asymmetry attracts one's attention. When Γ decreases, the resonance maximum is shifted towards the radiation frequency of an immovable particle. Under the transit time conditions, the influence of the SODE upon the resonance shape has been analysed in [26, 27]. The selection of slow particles strongly decreases the influence of the SODE. It has been experimentally shown in [28] that, in a low-pressure gas when the free-path length of the particles considerably exceeds the transverse-field dimensions, the main contribution to the saturation resonance is provided by the slow particles whose transverse velocity $v_r \simeq a\Gamma$ is much less than the average thermal one. Superhigh-resolution spectroscopy in methane with cold molecules was done in [28]. A record of the recoil doublet of the hyperfine magnetic 6-5 component of the $F_2^{(2)}$ line in methane is shown in fig. 11. The diameter of the laser beam was $\simeq 0.5$ cm. The main contribution to the saturation resonance is provided by molecules with effective temperature $\simeq 0.1$ K. The optical selection is particularly visible when using derivative resonances. At $\beta \gg 1$, the velocities of the particles making the main contribution to the resonance satisfy the condition $v = v_0$. Therefore, the contribution of the SODE to the shift and broadening of the resonance will be equal to about $\Delta\beta^2$. This confirms the fact that, at $\beta \ll 1$, the influence of the SODE upon the shift and broadening of the resonance becomes very weak. The effective temperature of the particles responsible for the resonance is $T_{\text{eff}} = T_0\beta^2$.

At $\beta \ll 1$, the resonance shape is described by

$$\alpha = \frac{\alpha_0}{4} G\beta^2 \int_0^\infty \int_0^\infty \frac{d\xi\, d\eta \exp[-\xi - \eta](A\cos(\Omega\xi/\Gamma) - \Delta\beta^2/\Gamma \sin(\Omega\xi/\Gamma))}{A^2 + (\Delta\beta^2\xi/\Gamma)^2}\,, \tag{8}$$

where $A = (\xi/2)^2 + (\xi/2 + \eta)^2 + \beta^2$.

Successful results are obtained when using the simple expression

$$\alpha = \alpha_0[1 - G\tau_0^2\Gamma^4/4] \int_0^\infty \frac{W(v)\, dv}{[\Omega + (v/c)^2\, \omega/2]^2 + (\Gamma + v/a)^2}\,. \tag{9}$$

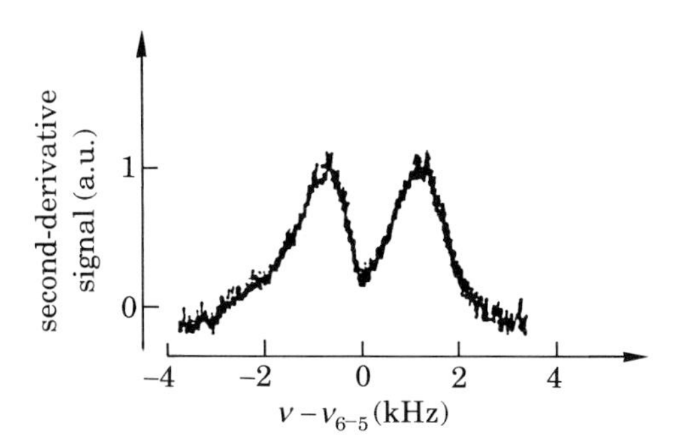

Fig. 11. – The record of the saturation resonance of the hyperfine magnetic 6-5 component of the $F^2_{(2)}$ line in methane by using cold CH_4 molecules ($a = 0.25$ cm, $P_{CH_4} = 4 \cdot 10^{-5}$ Torr, $P = 0.5$ μW, $T_{eff} \simeq 0.1$ K).

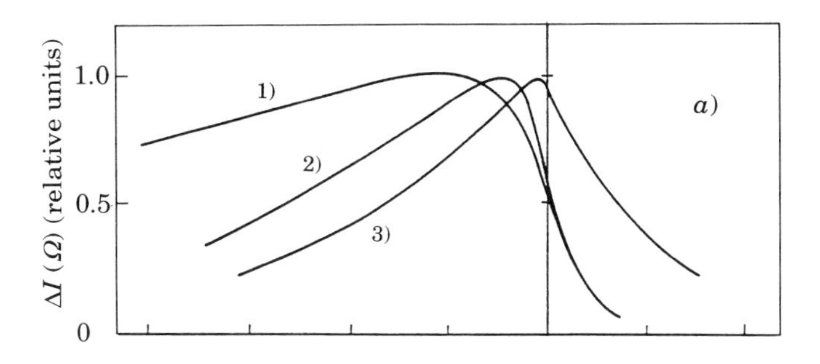

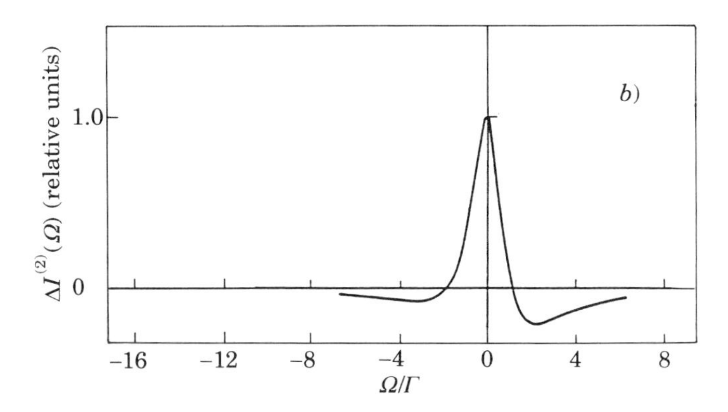

Fig. 12. – Calculated shape of the saturated resonance in methane taking into account the influence of second-order Doppler effect: *a*) $\Gamma/\Delta_0 = 0.025$, $\Gamma \gg \tau_0^{-1}$ (curve 1)), $\Gamma/\Delta_0 = 0.1$, $\Gamma \gg \tau_0^{-1}$ (curve 2)), and $\Gamma/\Delta_0 = 0.1$, $\Gamma\tau_0^{-1} = 0.1$ (curve 3)); *b*) second-derivative signal with $\Gamma\tau_0 = 0.1$, $\Gamma/\Delta_0 = 0.1$.

Figure 12 presents the results of the calculations using (9) and those of the precise ones obtained from (8). The calculations were made for the resonance in methane. In the case of transit time broadening, the shape of the saturated-absorption resonance changes compared with the homogeneous case (curves 1) and 2)). Also, the differences of the widths of the resonance and the second derivative are visible. The half-width of the second derivative is equal to 4 Hz; the shift is equal to 0.006 Hz; $T_{\text{eff}} = 0.02$ K.

Direct observation of the influence of the SODE on the shape of the resonance may be done at $\Gamma \simeq \Delta$. Under typical conditions, $\Delta \simeq 100$ Hz. Therefore, obtaining the resonances with such a small width requires a very low gas pressure, of about 10^{-6} Torr. To overcome inevitable difficulties connected with a strong decrease of the resonance intensity, a special spectrometer with a telescopic beam expander has been developed in [29] (fig. 13). The laser cavity was formed by six mirrors. The beam diameter in the cell was equal to 30 cm, and its length was of 800 cm. A narrow radiation line width of about 1 Hz and a

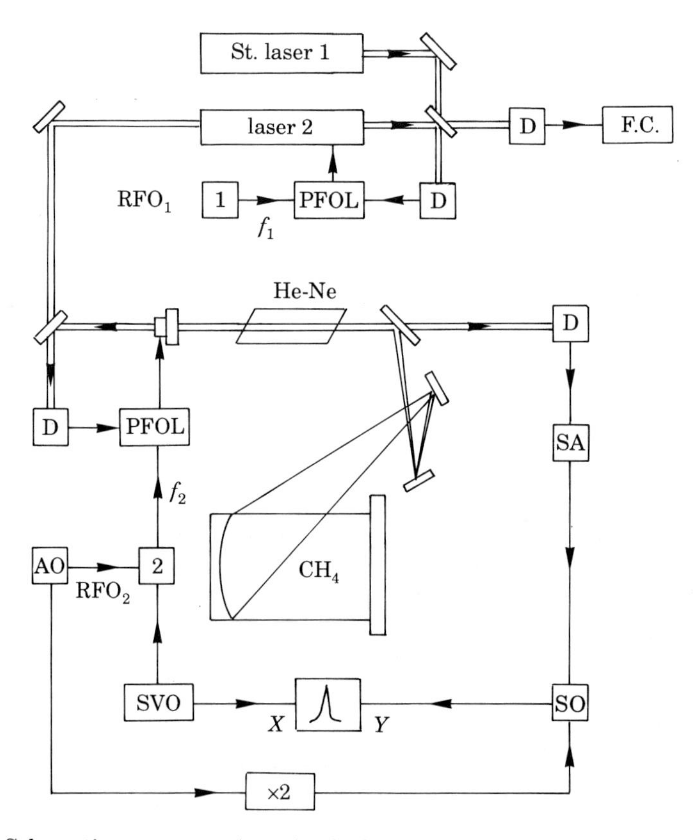

Fig. 13. – Schematic representation of a high-resolution laser spectrometer at 3.39 μm (D: photodetector; PFOL: electronic system of phase-frequency offset lock; SA: audio oscillator; SVO: sawtooth voltage oscillator; F.C.: frequency control; RFO: r.f. oscillator).

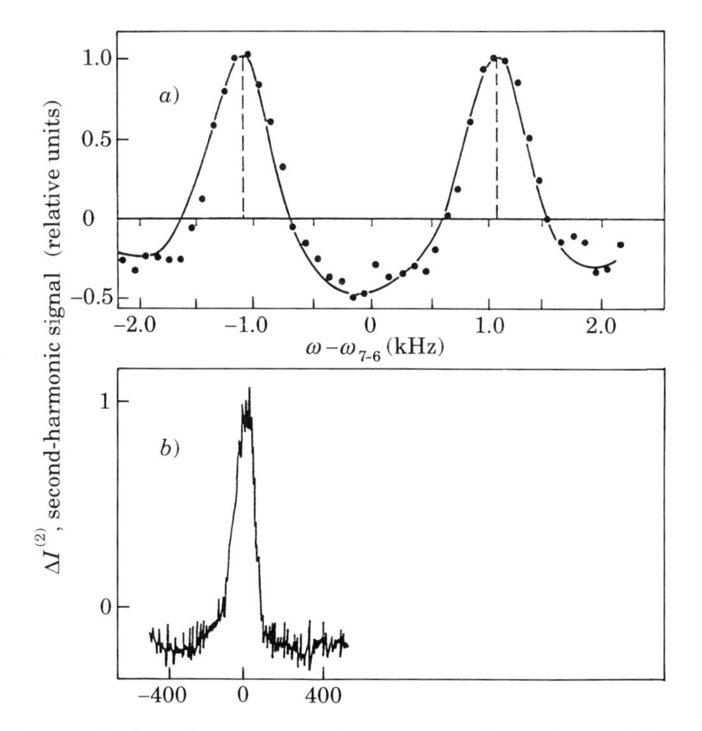

Fig. 14. – The record of recoil components in methane using a telescopic beam expander: *a*) methane pressure 10 μTorr, modulation frequency 230 Hz, deviation amplitude 200 Hz; *b*) methane pressure 5 μTorr, modulation frequency 60 Hz, deviation amplitude 100 Hz, record time 20 min.

long-term frequency stability of 10^{-14} enabled one to record a spectrum for a long time. Figure 14*a*) shows the record of the resonance at a pressure of 10^{-5} Torr. Also, the calculated shape of the second-derivative signal taking account of the SODE is shown there. The influence of the SODE resulted in the asymmetry of the resonance wings of the recoil doublet. The narrowest resonance has been obtained at a pressure of $\simeq 5 \cdot 10^{-6}$ Torr (fig. 14*b*)). The elimination of the amplitude radiation noise due to the misalignment of the cavity should enable one to obtain a resonance width of about 10 Hz.

Under the described conditions in which the recoil doublet of the $F_2^{(2)}$ line of methane was resolved and the influence of the second-order Doppler effect was strongly decreased, it was important to measure the absolute frequency of the centre of the methane line. The direct absolute frequency measurement of the central component was done using an optical-clock system [30] and a hydrogen standard (fig. 15). Few measurements have been done using the different regimes of operation of an He-Ne/CH_4 laser. The values $\nu_{CH_4} = (88\,376\,181\,600.7 \pm 0.5)$ kHz and $\nu_{CH_4} = (88\,376\,181\,600.4 \pm 0.1)$ kHz have been obtained with the help of the resonances which are shown in fig. 11 and in

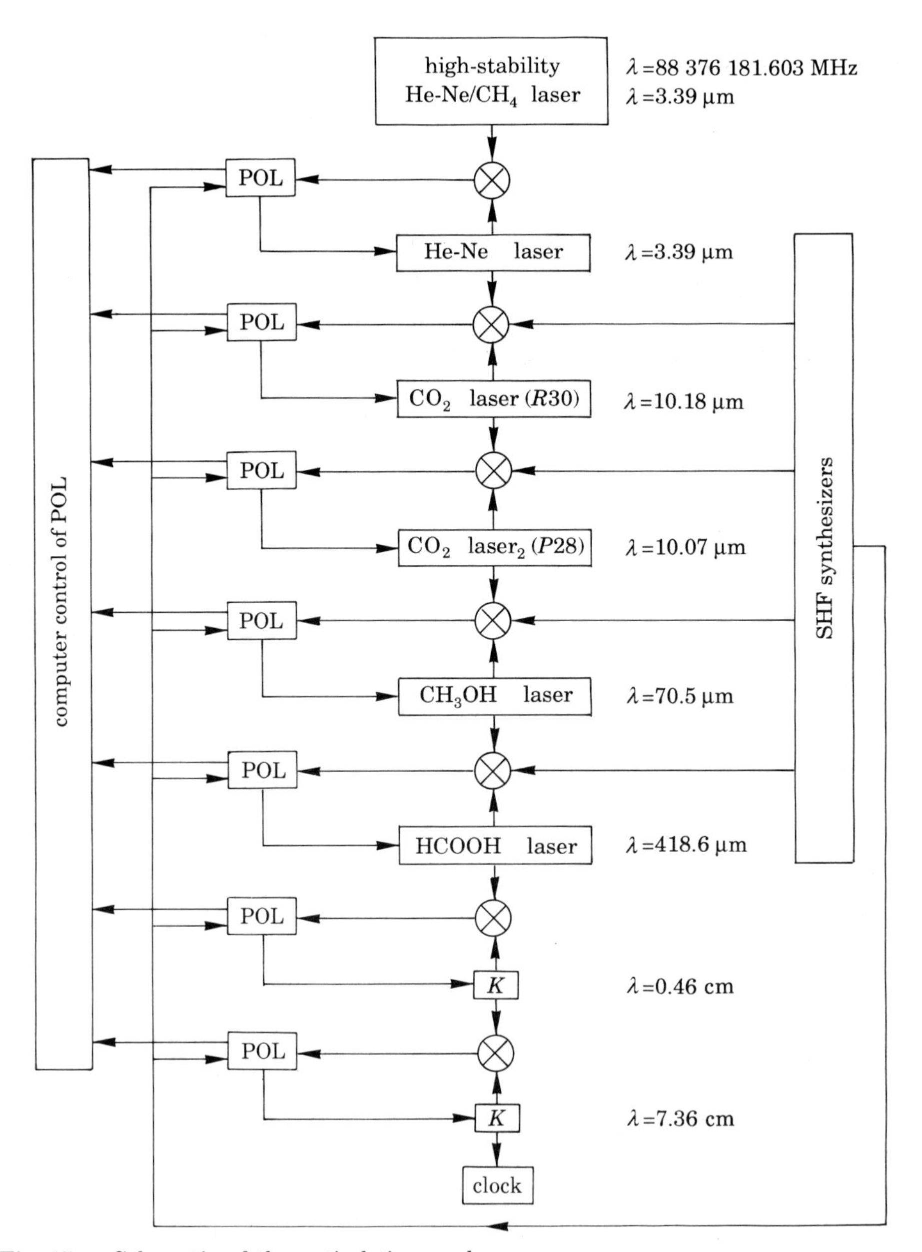

Fig. 15. – Schematic of the optical time scale.

fig. 14*a*), respectively. The histogram of the direct comparison of the frequencies of the He-Ne/CH_4 laser with the telescopic beam expander and an H maser is shown in fig. 16. The histogram shows the real possibility of absolute frequency measurement of the laser with an accuracy of $\simeq 10^{-14}$. In both cases the

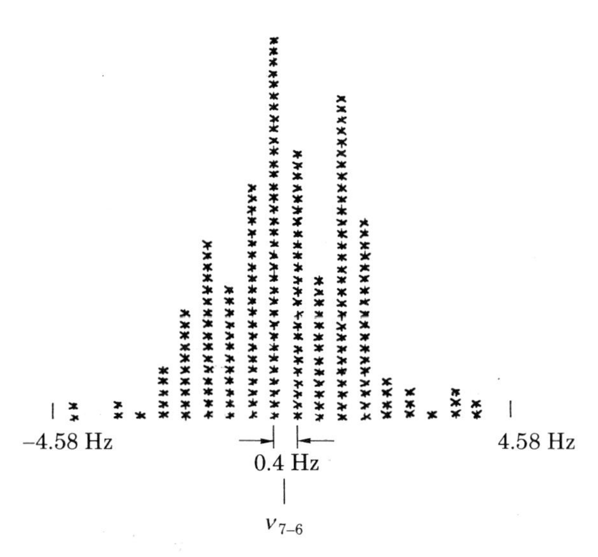

Fig. 16. – The histogram of the direct comparison of the frequencies of He-Ne/CH_4 laser with the telescopic beam expander and H maser, $\nu_{7\text{-}6} = (88\,376\,181\,600\,480 \pm 10)$ Hz.

measurement error was due to the accuracy of the definition of the absolute value of the frequency of the hydrogen standard.

3. – Laser stabilitron.

Recently theoretical attention has been given to the production of squeezed photon states [31] and a decrease in noise level below the quantum limit has been reported [32]. In the past a new method was considered which should not only provide a substantial reduction in the common sources of amplitude noise,

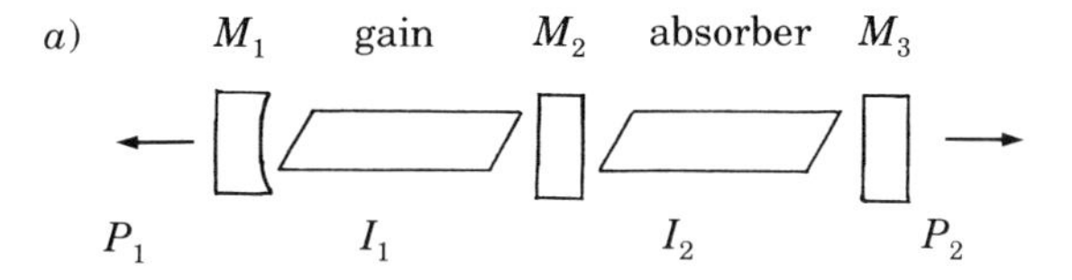

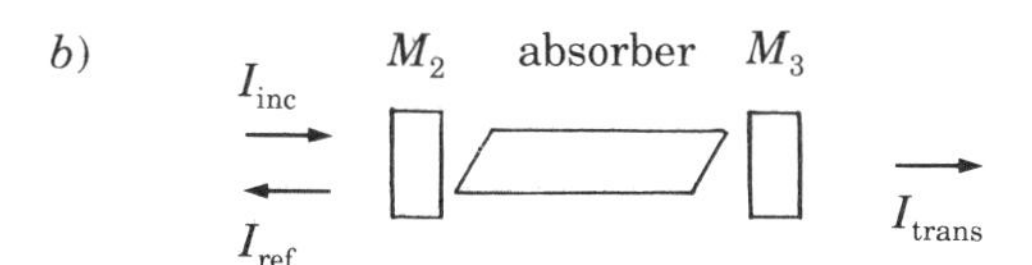

Fig. 17. – *a*) The coupled-cavity method to produce noise reduction, *b*) effective laser mirror composed of cavity 2.

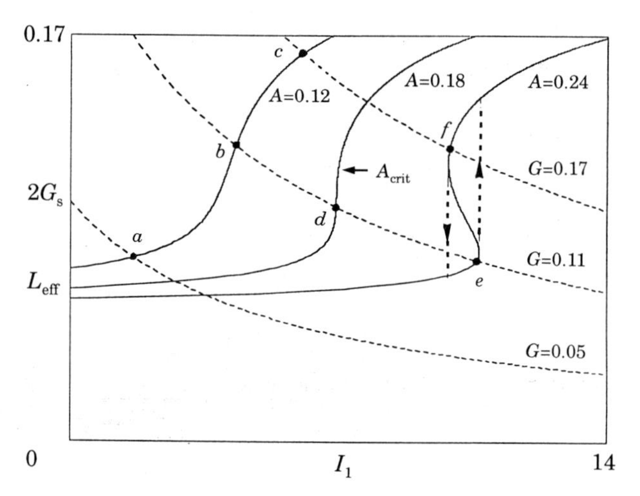

Fig. 18. – Effective loss L_{eff} (solid curves) and saturated gain $2G_s$ (dashed curves) as a function of intensity I_1 in the laser cavity for different values of the absorption A; $g_1 = 0.18$, $g_2 = 1$, $f = 0.021$, $T = 0.002$.

but also may reduce the shot noise below the level expected from the quantum effect. This type of device, which we will refer to as a laser stabilitron, involves a three-mirror system in which two coupled cavities are produced. One cavity contains an amplifier, the other contains a saturable absorber. Due to nonlinear effects, there are two regions of operation in the system: One of these is charac-

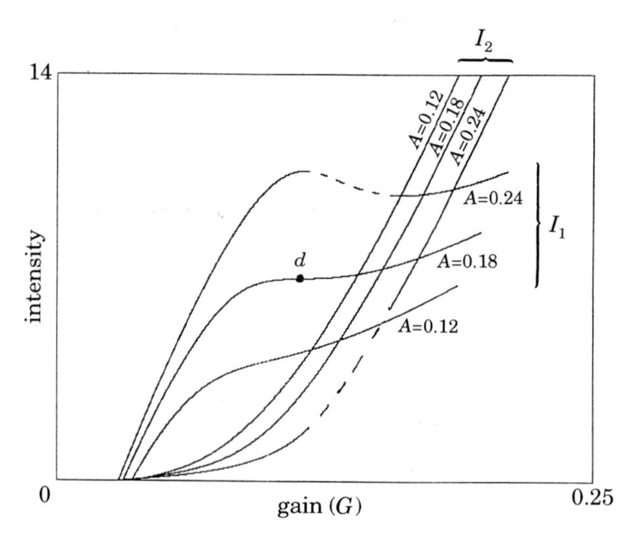

Fig. 19. – Intensities I_1, I_2 as a function of unsaturated gain for different values of A. The optimum low-noise condition (point d) corresponds to point d in fig. 18. The dashed region on the curves for $A = 0.24$ corresponds to the region of instability between points e and f in fig. 18; $g_1 = 0.18$, $g_2 = 1$, $f = 0.021$, $T = 0.002$.

terized by hysteresis and instability. The other is stable and should permit strong suppression of fluctuations in the laser intensity.

A schematic diagram of the method is shown in fig. 17. The amplifying and absorbing media are separated by a highly reflecting mirror M_2, which divides the system into two coupled resonators. If the resonant frequencies of two cavities are different, the coupling is small. However, when the frequencies of two resonators coincide, strong coupling occurs. Then saturation of the absorber by a strong field leads to an increase in output intensity through mirror M_3. The strong negative-feedback mechanism between the field intensity in cavity 1 and the absorption in cavity 2 works as follows: Increasing the gain and intensity in cavity 1 leads to increased field intensity in cavity 2 and, correspondingly, to a decrease in absorption. That, in turn, leads to increased transmission from cavity 1 to 2—hence, an increase in the loss of the laser and a decrease of the laser intensity in cavity 1. Figure 18 shows the dependence of the effective losses as a function of the intensity in cavity 1. The optimum low-noise condition corresponds to point d. The intensity dependences on the gain are shown in fig. 19. The dynamic and noise features of the stabilitron have been studied. The noise reduction in cavity 1 is efficient when the length of the first cavity is much longer than the length of cavity 2. The results of the quasi-classical consideration of the dispersion of photon numbers are shown in fig. 20.

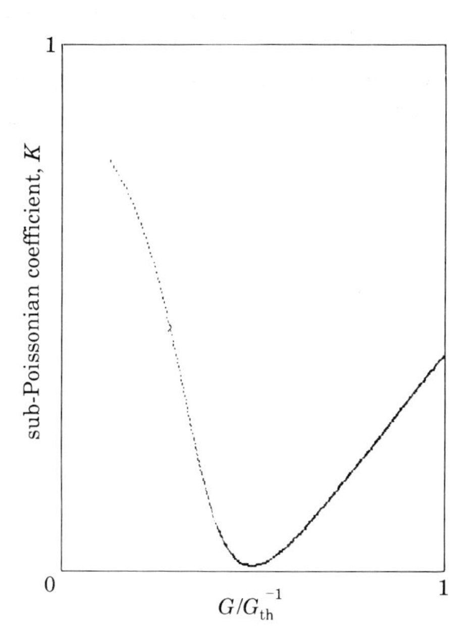

Fig. 20. – Dependence of the sub-Poissonian coefficient $K = \langle \Delta N^2 \rangle / N$, where $\langle \Delta N \rangle^2$ and N are the dispersion and the number of photons in the first cavity, on gain. The minimum of K corresponds to point d in fig. 19.

Near point d a strong reduction of the dispersion of photon distribution takes place. This method can be used to decrease the noise intensity of most known lasers.

REFERENCES

[1] J. R. Murray and A. Javan: *Mol. Spectrosc.*, **42**, 1 (1972).
[2] V. P. Chebotayev and V. A. Ulybin: *Appl. Phys. B*, **31**, 3 (1983).
[3] V. P. Chebotayev and V. A. Ulybin: *Opt. Spectrosk.*, **62**, 351 (1987).
[4] D. Kleppner, H. C. Berg, S. B. Crampton, N. F. Ramsey, R. F. C. Vessot, H. E. Peters and J. Vanier: *Phys. Rev.*, **138**, 972 (1965).
[5] V. P. Chebotayev and V. A. Ulybin: *Appl. Phys. B*, **49**, 361 (1989).
[6] N. F. Ramsey: *Molecular Beams* (Oxford University Press, New York, N.Y., London, 1955).
[7] E. V. Baklanov and V. P. Chebotayev: *Ž. Ėksp. Teor. Fiz.*, **60**, 551 (1971).
[8] V. P. Chebotayev: *Optical frequency standards*, in *Metrology and Fundamental Constants, Proc. S.I.F.*, LXVIII Course (North-Holland, Amsterdam, 1980), p. 623.
[9] E. V. Baklanov and V. P. Chebotayev: *Kvantovaya Electron.* (*Moscow*), **2**, 606 (1975).
[10] E. L. Hahn: *Phys. Rev.*, **80**, 580 (1950).
[11] E. V. Baklanov, B. Ya. Dubetsky and V. P. Chebotayev: *Appl. Phys.*, **9**, 201 (1976).
[12] E. V. Baklanov, B. Ya. Dubetsky and V. P. Chebotayev: *Appl. Phys.*, **9**, 171 (1976).
[13] V. P. Chebotayev and B. Ya. Dubetsky: *Appl. Phys.*, **18**, 217 (1979).
[14] V. P. Chebotayev: *Appl. Phys.*, **15**, 219 (1978).
[15] V. P. Chebotayev, R. Y. Dubetsky, A. P. Kasantsev and V. P. Yakovlev: *J. Opt. Soc. Am. B*, **2**, 1791 (1985).
[16] E. V. Baklanov *et al.*: *Ž. Ėksp. Teor. Fiz.*, **78**, 482 (1979).
[17] Ch. J. Borde: *Phys. Lett. A*, **140**, 10 (1989).
[18] M. Kasevich and S. Chu: *Appl. Phys. B*, **54**, 321 (1992).
[19] F. Riehle, A. Witte, T. Kisters and J. Helmcke: *Appl. Phys. B*, **54**, 333 (1992).
[20] V. P. Chebotayev: *Super high resolution spectroscopy*, in *Laser Handbook*, Vol. **5**, edited by M. Bass and M. L. Stitch (North-Holland, Amsterdam, Oxford, New York, N.Y., Tokyo, 1985).
[21] S. N. Bagayev, A. S. Dychkov and V. P. Chebotayev: *Appl. Phys.*, **15**, 209 (1978).
[22] D. W. Keith, M. L. Schattenburg, H. I. Smith and D. E. Pritchard: *Phys. Rev. Lett.*, **61**, 1580 (1988).
[23] D. W. Keith, C. R. Ekstrom, Q. A. Turchette and D. E. Pritchard: *Phys. Rev. Lett.*, **66**, 2693 (1991).
[24] S. N. Bagayev and V. P. Chebotayev: *Usp. Fiz. Nauk*, **148**, 143 (1981).
[25] V. P. Chebotayev: *Pis'ma Ž. Ėksp. Teor. Fiz.*, **49**, 429 (1989); V. P. Chebotayev and V. A. Ulybin: *Appl. Phys. B*, **50**, 1 (1990).
[26] E. V. Baklanov and B. Ya. Dubetsky: *Kvantovaya Electron.* (*Moscow*), **2**, 2041 (1975).

[27] S. N. BAGAYEV, A. K. DMITRIYEV, YU. V. NEKRASOV and B. N. SKVORTSOV: *Pis'ma Ž. Ėksp. Teor. Fiz.*, **50**, 173 (1989).
[28] S. N. BAGAYEV, A. E. BAKLANOV, V. P. CHEBOTAYEV and A. S. DYCHKOV: *Appl. Phys. B*, **48**, 31 (1989).
[29] S. N. BAGAYEV, V. P. CHEBOTAYEV, A. K. DMITRIYEV, A. E. OM, Y. V. NEKRASOV and B. N. SKVORTSOV: *Appl. Phys. B*, **52**, 63 (1991).
[30] E. GIACOBINO, T. DEBUISSCHERT, A. HEIDMANN, J. MERTZ and S. REYNAUD: in *Proceedings of the NICLOS, Bretton Woods, N.H., June 19-23, 1989* (Academic Press, New York, N.Y., 1989), p. 180.
[31] H. J. KIMBLE: in *Atomic Physics II*, edited by S. HAROSCHE, J. C. GAY and G. GRIMBERA (World Scientific, Singapore, 1989), p. 467.
[32] W. R. BENNETT jr. and V. P. CHEBOTAYEV: *Appl. Phys. B*, **54**, 552 (1992).

Frequency-Stabilized Lasers—a Driving Force for New Spectroscopies.

J. L. HALL (*)

Joint Institute for Laboratory Astrophysics
University of Colorado and National Institute of Standards and Technology
Boulder, CO 80309-0440

1. – Introduction and overview.

It is particularly appropriate to consider the evolution of stabilized lasers within the context of a Fermi School concerned with the *Frontiers of Laser Spectroscopy,* as the stabilization of lasers has clearly provided one of the stimulating influences for the development of new spectroscopic methods. It is hoped that this lecture will serve as a useful introduction to this art of stabilized lasers, especially for those who may be joining the sport from other fields. The remainder of this overview section may be regarded as an «executive summary» of the main text.

The first laser stabilization experiments with laser/absorber systems such as $HeNe/CH_4$ and $HeNe/I_2$ led to unprecedented laser stability performance, based on intracavity saturated absorption. The use of $HeNe/CH_4$ and CO_2/OsO_4, for example, yielded stability performance in the 10^{-13} domain. Later the added flexibility of experiments with an external cell led to further significant stability improvements. The natural extension of this work is to employ tunable lasers. This approach has dominated the past decade-and-a-half, with frequency stabilization performance ultimately limited by the same systematic shifts that trouble the gas laser work. However, in using broadly tunable sources—such as dye lasers, Ti:sapphire lasers and diode lasers—the first challenge concerns narrowing the laser spectrum and guiding the center frequency to explore interesting resonances. It has become clear that prestabiliza-

(*) Staff Member, Quantum Physics Division, National Institute of Standards and Technology.

tion of the laser on the nonsaturable resonance of a stable cavity is a good strategy: with adequate feedback system design, one can effectively replace the intrinsic noise of the laser with the measurement noise of the stabilizer system. By now sub-hertz frequency control and optical phase locking have been demonstrated with most of these tunable sources. The ready access to modulation of the diode laser can lead to a very simple but impressive source, while the external stabilizer approach [1] is attractive for dye and optically pumped solid-state sources. Current work on amplitude stabilization of the laser pump may lead to reduction of the «intrinsic» solid-state laser noise as well.

With the current explosion of interest in atom trapping techniques we can look forward to major progress in the narrow-line laser/super-sharp absorber high-resolution spectroscopy business. Applications range from atomic clocks to cold-atom collision physics to tests of special relativity. The combination of ultra-stable lasers with cold-atom interferometry will be especially powerful in offering new tests of atomic-charge neutrality and of time-reversal invariance via new limits on atomic electric-dipole moments. Remarkably, a practical instrument for oil and gas prospecting might be based on a laser-diode/atom-interferometric measurement [2] of local g.

2. – The gas laser epoch.

2·1. *Lamb-dip stabilization.* – One of the important approaches to laser stabilization came up early in the history of gas laser development, when the technical advances in reducing optical losses allowed saturation effects to play a larger role. This provided researchers with the sub-Doppler intracavity saturated-emission resonance which came to be known as Lamb's central tuning dip, in recognition of Lamb's fundamental contributions to the theory [3] of the effect. A clear physical picture was provided [4] by BENNETT jr., who discussed the saturation effect in terms of «holes» «burned» into the Gaussian velocity distribution by the strong fields. The saturation of the amplifying medium would be maximized at the line center tuning, as there both cavity running waves would appear to have the same frequency in the rest frame of a single-velocity group. These zero-axial-velocity atoms would then interact with increased strength, but still would yield less stimulated emission than in the detuned case where two separate velocity groups could contribute to the output power.

The width of the power dip at the central tuning was fixed in part by collisions (≈ 10 MHz/mbar), and in part by natural decay, augmented by the broadening effect of the intracavity laser transition rates themselves. Servo techniques enabled one to lock the laser frequency to the center of the power dip with an attainable frequency stability $\approx 10^{-10}$ or better, limited mainly by vi-

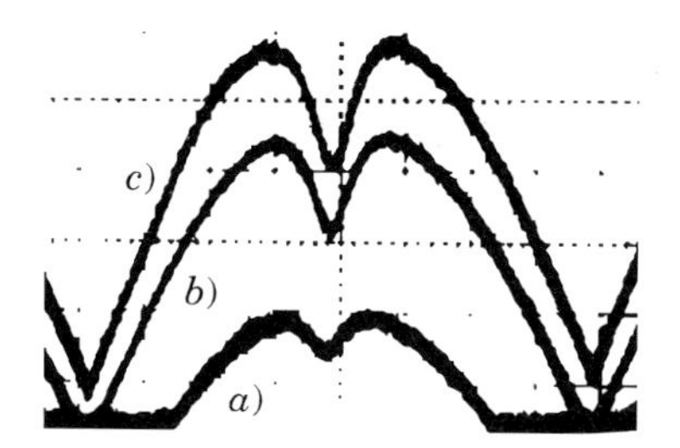

Fig. 1. – Lamb's (central tuning) dip in pure ^{20}Ne laser. Laser operated by r.f. discharge in 0.12 Torr (0.16 mbar) of pure ^{20}Ne. Frequency increases to right. Cavity fsr = 465 MHz. Powers increase from *a*) 1.4 μW, *b*) 4.2 μW, *c*) 5.6 μW. Weak asymmetry to high-frequency side probably due to small amount of ^{22}Ne impurity. Lamb dip widths are about *a*) 30 MHz, *b*) 36 MHz, *c*) 39 MHz.

brations. The long-term stability was compromised by the gradual pressure decrease of the HeNe gain cell, which reduced the pressure-induced blue shift associated with Ne-He collisions.

An important advance in this field was the development by BENNETT and KNUTSON [5] of a new, single-gas operating environment for the neon laser at 1.15 μm. The total pressure was reduced 15-fold to ~ 0.2 Torr which in turn provided a saturation feature of much higher contrast and sharpness (see fig. 1). The pressure shift of this line was reduced by the low-pressure operation, and removed as an important contributor to long-term drift by our easy ability to measure changes in the pressure of the one-component active gas. Interestingly, atomic physics also came to our advantage to make the shift intrinsically small, (-1.4 ± 0.8) MHz/mbar, as was eventually measured by MAGYAR in some careful experiments [6].

Certainly a stability and reproducibility in the $< 10^{-12}$ range would have obtained if this option had been strongly pursued [7]. CHEBOTAYEV [8] pointed out that adding hydrogen improved the power of the spectral doublet at 1.15 μm, and discussed measurement of the speed of light via interferometry and optical heterodyne detection of the 51 GHz beat frequency [9].

2·2. *Intracavity saturated absorption—atoms and molecules.* – In 1967 an important new idea was tried: LEE and SKOLNICK [10] and LISITSYN and CHEBOTAYEV [11] showed that it was attractive to use two separate cells in the laser cavity. One cell could contain a HeNe mix to give good gain and output power. The other would contain an absorber gas under conditions more favorable to serving as a sharp, stable reference transition. In these two experiments, pure neon was the absorbing reference gas. This choice ensured the needed wavelength coincidence, but required also a weak discharge to produce population in the neon absorbing level. Almost instantly many workers realized that molecules would be interesting as absorbers because their rich spectra helped ensure the necessary spectral overlap with the laser transition. Furthermore, molecules can offer sharp absorptions beginning from the ground state. The attractive system of HeNe oscillating at 3.39 μm, saturating the $P(7)$ line in methane (CH_4) was introduced at basically the same time by BARGER and HALL [12] in the U.S. and by CHEBOTAYEV and his associates in Novosibirsk [13]. Frequency reproducibility of 10^{-11} was obtained along with stabilities in the 10^{-13} range in the very early experiments! This HeNe-plus-methane system represents, even now, one of the best-performing optical frequency standards known, based on the advanced techniques to be discussed momentarily.

However, in keeping with our spectroscopic-frontiers theme, and noting that methane spectroscopy has served as a testing ground for spectroscopic ideas since long before the laser epoch, it seems a brief historical note would be appropriate here. In fig. 2 we show the progress over some 35 early years, beginning with the rotational resolution by NIELSEN and NIELSEN [14] in 1935 of the ν_3 fundamental band at 3.39 μm of CH_4. (The modern laser physics student, armed with sophisticated data acquisition hardware/software in his personal computer, may well notice the ordinate axis label in fig. 2*a*) («deflection») and reflect upon the efforts invested by earlier generations of students.) The line from $J'' = 7$ in the ground state to $J' = 6$ in the vibrationally excited state, $P(7)$ in usual notation, looks in fig. 2*a*) quite sharp and reasonable at this resolution level of ~ 1500. However, by the '60's people had become sensitized to the issue of symmetry in spectroscopy, perhaps from the ready observation of symmetry-dependent NMR and ESR spectra of solids. The CH_4 molecule, with its four equivalent hydrogen atoms, might initially be expected to exhibit perfect tetrahedral symmetry. But the nuclear spins, each of $I_i = 1/2$, can be combined in several ways, leading to a total I of 2, 1 and 0. Of course, an overall symmetry is required for the nuclear part of the product-type Born-Oppenheimer wave function, and so we recognize not all states should be available. Also, one might first suppose the energy splittings between these states should be only associated with hyperfine interactions, but HECHT [15] showed in 1960 that a Coriolis-type interaction could lower the symmetry of the excited state in a subtle way, resolving it into 6 components (see fig. 2*c*)). Basically, the $\nu_3 = 1$ triply degenerate vibration can give rise to a superposition in which a circulation is evident, designated by $L = 1$. Coriolis-type forces couple $L = 1$ and $\nu_3 = 1$, giving six levels within the excited state, as shown clearly in fig. 2*b*) by the high-resolution grating spectra (nearing the Doppler limit) by HENRY *et al.* [16]. As indicated, the Coriolis component which interacts with the unshifted HeNe laser is designated $F_2^{(2)}$, and corresponds to one of the $I = 1$ nuclear configurations. Nearby levels E ($I = 0$) and $A_2^{(1)}$ ($I = 2$) can be accessed with a Zeeman-shifted HeNe laser. When tunable lasers became available, PINE used infrared difference frequency generation in a nonlinear crystal ($LiNbO_3$) to make beautiful systematic measurements [17] of this spectrum.

After this digression, we return to fig. 2*d*) which shows the sub-Doppler

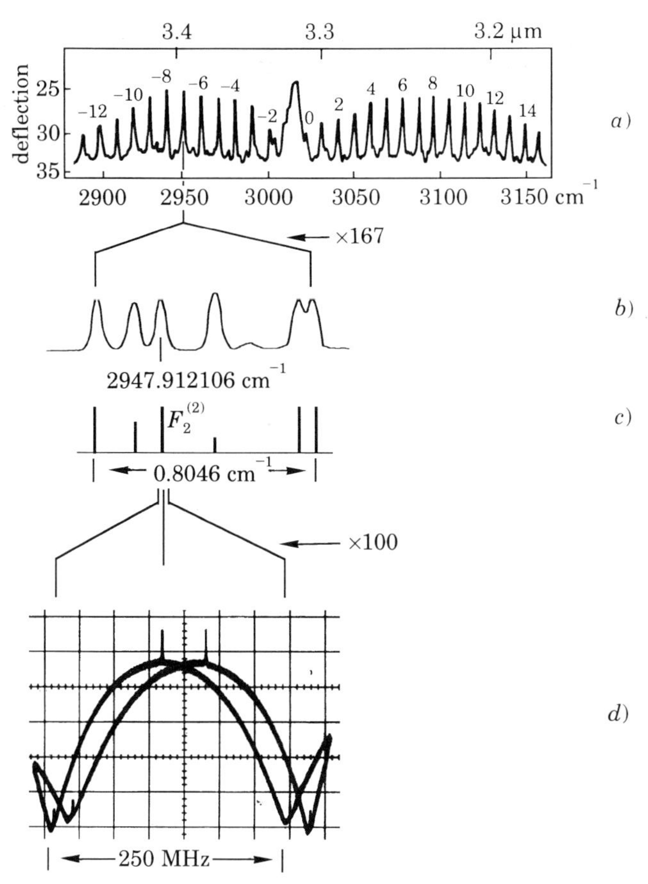

Fig. 2. – 35 years' progress in methane spectroscopy! *a*) shows the CH_4 spectrum obtained in 1935 by NIELSEN and NIELSEN (ref. [14]). The lines appear reasonable at this resolution (~ 1500) and give no hint of internal structure. *b*) shows the small splitting appearing between the different symmetry species, induced by a Coriolis interaction in the vibrationally excited state. From ref. [16]. *c*) shows the symmetry splitting predicted by HECHT (ref. [15]), although the designations follow the convention of Herzberg. *d*) Saturated-absorption peak in CH_4. HeNe laser at 3.39 μm is excited by r.f. discharge. CH_4 cell at 12 mTorr (16 mbar) is located inside laser cavity. HeNe pressure has been adjusted to bring laser frequency into near coincidence with the saturated-absorption feature. Power output is ~ 200 μW and peak contrast is ~ 12%. Peak width is ~ 270 kHz. Cavity free spectral range is 250 MHz. Note cross-over resonances in two-mode region near cusps (ref. [12] (1969)).

saturated-absorption peak obtained almost simultaneously in Boulder and Novosibirsk in 1968. In this figure, the peak shows a contrast of some 12% in the 200 μW output, which rises to 15% at maximum output power (≈ 0.8 mW). (Unfortunate hysteresis in the displacement of the PZT gives a doubled trace.) The peak width of ~ 270 kHz is controlled by power and pressure broadening of a basic linewidth fixed by the limited transit time of crossing the light beam.

Of course, other laser/molecular absorbing systems could be imagined. HANES and DAHLSTROM [18] were quick to introduce the HeNe/$^{127}I_2$ system in the red laser at 633 nm. KNOX and PAO [19] and DESLATTES, SWEITZER and their colleagues [20] pointed out the contrast advantages of a similar transition in the He^{22}Ne/$^{129}I_2$ system. In later times a relatively large number of other systems have been investigated, including HeNe/$^{127}I_2$ at 612 nm, 594 nm, 543 nm and 640 nm [21]. These systems work with the HeNe laser constrained to oscillate on one of several possible neon transitions, through the use of intracavity wavelength selectors such as prisms or special «edge» coatings. Another old favorite [22] is the Ar^+ ion laser operating with $^{127}I_2$. Because of the greater complexity of this laser, and especially its vastly higher power capability, it is typical for this combination to be operated with the absorber contained in a cell external to the laser cavity [23]. Other lasers and molecular absorbers have been found to give attractive performance, the most general system being CO_2 lasers operating with an intracavity cell containing CO_2 at low pressure [13]. The needed spectral coincidence is clearly ensured for every line. The associated 4.3 μm saturated fluorescence signals are widely used for stabilization [24]. Another excellent system uses the CO_2 laser with an external cell, employing molecules such as SF_6, SiF_4 and OsO_4 [25, 26]. The 3.39 μm HeNe laser was used to explore saturated-absorption spectroscopy and laser locking with a variety of methyl halide gases [27].

2'3. *Saturated-absorption wavelength/frequency standards—redefinition of c.* – The use of these various laser systems as reproducible standards of frequency/wavelength has been extensively studied: many variations and intercomparisons have been performed [28], mainly by the several national standards laboratories and a few other institutes. Progress has been monitored by an international committee of experts (Comité Consultatif pour la Définition du Mètre, or CCDM) which has recommended operating conditions within which frequency reproducibility is enhanced [29]. The good general agreement has made it possible also to offer recommended values for the frequency/wavelength under such conditions. In 1983 there was an international redefinition of the meter in terms of the frequency of a stable laser and an adopted value for the speed of light of 299 792 458 m/s, exactly [29]. This idea has worked out well, with the 1992 meeting of the CCDM being able to propose almost 10-fold more precise values for the absolute optical frequencies of these recommended transitions [30]. A number of laboratories now have specialized frequency synthesis chains operating to connect different of these optical frequency references to the primary standard of time/frequency, the cesium clock, which operates at 9.192 631 770 GHz, by definition. As an infrared standard, the CO_2/OsO_4 system in particular has served well, with many lines now being known in absolute frequency at the 50 Hz level [25]. A particularly good approach employs the OsO_4 reference molecules in an external resonator [31]. Also the HeNe/CH_4 system has been measured by many laboratories and an uncertainty in the ~ 200 Hz region seems realistic: $\nu = 88376181600$ kHz [30]. CLAIRON and his colleagues have recently given a much improved frequency [32] for the red 633 nm iodine-stabilized laser: (473 612 214 705 ± 12) kHz.

A significant difference exists between the iodine-based and other molecular systems with respect to the presence of internal structure within the saturated-absorption line. It happens that the main hyperfine energy arises in I_2 from an interaction of the nuclear electric-quadrupole moment with the electric-field gradient of the molecule. These splittings of tens of MHz are readily resolved in the optical sub-Doppler spectra, in contrast to other molecular cases where the absence of quadrupolar splitting leaves only magnetic energies in the tens of kHz range. For a time in the early 1970's there was a productive industry of laser-based measurements of these small hyperfine splittings, stimulating an improvement in the attainable optical resolution. Since the main broadening turned out to be associated with «time of flight» of the molecules as they crossed the interrogating laser beam, it was not surprising that absorption cells of larger transverse aperture were soon introduced. The JILA group pursued this direction, along with CHEBOTAYEV, BAGAEV and their colleagues in Siberia, and BORDÉ and his associates in Paris. It proved possible to resolve the methane hyperfine spectrum [33] optically and soon to observe the consequences of the recoil [34] when a photon was either absorbed or emitted. Figure 3*a*) represents a continuation of fig. 2*d*), where that single transit-limited peak has been explored with an absorption cell of 30 cm aperture. Note the frequency axis scale has been usefully magnified by 10^4. The magnetic hyperfine structure is clearly resolved and the pure of heart can appreciate the incipient spectral doubling associated with recoil in both absorption and emission. After this epoch, the Boulder group turned toward frequency stabilization of tunable lasers operating in the visible. However, CHEBOTAYEV and his associates improved their methane setup (see fig. 3*b*)) and pursued another extremely promising avenue.

2·4. *Optical selection of the slowest atoms.* – This idea [35] is based on enhancing the detection sensitivity to such a degree that one could afford to select molecules which have basically zero velocity along the axis (this is the usual group involved in saturation spectroscopy) but also have near-zero velocity transverse to the axis. Their absorption contributions can be relatively enhanced—perhaps it would be better to say they are not so disastrously *decreased* in amplitude—when the interrogating laser power is strongly reduced. Then only the slowest particles will experience the resonant drive for a time long enough to develop appreciable excitation. Maintaining a collision-free interaction will require severe reduction of the pressure, which also cuts the signal size. However, this velocity selection approach has been used successfully

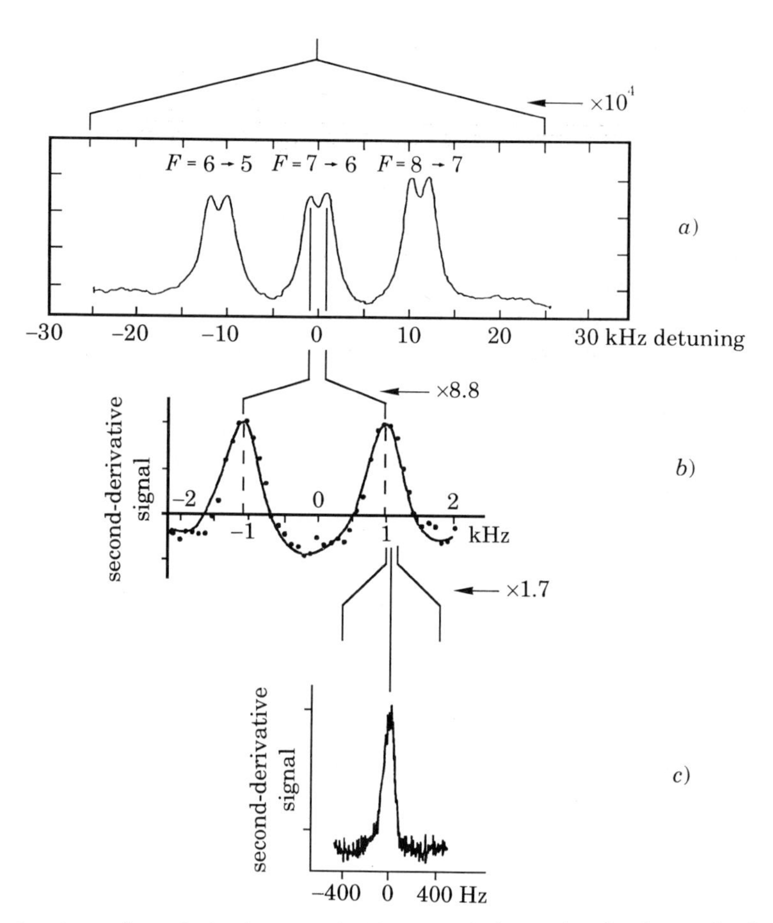

Fig. 3. – Another 10^5 resolution increase in 20 years. *a*) shows the clearly resolved magnetic hyperfine structure of the methane $P(7)$ line. Other, weaker transitions are observable in the next ~ 100 kHz lower-frequency interval. Each of the hfs peaks shows incipient resolution into doublet structure, due to presence of both absorption (red peak) and emission (bluer peak) in the saturated-absorption process (from ref. [34]). *b*) shows linewidths of ~ 700 Hz for the second derivative of the central methane hfs component. This resolution level clearly separates the two recoil components (from ref. [35]). *c*) Second-order Doppler-free spectral line. The «emission» recoil component is shown here with approximately 50 Hz linewidth, representing a line «Q» of $2 \cdot 10^{12}$, ~ 20-fold sharper than would be given by thermal molecules. The effective kinetic temperature of the resonating molecules is < 1 K, and the relativistic time dilation is below 1 Hz—at 88 THz! (From ref. [35].)

in Siberia (one wonders if their reduced ambient temperatures imply substantially larger fractions of slow molecules!) with the added advantage that the extended interaction time leads to surprisingly narrow resonances for a given size laser beam. A 1988 preprint by BAGAEV and CHEBOTAYEV presented a 1 kHz linewidth which they had obtained—at 88 THz!—, about 20-fold smaller than the scaling studies [36] would predict for a 5 mm diameter beam. This 20-fold reduced velocity means an effective temperature below 1 K, and a relativistic time dilation shift below 1 Hz. So the stability, reproducibility and accuracy capability of this laser system can be in the $\sim 10^{-14}$ region! This work has been further improved [35], leading to the data shown in fig. 3*c*): a 50 Hz optical resonance linewidth at 88 THz! The aperture here was increased to 30 cm, so the cooling effect is more than a factor of 20. This has been absolutely inspiring work—a clever mixture of brute-force scaling, careful engineering and audacious new physics ideas.

The entire laser spectroscopy community was saddened recently to learn of the sudden death of our colleague Veniamin P. CHEBOTAYEV on September 1, 1992, just shortly after our Fermi School meeting. He was 53 years of age. His contributions are many, varied and important, ranging from many of these methane spectroscopy tricks, to suggestions for measuring the speed of light [8], to Doppler-free two-photon spectroscopy [37], to optical Ramsey fringes [38], to optical clocks [39], frequency synthesis [40], and many, many other now widely employed techniques. We will all miss his cheerful style and his many good ideas in physics.

3. – The tunable laser arrives.

3·1. *Frequency stabilization is a severe problem for tunable lasers.* – Many spectroscopists welcomed the arrival in the early 1970's of tunable dye lasers and the rapid development of various additional methods of frequency extension, including new dyes, harmonic generation, frequency mixing, etc. Now one could tune to the real transition of physical interest! Of course, a laser that is free to be tuned to another color will itself freely tune to a *wrong* color: the pressure was on to learn how to control these potentially wonderful new sources. Single-mode operation became routine with the inclusion of a servo-controlled intracavity etalon to increase the losses for all but a selected mode. Commercial systems appeared with the capability of a 30 GHz tuning range. A stabilization system provided useful but relatively rough stabilization to a tunable cavity ($\approx$ MHz). Even now, it remains an interesting challenge to provide stabilization at the kHz level (or better) of a laser which can tune 50 THz in 30 GHz windows!

3·2. *Cavity stabilization of tunable lasers.* – Stabilization of a tunable laser to an optical resonator [41] has a number of advantages compared with the previous nearly universal scheme of stabilization of nearly untunable lasers to quantum absorptions. For one thing, a cavity has the interesting property that it has many uniformly spaced sharp resonances throughout a spectral region *vs.* just one, or at best a few, lines for a molecule. Having many potential lock points should be a useful advantage for a widely tunable laser. Another important advantage is the totally generic aspect of cavity resonances: they can be designed to occur at whichever wavelength we find useful. A more subtle, but crucial, difference arises from the fact that the cavity resonant system is basically linear. For our locking purpose, if the obtained signal/noise performance is insufficient, it can be increased by allocating a larger power for the task. With quantum absorbers this extra power would broaden the resonance and so degrade the useful frequency discriminant.

While a fringe side locking method [42] was in vogue first, as accuracy dreams became more demanding it was necessary to use the cavity resonance information in a better way. One method currently in use employs r.f. phase modulation of the laser source to produce optical sidebands on the laser signal [43]. The perfect antisymmetry of the sidebands produced by pure FM/ /phase modulation leads to a cavity frequency discriminator curve which is ideal for laser locking. Drever *et al.* [44] have discussed an interesting aspect of this discriminator when used in reflection: the stored field in the resonator can serve as a phase reference against which the present optical phase can be compared. This leads to the capability of an optical phase lock at short times, changing to a frequency lock for times long compared with the cavity storage time. The detection S/N ratio, of course, depends on the employed bandwidth. The interesting effect that occurs here is that we can broaden the servo bandwidth to large values ($\sim$ 10 MHz) with the extra bandwidth-induced noise leading to a slight reduction of the optical-phase measurement quality. However, from FM theory we know that, if the optical phase tracks the sinewave ideal to within 0.1 rad r.m.s., we will have only 1% of the power in the accompanying phase modulation sidebands [45]. So we paid only a little in noise performance—and that noise appears far from the carrier—in exchange for starting our servo locking curve at a much higher unity gain frequency. With the gain function rising toward lower frequency at the rate of $\approx$ 10 dB/octave, now when we reach the ugly range $\sim$ 200 kHz for Ar^+-pumped lasers, we just *may* have enough gain to reduce the parasitic frequency/phase noise to our desired low value ($\ll$ 1 rad). Eventually, below the cavity linewidth of perhaps tens of kHz, the cavity no longer stores a reference phase, so we begin to have a frequency measurement output. The measurement noise in the bandwidth from this frequency downward will be compared by the servo with the frequency discriminator output slope, and so we can see that the servo will impose this FM equivalent onto the laser as the servo drives toward a zero apparent error signal [46].

The bottom line of this story is that, even for dye lasers, one can expect to be able to suppress the intrinsic noise of the laser more-or-less totally with feedback, replacing it with measurement noise. The value of this noise is worth noting for its smallness: For an invested 1/4 mW one can readily engineer a subhertz bandwidth for the tunable laser, be it dye or Ti:sapphire! Recently Ohtsu's group has increased the speed of analog locking [47] of a simple single-frequency diode laser to achieve $\sim$ 10 Hz linewidth and phase locking. External cavity operation of diode lasers allows us to enter this domain also [48] with

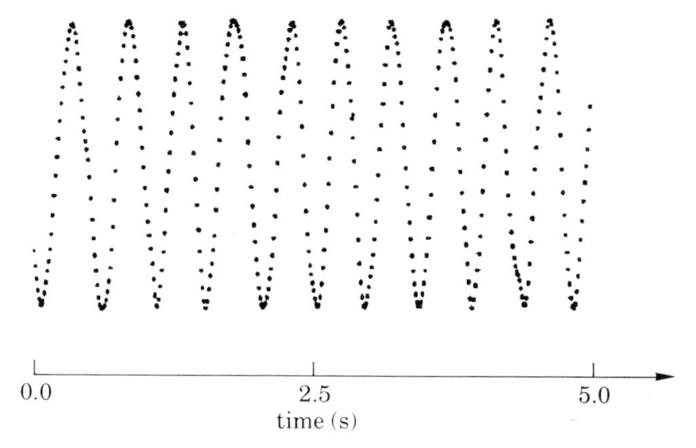

Fig. 4. – Wave form of optical heterodyne signal. Two lasers are independently locked onto adjacent axial orders of a stable interferometer leading to a beat frequency of ~ 500 MHz. This r.f. beat is heterodyned to near 1/2 Hz using a frequency synthesizer. The resulting wave is low-pass filtered ($f < 10$ kHz) and digitized at 60 samples/s. From least-squares fits to several such data sets, the apparent linewidth is found to be below 50 mHz per laser.

these attractive sources. It should be emphasized that this degree of locking is expressed relative to the resonator's frequency. If that frequency is being affected by vibrations or temperature changes, then surely our narrow-linewidth tunable laser will faithfully track these changes. So the problem has two parts: laser locking and cavity stabilization. Let us consider the laser locking part first.

3·3. *Precision of locking to a cavity resonance.* – In fig. 4 we show the beat between two HeNe lasers locked onto adjacent orders of a single Fabry-Perot resonator [49]. The optical frequencies differed by one unit of the free spectral range (~ 500 MHz), so the 500 MHz r.f. beat signal from the photodetector was heterodyned with a stable 500 MHz synthesized frequency to produce the difference frequency which was digitized in time to produce the wave form shown. The absence of fast phase noise shows that both of these independent optical-frequency lock circuits work extremely well, since any errors would show up in the difference. This phase-stable picture was obtained only after we used a special quartz oscillator for the frequency synthesizer, since, even though the radiofrequency is one million times lower than the locked optical frequency, the performance domain we are entering places almost unattainable demands on stability and spectral purity—of the r.f. oscillator! Obvious interest attaches to increasing the beat frequency to perhaps 10-fold higher values and selling this beat frequency wave form for r.f. synthesis tasks.

In fig. 5 we present data for the system of fig. 4 using the Allan variance presentation. Please remember that these curves relate to the quality of locking

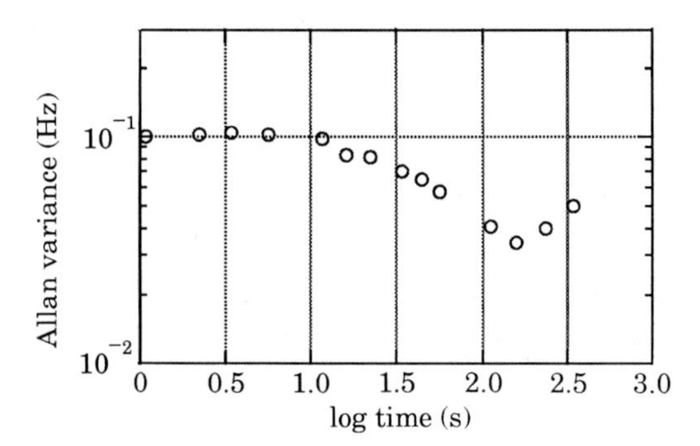

Fig. 5. – Allan variance of beat. Heterodyne beat between two HeNe lasers, independently locked to adjacent axial orders of a stable interferometer, similar to data of fig. 4. Note that the locking precision of 0.1 Hz is obtained with two oscillators at 473 THz, so the noise in locking is $\sim 2 \cdot 10^{-16}$. Of course, vibrations modulate the cavity frequency, giving ~ 5 Hz r.m.s. as the present optical linewidth.

lasers to a common cavity, which substantially reduces sensitivity to cavity instabilities: Here we are testing the locking.

As to the accuracy of locking to the optical resonances, we have shown in an earlier experiment that with some care one can reach the 1 Hz domain [50]. Some new experimental strategies to separate the cavity reflection from spurious reflection information are being explored at present. Experience in the r.f. domain has shown that with enormous and persistent effort one can hope to find a reproducible «center» of a resonance to within about 10^{-5} linewidths, assuming the line shape is free of asymmetry and other problems. For the current cavities, this criterion would be about $1/10$ Hz! Clearly there is room for major resourcefulness here because the shot-noise-defined linewidth is conveniently expressed in milli- or micro-hertz. It is worth re-emphasizing that basically any c.w. laser can in principle be stabilized at these levels: all it takes is lots of servo gain and the requisite short time delay to make it applicable [45, 46]. We turn now to the question of the cavity's intrinsic stability.

4. – Stability of the cavity resonance itself.

4·1. *Estimating environmental requirements.* – A useful calculation relative to stable cavities is to estimate the required isolation from vibration and changing temperature changes. Contemporary low-expansion materials [51] such as ULE and Zerodur have expansion coefficients in the neighborhood of $2 \cdot 10^{-8}$, which would correspond to about 10 MHz per degree. So to have stability in the kHz domain implies temperature control in the 0.1 mK range. Careful temperature servo design can approach this range, given favorable environmental stability. Still, considering the nice linewidth data presented above, why not try for 1 Hz? The associated temperature drift over the duration of the experiment must then be 0.1 μK. Double-shell dual-temperature regulators can bring one into the sub-mK range, but it is easy to see that soon we will want to return to locking the system to quantum absorption lines.

As for vibrations, it is also easy to estimate the scale of the sensitivity using the published values of elastic moduli, density and Poisson's ratio. One finds a compressional distortion of about + 10 MHz, assuming the cavity were to be supported vertically from the bottom. Holding it suspended from the top end would lead to a stretching, giving about − 10 MHz frequency shift. One could hope by symmetry of mounting to achieve something like 100 kHz/g as the vibration sensitivity. A typical laboratory has 1 milli-g vibration levels at the ~ 30 Hz frequency associated with mechanical imbalance of rotating electrical equipment. One should reach the sub-kHz level on an ordinary vibration-isolated optical table, while another 50 or 60 dB reduction would be necessary to reduce the phase modulation to sub-radian levels. This proves to be possible!

4·2. *Active vibration isolation.* – In fig. 6 we show some data on the performance of a 4′ × 8′ optical table which is servo-controlled vertically. The table employed the customary air legs, the vertical vibrations were sensed with a sensitive vertical accelerometer, and the vertical actuator was a low-cost «woofer» from a local stereo store. A time-proportioning air feed/exhaust of the air legs provided the long-range vertical control in preparation for future refinement, *i.e.* including a tilt servo as well. The seismometer has a simple resonance at 0.8 Hz associated with its suspended mass, and the first of its spurious resonances is at 200 Hz. With the aid of a simulation software package, it was feasible to design an effective 11th-order servo control loop for this system, as indicated in fig. 6.

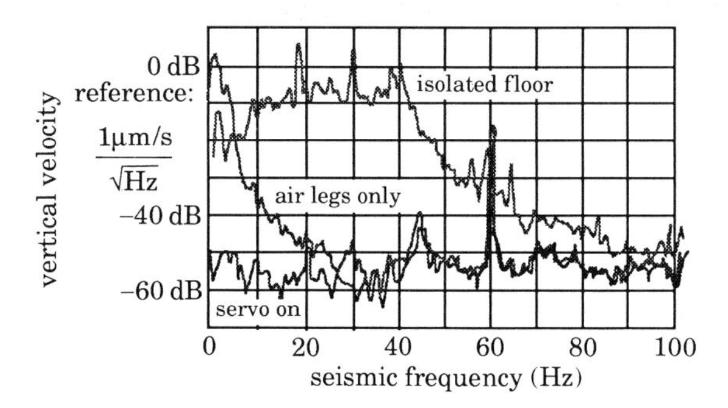

Fig. 6. – Influence of air legs with tilt and vertical servos. Vertical (velocity) noise on 4′ × 8′ optical table. Vertical sensor is a low-noise seismograph of moving-coil design, 0.8 Hz resonance. Servo «actuator» is electromagnetic type. Servo design compensates excess phase, is of eleventh-order design, but not highly optimized. The peak at 60 Hz is an artifact. Electronic «bubble» tiltmeter feeds time-proportioning pneumatic tilt servo to give 0.5 μrad r.m.s. residual tilt noise.

However, the problem with such an active reduction system is that the directional stability is heavily compromised by the air-legs' integrating character, and these tilts lead to unacceptable low-frequency length changes of the cavity frequency. Of course, we can plan to stabilize the laser table tilts as well.... The present cavity mounting is horizontal and shows a tilt-induced cavity frequency shift of 10 Hz/μrad (several times larger than expected). Our preliminary E/W tilt servo shows a noise residual of about 0.5 μrad r.m.s., which leads to a frequency excursion of the beat between our two independent systems of about $(5 \div 10)$ Hz as expected. An additional and faster tilt sensor is being considered to increase the tilt servo bandwidth well beyond the present ~ 1 Hz. The tilt motion of the table at present shows a spectrum peaked up around 6 s/cycle, the known center of the microseismic band associated with ocean wave loading on the sea floor.

4·3. *Long-term cavity length/frequency stability.* – Our earlier estimates have indicated that it will require some effort to obtain adequate long-term environmental conditions appropriate for hertz level stabilities, even assuming the cavity *per se* experiences no drift. Of course, all physical materials do show drift with time, so it is interesting to see what can be achieved experimentally. This information on drift and uniformity of drift will be useful for long-term experiments in space, and for relativity experiments which depend on a fixed length, as opposed to a fixed frequency. One can imagine the internal molecular vibrations as being of very high frequency ($\sim 10^{12}$/s), continually exploring the shape and extent of the intermolecular potential wells. In some cases a structural defect gives rise to another nearby potential minimum, but one which is inaccessible energetically. There are permanent intermolecular forces of chemical and polarization origin (van der Waals) that bias the specimen with their attractive force. It seems quite reasonable that occasionally the system somehow manages to pool different vibrational energies to jump into the preferred minimum-energy configuration. Perhaps this biased fluctuation process is the physical picture underlying the perpetual slow shortening we call «creep». In ULE [51] this process leads to shortening rates near 10 kHz/day asymptotically, after the larger response due to mirror substrate/spacer body mutual roughness accommodations have slowed down ($t \gtrsim 1$ month). These Fabry-Perot reference cavities employ mirrors fabricated with particular care for low optical loss ($L \lesssim 10$ p.p.m.) and are «optically contacted» onto the spacer. An initial shortening $\approx$ the residual surface roughness has been suggested by JACOBS *et al.* [52]. However, creep at the later times appears to be a homogeneous process which becomes very uniform in time. Some interesting photo-assisted aging effects can be seen also.

The low-expansion material «Zerodur» is a mixed-phase material in which the low expansion is assured by controlling the crystallite size growth by careful thermal annealing [51]. One could imagine that the creep rate might be asso-

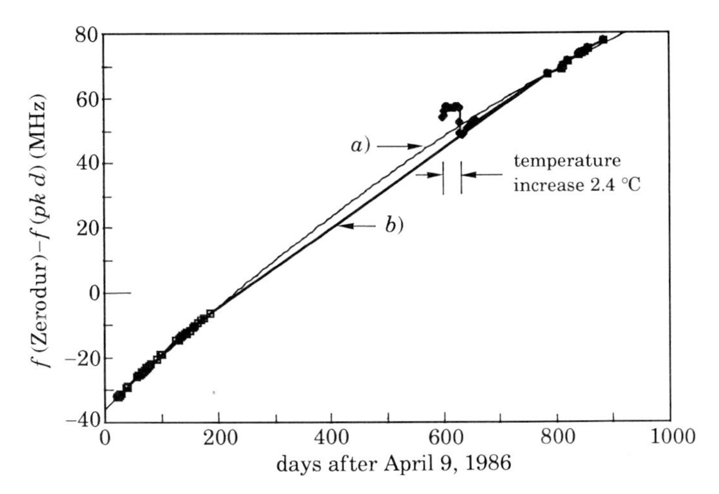

Fig. 7. – Long-term drift of Zerodur cavity. Zerodur rod is hanging at 24 °C constant temperature in vacuum: curve *a*) quadratic fit $y = -35.6073 + 0.167x - 4.497 \cdot 10^{-5} x^2$, curve *b*) standard line fit. Rapid initial drift after mirror contact not shown. Slow drift of length and deceleration are similar to that observed by BAYER-HELMS *et al.* Near day 600 we increased the stabilized temperature by 2.4 °C, resulting in basically zero drift during the one-month experiment, before restoring the original setpoint.

ciated with the gradual stress cracking of these crystalline zones under the internal compressive forces. Such a model would account naturally for the gradual deceleration of the creep as observed by BAYER-HELMS *et al.* [53]. Figure 7 shows our long-term data for one cylindrical spacer of 15 cm ∅ × 30 cm. The deceleration we observed is basically in accord with the value they observed for the gauge blocks.

We supposed that heating the Zerodur by several degrees would increase the creep rate by providing more available thermal energy. However, the data in the middle of fig. 7 show the physics to be vastly more complex. By increasing the storage temperature by 2.4 °C from 24 °C we found the creep rate changed more than 10-fold: the creep rate *decreased* to nearly zero at the higher temperature. An interesting and possibly related hysteresis of thermal expansion in the temperature range below ~ 30 °C has been observed by LINDIG [54] and by JACOBS [55]. A new formulation (Zerodur M) is now available and long-term tests are being prepared using this material [51].

5. – Tuning the laser relative to a fixed cavity.

5.1. *Using an acousto-optic modulator for tuning.* – The utility of a stable reference etalon in the optical domain is now quite clear, and we have indicated that sub-hertz linewidth and sub-hertz frequency predictability for perhaps one hour are now basically available from our stabilized laser. Unfortunately it is almost certain that our interesting atomic-physics resonances will not occur at one of the cavity's resonance frequencies. The required frequency offset can usefully be made up by a modulator which produces a frequency-offset output beam. Perhaps the most attractive such device is an acousto-optic modulator (AOM), which produces a (single-direction) input/output frequency shift equal to the r.f. drive frequency. So one can use a variable r.f. drive frequency, obtained from a computer-programmable source, to obtain the required tuning. A nice idea is to double-pass this AOM so that the angular offset intrinsic to the Bragg scattering can be cancelled in a second pass, while the frequency offset is doubled. The retroreflected beam can be produced with a curved mirror and isolated with polarization optics [56], or with a collimating lens and 90° roof prism. In the latter case the beam also emerges on the source side, but spatially resolved from the input beam. Wide-band modulators allow one to scan a full free spectral range of the reference cavity, and appropriate software allows the laser to be scanned many cavity fsr's by seamlessly joining sequential scans [57]. Some kind of gain normalization can be introduced in software or by a hardware intensity-levelling loop to overcome the change in Bragg-scattering efficiency as the frequency is scanned.

5·2. *Example 1: Line-narrowing and scanning a* Ti:*sapphire laser.* – This AOM tuning system has been implemented for our Ti:sapphire laser and is very effective in controlling the laser relative to the stable cavity frequency. For convenience we employ two AOM-based systems in cascade, the first using a tunable reference cavity and fast-locking system to line-narrow the laser. This error signal is derived with a Pound-Drever discriminator, and applied back to the laser to correct its «instantaneous» frequency. A fast PZT is used, but to obtain a more rapid control the high-frequency components are sent to the voltage-controlled oscillator which feeds the first AOM just outside the laser itself. This AOM has very small frequency excursions to deal with and so, in the interests of optical efficiency, is not doubly passed. This frequency-servo AOM serves the additional roles of optical isolator and AO intensity controller, in addition to its role as the fast actuator in locking the laser to the reference cavity [58]. The intensity noise is reduced by $> 30\,\mathrm{dB}$ at mid-kHz frequencies, and a linewidth in the $\sim 10\,\mathrm{kHz}$ range is obtained. This cavity uses an intracavity Brewster plate for tuning in the usual way, but its galvanometer is fitted with flex pivots and angular-position feedback to basically eliminate the troublesome hysteresis and «stick-tion» often observed otherwise. This line-narrowed and «civilized» tunable Ti:sapphire laser is then locked to the long-term reference cavity and scanned as described above [59].

Of course, if the cavity thermal environment is inadequate, the system may show excessive drift in frequency: we presently have about $1\,\mathrm{MHz/h}$ with a simple system. One is then glad to have the nearly inertia-less tuning capability, which allows a software subroutine to revisit a previous strong atomic/

/molecular resonance to confirm/re-measure the zero-crossing central frequency, which is separately recorded whenever it is measured. At present we time-stamp all data so that, by making use of this drift data vector, we can make *a posteriori* correction of drifts in the scan axis. With a foreground/background multitasking approach we hope to calculate the necessary corrected axis frequencies as we proceed. Line fitting accuracy and reproducibility in the range $\sim 10^{-3}$ linewidths is readily attainable with this approach [57].

5'3. *Example 2: Phase-locking laser diodes.* – Another approach to precise laser scanning is the tuning of one laser relative to a highly stabilized reference laser. With optical phase-locking techniques, one can transfer the full stability of the reference laser to the «tunable» laser, while enabling precise frequency scans via a tunable frequency offset in the locking loop. The most interesting applications of this stabilization technology may be for laser diode systems, as their reliability and relatively low cost facilitate the design of multilaser approaches to a physical experiment. Unfortunately, for most of the diode lasers now available, the injected charges we intended to correct the laser's frequency at time t are still present somewhere near the index-guided laser channel after an appreciable time. Taken with the large thermal tuning rate, the result is that the frequency change for an injected milliampere correction signal has a drastic phase shift with frequency. We have measured a $> 90°$ shift at 25 MHz for the common Sharp LT024 780 nm laser [51]. Such large phase changes within a feedback loop are, of course, unacceptable for stability reasons, so it is useful to build multiple phase compensators to develop a more pleasant transfer function [48]. Previous work in this compensator area has been done by Prof. Ohtsu's group [47]. With a five-stage compensator we locked $> 98\%$ of the emitted diode laser power into the phase-locked carrier. With a more favorable diode operating at 852 nm (STC-LT50A-03U) [51], a single compensator allowed a > 20 MHz unity gain frequency and > 99.8 power concentration into the carrier [48]. Work on the locking/acquisition strategy continues.

5'4. *Moving to higher/wider frequency scans.* – Locking with higher and higher beat frequency offsets is another worthwhile and important research area. Applications include bringing a full hyperfine spectrum into a single precise scan, establishing precise frequency intervals between different reference lines, and simplifying various schemes of optical-frequency synthesis such as the all-optical «Divide and Conquer!» scheme being developed by Hänsch and his associates [60]. Wong *et al.* [61] suggest another promising approach to tunability, frequency-locking the outputs of a nearly degenerate optical parametric oscillator.

5'4.1. R.f. harmonic mixing in photodetector. The straightforward approach is to obtain a fast photodetector—bandwidths up to 60 GHz are commercially available—and heterodyne its r.f. output in a microwave mixer to produce a beat in the comfortable range ≈ 100 MHz for the phase-locking loop. Unfortunately this approach requires an r.f. source output at the optical beat frequency. As the radiofrequency increases, such sources rapidly become extremely expensive! With a metal-insulator-metal diode as their mixer, DRULLINGER *et al.* used a FIR laser at 119 μm as the «local oscillator» source to detect beats at 2.5 THz [62]. Probably some microwave harmonic-locking approach will be the way to obtain useful bandwidth increases. For example, it has been shown [9] that r.f. drive of a Schottky-barrier detector serves to generate relatively high harmonics internally and recently even 4 THz beats [63] have been detected as the 60th harmonic with this approach! With the development of fast nonlinear transmission lines, one has available for the first time electrical pulses in the picosecond domain. It will be interesting to try to build «sampling-type» optical detectors, for example by putting the optical-signal pair into the last photodetector / Schottky-barrier diode in the transmission line [64].

5.4.2. Broad-band modulation: modulator-in-a-cavity approach. Another promising approach is based on strong r.f. modulation: a microwave modulator is enclosed in a low-loss cavity. The sideband produced on the first transit is used as the source for a second sideband, and the second for a third and.... KOUROGI, NAKAGAWA and OHTSU have reported [65] a spectral width of $\sim \pm 4$ THz, made up of individual lines spaced by the 5.6 GHz modulation frequency. Enhancements of this scheme certainly can be imagined: A nice improvement of the efficiency could be accomplished by «re-cycling» the light reflected back toward the source from the entrance mirror. (Since the in-coupled light is spectrally scattered to many frequencies, there is a painful loss of in-coupling efficiency as the internal cavity leakage field at the input frequency rapidly becomes too feeble to cancel the directly reflected input beam and thus to accomplish the desirable «optical impedance matching».) This same recycling cavity, if short enough, could be resonance-free until one reaches the desired high-order sideband, perhaps several THz away. The frequency-shifted power in this line would be coupled back toward the source and could be separated with a Faraday isolator system. Such schemes may make it feasible to transfer the stability of one optical source in a phase-coherent manner to another source located an appreciable frequency interval away. Perhaps a shift of 1% of the optical frequency could be accomplished in a single step! This would reduce the number of «Divide and Conquer» stages [60] to six or seven.

6. – Applications of stable lasers.

Of course, there are many applications of frequency-stabilized lasers in science and industry. One of the dreams is to build optical-frequency clocks with dramatically improved characteristics. BERGQUIST and his associates have discussed [66] some of their results about locking a stable laser to a single trapped Hg^+ ion. Trapped ions represent an excellent system since the ion is held basically without perturbation, for days if we like. Our JILA group is exploring an alternative idea, the so-called «atomic-fountain clock» [67, 68]. In this scheme we imagine bringing neutral atoms to near-zero velocity in a Zeeman-optical trap. One can capture easily 10^7 atoms in a mm^3. Now the magnetic fields are switched off, the molasses light further red-detuned and a few milliseconds long post-cooling interval begins. The product is ~ 3 million atoms within an 8 mm^3 ball. Switching the vertical frequencies of the molasses has a profound effect by forming molasses in a moving frame—one could say «walking-wave» molasses—where the captive atomic ball is now moving upward at the speed $v = \lambda\,\delta\nu$. These preparation fields are now switched off and the supersharp optical clock transition is irradiated for a few milliseconds. This excitation prepares a coherent admixture of ground and excited atomic states which forms the atomic-frequency coherence of interest. After the atoms reach apogee, fall and re-enter this clock excitation region, a second laser excitation pulse creates a second amplitude for being in the atomic excited state. Another laser can then interrogate the system to enjoy the Ramsey fringes between the two excitations, so as to judge the fraction of the atomic ball that underwent transitions into the excited state of the clock transition. Repeating this process with different detunings gives an appreciation of the location of the atomic center frequency. CHEBOTAYEV and his associates have considered this approach in some detail.

Several groups are busy investigating microwave clocks built along such lines [69]. Using the «cycling» optical resonance line transition, alkali metals can be cooled efficiently to the Doppler limit, $kT \sim \hbar\Gamma$. One then uses another cooling mechanism such as polarization gradient or «Sisyphus» cooling to reach a residual velocity of ≈ 3 recoil units. Unfortunately, these supercold alkali atoms do not appear to have an attractive optical clock transition. In fact, surprisingly, one of the problems for an optical standard is an apparent dearth of fully satisfactory atomic transitions, if we put on the additional restriction that the atom must also have a fast and basically closed two-level system for laser-cooling purposes.

The next idea is to use group II metals such as calcium. The initial cooling proceeds very efficiently as before to the Doppler limit. Now there is no hyperfine structure to employ for the second stage of cooling. One possibility is to switch to a weaker transition, such as the ${}^1S_0$-${}^3P_1$ intercombination line. The long lifetime assures a narrow linewidth (~ 400 Hz), but this is uncomfortably weak for a second stage of Doppler cooling, and uncomfortably broad for an atomic-fountain clock, where we might prefer a natural width of $(1 \div 10)$ Hz. Still, laser-cooled calcium atoms would give an extremely good optical-frequency reference possibility. HOLLBERG has shown that diode lasers will be suitable for this system, so one can see the advantages of portability and low-cost potential, along with an inaccuracy of probably $\sim 10^{-15}$ due to light shift effects.

Another possibility in calcium would be to use the $^1S_0$-$^3P_2$ magnetic-quadrupole transition as the clock. This line has a superlow natural linewidth [70] ($A \sim 6 \cdot 10^{-5}$/s), but even so we estimate a power of only some 10 mW would be sufficient to excite the Ramsey resonance using a build-up cavity around the atoms. This would be excited with a spectral comb to produce saturated-absorption signals from atoms with a number of distinct residual velocities. An alternative approach based on use of two-photon excitation is initially attractive since all atoms can contribute to the Doppler-free peak [37], but we are worried about problems with a.c. Stark shifts.

Frequency-doubled Nd laser output can give strong signals in the green on at least five different ro-vibronic transitions in the $I_2 B \Leftarrow X$ spectrum, although predissociation may limit the linewidths to ~ 100 kHz. These linewidths are $(10 \div 100)$-fold larger than one would like for a primary standard, but the signals are very strong and the laser intrinsic stability sufficient that a practical «good enough» standard can be envisioned, based on diode-laser-pumped Nd. Of particular value for this case of broad lines is the method of modulation-transfer spectroscopy [71], which provides good immunity to additive baseline/ /frequency shifts due to spurious AM produced by the phase modulator. Our experiments so far yield an Allan variance of ~ 200 Hz at one second, and a reproducibility from system to system of the same order. Both values represent more than two-orders-of-magnitude improvement over the standard 633 nm HeNe/I_2 laser system. So the future seems bright indeed!

7. – Conclusions.

Building on our pre-laser-era spectroscopic heritage, rather spectacular progress in laser stabilization has been made over the past 25 years. We saw first the development of methane-based stable lasers, whose accuracy capability was crippled by relativistic time dilation shifts of $\sim 10^{-12}$. These lasers are still very convenient and practical at this level. The development of external cavity resonances in OsO_4 are similar in principle but more attractive in the number of resonances available. What will be the ultimate system of preference in the visible remains to be seen. Ca with laser diode excitation at 657 nm will obviously be interesting. Maybe the magnetic fine-structure transitions between ground-state ^3P levels will be good. Diode-pumped, frequency-doubled Nd looks good for precision with simplicity. Maybe...

* * *

I thank the several members of my group at JILA, both current and past, for their skillful efforts, personal dedication and generous friendship. Particularly to be noted in this regard are Miao ZHU, and former colleagues L. HOLLBERG and J. BERGQUIST. A legion of postdocs and visitors have contributed their own special skills with enthusiasm and dedication. Especially I must thank D. HILS whose careful work over many years advanced our stabilization efforts so well. Finally I want to explicitly thank my wife Lindy for her friendship and for her patience and support during these long years, and for making things be fun.

The work at JILA is supported in part by the National Institute of Standards and Technology as part of its program of advanced research for possible applications in basic standards, and in part by the Office of Naval Research, the Air Force Office of Scientific Research and the National Science Foundation.

REFERENCES

[1] J. L. HALL and T. W. HÄNSCH: *Opt. Lett.*, **9**, 502 (1984).
[2] M. KASEVITCH and S. CHU: *Appl. Phys. B*, **54**, 321 (1992).
[3] W. E. LAMB: *Phys. Rev.*, **134**, 1429 (1964).
[4] W. R. BENNETT jr.: *Phys. Rev.*, **126**, 580 (1962).
[5] W. R. BENNETT jr. and J. W. KNUTSON jr.: *Proc. IEEE*, **52**, 861 (1964).
[6] J. A. MAGYAR: thesis University of Colorado (1974), and unpublished work.
[7] A brief look down this road was offered by J. L. HALL: *IEEE J. Quantum Electron.*, QE-4, 638 (1968).
[8] V. P. CHEBOTAYEV: *Radio Tekh. Elektron.*, **11**, 1712 (1966).
[9] J. L. HALL and W. W. MOREY: *Appl. Phys. Lett.*, **10**, 152 (1967).
[10] P. H. LEE and M. L. SKOLNICK: *Appl. Phys. Lett.*, **10**, 303 (1967).
[11] V. N. LISITSYN and V. P. CHEBOTAYEV: *Ž. Èksp. Teor. Fiz.*, **54**, 419 (1968); translation in *Sov. Phys. JETP,* **27**, 227 (1968).
[12] R. L. BARGER and J. L. HALL: *Phys. Rev. Lett.*, **22**, 4 (1969); J. L. HALL: *IEEE J. Quantum Electron.*, QE-4, 638 (1968).
[13] S. N. BAGAEV, YU. D. KOLOMNIKOV, V. N. LISITSYN and V. P. CHEBOTAYEV: *IEEE J. Quantum Electron.*, QE-4, 868 (1968).
[14] A. H. NIELSEN and H. H. NIELSEN: *Phys. Rev.*, **48**, 864 (1935).
[15] K. T. HECHT: *J. Mol. Spectrosc.*, **5**, 355, 390 (1960).
[16] L. HENRY, N. HUSSON, R. ANDIA and A. VALENTIN: *J. Mol. Spectrosc.*, **36**, 511 (1970).
[17] A. S. PINE: *J. Opt. Soc. Am.*, **66**, 97 (1976).
[18] G. R. HANES and C. E. DAHLSTROM: *Appl. Phys. Lett.*, **14**, 362 (1969).
[19] J. D. KNOX and Y.-H. PAO: *Appl. Phys. Lett.*, **18**, 360 (1971).
[20] W. G. SWEITZER, E. G. KESSLER jr., R. D. DESLATTES, H. P. LAYER and J. R. WHETSTONE: *Appl. Opt.*, **12**, 2927 (1973).
[21] The journal *Metrologia* has a significant fraction of these papers. The biennial Conferences on Precision Electromagnetic Measurements are published in the *IEEE Transactions on Instrumentation and Measurement.*
[22] S. EZEKIEL and R. WEISS: *Phys. Rev. Lett.*, **20**, 91 (1968); *IEEE J. Quantum Electron.*, QE-4, 367 (1968).

[23] Really impressive results have been obtained by C. J. BORDÉ, G. CAMY, B. DECOMPS, J.-P. DESCOUBES and J. VIGUE: *J. Phys. (Paris)*, **42**, 1391 (1981); *Phys. Rev. A*, **20**, 254 (1979).
[24] C. FREED and A. JAVAN: *Appl. Phys. Lett.*, **17**, 53 (1970).
[25] See A. CLAIRON, O. ACEF, C. CHARDONNET and C. J. BORDÉ: in *Frequency Standards and Metrology*, edited by A. DE MARCHI (Springer-Verlag, Heidelberg, 1989), p. 212 ff.
[26] B. BOBIN, C. J. BORDÉ, J. BORDÉ and C. BRÉANT: *J. Mol. Spectrosc.*, **121**, 91 (1987).
[27] J. L. HALL and J. A. MAGYAR: in *High Resolution Laser Spectroscopy*, edited by K. SHIMODA (Springer-Verlag, Heidelberg, 1976), p. 173.
[28] This type of work is usually reported in *Metrologia*. See, *e.g.*, L. R. PENDRILL, J. M. CHARTIER, M. FRENNBERG and L. ROBERTSSON: *Metrologia*, **28**, 95 (1991).
[29] CCDM documents, edited by BIPM authors, *Metrologia*, **19**, 163 (1984).
[30] The final form of these 1992 CCDM recommendations will be published in *Metrologia* in 1993 or early 1994.
[31] A. CLAIRON, B. DAHMANI, A. FILIMON and J. RUTMAN: *IEEE Trans. Instrum. Meas.*, IM-**34**, 265 (1985).
[32] O. ACEF, J. J. ZONDY, M. ABED, D. G. ROVERA, A. H. GÉRARD, A. CLAIRON, PH. LAURENT, Y. MILLERIOUX and P. JUNCAR: *Opt. Commun.*, **97**, 29 (1993); the quoted value includes some additional corrections by CCDM 92. See ref. [30].
[33] J. L. HALL and C. J. BORDÉ: *Phys. Rev. Lett.*, **30**, 1101 (1973).
[34] J. L. HALL, C. J. BORDÉ and K. UEHARA: *Phys. Rev. Lett.*, **37**, 1339 (1976).
[35] A recent summary of this work is given by S. N. BAGAEV, V. P. CHEBOTAYEV, A. K. DMITRIYEV, A. E. OM, YU. V. NEKRASOV and B. N. SKVORTSOV: *Appl. Phys. B*, **52**, 63 (1991).
[36] J. L. HALL: in *Fundamental and Applied Laser Physics*, edited by M. S. FELD, A. JAVAN and N. KURNITT (Wiley, New York, N.Y., 1973), p. 463.
[37] L. S. VASILENKO, V. P. CHEBOTAYEV and A. V. SHISHAEV: *JETP Lett.*, **12**, 113 (1970).
[38] YE. V. BAKLANOV, B. YA. DUBETSKII and V. P. CHEBOTAYEV: *Appl. Phys.*, **9**, 171 (1976); **11**, 201 (1976).
[39] S. N. BAGAEV, V. P. CHEBOTAYEV, V. M. KLEMENTYEV, B. A. TIMCHENKO and V. F. ZAKHARYSH: in *Frequency Standards and Metrology*, edited by A. DE MARCHI (Springer-Verlag, Heidelberg, 1989), p. 191.
[40] V. P. CHEBOTAYEV, V. M. KLEMENTYEV and YU. A. MATYUGIN: *Appl. Phys.*, **11**, 163 (1976).
[41] A. D. WHITE: *IEEE J. Quantum Electron.*, QE-**1**, 349 (1965).
[42] R. L. BARGER, M. S. SOREM and J. L. HALL: *Appl. Phys. Lett.*, **22**, 573 (1973).
[43] See, for example, G. C. BJORKLUND: *Opt. Lett.*, **5**, 15 (1980).
[44] R. W. P. DREVER, J. L. HALL, F. V. KOWALSKI, J. HOUGH, G. M. FORD, A. J. MUNLEY and H. WARD: *Appl. Phys. B*, **31**, 97 (1983).
[45] See, for example, M. ZHU and J. L. HALL: *J. Opt. Soc. Am. B*, **10**, 802 (1993); *Proc. S.I.F.*, Course CXVIII (North-Holland, Amsterdam, 1993), p. 671.
[46] See, for example, J. L. HALL: in *Quantum Optics IV*, edited by J. D. HARVEY and D. F. WALLS (Springer, Berlin, 1986), p. 273.
[47] M. OHTSU, C.-H. SHIN, H. FURUTA, M. KOUROGI, H. LUSUZAWA, S. JIANG and K. NAKAGAWA: in *QELS '91*, Baltimore, May 12-17, 1991, paper QThL1, p. 242; M. KOUROGI, C.-H. SHIN and M. OHTSU: *IEEE Photonics Technol. Lett.*, **3**, 496 (1991).
[48] S. SWARTZ, J. L. HALL, K. E. GIBBLE and D. S. WEISS: *Robust phase-locking of diode lasers, J. Opt. Soc. Am.*, submitted 1993.

[49] I would like to express my appreciation to D. HILS whose painstaking work has been the mainstay of the HeNe results.
[50] D. HILS, CH. SALOMON and J. L. HALL: *J. Opt. Soc. Am. B*, **5**, 1576 (1988).
[51] These trade names are used for technical communication purposes only, with no implied recommendation.
[52] J. W. BERTHOLD III, S. F. JACOBS and M. A. NORTON: *Metrologia*, **13**, 9 (1977).
[53] F. BAYER-HELMS, H. DARNEDDE and G. EXNER: *Metrologia*, **21**, 49 (1985).
[54] O. LINDIG and W. PANNHORST: *Appl. Opt.*, **24**, 3330 (1985).
[55] S. F. JACOBS: *Opt. Acta*, **33**, 1377 (1986).
[56] G. CAMY, D. PINAUD, N. COURTIER and H. C. CHUAN: *Rev. Phys. Appl. (Paris)*, **17**, 357 (1982).
[57] M. D. RAYMAN, C. G. AMINOFF and J. L. HALL: *J. Opt. Soc. Am. B*, **6**, 539 (1989); see also M. ANSELMENT, S. GÖRING and G. MEISEL: *Z. Phys. D*, **7**, 113 (1987).
[58] J. L. HALL, H. P. LAYER and R. D. DESLATTES: in *1977 IEEE/OSA Conference on Laser Engineering and Applications* (IEEE, New York, N.Y., 1977, Cat. No. 77CH-1207-0 Laser), p. 45; see also ref. [1].
[59] S. SWARTZ, B. E. KOHLER, H.-R. XIA and J. L. HALL: *A frequency-tuned stable Ti: sapphire laser*, in preparation.
[60] H. R. TELLE, D. MESCHEDE and T. W. HÄNSCH: *Opt. Lett.*, **15**, 532 (1990); R. WYNANDS, T. MUKAI and T. W. HÄNSCH: *Opt. Lett.*, **17**, 1749 (1992).
[61] N. C. WONG, D. LEE and L. R. BROTHERS: *International Conference on Quantum Electronics, Technical Digest Series*, 1992, p. 110; also see D. LEE and N. C. WONG: in *Frequency-Stabilized Lasers and their Applications*, edited by Y. C. CHUNG, *Proc. SPIE*, **1837**, 419 (1993).
[62] R. E. DRULLINGER, K. M. EVENSON, D. A. JENNINGS, F. R. PETERSEN, J. C. BERGQUIST, L. BURKINS and H.-U. DANIEL: *Appl. Phys. Lett.*, **42**, 137 (1983).
[63] Y. C. NI and C. O. WEISS: *Int. J. Infrared Millimeter Waves*, **11**, 1069 (1990).
[64] R. MARSLAND and D. BLOOM: private communication (1992); R. A. MARSLAND, M. S. SHAKOURI and D. M. BLOOM: *Electron. Lett.*, **26**, 1235 (1990).
[65] M. KOUROGI, K. NAKAGAWA and M. OHTSU: in *Frequency-Stabilized Lasers and their Applications*, edited by Y. C. CHUNG, *Proc. SPIE*, **1837**, 205 (1993).
[66] J. C. BERGQUIST, W. M. ITANO, F. ELSNER, M. G. RAIZEN and D. J. WINELAND: in *Light-Induced Kinetic Effects on Atoms, Ions, and Molecules*, edited by L. MOI, S. GOZZINI, C. GABBANINI, E. ARIMONDO and F. STRUMIA (E.T.S. Editrice, Pisa, 1991), p. 291.
[67] J. L. HALL, M. ZHU and P. BUCH: *J. Opt. Soc. Am. B*, **6**, 2194 (1989).
[68] K. GIBBLE and S. CHU: *Metrologia*, **29**, 201 (1992).
[69] K. GIBBLE and S. CHU: *Phys. Rev. Lett.*, **70**, 1771 (1993); A. CLAIRON, C. SALOMON and S. GUELLATI: *Europhys. Lett.*, **16**, 165 (1991).
[70] R. GARSTANG, JILA, Boulder: private communication (1992).
[71] L.-S. MA and J. L. HALL: *IEEE J. Quantum Electron.*, **26**, 2006 (1990).

FUNDAMENTAL EXPERIMENTS

Measurement of Parity Nonconservation in Atoms.

C. E. WIEMAN, S. GILBERT (*), CH. NOECKER (**), P. MASTERSON (***)
C. TANNER (***), CH. WOOD, DONGHYUN CHO and M. STEPHENS

Joint Institute for Laboratory Astrophysics
University of Colorado and National Institute of Standards and Technology
and Department of Physics, University of Colorado - Boulder, CO 80309-0440

1. – Introduction.

This lecture is an introduction to the subject of parity nonconservation in atoms. It is intended for the student or scientist who is not familiar with the field. The Colorado cesium experiment is described in detail, and we have attempted to present many of the technical details and considerations that led to the final experimental design.

The basic phenomenon discussed in this lecture is the parity-nonconserving (PNC) weak neutral-current interaction in an atom. As shown in fig. 1, this interaction arises from the exchange of a Z_0 boson between the electrons and the protons and neutrons in an atom. It can be contrasted with the much larger Coulomb interaction that arises from the exchange of a virtual photon between the electrons and the protons and predominantly determines the structure of the atom. Study of PNC in atoms began in earnest with the Weinberg-Salam-Glashow (WSG) electroweak theory which unified the weak and electromagnetic interactions and predicted a parity-violating neutral-current interaction.

In an atom, the parity-nonconserving Hamiltonian can be written as the sum of two parts [1]

$$H_{\mathrm{PNC}} = H_{1\mathrm{PNC}} + H_{2\mathrm{PNC}} \,, \tag{1}$$

$$H_{1\mathrm{PNC}} = \frac{G_{\mathrm{F}}}{2\sqrt{2}}\, Q_{\mathrm{w}}\, |\gamma_5| \,, \tag{2}$$

(*) Present address: National Institute of Standards and Technology, Boulder, CO.
(**) Present address: Harvard Smithsonian Institute for Astrophysics, Cambridge, MA.
(***) Present address: Melles Griot Inc., Boulder, CO.
(***) Present address: Department of Physics, University of Notre Dame, South Bend, IN.

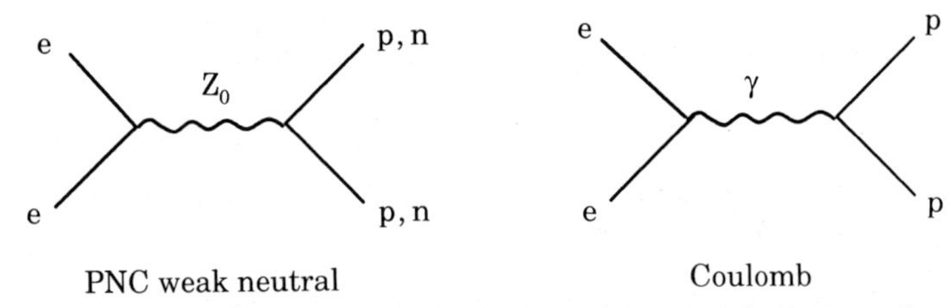

Fig. 1. – Parity-violating neutral-current interaction between electrons and nucleons predicted by the Weinberg-Salam-Glashow electroweak theory contrasted with the normal electromagnetic interaction.

where G_F is the Fermi constant and γ_5 is the fifth Dirac gamma matrix and operates on the electron. The largest part of the PNC Hamiltonian arises from the so-called weak charge, Q_W. This involves the axial-vector quark current where the contributions from all quarks simply add in similar manner to the way the electromagnetic charges add. The weak charge,

$$Q_W = 2[(2Z + N) C_{1u} + (2N + Z) C_{1d}], \tag{3}$$

involves the number of protons (Z) and neutrons (N), and the coupling constants C_{1u} and C_{1d} [2]. These are two of the four fundamental coupling constants which characterize the neutral-current interaction between electrons and quarks. They are multiplied by the appropriate factors involving the number of up and down quarks, respectively. The total Q_W is nearly proportional to the number of neutrons contained in the nucleus.

The H_2 PNC term, arising from the nuclear-spin-dependent or «weak isospin» contribution, is much smaller [3]. This contribution is about 1% of the total PNC, and, because the spins of the quarks in the nucleus tend to add up, they cancel. Actually, this nuclear-spin-dependent part also has two parts, the smallest of which is the weak neutral-current contribution involving the Z_0 exchange. The much larger part ($\sim \times 10$) is the nuclear anapole moment contribution [3]. This contribution comes from the more traditional weak interactions that take place in the nucleus which then couple to the electrons. We will discuss this in more detail in sect. 7 below.

The primary interest in studying these interactions is to test the «standard model» of the elementary-particle interactions. This standard model is made up of the WSG electroweak theory plus quantum chromodynamics to account for the strong interactions. In the standard model (neglecting radiative corrections), the constants C_{1u} and C_{1d} are given by [2]

$$C_{1u} = \frac{1}{2} - \frac{4}{3} \sin^2 \theta_W , \qquad C_{1d} = -\frac{1}{2} + \frac{2}{3} \sin^2 \theta_W . \tag{4}$$

The goal of atomic-parity-nonconservation work is to determine if these constants are accurately predicted by these formulae or if there are some corrections to these values which arise from «new» physics. By «new» physics, we mean physics that is not included in the standard model [2].

To be more specific, we test the standard model and hence look for this new physics using the following prescription. 1) Obtain the value of $\sin^2 \theta_w$ from the mass of the Z_0 boson, which is now measured very precisely at CERN. 2) Use this value to calculate Q_w as in eq. (3) plus the addition of small radiative corrections we neglected. 3) Compare the calculated value of Q_w with what one obtains from atomic-parity nonconservation.

Before beginning a discussion of the very lengthy and difficult experiments and atomic theory which have been undertaken to determine Q_w, the obvious question is: «Why bother?». What is the motivation for doing this? The answer to this is that the standard model is almost certainly not the final answer. It has many undesirable features; in particular, there are a number of important quantities that have to be put in on a somewhat artificial *ad hoc* basis. Notable among these are the coupling constants, the masses of all the particles and certain basic symmetries such as the handedness of the neutrinos. None of these different quantities arise naturally out of the model, as one would expect from the ultimate theory. In addition, the standard model runs into trouble at high energies. As the energies become large compared to the mass of the Z_0 boson, the theory becomes divergent.

These features indicate that there must be new physics beyond the standard model. We note that many aspects of the standard model have only been tested to about 2%. This is not like quantum electrodynamics where the theory has been tested to nine decimal places and one is asking whether there might be some deviation in the tenth. Because the standard model has only been tested relatively crudely, there is still a reasonable likelihood that small improvements in precision may lead to fundamental insight into new physics. Finally, the standard model does not occupy an unchallenged position. Many extensions and alternatives have been proposed, and atomic PNC measurements are uniquely sensitive to the physics that many of these would produce. The reason for this unique sensitivity is that the two coupling constants, C_{1d} and C_{1u}, are measured very poorly, or not at all, by the high-energy experiments. In addition, there are a number of radiative corrections to the standard model which have energy dependences. Since the atomic PNC is at low energy, the comparison of atomic-parity nonconservation with high-energy results is sensitive to such energy-dependent terms. All these features have combined to make atomic-parity nonconservation perhaps the only mainstream particle physics that is currently being done on a table top scale.

2. – PNC **neutral currents in atoms.**

Many of the early concepts of the effects of neutral currents in atoms were introduced by CURTIS-MICHEL in the 1960's [4]. However, activity in the field expanded rapidly after the papers of M. A. Bouchiat and C.

Bouchiat in 1974 [1, 5] in which they considered this in the context of the WSG model.

The principal effect of this parity-violating interaction in an atom is to mix the S and P parity eigenstates so that the S state is no longer a pure S state, but has a very small amount (δ_{PNC}) of P state mixed into it,

$$|S\rangle \rightarrow |S\rangle + \delta_{PNC}\,|P\rangle, \tag{5}$$

$$\delta_{PNC} = \frac{G_F}{2\sqrt{2}}\,Q_w \langle \gamma_5 \rangle_{nuc}\,, \tag{6}$$

$$\delta_{PNC} \approx (5 \cdot 10^{-18})(ZC_1)(Z^2) \approx 10^{-11}\,! \tag{7}$$

This quantity δ_{PNC} involves the Fermi constant of the weak interactions G_F, the weak charge, Q_w, mentioned before, and an atomic matrix element which is simply the matrix element of the γ_5 evaluated over the nucleus. The evaluation is only over the nucleus because the Z_0 boson is a massive particle and thus this is a short-range interaction. It is straightforward to estimate the approximate size of the mixing (eq. (7)). We have already indicated the weak charge is proportional to Z multiplied by constants which are on the order of one. The γ_5 matrix element, as pointed out by the Bouchiats, is proportional to Z^2. For a relatively heavy atom like cesium, the mixing then works out to be about 10^{-11}! It is this very tiny scale that makes the experiment so difficult and fraught with the possibility of error. To put this in perspective, a ratio of one part in 10^{11} is the same as the ratio of the diameter of a human hair to the diameter of the Earth.

In these experiments, the mixing of S and P states is observed as an electric-dipole transition amplitude between S states with different principal quantum number n. Or, alternatively, between P states with different n. In introductory quantum mechanics one learns that such electric-dipole transition amplitudes are absolutely forbidden by the parity selection rule and, therefore, if one measures this electric-dipole transition amplitude, it is a measure of how much P state is mixed into the S state.

The most direct approach to measuring this transition amplitude would be to simply take a laser, set its frequency in resonance with some $S \rightarrow S$ transition in an atom, and look for the excitation rate, trying to pick out the part that was due just to the electric-dipole contribution. In fact, in the days before the Bouchiats' paper, this approach provided the best test of the conservation of parity in atoms and set the limits at that time. However, the parity-nonconserving rate is proportional to the square of δ_{PNC},

$$R_{PNC} = |A_{PNC}|^2 \propto \delta^2_{PNC}\,. \tag{8}$$

Effectively, this means the oscillator strength for such a transition is about 10^{-22} and, therefore, the transition rate is 22 orders of magnitude smaller than a normal allowed electromagnetic transition. With any conceivable experiment, this is an impossibly small transition rate, and will always be lost in the noise. Therefore, this approach is clearly unsuitable for achieving the level of sensitivity needed for investigating the weak neutral-current effects.

As discussed by Michel and the Bouchiats, a much more intelligent approach is to use a «heterodyne» or interference approach. In this case, the transition rate, R_H, is equal to the square of the sum of two transition amplitudes, one being the parity-nonconserving amplitude and the other being a much larger parity-conserving transition amplitude between the S states:

$$R_H = |A_0 \pm A_{PNC}|^2 = A_0^2 \pm 2A_0 A_{PNC} + A_{PNC}^2 . \tag{9}$$

Here, A_{PNC} is the quantity we are interested in, and A_0 is the parity-conserving amplitude. Notice that there is now a term $2A_0A_{PNC}$, which is linear in the parity-nonconserving amplitude, and, therefore, can be large enough to measure. It is this interference term that we are interested in determining. This basic idea of mixing and detecting a small amplitude by having it mixed with a large amplitude is, of course, a very common one in physics and goes back to at least the early days of radio, if not before.

It is necessary to give some thought as to the phases of these amplitudes to make sure that such a parity-violating interference term can exist in any given experiment. One can work through all the mathematics and selection rules to find the appropriate conditions, but the basic point is that this is a parity-nonconserving term, and, therefore, one has to design an experiment which has a handedness built in. If the experiment has no such handedness, it will not be sensitive to a PNC effect. The second important question is, assuming the interference term exists, how does one isolate it from the much larger A_0^2 contribution to the rate? This is done by observing the modulation in the transition rate when the handedness is changed, or, in other words, a mirror reflection of the experiment is carried out. In that case there is a change ΔR_H in the transition rate due to the parity-violating $2A_0A_{PNC}$ term reversing sign. Thus all the experiments we are going to describe determine a fractional change in the transition rate $\Delta R_H/R_H$. A key point is that the transition rate itself, R_H, is a very weak transition by normal standards, and the fractional changes we are interested in are very small. For example, in our experiment R_H corresponds to an oscillator strength of 10^{-12} and the modulation ΔR_H is on the order of one part in 10^6. Thus the basic scale of the experiment is set by the fact that one has a very weak transition, and, in order to see small modulations on that transition, there must be a very high signal-to-noise ratio. It is this issue which makes these experiments quite difficult, and is behind most of the design considerations that we will discuss. There is also a continuing tradeoff between a large A_0, which makes ΔR_H larger, but $\Delta R_H/R_H$ smaller, and a small A_0, which does the opposite.

3. – Experimental approaches.

The various experimental approaches that have been tried can be categorized according to what they use as an interference amplitude, A_0. The first choice was to use an allowed magnetic-dipole ($M1$) amplitude for A_0. In this case, one drives a transition between two P states with the same principal quantum number. This experimental approach has been used to measure parity violation in bismuth, lead and thallium. In this work, the actual observable is the optical rotation of the plane of linearly polarized light. This corresponds to looking at a difference between the index of refraction for the left *vs.* right circularly polarized light. Obviously, such a difference reflects a handedness to the system. The basic set-up for such experiments is shown in fig. 2. The laser beam passes through a linear polarizer to ensure that its polarization is very clean. It then passes through a vapor cell which contains the atom of interest, and then finally goes into a second crossed polarizer. This second polarizer blocks out all the laser light unless its polarization has been rotated in the vapor, in which case some light passes through and can be seen at the detector. Of course, the actual experiments are somewhat more involved, but this is the basic idea. One then tunes the laser over the atomic transition and observes a small rotation in the polarization as an increase in the light at the detector when tuned on the transition.

Several steps have been taken to improve the signal-to-noise ratio and to test for potential systematic errors, which have always dominated the uncertainty. First, to improve the signal-to-noise ratio, the polarizers are rotated slightly from perfectly orthogonal, and the incident polarization is modulated using a Faraday rotator. This latter step reduces the noise by shifting the detection bandwidth away from d.c. To eliminate sources of potential systematic errors, some or all of the following steps have been taken in the various experiments: 1) alternating between an oven containing atomic vapor and an identical oven with no vapor, 2) reversing the direction of the light through the vapor, and 3) careful fitting to the atomic line shape. The parity nonconservation signal is dispersion shaped and thus has quite a different dependence on laser frequency from the absorption.

The second approach is to use a Stark-induced A_0 amplitude for the interference. In this approach, one simply applies a d.c. electric field to the atom. This electric field mixes S and P states by the Stark effect, in a parity-conserving

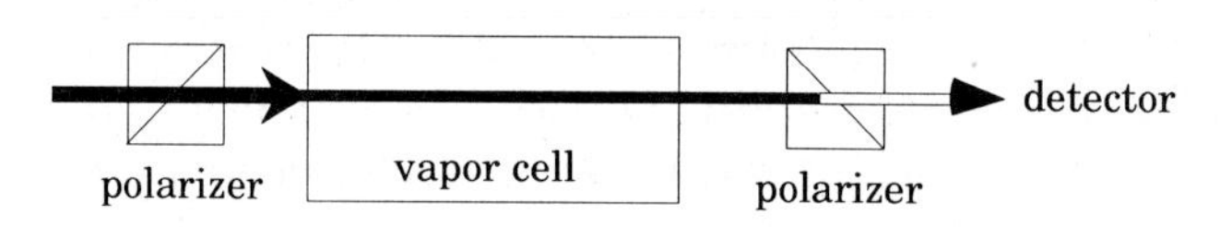

Fig. 2. – Basic experimental set-up for optical-rotation PNC experiments.

way as shown by

$$|S\rangle \to |S\rangle + \delta_{\mathrm{E}}\,|P\rangle + \delta_{\mathrm{PNC}}\,|P\rangle. \tag{10}$$

One now observes an interference between the δ_{E} and δ_{PNC} mixing terms. This approach has been used to measure parity violation in cesium by the groups at Paris and Boulder, and in thallium at Berkeley.

Each of these approaches (allowed $M1$ and Stark induced) has certain advantages and disadvantages. While these often get quite technical, we can summarize some of the more notable features. First, let us consider the $M1$ approach. It can be characterized by the fact A_0 is quite large, which makes $\Delta R/R$ small. However, since the transition rate itself is large, the statistical signal-to-noise ratio can be quite good. Another advantage of these experiments is that they are relatively simple. The drawback to this approach is that, because the fractional modulation is quite small ($\sim$ 1 part in 10^8) and there are relatively few reversals to isolate the PNC component, the systematic errors are major problems.

This can be contrasted with the Stark interference approach. Here A_0 is small and the problems generally are statistical signal-to-noise ratios. There are many reversals to isolate the parity-nonconserving effects and suppress systematic errors and the fractional modulation is relatively large. Therefore, systematic errors are much less of an issue. Because these experiments involve extra applied fields and more reversals, they tend to be rather more complicated, however.

The other important experimental issue is the choice of atom. Here there are a number of considerations that come into play. First, one wants a heavy atom because of the Z^3 dependence (eq. (7)) of the mixing. Second, one needs an atomic transition between states with the appropriate quantum numbers. If one is using the magnetic-dipole interference, this needs to be a nP-to-nP' transition. If one is using the Stark-induced approach, either an $nS \to n'S$ or an $nP \to$ $\to n'P$ will work. In this case it is desirable to have different initial and final n's to suppress the magnetic-dipole amplitude, which can introduce systematic errors and background noise. Also, in all cases it is desirable to have the transition at a wavelength where a good tunable laser is available to excite the atom. If one is using fluorescence detection (as in Stark-induced experiments), it is important that the fluorescing light be at a wavelength that is significantly different from the excitation light. Otherwise, the scattered excitation light will give an overwhelming amount of background. Another consideration is that the atoms must remain isolated rather than forming molecules, where the PNC effects are obscured. The final important issue, and at this stage of the field clearly the most important, is the accuracy with which the γ_5 matrix element for the atom can be calculated. This is important because one is now worried about precision tests of the standard model rather than simply detecting parity nonconservation. We will discuss this issue in more detail later. Here we will simply summarize that, for alkali atoms such as cesium, francium and rubidium, these calculations can be done with an accuracy on the order of 1%. In thallium calculations the accuracy is on the order of 5%, while for systems such as bismuth and lead, with many valence electrons, the accuracy of calculations of the γ_5 matrix element is relatively low.

Before discussing the experiments in more detail, we might mention a few other possibilities for PNC experiments which have been proposed but not carried out. Among other interference amplitudes that can be used, there really are very few that have been seriously discussed. There have been proposals by ANDERSON [6] to interfere a two-photon-allowed transition amplitude with the one-photon PNC amplitude. In this case, the interference is sensitive to relative optical phases. This can be both an advantage and a disadvantage, however, and this approach has not been seriously pursued up to this time.

A number of other experiments have been proposed, and a few attempted, involving other atomic species. Probably the most notable is the use of atomic hydrogen. In this experiment, the idea was to observe the mixing of the $2S$ and $2P$ states of hydrogen. Although Z is 1, and, therefore, the PNC matrix element is quite small, this is largely offset by the mixing energy denominator, which is nearly zero for these two states. Hence, the actual δ_{PNC} for $n = 2$ hydrogen is nearly the same as that for heavy atoms. Because of this there were a number of experimental programs initiated to study PNC in hydrogen. However, the great problem with the hydrogen case is systematic errors. Stray electric fields, which can also cause mixing of the $2S$ and $2P$ states, are amplified by the same near-zero energy denominator. As a result, the systematic errors, relative to the PNC signal, are enhanced compared to heavy atom by a Z^3 factor. Effectively, this means that, instead of needing to worry about mV/cm stray fields, one has to worry about nV/cm fields. This is a nearly impossible problem and, therefore, to our knowledge, all the experiments on hydrogen PNC have now been abandoned.

Another set of experiments which have been proposed and pursued to some degree involve the use of ions or muonic atoms for studying PNC. In these cases, the overlap of the electrons at the nucleus is much larger than for a normal atom, and, therefore, δ_{PNC} can be relatively large. However, this is more than offset by the fact that the sample size is very small. At the present time, the technology is not available to produce large enough samples to allow meaningful PNC measurements. However, this is very much a function of technology, which is likely to change.

We would now like to discuss the first generation of Stark interference experiments which were carried out in Paris [7] and Berkeley [8]. The basic schematic for these experiments is shown in fig. 3. A circularly polarized laser beam is sent into an atomic-vapor cell and excites the transitions of interest in the presence of a d.c. electric field. The transition rate is monitored by observ-

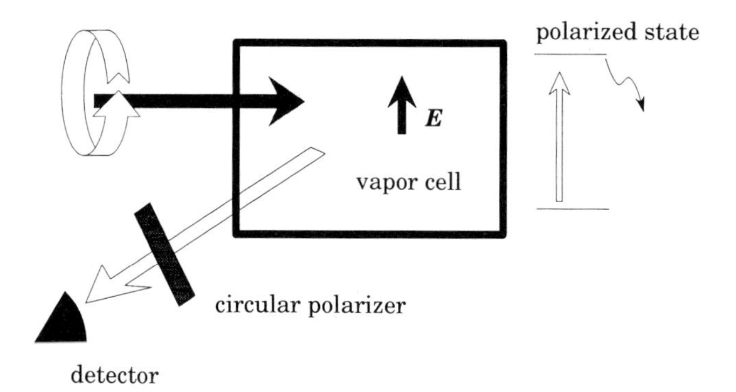

Fig. 3. – Basic experimental set-up for the first-generation Stark interference PNC experiments.

ing the fluorescence as the atom decays back to the ground state. Since the Zeeman transitions are not resolved, the parity-nonconserving interference term cannot be observed directly in the total atomic-transition rate, but it does cause a polarization of the excited state. This polarization is detected by looking at the degree of circular polarization of the fluorescence light. These experiments were successful at detecting a small circular polarization and hence a parity violation. It was measured with a fractional uncertainty at a level of $(10 \div 20)\%$.

Before beginning a lengthy discussion of the design considerations of the Colorado experiments, we would like to briefly review the lessons we have learned from these pioneering first-generation optical-rotation and Stark interference experiments. The two main lessons from the optical-rotation experiments are that the systematic errors are very serious and must be considered in great detail, and the atomic-structure issue is crucial if one is interested in a precision test of the standard model. These considerations made us decide to go with the Stark interference approach and use cesium, since Cs is a heavy alkali atom. However, there were also several important lessons from the Stark interference experiments. The first was that the signal-to-noise ratio was disappointingly low, predominantly due to problems from background signals. These signals can arise from sources such as blackbody radiation, molecules, collision-induced transitions and scattered light. Also, in these cell experiments there were a very large number of nonresonant atoms, atoms that were not being excited by the laser beam and did not contribute to the PNC signal. These atoms could give rise to background noise or systematic errors, which is clearly undesirable.

Finally, a nonobvious design feature is the issue of experimental flexibility. In such cell experiments, it is very difficult to build a new cell and, therefore, one is quite limited in how easily one can change the experiment in response to new data or ideas.

4. – Design concepts of Colorado experiments.

After reviewing the lessons from the work described in the previous section, we had three major concerns. The first was to have flexibility in the apparatus so that the experiment could be easily modified. The second concern was to increase the statistical signal-to-noise ratio (S/N), and the third was to control potential systematic errors.

To improve the S/N, we set out to achieve high detection efficiency and minimal background. One way to achieve this is to have a modulation directly in the transition rate rather than the polarization of the excited state. An applied magnetic field would allow one to do this by resolving the different m levels. It is particularly useful in conjunction with a collimated atomic beam because much weaker magnetic fields are required.

There are also several other desirable features to an atomic beam. First, it has rather few nonresonant atoms and molecules and allows high detection efficiency. Second, the beam can be turned off and the system opened up and changed quickly. The major drawback is that the actual number of resonant atoms one has is significantly lower than in a cell. However, we felt that the other factors would more than offset this, which has proved to be the case.

The primary feature for controlling potential systematic errors is to design an experiment with many mirror reversals. However, in addition, it is important to have ways to quickly measure potential systematic errors with high precision.

It is quite easy to understand how the addition of a magnetic field helps in these experiments. The A_{PNC} amplitude is proportional to the magnetic quantum number m, while the Stark interference amplitude A_{E} is proportional to the absolute value, $|m|$. Therefore, the sum over all m of the product $A_{\mathrm{PNC}}\, A_{\mathrm{E}}$ equals zero. This arises because of the requirement of time-reversal conservation. However, if one only looks at transitions between individual m levels, then the product $A_{\mathrm{PNC}}\, A_{\mathrm{E}}$ is not equal to zero, and will contribute to the transition rate. There are two ways one can experimentally observe transitions between individual m levels. The first way is to simply apply a large enough magnetic field and use the Zeeman effect to resolve the different m levels. Then the laser will only excite transitions from a single m. This approach was used in the Colorado 1985 and 1988 experiments [9, 10], and in the Drell and Commins thallium experiment [11] at Berkeley. The second option is to simply optically pump all the atoms into a single m level. This has the advantage of utilizing the atoms much more efficiently, but it does make the experiment more complicated. This approach is being used in the current experiment at Colorado.

The relevant energy levels of the cesium atom are shown in fig. 4. The $6S$ ground state has two hyperfine states with total angular-momentum quantum numbers $F = 3$ and $F = 4$. The $7S$ excited state also has $F = 3$ and $F = 4$ hyperfine states. In the presence of a magnetic field, each of these levels then splits

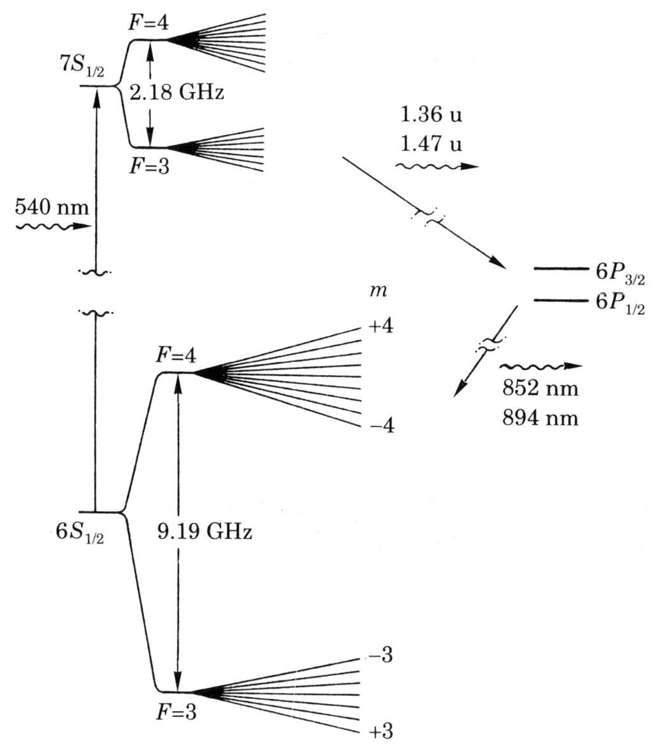

Fig. 4. – Cesium energy level diagram.

up into the different m levels, and one excites $6S \rightarrow 7S$ transitions between these independent m levels with 540 nm laser light. The excitation of the $7S$ state is monitored by observing the fluorescence produced primarily at 852 and 894 nm when the $7S$ state decays by an allowed transition to the $6P$ and then to the $6S$ states. In fig. 5, we show the theoretical spectrum of the $6S$, $F = 4 \rightarrow$

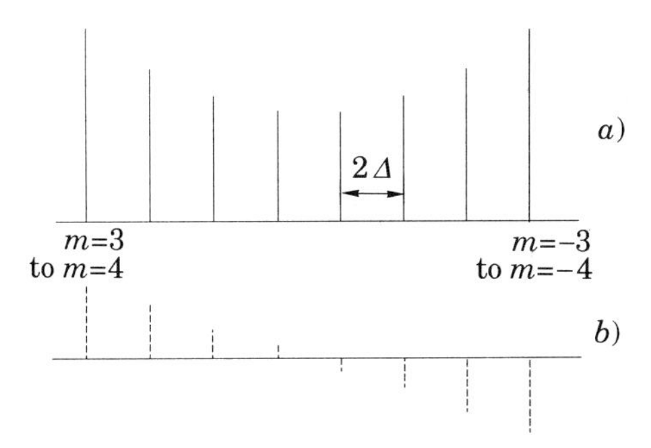

Fig. 5. – Theoretical spectrum of the $6S$, $F = 3 \rightarrow 7S$, $F' = 4$ transition of cesium in a weak magnetic field. *a*) The pure Stark-induced rate (A_{E}^2), *b*) the $A_{\mathrm{E}}\, A_{\mathrm{PNC}}$ interference terms multiplied by 10^6.

$\rightarrow 7S$, $F' = 3$ transition in a weak magnetic field. Figure 5*a*) represents the pure Stark-induced rate that is proportional to A_E^2; figure 5*b*) represents the $A_E A_{PNC}$ interference terms. This illustrates how the parity-nonconserving contribution adds to the rate for the $+m$ levels, while subtracting for the $-m$ levels.

A general layout of the experiment is shown in fig. 6. A collimated cesium beam is intersected by a laser beam which excites the $6S \rightarrow 7S$ transition. In the intersection region, there are three perpendicular vectors defining a coordinate system. These are a d.c. electric field, $\boldsymbol{E}$, a d.c. magnetic field, $\boldsymbol{B}$, and the angular momentum σ of the laser photons. The $6S \rightarrow 7S$ excitation rate is monitored by observing the subsequent fluorescence. In fig. 7, we show a typical spectrum observed when we sweep the laser over one of the hyperfine transitions. This shows eight lines of the Zeeman multiplet corresponding to the excitation of different m levels. It is quite obvious that there is a rather peculiar line shape. The explanation of this line shape is somewhat complicated. It involves the off-resonance a.c. Stark shift in combination with the small Doppler shifts. (This has been analyzed and explained in ref. [12].) The values of laser power and cesium beam divergence for this figure make the line shape look its most peculiar. For larger or smaller power, or less beam divergence, the lines become smoother and more systematic. In any case, this strange line shape is a minor anomaly and has no real effect on the parity nonconservation experiments.

To summarize the basic experiment, we set the laser frequency on one of the peaks of the Zeeman multiplet and look for a change, typically a part in 10^6, in the transition rate when the parity or handedness of the experiment is reversed. There are many different ways to reverse the handedness; most of them can be seen simply by considering the coordinate system defined by $\boldsymbol{E}$, $\boldsymbol{B}$ and σ. Anything that reverses the handedness of this coordinate system will change the sign of the parity-nonconserving term. Equivalently, this can be described as making a mirror reversal with the mirror oriented in three different planes. Thus the three reversals are $\boldsymbol{E} \rightarrow -\boldsymbol{E}$, $\boldsymbol{B} \rightarrow -\boldsymbol{B}$, $\sigma \rightarrow -\sigma$ (right circular polar-

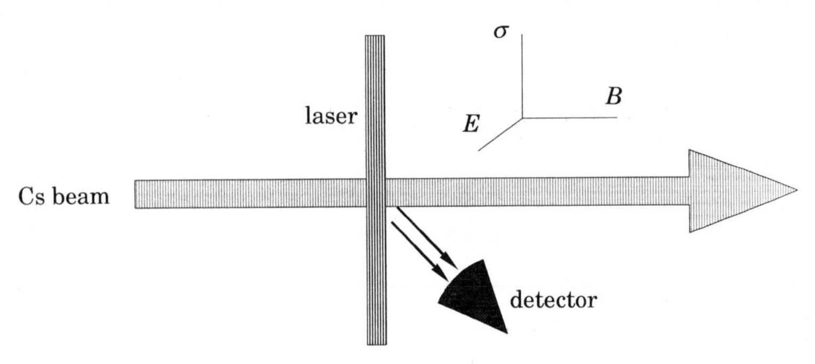

Fig. 6. – General layout of the Boulder cesium PNC experiment. The coordinate system in the interaction region is defined by a d.c. electric field, $\boldsymbol{E}$, a d.c. magnetic field, $\boldsymbol{B}$, and the angular momentum, σ, of the laser photons.

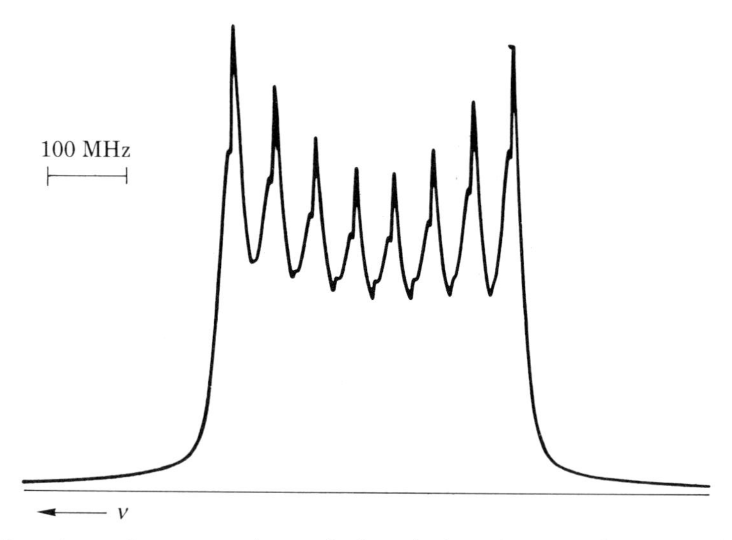

Fig. 7. – Experimental spectrum observed when the laser is scanned over the $6S$, $F = 3 \rightarrow$ $\rightarrow 7S$, $F' = 4$ transition of cesium in a 70 G magnetic field.

ized light to left circular polarized light), and, finally, $m \rightarrow -m$. This m reversal is carried out by simply changing the frequency of the laser so that it moves from one Zeeman peak to the symmetric Zeeman peak on the other side of the multiplet.

The presence of four independent reversals is crucial for suppressing systematic errors. Having many reversals provides a redundancy which tests that the signal observed truly is parity violating. It greatly suppresses potential systematic errors which might mimic PNC under one or two reversals, but are unlikely to pass the test of all four.

While the experiment is rather simple in principle, in fact a tremendous amount of time and effort has gone into optimizing the apparatus to achieve the necessary signal-to-noise ratio. The next section will discuss in detail these signal-to-noise issues, and how they have led to the construction of the apparatus in its present form.

5. – Details of apparatus.

5·1. *Signal and noise analysis.* – The majority of the time spent on this experiment has been used to achieve the necessary signal-to-noise ratio. The total detector current is given by

$$R \propto \{(\text{number of atoms})(\text{laser power})(\text{detection efficiency})\} \cdot \cdot [E^2 \pm 2E\,\delta_{\mathrm{PNC}}] + \text{background}, \tag{11}$$

$$S = \Delta R \propto \{\ \}[4E\delta_{\mathrm{PNC}}] \text{ and is } 10^{-5} \text{ to } 10^{-6} \text{ of } R \text{ in this experiment}.$$

The main signal depends on the number of atoms, the laser power and the detection efficiency multiplied by the term in brackets involving the electric field. Of course, the largest electric-field term is the E^2 pure Stark component. Then there is the much smaller PNC interference term which is linearly proportional to the electric field. The actual signal of interest, S, is the change in the detector current when the fields are reversed. This is proportional to the number of atoms $\times$ laser power $\times$ detection efficiency $\times\, 4E\delta_{\rm PNC}$. The noise, however, is made up of several parts which we will characterize according to their dependence on the electric field E as given by

$$N = N_{\rm bk} + \text{(detector noise + fluctuations in background)} \tag{12}$$

$$+N_{\rm sh} + \qquad \text{(shot noise on } R, \propto \sqrt{\{\ \}E^2}\text{)}$$

$$+N_{\rm tech} \qquad \text{(technical noise on } R, \propto \{\ \}E^2 f\text{)}.$$

First there is the background noise, which is independent of E. This can be the fluctuations in the signal due to the fundamental detector noise or any backgrounds due to scattered light, room lights, etc. The second term comes from the shot noise fluctuations on the total detector current. Since this noise is proportional to the square root of the current, it is linearly proportional to E. Finally, the third term, what we call the technical noise, is proportional to E^2 times some fluctuation fraction f. Here, f can be due to changes in a large number of things, for example, laser power, laser frequency and atomic-beam intensity. Because of this dependence of the noise on E, one has an interesting dependence of the signal-to-noise ratio of S on the electric field. This is shown in fig. 8.

For an ideal case, where ideal is rather low technical and background noise, one would have the signal-to-noise ratio rising linearly with electric field in the region where one is dominated simply by background noise. At higher field, shot noise, which is proportional to E, begins to dominate. At this point, the signal-to-noise ratio is independent of E since both the signal and the noise are

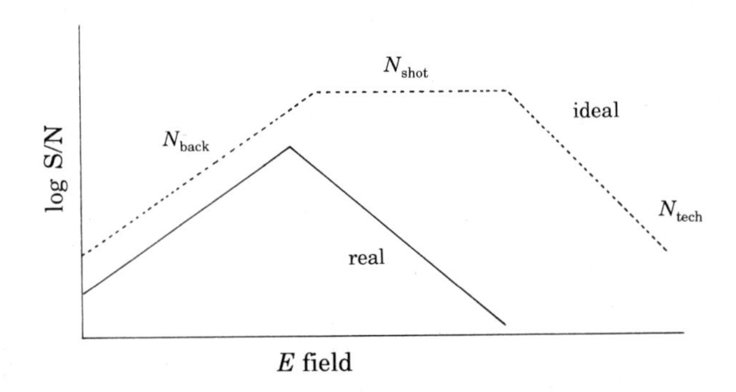

Fig. 8. – Dependence of the signal-to-noise ratio on the electric field.

proportional to E. Finally, at even larger E fields, the technical noise with its E^2-dependence will dominate. When this occurs, the signal-to-noise ratio then falls as $1/E$. In this ideal case one could then set the electric field anywhere in the region where the shot noise dominated and achieve the same signal-to-noise ratio. In fact, this is not the case in the real world.

In these experiments, we have always had a situation more like that shown in the lower curve of fig. 8 where the background and technical noise are larger. Here, the signal-to-noise ratio rises with the electric field until it reaches some peak, above which the technical noise dominates. There is no actual flat region where the shot noise dominates, and, therefore, there is always some optimum electric field. We work hard to push this peak up until it is fairly close to the shot noise limit, and then there is little to be gained by further improvements. At this field, typically the background noise and the technical noise are equal. The signal-to-noise ratio works out to be roughly proportional to the square root of the number of atoms times the laser power times the detection efficiency. In order to reach the shot noise signal-to-noise ratio, however, it is necessary to make the background noise small and to minimize the technical noise by making the quantity f small. The former means keeping the detector noise and scattered-light noise low, while the latter requires a number of parameters be highly stabilized. The primary feature that makes these experiments so difficult is the fact that achieving this to the level necessary to make high-precision PNC measurements requires pushing many different technologies to the state of the art or beyond the state of the art. This is different from many experiments where most of the experiment is based on well-developed technology and there are only one or at most two different technologies that one has to work on very hard.

Since the signal is proportional to the number of illuminated resonant atoms, the laser power and the detection efficiency, it is clearly important to make all three of these large. We will discuss efforts to maximize each of these factors separately.

5'2. *Atomic beam.* – In this experiment for an atom in the beam to be «useful» it has to meet slightly different requirements from most atomic beams, where the only requirement is either flux or density. In this case a long thin laser beam is exciting a narrow (2.8 MHz) transition. Thus atoms which have a divergence such that their velocity parallel to the laser beam gives a Doppler shift larger than 2.8 MHz are out of resonance with the laser and will not contribute to the signal. The divergence in the perpendicular direction affects the density but does not cause a Doppler shift. Thus the three basic considerations are: the density of atoms in the beam, the length of atomic beam intercepted by the laser and the velocity of the atoms parallel to the laser beam (the divergence in one direction).

If one starts with an atomic vapor and wants to make a collimated beam out

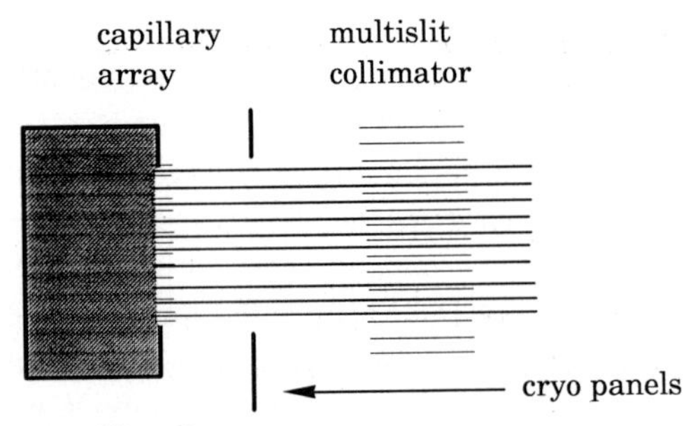

Fig. 9. – Cesium beam collimation.

of it, the simplest approach is to let it effuse out of a tube of length L and radius R. This tube is the output nozzle on an oven. In this case the intensity in the forward direction will increase as the pressure of the vapor is increased until the mean free path of the atoms in the vapor becomes less than L. At this point, the atoms coming out of the oven simply form a cloud, the divergence of the beam increases, and the number of atoms in the acceptable divergence angle will stay constant or decrease. To improve on this, we have used a technology that had been successful for helium, but which had never been successfully used for cesium. This technology is a glass capillary array nozzle. Such arrays are made up of thousands of very tiny tubes, typically 10 μm in diameter and a millimeter long. In order to increase the width of the beam, we use a section of array which is 0.4 cm by 2 cm so that it provides a collimated ribbon of Cs atoms. This provides a more intense beam with the appropriate divergence angle than provided by a single tube or slit.

We have found that operating a glass capillary array at the pressure where the mean free path is equal to L does not give the most intense possible beam. We have improved on this by using a glass capillary array as a nozzle on the oven which is operated at a vapor pressure which substantially degrades the collimation. About 2 cm from the array we place a multislit collimator made up of a large number of thin metal veins which provides collimation only along the laser beam direction (fig. 9). In between these two collimators, we have liquid-nitrogen-cooled panels which pump away the background Cs vapor. To break up the Cs_2 dimers, which cause a very noisy background signal, we keep the glass capillary array about 100 °C hotter than the rest of the oven. While we have been able to produce a state-of-the-art atomic beam in this manner, the beam intensity was lower than what we hoped for. The intensities we expected were based on results that had been achieved with helium, but we have achieved less than 1/10 of this goal.

5·3. *Laser power build-up.* – Fortunately, we have been able to compensate for lower than anticipated beam intensity by having more laser power than we had originally anticipated. The 540 nm laser light which excites the atoms

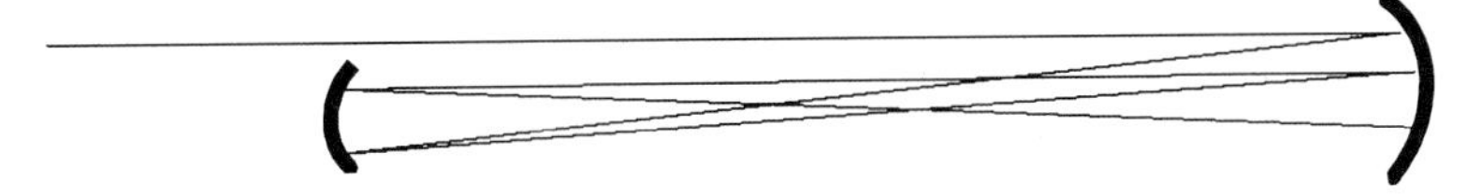

Fig. 10. – Multipass system for reusing laser light.

comes from a c.w. dye laser. The transition rate is very small, so only a small fraction ($< 10^{-11}$) of the light is absorbed as it passes through the atomic beam. This means that one can gain tremendously by reusing the light. There are two common ways to reuse light when looking at weak transitions. The first technique, which has been by far the mostly widely used, is to have a multipass system as shown in fig. 10. Here the light comes in and is reflected back and forth between two mirrors and then finally leaves without ever overlapping itself. This technique was used in the Bouchiat cesium experiment, and we initially tried to use it for our experiment, but we discovered it had some disadvantages. First, the number of bounces one can get is limited to the size of the mirrors divided by the size of the laser beam, since the beam occupies a different spot for each bounce. With a reasonably sized mirror, one has a limit of a few hundred bounces. Also, the interaction region becomes quite extended, which is a very serious limitation if one wants efficient detection. This is because imaging fluorescence from an extended region is always much less efficient than if the fluorescence is very localized. Finally, there is also a somewhat subtle effect involving systemic errors due to birefringence of the mirrors which is difficult to deal with in this configuration because one can only observe the average effect of many reflections.

As a result of these difficulties, we switched to a different configuration, shown in fig. 11. This is known as a «build-up cavity» because a resonant Fabry-Perot interferometer builds up the laser power between the two mirrors. The most important issues in a build-up cavity are the losses and transmissions of the two mirrors. To optimize the build-up, the transmission of the output mirror should be as low as possible. The transmission of the input mirror should be between one and two times the round-trip loss in the cavity. Of course, any optics book will provide the equations you need to calculate the exact optimum. Physically, what is happening is that, if the input transmission is much lower than the losses, the light will die out inside the cavity faster than it builds up. On the other hand, if the transmission is much larger than the losses, then the

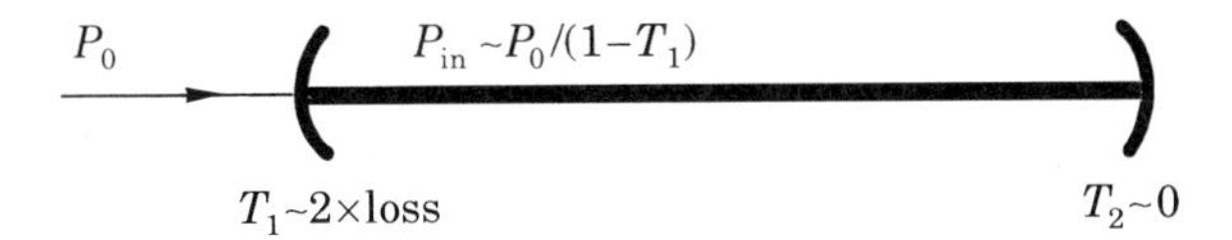

Fig. 11. – Laser power build-up cavity. T_1 is the transmission of the input mirror, T_2 is the transmission of the second mirror.

light will leak back out the way it came in before it can reflect back and forth many times. From this argument it is easy to see why the optimum transmission should be comparable or slightly more than the total losses due to absorption and scattering. In this case, one achieves a traveling-wave power inside the cavity which is equal to the incident power times the build-up factor $B = (1 - T_1)^{-1}$, where T_1 is the input transmission. This makes it clear that the build-up one obtains is limited entirely by the mirror losses. Over the course of this work, as we have obtained improved mirror coatings, we have progressed from a build-up factor of 100 to 1000 and, in our latest work, to 15000. With 0.3 W incident power, we have 4.5 kW of circulating power inside the cavity.

Unfortunately, this tremendous enhancement in the laser power, and thus in the signal, does not come without a price: laser stabilization. This is the primary disadvantage of the build-up cavity relative to a multipass cavity where there is no resonance. For a build-up cavity, the laser must be stabilized to the cavity resonance, which is only about 8 kHz wide for a build-up factor of 15000. However, it is not sufficient to simply stabilize the laser to 8 kHz. If the laser varied by 8 kHz, our $6S \rightarrow 7S$ signal would be varying by nearly 100%, while the desired measurement accuracy is about 1 part in 10^7. Therefore, it is necessary to have the laser stabilized to a small fraction of this 8 kHz cavity resonance width. Since the typical free-running dye laser has short-term frequency variation of about 10 MHz, this puts severe demands on laser stabilization. In addition, even if the laser is locked to the cavity, should the cavity move relative to the atoms' transition frequency, there will also be large changes in transition rate. Hence, the cavity must be stabilized to the atomic resonance to within a small fraction of the atom's natural linewidth. This requires considerable attention to the mechanical and thermal stability of the cavity. Ultimately all this stabilization is done with servo loops; we will discuss some of the concerns of servo systems used in this experiment in more detail in subsect. 5˙6.

5˙4. *Detection.* – The third important factor that determines the signal size is the detection efficiency of the $6P \rightarrow 6S$ fluorescence which is produced as a result of the $7S$ excitation. As mentioned earlier, the more localized the source, the more efficiently one can collect the emitted light. Although we do not have a point source, we do nearly as well by having a line source of fluorescence, which is defined by the region where the thin laser beam intersects the ribbon cesium beam (fig. 12). We use a cylindrical mirror to focus the light from this intersection region onto a long, narrow detector. Essentially we have a two-dimensional system, which allows us to obtain a rather high collection solid angle of nearly 2π sr.

A major difficulty is that the fluorescence photons must be collected from the same region where we must apply very carefully controlled electric fields. The only practical way to accomplish this is with parallel field plates, and hence the photons must pass through the electric-field plates. This required the de-

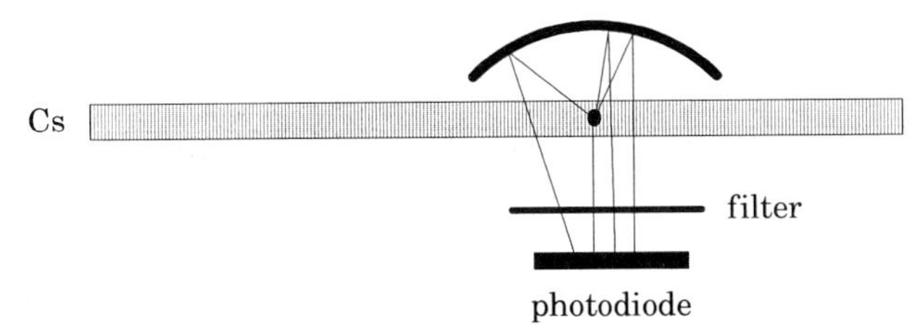

Fig. 12. – Fluorescence collection and detection.

velopment of appropriate transparent, highly conducting coatings for glass. The development and stability of these coatings has been a major headache in this experiment.

Silicon photodiodes operating at liquid-nitrogen temperature are used for the detection. Although it is generally believed that a photomultiplier is best for very sensitive detection, this is not true for relatively large signals. Silicon photodiodes have nearly 100% quantum efficiency at 850 nm, but a photomultiplier detection efficiency is only about 10% or less. Of course, a PIN photodiode has no gain, whereas a photomultiplier has gain which introduces very little noise. However, if the signal is large enough that the photon shot noise is comparable to the detector noise, then the photodiode is superior due to its quantum efficiency. Thus it is quite important to have the photodiode noise small. One way to accomplish this is to have a small diode, but this is incompatible with having a large detection solid angle. However, even with a large (1 cm^2) photodiode, we have been able to achieve a detector noise comparable to about 10^4 photons/$s^{1/2}$ by cooling it to liquid-nitrogen temperature. It is also necessary to use a very low noise op-amp with a large feedback resistor in the current-to-voltage amplifier so that amplifier noise is not a significant limitation. This means that the photodiode actually gives the best signal-to-noise ratio for signals larger than about 10^7 photons/s. Clearly, if one is interested in detecting effects of a part in 10^6, it is necessary to have far more than 10^7 photons/s.

Having efficient detection with low detector noise is only half the detection problem. The other half concerns the noise arising from background signals. The major source of background is the scattered laser light reaching the detector. This is a large problem because there are 10^{22} green laser photons in the power build-up cavity for every 10^9 infrared photons produced. Thus an isolation factor of 10^{13} is required! This isolation is achieved partly by geometry; simply putting in light baffles which absorb any green scattered light and shield the detector. Of course, this is never perfect, so we also use colored glass filters to block out the green light but pass the infrared photons. Colored glass filters are not as selective as interference filters, but they have a large acceptance angle, which allows much better detection efficiency. When we first looked at filter specifications, we thought there would be no problem in obtaining a filter that suppressed green light and transmitted near i.r. light. However, this was note the case. The specifications do not mention that green photons cause i.r. fluorescence in the filter at a part in 10^5 or 10^6, which is then transmitted through the filter. We spent considerable time studying this process and investigating many different filters. We found the conversion efficiency varies widely, but eventually obtained filters that worked quite well. It is interesting to note that the best filter was not one of the standard colored glass filters, but in fact was the red plastic in the back cover of the Schott filter book.

5·5. *Technical noise.* – These various signal and background issues led to the final design of the 1985 and '88 experiments shown in fig. 13. Light from a c.w. dye laser passes through optics, which control polarization very precisely, and then enters the power build-up cavity that is inside a vacuum chamber. It is necessary to have the cavity mirrors inside the vacuum chamber because there are no windows with low enough birefringence and loss to be put inside the cavity. The dye laser is frequency stabilized to the PBC with an electronic servo loop and its intensity is stabilized with another servo loop. The light inside the power build-up cavity intersects the atomic beam in a region of d.c. electric and magnetic fields. The details of the interaction region are shown in fig. 14.

Once this apparatus is assembled, the prime concern is reducing the technical noise and eventually reducing the size of the systematic errors. The technical noise must be reduced to a level where the modulation fraction, f, is less than or equal to $10^{-5}/\sqrt{\text{Hz}}$ ($N_{\text{tech}} = f \times R$, where R is the signal size). There are an astonishing number of processes that can produce technical noise at this level, including several that would not be encountered in a more typical experiment. For example, we have seen fluctuations in the atomic beam at this level. We found that the standard atomic beam is actually very quiet, but fluctuations of the pressure in the vacuum chamber of $\geqslant 10^{-8}$ Torr cause noise on our signal. We found that such pressure fluctuations come from the release of small bub-

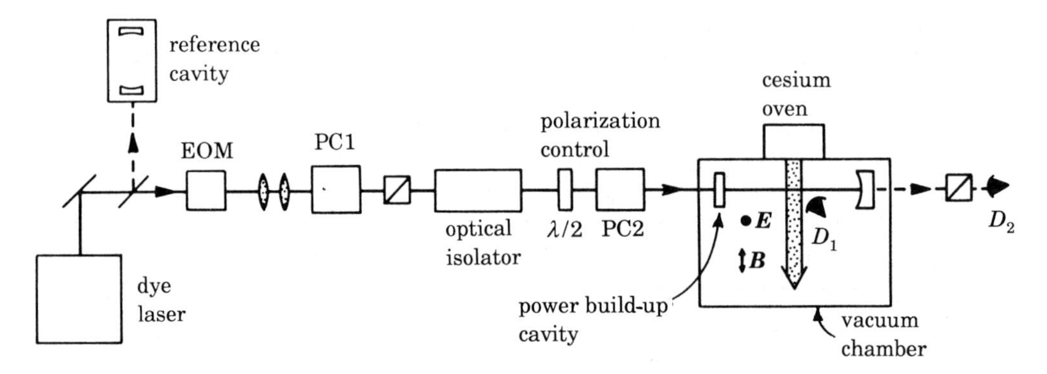

Fig. 13. – Overall schematic of the apparatus for the 1985 and 1988 Boulder cesium PNC experiment.

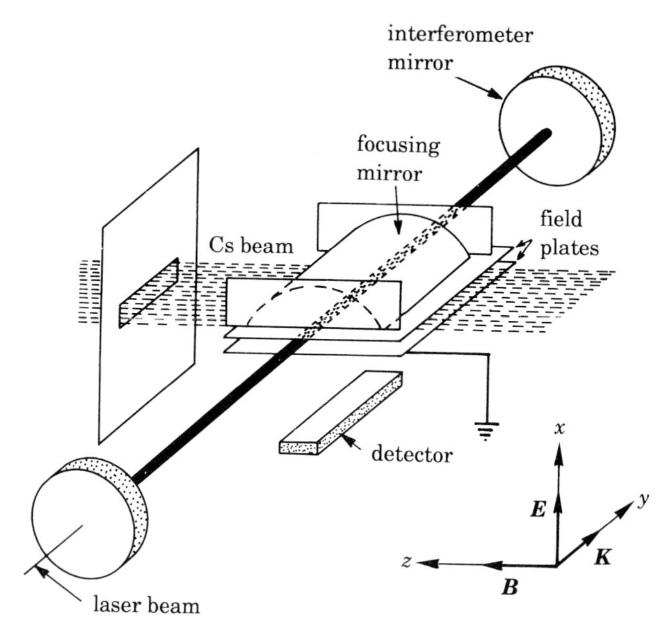

Fig. 14. – Details of the interaction region.

bles of gas. These can be eliminated with proper design of the vacuum and pumping system.

A more obvious source of technical noise is the laser power fluctuations inside the power build-up cavity. These can come from fluctuations in the laser power itself or fluctuations in the laser frequency. The latter is the more difficult to correct; as we mentioned earlier, an 8 kHz change in the laser frequency, relative to the cavity resonant frequency, leads to $f = 1$. Therefore, considerable effort must be made to stabilize both the frequency and the intensity of the laser. This is done with several servo loops. First, the laser is locked to the build-up cavity using the Pound-Drever frequency modulation technique. This lock has good signal-to-noise ratio and a rather fast response time, so that one can correct the errors in a time as short as 1 μs. However, the length and hence the resonant frequency of the build-up cavity can fluctuate because of inherent instabilities in the mechanical design and vibrations of the environment. To avoid this we lock the build-up cavity to a very stable reference cavity. Because this cavity does not have an experiment inside it, it can be much more rigid. The cavity-to-cavity lock also has good signal-to-noise ratio, but has a relatively slow response (400 Hz). This response is limited by the fact that we cannot move the build-up cavity mirrors very fast without introducing strains that cause birefringence and lead to systematic errors. Although the reference cavity has good short-term stability, it is subject to slow drifts and, therefore, it must be locked to the atomic-transition itself. This lock has relatively poor signal-to-noise ratio because it relies on the atomic-transition rate, which is low in this experiment. Fortunately, the drift rate of the reference cavity is small, so one does not require the high signal-to-noise ratio necessary for a very fast servo response. The signal-to-noise ratio limitation on this lock is the reason we use the reference cavity rather than locking the frequency of the build-up cavity directly to the atoms.

5·6. *Servo systems.* – Servo-control systems are very important to this and many other «frontier» experiments of the sort discussed elsewhere in this volume. Because they are often not part of a general physics education, we will provide a brief introduction to servo theory, which will allow one to grasp the key issues involved in the stabilization of systems. A basic servo loop is shown in fig. 15. The laser has some noise, for example, the table vibrations which move the mirrors and thus cause the output frequency to vary. One can compare this output frequency with some reference, such as a stable optical cavity, and from this obtain an error signal that indicates how far the frequency has moved away from what is desired. The basic error signal is then modified by electronics which provide gain and some type of compensation and the result is a feedback signal that is sent back to the laser. This feedback is negative, so it cancels the errors introduced by the table vibration and shifts the frequency back to where it is supposed to be. With this feedback, the error signal, $E_{\rm f}$, becomes $E_{\rm f} = E/\text{gain}$ in the limit of $\text{gain} \gg 1$, where E is the error signal with no feedback. Thus one wants to make this gain large in order to push the error signal as close to zero as possible, which is equivalent to forcing the laser frequency to be the same as the reference frequency.

The key point in designing any servo loop is that there are always time delays in the system, and much of the design is based on dealing with these time delays. This can be best understood by considering the laser mirror. The mirror is attached to a piezoelectric transducer (PZT) which will stretch it when voltage is applied. Although the PZT is a hard ceramic and the rest of the mount is a solid metal piece, on the scale of frequency errors and corresponding distances

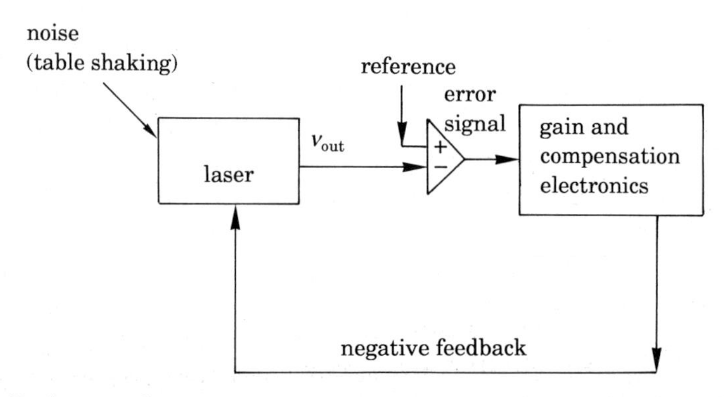

Fig. 15. – Basic servo loop.

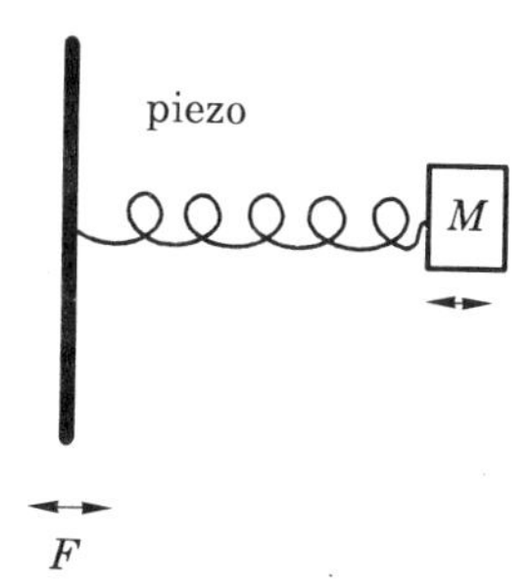

Fig. 16. – Spring analogy for PZT.

relevant here (atomic diameters or less), there is no such thing as a stiff mount. In fact, at these atomic scales the stiffest PZT is incredibly spongy and can compress many wavelengths. Therefore, a PZT can be best visualized as a spring with the mass of a mirror mounted on it, as shown in fig. 16. We then move the back end of the piezo to try to correct the position of the mirror and compensate for random vibrations. If one considers how a mass that is attached to a wall by a spring responds to sinusoidal motions of the wall as a function of frequency, this is just a driven harmonic oscillator. For frequencies below the resonant frequency, the response is in phase with the drive. However, above the resonant frequency of the mass-spring system, there is a 180° phase lag and the mirror moves in the opposite direction of the driving force. This shows the primary problem encountered in designing any servo system; if the amount of feedback is the same amplitude and phase relative to the error signal for all frequencies, the servo works fine for correcting for errors that fluctuate at frequencies lower than the resonant frequency. However, above the resonance frequency, this 180° phase shift causes positive feedback and the system becomes an oscillator if the feedback gain is greater than 1.

This is obviously unacceptable. A straightforward solution is to make the gain smaller than 1 for frequencies above the resonant frequency and make it larger than 1 for lower frequencies. This gain is produced simply by having the compensation electronics include a simple low-pass filter. The normal way to operate a servo system containing such a filter is to turn up the gain until it just starts to oscillate at the 180° phase shift point, and then reduce the gain slightly so that it stops oscillating. This provides a stable servo system. Notice that the gain at low frequencies is set by the response (phase lag) of the system at high frequencies. This is characteristic of any servo system.

Here we have presented the simplest possible compensation. In a more advanced servo design, one would put in more elaborate compensation involving electronic circuits that change the phase shifts and gain with frequency. With such systems one can optimize the gain at particular frequencies where the noise might be especially large or where one wants to have the system be particularly stable, such as at the frequency where the data are being acquired. Other reasons for more elaborate compensation are to improve the transient response so the system can recover more rapidly from a sudden shock. In the remainder of this section we provide a few more examples of relatively straightforward compensation and how one can deal with various kinds of phase lags in systems.

Several methods have been developed for designing servo systems—frequency response methods, the root locus method and state space methods. The method that one chooses depends on the requirements of the servo design, such as transient response and steady-state error. We will describe frequency response methods from an experimentalist's point of view. We have chosen this method because it is very easy to measure a system's frequency response with modern signal analyzers.

The basis of the frequency response method is the Bode plot. The Bode plot shows a system's gain and phase as a function of frequency. Both the system to be controlled and the compensation have characteristic Bode plots. When designing a servo, one first measures the frequency response of the system to be controlled and then designs a compensation circuit that tailors the open-loop frequency response to provide the desired control. It is important to keep in mind that high gain at frequencies where the phase is less than 180° provides good control, but low gain is required at frequencies where the phase is greater than or equal to 180°.

As an example of how to tailor the frequency response of a system with compensation, consider a simple harmonic oscillator. Many physical systems can be modeled as damped harmonic oscillators. The Bode plot of a harmonic oscillator is shown in fig. 17. There are two different goals one can have in trying to con-

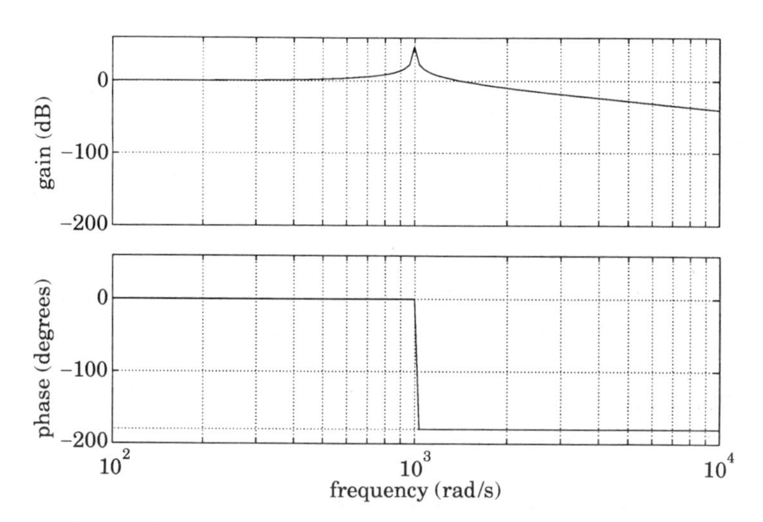

Fig. 17. – Bode plot for damped harmonic oscillator.

trol this system: i) to minimize the steady-state error, *i.e.* to have a large d.c. gain; or ii) to maximize the bandwidth of the servo, *i.e.* to provide damping of the resonant peak with feedback.

As we mentioned previously, the simplest way to prevent oscillation is to make the gain smaller than 1 at frequencies at which the phase shift is greater than 180°; a Bode plot of such compensation (an integrator, or a low-pass filter) is shown in fig. 18*a*). Figure 18*b*) shows the resultant frequency response for the oscillator-compensation system. Note that the gain at low frequencies has increased, but the phase shift has reached 180° at a lower frequency. We now have a larger d.c. gain but a smaller bandwidth. The controlled oscillator will

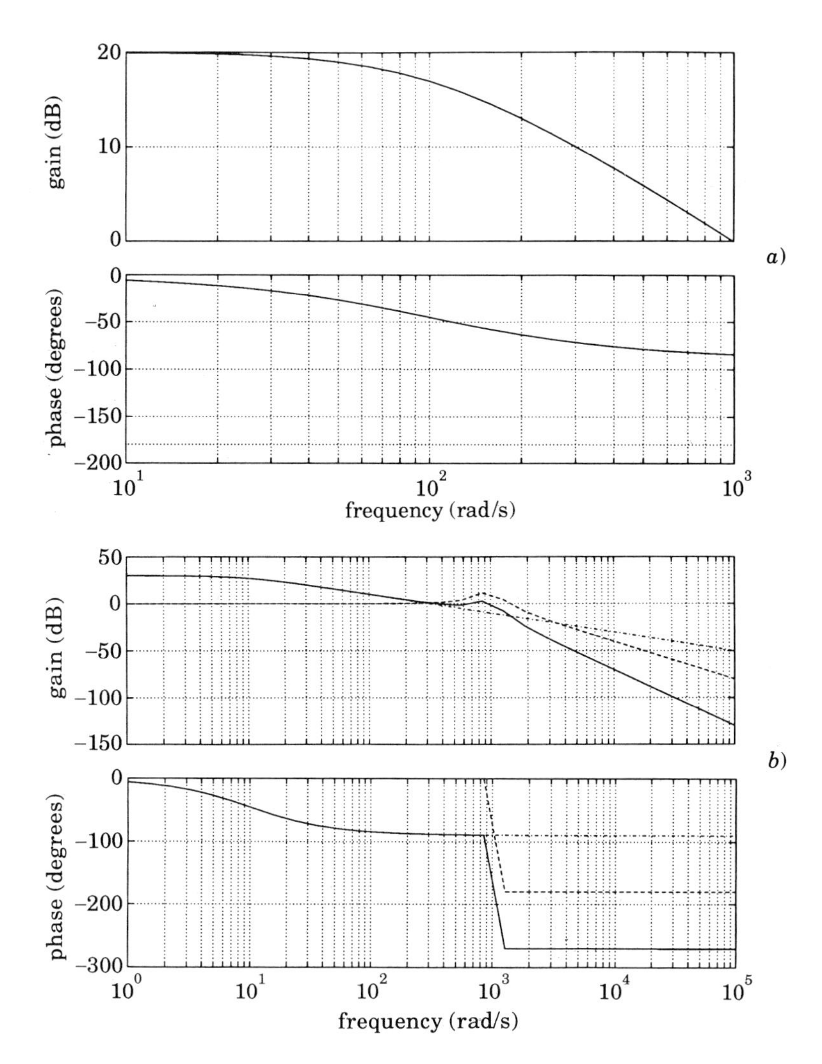

Fig. 18. – *a*) Bode plot for low-pass filter. *b*) Bode plot for harmonic oscillator plus low-pass filter compensation.

lock to the reference signal well at d.c., but will have a slow transient response and more noise at higher frequencies.

Suppose instead one compensates by adding a «phase lead» (*e.g.*, a differentiator) to the compensation so that the phase shift of the oscillator-compensation system has not yet reached 180° at the resonance. This allows the resonance to be artificially damped. Figure 19 shows the resultant frequency response for a harmonic oscillator compensated with a phase lead. Note that, compared to fig. 18, the gain near the resonant frequency is large, and the d.c. gain has decreased. This system will have a faster transient response, but more error in locking to the reference signal at d.c.

There is in general a trade-off between bandwidth (fast transient response) of a servo and d.c. gain (small steady-state error) of a servo. One way to think of it is in terms of integrators and differentiators. An integrator will generally provide less steady-state error because it has «memory» to make accurate adjustments at low frequencies. On the other hand, fluctuations that are fast compared to the integration time will be «washed out», and, as a result, the bandwidth of the system will be reduced. Differentiators predict the future performance of the system by looking at the slope of the error signal, and, therefore, increase the bandwidth. However, because a differentiator is compensating for future fluctuations it can slightly over or under compensate, leading to less steady-state accuracy.

Systems with more complicated frequency responses than that of a simple harmonic oscillator can be controlled by extending these ideas. A compensation circuit's phase and gain characteristics are tailored to provide the best compromise of bandwidth *vs.* d.c. response.

We now return to discuss how this stabilization is applied to the actual PNC

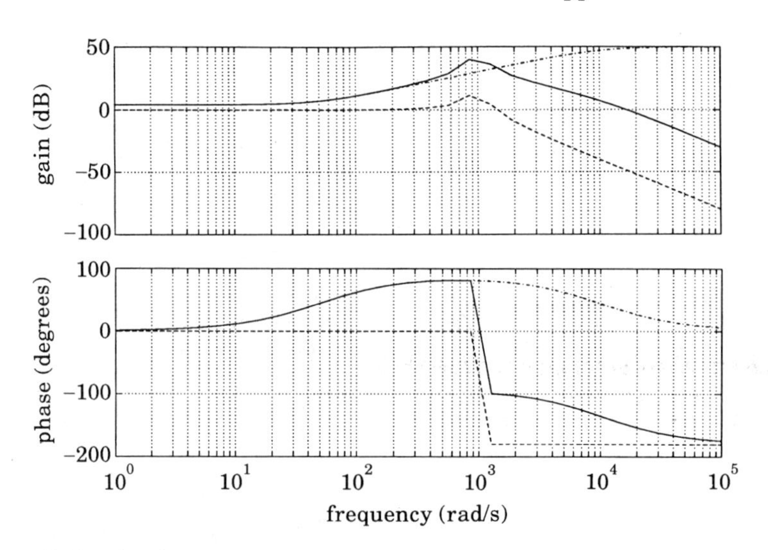

Fig. 19. – Bode plot for harmonic oscillator plus phase lead compensation.

experiment. The power build-up cavity is stabilized by moving the mirrors with piezoelectric transducers. Then laser frequency is locked to the power build-up cavity by a combination of elements, most of which are standard in c.w. dye lasers. First, we have a rotating plate on a galvanometer that changes the optical length of the laser cavity. This has a rather slow response, but a large dynamic range. Second, one of the mirrors is mounted on a PZT. This can change the cavity length with a frequency response extending to about 50 kHz. Third, the fastest feedback is provided by an electro-optic modulator. This has a unity gain frequency of about 2 MHz, but can only correct for rather small errors in the frequency. In addition to the frequency stabilization, to stabilize the optical power inside the power build-up cavity, we sense the light transmitted by the output mirror and hold it constant using acousto-optic or electro-optic modulators to control the incident laser power. One significant difficulty in this experiment is the fact that the power transmitted by this mirror does not seem to be exactly proportional to the power inside the cavity at the parts in 10^6 level. This discrepancy has been an ongoing problem which we do not yet fully understand.

5·7. *Field reversal and signal processing.* – With all the necessary frequencies, intensities and lengths stabilized, one then has to be concerned about reversing the various fields as precisely as possible, without upsetting the servo-control systems. The electric-field flip is accomplished by reversing the voltage applied to the electric-field plates. Initially we tried a sinusoidal reversal, but this gave unacceptably large electrical pickup on the detector. We then switched to a square-wave modulation with a few milliseconds of dead time after each reversal before taking data, to allow the transients to die away. The primary problem in obtaining a perfect electric-field reversal is the stray fields. There is considerable black magic we have learned for the preparation and handling of the plates which keeps the stray fields to a minimum, typically, a few tens of mV/cm.

For the actual voltage reversal, we have experimented with various solid-state and mechanical switches and have obtained the cleanest reversals when we use high-voltage relays. These have the minor annoyance that they are somewhat slow (less than ~ 40 Hz), but the reversal is much more exact than with any solid-state devices we have found. Mercury relays are faster but are limited in the voltage they can handle. To reverse the polarization, we use the same high-voltage relays to flip the voltage applied to the Pockels cell that provides a quarter-wave retardation. In this reversal, the major problem is the birefringence of the Pockels cell which drifts with temperature. However, with careful temperature stabilization this can be reduced to a reasonable level.

The magnetic-field reversal is the easiest; we simply reverse the currents flowing through coils using solid-state switches. One of the major concerns when doing any of these reversals is to avoid upsetting any of the servo loops.

This takes considerable care and involves the use of various sample-and-holds circuits and gates with precise timing to isolate the servos from the transients. We have succeeded in keeping everything stable enough that the noise while flipping the various fields is as low as when there are no reversals.

The signal processing is the final part of the apparatus, and it is fairly simple. The current from the photodiode is sent into a very-low-noise current-to-voltage converter, as mentioned above. The output voltage of this amplifier is monitored with two different systems. The first is relatively crude (1 part in 10^3), and simply monitors the overall d.c. signal level for normalization. The second system detects the small changes in the signal due to the PNC modulation. First, the signal passes through a low-noise amplifier which subtracts off a constant voltage so that the d.c. output is close to zero. This near-zero signal is sent into a gated integrator and the signal is integrated during the time between each reversal. At the end of each interval the output from the integrator is digitized and stored in a computer. Each of these numbers is stored with its appropriate label as to the state of E, B, m and polarization. Then the computer carries out the next reversal, resets the gated integrator, and the sequence is repeated. Using the offset and gated integrator in this manner, we avoid the dynamic-range problem encountered in trying to measure a very small modulation on top of a large signal.

6. – Systematic errors.

Most of the time spent taking data in these experiments is devoted to the study and reduction of potential systematic errors. Our approach to dealing with systematic errors follows the same general analysis used in the earlier cesium and thallium experiments. This procedure starts by considering the most general possible case of both d.c. and a.c. electric and magnetic fields which have components in the X, Y and Z directions. Thus we have 12 possible field components. Next we look at all the combinations of these fields that can give rise to a $6S \rightarrow 7S$ transition, either electric dipole or magnetic dipole. We allow each of these 12 field components to have both flipping and nonflipping (henceforth known as «stray» parts). We then go through the exhaustive list of combinations that produce terms that mimic the parity nonconservation by reversing with all of the possible various reversals. Then we measure the size of these 24 different field components and, in the process, try to make the stray and misaligned fields as small as possible. We have been able to reduce them to typically 10^{-4} to 10^{-5} of the main applied field. Using the measured sizes of the different components, we look at all the vast number of combinations that mimic PNC, and see which ones are significant. The 10^{-4} to 10^{-5} values effectively mean that any terms involving more than two stray or misaligned components are negligibly small compared to the true $(10^{-5} \div 10^{-6})$ PNC. At the end of

TABLE I. – *Potential systematic errors.*

Systematic contribution ([a])	Range ([b])	Average all data	Daily uncertainty
$(\Delta E_y/E)(B_x/B)$	$-0.3\% \rightarrow +1.1\%$	$+0.3\%$	0.4%
$(\Delta E_z/E)(E_y/E)$	$-1.3\% \rightarrow +0.4\%$	-0.1%	0.4%
$(E1M1\xi)(\Delta m = \pm 1)$	$-0.8\% \rightarrow +4.8\%$	$+1.7\%$	0.6% $(\Delta F = -1)$
	$-1.1\% \rightarrow +6.8\%$	$+2.4\%$	0.9% $(\Delta F = +1)$
$(\Delta m = 0)$	$-0.3\% \rightarrow +0.6\%$	$+0.04\%$	0.04% $(\Delta F = -1)$
	$-1.6\% \rightarrow +0.1\%$	-0.23%	0.06% $(\Delta F = +1)$

(a) ΔE_y and ΔE_z are nonreversing electric-field components, B_x and E_y are misaligned magnetic- and electric-field components, and ξ represents the birefringence of the coating on the output mirror.
(b) The range shows largest and smallest daily corrections.

this exercise we found there are three terms that contain two small components, and these are listed in table I. The first of these terms involves a stray electric field in the Y direction times the (misaligned) magnetic field in the X direction. The second term involves a stray electric field in the Z direction times a misaligned component of electric field in the Y direction. And the third term is a product of $E1$ and $M1$ transition amplitudes times a mirror birefringence factor.

We measure each of the fields and the birefringence involved in these terms while the experiment is running and subtract off their contributions. To do this we run a set of auxiliary experiments simultaneously, or interleaved with the PNC data acquisition. These auxiliary experiments involve observing the effects on the $6S \rightarrow 7S$ atomic-transition rate of different hyperfine transitions, different laser polarizations and application of additional E or B fields. Two points should be emphasized about dealing with systematic errors in this manner. First, it is important to use the atoms themselves so that the same region of space is sampled at nearly the same time as the PNC experiment. Second, the auxiliary experiments must be designed to allow systematic corrections to be measured with an uncertainty that is much less than the statistical uncertainty in the parity nonconservation experiment. It is highly desirable to have a measurement time much shorter than that required to take the parity violation data. If one fails to achieve this, then the uncertainty of an experiment increases because much of the running time is spent in taking data on systematic errors and little on the measurement itself.

In the experiments we have designed, achieving the necessary uncertainty requires a small fraction of the PNC integration time. In table I, we show the different sizes of the systematic uncertainties for our 1988 experiment, how much they vary from one run to another, and the average correction and uncertainty. It can be seen that the typical corrections are a few percent or less, and, most importantly, the uncertainties in all of these corrections in a given day are less than 1%, and thus much smaller than the statistical uncertainty.

An obvious question is, «Is this analysis foolproof, or did we miss something?». In fact we did miss something; there is no analysis that is absolutely foolproof. We overlooked a small correction the first time through, although we caught it well before we were ready to publish a result. However, it is educational to discuss the statistical-analysis procedure we used to discover this systematic error. This same type of analysis can (and probably should) be used in any precision measurement. It involves using a χ^2 test in a particular way to track down systematic errors. The Allan variance used to characterize frequency standards is related to this approach.

Our data consist of a large set of numbers, each number corresponding to a current which was integrated for 0.1 s. In the entire data set there are roughly 10^7 such numbers stored in the computer for analysis. The first step is to find the scatter in the numbers which is due purely to the statistics and has no contribution from any systematic source. This is accomplished by looking at the fluctuations on the shortest possible time scale where the statistical fluctuations are large. This gives us a standard deviation, σ, which is most likely to be purely statistical. In our case, we are doubly sure that is truly statistical because it corresponds to the shot noise limit for the signal.

Having found σ, we collect the data into various bin sizes, for example, the first million data points would be one bin, the second million would be the second bin, and so on for all the data. This produces 10 bins of data, and we can now predict how the average values in each bin should distribute based on σ, and the number of points in the bin ($\sigma_{\text{ave}} = \sigma\sqrt{N}$). This hypothetical distribution is then compared with the actual distribution. Specifically we find the value of χ^2 for the distribution using σ to obtain an uncertainty for the value in each bin, then we look up the probability for having that value of χ^2. If the resulting probability is 0.5 or larger, we are confident there was no systematic error that varied on a time scale of the length of a bin that would be significant relative to the statistical uncertainty.

Now, by choosing different bin sizes we probe for variations on different time scales. This is quite important because, if the bin size is much larger or much smaller than the time scale for the variations in some systematic error, the χ^2 will probably look reasonable. However, when one chooses a bin size that corresponds to the time scale of the variations, suddenly, the probability will be very low, indicating the presence of some unknown systematic error. This approach does not have to be limited to binning the data by time. It is equally useful to bin it according to any other factor that may lead to some systematic errors. For example, one might also bin the data according to room temperature to search for temperature-dependent systematic errors.

Of course, this approach only works if the systematic errors vary—if they are always constant you will never see them. However, one can make them vary by changing everything about the experiment that might be important, such as realigning all the optics or replacing critical components. Again, we binned the data corresponding to the different configurations and performed the χ^2 test. This is a remarkably sensitive test for potential systematic errors, and, although it is not generally taught, it is important to keep in mind in any precision experiment. In our case it revealed that we had neglected to consider the $E1M1$ interference correction associated with the off-resonance excitation of a m level. This excitation is forbidden at zero magnetic field but could occur because of the second-order Zeeman effect.

Having carried out all the detailed studies of systematic errors and χ^2 tests we finally achieved the result

$$\frac{\mathrm{Im}\,\mathcal{E}_{\mathrm{PNC}}}{\beta} = \begin{cases} -1.639(47)(08)\ \mathrm{mV/cm}\,, & F=4\rightarrow F'=3\,, \\ -1.513(49)(08)\ \mathrm{mV/cm}\,, & F=3\rightarrow F'=4\,, \\ -1.576(34)(08)\ \mathrm{mV/cm} & \text{(average)}. \end{cases} \tag{13}$$

The size of the parity-nonconserving mixing is given in terms of the equivalent amount of d.c. electric field that would be necessary to give the same mixing of S and P states. As shown, we have measured this mixing for two different hyperfine transitions, the $6S$, $F=4\rightarrow 7S$, $F'=3$ and the $6S$, $F=3\rightarrow 7S$, $F'=4$. In both cases, the amount of mixing corresponds to about 1.5 mV/cm. The average of these two is the most important quantity, as we will discuss below. We have measured this to an uncertainty of 2%, which is dominated by the 0.034 mV/cm statistical uncertainty. The systematic uncertainty is about 1/4 of the size of the statistical uncertainty. It should be noted that this systematic uncertainty is different from many systematic uncertainties, in that it is actually a true statistical uncertainty in the evaluation of the systematic correction. Therefore, if the statistical signal-to-noise ratio in the experiment is improved, this uncertainty will be reduced.

In fig. 20, we show a comparison of the different experimental measurements of parity nonconservation in cesium, the most thoroughly measured atom. On top are the two experimental results of the Paris group in '82 and '84, below is our 1985 result, and our 1988 result, with its 2% uncertainty. There is good agreement among all of these numbers. This gives one a certain amount of confidence that no tremendous systematic errors are being overlooked.

In table II, we show a summary of the results from all atomic parity nonconservation experiments. In the first section are the optical-rotation experiments which looked at the 648 nm line of bismuth. These results are somewhat controversial in that the results from the three groups showed substantial discrepancy, as did the theoretical calculations. In retrospect, the former was probably due to systematic errors that were not sufficiently controlled. More recent optical-rotation experiments have shown better consistency, and the uncertainties

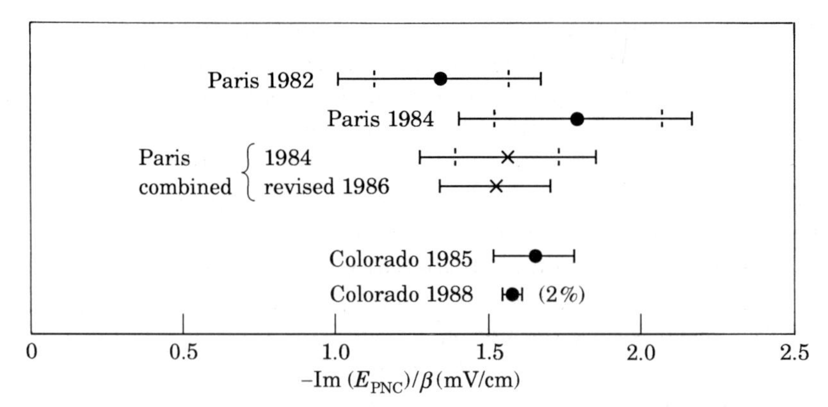

Fig. 20. – Comparison of the experimental measurements of PNC in cesium.

are mostly in the (15 ÷ 30)% range. The one exception is the recent Oxford measurement in bismuth which has an uncertainty of only 2%. The Stark-induced interference experiments are given at the bottom of this table. Most of these are the cesium measurements we have already mentioned, plus there is the one thallium result from Berkeley with an uncertainty of 28%.

Before we can consider what these measurements tell us about elementary-particle physics, we must return to the atomic-structure issue. The quantity that is experimentally measured is the δ_{PNC} mixing, which is equal to Q_w times the γ_5 matrix element. This matrix element is found by calculating the atomic

TABLE II. – *Summary of* PNC *results.*

		Experiment	Atomic theory
Ancient (controversial) history			
bismuth	Oxford [13]		
	University of Washington [14]	wide	wide
	Novosibirsk [15]	variations	variations
Modern civilized (?) era:			
Pb	(Washington '83) [16]	±28%	±10% (?) [17]
Bi	(Washington '81) [18]	±18%	±15% (?) [19]
Tl 1.3 μm	(Oxford '91) [20]	±2%	±15% (?) [19]
	(Oxford '91) [21]	±15%	±3% (?) [22]
Stark-induced interference:			
Tl	293 nm		
	(Berkeley '85) [11]	±28%	±6% (?) [22]
Cs	(Paris '84-'86) [23]	±12%	±1% [24]
	(Colorado '85) [9]	±12%	±1% [24]
⇒	(Colorado '88) [10]	±2%	±1% [24]
		all agree with the standard model	

structure. Thus, in order to obtain Q_W to 1%, both the experiment and the matrix element calculation must be accurate to better than 1%. As mentioned earlier, the calculation of γ_5 varies considerably in accuracy from one atom to another. In table II we have given the accuracy quoted for the best calculations for each atom.

Here we will limit our discussion to the cesium atom for which there have been the most abundant and most accurate calculations. Two basic approaches have been employed for these calculations. The first is the semi-empirical method which has been used in Paris, Oxford, Colorado and elsewhere. This approach uses experimental data to determine wave functions which are then used to find the matrix element. This technique is relatively easy. However, it is difficult to make a rigorous evaluation of the accuracy of the calculation, since all the relevant experimental data have already been incorporated into the calculation. The estimates for the uncertainty in these calculations are as small as 2%. The second approach is to use *ab initio* relativistic many-body perturbation theory. The need for accurate cesium PNC calculations has spurred major advances in this field, although the calculations are very long and difficult. The most recent and most accurate results have come from the Novosibirsk group of Flambaum, Sushkov *et al.*, who have achieved a 2% uncertainty, and the Notre Dame group of Blundell, Saperstein and Johnson, who have now reached 1% uncertainty. The advantage of this calculational approach is that there is a fairly clear prescription for evaluating the accuracy of the calculations. The most direct way is to simply use the same calculational technique to determine many properties of the atoms and compare these with experimental data. In this case, this means calculating hyperfine splittings, oscillator strengths between many transitions, energy levels and fine-structure splittings for cesium and other alkali atoms. Fortunately, a tremendous amount of experimental data is available for comparison. In all cases, the agreement between the calculations and the experiments has been within 1%. Another technique for estimating the uncertainty in these calculations is to estimate the size of the uncalculated higher-order terms in the perturbation series expansion. This approach also gives an uncertainty of about 1%.

7. – Implications.

In this section, we will consider the implications of the Colorado measurement of PNC in cesium. We will first discuss the significance of the comparison of the two different hyperfine transitions. This difference between the two numbers, $\Delta = 0.126(68)$ mV/cm, is probably not zero. More specifically, this value indicates a 97% probability that Δ is greater than zero. When we made this measurement, we did not anticipate a nonzero result at this level and, therefore, spent a considerable amount of time trying to determine what was wrong with the data. The result, however, stubbornly persisted. Only later did we discover that an effect of nearly this size had been predicted.

The primary difference between these two transitions is that the nuclear spin is reversed relative to the electron spin. Thus Δ is a measure of the nuclear-spin-dependent contribution to the PNC signal. Two processes have been discussed which would cause a nuclear-spin-dependent parity nonconservation. The first is simply the electron-quark portion of the weak neutral current which depends on the spin of the quarks. This interaction is characterized by the C_{2u} and C_{2d} coefficients. As we mentioned earlier, because of the size of these coefficients and the fact that the effect is proportional to the total nuclear spin (and not proportional to the number of quarks), this contribution is much smaller than the weak-charge contribution. However, it has also been pointed out [3] that there is a substantially larger contribution, called the nuclear anapole moment, which arises from weak interactions within the nucleus. The effects of these weak interactions (both charged and neutral) are to mix the parity eigenstates of the nucleus, leading to a parity-nonconserving electromagnetic current in the nucleus. This current takes the form of a toroidal helix, and, therefore, has no long-range electric or magnetic fields. Thus it gained the name «anapole moment». This phenomenon was first proposed by ZEL'DOVICH in 1957 in the general context of parity violation in charged systems [25]. It is not well known because people shortly thereafter decided such an effect could never be measured. However, because the cesium electrons penetrate the nucleus, they spend some time inside the toroidal helix and thereby detect its existence. The coupling to the electrons is purely electromagnetic, but, because the underlying nuclear currents are parity violating, it leads to parity violation in the electronic transition.

There has been a significant amount of interest in this nuclear anapole moment by the nuclear-physics community and several authors have calculated the expected size. The first calculations were by KHRIPLOVICH and FLAMBAUM [3] and their estimates are consistent with our observations. HAXTON *et al.* [26] have also made similar calculations, but have treated the nuclear physics rather differently. Finally, BOUCHIAT and PIKETTY have done a calculation which is not consistent with our result [27]. We have been told by FLAMBAUM that the differences in these calculations are not due to any fundamental difference in the theory, but are a problem of the basic interpretation of nuclear PNC from other experiments. Depending on how one chooses to interpret the other experiments, it is possible to obtain very different constants which characterize PNC interactions in the nucleus. This emphasizes the need for more accurate data in this field. There is hope that these nuclear-anapole-moment measurements can provide these data. The nuclear anapole moment is unique in that it is a PNC distortion of the nuclear ground state. The previous measurements on nuclear PNC have observed parity mixing of excited, and often rather distorted, nuclear states where there is considerable uncertainty about the nuclear wave functions. Thus it is clear that future improvements in atomic PNC precision should substantially improve the understanding of nuclear parity nonconservation.

Obviously, the uncertainty due to the nuclear physics is a serious issue in the interpretation of atomic parity nonconservation. If we had measured only a single transition, it would seriously compromise our ability to test the standard model. Fortunately, if we take the average of the measurements on the two hyperfine transitions, as opposed to the difference, the nuclear-spin-dependent part cancels out. In this way, we also cancel out any questions involving the nuclear structure, which is critical in allowing a precision test of the standard model.

From this average and the Notre Dame matrix element calculation, we obtain a weak charge, $Q_W = 71.0 \pm 2\% \pm 1\%$. If one assumes the standard model is correct, one can then from this extract a value of $\sin^2\theta_W$ which is equal to

$$\sin^2\theta_W = 0.223 \pm 0.007 \text{ (experimental)} \pm 0.003 \text{ (theoretical)}. \tag{14}$$

This value of $\sin^2\theta_W$ can now be compared with values obtained from other experiments such as the measurement of the Z_0 mass or the neutrino scattering results, as shown in table III. In addition to these two measurements, there are many other measurements from high-energy experiments which can be used, but which have lower precision or involve other properties of the Z_0. We have omitted the latter group because, while these are reputed to be independent measurements, the variations in the values of $\sin^2\theta_W$ obtained are much smaller than the quoted uncertainties. This leads to unrealistic χ^2 probabilities and suggests that these measurements are not truly independent.

The comparison of the values of $\sin^2\theta_W$ provides a precise test of the standard model. It is worth noting that in other tests of the standard model, particularly those involving the comparison of neutrino scattering and Z_0 data, the uncertainty in the mass of the top quark introduces an uncertainty of 0.003 in the relative values of $\sin^2\theta_W$ [2]. However, the comparison of the atomic and the Z_0 mass values is unique in that dependence on the top quark mass is essentially identical in the two cases. Because the atom is sensitive to a different set of electron-quark couplings and a different energy scale, this comparison, however, is very sensitive to new physics which is not contained in the standard

TABLE III. – $\mathrm{Sin}^2\theta_W$ *values.*

δ_{PNC} (experiment) + $\langle\gamma_5\rangle$ (theory) $\Rightarrow Q_W = 71.0 \pm 2\% \pm 1\%$ $\Rightarrow \sin^2\theta_W = 0.223 \pm 0.007 \pm 0.003$ W mass $\Rightarrow 0.2320 \pm 0.0007$
neutrino scattering $\Rightarrow 0.233 \pm 0.003 \pm 0.005$

model. Proposed examples of such new physics include technicolor and extra Z bosons, which occur in many models.

Figure 21 shows the values of C_{1d} *vs.* C_{1u} as determined by a model-independent analysis of the experiments, along with the standard model prediction. The hatched area is the constraint from the SLAC deep inelastic electron scattering experiments. In contrast, the constraint set by our cesium experiment and the Notre Dame theory is the narrow solid line which is nearly orthogonal and, in particular, constrains the value of C_{1d} much more severely. The crossed line shows the values allowed by the $SU_2 \times U_1$ Weinberg/Salam//Glashow theory. The point on that line is determined by the value chosen for $\sin^2 \theta_w$. This figure shows quite clearly that any new physics that would cause a change in the value of C_{1d}, but not affect the other coupling constants, would

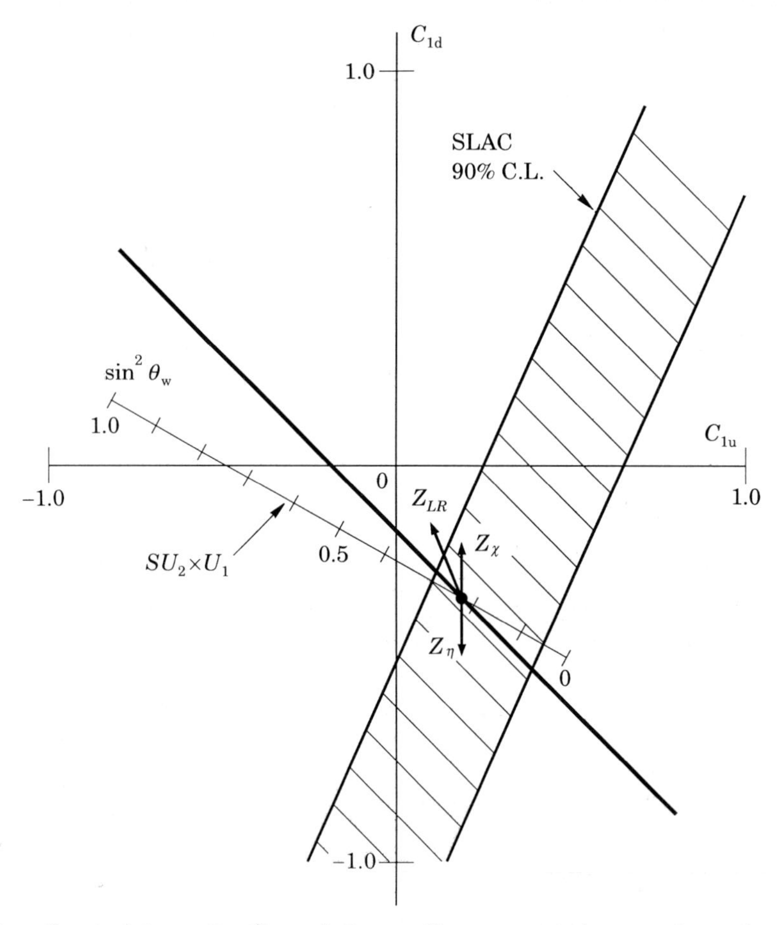

Fig. 21. – Constraints on the C_{1d} and C_{1u} coupling constants by experimental measurements. The hatched region is from SLAC deep-inelastic-scattering data, while the solid line is from atomic PNC. The $SU_2 \times U_1$ line is the standard model value as a function of $\sin^2 \theta_w$.

only be revealed by the atomic-parity-nonconservation measurements. The arrows on this figure show how a few popular proposed models would shift the values of these two coupling constants to a different place in the plane. Because the current atomic-physics line passes through the standard model point, there is no indication of the existence of new physics. However, this does put constraints, in some cases quite severe, on the parameters of models that propose such new physics.

One example which has drawn considerable attention in the last few years is how atomic PNC results constrain the proposed mechanism known as technicolor, or, more generally, dynamical symmetry breaking involving heavy particles. This type of new physics has been characterized in terms of S and T parameters which enter directly into Q_W [28]. Generic technicolor models predict the S parameters should be around $+2$ or somewhat larger [28]. From the comparison of results just mentioned, one finds that the atomic PNC yields a value of S which is $-2.7 \pm 2 \pm 1.1$ as given in ref. [29]. Thus one finds that technicolor is on somewhat shaky ground, although the atomic PNC experimental results are not good enough to completely rule it out.

Finally, atomic PNC provides the best constraints on many models that involve additional neutral Z bosons. While there are many papers on this subject (see references in ref. [2]), we note particularly the paper by MAHANTHAPPA and MOHAPATRA [29]. They consider 11 different models with additional Z's which have been proposed, and they find that, in 8 of these 11 cases, atomic PNC provides the most severe constraints. Thus it is clear that the cesium PNC results are providing information on elementary-particle interactions that is not available from any other source at the present time.

8. – Future improvements.

8·1. *Near term.* – While atomic PNC experiments are providing useful information, it is clear that more precise results would be desirable and useful. The mass of the Z_0 is now known to around 1 part in 10^3. If atomic PNC results could be improved to that level, we would have a 10-fold improvement in the test of the standard model and correspondingly improved sensitivity to possible new physics. With this in mind, we would like to discuss our efforts to improve the cesium PNC results. Work is also under way to improve atomic-parity-nonconservation measurements by several other groups: In Paris, the Bouchiat group is building a new experiment that involves stimulated-emission probing of the excited state in cesium. At Oxford and Washington, experiments are under way to obtain more precise optical-rotation measurements in thallium. At Berkeley, efforts continue to obtain a more precise Stark interference measurement in thallium. All of these experiments have been under development for a number of years, and we hope to have results in the not too distant future.

In our efforts to improve the Colorado 1988 experiment, our primary focus has been on improving the signal-to-noise ratio. This is clearly the major limitation of our experiment since the statistical uncertainty was much larger than the systematic uncertainties. We also took into account the fact that the scattered laser light was a major nuisance requiring frequent realignment of the optics, and the transparent conducting coatings were a substantial problem. The coatings deteriorated under exposure to the cesium vapor, and lead to frequent interruptions in the experiment while they were replaced. With these issues in mind we built a new apparatus that uses an optically pumped atomic beam which, in principle, should provide 16 times more atoms since there are 16 possible m levels of the $6S$ state. We now have better mirrors for our power build-up cavity; these increase the build-up by about a factor of 10, resulting in a total build-up of 15 000. A third improvement is using downstream detection of the $6S \rightarrow 7S$ excitation. This concept is illustrated in fig. 22, which shows the schematic of the new apparatus.

After leaving the oven, the cesium atomic beam is optically pumped into a single F and m_F level by light from two diode lasers which drive two hyperfine transitions of the $6S \rightarrow 6P_{3/2}$ transition. The atoms in the single m level then propagate down the atomic beam and intersect the power build-up cavity beam where they are excited to the $7S$ state. They then have a 70% probability of decaying back down into the $6S$ hyperfine level which was previously depleted. The atoms continue down the optical beam in this state until they reach the probe region. In this region, light from another diode laser again excites the

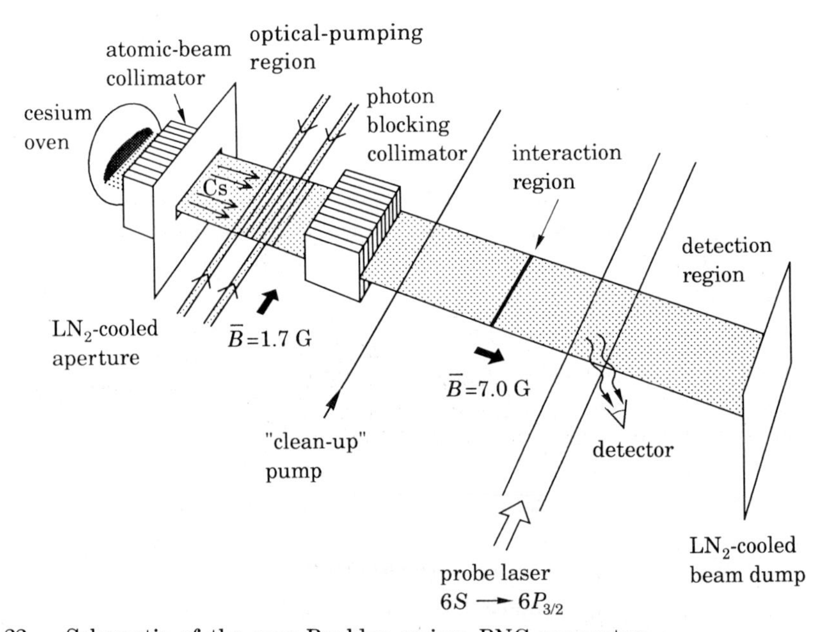

Fig. 22. – Schematic of the new Boulder cesium PNC apparatus.

$6S \rightarrow 6P_{3/2}$ transition. However, here we excite a cycling transition ($F = 4 \rightarrow$ $\rightarrow F' = 5$ or $F = 3 \rightarrow F' = 2$). On a cycling transition the atom returns only to the same initial state and hence can be excited many times. Typically 1000 i.r. photons are scattered for each $6S \rightarrow 7S$ excitation. We detect this fluorescence to determine the $6S \rightarrow 7S$ excitation rate. This detection scheme provides a substantial amount of amplification, yielding a detection of about 200 photons per $6S \rightarrow 7S$ transition, instead of the 0.3 detected in the previous apparatus.

In addition, since the detection takes place at a different region from the excitation region, we can now construct our electric-field plates out of any material. This greatly simplifies their construction and increases their longevity. Also, scattered light from the green laser light is now negligible, as is detector noise, because the signal size is much larger. All of these improvements would suggest that the experiment should be much easier. In fact, there have been major headaches and delays with this approach, and it is educational to consider what has gone wrong and what lessons can be learned about doing experiments at the frontier of laser spectroscopy.

We will now discuss the unexpected problems we encountered in making this «improved» experiment work. The first problem was noise in the optical-pumping and resonance fluorescence detection regions due to the fact that we were using diode lasers. Diode lasers have very rapid (ns) fluctuations in the optical phase. Through a somewhat obscure process, this leads to very-low-frequency fluctuations in the atomic-transition rate. This was quite puzzling when we first observed it, and has now been explained in a series of papers by ZOLLER and collaborators [30]. While this has become interesting atomic/optical physics to a number of people, to us it is a major experimental problem. To avoid the problem one must have a feedback system capable of providing gigahertz bandwidth correction signals in order to eliminate the noise at $(10 \div 20)$ Hz, where we detect our signals. In the first attempt we used optical feedback from narrow-band resonators, as demonstrated by HOLLBERG and co-workers [31]. This approach gave low-noise signals, but the locking was not reliable enough to allow three diode lasers to operate for reasonable periods of time. After considerable additional work, we settled on using optical feedback from diffraction gratings [32]. These gave much more reliable performance, but the noise levels were still unacceptably high. We solved this problem by finding a laser manufacturer whose instruments gave superior performance when operated with grating feedback. Thus we have finally succeeded in producing a very reliable source of diode laser light which provides very good signal-to-noise ratio in excitation of narrow-band atomic transitions. Specifically, we can now achieve a noise-to-signal ratio of $3 \cdot 10^{-6}/\text{Hz}^{1/2}$ when exciting a 10 MHz wide atomic transition. The necessary laser has a combination of optical, mechanical and current feedback, as shown in fig. 23. The grating which provides the optical feedback is mounted on a piezoelectric transducer which allows mechanical adjustment of the grating position. This holds constant the length, and hence the frequency,

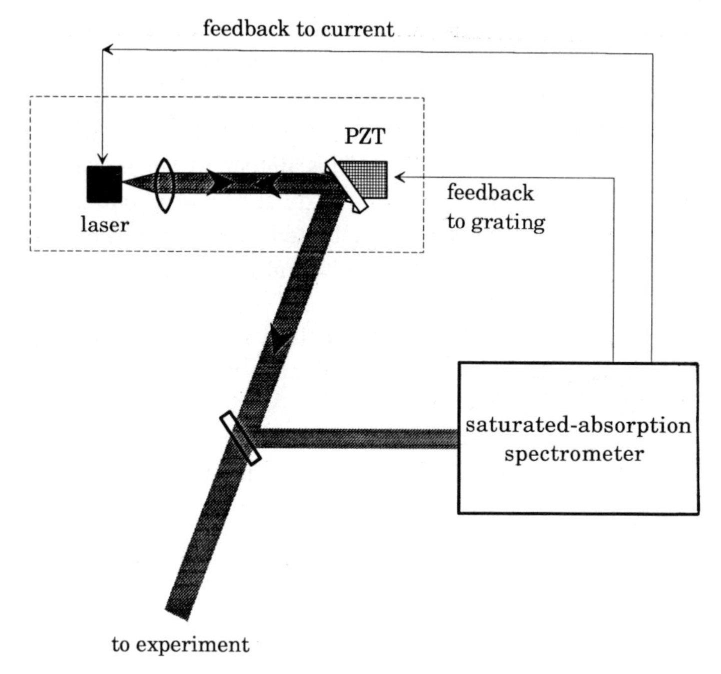

Fig. 23. – Schematic of the diode laser control system.

of the optical cavity. For faster corrections to the cavity resonant frequency we servo the laser current to achieve the highest level of stabilization.

The second major problem in the new experiment was background atoms in the supposedly empty F state. It is relatively easy to deplete one F level of a low-intensity atomic beam very well ($< 10^{-4}$). However, with a more intense beam there are a number of mechanisms that can repopulate the empty level. For example, collisions between atoms in the beam and surfaces or other atoms (particularly oxygen) is the first mechanism. We have eliminated this source by improving the vacuum and carefully positioning the collimating surfaces. A second contribution to the background, which appears to be from atoms in the wrong state, actually comes from the excitation of the other hyperfine line by the tail of the spectral distribution of light in the probe beam. We have eliminated this source by sending the probe light through an interferometer filter cavity which blocks out the tails of the spectral distribution. The third and most serious source of atoms in the wrong F state has been the multiple scattering of the optical-pumping light. The optical-pumping process produces fluorescence that can travel down the atomic beam and re-excite the atoms out of the desired state. The number of atoms pumped back into the empty state scales as the square of the atomic-beam intensity. We have found several ways to reduce this background: multiple pumping beams (the «clean-up» beam in fig. 22), picking the optical-pumping transition which minimizes the scattered fluorescence, and using the photon blocking collimator. This is a collimator with very thin black vanes which allows only the highly collimated photons to pass through. Finally, even with all these steps, the background was still too large, and we ultimately had to reduce our atomic-beam intensity. In spite of all these setbacks and delays, this new, improved experiment is now operational and we are taking data with a signal-to-noise ratio several times better than that of the 1988 experiment.

A very painful lesson has been brought home to us in carrying out this improved experiment. When one is probing a region of technology and physics which is unexplored, it is important to step warily, and to keep all your options open. In terms of an experiment, this means you should keep the apparatus flexible and be ready to adapt, as we mentioned earlier. In this experiment we were somewhat seduced by the fact that this approach seemed to solve all our old problems, and we committed ourselves to a design that turned out to be filled with major unexpected difficulties.

8'2. *Long term.* – As the experimental accuracy improves beyond 1%, the principal limitation on the usefulness of atomic PNC will become the atomic theory. There have been credible speculations that it will be possible to calculate the theory in cesium to a part in 10^3. However, it is not clear when these calculations will be completed, and the question of how to check their accuracy becomes a major issue.

We have begun a longer-term experimental project that avoids the atomic-theory question. The basic idea is to compare precise measurements of atomic PNC for different isotopes of cesium. The weak charge is sensitive to the number of neutrons, and hence will change for different isotopes. The atomic matrix element, however, depends on the electronic structure and is almost independent of the number of neutrons. If one then looks at appropriate combinations and ratios of experimental results, for example

$$\frac{\delta_{\mathrm{PNC}}^{\mathrm{Cs}^{130}} - \delta_{\mathrm{PNC}}^{\mathrm{Cs}^{150}}}{\delta_{\mathrm{PNC}}^{\mathrm{Cs}^{130}} + \delta_{\mathrm{PNC}}^{\mathrm{Cs}^{150}}} = \frac{Q_{\mathrm{W}}^{130} - Q_{\mathrm{W}}^{150}}{Q_{\mathrm{W}}^{130} + Q_{\mathrm{W}}^{150}},$$

the atomic matrix element will drop out, leaving a ratio of weak charge which can be directly compared with standard model predictions. In this manner we hope to achieve measurements that can be compared with the standard model predictions at the part in 10^3 level. There are two major obstacles to carrying out these experiments. First is the need for even better signal-to-noise ratios. Second, and most critical, is the need to carry out PNC measurements with small atomic samples, rather than the many grams used in the atomic-beam measurements. This requirement is necessary because all the other isotopes of cesium are radioactive and can only be obtained and used in small quantities. We propose to overcome both of these obstacles by using the new technology of

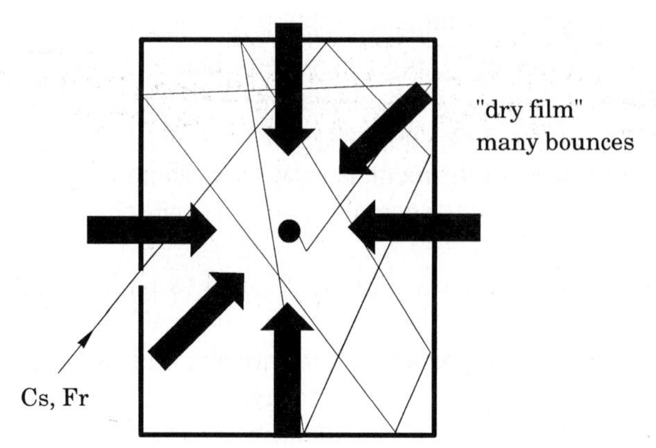

Fig. 24. – Laser trap cell.

laser trapping. This will improve the signal-to-noise ratio because it is possible, even easy, to obtain optical thicknesses in trapped-atom samples 10 or 100 times larger than can be achieved in our atomic beam.

It is more difficult to show that optical trapping will allow the experiments to be done with very small atomic samples (10^{10} atoms). We are currently working on this problem. The approach we are using starts with a very small sample of a given isotope (short-lived isotopes will be produced at an accelerator, while longer-lived isotopes can be brought to our laboratory), which is injected into a special cell where the atoms will be efficiently captured by a laser trap (fig. 24). We have carried out detailed studies on capturing atoms from a vapor and we are currently developing wall coatings which will allow the cesium atoms to bounce around inside the cell without sticking until they are captured. Preliminary work with silane coatings has been quite encouraging.

Once the atoms are captured, PNC measurements can then be carried out in the cold dense samples. If all goes according to plan, the next decade will see high-precision measurements of PNC in a number of cesium isotopes. This will provide detailed information on the nuclear anapole moment and a very precise test of the standard model.

* * *

This work has been supported by the National Science Foundation.

REFERENCES

[1] M. A. Bouchiat and C. Bouchiat: *J. Phys. (Paris)*, **35**, 899 (1974).
[2] P. Langacker, M.-X. Luo and A. Mann: *Rev. Mod. Phys.*, **64**, 87 (1992).

[3] V. V. FLAMBAUM and I. B. KHRIPLOVICH: *Sov. Phys. JETP*, **52**, 835 (1980); V. V. FLAMBAUM, I. B. KHRIPLOVICH and O. P. SUSHKOV: *Phys. Lett. B*, **146**, 367 (1984).
[4] F. CURTIS-MICHEL: *Phys. Rev.*, **138B**, 408 (1965).
[5] M. A. BOUCHIAT and C. C. BOUCHIAT: *Phys. Lett. B*, **48**, 111 (1974).
[6] D. Z. ANDERSON, University of Colorado: private communication.
[7] M. A. BOUCHIAT, J. GUENA, L. HUNTER and L. POTTIER: *Phys. Lett.*, **1178**, 358 (1982).
[8] R. CONTI, P. BUCKSBAUM, S. CHU, E. COMMINS and L. HUNTER: *Phys. Rev. Lett.*, **42**, 343 (1979); E. COMMINS, P. BUCKSBAUM and L. HUNTER: *Phys. Rev. Lett.*, **46**, 640 (1981); P. H. BUCKSBAUM, E. D. COMMINS and L. R. HUNTER: *Phys. Rev. D*, **24**, 1134 (1981).
[9] S. L. GILBERT, M. C. NOECKER, R. N. WATTS and C. E. WIEMAN: *Phys. Rev. Lett.*, **55**, 2680 (1985).
[10] M. C. NOECKER, B. P. MASTERSON and C. E. WIEMAN: *Phys. Rev. Lett.*, **61**, 310 (1988).
[11] P. S. DRELL and E. D. COMMINS: *Phys. Rev. Lett.*, **53**, 968 (1984).
[12] C. E. WIEMAN, M. C. NOECKER, B. P. MASTERSON and J. COOPER: *Phys. Rev. Lett.*, **58**, 1738 (1987).
[13] P. E. G. BAIRD, M. W. S. M. BRIMICOMBE, R. G. HUNT, G. J. ROBERTS, P. G. H. SANDARS and D. N. STACEY: *Phys. Rev. Lett.*, **39**, 798 (1977).
[14] L. L. LEWIS, J. H. HOLLISTER, D. C. SOREIDE, E. G. LINDAHL and E. N. FORTSON: *Phys. Rev. Lett.*, **39**, 795 (1977).
[15] L. M. BARKOV and M. S. ZOLOROTEV: *Sov. Phys. JETP*, **27**, 357 (1978).
[16] T. P. EMMONS, J. M. REEVES and E. N. FORTSON: *Phys. Rev. Lett.*, **51**, 2089 (1983).
[17] V. A. DZUBA, V. A. FLAMBAUM, P. G. SILVESTROV and O. P. SUSHKOV: *Europhys. Lett.*, **7**, 413 (1988).
[18] J. H. HOLLISTER, G. R. APPERSON, L. L. LEWIS, T. P. EMMONS, T. G. VOLD and E. N. FORTSON: *Phys. Rev. Lett.*, **46**, 642 (1981).
[19] V. A. DZUBA, V. A. FLAMBAUM and O. P. SUSHKOV: *Phys. Lett. A*, **141**, 147 (1989).
[20] M. J. D. MACPHERSON, K. P. ZETIE, R. B. WARRINGTON, D. N. STACEY and J. P. HOARE: *Phys. Rev. Lett.*, **67**, 2784 (1991).
[21] T. WOLFENDEN, B. BAIRD and P. SANDERS: *Europhys. J.*, **15**, 731 (1991).
[22] V. A. DZUBA, V. A. FLAMBAUM, P. G. SILVESTROV and O. P. SUSHKOV: *J. Phys. B*, **20**, 3297 (1987).
[23] M. A. BOUCHIAT, J. GUENA, L. HUNTER and L. POTTIER: *J. Phys. (Paris)*, **47**, 1709 (1986).
[24] S. A. BLUNDELL, W. R. JOHNSON and J. SAPERSTEIN: *Phys. Rev. Lett.*, **65**, 141 (1990).
[25] YA. B. ZEL'DOVICH: *Ž. Ėksp. Teor. Fiz.*, **33**, 1531 (1958) (*Sov. Phys. JETP*, **7**, 1184 (1957)).
[26] W. C. HAXTON, E. M. HENLEY and M. J. MUSOLF: *Phys. Rev. Lett.*, **63**, 949 (1989).
[27] C. BOUCHIAT and C. PIKETTY: *Z. Phys. C*, **49**, 91 (1991).
[28] W. MARCIANO and D. ROSNER: *Phys. Rev. Lett.*, **65**, 2963 (1990).
[29] K. T. MAHANTHAPPA and P. K. MOHAPATRA: *Phys. Rev. D*, **43**, 3093 (1991).
[30] T. HASLWANTER, H. RITSCH, J. COOPER and P. ZOLLER: *Phys. Rev. A*, **38**, 5652 (1988).
[31] B. DAHMANI, L. HOLLBERG and R. DRULLINGER: *Opt. Lett.*, **12**, 876 (1987).
[32] K. MACADAM, A. STEINBACH and C. WIEMAN: *Am. J. Phys.*, **60**, 1098 (1992).

High-Resolution Spectroscopy of the Hydrogen Atom; Measurement of the Rydberg Constant.

L. JULIEN, F. NEZ, M. D. PLIMMER, S. BOURZEIX
R. FELDER (*) and F. BIRABEN

Laboratoire de Spectroscopie Hertzienne de l'ENS, Université Pierre et Marie Curie
4 place Jussieu, BP 74, 75252 Paris Cedex 05, France

1. – The Rydberg constant.

The Rydberg constant is the scaling factor for the energy levels of simple atomic systems. In the hydrogen atom and to a rough approximation, the energy depends only on the principal quantum number n and is given by the elementary formula

$$E_n = -hcR_{\mathrm{H}}\left(\frac{1}{n^2}\right), \tag{1}$$

where R_{H} is the Rydberg constant for the hydrogen atom.

From the above expression, one can derive the Balmer-Rydberg formula which gives the transition wavelength between two different levels n and p:

$$\frac{1}{\lambda} = R_{\mathrm{H}}\left(\frac{1}{n^2} - \frac{1}{p^2}\right). \tag{2}$$

The hydrogen atom has transitions in a wide domain of wavelengths from the UV to the microwave range and the Rydberg constant R_{H} can be deduced from any wavelength or frequency measurement between two levels having different principal quantum numbers.

A knowledge of the electron-to-proton mass ratio allows one to deduce the Rydberg constant R_∞ for the case of an infinite nuclear mass: $R_\infty = R_{\mathrm{H}}(1 + m/m_{\mathrm{p}})$.

(*) Bureau International des Poids et Mesures, Pavillon de Breteuil, 92312 Sèvres Cedex, France.

R_∞ is a fundamental constant related to the electronic mass m, the electronic charge e and Planck's constant h:

$$R_\infty = \frac{me^4}{8\varepsilon_0^2 h^3 c} . \tag{3}$$

In fact, in order to obtain the exact energy levels, some corrections must be added to this very simple treatment:

relativistic corrections, calculated by the Dirac theory, which are responsible for the fine structure of the levels;

hyperfine corrections, due to the magnetic momentum of the proton, which split all the levels into two components;

radiative corrections which result from the interaction of the electron with the quantized electromagnetic field and are calculated using the theory of quantum electrodynamics (QED);

the nuclear-size effect corrections arising from the finite volume of the proton.

The two latter corrections are especially important for the S levels: they give together the Lamb shift. This Lamb shift is about 8 GHz for the ground-state $1S$ level and is known theoretically with a precision of a few parts in 10^{11}.

2. – Spectroscopic measurements in atomic hydrogen.

In atomic hydrogen two types of spectroscopic measurements can be distinguished:

1) *Absolute measurements of a given transition frequency.* If the various corrections to the energy are well known, these measurements allow one to deduce the Rydberg constant. The difficulty of this type of measurement is the need of a good frequency standard and of a frequency chain to connect the studied transition to the standard.

2) *Comparisons of two different transition frequencies,* one involving the ground state. Such a measurement provides a determination of the $1S$ Lamb shift. The most convenient scheme, proposed by T. W. Hänsch, is the comparison of two frequencies lying in an almost integer ratio, as the $1S$-$2S$ and $2S$-$4S/4D$ two-photon transition: this has been done recently in Garching [1]. A similar experiment is currently under way prepared in Paris based on the $1S$-$3S$ and $2S$-$6S/6D$ two-photon transitions at 205 and 820 nm.

In the following, we will concentrate on the first type of measurement only.

The development of tunable lasers has allowed the conquest of three orders of magnitude in the precision of the Rydberg constant in less than twenty years [2] and the experimental precision is now challenging the theoretical calculations. This rapid improvement is of great interest for several reasons:

The Rydberg constant plays a key role in the least-squares adjustment of the fundamental constants where it is used as an auxiliary constant (that is, with a fixed value). It then gives a test of the consistency between various domains of physics. It can also be used to deduce other important constants, such as, for example, the fine-structure constant.

As it is the scaling factor of energy levels of simple atomic systems, a precise value of this constant is needed to test theoretical predictions for these systems; among them the positronium atom is of special interest because it is a purely leptonic system.

Finally, it gives a means of connecting microwave and optical frequencies: the hydrogen atom itself can be used to check frequency chains between these two domains and could possibly be used in the future to realize the metre in the visible range.

Various methods have been used recently to measure the Rydberg constant.

The study of transitions between two adjacent circular states ($|m| = l = n - 1$) around $n = 25$ is carried out in Paris in the Li atom [3] and at the MIT in hydrogen. These transitions have the advantage of lying in the microwave range where good frequency standards are easily available. Because of the long lifetimes of the circular states leading to natural widths of the order of 10^{-9}, they will probably give soon the Rydberg constant with a precision of 1 part in 10^{10} or better.

In the optical range, the Balmer single-photon transitions in atomic hydrogen have been studied extensively. The most precise result, with a precision of 3 parts in 10^{10} [4], was obtained at Yale on the $2S$-$4P$ transition. It was limited by the natural width of the P levels which is responsible for a relative linewidth of a few parts in 10^8. Indeed in hydrogen the P levels are the broadest ones since they can radiate directly to the ground state.

The advantage of the two-photon transitions is to avoid these P states. The $1S$-$2S$ transition has an extremely small natural width and lies in the UV range. It has been studied by the Oxford group and by the Stanford group, now in Münich, as discussed in the lecture given by T. W. Hänsch. The $2S$-nS/nD ($n \geqslant 6$) transitions are studied in Paris. They lie in the near infrared and have relative linewidths of a few parts in 10^{10}. Because the $2S$ Lamb shift has been measured precisely in the radiofrequency range, they can lead by themselves to a determination of the Rydberg constant with a precision better than 2 parts in 10^{11}.

3. – The Paris experiment.

In order to obtain very narrow signals, we use the following experimental conditions [5]:

Metastable atoms are produced in an atomic beam to avoid collisional broadening.

They are optically excited by two counterpropagating laser beams so that the first-order Doppler effect is cancelled.

In the interaction region, the atomic beam is collinear with the two laser beams so that the transit time broadening is reduced.

The metastable atoms are produced by dissociation of molecular hydrogen in a radiofrequency discharge followed by atomic excitation to the $2S$ state by electronic impact. The deviation of the atomic beam due to the inelastic collisions with electrons is used to align the $2S$ beam with the laser beams. After passing through the interaction region (length ≈ 50 cm), the metastable atoms are detected by quenching in an applied electric field and measurement of the Lyman-α fluorescence induced by two photomultipliers.

The light source in the range $(730 \div 820)$ nm was previously a dye laser and is now a Ti-sapphire laser. To enhance the two-photon transition probability, the whole apparatus is placed inside a Fabry-Perot cavity. The length of this

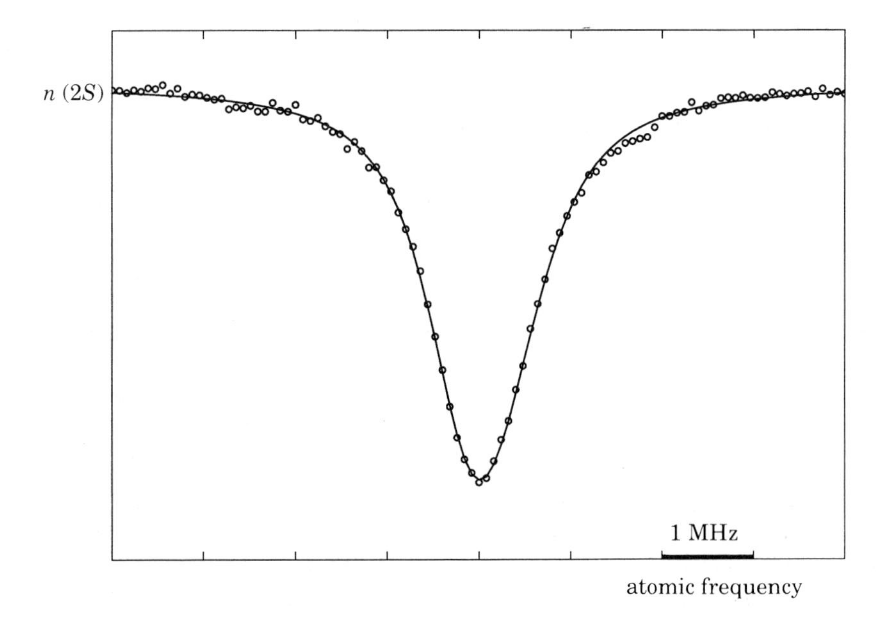

Fig. 1. – Experimental profile (circles) and theoretical fit (solid line) for the two-photon $2S_{1/2}$-$8D_{5/2}$ transition in hydrogen. The amplitude of the signal corresponds to a decrease of 19% of the metastable yield.

cavity is locked to the laser wavelength so that the light power experienced by the atoms can be up to 100 W in each direction of propagation. After optical excitation to the nS or nD states, most of the atoms decay to the ground state: the two-photon absorption signal can then be detected through the associated decrease of the $2S$ beam intensity.

A very reliable system allows us to bring the laser frequency to any desired value and sweep it with a reproducibility better than 1 part in 10^{11}. It makes use of an acousto-optic device driven by a computer-controlled synthesizer and of a Fabry-Perot reference cavity locked to a 633 nm standard laser. During a typical signal recording, which lasts 20 min, the laser frequency is scanned 10 times across the atomic resonance. The quenching voltage used for the metastable detection is square-wave modulated and a lock-in amplifier demodulates the signal given by the photomultipliers. Figure 1 shows a signal obtained for the $2S_{1/2}$-$8D_{5/2}$ transition. The circles are the experimental data and a calculated profile is superimposed.

4. – Study of the signals.

A detailed comparison of the line profiles with theoretical ones has been performed. The numerical calculation of these profiles is done taking into account the combined effect of natural width, light shift and saturation [6] along each atomic trajectory inside the interaction region and summing the contributions of all trajectories. The features of the calculated curves are a broadening and an asymmetry of the line due to both the saturation and inhomogeneous light shift and a global shift of the signal. Because of the saturation, this shift does not vary linearly with the light power.

The other broadening effects which are not considered in the above calculation (laser linewidth, residual transit time broadening, Stark and Zeeman effects due to residual stray electric or magnetic fields, ...) are simulated by making a convolution of the calculated line profile with a Gaussian curve.

For each run, the theoretical curve is fitted to the experimental one by adjusting four parameters: the metastable yield when the laser is off resonance, the light power, the Gaussian broadening and the line position without light shift with respect to the reference Fabry-Perot cavity. An example of a fit is shown in fig. 1. This treatment has been carried out for various light powers and different transitions. In all cases, we have found a very good agreement between experimental and calculated profiles and we have checked the consistency between the results deduced from the S and D signals [7]. Finally, for a suitable distribution of the atomic trajectories, we have obtained power-independent light-shift-corrected line positions for all the studied transitions.

5. – Wavelength measurement.

In our previous measurement of the Rydberg constant, performed in 1988, the wavelength calibration of the transitions under investigation was carried out very carefully by an interferometric method. The laser wavelength was compared to that of an I_2-stabilized standard He-Ne laser at 633 nm using a non-degenerate high-stability etalon built with two silver-coated mirrors optically adhered to a Zerodur rod. The length of this cavity was measured by beating an auxiliary He-Ne laser with the standard laser.

During the measurement, the auxiliary He-Ne laser and the excitation laser were both locked to peaks of the etalon cavity. Thus both their frequencies ν satisfied a resonance condition:

$$\nu = (c/2L)(N + \phi + \psi),$$

where N is an integer number easily determined, ϕ is the reflective phase shift due to the mirror coatings and ψ is the Fresnel phase shift due to the wave front curvature inside the cavity. The phase shift ϕ was eliminated by the method of virtual mirrors [8] using alternately two rods of different lengths (50 and 10 cm). The other phase shift ψ was precisely deduced from the frequency interval between the fundamental mode and the first transverse mode of the cavity [9].

By this method we measured the wavelengths of 6 transitions: the $2S_{1/2}$-$nD_{5/2}$ transitions with $n = 8$, 10 and 12 in hydrogen and deuterium. Six independent values of the Rydberg constant were deduced and our average result was [10]

$$R_\infty = 109\,737.315\,709\ (18)\ \text{cm}^{-1}$$

in good agreement with the results obtained by other groups between 1986 and 1988 from the Balmer lines [4] or the $1S$-$2S$ transition [11, 12]. For our result we assumed for the frequency of the standard laser at 633 nm the recommended value of the CCPM [13].

In the overall uncertainty of this measurement (1.7 parts in 10^{10}), the main contribution (1.6 parts in 10^{10}) came from the calibration of the standard laser itself performed at the NBS [14]. Very recently, this cause of uncertainty has been reduced, thanks to a new measurement of the standard frequency performed at the Laboratoire Primaire des Temps et Fréquences (LPTF) in Paris by A. CLAIRON and co-workers. The starting point of their frequency chain is a CO_2 laser stabilized to the OsO_4 molecule. Two intermediate sources are used: a colour centre laser at 2.6 μm and a laser diode at 1.3 μm.

After a preliminary result presented during the Enrico Fermi school, a final value for the standard frequency was obtained in July 1992 for the standard frequency. This value differs from the NBS value by −133 kHz. As a result, we can correct the value for the Rydberg constant obtained previously [15]:

$$R_\infty = 109\,737.315\,681\ (5)\ \mathrm{cm}^{-1}\ .$$

In the residual uncertainty of this corrected value, one can distinguish two types of contributions:

The first ones are not inherent in our experiment and arise from other data needed to deduce the Rydberg constant from our measurements, for example the electron-to-proton mass ratio, the $2S$ Lamb shift and QED calculations.

The second ones are due to our experimental method itself and give altogether a contribution of 4.2 parts in 10^{11}; among them, some come from our signal and its study and have been reduced by various improvements of our experimental set-up [7] and others are due to the interferometric method itself.

We estimate the limit of the interferometric method to be about 2 parts in 10^{11}. The physical reason is that in the interferometer we compare two beams having different wavelengths and hence different spatial extensions: a small imperfection on a mirror surface can be «seen» differently by the two beams and so induce an error of this magnitude in the wavelength comparison.

6. – Frequency measurement.

In order to improve our precision, we have recently replaced our wavelength measurement by a direct frequency comparison of the $2S$-$8S_{1/2}$, $8D_{3/2}$ and $8D_{5/2}$ lines with two optical standards. The scheme used for this comparison, suggested by A. Clairon, is shown in fig. 2: it takes advantage of a near coincidence (89 GHz) between the frequency of the I_2-stabilized He-Ne standard at 633 nm and the frequency sum of the CH_4-stabilized He-Ne laser at 3.39 μm and the excitation laser at 778 nm. In fact, it is convenient to use two Ti-sapphire lasers, one for the excitation of the atomic transitions (TiS1) and the other for the frequency sum (TiS2) and an auxiliary He-Ne at 3.39 μm.

The outputs of the auxiliary laser and of TiS2 are mixed in a $LiIO_3$ crystal. The red beam generated at 633 nm has a power of 250 nW for incident powers 20 mW at 3.39 μm and 1 W at 778 nm.

The measurement is performed in two steps. First we record the hydrogen spectrum with TiS1 which is scanned with respect to a given peak of the Fabry-Perot reference cavity. In the second step, we reduce the TiS1 frequency by an integer number of the free spectral range of this cavity and measure simultaneously the following beat frequencies:

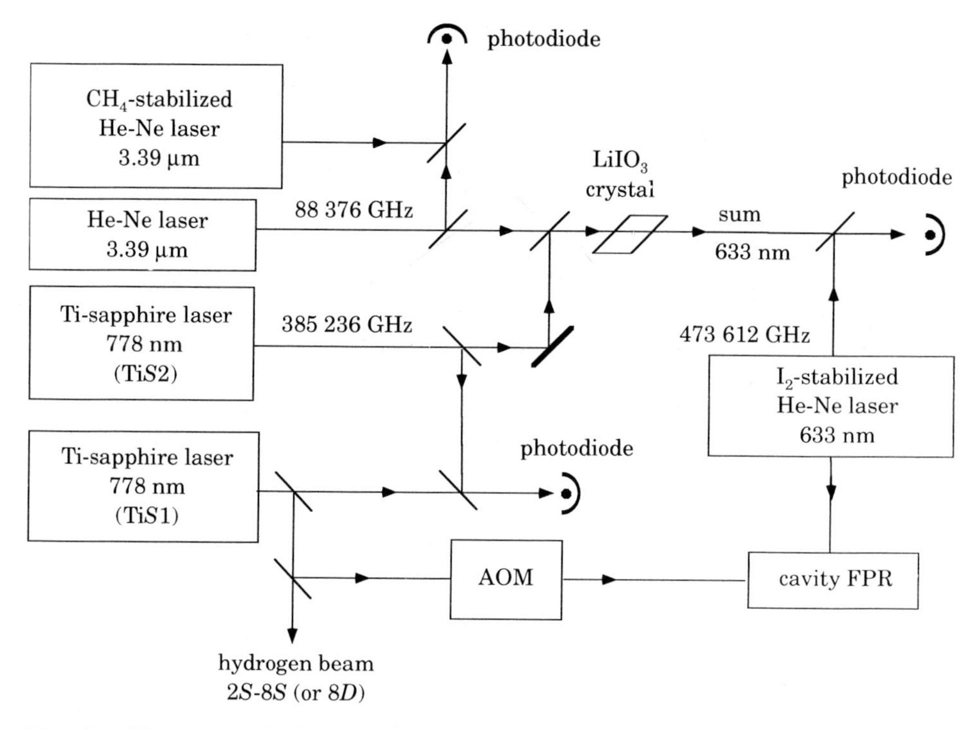

Fig. 2. – Experimental scheme for the calibration of the atomic-transition frequency.

between the auxiliary laser and the standard laser on loan from the BIPM at 3.39 μm,

between the two Ti-sapphire lasers at 778 nm,

between the synthesized radiation and our standard laser at 633 nm; this standard laser has been calibrated by direct comparison with another He-Ne laser whose frequency was measured at the LPTF.

From all these frequencies, we can deduce the hydrogen transition frequencies and hence determine the Rydberg constant.

7. – Result and perspectives.

We have measured the frequencies of three two-photon transitions in atomic hydrogen: the $2S_{1/2}$-$8S_{1/2}$, $2S_{1/2}$-$8D_{3/2}$ and $2S_{1/2}$-$8D_{5/2}$ transitions. The three independent determinations of the Rydberg constant which each one provides are in good agreement. Our weighted mean result is

$$R_\infty = 109\,737.315\,683\,0(31)\ \text{cm}^{-1}\,.$$

This result is in very good agreement with the value recently obtained by T.

W. HÄNSCH and his collaborators from the frequency measurement of the $1S$-$2S$ transition as well as being slightly more precise. The apparent discrepancy between the results of the two groups, which was discussed during the Enrico Fermi school, was due to a systematic error in the calibration of the LPTF frequency chain and was eliminated just after the school has ended.

Our global uncertainty is 2.9 parts in 10^{11}. The various contributions to this uncertainty are listed in table I. The electron-to-proton mass ratio, the $2S$ Lamb shift and QED calculations give together an uncertainty of 1.7 parts in 10^{11}. This is the current limitation of our method.

The uncertainty due to the measurement and the comparison of the frequency standards at 633 nm could be reduced, for example by a direct link between our laboratory and the LPTF by means of an optical fibre.

Finally, we plan to reduce soon the uncertainty of the measurement of the 89 GHz interval. In this work, to determine this interval, we used a Fabry-Perot cavity whose free spectral range was known very precisely. This contrasts with our previous measurement in which we employed cavities to compare two very different wavelengths and where uncertainties due to phase shifts were consequently of greater importance. In the future we plan to eliminate altogether the need for a cavity for the frequency comparison. All the beat frequencies mentioned above will be recorded simultaneously with the hydrogen spectra. The frequencies of the two Ti-sapphire lasers will be compared using a MIM diode, with a Gunn diode at 89 GHz as an auxiliary microwave source. It should then be possible to improve our precision to 2 parts in 10^{11} or better.

TABLE I. – *Uncertainty budget.*

Contribution	Parts in 10^{11}
Lamb shift and QED calculations	1.3
electron-to-proton mass ratio	1.1
statistical uncertainty	0.9
fits to theoretical profile	0.5
standard He-Ne/I_2 lasers	1.8
standard He-Ne/CH_4 laser	0.3
89 GHz measurement	0.8
second-order Doppler effect	0.4
Stark effect	0.3
uncertainty in the Rydberg constant	2.9

REFERENCES

[1] M. WEITZ, F. SCHMIDT-KALER and T. W. HÄNSCH: *Phys. Rev. Lett.*, **68**, 1120 (1992).

[2] L. JULIEN, F. BIRABEN and M. ALLEGRINI: *Comments At. Mol. Phys.*, **26**, 219 (1991).

[3] A. NUSSENZWEIG, J. HARE, A. M. STEINBERG, L. MOI, M. GROSS and S. HAROCHE: *Europhys. Lett.*, **14**, 755 (1991).

[4] P. ZHAO, W. LICHTEN, H. P. LAYER and J. C. BERGQUIST: *Phys. Rev. Lett.*, **58**, 1293 (1987); *Phys. Rev. A*, **39**, 2888 (1989).

[5] J. C. GARREAU, M. ALLEGRINI, L. JULIEN and F. BIRABEN: *J. Phys. (Paris)*, **51**, 2263 (1990).

[6] J. C. GARREAU, M. ALLEGRINI, L. JULIEN and F. BIRABEN: *J. Phys. (Paris)*, **51**, 2275 (1990).

[7] L. JULIEN and F. BIRABEN: in *Atomic Physics 12*, edited by J. C. ZORN and R. R. LEWIS (American Institute of Physics, New York, N.Y., 1991), p. 381.

[8] H. P. LAYER, R. D. DESLATTES and W. G. SCHWEITZER jr.: *Appl. Opt.*, **15**, 734 (1976).

[9] J. C. GARREAU, M. ALLEGRINI, L. JULIEN and F. BIRABEN: *J. Phys. (Paris)*, **51**, 2293 (1990).

[10] F. BIRABEN, J. C. GARREAU, L. JULIEN and M. ALLEGRINI: *Phys. Rev. Lett.*, **62**, 621 (1989).

[11] R. G. BEAUSOLEIL, D. H. MCINTYRE, C. J. FOOT, E. A. HILDUM, B. COUILLAUD and T. W. HÄNSCH: *Phys. Rev. A*, **35**, 4878 (1987).

[12] M. G. BOSHIER, P. E. G. BAIRD, C. J. FOOT, E. A. HINDS, M. D. PLIMMER, D. N. STACEY, J. B. SWAN, D. A. TATE, D. M. WARRINGTON and G. K. WOODGATE: *Nature (London)*, **330**, 463 (1987); *Phys. Rev. A*, **40**, 6169 (1989).

[13] *Documents concerning the New Definition of the Metre, Metrologia*, **19**, 163 (1984).

[14] D. A. JENNINGS, C. R. POLLOCK, F. R. PETERSEN, R. E. DRULLINGER, K. M. EVENSON, J. S. WELLS, J. L. HALL and H. P. LAYER: *Opt. Lett.*, **8**, 136 (1983).

[15] F. NEZ, M. D. PLIMMER, S. BOURZEIX, L. JULIEN, F. BIRABEN, R. FELDER, O. ACEF, J. J. ZONDY, P. LAURENT, A. CLAIRON, M. ABED, Y. MILLERIOUX and P. JUNCAR: *Phys. Rev. Lett.*, **69**, 2326 (1992).

Laser Spectroscopy of Atomic Hydrogen.

T. W. HÄNSCH

Sektion Physik, Universität München - Schellingstrasse 4, 8000 Munich 40
Max-Planck-Institut für Quantenoptik - 8046 Garching, B.R.D.

1. - Introduction.

As the simplest of the stable atoms, hydrogen permits unique confrontations between spectroscopic experiment and fundamental theory. Spectroscopy of hydrogen has played a central role in the development of quantum mechanics and atomic physics [1-3]. The interpretation of the regular visible Balmer spectrum of hydrogen has inspired several conceptual breakthroughs, from BOHR and the old quantum physics to the theories of Sommerfeld, de Broglie, Schrödinger and Dirac to the discovery of the Lamb shift and the development of modern quantum electrodynamics (QED).

Despite persisting conceptional difficulties, QED is undisputedly the most successful theory in physics. With its help we can calculate the energy levels and transition frequencies of the hydrogen atom with exquisite precision. A comparison of measured hydrogen transition frequencies with such calculations can yield accurate values of important quantities, such as the Rydberg constant, the electron/proton mass ratio, or the charge radius of the proton. There is also the much more intriguing prospect that very precise spectroscopic studies of hydrogen and other hydrogenlike atoms might eventually discover conceivable limits or flaws in the theory of QED. Despite the impressive successes of QED, several small and unexplained discrepancies between experiment and theory seem to persist [2, 3]. Examples include the lifetime of positronium or the magnetic moment of the electron. For these reasons, the spectroscopist faces the intriguing challenge of studying the spectrum of atomic hydrogen with the most advanced experimental tools available in order to reach new and untested levels of resolution and measurement accuracy.

The most serious problem in classical optical spectroscopy of hydrogen has been the large Doppler broadening of the spectral lines due to the random thermal motion of the light atoms which masks the intricate details of line shape and structure. Major advances became possible only with the advent of tunable lasers and coherent light techniques. About twenty years ago, HÄNSCH and coworkers at Stanford [4] first succeeded in resolving single fine-structure components of the prominent red Balmer-α line of hydrogen with the help of a highly monochromatic tunable dye laser and the newly developed method of saturation spectroscopy. In this technique, Doppler broadening is eliminated by sending a strong saturating beam and a weak probe beam in opposite directions through an absorbing gas sample. If the laser is tuned near the centre of a Doppler-broadened line, the saturating beam can bleach a path for the probe because both beams are now interacting with the same atoms, those with zero axial velocity. Figure 1 shows at the bottom one of the first Doppler-free spectra of the Balmer-α line recorded in this way. Shown above is the Doppler profile at room temperature and the seven fine-structure components as predicted by QED. For the first time, the «classical» $2S$ Lamb shift could be directly observed in the optical absorption spectrum. (Experiments of this kind can today be real-

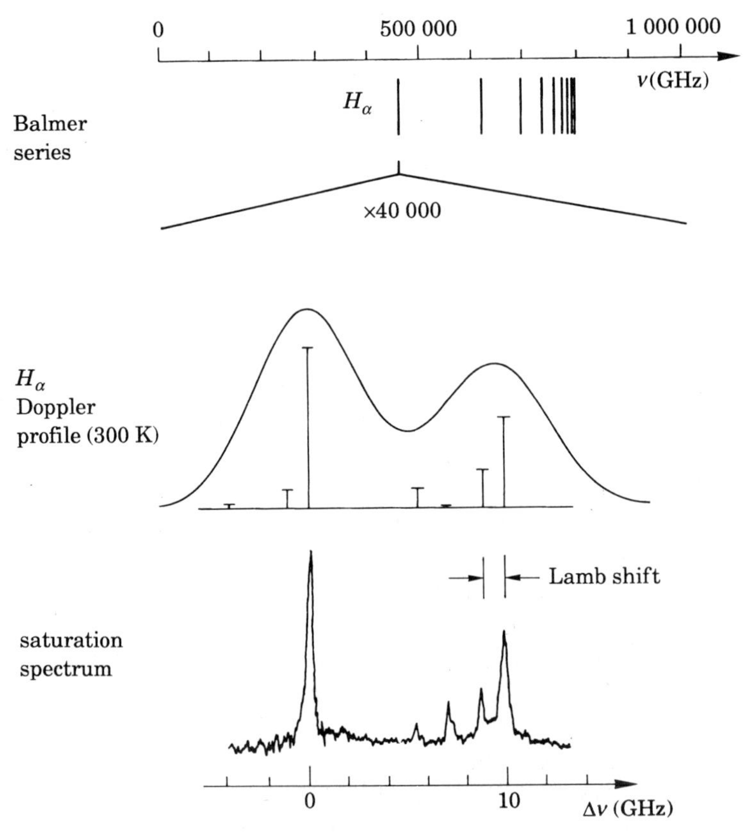

Fig. 1. – Doppler-free saturation spectrum of the red Balmer-α line of atomic hydrogen with resolved fine-structure components, as recorded around 1972 at Stanford (ref. [4]). Shown above for comparison is the Doppler-broadened line profile at room temperature and the seven fine-structure components predicted by QED.

ized even in a student laboratory with the help of an inexpensive visible laser diode.) In 1974, an interferometric determination of the absolute wavelength of the strongest $2P_{3/2}$-$3D_{5/2}$ component at Stanford [5] yielded the first laser measurement of the Rydberg constant R_∞ with an improvement over the best conventional measurements of almost one order of magnitude.

Since then, the resolution and measurement accuracy of laser spectroscopy of hydrogen have been advanced by almost 5 orders of magnitude, and the rate of improvement is still growing exponentially. New spectroscopic techniques and instruments have been developed during this pursuit, which are finding important applications beyond their original purpose. Examples include the technique of polarization spectroscopy [6], the Hänsch-Couillaud method of frequency stabilization [7], or even the original proposal for laser cooling of atoms [8]. At the same time, continuous rapid advances in the technology of lasers, nonlinear optics and optoelectronics, fuelled by technical applications such as telecommunications, are providing a rich source of new experimental tools. For the future, one can foresee an increasing role of sophisticated methods for cooling, manipulating and trapping neutral atoms with electromagnetic fields, as well as techniques of atom interferometry, which are being developed and investigated in many laboratories [9]. By their very nature, such experiments at the forefront of the state of the art tend to be technically very demanding, and hydrogen almost inevitably adds particular technical difficulties of its own.

2. – Doppler-free two-photon spectroscopy of the 1*S*-2*S* transition.

The resolution of the visible hydrogen Balmer lines cannot be reduced below the natural linewidth limits of at least several MHz imposed by the short radiative lifetimes of the excited levels. One can find, however, other transitions between longer-living states which can yield much higher resolution. Figure 2 shows a simplified energy level diagram of hydrogen. Perhaps the most fascinating resonance is the transition from the $1S$ ground state to the metastable $2S$ state, because the $1/7$ s lifetime of the upper level implies a natural linewidth of only 1.3 Hz. The magnified sections at the right-hand side of fig. 2 show the level energies as predicted by the relativistic Dirac theory, the energy shifts due to QED effects (Lamb shifts) and the hyperfine splitting due to interaction of the magnetic moments of electron and proton.

Since the charge distribution in both S states is spherically symmetric, there is no allowed electric-dipole transition. The $2S$ level can be excited, however, by simultaneous absorption of two photons, which jointly provide the required energy, as first demonstrated at Stanford in 1975 [10]. The selection rules for two-photon transitions without intermediate resonance ($\Delta F = 0$, $\Delta m = 0$) permit two hyperfine components, as indicated in fig. 2.

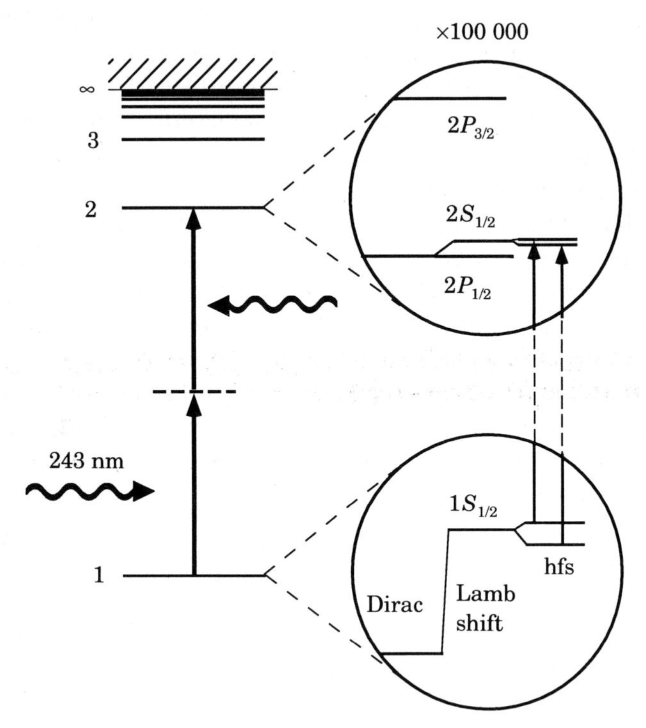

Fig. 2. – Simplified energy level diagram of hydrogen with $1S$-$2S$ two-photon transition.

Linear Doppler broadening in two-photon spectroscopy can be simply eliminated without any need for velocity selection by exciting the atoms with two counterpropagating laser beams of equal frequencies. From a moving atom, one beam will appear red-shifted, the other blue-shifted, and these two Doppler shifts cancel to first order. This elegant method of Doppler-free two-photon spectroscopy was first proposed in 1970 by CHEBOTAEV and co-workers [11]. If the natural linewidth of the hydrogen $1S$-$2S$ transition could be approached experimentally, the resolution $\Delta\nu/\nu$ would be better than 10^{-15}, and a measurement uncertainty smaller than 10^{-18} appears conceivable.

The most serious experimental obstacle in Doppler-free two-photon spectroscopy of the hydrogen $1S$-$2S$ transition has long been the lack of a suitable intense and highly monochromatic tunable laser source near 243 nm. Early experiments with frequency-doubled pulsed dye lasers [10, 12, 13] reached at best line widths of a few 100 MHz. Nonetheless, these experiments led to the first accurate measurements of the Lamb shift of the $1S$ ground state [12, 13]. Around 1986, BEAUSOLEIL and MCINTYRE at Stanford [14, 15] recorded the first continuous-wave spectra with much improved resolution by exciting hydrogen atoms in a gas cell with an ultraviolet standing wave produced by sum frequency mixing of an Ar^+ laser at 365 nm and a dye laser at 780 nm in a nonlinear crystal of KDP (= potassium dihydrogen phosphate). The remaining line broadening of a few MHz was due to collisions, the short transit time of the atoms passing through the focussed light field and laser frequency fluctuations. Similar experiments have been performed afterwards by STACEY and co-workers at Oxford [16], who produced 243 nm light more simply by doubling the frequency of a c.w. dye laser operating near 486 nm in a nonlinear crystal of BBO (= beta barium borate) which had just become commercially available.

A new generation of $1S$-$2S$ two-photon spectrometers has been developed since 1987 in our laboratory at the Max-Planck-Institut for Quantum Optics in Garching [17]. Considerable efforts have been devoted to the development of a suitable ultrastable laser source. To generate several milliwatt near 243 nm, the frequency of a modified commercial c.w. dye laser at 486 nm is doubled in a BBO crystal inside an external build-up ring cavity. Laser frequency fluctuations such as those caused by optical-path variations in the dye jet are compensated with the help of an electro-optic phase modulator inside the laser cavity. A fast servo system locks the laser frequency to a mode of a stable external reference cavity. The error signal is generated by the r.f. sideband technique of Pound and Drever, *i.e.* by introducing a phase modulation with an electro-optic modulator and by observing a conversion to amplitude modulation in the light reflected by the resonator. The spacer of the reference cavity consists of a 45 cm long massive bar of the glass-ceramic material Zerodur with low thermal-expansion coefficient. Optically contacted gyroscope quality mirrors provide a finesse of about 57 000 or a resonance linewidth of about 5 kHz. The entire assembly is suspended from soft springs inside a temperature-stabilized and vibration-iso-

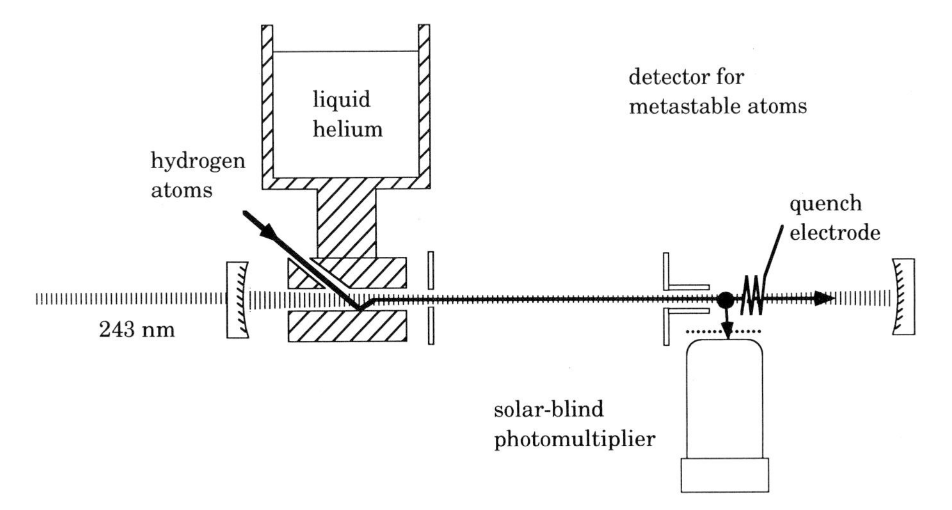

Fig. 3. – Set-up for Doppler-free two-photon spectroscopy of the hydrogen $1S$-$2S$ transition in a cold atomic beam.

lated vacuum tank. Remaining tiny vibrations limit the achievable laser line width to a few 100 Hz at the present time.

To reduce line broadening due to collision and transit time effects, the hydrogen atoms are now observed by longitudinal excitation of a cold atomic beam, as sketched in fig. 3. The atoms are produced by microwave dissociation of molecular H_2 and guided with a Teflon tube to a nozzle which is mounted at the bottom of a cryostat so that the atoms can be cooled by collisions with the walls. At the temperature of liquid helium, the mean velocity is reduced to about 300 m/s. The ultraviolet light is coupled into a build-up cavity inside the vacuum apparatus, producing a standing-wave field of (20 ÷ 50) mW circulating power coaxial to the atomic beam. The 15 cm long interaction region is screened from electric stray fields with a graphite-coated wire mesh. Finally, an electric quenching field mixes the $2S$ and $2P$ states and forces the atoms to emit vacuum ultraviolet Lyman-α photons which are counted with a photomultiplier.

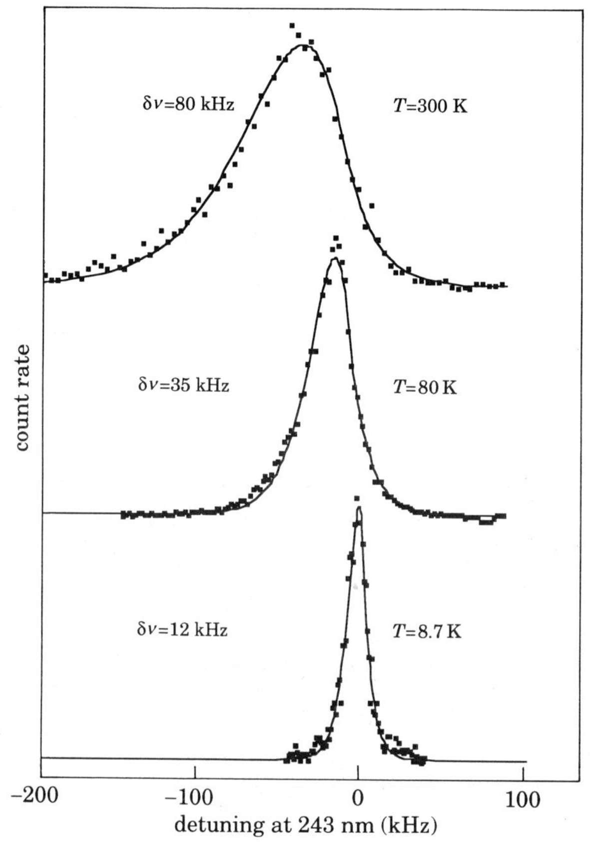

Fig. 4. – Doppler-free two-photon spectra of the $F = 1$ hyperfine component of the hydrogen $1S$-$2S$ transitions, recorded by coaxial excitation of an atomic beam at three different nozzle temperatures. At 8.7 K, the resolution reaches 1 part in 10^{11}.

Figure 4 shows spectra of the $F = 1$ hyperfine component of the hydrogen $1S$-$2S$ transition which have been recorded in this way at different nozzle temperatures. Near room temperature, the line shows a clear asymmetry which can be explained by the second-order Doppler effect. Due to the time dilation of special relativity, moving atoms are oscillating more slowly than atoms at rest, and the thermal-velocity distribution leads to a red shift and asymmetric line broadening proportional to the temperature. This effect is much reduced at liquid-nitrogen temperature, and it is negligible near liquid-helium temperature where the symmetric resonance line shape is almost totally determined by the finite transit times of the atoms passing through the light field. The line width of 12 kHz in the ultraviolet corresponds to a resolution of one part in 10^{11}, *i.e.* the ratio of line width to frequency is the same as that of the diameter of a human hair to the diameter of the Earth. For the heavier deuterium isotope, we observe an even narrower line of 9 kHz.

This extremely narrow $1S$-$2S$ two-photon resonance has recently been used at Garching for three different precision measurements. New techniques of optical frequency metrology had to be developed for these experiments, since the spectral resolution already exceeds the precision of available optical wavelength standards and established interferometric measuring methods.

3. – Precision measurement of the $1S$ Lamb shift.

In the first of these experiments, the Lamb shift of the hydrogen $1S$ ground state has been measured to within 1 part in 10^5 [18]. The discovery of the «classic» Lamb shift in the excited $n = 2$ level after the second world war [19] has ushered in the development of modern QED by FEYNMAN, SCHWINGER and TOMONAGA. Using radiofrequency spectroscopy of a metastable hydrogen beam, LAMB and RETHERFORD were able to show that the $2S_{1/2}$ state does not coincide with the $2P_{1/2}$ state, as expected according to the theory of Dirac, but that the $2S$ level is actually shifted upwards by about 1000 MHz. The largest contribution to this Lamb shift is caused by the interaction of the electron with the fluctuating quantized radiation field of the vacuum, *i.e.* the «self-energy». Measurements of the $2S$ Lamb shift still provide one of the most critical tests of QED [20]. However, they have already been pushed to the accuracy limit of about 10^{-5}, imposed by the large 100 MHz natural linewidth of the r.f. transition due to the short lifetime of the neighbouring $2P$ state.

Since the Lamb shift scales with the inverse cube of the principal quantum number n, the Lamb shift of the $1S$ ground state is 8 times larger than that of the $2S$ state. However, it cannot be measured directly by microwave spectroscopy, since there is no nearby P level that could serve as a reference. For this reason, two different optical transitions have been accurately compared in the Garching experiment, similar to earlier and much less accurate measure-

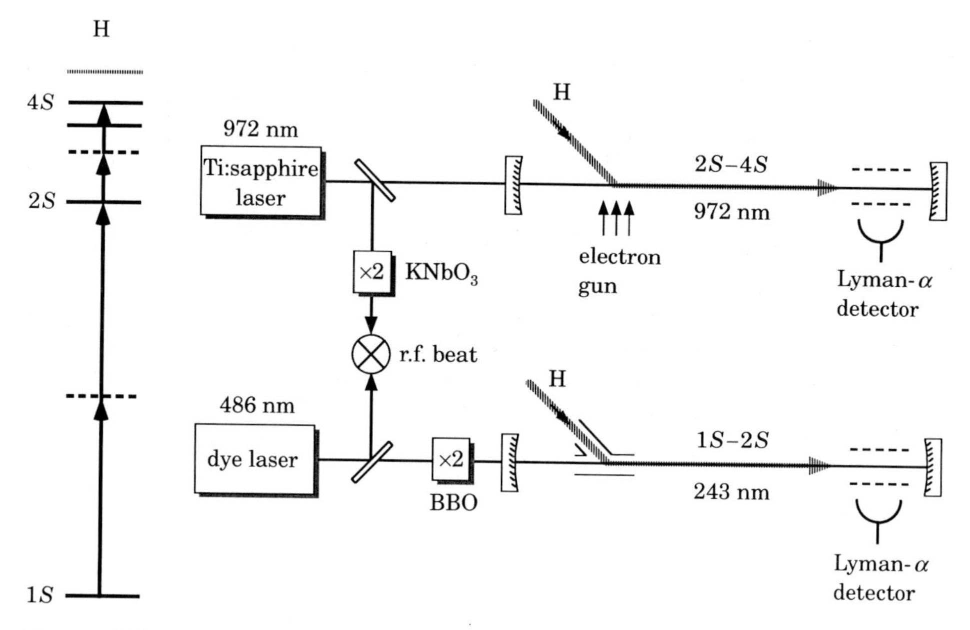

Fig. 5. – Scheme for a precision measurement of the hydrogen 1S Lamb shift. Two separate atomic beams are used to compare the frequencies of the two-photon transitions 2S-4S and 1S-2S via a microwave beat signal.

ments of the 1S Lamb shift at Stanford [12, 13]. In the simple Bohr or Schrödinger model of the hydrogen atom, the 1S-2S interval is precisely 4 times larger than the 2S-4S interval. Any deviation from this integer ratio is determined by the Lamb shifts of the participating levels in addition to well-understood relativistic effects. Since the smaller Lamb shifts of the 2S and 4S states are rather accurately known, the Lamb shift of the 1S state can be determined from a precise comparison of the two transition frequencies.

A scheme of the new Garching experiment is shown in fig. 5. Two separate atomic beams are used to observe and compare the 1S-2S and the 2S-4S (or 2S-4D) two-photon transitions. In the 2S-4S spectrometer (top), metastable hydrogen 2S atoms are produced by bombarding ground-state hydrogen atoms with electrons, similar to the scheme of Biraben *et al.* [21, 22]. The 2S-4S (or 2S-4D) two-photon resonance is excited with a collinear infrared standing wave of about 50 W circulating power near 972 nm, produced by injecting the output from a highly stable Ti : sapphire laser into a standing-wave build-up cavity. Since the 4S atoms decay with high probability to the 1S ground state via radiative cascades, the 2S-4S excitation can be detected by observing a reduction in the count rate of metastable 2S atoms.

To compare the two transition frequencies, a nonlinear crystal produces some blue light at the second-harmonic frequency of the infrared laser, and a fast photodiode monitors a beat signal with the output of the dye laser which is

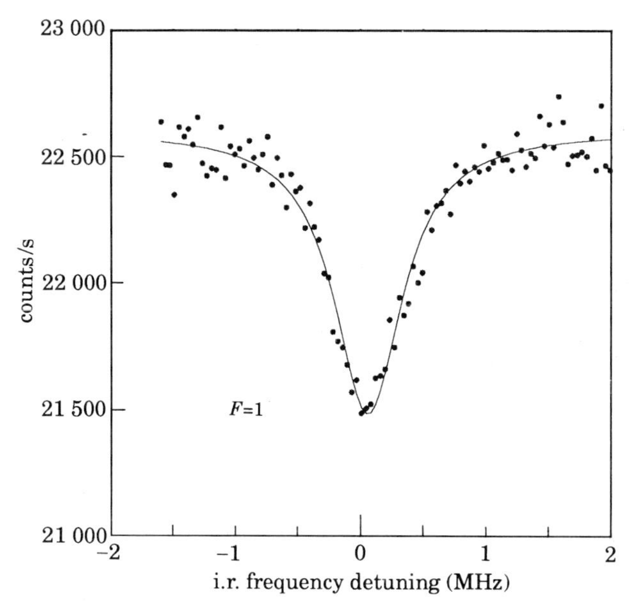

Fig. 6. – Doppler-free two-photon spectrum of the hydrogen 2*S*-4*S* transitions. The resonance appears as a dip in the count rate of metastable 2*S* atoms.

locked to the maximum of the 1*S*-2*S* resonance. Without Lamb shifts or relativistic effects, this beat frequency at resonance would be exactly zero. In reality one observes a beat frequency near 5 GHz when the infrared laser is tuned to the centre of the 2*S*-4*S* transition.

Figure 6 shows a Doppler-free two-photon spectrum of the 2*S*-4*S* resonance. Forty scans have been averaged during a total measuring time of 20 min, since the drop of the 2*S* count rate at resonance amounts to only a few percent. Nonetheless, a line of sub-MHz width is recorded, since the beat signal allows a completely drift-free and reproducible frequency tuning of the Ti : sapphire laser. To find the true resonance frequency it was necessary to carefully correct for light shifts due to the a.c. Stark effect, while accounting for the saturation of the transition rate in the strong laser field with the help of computer codes kindly made available to us by JULIEN and BIRABEN [13].

In this way, a 1*S* Lamb shift of 8172.82(11) MHz has been measured with an uncertainty near 1 part in 10^5. Figure 7 illustrates the improvement over the best previous measurements. Following tradition, the Lamb shift includes the effects due to the finite size of the nucleus. As indicated in the lower part of fig. 7, this causes some problems for a comparison with theory, since scattering experiments with fast electrons from high-energy accelerators have so far yielded two somewhat contradictory values of the r.m.s. charge radius of the proton. The measured 1*S* Lamb shift is in better agreement with the theoretical value of 8172.94(9) which is predicted if the proton charge radius is taken to be

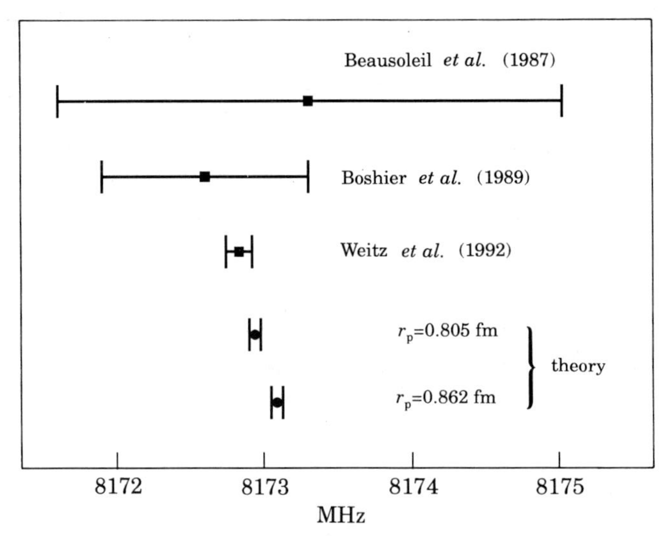

Fig. 7. – Garching measurement of the hydrogen $1S$ Lamb shift in comparison with the best previous experiments.

0.805(11) fm [23]. With a more recent and probably more reliable proton radius of 0.862(12) fm [24], however, the agreement is less convincing.

Very recently, we have recorded even narrower $2S$-$4S$ resonances with better signal-to-noise ratio by observing the blue $4S$-$2P$ fluorescence rather than the drop in metastable count rate [25]. These experiments should give an almost twofold further improvement in accuracy, so that the new measurement of the $1S$ Lamb shift will surpass the r.f. measurements of the $2S$ Lamb shift and provide the most stringent test of QED for a bound system. In the future it will also become possible to compare the $1S$-$2S$ frequency with even narrower two-photon resonances from $2S$ to higher S or D states. Once the measurement accuracy exceeds the accuracy of radiofrequency Lamb shift measurements it will be advantageous to compare the observed frequency ratio with theory directly.

As long as QED is correct, such experiments can provide a new value of the proton charge radius. Independent measurements of this charge radius are required, however, to test the validity of QED. Although improved electron scattering experiments are clearly desirable, experiments with muonic hydrogen ($p^+ \mu^-$) may provide even better access to the size of the proton. The muon is 200 times heavier than the electron and moves in much closer vicinity to the nucleus. The Lamb shift of the $2S$ state of muonic hydrogen is, therefore, expected in the infrared (near 6 μm). Unlike the shift in ordinary hydrogen, it is dominated by the polarizability of the vacuum and nuclear-size effects are much more important. Measurements of this shift could probably determine the proton charge radius to better than 1 part in 10^3.

The prospects of such future precise measurements are rekindling an interest in more precise QED calculations of hydrogen energy levels. Very recently, PACHUCKI has cut the uncertainty of the predicted $1S$ self-energy almost in half using a novel semi-analytical approach [26]. It is perhaps worth noting that the symbolic manipulation and record keeping of a large number of terms were much facilitated with the help of the program MATHEMATICA on a personal computer. More accurate calculations of higher-order two-loop corrections are now well under way.

4. – Hydrogen-deuterium isotope shift of the $1S$-$2S$ frequency.

In a recent experiment which takes even better advantage of the very high resolution of the $1S$-$2S$ resonance, the difference of the $1S$-$2S$ transition frequencies for hydrogen and deuterium (672 GHz) has been measured to within 3.7 parts in 10^8 [27]. QED effects cancel almost completely in this isotope shift, and the frequency difference is essentially due to the different masses of the nuclei. Since the proton/deuteron mass ratio is known very accurately [28], a precise measurement of the isotope shift can yield an accurate value of the electron/proton mass ratio, if the nuclear-size corrections are known, *e.g.* from measurements of the $1S$ Lamb shift.

Figure 8 illustrates the scheme of the isotope shift measurement. To produce accurately known marker frequencies near the hydrogen and the deuterium resonance, a second dye laser is locked to a mode of the stable reference resonator so that its frequency is about half-way in between the two resonances. Two marker frequencies spanning the isotope shift (168 GHz at 486 nm) are generated as modulation sidebands with the help of a novel very fast electro-optic phase modulator [29], operating near 84 GHz. The modulator employs a thin plate of lithium tantalate which is inserted like an etalon into an open Fabry-Perot-type microwave cavity. The light propagates inside the crystal along a zig-zag path under total internal reflection. Phase matching and thus the modulation index can be optimized by proper choice of the propagation angle.

To measure the isotope shift, the dye laser of the $1S$-$2S$ spectrometer is alternatingly locked to the line maximum of the hydrogen and the deuterium resonance, and its frequency is compared with the nearby sideband marker frequency by counting the frequency of a radiofrequency beat signal. The separation of the two markers is calibrated by comparing the frequencies of the microwave oscillator and a rubidium atomic clock.

So far, we have measured an isotope shift of 670 994 337(22) kHz, which is about 25 times more accurate than the best previous measurement. A further improvement by one or two orders of magnitude should be achievable, if slow drifts of the reference cavity are eliminated. Theoretically, one expects a shift

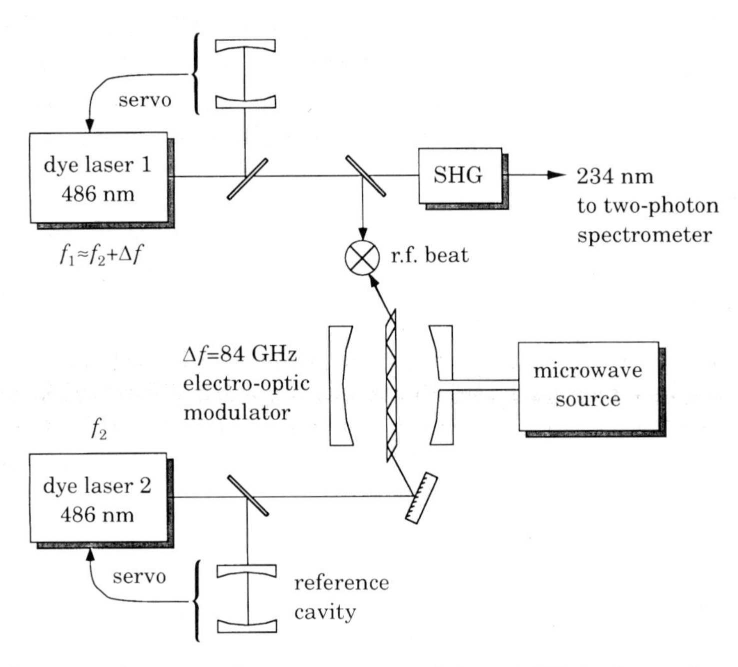

Fig. 8. – Apparatus for a precision measurement of the 672 GHz hydrogen-deuterium isotope shift of the $1S$-$2S$ transition. A novel fast electro-optic modulator produces marker sidebands spanning the frequency difference of 168 GHz at 486 nm.

of 670 994 414(22) kHz, taking into account the measured charge radii of proton and deuteron. The discrepancy may be associated with the polarizability of the deuteron, which has so far been ignored. Using a rather simplified model [30], we have estimated that the deuteron polarizability reduces the deuterium $1S$-$2S$ interval by about 20 kHz. Tabletop spectroscopic experiments are thus beginning to reach to effects of nuclear structure and dynamics in a regime of low energies and momentum transfer which is inaccessible to experiments with large colliders.

5. – Absolute frequency of hydrogen $1S$-$2S$ and a new Rydberg constant.

In the most elaborate experiment so far we have compared the absolute frequency of the hydrogen $1S$-$2S$ transition with the 9 GHz microwave frequency of a cesium atomic clock, using a phase-locked harmonic laser frequency chain [31]. This experiment is the first of its kind which is not limited by the systematic errors of interferometric wavelength metrology and the first which overcomes the uncertainty limit of 1.6 parts in 10^{10}, imposed in the past by the internationally accepted visible wavelength standard, an I_2-stabilized He-Ne laser near 633 nm. From the measured frequency, together with the observed 1*S* Lamb shift, we derive an accurate new value of the Rydberg constant.

As a secondary frequency standard we use a transportable CH_4-stabilized infrared He-Ne laser at 3.39 μm which has been built in the laboratory of Chebo-

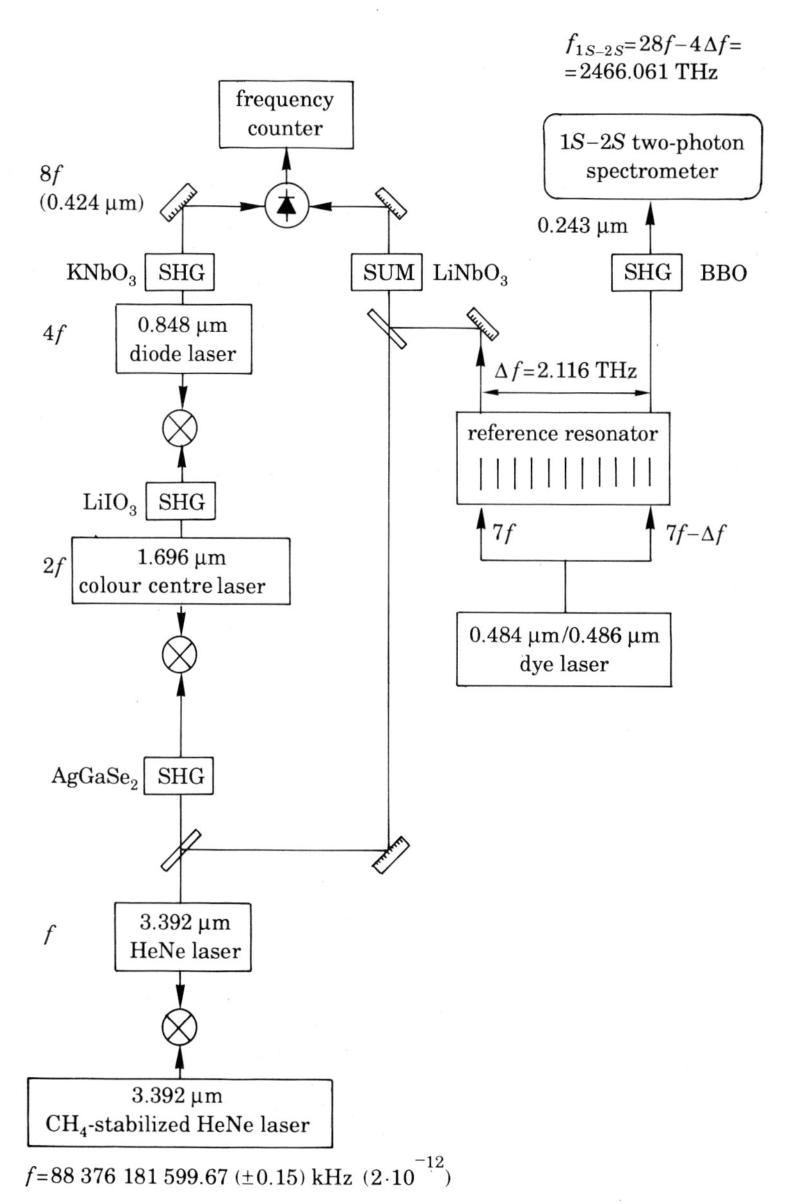

Fig. 9. – Garching laser frequency chain for a precision measurement of the absolute frequency of the hydrogen 1*S*-2*S* transitions. An infrared CH_4-stabilized He-Ne laser at 3.39 μm serves as a secondary frequency standard with an uncertainty of 10^{-12}.

taev in Novosibirsk and calibrated at the PTB in Braunschweig in a direct comparison with a Cs frequency standard. The measured absolute reproducibility was $1.8 \cdot 10^{-12}$ and the stability 10^{-13} at $(10 \div 100)$ s integration time. As illustrated in fig. 9, we synthesize a violet reference frequency (424 nm) as the 8th harmonic of the CH_4 standard in three steps of second-harmonic generation. Since the efficiency of c.w. frequency doubling in the chosen nonlinear crystals is very low, we boost the power with the help of three transfer oscillators: a He-Ne laser at 88 THz, a NaCl:OH^- colour centre laser at 176 THz and a diode laser at 352 THz. By electronically phase-locking all these oscillators simultaneously, we transfer the full accuracy of the He-Ne/CH_4 standard to the violet reference at 707 THz. This visible light represents the highest optical frequency that has been synthesized with an accuracy of 10^{-12}.

The dye laser of the hydrogen spectrometer is operating near the 7th harmonic of the He-Ne/CH_4 standard. To compare its frequency with the violet reference, we generate the sum frequency of this dye laser and the He-Ne/CH_4 standard in another nonlinear crystal and observe a beat signal on a photodiode. A remaining frequency mismatch of about 2 THz at 486 nm has so far been bridged with the help of 6372 axial modes of the dye-laser reference cavity. The free spectral range of this cavity was known to better than 1 Hz after a calibration with a fast electro-optic modulator [29]. However, we observe residual erratic drifts, perhaps due to recrystallization processes in the cavity spacer, which dominate the final error budget.

In this way, we have measured a frequency $f(1S\text{-}2S) = 2\,466\,061\ 413.182(45)$ MHz with an 18-fold improvement in accuracy over the best previous measurement. From a comparison with theory, together with the measured 1S Lamb shift, we derive a new and much improved value of the Rydberg constant, $R_\infty = 109\,737.315\,684\,1(42)\ \text{cm}^{-1}$. The relative uncertainty of 3.8 parts in 10^{11} is somewhat larger than that of the 1S-2S frequency, but it will be reduced when the 1S Lamb shift with its troublesome nuclear-size corrections is measured more accurately.

The Garching result represents the most accurate measurement of any fundamental constant. The new Rydberg value is higher than the value of $109\,737.315\,34(13)\ \text{cm}^{-1}$, chosen in the 1986 *Adjustment of the Fundamental Constants* [28], but below the preceding most accurate measurement of $109\,737.315\,709(18)\ \text{cm}^{-1}$ by BIRABEN and co-workers in Paris [22]. Figure 10 shows the new result in a comparison with several previous measurements. (About a month after this Varenna Course, the Garching experiments have been fully confirmed by new measurements in Paris, based on an improved calibration of the 633 nm He-Ne laser frequency reference [32].)

The Rydberg constant is an important cornerstone in the system of fundamental constants, and the principal scaling factor for any spectroscopic transition. The new Rydberg measurement improves our knowledge about the relationship between the arbitrarily chosen unit of time, the second, and the system

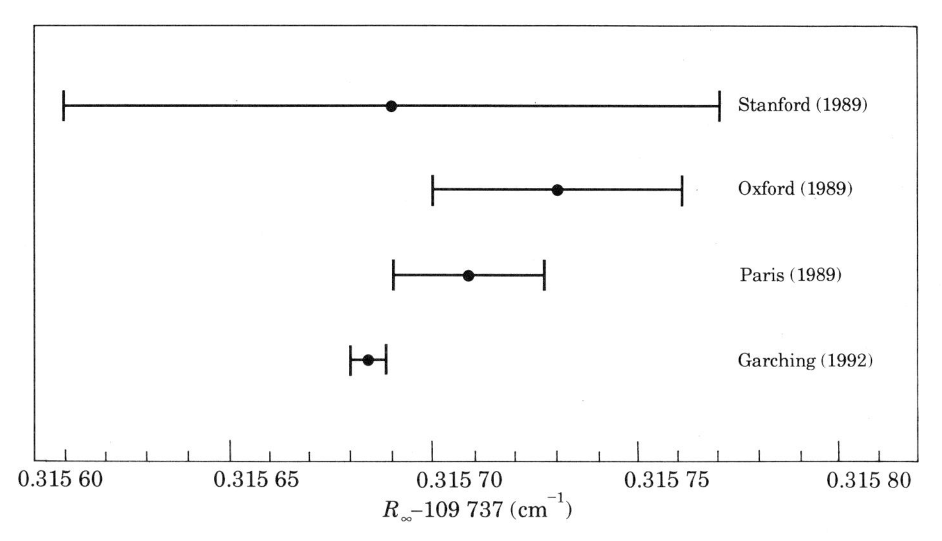

Fig. 10. – 1992 Garching measurement of the Rydberg constant in comparison with the best preceding measurements.

of fundamental constants. At the same time it establishes an entire system of accurately known reference frequencies, from the vacuum ultraviolet to the visible, infrared and microwave regions, because the frequencies of all other transitions in atomic hydrogen can now be predicted with improved accuracy, provided the theory of QED is correct.

For future fully phase-coherent measurements of hydrogen transition frequencies we are perfecting a new method for bisecting optical frequency intervals [33, 34]. The operating principle of a basic building block is rather simple. The frequency of a laser f_3 is phase-locked to the centre of two given laser frequencies f_1 and f_2 by observing a beat signal between the second harmonic $2f_3$ and the sum frequency $f_1 + f_2$. By linking several such optical-divider stages, it becomes possible to divide arbitrary optical frequencies until they become accessible to direct frequency counting in the microwave region. It is also possible to phase-coherently synthesize an arbitrary optical frequency, starting from a given optical frequency standard. In this way, one could realize an «artificial hydrogen atom» whose frequencies can be compared with Nature [35]. It is conceivable that a precise comparison of different atomic clocks with the help of such optical frequency dividers or synthesizers might reveal possible slow changes of fundamental constants during the continuing evolution of the Universe. To give an example, the $1S$-$2S$ transition in hydrogen is essentially determined by electromagnetic interactions, while the Cs atomic frequency standard is based on a hyperfine transition and thus depends on the nuclear magnetic moment. The two clocks would be affected at different rates, if the relative strengths of electromagnetic and strong interactions should slowly change.

6. – Outlook.

Advances in the state of the art of high-resolution laser spectroscopy are allowing measurements and comparisons of hydrogen frequencies with ever higher precision. Future studies of the $1S$-$2S$ two-photon transitions might take advantage of the fact that cold hydrogen atoms in low-field-seeking spin states can be suspended and confined for extended periods in a magnetic trap, as first demonstrated by KLEPPNER and co-workers at MIT [36]. Temperatures as low as 100 μK have since been achieved with the help of advanced cryogenic technology together with evaporative cooling of the trapped atoms. Very recently, the group of Walraven at Amsterdam has succeeded in laser spectroscopy [37] and even laser cooling [38] of magnetically trapped hydrogen atoms with the help of a pulsed source of Lyman-α radiation. For ultrahigh-resolution spectroscopy without perturbing external fields, one could try to observe cold atoms which are freely falling in an «atomic fountain» [39]. A fountain of a few cm height should be sufficient to approach the natural linewidth of 1.3 Hz of the $1S$-$2S$ transition with Ramsey-type two-photon spectroscopy, *i.e.* with the atoms passing twice through the same standing-wave light field.

Future improved Lamb shift measurements could take advantage of very sharp two-photon transitions from the metastable $2S$ level to high Rydberg states, if perturbing external fields can be sufficiently controlled. The frequency ratio of two different such transitions from the $n = 2$ state upwards could provide a more accurate value for the $2S$ Lamb shift.

To measure the Rydberg constant or the electron/proton mass ratio without the present limits due to the uncertainty of the proton radius, one could choose linear combinations of observable transition frequencies so that nuclear-size effects and uncalculated higher-order QED corrections cancel to first order. As an example we consider the combination $8f(2S\text{-}4S)\text{-}f(1S\text{-}2S)$ which no longer contains energy shifts which scale with the inverse cube of the principal quantum number. This frequency combination or its isotope shift can be predicted much more accurately than the two individual frequencies. Microwave spectroscopy of transitions between circular Rydberg levels [40] promises an alternative path to measurements of the Rydberg constant without troublesome nuclear-structure corrections.

A different and rather intriguing possibility would be a precise measurement of the inertial mass of the hydrogen atom by observing the small line shifts caused by radiative recoil. If an atom absorbs or emits a photon, total momentum has to be conserved, and the atom changes its kinetic energy. Atom interferometers can be designed where the time-integrated kinetic energy is different along two possible paths, so that the recoil energy can be measured precisely via the associated phase shift of the de Broglie wave [41, 42]. Since the recoil momentum is very well known, such an experiment gives a direct measure for the hydrogen mass m_{H}, or, more correctly, for the ratio m_{H}/h, where h

is Planck's constant. Such an experiment could thus establish an accurate atomic standard for the unit of mass, the kilogram. At the same time, it could yield an accurate new value for the fine-structure constant α, since $\alpha^2 = (h/m_H)(m_H/m_e)\, 2R_\infty/c$. Because of its light mass, hydrogen appears as an ideal candidate for such experiments. As long as no good laser source for the hydrogen Lyman-α transition exists, one might transfer recoil momentum to the metastable hydrogen 2*S* atoms by inducing Raman transitions between Zeeman or hyperfine sublevels of 2*S* with the help of two counterpropagating laser beams. Such an approach is already being used very successfully by the group of Chu at Stanford for measuring the recoil energy and thus the mass of cesium atoms [42, 43].

High-resolution laser spectroscopy can also be applied to other hydrogenlike atoms. The exotic atoms positronium (e^+e^-) and muonium (μ^+e^-) are particularly intriguing, because these purely leptonic systems do not involve troublesome large hadronic-structure corrections. On the other hand, the theory for the relativistic two-body system positronium is considerably more difficult than for hydrogen, and has not yet been advanced to the same level of sophistication. Doppler-free two-photon spectroscopy of the 1*S*-2*S* transition has been successfully demonstrated both for positronium [44] and for muonium [45, 46]. Even though the achievable resolution is ultimately limited by the short annihilation lifetime of positronium or by the decay of the muon, experiments of this kind provide valuable tests of QED.

Antihydrogen (p^+e^-) is a different and most fascinating candidate for high-resolution laser spectroscopy. The prospects for producing such atoms in the laboratory have advanced from impossible to merely very difficult, since GABRIELSE and co-workers have succeeded in slowing and trapping antiprotons at the LEAR storage ring of CERN [47]. It should be possible to observe antihydrogen atoms with very high spectroscopic resolution, if they can be suspended in a magnetic trap so that they do not annihilate on contact with ordinary matter. With this goal in mind we are developing novel electromagnetic traps which can also capture charged electrons and protons or their antiparticles. Such a trap should make it possible to produce slow antihydrogen atoms by laser-induced recombination of cold positrons and antiprotons.

We are also developing a rugged and compact source of tunable ultraviolet light for spectroscopy of the 1*S*-2*S* transition, starting with a frequency-doubled high-power semiconductor laser, to facilitate future experiments in the hostile environment of an accelerator laboratory. Even the slightest spectroscopic difference between matter and antimatter would be a revolutionary discovery. It might indicate a breakdown of one of the most basic symmetry laws of physics. In any case, such experiments could provide stringent new tests of such basic assumptions as CPT symmetry or the equivalence principle. It has been speculated that the gravitational force between matter and antimatter might differ from the force between ordinary matter [48]. Even a tiny difference might lead to a detectable difference in the gravitational red shift of the $1S$-$2S$ resonance.

High-resolution spectroscopy of atomic hydrogen continues to hold fascinating challenges. Perhaps the biggest surprise in this endeavour would be if we found no surprise.

* * *

Many dedicated researchers of our laboratories in Garching and Munich have made important contributions to the work presented here. They include (in alphabetical order) T. ANDREAE, T. FREEGARDE, A. HEMMERICH, M. HUBER, R. KALLENBACH, W. KÖNIG, D. LEIBFRIED, D. MESCHEDE, T. MUKAI, K. PACHUCKI, M. PREVEDELLI, L. RICCI, F. SCHMIDT-KALER, W. VASSEN, V. VULETIC, J. WALZ, M. WEITZ, R. WYNANDS, C. ZIMMERMANN and S. ZIMMERMANN. We owe special gratitude to V. BAGAEV, F. BIRABEN, C. BORDÉ, V. CHEBOTAEV, S. CHU, G. GABRIELSE, J. L. HALL, M. INGUSCIO, S. HAROCHE, R. J. HUGHES, L. JULIEN, D. KLEPPNER, G. KRAMER, M. KASEVICH, A. L. SCHAWLOW, D. TAQQU, H. TELLE and J. T. M. WALRAVEN for stimulating discussions. This work has been supported in part by the Deutsche Forschungsgemeinschaft.

REFERENCES

[1] T. W. HÄNSCH, G. W. SERIES and A. L. SCHAWLOW: *Sci. Am.*, **240**, 94 (1979).
[2] G. W. SERIES, Editor: *The Spectrum of Atomic Hydrogen, Advances* (World Scientific, Singapore, 1988).
[3] *The Hydrogen Atom, Proceedings of the Symposium held in Pisa, June 30-July 2, 1988,* edited by G. F. BASSANI, M. INGUSCIO and T. W. HÄNSCH (Springer-Verlag, Heidelberg, 1989).
[4] T. W. HÄNSCH, I. S. SHAHIN and A. L. SCHAWLOW: *Nature (London)*, **235**, 63 (1972).
[5] M. H. NAYFEH, S. A. LEE, S. M. CURRY and I. S. SHAHIN: *Phys. Rev. Lett.*, **32**, 1396 (1974).
[6] C. WIEMAN and T. W. HÄNSCH: *Phys. Rev. Lett.*, **36**, 1170 (1976).
[7] B. COUILLAUD and T. W. HÄNSCH: *Opt. Commun.*, **35**, 441 (1981).
[8] T. W. HÄNSCH and A. L. SCHAWLOW: *Opt. Commun.*, **13**, 68 (1975).
[9] *Laser Cooling and Trapping of Atoms*, special issue of *J. Opt. Soc. Am. B*, **6**, 2023ff (1989).
[10] T. W. HÄNSCH, S. A. LEE, R. WALLENSTEIN and C. WIEMAN: *Phys. Rev. Lett.*, **34**, 307 (1975).
[11] L. S. VASILENKO, V. P. CHEBOTAEV and A. V. SHISHAEV: *JETP Lett.*, **12**, 113 (1970).
[12] S. A. LEE, R. WALLENSTEIN and T. W. HÄNSCH: *Phys. Rev. Lett.*, **35**, 1262 (1975).
[13] C. WIEMAN and T. W. HÄNSCH: *Phys. Rev. A*, **22**, 192 (1980).

[14] R. G. BEAUSOLEIL, D. H. MCINTYRE, C. J. FOOT, B. COUILLAUD, E. A. HILDUM and T. W. HÄNSCH: *Phys. Rev. A, Rapid Communications*, **36**, 4115 (1987).
[15] D. H. MCINTYRE, R. G. BEAUSOLEIL, C. J. FOOT, E. A. HILDUM, B. COUILLAUD and T. W. HÄNSCH: *Phys. Rev. A*, **39**, 4591 (1989).
[16] M. G. BOSHIER, P. E. G. BAIRD, C. J. FOOT, E. A. HINDS, M. D. PLIMMER, D. N. STACEY, J. B. SWAN, D. A. TATE, D. M. WARRINGTON and G. K. WOODGATE: *Phys. Rev. A*, **40**, 6169 (1989).
[17] C. ZIMMERMANN, R. KALLENBACH and T. W. HÄNSCH: *Phys. Rev. Lett.*, **65**, 571 (1990).
[18] M. WEITZ, F. SCHMIDT-KALER and T. W. HÄNSCH: *Phys. Rev. Lett.*, **68**, 1120 (1992).
[19] W. E. LAMB and R. C. RETHERFORD: *Phys. Rev.*, **72**, 241 (1947).
[20] S. R. LUNDEEN and F. M. PIPKIN: *Phys. Rev. Lett.*, **46**, 232 (1981).
[21] F. BIRABEN, J. C. GARREAU, L. JULIEN and M. ALLEGRINI: *Rev. Sci. Instrum.*, **61**, 1468 (1990).
[22] J. C. GARREAU, M. ALLEGRINI, L. JULIEN and F. BIRABEN: *J. Phys. (Paris)*, **51**, 2263, 2275, 2293 (1990).
[23] L. N. HAND, D. J. MILLER and R. WILSON: *Rev. Mod. Phys.*, **35**, 335 (1963).
[24] G. G. SIMON, CH. SCHMITT, F. BOROWSKI and V. H. WALTHER: *Nucl. Phys. A*, **333**, 381 (1980).
[25] M. WEITZ, A. HUBER, F. SCHMIDT-KALER, D. LEIBFRIED and T. W. HÄNSCH: *Phys. Rev. Lett.*, **72**, 328 (1994).
[26] K. PACHUCKI: *Phys. Rev. A*, **48**, 120 (1993).
[27] F. SCHMIDT-KALER, D. LEIBFRIED, M. WEITZ and T. W. HÄNSCH: *Phys. Rev. Lett.*, **70**, 2261 (1993).
[28] E. R. COHEN and B. N. TAYLOR: *Codata Bull.*, **63** (November 1986) (Pergamon Press).
[29] D. LEIBFRIED, F. SCHMIDT-KALER, M. WEITZ and T. W. HÄNSCH: *Appl. Phys. B*, **56**, 65 (1993).
[30] K. PACHUCKI, D. LEIBFRIED and T. W. HÄNSCH: *Phys. Rev. A*, **48**, R1 (1993).
[31] T. ANDREAE, W. KÖNIG, R. WYNANDS, D. LEIBFRIED, F. SCHMIDT-KALER, C. ZIMMERMANN, D. MESCHEDE and T. W. HÄNSCH: *Phys. Rev. Lett.*, **69**, 1923 (1992).
[32] F. NEZ, M. D. PLIMMER, S. BOURZEIX, L. JULIEN, F. BIRABEN, R. FELDER, O. ACEF, J. J. ZONDY, P. LAURENT, A. CLAIRON, M. ABED, Y. MILLERIOUX and P. JUNCAR: *Phys. Rev. Lett.*, **69**, 2326 (1992).
[33] H. R. TELLE, D. MESCHEDE and T. W. HÄNSCH: *Opt. Lett.*, **15**, 532 (1990).
[34] R. WYNANDS, T. MUKAI and T. W. HÄNSCH: *Opt. Lett.*, **17**, 1749 (1992).
[35] T. W. HÄNSCH: in *The Hydrogen Atom, Proceedings of the Symposium held in Pisa, June 30-July 2, 1988*, edited by G. F. BASSANI, M. INGUSCIO and T. W. HÄNSCH (Springer-Verlag, Heidelberg, 1989), p. 93.
[36] H. F. HESS, P. GREG, G. P. KOCHANSKI, J. M. DOYLE, N. MASUHARA, D. KLEPPNER and T. J. GREYTAK: *Phys. Rev. Lett.*, **59**, 672 (1987).
[37] O. J. LUITEN, H. G. C. WERIJ, I. D. SETIJA, M. W. REYNOLDS, T. W. HIJMANS and J. T. M. WALRAVEN: *Phys. Rev. Lett.*, **70**, 544 (1993).
[38] J. T. M. WALRAVEN: private communication.
[39] R. G. BEAUSOLEIL and T. W. HÄNSCH: *Opt. Lett.*, **10**, 547 (1985); *Phys. Rev. A*, **33**, 1661 (1986).
[40] D. KLEPPNER: private communication.
[41] C. BORDÉ: private communication.
[42] S. CHU and M. KASEVICH: private communication.

[43] M. Kasevich, D. S. Weiss, E. Riis, K. Moler, S. Kasapi and S. Chu: *Phys. Rev. Lett.*, **66**, 2297 (1991).

[44] S. Chu and A. P. Mills: *Phys. Rev. Lett.*, 48, 1333 (1982); M. S. Fee, A. P. Mills, S. Chu, E. D. Shaw, K. Danzmann, R. J. Chichester and D. M. Zuckerman: *Phys. Rev. Lett.*, **70**, 1397 (1993).

[45] S. Chu, A. P. Mills, A. G. Yodh, K. Nagamine, Y. Miyake and T. Kuga: *Phys. Rev. Lett.*, **60**, 101 (1988).

[46] K. Jungmann, C. Bresler, H. Geerds, J. Kenntner, F. Maas, G. Zu Putlitz, W. Schwarz, L. Zhang, Z. Zhang, P. E. G. Baird, S. N. Sea, P. G. H. Sandars, G. Woodman, J. R. M. Barr, P. F. Curley, A. I. Ferguson, M. A. Persaud, R. Dixon, V. W. Hughes, G. H. Eaton, W. T. Toner and M. Towrie: *Z. Phys. D*, **21**, 241 (1991).

[47] G. Gabrielse, X. Fei, L. A. Orozco, R. L. Tjoelker, J. Hass, H. Kalinowsky, T. A. Trainor and W. Kells: *Phys. Rev. Lett.*, **65**, 1317 (1990).

[48] T. Goldman, R. J. Hughes and M. M. Nieto: *Sci. Am.*, **258**, 48 (1988).

Precision Atom Interferometry and an Improved Measurement of the $1\,^3S_1$-$2\,^3S_1$ Transition in Positronium.

S. CHU

Physics Department, Stanford University - Stanford, CA 94305

1. – Stimulated Raman transitions and atom interferometry.

The first two sections describe how stimulated Raman transitions have been used to manipulate laser-cooled atoms. In particular, they give a summary of how these transitions have been used for velocity selection and in the construction of atom interferometers. The interferometers described in this lecture depend on the interference of two internal quantum states of the atom as well as on the external degrees of freedom common with other types of interferometers. Their design is an extension of ideas that originated with nuclear magnetic resonance and the separated-oscillatory-field method of measuring energy level splittings. The application of these types of interferometers to the measurement of the acceleration due to gravity and the velocity recoil an atom experiences when it absorbs a photon is described. For a discussion of other types of atom interferometers that have been recently demonstrated, the reader is referred to some of the other contributions to this volume and to the original papers [1-3].

Stimulated Raman transitions have also been used by KASEVICH and CHU [4] to cool sodium atoms along one dimension to an effective temperature of less than 100 nK, more than a factor of ten below the photon recoil limit. This method of cooling has been recently extended to 2 and 3 dimensions [5]. This work will not be covered in this lecture.

The third section gives an account of a recent measurement of the $1\,^3S_1$-$2\,^3S_1$ interval in positronium with a 2.6 p.p.b. uncertainty. This result corresponds to a $3.5 \cdot 10^{-5}$ measurement of the $\alpha^2 R_\infty$ QED contribution to the energy level

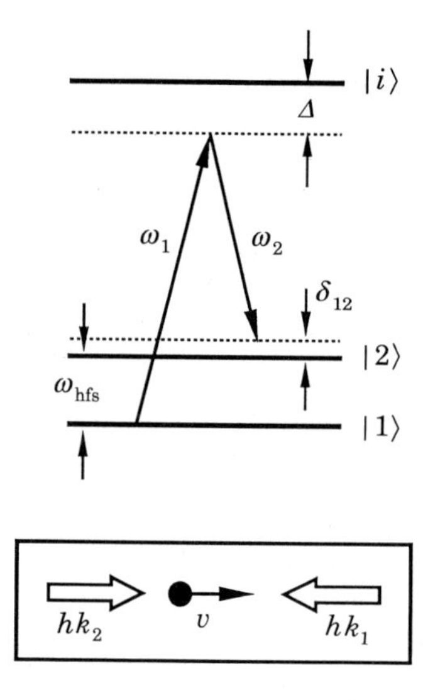

Fig. 1. – Energy diagram for the Raman transitions. If the beams are oriented as shown in the inset, the atom momentum changes by $\hbar(\boldsymbol{k}_1 - \boldsymbol{k}_2)$ in making the transition from state $|1\rangle$ to state $|2\rangle$.

splitting and is sufficiently accurate to test the as yet uncalculated $\alpha^4 R_\infty$ QED corrections.

1·1. *Stimulated Raman transitions.* – The use of stimulated Raman transitions for velocity selection and manipulation has been discussed rigorously in a number of papers [6-8].

Consider an atom with two ground states $|1\rangle$ and $|2\rangle$ separated by a hyperfine splitting ω_{hfs} and an excited state $|i\rangle$. If two laser beams with frequencies ω_1 and ω_2 are tuned so that their frequencies are close to the excited-state resonances and their difference is near ω_{hfs} as shown in fig. 1, the two ground states will be coupled to each other. If the detuning Δ is large enough so that the spontaneous emission from the excited state is negligible, the intermediate level can be «adiabatically eliminated» and the system can be reduced to an effective two-level system [7].

Adiabatic elimination of level $|i\rangle$ results in the familiar equations for a two-level atom in an external driving field:

$$(1) \quad \begin{cases} \dot{a}_{1,\,p} \approx -\dfrac{i}{2}\,\Omega_1^{AC}\,a_{1,\,p} - \dfrac{i}{2}\exp[i\delta_{12}t]\,\Omega_{\rm eff}\,a_{2,\,p+\hbar k_1-\hbar k_2}\,, \\[2ex] \dot{a}_{2,\,p+\hbar k_1-\hbar k_2} \approx -\dfrac{i}{2}\,\Omega_2^{AC}\,a_{2,\,p+\hbar k_1-\hbar k_2} - \dfrac{i}{2}\exp[-i\delta_{12}t]\,\Omega_{\rm eff}^*\,a_{1,\,p}\,, \end{cases}$$

where

$$(2)\qquad \begin{cases} \Omega_{\text{eff}} = \dfrac{\Omega^*_{1i}\Omega_{i2}}{2\Delta}\,, \\ \Omega_j^{\text{AC}} = \dfrac{|\Omega_{ji}|^2}{2\Delta}\,, \\ \delta_{12} = (\omega_1 - \omega_2) - \left(\omega_{\text{hfs}} + \nu_x(k_1 - k_2) + \hbar\,\dfrac{(k_1 - k_2)^2}{2m}\right). \end{cases}$$

Since the equations reduce to those of a two-level system, the NMR language used to describe simple spin systems applies equally well to those which employ two-photon Raman transitions. Momentum recoil explicitly shows up in the coefficients of eqs. (1), where there is a one-to-one correlation between the atom's internal state and its momentum. The first terms on the right of eqs. (1) lead to AC Stark shifts of levels $|1\rangle$ and $|2\rangle$. The second terms lead to Rabi flopping between the two levels. Note that the effective Rabi frequency Ω_{eff} depends on the initial phase of the light fields. This dependence eventually leads to the atom's sensitivity to inertial forces during a $\pi/2$-π-$\pi/2$ pulse sequence and to the precision measurement of the recoil of an atom due to the absorption of a discrete number of photons.

The effective detuning δ_{12} from the Raman resonance contains three terms: AC Stark shifts, Doppler shifts and the recoil shift. The AC Stark shifts arise from the weak coupling of levels $|1\rangle$ and $|2\rangle$ to level $|i\rangle$. They are typically on the order of the effective Rabi frequency Ω_{eff}. The Doppler shifts arise from the atom's motion along the laser beams. In the atom's frame of reference, the frequency of each light beam is Doppler shifted by an amount $k_i v_x$. The Raman resonance, therefore, shifts by $(k_1 - k_2)v_x \sim 2kv_x$ when the beams are counterpropagating and of nearly the same frequency.

1.2. *Velocity selection with Raman transitions.* – There are two important points worth noting in the above treatment. First, the transition is between stable ground states of the atom so that the spectral width of the transition will be governed by the Rabi frequency Ω_{eff} of the transition. Second, both the velocity recoil and Doppler sensitivity of these transitions are twice those of a single-photon transition. These two features enabled us to select a very narrow velocity slice out of an ensemble of atoms as shown schematically in fig. 2 [4].

Atoms from an atomic beam were slowed, collected in a magneto-optic trap, and then further cooled before launched upwards in an atomic-fountain trajectory. The atoms are launched by sweeping the frequencies of the downward propagating molasses laser beams by -2.9 MHz and the upward going beams by $+2.9$ MHz in 250 μs and then leaving the molasses at the shifted frequencies for an additional 750 μs. The new frequencies create polarization gradients mov-

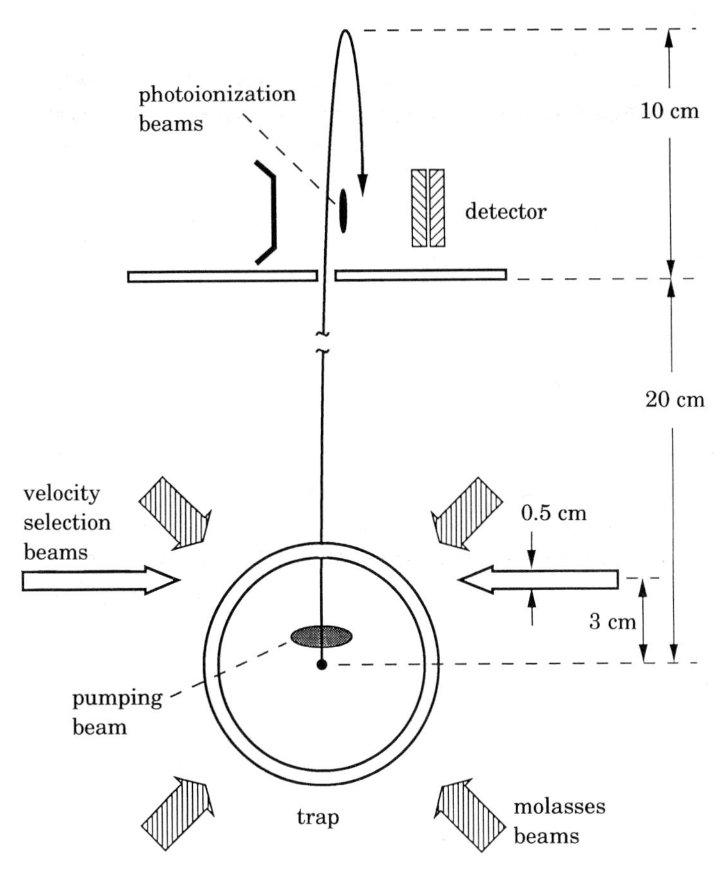

Fig. 2. – Schematic of the apparatus used to demonstrate velocity selection with stimulated Raman pulses. After trapping and cooling the atoms were launched upwards before the velocity selection is made. A photoionization beam was scanned laterally in order to measure the spread of the fountain during the upward and downward going portions of the atomic trajectories.

ing upwards at a velocity of 250 cm/s. Atoms in this molasses are then dragged upwards by the moving molasses.

The atoms were then optically pumped into the $F = 1$ hyperfine state and then allowed to pass through the Raman velocity selection beams which induced transitions between the $m_F = 0 \rightarrow 0$ and $m_F = -1 \rightarrow +1$ magnetic-field-insensitive transitions. Figure 3 shows the spatial distribution of the atoms at times of 140 ms and 370 ms after launch as measured by 1 mm wide resonant photoionization laser beams. The times correspond to atoms immediately after passing through a 1.6 mm aperture and atoms that have gone 10 cm above the aperture and have returned to the detection region. There was no measurable spread in the atomic distribution, indicating that the width of the velocity-selected atoms is less than 1 mm/s. The transit-time-limited linewidth of the Ra-

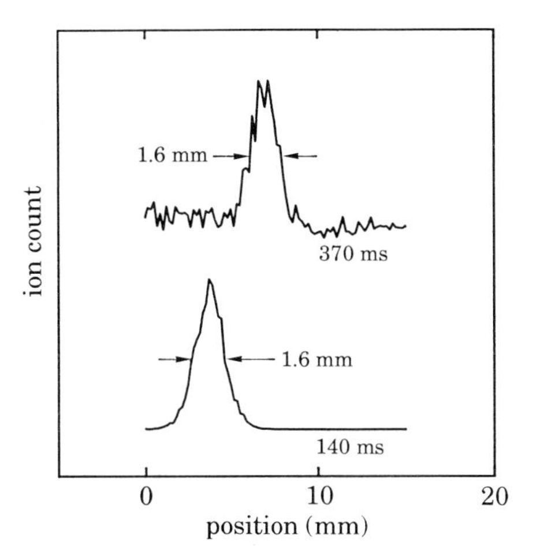

Fig. 3. – Time-of-flight data of the velocity selection experiment. There was no measurable spread in the width of the ensemble of atoms. The inferred velocity width of 270 μm/s corresponds to an effective temperature of $24 \cdot 10^{-12}$ K.

man transition was measured to be 900 Hz and from this measurement one can infer a velocity width of 270 μm/s.

During the velocity selection process, there is a delocalization of the wave packet that must be consistent with the Heisenberg uncertainty principle. Consider a wave packet well localized along the direction of the laser beams as shown in fig. 4. The amplitude of the atom in the other momentum state begins to grow and separate from the initial state. As this happens, the excited-state amplitude becomes a new source of amplitude for the atom in the original state. After the end of the pulse, both wave packets have become significantly delocalized. The inherent fuzziness of the wave packets is an important consideration when analyzing the behavior of the Raman pulse atom interferometers.

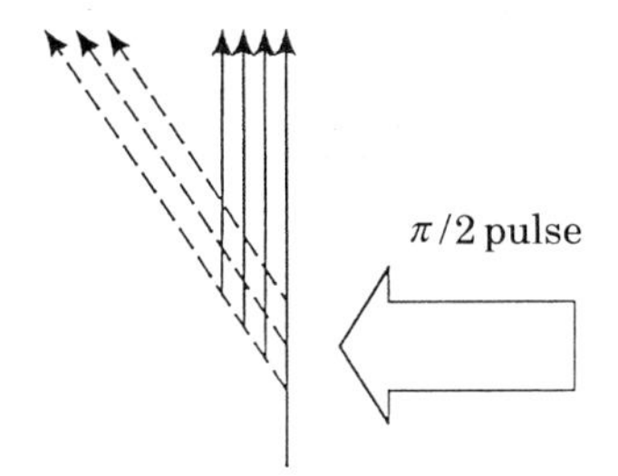

Fig. 4. – An initially transversely well-localized wave packet is excited by a $\pi/2$ pulse of light. If the light pulse is extended in time, the atom will necessarily become delocalized as the amplitudes of the two states evolve.

1˙3. *Optical Ramsey fringes.* – A major systematic effect in measurements using Ramsey's separated-oscillatory-field technique is the so-called «phase shift» problem. In fig. 5, an atom that travels along path 1 traverses the two microwave cavities at exactly the same position with respect to the geometry of the cavities. For this case, the electromagnetic fields in the two cavities are in phase with each other and Ramsey's method correctly measures the phase difference between the atomic phase and the phase of the local oscillator. However, if the atom takes path 2, it experiences an additional phase shift of $\sim \pi$ radians as it traverses the second cavity. For typical microwave wavelengths, the atoms can be easily localized so that the distribution of atoms is much less than the microwave wavelength. However, if one wants to generalize this idea to the optical regime (*e.g.*, to make a frequency standard based on an optical transition), the phase reversal problem is severe.

A solution to the problem was first pointed out by BAKLANOV *et al.* [9] with standing waves and by BORDÉ in the case of traveling waves [10]. More recent accounts are given in papers by BORDÉ and co-workers [11] and CHEBOTAYEV *et al.* [12]. The essential idea of the traveling-wave scheme can be seen by referring to fig. 6. After the first $\pi/2$ pulse at position a, the atom is divided into two parts with different trajectories. The part of the atom in path 2 sees a phase loss relative to path 1 at point b of $\Delta\phi = k\Delta x$, where k is the wave vector of the light field and Δx is the spatial separation of the two arms along the direction of the light beams. However, at point c, the part of the atom in path 2 sees a phase gain relative to path 1 because the direction of the beams is reversed. Moreover, an atom that is on another trajectory such as shown by the dashed line experiences the identical phase compensation, so that all the atoms in an ensemble with different trajectories will contribute constructively to the interference signal.

Phase compensation also occurs automatically in a Ramsey transition with two-photon Doppler-free excitation in an atomic fountain [13]. The excitation

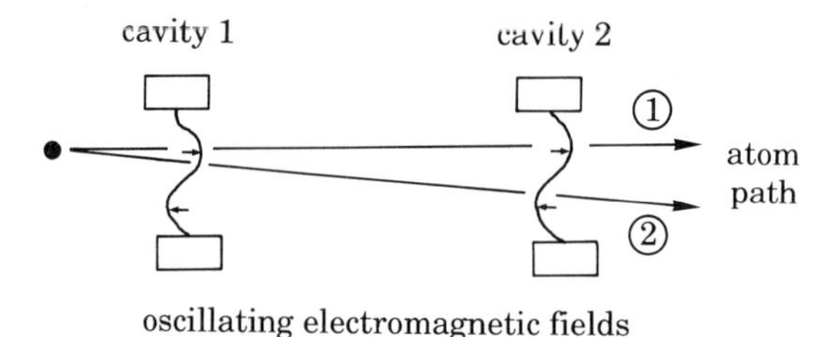

Fig. 5. – A phase shift problem in the Ramsey spectroscopy. Atoms that take path 1 experience no phase shift since the fields in the two regions are in phase with each other. Atoms that take path 2 see a net phase shift. For microwave frequencies $\leqslant 10$ GHz, this phase shift can be managed. For optical frequencies, tiny lateral displacements give phase reversals.

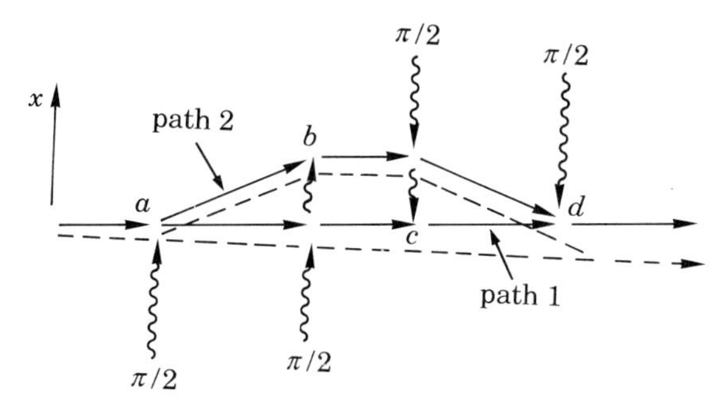

Fig. 6. – A sequence of four $\pi/2$ pulses eliminates the phase shift problem for optical Ramsey spectroscopy. Each path is phase compensated so that atoms that take different paths 1 and 2 experience no net phase shift. The part of the atoms in path 2 sees a phase *loss* relative to path 1 at point b. However, at point c, path 2 sees a phase gain relative to path 1. Moreover, for another atom (dotted line) the phase compensation is the same.

requires contributions from the two counterpropagating beams. If the atom does not return exactly to the same position in space, there will be a phase shift associated with each light beam, but these phase shifts will have equal and opposite signs and the overall phase shift will be zero.

It is interesting to note that the pulse scheme that was used to cure the phase shift problem for optical Ramsey fringes also created an atom interferometer. However, it took fifteen years to realize that the photon recoil kicks experienced by the atom can cause a physical separation and the resulting interference is due to both an interference of the internal degrees of freedom of the atom, as in the case of the Ramsey technique, and the external degrees of freedom, as in conventional particle and light interferometers. In an atomic fountain with a long separation time between $\pi/2$ pulses, the original Ramsey technique would not give interference because the two parts of the atom wave packet would not overlap in space.

1·4. *Atom interferometry.* – A simpler three-pulse scheme to atom interferometry was introduced by KASEVICH and CHU [6] at the same time it was realized that the 4-pulse scheme would produce coherent but physically separated atomic paths. The Kasevich/Chu scheme differed from the other work on a number of counts. 1) The light pulses used were Raman pulses that induced transitions between ground states of the atom. As a result, the finite lifetime of the excited state was no longer an issue and the long drift times made possible in an atomic-fountain time could be used to full advantage. 2) The absolute frequency of the light pulses did not have to be as stable as that required for the single-photon interferometers. If the time between light pulses is on the order of 0.1 s, the frequency stability of the optical source in the single-photon transition interferometer must be on the order of 10 Hz. (This is of, course, why these pulse

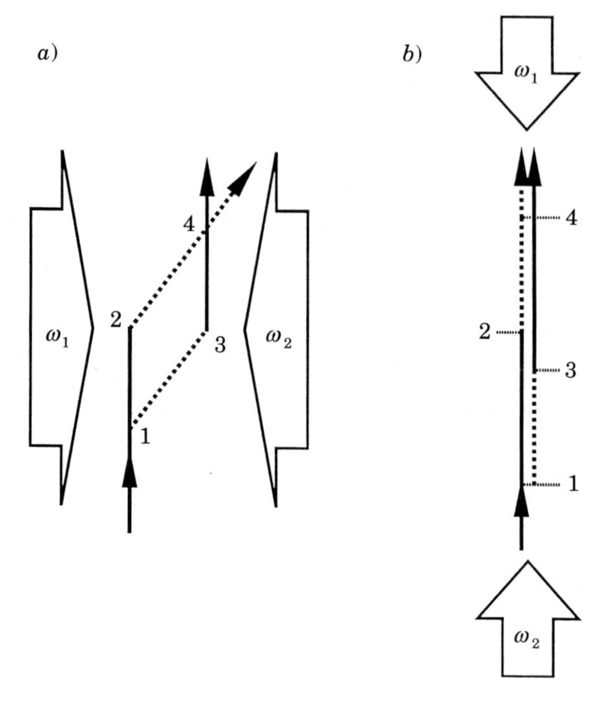

Fig. 7. – *a*) An atom interferometer using the $\pi/2$-π-$\pi/2$ pulse sequence. This geometry is analogous to the Mach-Zender geometry used in optical interferometry. *b*) A vertical geometry, where the «area» of the interferometer is zero.

schemes were first considered in connection with the construction of an optical frequency standard.) In the Raman pulse scheme, only the relative frequency between the two laser beams has to be stable. There are now techniques such as electro-optic sideband modulation and direct phase locking of one independent laser to another that can produce a relative frequency uncertainty between the two frequencies of less than one mHz. 3) Finally, the recoil is double that of a single-photon transition.

The price one pays for using Raman transitions is that higher laser power (but not out of the range of diode lasers) is required and the AC Stark shifts become a significant problem. As we shall see later, there are methods that have been used to minimize the AC Stark shift problem.

The geometry used in the first Raman pulse interferometer is shown in fig. 7*a*). The first $\pi/2$ pulse acts as a beam splitter, creating a coherent superposition in the two ground states of the atom. The π pulse acts as a mirror in that the atomic trajectories are reversed so that the wave packets will overlap by the time the second $\pi/2$ pulse is applied. In the actual experiment, a single large-diameter beam was used and the three pulses were applied to the atom by gating an acousto-optic modulator on and off. A vertical geometry was also used (fig. 7*b*)) where the separation of the atom was along the direction of its motion. This geometry allows a longer pulse separation time since the atomic trajectory can be contained in the beam for the entire duration of the atomic fountain.

1·5. *Interferometer phase shifts.* – We will summarize the results of an analysis of the interferometer phase shifts in terms of the position of the wave packet. In this approach, contributions to the total phase shift $\Delta\phi$ fall into two classes: those arising from the atom's interaction with the light ($\Delta\phi_\lambda$) and those arising during the atom's free evolution between light pulses ($\Delta\phi_f$). The phase change due to the interaction with the light field takes the form

$$\Delta\phi_l = \phi_R(x_1^1, t_1) - \phi_R(x_2^1, t_2) - \phi_R(x_2^2, t_2) + \phi_R(x_3^1, t_3), \quad (3)$$

where $\phi_R(\boldsymbol{x}, t) = (\boldsymbol{k}_1 - \boldsymbol{k}_2)\cdot\boldsymbol{x} - \Delta\omega t$ is the phase of the Raman field at position $\boldsymbol{x}$ and time t. The position vectors $\boldsymbol{x}_i^\alpha$ represent the mean position of the wave packet correlated with internal state α at the time t_i of the i-th pulse [6].

As with the 4 $\pi/2$ pulse interferometer, $\Delta\phi_\lambda$ does not vary with the initial momentum of the atom. Thus a large spread in initial velocities can coherently contribute to the interference signal, and the phase shift for a spatially localized wave packet, which is a coherent sum over momentum states, is also given by eq. (3).

Next consider an atom which is falling in a gravitational field. In the frame falling with the atom the Raman frequency changes linearly with time at the rate of $(\boldsymbol{k}_1 - \boldsymbol{k}_2)\cdot\boldsymbol{g}t$. If we change $\omega_1(t) - \omega_2(t)$ as $\omega_1(t) - \omega_2(t) = \omega_1 - \omega_2 - (\boldsymbol{k}_1 - \boldsymbol{k}_2)\cdot\boldsymbol{g}t$, one can show that the total phase change is given by

$$\Delta\phi_l = -(\boldsymbol{k}_1 - \boldsymbol{k}_2)\cdot\boldsymbol{g}T^2 . \quad (4)$$

For sodium, with $\boldsymbol{k}_1, \boldsymbol{k}_2$ parallel to $\boldsymbol{g}$, $(\boldsymbol{k}_1 - \boldsymbol{k}_2) g \sim 2\pi \times 3.3 \cdot 10^7\,\text{s}^{-2}$.

In the experiment the effective Rabi frequency was ~ 50 kHz, whereas the change in the Doppler shifts, $\sim 2k_1 gT$, was greater than 3 MHz. Thus a momentum component in resonance with the first $\pi/2$ pulse will be far out of resonance with the remaining pulses. By actively changing the frequency difference $\Delta\omega(t) = \omega_1(t) - \omega_2(t)$, we compensated for the atom's deceleration. We changed $\omega_1(t) - \omega_2(t)$ so that, for each Raman pulse, the detuning from resonance was less than the Rabi frequency. The atom is resonantly driven for all three pulses when $\Delta\omega = \omega_0$ for the first $\pi/2$ pulse, $\Delta\omega = \omega_0 + \omega_m$ for the π pulse, and $\Delta\omega = \omega_0 + 2\omega_m$ for the final $\pi/2$ pulse, with $\omega_m \sim (\boldsymbol{k}_1 - \boldsymbol{k}_2)\cdot\boldsymbol{g}T$. In the falling frame, the phase of the Raman field ϕ_R before each Raman pulse is $\omega_0 t_1 + \phi_1^0$, $(\omega_0 + \omega_m)t_2 - (\boldsymbol{k}_1 - \boldsymbol{k}_2)\cdot\boldsymbol{g}t_2^2 + \phi_2^0$ and $(\omega_0^2 + \omega_m)t_3 - (\boldsymbol{k}_1 - \boldsymbol{k}_2)\cdot\boldsymbol{g}t_3^2 + \phi_3^0$, respectively. In this case, $\Delta\phi = 2\omega_m T - (\boldsymbol{k}_1 - \boldsymbol{k}_2)\cdot\boldsymbol{g}T^2 + \Delta\phi^0$. The term $\Delta\phi^0 = \phi_1^0 - 2\phi_2^0 + \phi_3^0$ represents the initial phase relationship between the three pulse frequencies. If each of the three frequencies is derived from an independent synthesizer, the ϕ_i^0's are interpreted as the phase of the i-th synthesizer at time $t = 0$ (*e.g.*, $\phi_i(t) = \omega_i t + \phi_i^0$).

The free evolution of the atomic wave packet can be obtained from Feynman's path integral formulation of quantum mechanics [14]. He shows that the wave function is proportional to $\Psi \sim \exp[i \sum (S/\hbar)]$, where the action $S = \int_\Gamma L\, \mathrm{d}t$, L is the Lagrangian for the atom, and the sum is over all possible paths. If S is much greater than $\hbar$, the free-evolution contribution is

$$\Delta f_\mathrm{f} = \int_\Gamma \boldsymbol{k}_a \cdot \mathrm{d}x - \omega_a\, \mathrm{d}t\,, \tag{5}$$

where Γ is the circuit describing the classical path of the wave packet, $\boldsymbol{k}_a =$ (wave packet's mean momentum $\boldsymbol{p})/\hbar$, and $\omega_a =$ (total energy of wave packet)$/\hbar$.

The free-evolution contribution from eq. (5) turns out to be zero if there is no violation of the equivalence principle: the contribution from the asymmetry in de Broglie wavelengths is canceled by the changes in the atom's total energy. This is in contrast to single-crystal neutron interferometers, where the free-evolution contribution *is* the dominant phase shift [15]. The distinguishing feature between the two cases is that the energy of the particle is not conserved during the light pulses of our interferometer while it is conserved during the Bragg reflection processes of neutron interferometers.

1·6. *Fringe visibility.* – Fringe visibility $V = (P_\mathrm{max} - P_\mathrm{min})/(P_\mathrm{max} + P_\mathrm{min})$ quantifies the extent to which wave packet interference is experimentally observable. For our interferometer P_max and P_min are the maximum and minimum probabilities of finding the atom in state $|2\rangle$ after the $\pi/2$-π-$\pi/2$ pulse sequence. If the spectral content of the $\pi/2$ and π pulses used in the interferometer exceeds the Doppler-broadened width of the atoms in the atomic fountain, the fringe visibility of the interferometer can be 100%. However, limitations to the laser power may produce pulses with Rabi frequencies less than the inhomogeneous linewidth. In our experiment, the Doppler width of the laser-cooled source of atoms is 600 kHz while the maximum Rabi frequency used was 50 kHz. Consequently, not all atoms see a $\pi/2$-π-$\pi/2$ pulse sequence because some are Doppler shifted out of resonance with the Raman beams. Contributions from off-resonant atoms are calculated by using the exact solutions to the three-level problem. Figures 8*a*), *b*) show the probability of excitation to level 2 following a $\pi/2$-π-$\pi/2$ sequence as a function of the atom's detuning from the resonance. In fig. 8*b*) the phase of the final $\pi/2$ pulse is shifted by π radians with respect to the first two pulses. In the limit where the Doppler width of the source of atoms is much larger than the Rabi frequency, P_max and P_min are obtained by integrating the excitation probabilities over detuning (velocity). The resulting visibility is 27% in the limit where $T \gg \tau$.

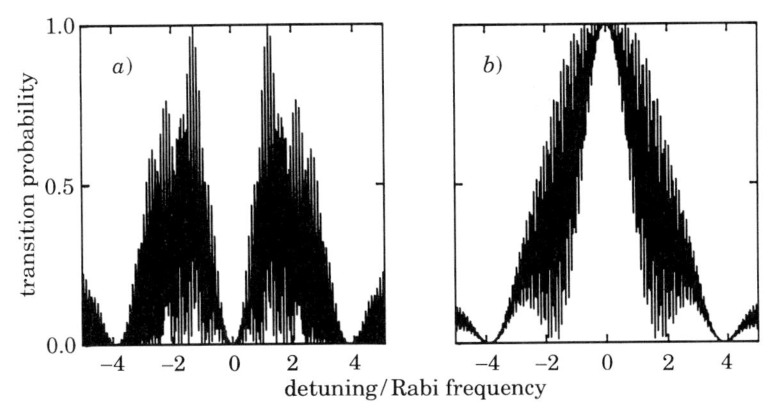

Fig. 8. – The transition probability *vs.* detuning from resonance for the $\pi/2$-π-$\pi/2$ pulse interferometer. *a*) The initial phase of all pulses is the same. *b*) The phase of the final $\pi/2$ pulse was shifted by 180° with respect to the first two pulses. When the Doppler width is much larger than the Rabi frequency, the fringe visibility is obtained by integrating *a*) or *b*) to find the velocity-averaged transition probabilities.

There are other factors that can further degrade the fringe visibility of the interferometer. For example, if there are magnetic-field gradients, atoms that take different trajectories will experience phase shifts. Our interferometer uses only the $m_F = 0 \rightarrow 0$ transitions which are magnetic-field insensitive to first order. The frequency shift for this transition in sodium is $\sim (2.2\,\text{MHz/mG}^2) \times B^2$, where B is the strength of the bias field.

A bias field was needed to Zeeman shift the field-sensitive transitions out of resonance with the Raman beams. Off-resonant excitation of magnetic-field-sensitive transitions produces unwanted background counts. To avoid off-resonant excitation the Zeeman shifts between field-sensitive and field-insensitive transitions must be greater than the frequency linewidth of an individual Raman pulse. The magnetic-field-sensitive transitions Zeeman shift by $\sim 1.4\,\text{MHz/G}$. For a $10\,\mu\text{s}$ $\pi/2$ pulse, the transition linewidth is $\sim 100\,\text{kHz}$, so a bias field of at least $100\,\text{kHz}/(1.4\,\text{MHz/G}) \sim 80\,\text{mG}$ is needed. Interference fringes were observed to be insensitive to bias field values over a range from 100 mG to 300 mG.

The greatest loss of fringe contrast for long times between pulses in our interferometer is the vibration of the mirror used to retroreflect the Raman beams. This problem will be discussed after a brief description of the apparatus is given.

1·7. *The atomic-fountain apparatus.* – Figure 9 gives a schematic diagram of the apparatus. The atoms were launched at 5 Hz. Each pulse of atoms was generated from a three-step sequence which consisted of loading a magneto-optic atom trap [16], then cooling the trapped atoms in a polarization gradient op-

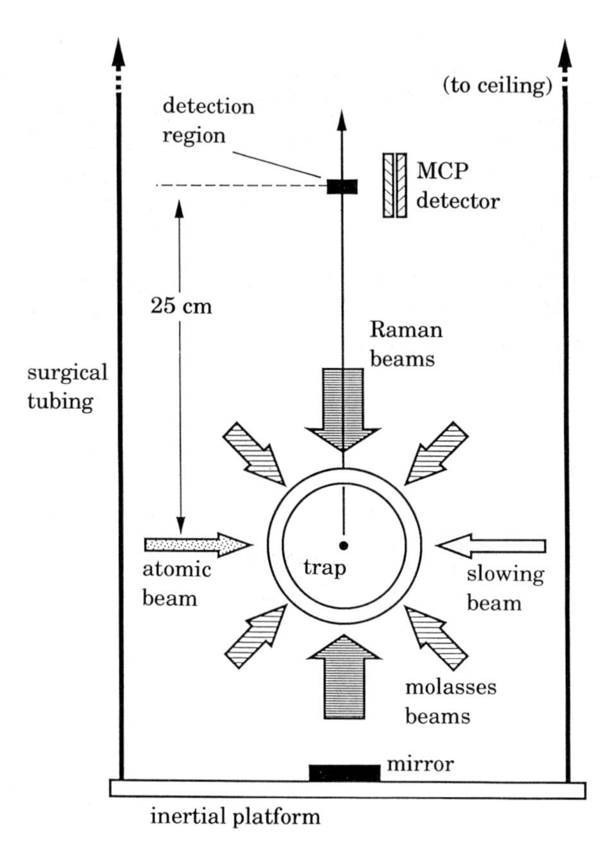

Fig. 9. – Schematic diagram of the interferometer apparatus used to measure g. An atomic beam was trapped, cooled and launched before interacting with the Raman beams. Atoms that were transferred from the $F = 1$ to the $F = 2$ state by the interferometer were detected by resonant photoionization.

tical molasses [17], and finally launching the cooled atoms vertically [18]. Atoms in a thermal atomic beam were initially slowed by the scattering force from a resonant, counterpropagating laser beam. The frequency of this beam was chirped by an electro-optic modulator to stay in resonance with the atom as it decelerated to near-zero velocity [19]. An electro-optic modulator was used to impose 1.712 GHz sidebands on the trapping beams. The high-frequency sideband re-excited atoms which were optically pumped into the $F = 1$, $3S_{1/2}$ level.

The trapping magnetic field was turned off after $\sim 1.5 \cdot 10^7$ atoms had been collected in the trap. This left the atoms in a polarization gradient molasses light field which had the effect of further cooling the sodium atoms to temperatures of $\sim 30\,\mu$K. At 30 μK, the r.m.s. velocity spread of the sample is $\sim 20\,\text{cm/s}$.

Following this post-cooling period, the atoms were launched vertically on a ballistic trajectory by creating a molasses with polarization gradients moving vertically with respect to the laboratory frame. After the atoms reached their equilibrium temperature in the moving molasses, the molasses beams were extinguished by blocking them with a mechanical shutter. Just prior to closing the shutter, atoms were optically pumped into the $F = 1$ ground state by turning off the re-pumping sideband for the optical molasses.

Atoms were detected by resonant photoionization where a microchannel plate was used to detect the ions. In the photoionization process, atoms in the $F = 2$ ground state were resonantly excited into the $3P_{3/2}$ state by a ~ 500 ns pulse of light from the trapping laser before being ionized by a co-propagating 15 ns pulse of 355 nm laser light.

The light used to induce the Raman transition was generated from a second dye laser. The light pulses were generated by switching the light through an electro-optic modulator with an acousto-optic modulator located at the output of the laser. The frequency-modulated light was then coupled into a ~ 2 m long polarization-preserving optical fiber. The output beam from the fiber was expanded and collimated before being sent into the vacuum can through a window mounted on its top flange. The light exited the vacuum chamber through another window located on the bottom of the can. The beam was then retroreflected back into the vacuum by a mirror mounted on a stable «inertial» reference just outside the lower window.

The effective Rabi frequency of the Raman transition is determined by the power output of the interferometer laser, the detuning of this laser from the optical transition and the modulation index of the electro-optic modulator. Typically ~ 20 μs were required to drive a π pulse when the laser carrier frequency was detuned 2.5 GHz from the $F = 2$, $3S_{1/2} \rightarrow 3P_{3/2}$ resonance. We used circularly polarized light to drive the transition. The details of how the Raman frequency differences were generated have been detailed previously, and will not be given here [6].

1·8. *Mechanical vibrations.* – Since all frequency components of the Raman beams follow the same optical path into the vacuum can, the Raman frequency is insensitive to mechanical vibrations of the optical elements used to steer the beams into the can. Thus the net shift to the Raman beat frequency was negligible when the beams were being directed into the vacuum can (on the order of $(1\,\text{kHz}) \times (k_{\text{rf}}/k_1) \sim 10^{-3}$ Hz). On the other hand, vibrations of the retroreflecting mirror beneath the vacuum can will directly Doppler shift the Raman frequency, since one frequency component from each of the counterpropagating beams was used to excite the atoms. The Doppler shift due to the random motion of this mirror, in turn, randomizes the phase of the Raman beams between the pulses and from launch to launch. From another point of view, the retroreflecting mirror is providing an inertial reference for the measurement of the acceleration of the atom. It is, therefore, essential that this mirror not vibrate.

We passively vibration isolate the mirror by mounting it on a platform which was suspended from the ceiling of our laboratory with ~ 3 m long surgical tubing. The platform was damped by coupling it with vacuum grease to aluminum posts resting on rubber isolation pads on the floor of the laboratory. A Kinemetrics SS-1 seismometer [20] was used to characterize the platform's motion. The platform's natural-oscillation frequency was ~ 0.5 Hz. At frequencies above 3 Hz the peak-to-peak displacement of the position of the platform parallel to the Raman beams was less than 20 nm. Since the amplitude of these high-frequency vibrations was much less than the effective wavelength of the Raman beams (295 nm), they were not large enough to wash out interference fringes. The dominant platform vibrations were large-amplitude ($\sim 2.5\,\mu$m) oscillations near the platform's resonance frequency, which corresponded to platform accelerations of $\sim 2.5 \cdot 10^{-6}\,g$. At longer drift times between pulses, these vibrations destroyed fringe visibility. Despite our efforts at vibration isolation, our best data were taken late at night after the theorists on the floor above us had left the building and the elevator of the building was shut off.

1·9. *Results.* – Data were taken by scanning the phase of the final $\pi/2$ pulse and recording the number of ions detected in the $F = 2$ state after the interferometer pulse sequence. Figure 10 shows a fringe for a drift time of $T = 50$ ms between pulses. Each data point in fig. 10 represents an average of 251 launches. There were 5 launches per second giving a total scan time of $\sim 2 \cdot 10^3$ s. A least-squares fit to a sine wave gives an uncertainty in phase of $3 \cdot 10^{-3}$ cycles.

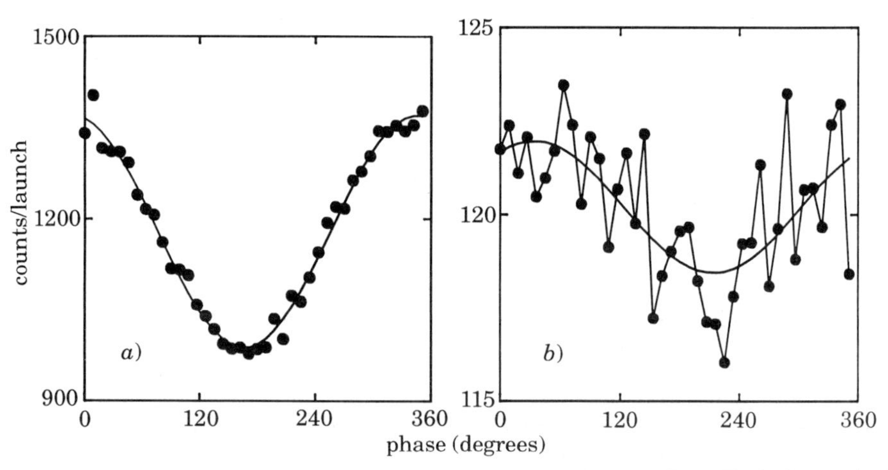

Fig. 10. – *a*) An interference fringe for $T = 50$ ms delay between pulses. Each data point is an average of 251 pulses (~ 25 s). The fit to the data has an uncertainty that corresponds to a resolution of 3 parts in 10^8. *b*) An interference fringe for $T = 100$ ms delay between pulses. Each data point is an average of 22 shots.

With 50 ms between pulses, phase will shift by one fringe if g changes by $1.2 \cdot 10^{-5}$. Therefore, the sensitivity to changes in g is $\Delta g/g = 3 \cdot 10^{-8}$.

Fringe visibility was 18%, with ~ 500 counts per launch (or ~ 2500 counts per second) contributing to the interference signal. Vibration of the inertial platform was most likely responsible for degrading visibility from the 27% theoretical maximum. For $T = 10$ ms we observed the maximum allowed fringe visibility. At longer drift times, the effects of vibrations become more pronounced. Figure 10*b*) shows data for $T = 100$ ms. At $T = 100$ ms between pulses the estimated phase noise from platform vibrations was 0.8 cycles. Due to these large phase variations, visibility was degraded to a few percent, which is consistent with simulations of the signal. Each data point is an average of 22 launches, with a launch rate of 1 launch/s. Even with the poor visibility, the sensitivity to changes in g was $\Delta g/g = 7 \cdot 10^{-8}$.

1·10. *Limits to resolution and accuracy.* – The ultimate resolution and accuracy of the accelerometer hinge on sources of short-term fluctuations and long-term stability of systematic offsets.

Our present short-term sensitivity to changes in g is limited by vibrations, fluctuations in the number of trapped atoms and fluctuations in detection efficiency. Neither of these is an insurmountable obstacle towards achieving better resolution.

Noise arising from low-frequency vibrations can be significantly reduced with an active isolation system: state-of-the-art «supersprings» routinely achieve r.m.s. residual accelerations of $10^{-9} g$ [21]. Towards this end, we have mounted a sensitive, low-frequency seismometer and retromirror on a platform suspended by a spring. A feedback system was constructed that minimized the output signal of the seismometer. In the spectral range between 0.1 and 10 Hz, the error signal indicates that the platform is being held to an acceleration of less than $\sim 10^{-8}\ g/\sqrt{\text{Hz}}$. This improvement should enable the use of drift times ~ 0.3 s. For $T = 0.3$ s, a change in g by 3 parts in 10^7 leads to a phase shift $\Delta\phi$ of one cycle.

Noise sources associated with fluctuations in numbers of trapped atoms or from fluctuations in detection efficiency can be reduced by techniques now being used for atomic-fountain-based atomic clocks. Fluctuations in detection efficiency and numbers of trapped atoms are minimized by «normalization» detection techniques, where the number of atoms in the $|1\rangle$ state and $|2\rangle$ state are detected simultaneously [22, 23].

The number of atoms contributing to the interference can be significantly improved by using cesium, rather than sodium, atoms, and such an experiment is under way in our laboratory. The r.m.s. velocity spread of a sample of laser-cooled cesium atoms is ~ 1 cm/s. Thus fewer atoms will be lost due to ballistic spreading of the sample during the ~ 600 ms flight time of the atoms. A 1 mm diameter trap contains nearly 10^8 atoms. Assuming 10% of the initially trapped atoms ultimately contribute to the detected interference signal, the anticipated count rate is $\sim 10^7$ atoms/s. If the measurement were shot noise limited, these statistics would enable a fringe to be split by better than 1 part in 10^3 after 1 s. If the measurement were shot noise limited, the estimated resolution is $\sim 10^{-10}\, g\sqrt{\text{Hz}}$.

Other uncertainties which will limit the resolution of a measurement of g include those arising from the initial spread in position and velocity of the atoms, from spurious electromagnetic forces on the atoms and from imperfections in the light beams. Coriolis forces and gravitational gradients produce inertial forces which are velocity and position dependent. Magnetic-field gradients are the dominant electromagnetic perturbation. Finally, AC Stark shifts, Raman beam alignment and wave front distortion, laser frequency stability and multiple sideband interference are sources of laser-dependent phase shifts. Many of these systematic effects have been discussed in great detail in earlier work [6, 24-26] and will not be discussed here.

1·11. *Speculations.* – Potential applications of a sensitive absolute g-meter include geophysics and oil and mineral exploration. As an example of a geophysical application, we note that the fractional change in g is $\Delta g/g \sim 10^{-9}$ if the location of the atom interferometer is raised by 3 mm with respect to the center of the Earth. The long-term vertical change of sea level, due to the possible melting of the polar icecaps, could be measured by using laser ranging to measure the distance between a platform and the sea height and a g-meter to measure the distance of the platform relative to the center of the Earth. Density changes in the water due, for example, to temperature and salinity changes would also have to be monitored.

The measurement of local changes in g is used in oil exploration. The instruments are based on mechanical pendulums which requires approximately 15 min of set-up time and an additional 15 min to get a $\Delta g/g$ reading of 10^{-8} sensitivity. An atom interferometer based on a cell magneto-optic trap and diode lasers has the potential of higher sensitivity with more than two orders of magnitude less time to make the measurement. If the sensitivity in g reaches $\Delta g/g \leqslant 10^{-12}$, equivalence principle and «5th force» measurements may become possible.

The $\pi/2$-π-$\pi/2$ interferometer can also be used to construct a sensitive gyroscope. The phase shift due to the Coriolis force is $\Delta\phi_{\text{Cor}} = (\boldsymbol{k}_1 - \boldsymbol{k}_2) \cdot (\boldsymbol{\Omega}_{\text{Earth}} \times \times \boldsymbol{v})\, \Delta t^2$. For this application, the area of the interferometer is important. Thus the velocity of the atoms used in this type of interferometer should be higher than for the measurement of g. Also, a «beamsplitter» that generates a larger momentum difference between the two arms of the interferometer is also desirable. Without going into the design of an appropriate beamsplitter and mirror, suppose that the transverse-momentum difference between the two arms of the interferometer is on the order of $40\hbar k$, that the velocity is 5 m/s, $\Delta t = 0.3$ s, and we can split the interferometer fringe by 10^4. Then the resolution of the gyroscope is on the order of $\Delta\Omega \sim 5 \cdot 10^{-13}$ rad/s. Other systematic effects including geological noise may prevent one from achieving this resolution.

2. – An interferometer measurement of the photon recoil velocity.

The first observation of the effect of the photon recoil was made by COMPTON when he observed the scattering of electrons and X-rays [27]. In 1933 Otto FRISCH observed sodium recoil from the absorption of a single photon [28]. Finally, the recoil of a visible photon was first observed spectroscopically in the doubling of certain spectral peaks in saturation spectroscopy [29]. The importance of measuring the recoil with visible photons is due to the fact that optical frequencies can now be related directly to the cesium frequency standard with an uncertainty of less than one part in 10^{11}.

A measurement of the photon recoil leads directly to a measurement of $\hbar/M$, where M is the mass of the atom used in the experiment. The significance of $\hbar/M$ lies in the fact that mass appears in quantum-mechanical equations only in that ratio, so that tests of quantum theories typically require only knowledge of $\hbar/M$, and not of M. As a particularly important example, the fine-structure constant α is determined by combined measurements of $\hbar/m_e$ and the Rydberg constant, R. Many mass ratios can be measured to high precision through differences in their cyclotron frequencies in a Penning trap, so that the particular mass in the ratio $\hbar/M$ is of secondary importance [30]. Recently, $\hbar/M_n$ has been measured with an accuracy of $8 \cdot 10^{-7}$ by diffracting neutrons with a silicon crystal [31].

We have made new measurement of $\hbar/M_{Cs}$ based on recent advances in laser cooling and atom interferometry with Raman transitions. Our current precision is at the level of 10^{-7} after two hours of integration time. It is hoped that improved versions of this experiment will lead to an absolute measurement of $\hbar/M_{Cs}$ on the order of 10^{-9}, which will then lead to a better value of α. Many of the key tests of QED (which properly should be viewed as constraints on possible extensions of the standard model) are impeded by the uncertainty of this coupling constant [32].

Laser-cooled atoms and atomic interferometers have great promise for a broad range of sensitive measurements, but no previous experiment using either of these new technologies has measured a fundamental constant to high precision. Because $\hbar/M_{Cs}$ is already known to high accuracy, our measurement is well suited for studying systematic errors in light pulse interferometers.

2'1. *The interferometric method of measuring the photon recoil.* – The basic physical principle of this atomic recoil velocity measurement does not depend on atomic interference. Figure 11*a*) shows the center of mass of an atom with zero horizontal velocity in the laboratory frame which absorbs a photon from a

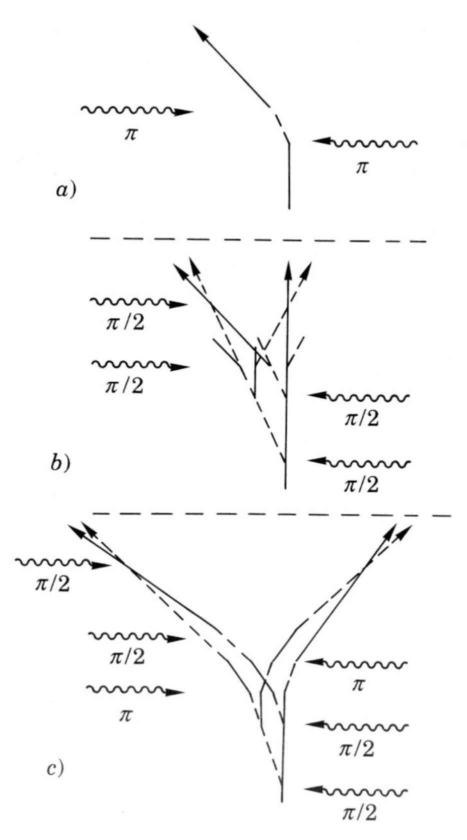

Fig. 11. – *a*) A simplified measurement of the photon recoil. The solid and dashed lines indicate different internal states of the atom. *b*) The double atom interferometer that results from the application of four $\pi/2$ pulses. The four paths that do not interfere have been truncated for clarity. *c*) The double interferometer with two additional π pulses inserted in between the two sets of $\pi/2$ pulses. Up to $15\,\pi$ pulses have been used in the experiment.

rightward propagating laser beam tuned on resonance. To conserve momentum its velocity changes by $\hbar k/m$. It is then de-excited by a photon from a leftward propagating beam. The de-excitation and excitation frequencies are separated by $\Delta\omega = 2\hbar k^2/M$, so that accurate measurements of $\Delta\omega$ and the photon wavelength determine $\hbar/M$.

This experiment can be improved by adopting the ideas of Ramsey spectroscopy. The optical Ramsey spectroscopy version of the experiment is shown in fig. 11*b*), where each π pulse is replaced with two $\pi/2$ pulses [3, 9]. This simple change gives increased resolution, larger signal and reduced systematic shifts. There are two pairs of trajectories that interfere when the spacing within the $\pi/2$ pulse pairs is equal. The crucial difference between the two interfering pairs is that their velocities differ by two photon recoils, so that the center frequencies of the sets of Ramsey fringes they produce are displaced by $2\hbar k^2/M$. The atom's initial velocity, the acceleration due to gravity and all position-independent frequency shifts do not affect the recoil shift measurement.

As with our measurement of the acceleration of gravity, our interferometer is based on stimulated Raman pulses instead of light pulses between the ground state and a metastable state. Another major improvement we made to the basic four $\pi/2$ pulse configuration is the addition of up to 15 π pulses with alternating propagation directions, sandwiched between the middle two $\pi/2$ pulses (see fig. 11*c*)). Each π pulse adds an ultraviolet-photon recoil to each atomic path, so that for N π pulses the separation between the two sets of interference fringes is multiplied by $N+1$. In this manner we have separated the two interferometer endpoints by over 4 mm.

2·2. *Experimental apparatus.* – The apparatus used in the $\hbar/M$ experiment is shown in fig. 12. A beam of thermal Cs atoms is slowed and $\sim 5\cdot 10^8$ atoms are loaded into a magneto-optic trap. The atoms are then further cooled in polarization gradient molasses before being launched upwards with the methods described in sect. **1** of this lecture. The repetition rate of the experiment is 2 Hz.

On their way up most of the atoms are optically pumped into the $F=4$, $m_F=0$ magnetic sublevel. Then a velocity-selective Raman beam pair, which propagates along the 85 mG bias magnetic-field axis, transfers a group of atoms with a velocity width of ~ 500 mm/s into the $F=3$, $m_F=0$ level. A beam resonant with the $6S_{1/2}$, $F=4$ to $6P_{3/2}$, $F=5$ cycling transition pushes away the remaining $F=4$ atoms, while the selected atoms continue upward. With this velocity filter, the contrast of the interference fringes is improved at the cost of a loss of count rate.

Near the top of the trajectory, a single pair of Raman beams, parallel to the previous Raman beams, delivers all of the interferometer pulses. The Raman beams are generated by two diode lasers which are phase-locked to each other with a 9.2 GHz frequency difference [33]. The lock reference frequency is the frequency sum of a fixed oscillator and another digital synthesizer which controls the Raman beam difference frequency. Both oscillators are locked to a SRS Loran C receiver.

The beams are combined, sent through an acousto-optic modulator for intensity control, and coupled into a single-mode optical fiber to ensure precise overlap. The final beams have 15 mW total power in a 2.3 cm Gaussian diameter, and are frequency-locked within $(0.5 \div 2.5)$ GHz from the $6P_{3/2}$ state. The maximum Rabi frequency of the interferometer pulses is ~ 10 kHz. The interferometer Raman beams are retroreflected from a mirror mounted on a precision air rail, which isolates it from residual horizontal vibrations on the vibration-isolated table. Both Raman beam frequencies travel in both directions, but the velocity

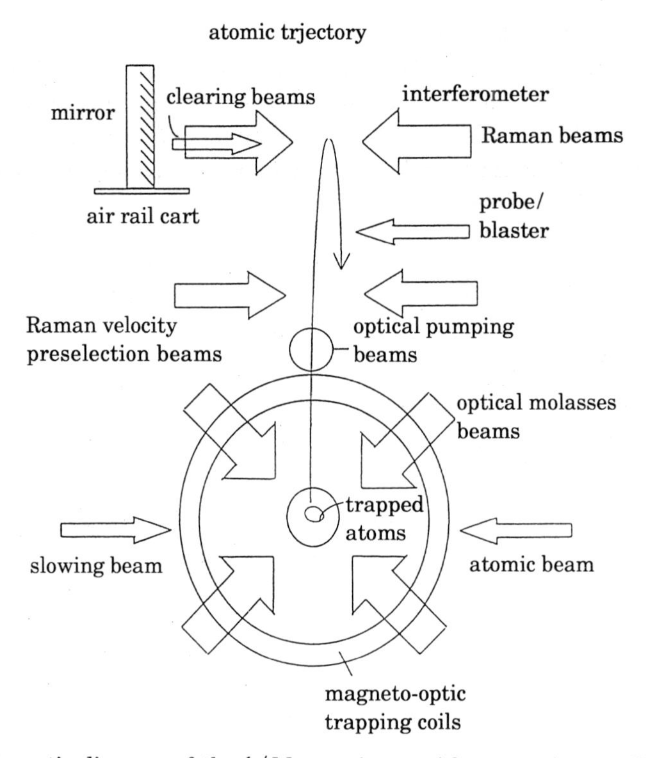

Fig. 12. – Schematic diagram of the $\hbar/M$ experiment. After trapping, cooling, launching and optically pumping the atoms into particular hyperfine state, only atoms within a narrow velocity spread were allowed to continue along the upward trajectory. The four $\pi/2$ pulses were applied at the apogee of the ballistic trajectory.

along the beam is made large enough so that only one of the two Doppler-sensitive pairs is resonant at any time, and the Doppler-free pairs and standing-wave pairs are not resonant with the atoms.

The frequency width of each interferometer pulse is narrower than the recoil shift, so a given sequence of pulse frequencies is only resonant with one pair of interfering paths. Atoms which branch off into nonresonant paths, either because they are in the other interferometer paths or because of imperfect π pulses, contribute a background which reduces fringe contrast. Most of these atoms can be removed because both paths in any given interferometer are in the same atomic state during the time between the $\pi/2$ pulse pairs. Therefore, before and after the last π pulse, we pulse on one of two clearing beams aligned nearly parallel to the interferometer beams. $F = 3$ atoms are cleared by a linearly polarized beam resonant with the $F = 2$ excited state, and $F = 4$ atoms by a beam resonant with the $F = 5$ excited state.

After the interferometer sequence, the remaining atoms fall back down through the probe/blasting beam. Fluorescence from $F = 4$ atoms is detected with a photomultiplier tube, as the atoms are pushed out of resonance by the unidirectional beam. In 2 ms, fluorescence from $F = 4$ atoms nearly vanishes, while the $F = 3$ atoms have fallen only a fraction of a probe beam diameter. A beam resonant with the $F = 3$ to $F = 4$ transition then optically pumps the $F = 3$ atoms into the $F = 4$ state and the probe excites them with nearly equal efficiency as the original $F = 4$ atoms. The fluorescence ratio allows us to normalize the signal [22, 23].

2'3. *Experimental data.* – Illustrative data with 1 ms between $\pi/2$ pulses (within each pair) are shown in fig. 13. The $\pi/2$ pulse separation was typically 15 to 20 ms. The two sets of fringes correspond to the two interferometers in fig. 11*c*). The fringes at the Rabi peak do not extend to the baseline because half of the atoms that survive the clearing are necessarily in two noninterfering trajectories. Although the broad structure

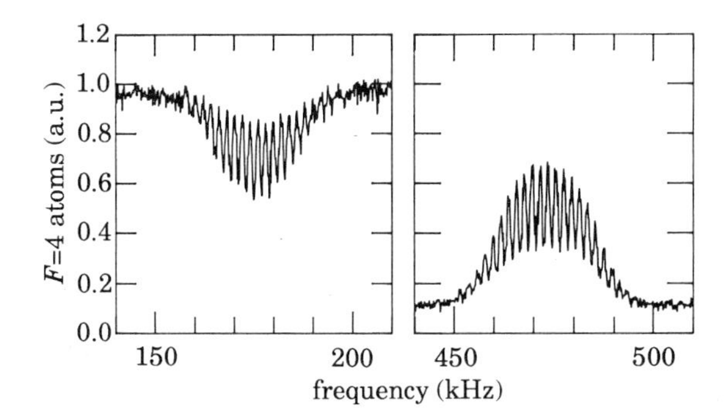

Fig. 13. – The Ramsey fringe recoil doublet for the two arms of the interferometers. The separation between $\pi/2$ pulses was reduced to 1 ms to show the overall envelope structure.

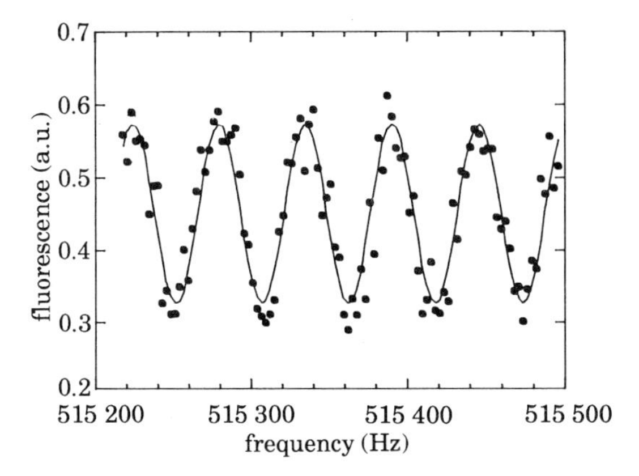

Fig. 14. – A typical interference fringe pattern after one minute of data collection. For these data, the time between $\pi/2$ pulses was 18 ms and there were 10 π pulses in between the two $\pi/2$ pulse sets.

of the signal is inverted for the two interferometers, the fringe pattern is not.

Alternate data points are taken from the two interferometers as the frequency of the final $\pi/2$ pair is scanned. Least-squares fits of the interference patterns to sinusoids yield the recoil splitting to within an integer multiple of the fringe frequency. Scans with different $\pi/2$ pulse separations determine the correct fringe number. Figure 14 shows a fringe pattern taken in one minute with 18 ms between $\pi/2$ pulses and 10 π pulses. In two hours we can measure the photon recoil with a relative precision of $\sim 10^{-7}$.

2·4. *Systematic effects.* – We have studied the systematic errors in this experiment in some detail. A more lenghty discussion of anticipated and measured limits on systematic errors appears in ref. [25, 26], and will be published in a longer paper. The most important experimental handles on systematic errors are the time between $\pi/2$ pulses (T), number of π pulses (N), Raman laser detuning from the excited state (Δ) and Rabi frequency (Ω). Using these experimental «knobs», we have uncovered and removed several errors at the 10^{-6} level. The largest systematic error was caused by optical standing waves from retroreflected, linearly polarized Raman beams. These standing waves shift the measurement as a function of the distance of the interferometer from the retroreflecting mirror. Switching to $\sigma^+\sigma^-$ polarizations eliminated the standing waves and this shift.

Magnetic-field gradients shift the phases of the interferometers differently, and will alter the measured recoil. We monitor this shift in two ways. First, we reverse the direction of all the Raman pulses, which creates interference paths that are mirror symmetric to the original ones. Since the two paths in each interferometer flip sides, the shift from any linear gradient reverses. Second, we change the parity of the number of π pulses, which changes the internal state during the time between the last two $\pi/2$ pulses for all the paths. Increasing the bias magnetic field fourfold causes a 1 p.p.m. shift in the measured recoil that reverses sign in just this manner. With our typical bias field, these reversals do not affect our measurement.

We have also looked for dependences on beam alignment (D), the relative AC Stark shifts of the hyperfine levels and clearing beam power. Our current empirical limits on shifts due to these effects are at the 0.25 p.p.m. level. Although we do not see a clear dependence on T, D or N, our measured value is $8.5 \cdot 10^{-7}$ below the accepted value, which has a relative uncertainty of $7.7 \cdot 10^{-8}$.

We believe the cause of this presumed error is wave front distortion from imperfect optics. In the past we have seen repeatable changes in the measured recoil of up to 10 p.p.m. with our routine reversals, at about the same level as the disagreement with the accepted value. Successive improvements in the optics have led to measured recoils progressively closer to the accepted value with progressively smaller internal inconsistencies. Still, in the current configur-

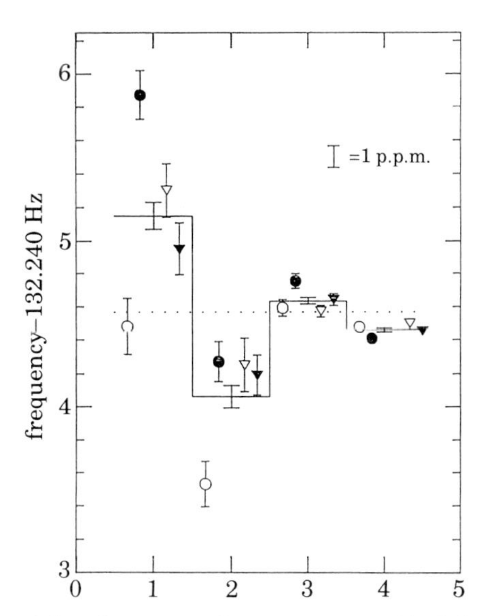

Fig. 15. – The evolution of the photon recoil measurement. Each horizontal line for a given «date» is the mean value of the measured recoil at a different time. The shift in the mean value is an indication of a change in the magnitude of systematic effects as the apparatus was changed. Circles correspond to even number of π pulses and the triangles to odd number. The open and closely symbols correspond to parity-reversed interferometers. The dotted line corresponds to the accepted value for $\hbar/M_{\mathrm{Cs}}$.

ation it is difficult to ensure that all effects due to wave front distortion have been eliminated.

Figure 15 shows the historical evolution of the recoil measurement after the precision was better than 1 part per million in an hour of integration time. Each change of «date» is accompanied with specific experimental improvements that affected the systematic shifts. Adjacent «dates» are separated by roughly a month. The dotted line shows the accepted value of the photon recoil for our experimental conditions. Between date 1 and 2, the Raman beams were spatially filtered. Between date 2 and 3, the vacuum chamber windows and retroreflection mirror were changed to $\lambda/10$ quality optics. Finally, between date 3 and 4, a combination of standing-wave effects and AC Stark shifts was eliminated by changing the polarization of the Raman beams so that the counterpropagating ω_1 and ω_2 beams form a corkscrew polarization instead of standing waves. Details of these changes and the accompanying analysis are given in other publications [25, 26].

2·5. *Future work.* – Various improvements to this experiment have been discussed [26, 34]. The next improvement in this experiment is to orient the Raman beams vertically. The effects of wave front distortions will be easier to study in this configuration because the transverse motion of atoms in the beams can be varied. Also, we will be able to easily vary where along the beam atoms receive interferometer pulses, and the $\pi/2$-π-$\pi/2$ pulse interferometer will provide extra information on the effects of wave front distortion. The frequencies of the Raman beams will have to be adjusted between each pulse to account for gravitational acceleration, but the adjustment is the same for the two interferometers apart from a slight difference in the local gravitational acceleration. Furthermore, the vertical launch should permit a 200 ms $\pi/2$ pulse separation and give time for additional pulses. We anticipate a resolution of 10^{-9} in one hour, but it is difficult to now predict the severity of the associated systematic errors, particularly those related to the quality of the Raman beams. However, the simplicity of this system, which fundamentally depends only on simple interactions between free atoms and photons, makes the prospects for accuracy in this and similar measurements promising.

Finally, we note that π pulses sandwiched in between the two sets of $\pi/2$ pulses enhances the resolution of the experiment by a factor of N, where N is the number of photon recoil kicks the atom receives during the measurement. By introducing π pulses in between each of the $\pi/2$ pulse sequences so that one arm of each of the interferometers is widened, the resolution can increase as N^2 [35, 36].

3. – Positronium spectroscopy with a c.w. dye laser.

Positronium (Ps) is a quasi-stable atom composed of an electron and its antiparticle, the positron. It is one of a family of leptonic hydrogenlike systems that include muonium ($\mu^+ e^-$) and tauonium ($\tau^+ e^-$). Ps is also one of a family of more exotic particle-antiparticle systems such as $\mu^+ \mu^-$ and $p^+ p^-$, the proton-antiproton bound state whose Lyman and Balmer series have recently been observed in the X-ray region [37]. Of these, only Ps, muonium and tauonium have energy levels accessible to laser excitation, and only Ps and muonium have been produced in sufficient quantity to do laser spectroscopy [38-40].

A nearly complete description of Ps is possible with the electroweak interaction, and to high accuracy by the electromagnetic interaction alone, affording an ideal test of quantum electrodynamics. In contrast, hydrogen is composed of an electron and three quarks interacting through the strong, weak and electromagnetic forces. Unfortunately, the simplicity of the forces in positronium does not carry over to simplicity in the calculations of the energy levels of the atom. In a nonrelativistic system, the interaction of two equal-mass particles can be reduced to a one-body problem in a center-of-mass frame through a simple Galilean transformation. The Hamiltonian is then written using the reduced mass of the particles. The Ps atom, however, is relativistic at the level of $E_I/2m_e c^2 \sim 10^{-5}$, where $E_I = 6.8$ eV is the Ps binding energy, and cannot be reduced in a simple way to a one-body problem.

In contrast, although H also has relativistic corrections, the large proton-to-electron mass ratio allows one to calculate the fine structure to first order with the Dirac equation, by considering the electron in a fixed Coulomb potential. Nonrelativistic reduced-mass corrections are included as a modification of the value of the Rydberg constant, and relativistic reduced-mass, or relativistic «nuclear recoil», corrections may subsequently be applied as perturbations to the Dirac energy levels. In Ps, however, the Dirac equation fails completely, and even to first order one must resort to a relativistic two-particle equation to obtain the fine structure.

3·1. *Status of theory.* – BETHE and SALPETER [41] and SCHWINGER [42] developed the basic field-theoretic equation describing the relativistic two-body problem; unfortunately the Bethe-Salpeter equation has no exact solutions, even in lowest order. A two-body formalism applicable to positronium and muonium has more recently been introduced by LEPAGE and CASWELL [43] and REMIDDI and BARBIERI [44]. They have succeeded in transforming the Bethe-Salpeter equation to an equivalent one-body Dirac equation, resulting in analytical solutions to lowest order. The higher-order solutions to this transformed equation are essentially nonperturbative. Although they involve integrals, called kernels, that can be expressed in terms of Feynman diagrams, the fine-stucture constant does not enter the wave functions and propagators in a perturbative way as it does for other precision calculations such as the magnetic moment of the electron. Despite this difficulty, theorists are still able to use small parameters such as α to develop expressions of decreasing magnitude for succeeding corrections. The nonperturbative character of the expansion is revealed by the appearance of logarithmic factors, resulting in an expansion such as R_∞, $R_\infty \alpha^2$, $R_\infty \alpha^4 \ln \alpha$, $R_\infty \alpha^3$, ... where the Rydberg constant is $R_\infty = \alpha^2 mc/2h$. This approach allows a rough estimate to be made of the magnitude of the neglected terms, and, therefore, of the accuracy of the calculation. Unfortunately, there is no reliable means of estimating the size of the coefficient associated with the uncalculated term.

The particle-antiparticle nature of positronium has a clear impact on the difficulty of the calculations at each order of the calculation. The calculation in H has a contribution from only two Feynman diagrams, as compared to ten for Ps, seven of which are due to virtual annihilations. Two-photon interaction and two-photon virtual-annihilation terms do not contribute to the energy of the Ps triplet states by charge conjugation symmetry. Multiphoton interactions first appear in the corrections for the triplet states. The calculation of corrections involves hundreds of Feynman graphs and is clearly difficult, but this must be offset by the fact that positronium is the only experimentally accessible system in which self-annihilation terms may be tested by precision measurements. Because the positron and electron have equal masses, Ps is also the most sensitive system in which to test the QED relativistic two-body formalism.

Currently, the coefficients for the terms in the expansion given above have been calculated for the triplet states through $R_\infty \alpha^4 \ln \alpha$, although the coefficient for the latter is in dispute. The expansion for the n^3S states has the following form:

$$(6)\quad \frac{E(n^3S)}{h} = -\frac{1}{2n^2} c(R_\infty) + \qquad \text{(nonrelativistic Schrödinger equation)}$$

$$+2\frac{1}{n^3} c(R_\infty \alpha^2)\left[\frac{11}{64}\frac{1}{n} - \frac{1}{2}\right] - \qquad \text{(Ferrel, 1951)}$$

$$-\frac{6}{4\pi}\frac{1}{n^3} c(R_\infty \alpha^3 \ln \alpha) + \qquad \text{(Fulton, 1954)}$$

$$+\frac{1}{4\pi}\frac{1}{n^3} c(R_\infty \alpha^3)\left\{\frac{14}{3}\left[\frac{7}{15} + \ln\frac{2}{n} + \frac{n-1}{2n} + \sum_{k=1}^{n} k + \ln 2\right] - \frac{16}{3}\ln k_0(n)\right\} -$$

$$\text{(Fulton, 1982, and Gupta, 1989)}$$

$$-\frac{1}{6}\frac{1}{n^3} c(R_\infty \alpha^4 \ln \alpha) \qquad \text{(Fell, 1992)}$$

or

$$-\frac{5}{12}\frac{1}{n^3} c(R_\infty \alpha^4 \ln \alpha) \qquad \text{(Khriplovich, 1992)},$$

where $\ln k_0(n)$ are the Bethe logarithms. The first term in eq. (6) is simply the result of the nonrelativistic Schrödinger equation, using the Ps reduced mass as discussed earlier. The term was calculated by FERRELL in 1951 [45]. The $R_\infty \alpha^3 \ln \alpha$ and $R_\infty \alpha^3$ terms were calculated together for the $n = 2$ levels by FULTON and MARTIN [46] and later generalized by FULTON to the $n = 1$ levels with the erroneous assumption of a $1/n^3$ scaling law for this term [47], as noted and corrected by several authors [48]. Finally, conflicting results have been reported for the correction by FELL [49] and KHRIPLOVICH, MILSTEIN and YELKHOVSKY [50]. Table I shows the frequency contribution of each of these terms to the $1\,^3S$ and $2\,^3S$ levels, and to the $1S$-$2S$ transition. The final frequency of the $1S$-$2S$ transition is $(1\,233\,607\,212.7 \pm O(10))$ MHz and $1\,233\,607\,211.7 \pm O(10)$ for the work of Fell and Khriplovich.

3·2. *Earlier optical spectroscopy of positronium.* – CHU and MILLS [52] used the two-photon Doppler-free transition to observe the $1\,^3S_1$-$2\,^3S_1$ transition in Ps in 1982 with a resolution of 1 p.p.m. In 1984, CHU, MILLS and HALL [38] measured this transition to be 1 233 607 185.15 MHz. Roughly half of the 12 p.p.b. error estimate was due to the determination of the positronium line center rela-

TABLE I. – QED *corrections to* $1\,^3S_1$ *and* $2\,^3S_1$ *levels and* $1\,^3S_1$-$2\,^3S_1$ *transition frequency, in* MHz (from ref. [51]).

Order	$\Delta\nu(1\,^3S_1)$	$\Delta\nu(2\,^3S_1)$	$\Delta\nu(2\,^3S_1$-$1\,^3S_1)$
R_∞	– 1 644 920 980.6	– 411 230 245.1	1 233 690 735.41
$R_\infty\,\alpha^2$	89 419.2	7413.6	– 82 005.59
$R_\infty\,\alpha^3 \ln\alpha + R_\infty\,\alpha^3$	1733.1	231.7	– 1501.44
$R_\infty\,\alpha^4 \ln\alpha$	7.6	1.0	– 6.69
total	– 1 644 829 820.6	– 411 222 598.9	1 233 607 221.7

tive to their secondary frequency standard, a line in tellurium vapor. The remainder of the error estimate was attributed to the calibration of the tellurium standard to the deuterium $2S_{1/2}$-$4P_{3/2}$ Balmer line. The 1984 $1S$-$2S$ measurement was found to be in good agreement with the R^3 QED corrections [43], the two differing by 1315 MHz.

However, a subsequent recalibration of the Ps reference line in Te_2 by MCINTYRE and HÄNSCH [53] corrected the Ps transition frequency to the red by 42 MHz to (1 233 607 142.9 ± 10.7) MHz, a value (56.4 ± 10.7) MHz smaller than Fulton's 1982 theoretical value. Then, an error was discovered in the sign of the pulsed-laser frequency offset in the 1984 measurement, and published in 1989 by DANZMANN, FEE and CHU [54], moving the value of the transition frequency of (1 233 607 218.9 ± 10.7) MHz, (16.6 ± 10.7) MHz above the corrected theoretical value. Finally, Fell's $R_\infty\,\alpha^4 \ln\alpha$ QED corrections once again eliminated discrepancy between theory and experiment. There remainded, however, the clear need for further advances in the measurement and theory of the $1S$-$2S$ interval in positronium.

The large intermediate-state detuning of the $1S$-$2S$ two-photon transition requires high intensities to drive the transition. Thus the first precision measurements in hydrogen and positronium were made using narrow-band pulsed lasers produced by amplifying a c.w. «seed» laser with a series of traveling-wave amplifiers. Pulsed lasers are usually characterized in terms of their temporal and spectral intensity profiles. Any frequency fluctuations appear as a spectral broadening of the laser output, which is expressed by the degree to which the laser is not Fourier-transform limited. Apart from picosecond and femtosecond lasers, where the pulses can actually be very close to the Fourier-transform limit, most lasers fall short of achieving the minimum uncertainty in $\Delta\nu\,\Delta t$.

The process of amplifying a pulsed laser inherently involves rapid changes in the gain. By the Kramers-Kronig relations, the index of refraction of the amplifying medium is rapidly modulated, resulting in fluctuations of the laser frequency within a single pulse. Standard measurements of the spectral broadening from the induced frequency modulation do not tell us the instantaneous frequency (the time derivative of the instantaneous phase of the complex amplitude) of the pulse. The time evolution of the phase is essential to thoroughly predict or interpret the results of several types of experiments, particularly the spectroscopy of nonlinear transitions. This limitation was recognized by WIEMAN and HÄNSCH [55], and was the principal uncertainty in their measurement of the hydrogen $1S$ Lamb shift.

In general, one would not attempt to make a precision measurement with a pulsed laser as it appears at the output of the amplifier, for exactly the reasons outlined above. In the pulsed measurements of H and Ps, the laser light was filtered in a Fabry-Pérot interferometer before being used to excite the atoms, the assumption being that the filtered light had sufficiently small frequency chirps as to be neglected.

This assumption was shown to be incorrect by an experimental measurement of the instantaneous frequency behavior of an excimer-pumped dye-laser system [54]. The output of the laser amplifier was Fabry-Pérot filtered and heterodyned with a frequency-shifted sample of the c.w. seed light, and the resulting beat note measured. The instantaneous frequency behavior was then extracted from the beat signal using Fourier-transform techniques. We found that pronounced frequency chirps are possible even after the filtering process. The effect of such frequency chirps on the resulting two-photon line shape was modeled, indicating that laser frequency chirps contributed roughly 2 MHz uncertainty to the 1984 measurement of the $1S$-$2S$ interval in positronium.

By directly measuring the phase behavior of traveling-wave pulsed amplifiers, precision pulsed measurements of nonlinear transitions to the level of 1 MHz may be possible. However, several factors led us to abandon this approach in favor of a continuous-wave measurement. The possibility of an intense, cryogenic (50 K) positronium source raised the c.w. excitation rate (proportional to $1/T$) to an acceptable level. Additionally, the availability of ultra-low-loss laser mirrors at 486 nm prompted us to construct the required apparatus to frequency stabilize our Coherent 699 dye laser to the kilohertz wide resonance of an ultra-high-finesse cavity, eventually allowing us the generate more than 2.5 kW c.w. circulating laser power.

3·3. *Experimental apparatus.* – A schematic overview of the experimental apparatus is shown in fig. 16. Positrons generated at the beam dump of a free-electron laser are collected in an electromagnetic bottle and bunched into a 25 ns pulse at the positronium source [56]. Ps atoms are then generated in vacuum at the atomically clean surface of a pure aluminum crystal (111 plane). Approximately 30% of the incident positrons are thermally desorbed from the 576 K sur-

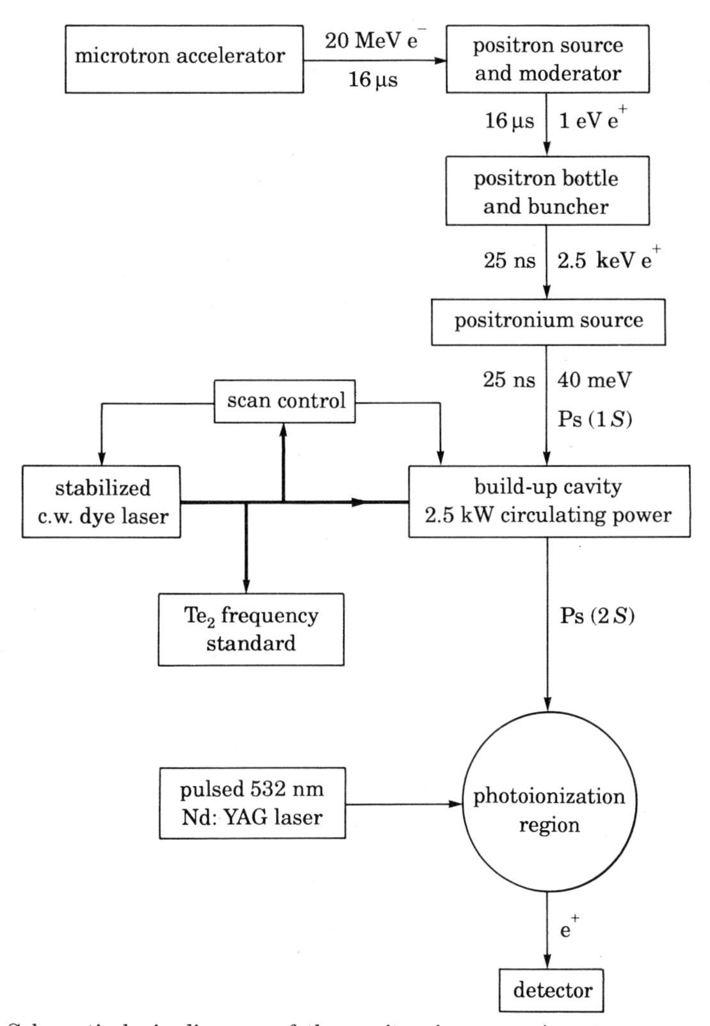

Fig. 16. – Schematic logic diagram of the positronium experiment.

face as positronium atoms with a beam Maxwellian velocity distribution. Unfortunately, experiments run with a target of physisorbed oxygen on aluminum failed to produce sufficient amounts of cryogenic positronium.

A c.w. laser beam, built up to a circulating power of 2.5 kW by a resonant Fabry-Pérot cavity, was placed 1.5 mm in front of the Al target to excite the $1S$-$2S$ transition, as shown in fig. 17 and 18. After a variable time delay between 50 and 200 ns during which the Ps atoms drift ballistically through the c.w. excitation beam, excited-state atoms were photoionized a few millimeters from the target using a YAG pulsed laser. The e^+ photoionization fragments were accelerated out of the laser interaction region with a $\sim 2\,\mathrm{V/cm}$ electric

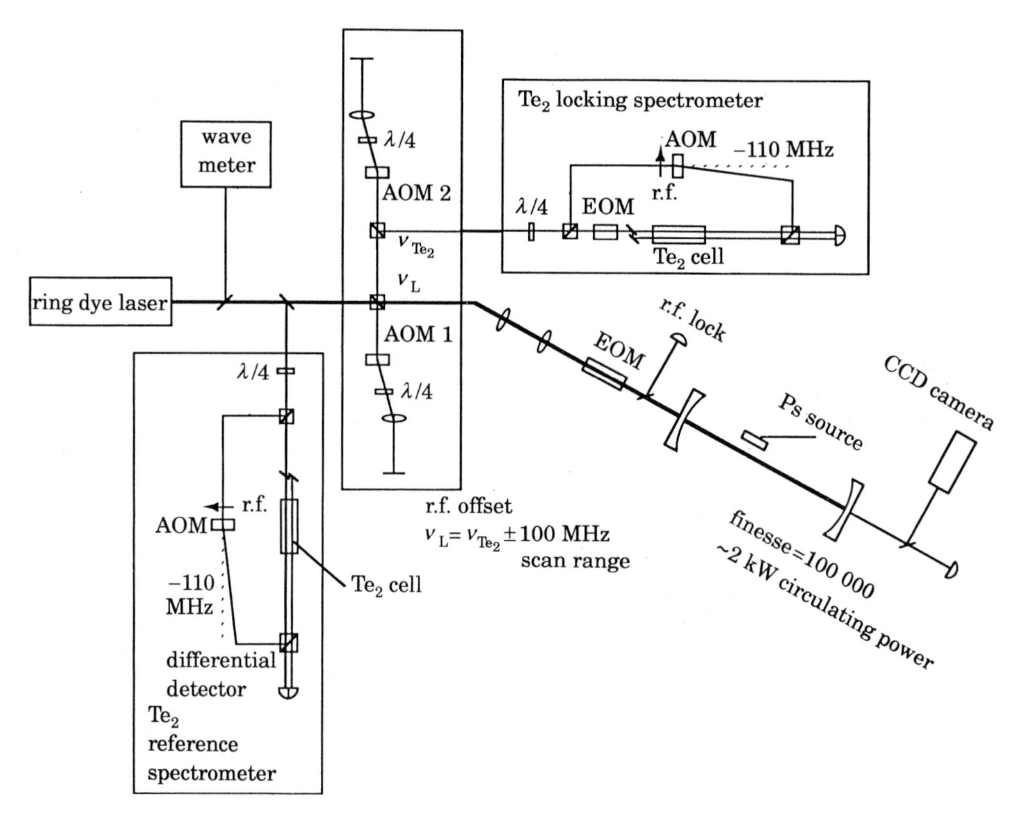

Fig. 17. – Schematic diagram of the optical configuration. A Te_2 reference spectrometer was included as a check of the accuracy of the Te_2 locking spectrometer and the performance of the frequency off-set lock. Most of the laser power was directed into the high-finesse Fabry-Pérot build-up cavity.

field, through a constant-potential drift tube, to a channel electron multiplier array (CEMA). Detected positrons were counted as the laser is tuned relative to the Ps line center. Because of the small excitation probability of Ps in the c.w. beam, the re-emitted positrons constitute an overwhelming background if allowed to contaminate the photoionized e^+ signal. Thus signal positrons were collected only from the region below the target while the re-emitted positrons were trapped along magnetic-field lines and cannot be accelerated to the detector.

The majority of the c.w. laser power was delivered to the build-up cavity in a small bandwidth using a frequency modulation locking scheme [57] (see fig. 17). The laser is locked to the cavity resonance by rapidly correcting the laser frequency with an intracavity phase modulator inside the ring dye laser. With a cavity finesse of 10^5, we achieved an intensity build-up in the cavity of $7.6 \cdot 10^3$ and a maximum circulating power of 2.5 kW. A peak intensity of

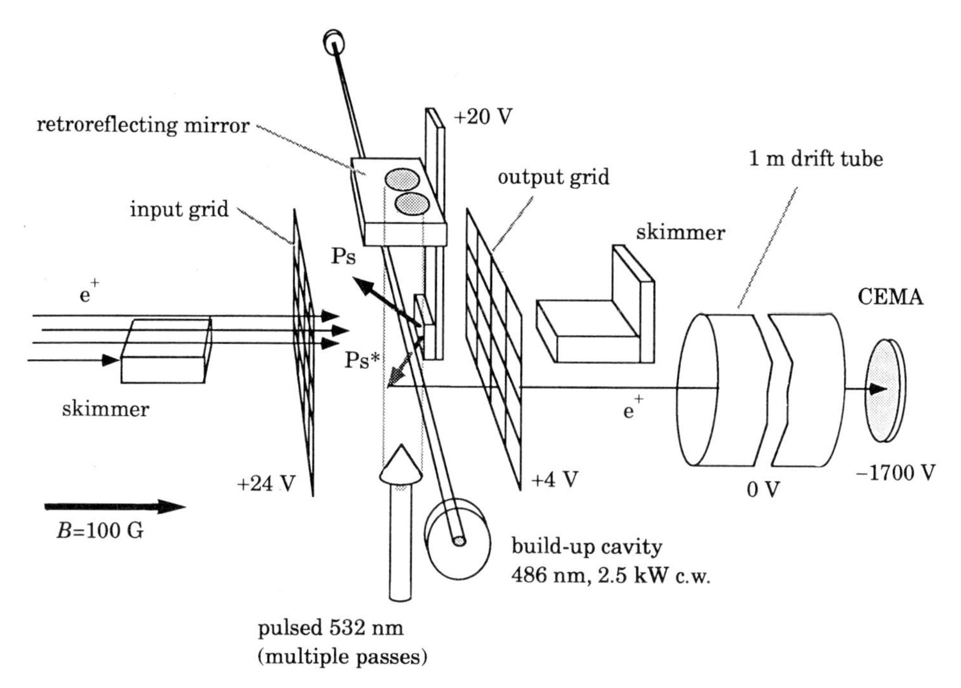

Fig. 18. – Schematic view of the positronium production target with the grids, skimmers and baffles used to isolate the photoionized positronium from the positrons that originate from the incident beam. The skimmers have a finite length along the magnetic-field axis in order to eliminate those positrons that have helical orbits with a large pitch that may orbit past a thin baffle.

1.7 MW/cm^2 was obtained before thermal distortion of the mirrors prevented effective coupling into the TEM_{00} mode of the cavity. Within a few hours, permanent optical damage to the mirrors degraded the performance of the build-up cavity enough so that data could no longer be taken. The details of the build-up cavity and locking scheme are discussed in the Ph.D. thesis of M. Fee [58] and in the *Physical Review.*

A small fraction (25 mW) of the c.w. laser power was frequency offset using r.f.-driven acousto-optic modulators. The frequency-shifted light was then sent to a frequency-modulated saturated-absorption spectrometer [59, 60]. Another small fraction of the laser power (10 mW) is used for simultaneous saturation spectroscopy of a recently calibrated reference line in tellurium molecular vapor. We use the e_3 component of ref. [53], which has a frequency approximately 65 MHz below the Ps 1S-2S transition. The laser frequency was measured as a radiofrequency offset relative to the zero crossing of the e_3 line center of a second, frequency-modulated Te_2 spectrometer, and scanned in 5 MHz steps. The transmitted laser power through the build-up

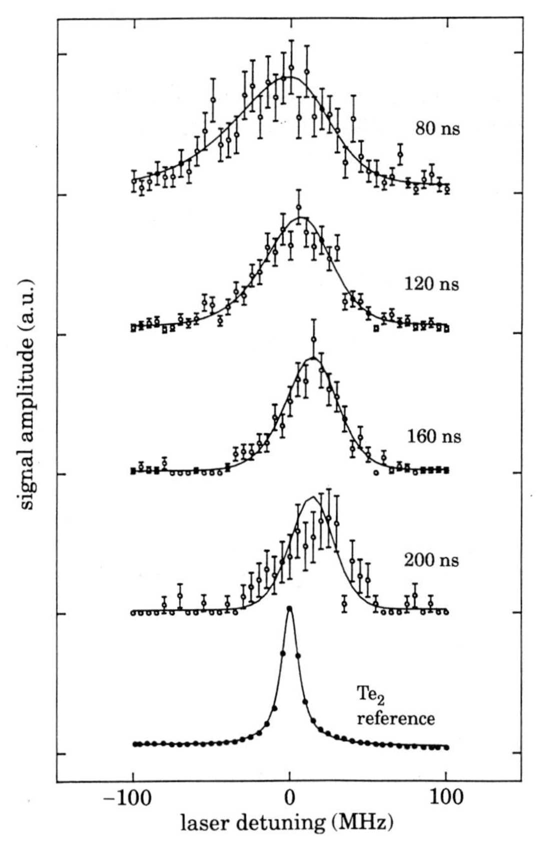

Fig. 19. – Five positronium resonances taken at different YAG photoionization delays. The solid lines are fits to the data based on Monte Carlo models of the experiment with only the amplitude and frequency off-set as the free parameters. The data were taken at a target temperature of 600 K and a circulating c.w. laser power in the build-up cavity of $(2.0 \div 2.5)$ kW.

cavity was continuously monitored by a photodiode and calibrated to reflect the absolute power in the build-up cavity.

3·4. *Experimental results and analysis.* – In fig. 19 we show the Ps resonance taken at five different YAG pulse delays. The major systematic effects in determining the positronium $1S$-$2S$ line center relative to the Te_2 reference line are the second-order Doppler shift due to the motion of the Ps atoms in the laser reference frame and the AC Stark effect due to the nearly saturating c.w. laser beam. For the majority of the data taken, both systematic corrections are less than 20 MHz. Scans were taken at several time delays of the photoionization pulse and at several settings of the build-up cavity power to facilitate the understanding and correction of these systematic effects.

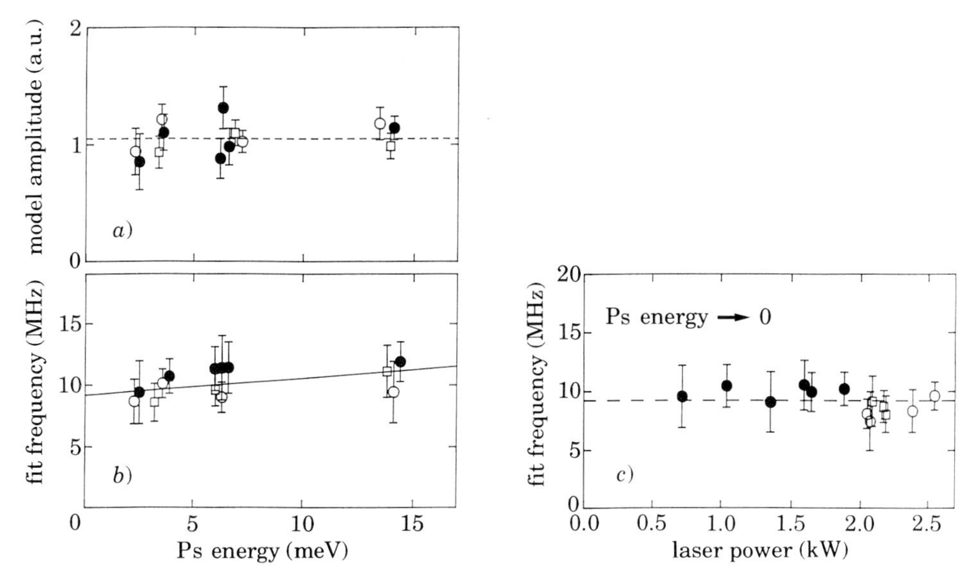

Fig. 20. – *a*) Signal amplitude parameter as a function of Ps energy. *b*) The line center fit relative to the Te_2 reference line for a set of model parameters that describe the experimental geometry. *c*) The line center fit as a function of the measured circulating power in the build-up cavity. The different symbols correspond to different data runs.

The precision to which we can determine the Ps line center is limited by both counting statistics and our understanding of the line shape. A computer model of our experimental geometry allows us to integrate the Schrödinger equation for a large number of positronium trajectories chosen randomly from a beam Maxwellian distribution at the temperature of the Al sample. The calculation includes the AC Stark shift, ground-state annihilation and 486 nm photoionization. Calculated line shapes at each YAG delay are multiplied by a signal amplitude parameter and shifted to a frequency relative to the measured Te_2 reference line. The value of the amplitude parameter is shown in fig. 20*a*) to be independent of the Ps energy, thus supporting our use of a beam Maxwellian velocity distribution. The error bars are from a Monte Carlo calculation in which the model was fitted to 2000 randomly generated data sets to determine the variance of the fit amplitude and line center for each resonance.

Figure 20*b*) shows that the fit frequencies as a function of estimated Ps energy have a small residual slope 1.13 standard deviations above zero. If the modeling of the experiment were accurate, the slope in fig. 20*b*) would be zero. To account for possible overcorrection of the second-order Doppler shift, we extrapolate linearly to zero Ps energy to find the Ps resonance frequency. We note that fit line centers are also perfectly consistent with their weighted mean, yielding a reduced chi-square parameter of $\chi^2 = 0.5$ indicating that the scatter in the line centers is actually $\sqrt{2}$ smaller than expected based on our Monte Carlo statistical-error estimate. The extrapolation is a more conservative approach, giving a statistical uncertainty of 0.90 MHz, compared to a 0.46 MHz uncertainty in the weighted mean.

The uncertainty in the determination of the Ps transition frequency due to uncertainty in the model parameters is determined by varying each parameter and noting its effect on the extrapolated frequency. The values of the fit frequency for the 13 resonances are shown in fig. 20*c*) as a function of circulating laser power. The slope of the data is consistent with zero, indicating that our model has correctly accounted for the AC Stark effect. The uncertainty in our laser power measurements contributes less to the uncertainty in the slope of these data than the statistical errors. It is, therefore, not necessary to extrapolate these data to zero laser intensity in the determination of the Ps line center.

The reduction of our measurements to an absolute value for the $1\,^3S_1$-$2\,^3S_1$ transition frequency is detailed in table II. An acousto-optic frequency shift produces a $+55$ MHz offset in the observed position of the Te_2 reference line. Contributions to the ± 1.5 MHz uncertainty in the extrapolated Ps line center are the statistical uncertainty and the uncertainty of the second-order Doppler and AC Stark shift corrections. The dimensions of the YAG beam affect the velocity distribution of the sampled atoms, and, therefore, the size of the second-order Doppler shift correction. The c.w. beam diameter and laser power calibration determine the c.w. laser intensity and thus the AC Stark shift correction. All contributions to the final uncertainty are assumed to be uncorrelated and are added in quadrature. The motional Stark shift in the 150 G magnetic field is less than 1 MHz and is eliminated by our extrapolation procedure. The calculated Zeeman shift of the $1\,^3S$-$2\,^3S$, $m_F = 0 \rightarrow 0$ transition produces a $+1.14$ MHz shift in the observed Ps resonance when averaged over the 3 magnetic sublevels. With these corrections, the absolute transition frequency is $(616\,803\,608.2 \pm 1.6)$ MHz, and the full $1\,^3S_1$-$2\,^3S_1$ interval in positronium is $(1\,233\,607\,216.4 \pm 3.2)$ MHz.

The agreement between our experimental result and theory represents an improved measurement of the positron-electron mass ratio [52, 61]. In the comparison of the QED prediction with experiment, one implicitly assumes that the positronium Rydberg is exactly half the hydrogen Rydberg. The agreement between theory and our experiment at the level of ± 10 MHz places a limit on the positron-electron mass difference on the order of $\Delta m/m = 8 \cdot 10^{-9}$. If agreement with theory persists with the $R_\infty \alpha^4$ QED correction in progress, our present measurement will place a constraint of $\Delta m/m < 10^{-9}$ on the e^+-e^- mass ratio. This measurement places the best limit on a particle-antiparticle mass ratio apart from the K^0-K^0 bar system. HUGHES and DEUTCH [62] point out that the measurement of the Ps $1S$-$2S$ interval also constrains the positron-electron charge ratio. They use the measured ratio of the positron and electron cyclotron frequencies to eliminate the e^+-e^- mass ratio in the expression for the ratio of

TABLE II. – *Summary of uncertainties and results* (from ref. [51]).

Te_2 reference line			(616 803 544.4 ± 0.6) MHz
AOM offset			255.0 ± 0.0
model fit to line center			210.0 ± 1.5
statistical uncertainty		± 0.9	
YAG parameters			
diameter a:	(6.0 ± 0.5) mm	± 0.3	
eccentricity b/a:	1.3 ± 0.2	± 0.5	
distance from target:	(1.5 ± 0.5) mm	± 0.1	
height:	(6.0 ± 1.0) mm	± 0.8	
Ps temperature:	(576 ± 30) K	± 0.2	
c.w. beam diameter:	(0.31 ± 0.01) mm	± 0.3	
c.w. power measurement:	(761 ± 38) W/V	± 0.5	
d.c. Stark shift	− 0.07 ± 0.1		
quadratic Zeeman shift	− 1.14 ± 0.0		
$1S$-$2S$ transition frequency			(616 803 608.2 ± 1.6) MHz
$1S$-$2S$ interval			1 233 607 216.4 ± 3.2
previous measurement CHU, MILLS jr. and HALL [38], DANZMANN, FEE and CHU [54].			1 233 607 218.9 ± 10.7
theory FELL [49]			1 233 607 221.7 ± $O(10)$

the positronium and hydrogen Rydberg constants. The result is an expression for the uncertainty in the charge ratio in terms of the uncertainty in the ratio of cyclotron frequencies and the uncertainty in the Rydberg ratio. The result, $q_e/q_{e^+} = 1 \pm 2.5 \cdot 10^{-8}$, is limited entirely by the 130 p.p.b. uncertainty in the cyclotron frequency ratio. With a factor of 100 improvement in the measurement of the cyclotron ratio, the result would be limited by the positronium $1S$-$2S$ measurement at the level of $1 \cdot 10^{-9}$.

3·5. *Future improvements.* – Further improvements over new measurement of the $1\,^3S_1$-$2\,^3S_1$ are clearly possible and have been previously discussed [51]. The use of cold Ps would greatly reduce the second-order Doppler shift and the transit time broadening of the observed resonances. Since the excitation probability scales roughly as the square of both the interaction time and the laser intensity, a factor of 10 reduction in temperature and a factor of 3 lower laser intensity would yield the same signal. The AC Stark shift would be concomitantly reduced to roughly 6 MHz from the 18 MHz of the present measurement. It is clear that a measurement of the $1S$-$2S$ transition might reach a precision better than the 1.3 MHz natural linewidth with apparatus similar to that described in this lecture.

It is possible that a future measurement of the Ps $1S$-$2S$ transition can be made using a ^{22}Na β^+ source. Beam time constraints at an accelerator facility are contrary to the requirements of making a precision measurement. Recently a positron flux $5 \cdot 10^6$ slow positrons per second has been obtained using a solid-Ne moderator and a 70 mCi ^{22}Na source [63]. The improved efficiency of the solid-Ne moderator makes possible yields larger by a factor of 10 than with a tungsten moderator. Various bottling and trapping schemes must be considered to convert the c.w. source of positrons into a source suitable for laser spectroscopy with photoionization detection.

The ultimate reduction in the positronium temperature would, of course, be achieved by laser cooling the atoms. Sub-kelvin positronium temperatures may be possible. The laser cooling of positronium presents an unusual challenge since i) the ground-state atom annihilates in 140 ns, ii) the Doppler width at room temperature is ~ 500 GHz, and iii) the single-photon recoil velocity corresponds to a Doppler shift of 6.1 GHz. Thus the «chirped slowing» commonly used for heavier atoms is not feasible. Even if it were possible to generate a frequency chirp of 500 GHz in 140 ns, the random recoil due to the spontaneous emission could easily give the atom a velocity that would Doppler-shift the laser to the blue side of the resonance.

We propose a white-light cooling scheme that is a modification of a scheme first proposed by HOFFNAGLE [64], and discussed for Ps by LIANG and DERMER [65]. The theory of broad-band laser cooling on narrow transitions was explored in detail by WALLIS and ERTMER [66]. The idea is to irradiate the atoms with a quasi-continuum radiation, that is to the low-frequency side of the $1S$-$2P$ frequency. The radiation spectrum is designed to drop off sharply (relative to the recoil Doppler shift) at the resonance frequency. Atoms moving toward any particular laser beam will be Doppler-shifted into resonance and will scatter radiation until its velocity is reversed. A counterpropagating laser beam that will then be Doppler-shifted into resonance as the atom reverses direction will also be included. Thus the cooling works in a similar fashion to the usual optical molasses, except that the cooling limit will now be determined ultimately by the recoil velocity (~ 6.1 GHz), which is much larger than the transition linewidth of 50 MHz.

We propose to create the «white light» by constructing an FM laser [67], where modulation depths of 1000 are easily produced [68]. A modulation frequency of 150 MHz will result in a laser frequency varying sinusoidally over a range of 150 GHz. The FM laser beam would then be amplified in a series of traveling-wave amplifiers and then frequency doubled in a beta barium borate crystal which can phase-match to the 150 GHz simultaneously. An inherent characteristic of the FM laser technique is that the edge of the resulting spectral distribution falls off in ~ 3 GHz, less than the 6 GHz single-photon recoil shift. The lifetime of the $2P$ state is 3.2 ns, so with a modulation frequency of 150 MHz, the laser frequency sweeps over the $1S$-$2P$ transition roughly once per $2P$ lifetime.

Beginning with atoms at 50 K, we will need only ~ 18 photon kicks to cool the atoms to the recoil limit. We estimate a one-dimensional cooling time of 130 ns. In three dimensions, the situation is more complicated. The stimulated redistribution of photons among the three resonant beams will increase the cooling time by perhaps a factor of three to roughly 400 ns. For rapid slowing, however, the $1S$-$2P$ transition must be nearly saturated, increasing the Ps lifetime to ~ 250 ns because the atom spends half of its time in the nonannihilating $2P$ state. At shorter cooling times, one slows a correspondingly smaller fraction of the initial thermal distribution, emphasizing the importance of starting with as cold an initial thermal distribution as possible.

Due to annihilation, there is clearly an optimum duration of the cooling pulse after which the number of cooled atoms remaining does not increase. However, the long cooling time emphasizes the importance of starting with as cold a Ps source as possible. The long cooling pulses required are not trivial to generate. One possibility is to frequency-double the 20 ns pulse of an excimer pumped amplifier and stretch the doubled light by a factor of 16 using several stages of optical delay lines.

We estimate a final recoil-limited temperature of < 0.4 K, which gives a second-order Doppler width of only ~ 50 kHz, and a transit time broadening in a c.w. laser beam similar to the one used in the present measurement of ~ 300 kHz. Of course, the linewidth of the transition would be dominated by the 140 ns ground-state lifetime. The 20-fold increase in interaction time over the present experiment would allow a c.w. beam intensity of only 100 kW/cm^2 to be used, resulting in an AC Stark shift of the Ps line center of only 1 MHz. It is possible that the 1 MHz resonance width could then be split by a factor of 10 to 100 by a careful study of the systematic effects.

REFERENCES

[1] O. Carnal and J. Mlynek: *Phys. Rev. Lett.*, **66**, 2689 (1991).
[2] D. W. Keith, C. R. Ekstrom, Q. A. Turchette and D. E. Pritchard: *Phys. Rev. Lett.*, **66**, 2693 (1991).
[3] F. Riehle, T. Kisters, A. White, J. Helmecke and C. J. Bordé: *Phys. Rev. Lett.*, **67**, 181 (1991).
[4] M. Kasevich and S. Chu: *Phys. Rev. Lett.*, **69**, 1741 (1992).

[5] N. DAVIDSON, H. J. LEE, M. KASEVICH and S. CHU: *Phys. Rev. Lett.*, in press (1994).
[6] M. KASEVICH, D. S. WEISS, E. RIIS, K. MOLER, S. KASAPI and S. CHU: *Phys. Rev. Lett.*, **66**, 2297 (1991).
[7] K. MOLER, D. S. WEISS, M. KASEVICH and S. CHU: *Phys. Rev. A*, **45**, 342 (1992).
[8] M. KASEVICH and S. CHU: *Phys. Rev. Lett.*, **67**, 181 (1991); *Appl. Phys. B*, **54**, 321 (1992).
[9] Y. V. BAKLANOV, V. P. CHEBOTAYEV and B. Y. DUBETSKI: *Appl. Phys.*, **9**, 201 (1976).
[10] CH. J. BORDÉ: *C. R. Acad. Sci. Ser. B*, **284**, 101 (1977).
[11] CH. J. BORDÉ, CH. SALOMON, S. AVRILLIER, A. VAN LERBERGHE, CH. BRÉANT, D. BASSI and G. SCOLES: *Phys. Rev. A*, **30**, 1836 (1984), plus earlier references contained in the paper.
[12] V. P. CHEBOTAYEV, B. YA. DUBETSKY, A. P. KASANTSEV and V. P. YAKOVLEV: *J. Opt. Soc. Am. B*, **2**, 1791 (1985), and early references in the paper.
[13] Y. V. BAKLANOV, V. P. CHEBOTAYEV and B. Y. DUBETSKI: *Appl. Phys.*, **11**, 201 (1976). See also R. G. BEAUSOLEIL and T. W. HÄNSCH: *Phys. Rev. A*, **33**, 1661 (1986).
[14] R. P. FEYNMAN and A. R. HIBBS: *Quantum Mechanics and Path Integrals* (McGraw-Hill, New York, N.Y., 1965).
[15] A. V. OVERHAUSER and R. COLELLA: *Phys. Rev. Lett.*, **33**, 1237 (1974).
[16] E. L. RAAB, M. PRENTISS, A. CABLE, S. CHU and D. E. PRITCHARD: *Phys. Rev. Lett.*, **59**, 2631 (1987).
[17] J. DALIBARD and C. COHEN-TANNOUDJI: *J. Opt. Soc. Am. B*, **6**, 2023 (1989); P. J. UNGAR, D. S. WEISS, E. RIIS and S. CHU: *J. Opt. Soc. Am. B*, **6**, 2058 (1989).
[18] D. S. WEISS, E. RIIS, M. KASEVICH, K. MOLER and S. CHU: in *Light Induced Kinetic Effects*, edited by L. MOI, S. GOZZINI, C. GABBANINI, E. ARIMONDO and F. STRUMIA (ETS Editrice, Pisa, 1991), p. 35.
[19] W. ERTMER, R. BLATT, J. L. HALL and M. ZHU: *Phys. Rev. Lett.*, **54**, 996 (1985).
[20] The natural-resonance frequency of the seismometer is 1 Hz.
[21] P. NELSON: *Rev. Sci. Instrum.*, **62**, 2069 (1991); P. SAULSON: *Rev. Sci. Instrum.*, **55**, 1315 (1984).
[22] C. MONROE, H. ROBINSON and C. WIEMAN: *Opt. Lett.*, **16**, 50 (1991).
[23] K. GIBBLE and S. CHU: *Phys. Rev. Lett.*, **70**, 1771 (1993).
[24] M. KASEVICH: Ph. D. thesis, unpublished (1992).
[25] D. S. WEISS: unpublished Ph. D. thesis (1993).
[26] D. S. WEISS, B. C. YOUNG and S. CHU: submitted to *Appl. Phys. B*.
[27] A. H. COMPTON: *Phys. Rev.*, **22**, 409 (1923).
[28] O. R. FRISCH: *Z. Phys.*, **86**, 42 (1933).
[29] J. L. HALL, CH. BORDÉ and K. UEHARA: *Phys. Rev. Lett.*, **37**, 1339 (1976).
[30] See, for example, J. R. S. VAN DYCK, F. L. MOORE, D. L. FARNHAM and P. B. SCHWINBERG: *Int. J. Mass Spectrom. Ion Proc.*, **66**, 327 (1985); A. H. WAPSTRA, G. AUDI and R. HOEKSTRA: *At. Data Nucl. Data Tables*, **38**, 290 (1988).
[31] A. G. KLEIN, G. I. OPAT and W. A. HAMILTON: *Phys. Rev. Lett.*, **50**, 563 (1983).
[32] See T. KINOSHITA, Editor: *Quantum Electrodynamics* (World Scientific, Singapore, 1990).
[33] S. SWARTZ, J. L. HALL, K. E. GIBBLE and D. S. WEISS: to be published.
[34] D. S. WEISS, B. YOUNG and S. CHU: *Phys. Rev. Lett.*, **70**, 2706 (1993).
[35] C. J. BORDÉ, M. WEITZ and T. W. HÄNSCH: in *Laser Spectroscopy XI*, edited

by L. BLOOMFIELD, T. GALLAGHER and D. LARSON (American Institute of Physics, New York, N.Y., 1944), p. 76.
[36] M. KASEVICH: private communication.
[37] E. KLEMPT: in *The Hydrogen Atom*, edited by G. F. BASSANI, M. INGUSCIO and T. W. HÄNSCH (Springer-Verlag, Pisa, 1988), p. 211.
[38] S. CHU, A. P. MILLS jr. and J. L. HALL: *Phys. Rev. Lett.*, **52**, 1689 (1984).
[39] S. CHU, A. P. MILLS jr., A. G. YODH, K. NAGAMINE, H. MIYAKE and T. KOGA: *Phys. Rev. Lett.*, **60**, 101 (1988).
[40] M. S. FEE, A. P. MILLS, S. CHU, E. D. SHAW, K. DANZMANN, R. J. CHICHESTER and D. M. ZUCKERMAN: *Phys. Rev. Lett.*, **70**, 1397 (1993).
[41] E. E. SALPETER and H. A. BETHE: *Phys. Rev.*, **84**, 1232 (1951).
[42] J. SCHWINGER: *Proc. Natl. Acad. Sci. U.S.A.*, **37**, 452 (1951).
[43] W. E. CASWELL and G. P. LEPAGE: *Phys. Rev. A*, **18**, 810 (1978).
[44] R. BARBIERI and E. REMIDDI: *Nucl. Phys. B*, **141**, 413 (1978).
[45] R. A. FERREL: *Phys. Rev.*, **84**, 858 (1951).
[46] T. FULTON and P. C. MARTIN: *Phys. Rev.*, **95**, 811 (1954).
[47] T. FULTON: *Phys. Rev. A*, **26**, 1794 (1982).
[48] S. N. GUPTA, W. W. REPKO and C. J. SUCHYTA III: *Phys. Rev. D*, **40**, 4100 (1989); J. R. SAPERSTEIN and D. R. YENNIE: in *Quantum Electrodynamics*, edited by T. KINOSHITA (World Scientific, Singapore, 1990), p. 560.
[49] R. N. FELL: *Phys. Rev. Lett.*, **68**, 25 (1992).
[50] I. B. KHRIPLOVICH, A. I. MILSTEIN and A. S. YELKHOVSKY: *Phys, Lett. B*, **282**, 237 (1992). An initial disagreement with ref. [49] has been resolved in favor of ref. [49] (R. N. FELL: private communication).
[51] M. FEE, S. CHU, A. P. MILLS, R. J. CHICHESTER, D. M. ZUCKERMAN, E. D. SHAW and K. DANZMANN: *Phys. Rev. A*, **48**, 192 (1993).
[52] S. CHU and A. P. MILLS jr.: *Phys. Rev. Lett.*, **48**, 1333 (1992).
[53] D. H. MCINTYRE and T. W. HÄNSCH: *Phys. Rev. A*, **22**, 192 (1986).
[54] M. S. FEE, K. DANZMANN and S. CHU: *Phys. Rev. A*, **45**, 4911 (1992).
[55] C. WIEMAN and T. W. HÄNSCH: *Phys. Rev. A*, **22**, 5791 (1991).
[56] A. P. MILLS jr., E. D. SHAW, R. J. CHICHESTER and D. M. ZUCKERMAN: *Rev. Sci. Instrum.*, **60**, 825 (1989).
[57] R. W. P. DREVER, J. L. HALL, F. V. KOWALSKI, J. HOUGH, G. M. FORD, A. J. MUNLEY and H. WARD: *Appl. Phys. B*, **31**, 97 (1983).
[58] M. FEE: unpublished Ph. D. thesis.
[59] G. C. BJORKLUND: *Opt. Lett.*, **5**, 15 (1980).
[60] J. L. HALL, L. HOLLBERG, T. BAER and H. G. ROBINSON: *Appl. Phys. Lett.*, **39**, 680 (1981).
[61] E. W. WEBER: in *Present Status and Aims of Quantum Electrodynamics*, edited by G. GRÄFF, E. KLEMPT and G. WERTH, Vol. **143** (Springer-Verlag, Berlin, 1981), p. 146.
[62] R. J. HUGHES and B. I. DEUTCH: *Phys. Rev. Lett.*, **69**, 578 (1992).
[63] A. P. MILLS jr.: personal communication.
[64] J. HOFFNAGLE: *Opt. Lett.*, **13**, 102 (1988).
[65] E. P. LIANG and C. D. DERMER: *Opt. Commun.*, **65**, 419 (1988).
[66] H. WALLIS and W. ERTMER: *J. Opt. Soc. Am. B*, **6**, 2211 (1989).
[67] S. E. HARRIS and R. TARG: *App. Phys. Lett.*, **5**, 202 (1964).
[68] D. M. KANE, S. R. BRAMWELL and A. I. FERGUSON: in *Laser Spectroscopy VII*, edited by T. W. HÄNSCH and Y. R. SHEN (Springer-Verlag, Berlin, 1985), p. 362.

ION TRAPS

Laser Stabilization to a Single Ion.

J. C. BERGQUIST, W. M. ITANO and D. J. WINELAND

Time and Frequency Division, National Institute of Standards and Technology
Boulder, CO 80303

I. – Spectrally narrow Optical Oscillators.

I.1. – Frequency references.

An unperturbed optical resonance in an atom or molecule provides a good absolute frequency reference, but practically there are limitations. The signal-to-noise ratio is limited by the number of atoms in the interrogation region and by saturation of the resonance. If the probe laser is spectrally broad, then the measured atomic line is broadened, which degrades the stability. Furthermore, atoms recognize interactions and collisions with neighboring atoms, usually with shifts to the internal energy level structure. The motion of the atoms also produces Doppler shifts and broadening. A single laser-cooled atom removes many of these problems but only with a severe penalty in the signal-to-noise ratio [1]. Even so, by detecting each transition in the single atom, it should be possible to stabilize the frequency of a laser oscillating in the visible to better than $10^{-15}\,\tau^{-1/2}$ with an accuracy approaching 10^{-18} [2], if the laser were sufficiently stable (or spectrally narrow) for times that are comparable to the interrogation time of the transition in the atom.

A two-step approach might then be appropriate; spectrally narrow the laser by some scheme that offers good short- to medium-term stability (*e.g.*, 1 ms to 10 s), then stabilize the frequency of the laser for longer times to a narrow resonance in a single atom. The reference for the short- to medium-term stabilization of the laser not only needs good stability in this time frame, but also low phase and frequency noise. If a Fabry-Perot cavity is used as the frequency discriminator, then the response of the frequency discriminator can be nearly linear as a function of power and the signal-to-noise ratio can be high [3]. Frequency shifts and fluctuations arise due to thermal distortion in the mirror coatings caused by absorption of light in a small volume in the dielectric stack and due to photochemical processes at the surfaces of the coatings. Also, practical limits are reached at powers that saturate the detector or at powers (circulating in the cavity) that cause radiation pressure noise, but the signal-to-noise ratio from a cavity can be many orders of magnitude larger than that obtained with atoms. Although there are other types of optical frequency references, the most widely used is the Fabry-Perot interferometer, principally because the frequency excursions to error-voltage can be extremely large in a high-finesse cavity. We will spend time in the next few sections discussing some of the details and limits of a suitably stable reference cavity, and then turn our attention to the single atom.

I.2. – Reference cavity limitations.

We can begin with a brief look at the demands that a spectral purity of 1 Hz places on the physical stability of the reference cavity. If the cavity has a length of 30 cm and the optical wavelength is 500 nm, then the optical-path length between the mirrors must not change by more than 10^{-15} m, the approximate size of the nucleus of any of the constituent atoms in the dielectric coatings. Researchers who study parity-nonconserving interactions in atomic systems sometimes use the analogy that a human hair added to the radius of the Earth is equivalent to the distortion in an atom caused by the parity-nonconserving part of the Hamiltonian. By the same taken, if the spacer for an optical cavity were the Earth, a human hair added to the diameter would cause a frequency shift of about 300 Hz! In the first part of this section, we investigate some of the fundamental limits to the attainment of an average spacing between two mirrors that is stable to better than $1 \cdot 10^{-15}$ and to the achievement of a laser that is spectrally narrowed to better than 1 Hz. We will also address some of the limitations imposed by various environmental factors. We will see that, although fundamental limits come from quantum mechanics and thermodynamics, important practical limitations come from mechanics and gravitational coupling to the noisy terrestrial environment. In particular, at low Fourier frequencies (< 100 Hz), seismic noise and pendulum motion dominate the noise budget.

At the quantum level, the measurement of the length of the cavity to which the laser is locked brings about the inevitable competition between the measurement precision and the perturbation of the measurement to the system. The measurement precision can be improved by a factor of $1/\sqrt{N}$ by increasing the number N of (signal) photons in the measurement interval, whereas the shot noise of the radiation pressure on the mirrors increases as $\sqrt{N}$. The optimum flux of signal photons, or input power (assuming 100% of the light is coupled into the cavity on resonance and that the detection efficiency is unity), is given when both effects are equal in magnitude [4]. For the laser interferometers used in gravity wave detection, the mirrors are suspended as pendulums and the optimum power is calculated for frequencies $\omega/2\pi$ well above the resonance frequency $\omega_0/2\pi$ of the pendulum support. As a function of frequency, this limit length uncertainty is given by [5]

$$\Delta x \approx (4\hbar \Delta f/m)^{1/2}/\omega \,, \tag{1}$$

where m is the mass of suspended mirrors and Δf is the measurement bandwidth. Interestingly, this is the same measurement precision allowed by the standard quantum limit [4-7]. In our case the resonator is composed of a single bar, or spacer, to which the mirrors are rigidly attached. The resonator is then suspended, often by small-diameter wires, inside a temperature-regulated, evacuated housing. The resonator bar can also be treated as a harmonic system, but now the frequency of the lowest mechanical resonance frequency $\omega_0/2\pi$ is typically greater than a few kHz. The noise spectrum of length fluctuations that is of particular importance to us is at Fourier frequencies that are below the lowest resonance frequency of the bar. At frequencies lower than $\omega_0/2\pi$ and for optimum power, the measurement precision limit is independent of Fourier frequency

$$\Delta x \approx (4\hbar \Delta f/m)^{1/2}/\omega_0 \,. \tag{2}$$

For $\omega_0/2\pi = 10\,\text{kHz}$, a cavity finesse of 100000 at $\lambda = 500\,\text{nm}$ and a near optimum input power of about 3 W, this limit corresponds to a fractional length uncertainty of less than $(5 \cdot 10^{-22}\ m/\sqrt{\text{Hz}})(\Delta f)^{1/2}$ for a typical bar mass of 4 kg. Thus, with as little as 100 μW incident on the cavity, the quantum fluctuations in the radiation pressure acting on the mirrors are negligible, yet there is sufficient signal-to-noise ratio that the frequency of a laser can be made to track the resonance of the cavity to well below 1 Hz [8] (if only limited by the shot noise of the detected signal).

The technical fluctuations in the laser light that is coupled into the cavity must also be considered. If the finesse is 100000 and the corresponding power enhancement as high as 30000, then 100 μW of input power translates into 3 W of *circulating* power when the laser is resonant with the cavity. This in turn gives a force W on each mirror of about $2 \cdot 10^{-8}$ N. If the mirrors are treated as clamped disks of thickness t, the deflection of each mirror, δ, is given by [9]

$$\delta = 3Wr^2(1-\rho^2)/4\pi E t^3 \,, \tag{3}$$

where the light force is assumed to act uniformly over a concentric area much less than the area of the mirror. The radius r of the mirror is measured from the center to the clamped edge, ρ is Poisson's ratio and E is Young's modulus of elasticity. For a mirror substrate with a material composition similar to fused silica, ρ is about 0.17 and E is about $7 \cdot 10^{10}\ \text{N/m}^2$. If t is 5 mm and r is 10 mm, then the deflection at the center of the mirror is about $1.7 \cdot 10^{-16}$ m for a radiation mode size (w_0) of 200 μm. For a cavity that is 30 cm long, the corresponding fractional length change ($2\delta/L$) is about $1 \cdot 10^{-15}$, or about 0.5 Hz. The dimensions and physical properties of the resonator used in this example are typical. Cavities have been constructed that have been shorter, that have used thinner mirrors, and that have coupled in more light; all of these conspire to degrade the frequency stability through fluctuations of the intracavity light intensity. In our example, if 1 mW of power is used to stabilize the laser to the cavity, then the amplitude of the circulating light must be constant to about 10% to achieve a laser stability better than 1 Hz. The radiation pressure also works to stretch the entire bar. The strain, or fractional length change, induced by a force acting normal to the end of a cylindrical bar of cross-sectional area A is given by [9]

$$\Delta L/L = F/AE \,. \tag{4}$$

The fractional length change for 3 W of circulating power is about $2.5 \cdot 10^{-17}$. This is smaller than the elastic distortion of our typical mirror, and even power fluctuations as large as 10% would cause only millihertz frequency fluctuations through this term. The clear indication is that the fluctuations in the circulating power must be controlled if the laser frequency is to be stabilized to much better than 1 Hz. Cavity power fluctuations are caused by intensity fluctuations of the laser light external to the cavity and by variations in the amount of light that is coupled into the cavity. The latter fluctuations are caused, for example, by any motion of the resonator with respect to the input beam.

Another problem is the local heating of the dielectric mirror coating from the light circulating in the cavity. With high-finesse low-loss mirror coatings, one might assume that this would not be an important concern. However, as we have seen, even with as little as 100 μW of power coupled into the cavity that has a finesse of 100 000, the circulating power inside the cavity can exceed 3 W. If the absorption losses are as little as a few p.p.m., $(5 \div 10)$ μW are absorbed in the coating. Most of this power is absorbed in the first few layers of the dielectric stack where the light intensity is the highest. When the light amplitude fluctuates, there is a transient response followed by relaxation to a steady-state condition. For a radiation mode size of 200 μm, we have measured a 2 Hz/μW shift of the cavity resonance to higher frequencies with increased power. The magnitude has been measured to be as much as 20 Hz per μW change in the power of the input coupled light [10]. Both the thermal distortion of the mirror and the light pressure problem could be reduced by adjusting the mirror radii and the cavity length so that the mode size is larger at the mirrors (for example, by using a near-spherical resonator).

The thermal noise in the bar or spacer must also be considered. We can think of this as the weak coupling of the fundamental mode of the spacer to its environment, *e.g.*, the residual background gas, the wire suspension, radiation from the walls of the vacuum vessel, etc. If the bar is thermalized to this background or thermal bath at some physical temperature, T, then the weak coupling to the thermal bath causes the mode's amplitude to execute a random walk in the domain $|\Delta x| \leqslant \chi_{\text{r.m.s.}}$. $\chi_{\text{r.m.s.}}$ is the average deviation of one end of the spacer from the equilibrium position assuming the other end is fixed. The magnitude of this deviation can be found by equating the energy in the harmonic motion of the fundamental mode to $k_B T$. After rearranging, this gives

(5) $$\chi_{\text{r.m.s.}} = \{2k_B T / M_{\text{eff}} \omega_0^2\}^{1/2},$$

where k_B is Boltzmann's constant and M_{eff} is the effective mass of the spacer. The fluctuation-dissipation theorem states that the time scale on which this random walk produces changes of order $\chi_{\text{r.m.s.}}$ is the same as the time scale given by the decay time of the fundamental mode [11]. The decay time τ_0 is related to the quality factor of the fundamental mode by $\tau_0 = 2Q/\omega_0$. Q is inversely proportional to the fractional energy loss per cycle. For times τ shorter than τ_0, the mean-square change in the mode's amplitude is $\chi_{\text{r.m.s.}}$ reduced by the ratio τ/τ_0, $\Delta x = \chi_{\text{r.m.s.}}\ \tau/\tau_0$ [12]. Physically, this expresses the fact that a harmonic oscillator is a tuned system that responds to noise and other perturbations in a narrow range of frequencies. So, while the noise is proportional to temperature, the fluctuating forces cannot appreciably change the mode amplitude in times short compared to the decay time.

To reach an appreciation of the size of the length fluctuations due to thermal noise, we can calculate the frequency of the lowest mode of the spacer and solve for $\chi_{\text{r.m.s.}}$. Alternatively, since M_{eff} is only estimated, we can equate the work necessary to stretch or compress a cylindrical bar by $\chi_{\text{r.m.s.}}$ to the energy $k_B T$. The force necessary to elastically stretch a bar by a small amount x is given by eq. (4),

(6) $$F = EAx/L.$$

$F\,\mathrm{d}x$ is the incremental work done by this force in going from x to $x + \mathrm{d}x$. Integrating from $x = 0$ to $x = \chi_{\text{r.m.s.}}$, the work done is

(7) $$W = EA(\chi_{\text{r.m.s.}})^2/2L.$$

When this is equated to $k_B T$, $\chi_{\text{r.m.s.}}$ is simply related to the temperature of the spacer and to its physical properties,

(8) $$\chi_{\text{r.m.s.}} = \{2Lk_B T/EA\}^{1/2}.$$

Recall that, for a spacer made from a material comparable in its properties to fused silica, E is about 10^{11} N/m^2. If the spacer is 30 cm long and 10 cm in diameter, then $\chi_{\text{r.m.s.}} \approx 1.8 \cdot 10^{-15}$ m at a thermal-bath temperature of 300 K. The frequency of a laser in the mid-optical locked to this cavity would move by about 3 Hz for a length change of this magnitude. However, this excursion occurs dominantly at the vibrational frequency of the lowest fundamental (mechanical) mode of the spacer. For a fused-silica spacer of this size, the resonance frequency of its lowest mode is about 10 kHz. Hence, there is little power in the thermally induced sidebands at ± 10 kHz since the modulation index is so small [13] ($\approx 3 \cdot 10^{-4}$). It is worth noting that the resonant Q in a fused-silica bar at room temperature is only about 10^5 [14]. Therefore, the decay time for the lowest mode is about 10 s, and, unlike our colleagues looking for gravity waves with high-Q resonant-bar detectors, we are sensitive to the full change in the mode amplitude on time scales of critical interest to us. However, a few hertz at 10 kHz driven by the Brownian motion of the cavity is not a limitation to the frequency stability nor spectral purity of a 1 Hz laser.

The temperature sensitivity of low-expansion materials suitable for spacers can be better than 10^{-8}/K, which implies μK control at the cavity in order to achieve stabilities of a few hertz. However, if the cavity is suspended in a thermally massive, evacuated chamber, then the thermal coupling to the environment is primarily radiative. The time constant can exceed a day. Therefore, if the temperature fluctuations of the walls of the vacuum vessel never exceed 10 mK, then the frequency fluctuation rate of the laser will be less than 1 Hz/s, independent of the time rate of charge of the wall temperature.

Pressure fluctuations in the gas between the mirrors produce density variations which in turn cause refractive-index changes. This causes an effective optical-length change to the cavity, nL, where n is the index of refraction for air. Near room temperature, the index of refraction of dry air is linearly related to its pressure, P, by

$$n - 1 \approx 3 \cdot 10^{-9}\, P, \tag{9}$$

where P is in Pa (133 Pa = 1 Torr). Thus, even if the cavity is evacuated to 10^{-5} Pa, the absolute shift in the cavity resonance for optical frequencies near $\lambda \approx 500$ nm is about 15 Hz (from that of 0 Pa). 10% fluctuations in this pressure cause frequency excursions of the laser of approximately 1.5 Hz. The air pressure at 10^{-5} Pa also causes a strain in the bar but the length compression is negligible. The pressure at 10^{-5} Pa is less than 10^{-5} N/m^2, which produces an axial strain in the spacer of about $1 \cdot 10^{-16}$. The fractional length change is not fully this magnitude since the pressure-induced stress in the axial direction is somewhat compensated by the radial stress. Consequently, in a reasonably stiff spacer, fluctuations in the cavity length due to fluctuations in the pressure are dominated by index-of-refraction changes.

Additional limitations to the stability of a laser locked to a Fabry-Perot resonator come from technical problems such as optical feedback, intensity fluctuations, beam pointing stability, etc. [15], but seismic noise or ambient vibrations that cause changes in the cavity length are the most important practical problems limiting the frequency stability for times longer than a few ms. Generally, there are two distinct effects: high-frequency vibrations, which may excite fundamental mechanical resonances of the bar, and low-frequency vibrations, which tend to produce nonresonant distortions of the bar. The first effect typically occurs at frequencies in the range of $(100 \div 1000)$ Hz and often can be effectively eliminated, for example, by mounting the system on alternating layers of shock-absorbing material and passive masses, such as thin rubber and cinder block. Low-frequency vibrations ($(0.1 \div 100)$ Hz), which are typically driven by ground noise and building vibrations, are much harder to eliminate. Some sophisticated vibration isolation systems, both active and passive, have been developed to reduce noise in this frequency interval [16]. While an active vibration isolation system may ultimately be necessary for sufficient attenuation of seismic noise to reach laser spectral purity below 1 Hz, we, for the moment, isolate only with passive systems. The simplest method of passive vibration isolation consists of mounting the device to be isolated on a resilient support, such as a pendulum or spring. A pendulum isolates in the horizontal plane, using gravity as its spring constant (note that this spring can be nearly lossless). A spring can isolate horizontally and vertically simultaneously. It is relatively simple to show that the attenuation in the amplitude of motion between the support and the bar increases as ω^2 for motion at frequencies higher than the resonance frequency ω_0 ($\omega_0^2 = k/m$, spring; $\omega_0^2 = g/l$, pendulum). Damping must be included to limit the amplitude of motion, a_1, at resonance. The response of a damped system is

$$a_1 = a_2(1 + i\Gamma\omega)/\{1 - (\omega/\omega_0)^2 + i\Gamma\omega\}, \tag{10}$$

where a_1 is the amplitude of the motion at the bar, a_2 is the amplitude of motion of the support at frequency ω, and Γ is the coefficient of damping. The amplitude a_1 is complex (phase shifted) and everywhere bounded. In practice the choice of damping is a compromise between low resonant amplitude and sufficient high-frequency isolation. There are more complex passive systems that offer better high-frequency isolation while at the same time reduce the resonance peaking to a factor of 2 or less [16]. Calculations of the vibration isolation produced by various mechanical suspensions have been driven by the work done on gravity wave detectors; a good treatment is that of Veitch [17].

The most important function of the isolation system is to reduce fractional length changes of the reference cavity to below (ideally, well below) $1 \cdot 10^{-15}$. If the cavity is suspended with its axis horizontal by wires that act as vertically stiff pendulums, then it has been demonstrated that the isolation from horizontal vibrations in the direction of the cavity axis can be adequate to attain a stability approaching 1 Hz [18]. However, a suspended bar is subjected to a distributed load resulting from the pull of gravity, which produces considerable stress to the support structure and to the bar. In addition, the bar is not perfectly rigid and must distort at some level under its weight and this causes additional stress. All these stresses can produce sudden acoustical emission [19] at the rate of up to several per second and at frequencies that may or may not coincide with the eigenfrequencies of the bar. Also, since the wires are essentially stiff to vertical vibrations, these perturbations can be coupled into the bar. If, though, the wires were connected to the nodal points of the bar, then the external force would be unable to excite that mode (or modes) through that nodal position. Unfortunately, there are many bending modes and vibrational modes with disparate nodal positions, so, although it is possible to reduce some of the cavity sensitivity to vertical perturbations, single suspension points are not sufficient to eliminate excitation of all modes. Further improvement can be achieved if the cavity is isolated vertically as well. We now turn our attention to some of the experimental studies pursued over the past few years in the ion storage group at NIST.

I.3. – **Experimental results: cavity comparisons.**

In some of our studies, a direct heterodyne comparison of two independently frequency-stabilized light beams was made. The linewidths and frequency stabilities were analyzed in various ways. The beat note was recorded in an r.f. or microwave spectrum analyzer, thereby directly displaying the combined linewidth of the two sources. The signal-to-noise ratio was improved and the long-term relative frequency drift was studied by averaging many successive scans of the beat note. It was also possible to use two spectrum analyzers in tandem to identify the specific frequency noise components that contribute to the laser linewidth. This was particularly useful toward unraveling the noise sources that cause fluctuations to the length of the cavities. For instance, if the seismic noise was independently studied with a seismometer, a correspondence between the seismic-noise terms and the dominant contributors to the laser linewidth could be made. If the length fluctuations of the two cavities were similar in frequency and amplitude, but otherwise independent, the linewidth of either frequency-stabilized system is smaller than the recorded beat note by $\sqrt{2}$. If the effective length stability of one cavity had been worse than that of the other, the measured linewidth would have been be dominated by the frequency fluctuations of the laser locked to the noisier cavity.

Either the transmitted light beam or the reflected beam that interferes with the light re-emerging from the cavity can be used to stabilize the frequency of the laser. In our experiments, we used the beam reflected from the cavity. The error signal can be derived either near zero frequency or at some higher frequency. Since technical-noise terms are present at low frequencies, it is better to modulate and detect at a frequency at which the signal-to-noise ratio is limited by the shot noise in the detected light beam. This is the Pound-Drever-Hall reflected-sideband technique that has been treated in detail elsewhere [8, 20]. Attention to the optical layout and to the electronic-noise terms is critical to achieving a spectral linewidth that is smaller than a few hertz. This also has been discussed in papers by HOUGH *et al.* [21] and by SALOMON *et al.* [15]. By several separate measurements, the electrical problems in our work were determined to be unimportant to the attainment of a laser linewidth below 1 Hz; the dominant contribution to the laser linewidth were mechanical (and perhaps optical) perturbations that caused length changes in the reference cavity.

The frequency-stabilized light beams were derived from a home-built ring dye laser oscillating at 563 nm. The wavelength of the laser was chosen because its frequency would eventually be doubled into the ultraviolet to probe a narrow transition in $^{199}Hg^+$. Historically, the dye laser was locked in a two-step process. The laser was pre-stabilized to the order of a few hundred Hz by locking the laser to a lower-finesse (about 800) cavity. The cavity resonance was probed by the reflected-sideband technique using a modulation frequency of about 10 MHz. Rapid frequency fluctuations of the laser were removed by a fast (bandwidth > 2 MHz) servo driving an intracavity E/O modulator, and the lower-frequency fluctuations were corrected by a second-order servo-loop driving an intracavity PZT-mounted mirror. The frequency-stabilized light was then transmitted through optical fibers to the high-finesse cavities. Since the laser linewidth was less than 1 kHz, a smaller-bandwidth (100 kHz), lower-noise servo could be used to strip the remaining noise from the laser beams. Again, the reflected-sideband technique was used to probe the resonance of the high-finesse cavities. In the second stage, the frequency-correcting element was an acousto-optic modulator through which the laser beam was singly or doubly passed. The frequency corrections were simply written onto the acousto-optically shifted beam by the servo. Long-term corrections were fed back to the low-finesse cavity to maintain frequency alignment of the laser with one of the high-finesse cavities. The laser power was also stabilized by adjusting the r.f. power applied to the acousto-optic modulator. The overall intensity regulation was better than 0.1%, but this was applied to the beam before it entered the cavity. From the arguments made in sect. I.2, intensity variations in the light circulating in the cavity cause length fluctuations by light pressure changes and by heat variations in the mirror coatings. Therefore, although we have not done so yet, it may be better to stabilize the intensity of the light circulating in the resonator by using the light transmitted from the cavity. This should give a first-order insensitivity of the frequency of the laser to power fluctuations caused by relative motion between the cavity and the injected light beam.

The cavities were constructed from a cylindrical, Zerodur [22] rod that was cut and rough ground to a diameter of about 10 cm and a length of about 27 cm. A 1 cm round hole was bored along the axis of the spacers, and a smaller one was bored through the center of the rod midway from the ends and normal to its axis. The smaller bore permitted evacuation of the space between the mirrors. The length of the cavities and the mirror radii of curvature were chosen so that the cavities were highly nondegenerate for the spatially transverse modes. In particular, the frequency of the TEM_{01} and TEM_{10} modes are separated from the lowest-order TEM_{00} mode by approximately 30% of the free spectral range.

Even at the 12th transverse order, the frequencies are still separated from the lowest-order fundamental mode (modulo $c/2L$) by several percent of the free spectral range. This gives good immunity to any line pulling if any light power is coupled into the higher-order modes. Even so, great care is taken to mode match into the cavities. Note that this high immunity to line pulling would be somewhat compromised by going toward a near spherical resonator as suggested in sect. I.2.

Each cavity is suspended by two thin (250 μm) molybdenum wires inside a thick-walled (19.1 mm) aluminum vacuum vessel that has an inner diameter of 26.7 cm. The wires are placed as slings under either end of the spacer, about 1/5 the cavity length from each end, in an attempt to support at the nodal positions for the lowest-order bending mode. The ends of each wire are attached to the inner wall of the vacuum vessel by small clamps. Each wire either travelled vertically upward from either side of the spacer to the wall («U» shaped), or opened slightly away from the bar («V» shaped). This allowed free movement of the cavity along the direction of its axis and restricted, but not rigidly, the motion perpendicular to the axis. Damping of the pendulum motion was weak and dominantly into the table and its padding through the aluminum housing. The aluminum vessels are thermally insulated and temperature regulated to the order of a few mK. The temperature coefficient of the spacers is approximately $6 \cdot 10^{-9}\,K^{-1}$ at $T = 300$ K. The thermal time constant from the walls of the evacuated aluminum housing to the spacer is on the order of one day, giving adequate isolation to small temperature fluctuations at the vessel walls. Since the variations in the wall temperature were controlled to less than 10 mK for any time period, the maximum rate of change in the temperature of the bar did not exceed 100 pK/s. This corresponds to a frequency fluctuation rate of about 0.3 Hz/s and a maximum frequency change of about 50 kHz. A pressure of $1 \cdot 10^{-6}$ Pa ($8 \cdot 10^{-9}$ Torr) is maintained in each vacuum vessel by an ion pump that is rigidly attached to the vessel. This is adequate to give frequency fluctuations of less than 1 Hz for pressure fluctuations of 10%.

The length of the longer rod corresponds to a free spectral range of 622 MHz and the shorter rod has a free spectral range of 562 MHz. The mirrors are highly polished Zerodur substrates that are coated to give high finesse and good efficiency and then optically contacted to the polished ends of the spacer. The finesses of the cavities are about 60 000 and 90 000, respectively, and, for both cavities, the transmission on resonance exceeded 30%. The high finesse F, or the long optical storage time, translates into a narrow fringe whose HWHM is given by $c/2LF$. Consequently, the ratio of error voltages to frequency excursions can be very high, even for short cavities. Shorter cavities have been constructed from ULE [22] that have measured finesses that exceed 130 000 (and, recently, finesses exceeding 10^6 have been reported [23]). The frequencies of the mechanical resonances of these shorter, stiffer bars will be about a factor of 3 higher than those of the Zerodur bars. As long as the frequencies of the mechanical vibrations are high enough, a narrow optical resonance can be probed with high resolution by the unperturbed carrier of the laser spectrum.

The aluminum vacuum vessels rested on Viton [22] rubber strips attached to v-blocks made of aluminum. The v-blocks were secured to a rigid acrylic plastic plate. Each reference cavity system was mounted on a separate optical table that (initially) consisted of surface plates that were deadened by damping their internal vibrations into sand. The sandbox sat on soft rubber pads and cinder blocks in one case and on low-pressure inner tubes and cinder block legs in the second. Noise vibrations from the floor and on the table tops were monitored with moving-coil seismometers that had a sensitivity of 629 V·s/m from approximately (1 ÷ 100) Hz. Measurements of the floor motion in our laboratory revealed bright resonances at 14.6 Hz, 18.9 Hz and 29.2 Hz on top of a broad pedestal from about 1 to 40 Hz. The average amplitude of the resonant motion was greater than 10^{-6} m/s and the pedestal peak was about 10^{-7} m/s. Isolation from mechanical vibrations began at frequencies above about 5 Hz for both table tops. By 100 Hz the isolation from floor noise for either table had improved by a factor of 40 or better.

For heterodyning, a small fraction of the frequency-stabilized light from each cavity was combined on a beam splitter and detected with a fast diode. The heterodyned signal was amplified and analyzed with two spectrum analyzers used in series. This allowed us to look directly at the beat note and also Fourier-analyze the noise terms that contributed to its linewidth. The first analyzer could be used to observe the beat signal, or as a frequency discriminator. As a frequency discriminator, the scan was stopped and the analyzer was used as a tunable r.f. filter/receiver with a variable bandwidth. The center frequency of the analyzer was then shifted so that the heterodyned signal lay at the half-power point of the response curve. The bandwidth was adjusted so that the frequency excursions of the beat signal were much less than the bandwidth. This produced a one-to-one map of frequency excursions-to-voltage excursions whose Fourier power spectrum could be analyzed in the second low-frequency ((0 ÷ 100) kHz) spectrum analyzer. The noise power spectrum in the second analyzer helped to reveal the nature and origin of the vibrational noise that contributes to the linewidth of the stabilized lasers.

The width of the heterodyne signal between the two stabilized lasers was less than 50 Hz. The noise power spectrum disclosed that low-frequency fluctuations in the range from near zero to 30 Hz dominated this linewidth. The vibrational-noise spectrum measured by the seismometers on the table tops matched the largest noise components of the beat note. The frequencies of the pendulum motion of the suspended cavities were about 1.4 Hz and 1.48 Hz which gave FM at these frequencies. There were also bright features in the laser power spectra that came from the floor motion at 14.6 Hz, 18.9 Hz and 29.2 Hz. The modulation indices of the latter three noise components were about one, so they all had enough power to contribute to the beat note linewidth. When the pendulum motions of the bars were quiet, their FM contributed little to the laser linewidths. However, the integral of the nearly featureless noise power spectrum from about 0 Hz to 10 Hz contributed about 15 Hz to the combined spectral purity of the lasers. Some, if not all, of this noise was mechanical, but it was not clear how it coupled to the suspended cavity. To help elucidate the connection, we drove one of the table tops in either the horizontal or vertical direction with a small loudspeaker connected to an audio signal generator. The motion of the speaker diaphragm was coupled to the table by a rod glued to the diaphragm and gently loaded against the table. The table could be driven at frequencies from a few hertz to about 100 Hz with enough power to be 40 dB above background noise. When the loudspeaker drove the table in the horizontal plane in a direction parallel to the axis of the cavity, the isolation of the suspended cavity was sufficiently good that the beat signal showed little evidence of the perturbation even at the high drive levels. However, when the drive was applied vertically at a level barely perceptible above the vertical background noise, the heterodyne signal showed added noise power at the drive frequency. The stiff support in the vertical direction strongly coupled vertical motion into effective cavity length changes. The sensitivity of the Fabry-Perot cavity to vertical motion was orders of magnitude higher than for horizontal motion parallel to the

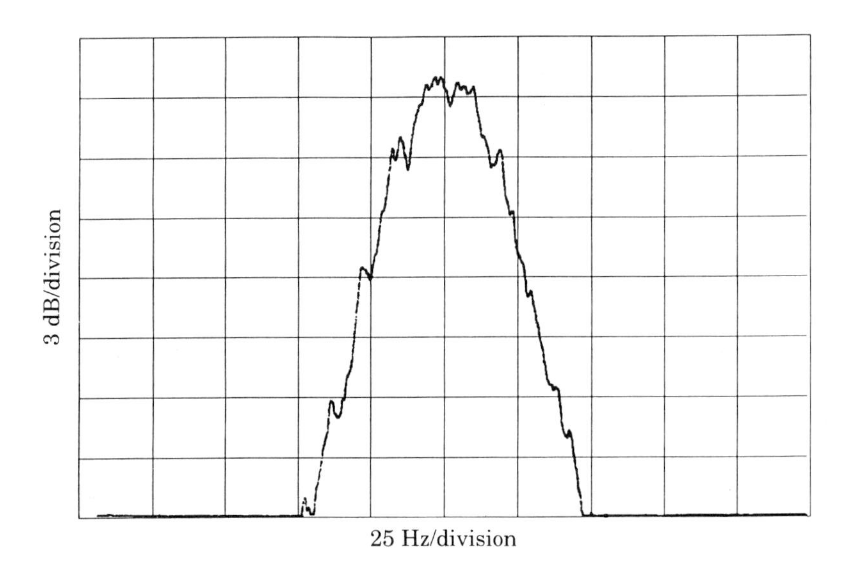

Fig. 1. – Spectrum of the beat frequency between the two independent, cavity-stabilized lasers discussed in the text. The resolution bandwidth is 30 Hz. Total integration time for these data is about 70 s (the relative linear drift between the two cavities is removed by mixing the beat note with the frequency from a synthesizer that is swept in time). The partially resolved sidebands at 14.6 Hz are due to a resonant floor vibration. The apparent linewidth is about 30 Hz; the linewidth of the better-stabilized laser is at least $\sqrt{2}$ narrower.

cavity axis. (For practical reasons, we did not drive the table in the horizontal plane in a direction perpendicular to the cavity axis.)

In order to improve the vertical isolation, one table was suspended just above the floor with latex-rubber tubing attached to the ceiling. The resonance frequencies for both the vertical motion and the horizontal pendulum motion of the suspended table were near 0.33 Hz. These were damped to the floor with two small dabs of grease, but the damping did not significantly change the isolation afforded by the latex tubing at higher frequencies. The isolation from vibrational noise above 1 Hz was more than an order of magnitude better than that of the quietest sandbox table. This was partially reflected in the linewidth of the heterodyne signal between the laser radiation stabilized to the cavity supported on this table and the laser radiation stabilized to the cavity supported on the best sandbox table; it dropped from about 50 Hz to less than 30 Hz. In fig. 1 the spectrum of the beat frequency is shown. Zerodur is known to temporally contract with a time constant of years [8, 24]. The creep rate for both cavities corresponds to a linear frequency drift of $(3 \div 5)$ Hz/s, but the rates are not identical. The linear, relative cavity drift is removed by mixing the beat frequency with the frequency from a synthesizer that was swept in time. The spectrum of fig. 1 represents an integration time of 70 s. The noise vibrational sidebands at 14.6 Hz are partially resolved. Note that the linewidth of the best stabilized laser is at least $\sqrt{2}$ narrower. The Fourier noise power spectrum from 0 to 10 Hz is shown in fig. 2. The beat note linewidth obtained by integrating the power spectrum is about 15 Hz. We suspect that the laser stabilized to the cavity on the sandbox table is the dominant contributor to the width of the heterodyne signal since the vibrational noise measured on this table is greater. Perhaps the linewidth of the laser stabilized to the cavity on

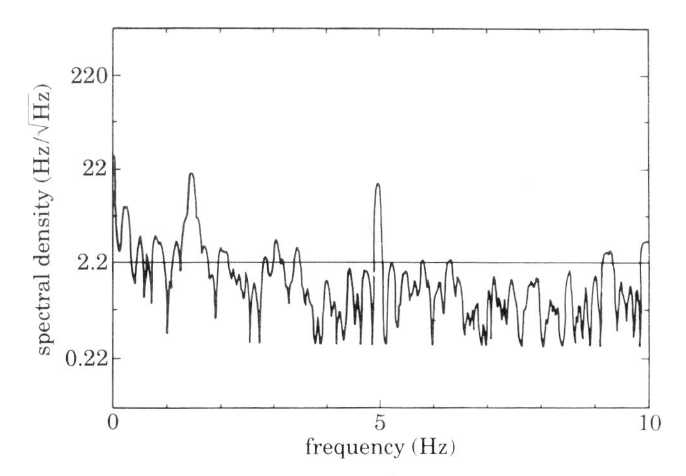

Fig. 2. – Fourier noise power spectrum of the laser heterodyne signal (shown in fig. 1) from 0 to 10 Hz.

the table suspended by the latex tubing is below 1 Hz. We are working to verify this and to build better cavities.

II. – Single-atom Spectroscopy.

II.1. – Single-ion results.

The ion trapping and laser cooling relevant to our experiments have been described elsewhere [25,26]. A ^{199}Hg atom is ionized and trapped in the harmonic pseudopotential well created by an r.f. potential applied between the electrodes of a miniature Paul trap. The separation between the endcap electrodes ($2z_0$) is about 650 μm. The frequency of the r.f. potential is about 21 MHz. Its amplitude can be varied up to 1.2 kV; at the maximum r.f. amplitude, the quadratic pseudopotential is characterized by a secular frequency of nearly 4 MHz. The ion is laser-cooled to a few millikelvin by a few microwatt of radiation from two 194 nm sources. One source drives transitions from the $5d^{10}6s\ ^2S_{1/2}(F=1)$ to the $5d^{10}6p\ ^2P_{1/2}(F=0)$ level (see fig. 3). This is essen-

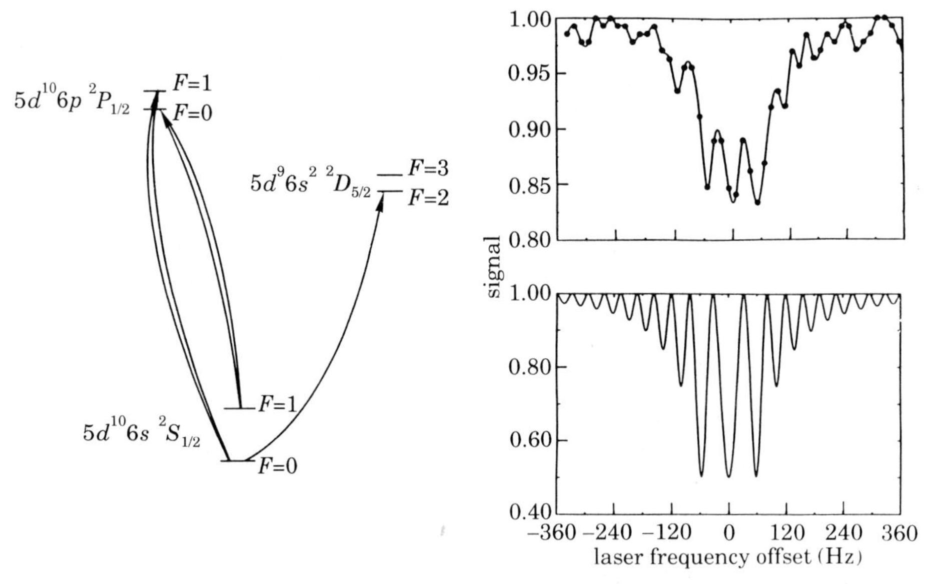

Fig. 3. – On the left is a simplified energy level diagram for $^{199}\mathrm{Hg}^+$ at zero field. Shown in the upper figure on the right is the power-broadened lineshape obtained by scanning through the Doppler-free resonance of the $^2S_{1/2}(F=0,\ m_F=0)$-$^2D_{5/2}(F=2,\ m_F=0)$ transition in a single laser-cooled $^{199}\mathrm{Hg}^+$ ion. A 563 nm laser that is stabilized to a high-finesse reference cavity, which in turn is long-term stabilized to the ion, is frequency-doubled and stepped through the resonance for 138 consecutive sweeps. The step size is 15 Hz at 563 nm (30 Hz at 282 nm). The lower right figure shows the lineshape calculated for conditions similar to the experimental conditions for the upper figure, except that the ion is assumed to have zero temperature and the laser is assumed to have zero linewidth.

tially a two-level system suitable for laser cooling, except for weak off-resonance pumping into the $^2S_{1/2}(F=0)$ state. The second 194 nm source, tuned to the $^2S_{1/2}(F=0)$ to $^2P_{1/2}(F=1)$ transition, returns the ion to the ground-state $F=1$ hyperfine level. The frequency separation between the two radiation sources is equal to the sum of the ground- and excited-state hyperfine splittings (about 47 GHz). The two 194 nm beams propagate collinearly and irradiate the ion at an angle of 55° with respect to the symmetry (z) axis of the trap. In this way, all motional degrees of freedom are cooled to near the Doppler-cooling limit of 1.7 mK. 194 nm fluorescence from the ion, collected in a solid angle of about $5 \cdot 10^{-3}\, 4\pi$ sr, is detected with an efficiency of 10% to give a peak count rate on resonance of about 25 000/s. The complication of laser cooling with two lasers is brought about by the hyperfine structure of $^{199}Hg^+$. Only an isotope with nonzero nuclear spin can have first-order, field-independent transitions, which give great immunity to magnetic-field fluctuations. In $^{199}Hg^+$, the nuclear spin is 1/2. Near $B=0$, the narrow $5d^{10}6s\ ^2S_{1/2}$-$5d^9 6s^2\ ^2D_{5/2}$ transition at 282 nm is first-order field-independent. The decay rate of the metastable $^2D_{5/2}$ state corresponds to an optical linewidth of less than 2 Hz—certainly, a suitably challenging test for the stabilized dye laser.

The 282 nm radiation is obtained by frequency doubling the radiation from the dye laser that is stabilized to the Fabry-Perot cavity on the sandbox table. (The cavity comparisons were done subsequent to the single-ion studies and we had not yet suspended a table with latex tubing.) Prior to being frequency-doubled, the 563 nm radiation (beam 1) is passed through an acousto-optic modulator (A/O-1) so that its frequency can be tuned through the *S-D* resonance. We also used A/O-1 to suppress the linear drift of the cavity and the frequency fluctuations caused by relative motion between the cavity and the ion trap (which are supported on different tables separated by 3 m). These Doppler effects can be removed in a fashion similar to that used by VESSOT to remove Doppler frequency shifts between a ground-based microwave source and a rocket-borne microwave oscillator [27]. Another acousto-optic modulator (A/O-2) is placed in an auxiliary laser beam (beam 2) near the ion trap. The frequency of beam 2 need not be stabilized. A/O-2 generates a frequency-shifted beam (beam 3) that is sent to the cavity table and returned on a path very close (< 2 cm) to that followed by beam 1. The light paths need not be overlapping in order to reach a frequency stability of 1 Hz. Beam 3 is recombined with its carrier to produce a beat note at the r.f. frequency of A/O-2. However, because the shifted beam traveled over to the cavity and back to the trap, the frequency fluctuations caused by relative motion between the tables and atmospheric turbulence are impressed on the beat note. Dividing the beat frequency by 2 gives the one-way path noise information carried at half the radiofrequency of A/O-2. If this frequency is summed with the right quadrature to the frequency that sweeps the stabilized laser through the *S-D* resonance, then path noise is eliminated. This is equivalent to bringing the cavity and trapped ion together. Stepping the frequency of the stabilized laser through the S-D resonance and removing the linear cavity drift are accomplished with an r.f. drive frequency to A/O-1 obtained by summing the output of two synthesizers. The frequency of one synthesizer sweeps opposite to the cavity drift and the frequency of the second synthesizer is stepped back and forth, sweeping the frequency of the stabilized laser through the narrow atomic resonance.

The 282 nm radiation and the two-frequency 194 nm source are turned on and off sequentially using shutters and the acousto-optic modulator. This prevents any broadening of the narrow S-D transition due to the 194 nm radiation. Electron shelving [25, 28] is used to detect each transition made to the metastable D state as a function of the frequency of the 282 nm laser. At the beginning of each cycle, both 194 nm lasers irradiate the ion. The fluorescence counts in a 10 ms period must exceed a minimum threshold (typically 20 counts) before the interrogation sequence can continue. The 194 nm beams irradiate the ion for sequential 10 ms periods until the threshold is met. The 194 nm radiation tuned to the $^2S_{1/2}(F=0)$-$^2P_{1/2}(F=1)$ transition is chopped off for 5 ms. During this time, the 194 nm radiation tuned to the $^2S_{1/2}(F=1)$-$^2P_{1/2}(F=0)$ transition optically pumps the ion into the $^2S_{1/2}(F=0)$ ground state. Then this 194 source is turned off. One millisecond later, the 282 nm radiation, tuned to a frequency resonant or nearly resonant with the $^2S_{1/2}(F=0,\ m_F=0)$-$^2D_{5/2}(F=2,\ m_F=0)$ transition, irradiates the ion for an interrogation period that is varied up to 25 ms. At the end of this period, the 282 nm radiation was turned off and both 194 nm sources were turned on. Another 10 ms detection period was used to determine whether a transition to the D state had been made (fluorescence counts > threshold, no; fluorescence counts < threshold, yes). The result was recorded as a 1 or 0 (no or yes) and averaged with the previous results at this frequency. Then the frequency of the 282 nm radiation was stepped and the measurement cycle repeated.

Since the frequency drift of the 282 nm laser depended not only on the reference cavity contraction rate, but also on small pressure and temperature changes, on laser power variations, and so on (as discussed in sect. I.2), we locked the frequency of the laser to the narrow S-D transition to remove long-term frequency drifts. To do this, we modified the measurement cycle to include a locking cycle. We began each measurement cycle by stepping the frequency of the 282 nm radiation to near the half maximum on each side of the resonance N times (N varied from 8 to 32). At each step, we probed for 5 ms and then looked for any transition with the electron-shelving technique. We averaged the N results from each side of the resonance line, took the difference and corrected the frequency of the synthesizer used to compensate the cavity drift. The gain of this lock needed to be properly adjusted to avoid adding frequency noise to the laser. In this way, variations in the frequency of the 282 nm laser for time periods exceeding a few seconds were reduced.

In fig. 3, we show the spectrum obtained by scanning in this drift-free way through the Doppler-free resonance of the $^2S_{1/2}(F=0,\ m_F=0)$-$^2D_{5/2}(F=$ $=2,\ m_F=0)$ transition. The lineshape shown is the result of 138 consecutive scans, each of which included a locking cycle. The probe period was 15 ms, and the step size was 15 Hz at 563 nm (30 Hz at 282 nm). The resonance shows a clearly resolved triplet with the linewidth of each component less than 40 Hz ($<$ 80 Hz at 282 nm). We first thought that the triplet structure might be due to 60 Hz modulation of the frequency of the 563 nm laser either due to grounding problems, line pickup or inadequate servo gain. However, when the radiation from two independently stabilized laser beams was heterodyned together, the 60 Hz modulation index was far too small to account for the sideband structure observed on the *S-D* resonance. In addition, the frequency separation of the peaks is nearer to 50 Hz, not to 60 Hz. We now think that, most likely, the triplet structure is caused by Rabi power broadening. The 282 nm radiation is focussed to a spot size of about 25 μm; therefore, on resonance, fewer than 10^6 photons/s ($<$ 1 pW) will saturate the transition. Below, the data is a theoretical lineshape calculated for an ion a rest, for no broadening due to collisions or laser bandwidth, for a pulse length of 15 ms and for sufficient power at resonance to give a 3.5π pulse (which roughly corresponds to the power used).

Qualitatively, the figures compare well. The fluctuations from measurement cycle to measurement cycle in the quantum occupation number of the ion in the harmonic well of the trap cause variations in the transition probability of the ion. This, and the finite laser linewidth, likely cause the general broadening and weakening of the signal. Current efforts are devoted to measuring the narrow *S-D* transition using the laser stabilized to the cavity on the suspended table. A cryogenic, linear r.f. Paul trap has been constructed and will soon be tested. With this trap, it should be possible to laser-cool many ions and to store them without attrition for days. The increased numbers of trapped ions will give a better signal-to-noise ratio (thereby better stability), but it will still be possible to have a small second-order Doppler shift. We also plan to investigate the lineshape and the effects of power broadening in more detail in future experiments.

* * *

The authors gratefully acknowledge the contributions of the colleagues who participated in the work reported here: F. DIEDRICH, F. ELSNER and M. RAIZEN. We also acknowledge the support of the Office of Naval Research.

REFERENCES

[1] D. J. WINELAND, W. M. ITANO, J. C. BERGQUIST and F. L. WELLS: in *Proceedings of the 35th Annual Symposium on Frequency Control, Philadelphia, Pa., May 1981* (copies available from Electronic Industries Assoc., 2001 Eye St., Washington, DC, 20006), p. 602.

[2] D. J. WINELAND, J. C. BERGQUIST, J. J. BOLLINGER, W. M. ITANO, D. J. HEINZEN, S. L. GILBERT, C. H. MANNEY and M. G. RAIZEN: *IEEE Trans. Ultrason., Ferroelectr. Frequency Control*, **37**, 515 (1990).
[3] MIAO ZHU and J. L. HALL: *J. Opt. Soc. Am. B.*, **10**, 802 (1993).
[4] C. M. CAVES: *Phys. Rev. Lett.*, **45**, 75 (1980).
[5] W. A. EDELSTEIN, J. HOUGH, J. R. PUGH and W. MARTIN: *J. Phys. E*, **11**, 710 (1980).
[6] C. M. CAVES: *Phys. Rev. Lett.*, **54**, 2465 (1985).
[7] R. LOUDON: *Phys. Rev. Lett.*, **47**, 815 (1981).
[8] D. HILS and J. L. HALL: in *Frequency Standards and Metrology*, edited by A. DEMARCHI (Springer-Verlag, Berlin, 1989), p. 162.
[9] R. J. REARK and W. C. YOUNG: in *Formulas for Stress and Strain* (McGraw-Hill, New York, N.Y., 1975).
[10] N. SAMPRAS, Stanford University: private communication.
[11] H. B. CALLEN and R. F. GREENE: *Phys. Rev.*, **86**, 702 (1952); H. B. CALLEN and T. A. WELTON: *Phys. Rev.*, **83**, 34 (1951).
[12] G. W. GIBBONS and S. W. HAWKING: *Phys. Rev. D*, **4**, 2191 (1971).
[13] P. F. PANTER: *Modulation, Noise and Spectral Analysis* (McGraw-Hill, New York, N.Y., 1965).
[14] V. B. BRAGINSKY, V. P. MITROFANOV and V. I. PANOV: *Systems with Small Dissipation* (University of Chicago Press, Chicago, Ill., 1985).
[15] CH. SALOMON, D. HILS and J. L. HALL: *J. Opt. Soc. Am. B*, **5**, 1576 (1988).
[16] D. L. PLATUS: *SPIE Proceedings*, **1619**, 44 (1991); P. R. SAULSON: *Gravitational Astronomy; Instrument Design and Astrophysical Prospects*, edited by D. E. MCCLELLAND and H.-A. BACHOR (World Scientific, Singapore, 1991), p. 248.
[17] P. J. VEITCH: *Rev. Sci. Instrum.*, **62**, 140 (1991).
[18] J. C. BERGQUIST, W. M. ITANO, F. ELSNER, M. G. RAIZEN and D. J. WINELAND: in *Light Induced Kinetic Effects on Atoms, Ions, and Molecules*, edited by L. MOI, S. GOZZINI, C. GABBANINI, E. ARIMONDO and F. STRUMIA (ETS Editrice, Pisa, 1991), p. 291.
[19] F. LARSEN: *Acoustic Emission* (IFI/Plenum, New York, N.Y., 1979).
[20] R. W. P. DREVER, J. L. HALL, F. V. KOWALSKI, J. HOUGH, G. M. FORD, A. J. MUNLEY and H. WARD: *Appl. Phys. B*, **31**, 97 (1983).
[21] J. HOUGH, D. HILS, M. D. RAYMAN, L.-S. MA, L. HOLLBERG and J. L. HALL: *Appl. Phys. B*, **33**, 179 (1984).
[22] Mention of a commercial product is for technical communication only.
[23] G. REMPE, R. J. THOMPSON, H. J. KIMBLE and R. LALEZARI: *Opt. Lett.*, **17**, 363 (1992).
[24] F. BAYER-HELMS, H. DARNEDDE and G. EXNER: *Metrologia*, **21**, 49 (1985).
[25] J. C. BERGQUIST, W. M. ITANO and D. J. WINELAND: *Phys. Rev. A*, **36**, 428 (1987).
[26] J. C. BERGQUIST, D. J. WINELAND, W. M. ITANO, H. HEMMATI, H.-U. DANIEL and G. LEUCHS: *Phys. Rev. Lett.*, **55**, 1567 (1985).
[27] R. F. C. VESSOT, M. E. LEVINE, E. M. MATTISON, E. L. BLOMBERG, T. E. HOFFMAN, G. U. NYSTROM, B. F. FARREL, R. DECHER, P. B. EBY, C. R. BAUGHER, J. W. WATTS, D. L. TEUBER and F. D. WILLS: *Phys. Rev. Lett.*, **45**, 2081 (1980).
[28] H. DEHMELT: *Bull. Am. Phys. Soc.*, **20**, 60 (1975); *J. Phys. (Paris) Colloq.*, **42**, C8-299 (1981).

Single-Atom Experiments and the Test of Quantum Physics.

H. WALTHER

Sektion Physik der Universität München
and Max-Planck-Institut für Quantenoptik - 85748 Garching, B.R.D.

1. – Introduction.

In recent years single-atom experiments have become possible. In this lecture two of those experiments will be reviewed with special emphasis on applications to study quantum phenomena. The first one deals with the one-atom maser and the second one with trapped ions [1]. In recent years there has also been considerable progress in trapping neutral atoms [2]. These techniques are very promising and are undergoing rapid development. Experiments using these methods will be discussed in other lectures of these proceedings.

2. – Review of the one-atom maser.

The most promising avenue to study the generation process of radiation in lasers and masers is to drive a maser consisting of a single-mode cavity by single atoms. This system, at first glance, seems to be another example of a Gedanken experiment, but such a one-atom maser [3] really exists and can in addition be used to study the basic principles of radiation-atom interaction. The advantages of the system are:

1) it is the first maser which sustains oscillations with less than one atom on the average in the cavity,

2) this setup allows one to study in detail the conditions necessary to obtain nonclassical radiation, especially radiation with sub-Poissonian photon statistics in a maser system directly, and

3) it is possible to study a variety of phenomena of a quantum field including the quantum measurement process.

What are the tools that make this device work? It is the enormous progress in constructing superconducting cavities together with the laser preparation of highly excited atoms—Rydberg atoms—that have made the realization of such a one-atom maser possible. Rydberg atoms have quite remarkable properties [4] which make them ideal for such experiments: The probability of induced transitions between neighbouring states of a Rydberg atom scales as n^4, where n denotes the principal quantum number. Consequently, a few photons are enough to saturate the transition between adjacent levels. Moreover, the spontaneous lifetime of a highly excited state is very large. We obtain a maser by injecting these Rydberg atoms into a superconducting cavity with a high quality factor. The injection rate is such that on the average there is less than one atom present inside the resonator at any time. A transition between two neighbouring Rydberg levels is resonantly coupled to a single mode of the cavity field. Due to the high quality factor of the cavity, the radiation decay time is much longer than the characteristic time of the atom-field interaction, which is given by the inverse of the single-photon Rabi frequency. Therefore, it is possible to observe the dynamics [5] of the energy exchange between atom and field mode leading to collapse and revivals in the Rabi oscillations [6, 7]. Moreover a field is built up inside the cavity when the mean time between the atoms injected into the cavity is shorter than the cavity decay time.

The detailed experimental setup of the one-atom maser is shown in fig. 1. A highly collimated beam of rubidium atoms passes through a Fizeau velocity selector. Before entering the superconducting cavity, the atoms are excited into the upper maser level $63p_{3/2}$ by the frequency-doubled light of a c.w. ring dye laser. The laser frequency is stabilized onto the atomic transition $5s_{1/2} \rightarrow 63p_{3/2}$, which has a width determined by the laser linewidth and the transit time broadening corresponding to a total of a few MHz. In this way, it is possible to prepare a very stable beam of excited atoms. The ultraviolet light is linearly polarized parallel to the electric field of the cavity. Therefore, only $\Delta m = 0$ transitions are excited by both the laser beam and the microwave field. The superconducting niobium maser cavity is cooled down to a temperature of 0.5 K by means of a ^{3}He cryostat. At this temperature the number of thermal photons in the cavity is about 0.15 at a frequency of 21.5 GHz. The cryostat is carefully designed to prevent room temperature microwave photons from leaking into the cavity. This would considerably increase the temperature of the radiation field above the temperature of the cavity walls. The quality factor of the cavity is $3 \cdot 10^{10}$ corresponding to a photon storage time of about 0.2 s. The cavity is carefully shielded against magnetic fields by several layers of cryoperm. In addition, three pairs of Helmholtz coils are used to compensate the Earth's magnetic field to a value of several mG in a volume of $(10 \times 4 \times 4)\,\mathrm{cm}^3$. This is necessary in order to achieve the high quality factor and prevent the different magnetic substates of the maser levels from mixing during the atom-field interaction time. Two maser transitions from the $63p_{3/2}$ level to the $61d_{3/2}$ and to the $61d_{5/2}$ level are studied.

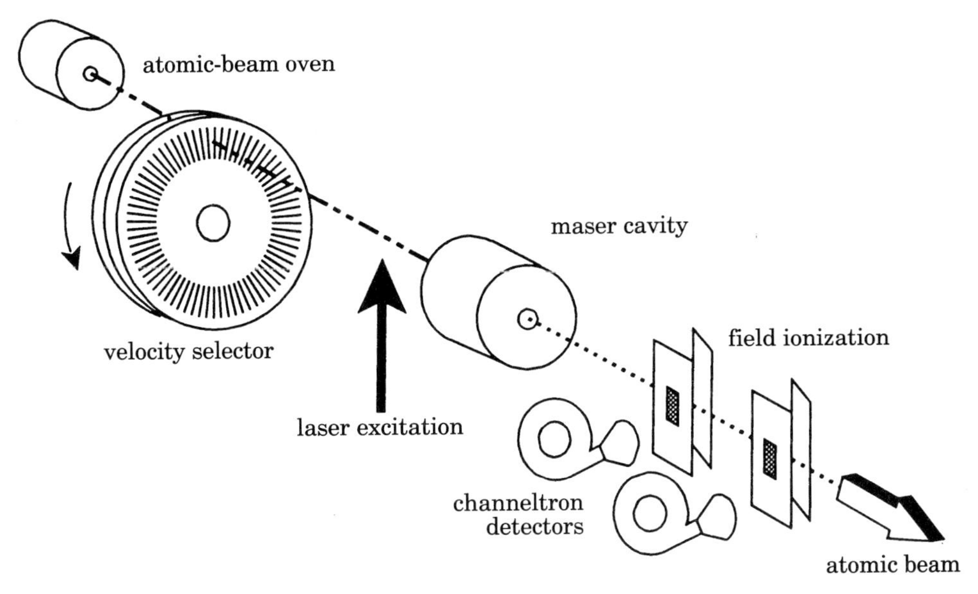

Fig. 1. – Scheme of the one-atom maser. To suppress blackbody-induced transitions to neighbouring states, the Rydberg atoms are excited inside the liquid-helium-cooled environment.

The Rydberg atoms in the upper and lower maser levels are detected in two separate field ionization detectors. The field strength is adjusted so as to ensure that in the first detector the atoms in the upper level are ionized, but not those in the lower level. The lower-level atoms are then ionized in the second field.

To demonstrate maser operation, the cavity is tuned over the $63p_{3/2}$-$61d_{3/2}$ transition and the flux of atoms in the excited state is recorded simultaneously. Transitions from the initially prepared $63p_{3/2}$ state to the $61d_{3/2}$ level (21.506 58 GHz) are detected by a reduction of the electron count rate.

In the case of measurements at a cavity temperature of 0.5 K, shown in fig. 2, a reduction of the $63p_{3/2}$ signal can be clearly seen for atomic fluxes as small as 1750 atoms/s. An increase in flux causes power broadening and a small shift. This shift is attributed to the a.c. Stark effect, caused predominantly by virtual transitions to neighbouring Rydberg levels. Over the range from 1750 to 28 000 atoms/s the field ionization signal at resonance is independent of the particle flux which indicates that the transition is saturated. This and the observed power broadening show that there is a multiple exchange of photons between Rydberg atoms and the cavity field.

For an average transit time of the Rydberg atoms through the cavity of 50 μs and a flux of 1750 atoms/s we estimate that approximately 0.09 Rydberg atoms are in the cavity on the average. According to Poisson statistics this implies that more than 90% of the events are due to single atoms. This clearly demon-

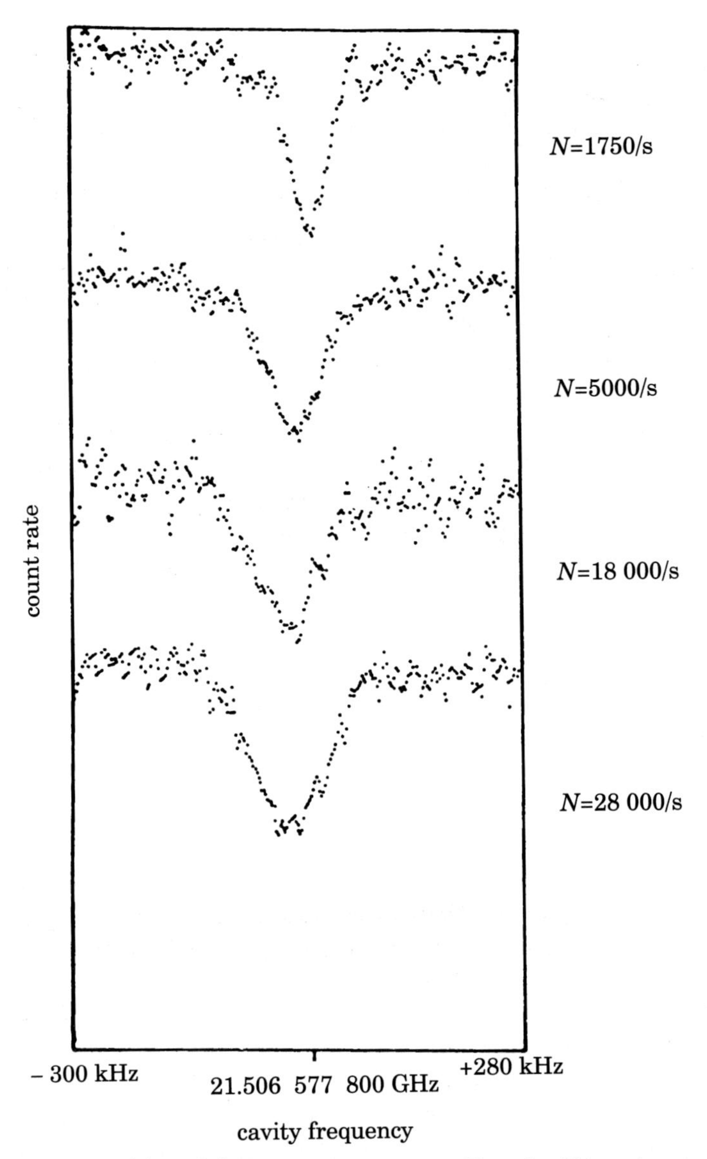

Fig. 2. – A maser transition of the one-atom maser manifests itself in a decrease of atoms in the excited state; ^{85}Rb, $63p_{3/2} \leftrightarrow 61d_{3/2}$, $T = 0.5$ K. The flux of excited atoms N governs the pump intensity. Power broadening of the resonance line demonstrates the multiple exchange of a photon between the cavity field and the atom passing through the resonator.

strates that single atoms are able to maintain a continuous oscillation of the cavity with a field corresponding to a mean number of photons between unity and several hundred.

3. – The generation of nonclassical light in the one-atom maser

One of the most interesting questions in connection with the one-atom maser is the photon statistics of the electromagnetic field generated in the superconducting cavity. This problem will be discussed in this section.

Electromagnetic radiation can show nonclassical properties [8, 9], that is, properties that cannot be explained by classical probability theory. Loosely speaking we need to invoke «negative probabilities» to get deeper insight into these features. We know of essentially three phenomena which demonstrate the nonclassical character of light: photon antibunching [10], sub-Poissonian photon statistics [11] and squeezing [12]. Mostly methods of nonlinear optics are employed to generate nonclassical radiation. However, the fluorescence light from a single atom caught in a trap also exhibits nonclassical features [13, 14].

Another nonclassical light generator is the one-atom maser. We recall that the Fizeau velocity selector preselects the velocity of the atoms: Hence the interaction time is well defined, which leads to conditions usually not achievable in standard masers [15-20]. This has a very important consequence when the intensity of the maser field grows as more and more atoms give their excitation energy to the field: Even in the absence of dissipation this increase in photon number is stopped when the increasing Rabi frequency leads to a situation where the atoms reabsorb the photon and leave the cavity in the upper state. For any photon number, this can be achieved by appropriately adjusting the velocity of the atoms. In this case the maser field is not changed any more and the number distribution of the photons in the cavity is sub-Poissonian [15, 16], that is, narrower than a Poisson distribution. Even a number state that is a state of well-defined photon number can be generated [17, 18] using a cavity with a high enough quality factor. If there are no thermal photons in the cavity—a condition achievable by cooling the resonator to an extremely low temperature—very interesting features such as trapping states show up [19]. In addition, steady-state macroscopic quantum superpositions can be generated in the field of the one-atom maser pumped by two-level atoms injected in a coherent superposition of their upper and lower states [20].

Unfortunately, the measurement of nonclassical photon statistics in the cavity is not that straightforward. The measurement process of the field invokes coupling to a measuring device, with losses leading inevitably to a destruction of the nonclassical properties. The ultimate technique to obtain information about the field employs the Rydberg atoms themselves: measure the photon statistics via the dynamic behaviour of the atoms in the radiation field, *i.e.* via the collapse and revivals of the Rabi oscillations. That is one possibility. However, since the photon statistics depend on the interaction time which has to be changed when collapse and revivals are measured, it is much better to probe the population of the atoms in the upper and lower maser levels when they leave the cavity. In this case, the interaction time is kept constant. Moreover, this measurement is relatively easy since electric fields can be used to perform selective ionization of the atoms. The detection sensitivity is sufficient so that the atomic statistics can be investigated. This technique maps the photon statistics of the field inside the cavity via the atomic statistics.

In this way, the number of maser photons can be inferred from the number of atoms detected in the lower level [3]. In addition, the variance of the photon number distribution can be deduced from the number fluctuations of the lower-level atoms [21]. In the experiment, we are, therefore, mainly interested in the atoms in the lower maser level. Experiments carried out along these lines are described in the following section.

4. – Experimental results—sub-Poissonian statistics of atoms and photons.

Under steady-state conditions, the photon statistics of the field is essentially determined by the dimensionless parameter $\Theta = (N_{ex} + 1)^{1/2}\,\Omega t_{int}$, which can be understood as a pump parameter for the one-atom maser [15]. Here, N_{ex} is the average number of atoms that enter the cavity during the lifetime of the field T_c, t_{int} the time of flight of the atoms through the cavity, and Ω the atom-field coupling constant (one-photon Rabi frequency). The one-atom maser threshold is reached for $\Theta = 1$. At this value and also at $\Theta = 2\pi$ and integer multiples thereof, the photon statistics is super-Poissonian. At these points, the maser field undergoes first-order phase transitions [15]. In the regions between these points sub-Poissonian statistics are expected. The experimental investigation of the photon number fluctuation is the subject of the following discussion.

In the experiments [22], the number N of atoms in the lower maser level is counted for a fixed time interval T roughly equal to the storage time T_c of the photons. By repeating this measurement many times the probability distribution $p(N)$ of finding N atoms in the lower level is obtained. The normalized variance [23] $Q_a = [\langle N^2\rangle - \langle N\rangle^2 - \langle N\rangle]/\langle N\rangle$ is evaluated and is used to characterize the deviation from Poissonian statistics. A negative (positive) Q_a value indicates sub-Poissonian (super-Poissonian) statistics, while $Q_a = 0$ corresponds to a Poisson distribution with $\langle N^2\rangle - \langle N\rangle^2 = \langle N\rangle$. The atomic Q_a is related to the normalized variance Q_f of the photon number by the formula

$$Q_a = \varepsilon P Q_f (2 + Q_f), \tag{1}$$

which was derived by Rempe and Walther [21] with P denoting the probability of finding an atom in the lower maser level. It follows from eq. (1) that the nonclassical photon statistics can be observed via sub-Poissonian atomic statistics. The detection efficiency ε for the Rydberg atoms reduces the sub-Poissonian character of the experimental result. The detection efficiency was 10% in our

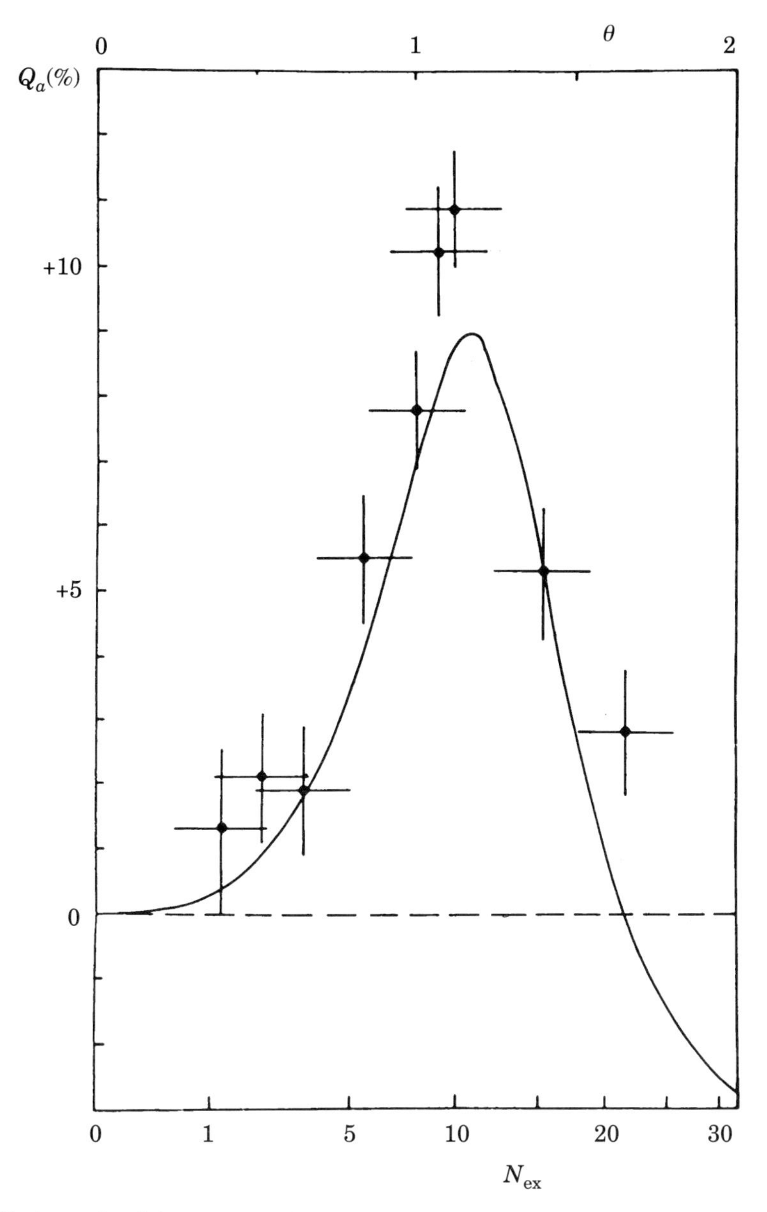

Fig. 3. – Variance Q_a of the atoms in the lower maser level as a function of flux N_{ex} near the onset of maser oscillation for the $63p_{3/2} \leftrightarrow 61d_{3/2}$ ^{85}Rb transition at 0.5 K (see also ref. [22]).

experiment; this includes the natural decay of the Rydberg states between the cavity and field ionization. It was determined by both monitoring the power-broadened resonance line as a function of flux [3] and observing the Rabi oscillation for constant flux but different atom-field interaction times [7]. In addition,

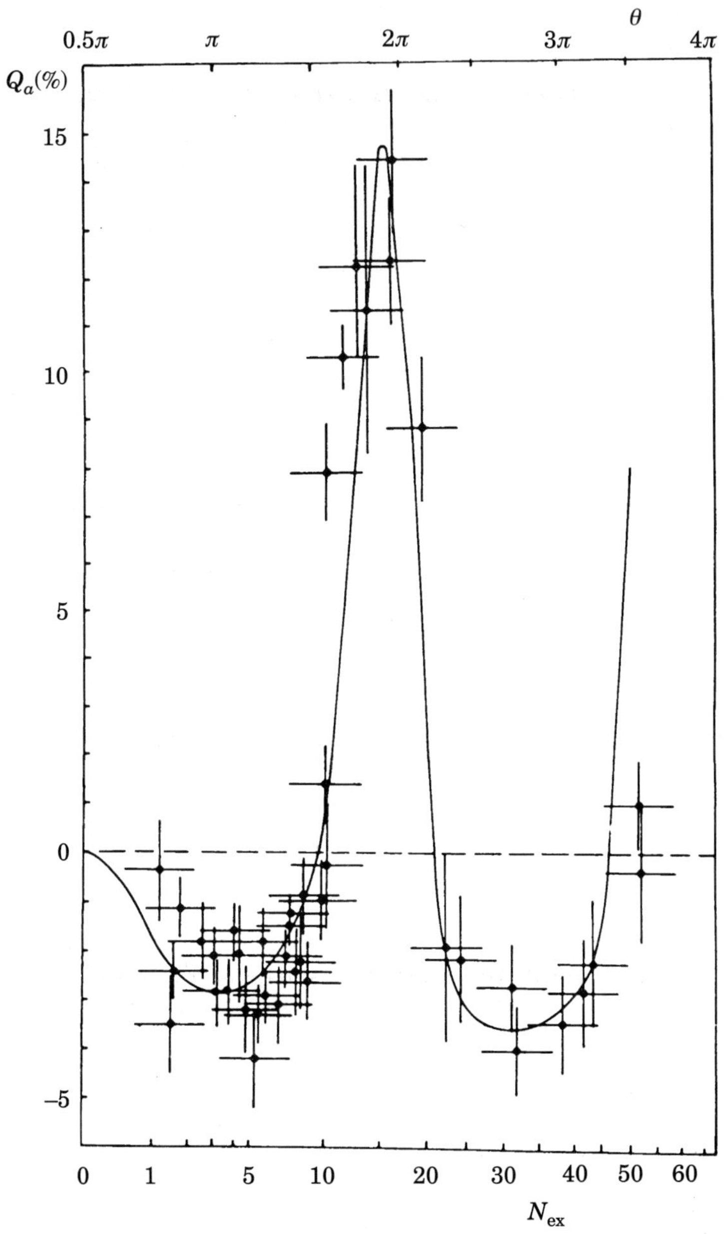

Fig. 4. – Same as fig. 3, but above threshold for the $63p_{3/2} \leftrightarrow 61d_{5/2}$ ^{85}Rb transition at 0.5 K (see also ref. [22]).

this result is consistent with all other measurements described in the following, especially with those on the second maser phase transition.

Experimental results for the transition $63p_{3/2} \leftrightarrow 61d_{3/2}$ are shown in fig. 3.

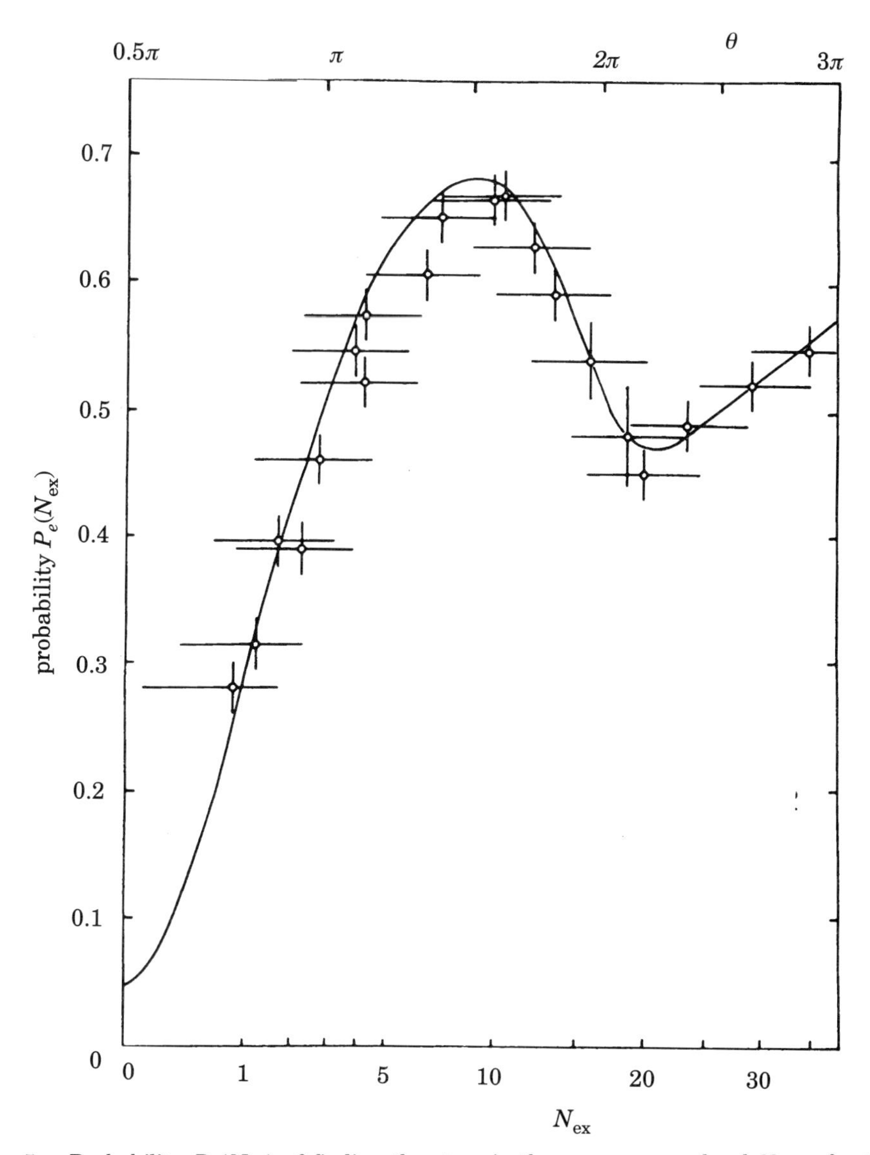

Fig. 5. – Probability $P_e(N_{ex})$ of finding the atom in the upper maser level $63p_{3/2}$ for the $63p_{3/2} \leftrightarrow 61d_{5/2}$ ^{85}Rb transition at 0.5 K as a function of the atomic flux.

The measured normalized variance Q_a is plotted as a function of the flux of atoms. The atom-field interaction time is fixed at $t_{int} = 50\,\mu s$. The atom-field coupling constant Ω is rather small for this transition, $\Omega = 10$ kHz. A relatively high flux of atoms $N_{ex} > 10$ is, therefore, needed to drive the one-atom maser above threshold. The large positive Q_a observed in the experiment proves the large intensity fluctuations at the onset of maser oscillation at $\Theta = 1$. The solid curve is plotted according to eq. (1) using the theoretical predictions for Q_f of the photon statistics [15, 16]. The error in the signal follows from the statistics of the counting distribution $p(N)$. About $2 \cdot 10^4$ measurement intervals are needed to keep the error of Q_a below 1%. The statistics of the atomic beam are measured with a detuned cavity. In this case the regular statistics of the atomic beam are measured which corresponds to Poissonian statistics. The error bars of the flux follow from this measurement. The agreement between theory and experiment is good.

The nonclassical photon statistics of the one-atom maser are observed at a higher flux of atoms or a larger atom-field coupling constant. The $63p_{3/2} \leftrightarrow$ $\leftrightarrow 61d_{5/2}$ maser transition with $\Omega = 44$ kHz is, therefore, studied. Experimental results are shown in fig. 4. Fast atoms with an atom-cavity interaction time of $t_{int} = 35\,\mu s$ are used. A very low flux of atoms of $N_{ex} > 1$ is already sufficient to generate a nonclassical maser field. This is the case since the vacuum field initiates a transition of the atom to the lower maser level, thus driving the maser above threshold.

The sub-Poissonian statistics can be understood from fig. 5, where the probability of finding the atom in the upper level is plotted as a function of the atomic flux. The oscillation observed is closely related to the Rabi nutation induced by the maser field. The solid curve was calculated according to the one-atom maser theory with a velocity dispersion of 4%. A higher flux generally leads to a higher photon number, but for $N_{ex} < 10$ the probability of finding the atom in the lower level decreases. An increase in the photon number is, therefore, counterbalanced by the fact that the probability of photon emission in the cavity is reduced. This negative feedback leads to a stabilization of the photon number [21]. The feedback changes sign at a flux $N_{ex} \approx 10$, where the second maser phase transition is observed at $\Theta = 2\pi$. This is again characterized by large fluctuations of the photon number. Here the probability of finding an atom in the lower level increases with increasing flux. For even higher fluxes, the state of the field is again highly nonclassical. The solid curve in fig. 4 represents the result of the one-atom maser theory using eq. (1) to calculate Q_a. The agreement with experiment is very good. The sub-Poissonian statistics of atoms near $N_{ex} = 30$, $Q_a = -4\%$ and $P_e = 0.45$ (see fig. 5) is generated by a photon field with a variance $\langle n^2 \rangle - \langle n \rangle^2 = 0.3 \langle n \rangle$, which is 70% below the shot noise level. Again, this result agrees with the prediction of the theory [15, 16]. The mean number of photons in the cavity is about 2 and 13 in the regions $N_{ex} \approx 3$ and $N_{ex} \approx 30$, respectively. Near $N_{ex} \approx 15$, the photon number changes abruptly between these two values. The next maser phase transition with a super-Poissonian photon number distribution occurs above $N_{ex} \approx 50$.

Sub-Poissonian statistics are closely related to the phenomenon of antibunching, for which the probability of detecting the next event shows a minimum immediately after a triggering event. The duration of the time interval with reduced probability is of the order of the coherence time of the radiation field. In our case this time is determined by the storage time of the photons. The Q_a value, therefore, depends on the measuring interval T. The measured Q_a value approaches a time-independent value for $T > T_c$. For very short sampling intervals, the statistics of atoms in the lower level show a Poisson distribution. This means that the cavity cannot stabilize the flux of atoms in the lower level on a time scale which is short in relation to the intrinsic cavity damping time.

We emphasize that the reason for the sub-Poissonian atomic statistics is the following: A changing flux of atoms changes the Rabi frequency via the stored photon number in the cavity. By adjusting the interaction time, the phase of the Rabi nutation cycle can be chosen so that the probability for the atoms leaving the cavity in the upper maser level increases when the flux and, therefore, the photon number are raised or *vice versa*. We observe sub-Poissonian atomic statistics in the case in which the number of atoms in the lower state is decreasing with increasing flux and photon number in the cavity. The same argument can be applied to understand the nonclassical photon statistics of the maser field: Any deviation of the number of light quanta from its mean value is counterbalanced by a correspondingly changed probability of photon emission for the atoms. This effect leads to a natural stabilization of the maser intensity by a feedback loop incorporated into the dynamics of the coupled atom-field system.

The experimental results presented here clearly show the sub-Poissonian photon statistics of the one-atom maser field. An increase in the flux of atoms leads to the predicted second maser phase transition. In addition, the maser experiment leads to an atomic beam with atoms in the lower maser level showing number fluctuations which are up to 40% below those of a Poissonian distribution found in usual atomic beams. This is interesting, because atoms in the lower level have emitted a photon to compensate for cavity losses inevitably present under steady-state conditions. But this is a purely dissipative phenomenon giving rise to fluctuations. Nevertheless the atoms still obey sub-Poissonian statistics.

5. – A new probe of complementarity in quantum mechanics—the one-atom maser and atomic interferometry.

The preceding section discussed how to generate a nonclassical field inside the maser cavity. But this field is extremely fragile because any attenuation causes a considerable broadening of the photon number distribution. Therefore, it is difficult to couple the field out of the cavity while preserving its nonclassical character. But what is the use of such a field? In the present section we want to propose a new series of experiments performed inside the maser cavity to test the «wave-particle» duality of nature or, better said, «complementarity» in quantum mechanics.

Complementarity [24] lies at the heart of quantum mechanics: Matter sometimes displays wavelike properties manifesting themselves in interference phenomena, and at other times it displays particlelike behaviour thus providing «which-path» information. No other experiment illustrates this wave-particle duality in a more striking way than the classic Young double-slit experiment [25, 26]. Here we find it impossible to tell which slit light went through while observing an interference pattern. In other words, any attempt to gain «which-path» information disturbs the light so as to wash out the interference fringes. This point has been emphasized by BOHR in his rebuttal to Einstein's ingenious proposal of using recoiling slits [26] to obtain «which-path» information while still observing interference. The physical positions of the recoiling slits, BOHR argued, are only known to within the uncertainty principle. This error contributes a random phase shift to the light beams which destroys the interference pattern.

Such random-phase arguments, illustrating in a vivid way how the «which-path» information destroys the coherent wavelike interference aspects of a given experimental setup, are appealing. Unfortunately, they are incomplete: In principle, and in practice, it is possible to design experiments which provide «which-path» information via detectors which do not disturb the system in any noticeable way. Such «Welcher Weg» (German for «which-path») detectors have been recently considered within the context of studies involving spin coherence [27]. In the present section we describe a quantum optical experiment [28] which shows that the loss of coherence occasioned by «Welcher Weg» information, that is, by the presence of a «Welcher Weg» detector, is due to the establishment of quantum correlations. It is in no way associated with large random-phase factors as in Einstein's recoiling slits.

The details of this application of the micromaser are discussed [29]. Here only the essential features are given. We consider an atomic interferometer where the two particle beams pass through two maser cavities before they reach the two slits of the Young's interferometer. The interference pattern observed is then also determined by the state of the maser cavity. The interference term is given by

$$\langle \Phi_1^{(\mathrm{f})}, \Phi_2^{(\mathrm{i})} | \Phi_1^{(\mathrm{i})}, \Phi_2^{(\mathrm{f})} \rangle ,$$

where $|\Phi_j^{(\mathrm{i})}\rangle$ and $|\Phi_j^{(\mathrm{f})}\rangle$ denote the initial and final states of the maser cavity.

Let us prepare, for example, both one-atom masers in coherent states $|\Phi_j^{(\mathrm{i})}\rangle = |\alpha_j\rangle$ of large average photon number $\langle m \rangle = |a_j|^2 > 1$. The Poissonian photon number distribution of such a coherent state is very broad, $\Delta m \approx \alpha \gg 1$. Hence the two fields are not changed much by the addition of a single photon associated with the two corresponding transitions. We may, therefore, write

$$|\Phi_j^{(\mathrm{f})}\rangle \approx |\alpha_j\rangle,$$

which to a very good approximation yields

$$\langle \Phi_1^{(\mathrm{f})}, \Phi_2^{(\mathrm{i})} | \Phi_1^{(\mathrm{i})}, \Phi_2^{(\mathrm{f})} \rangle \approx \langle \alpha_1, \alpha_2 | \alpha_1, \alpha_2 \rangle = 1 .$$

Thus there is an interference cross-term different from zero.

When, however, we prepare both maser fields in number states [17-19] $|n_j\rangle$, the situation is quite different. After the transition of an atom to the d-state, that is, after emitting a photon in the cavity the final states read

$$|\Phi_j^{(\mathrm{f})}\rangle = |n_j + 1\rangle$$

and hence

$$\langle \Phi_1^{(\mathrm{f})}, \Phi_2^{(\mathrm{i})} | \Phi_1^{(\mathrm{i})}, \Phi_2^{(\mathrm{f})} \rangle = \langle n_1, n_2 | n_1, n_2 + 1 \rangle = 0 ,$$

that is the coherence cross-term vanishes and no interference is observed.

At first sight this result might seem a bit surprising when we recall that in the case of a coherent state the transition did not destroy the coherent cross-term, *i.e.* did not affect the interference fringes. However, in the example of number states we can, by simply «looking» at the one-atom maser state, tell which «path» the atom took.

It should be pointed out that the beats disappear not only for a number state. For example, a thermal field leads to the same result. In this regard, we note that it is not enough to have an indeterminate photon number to ensure interferences. The state $|\Phi_j^{(\mathrm{f})}\rangle$ goes as $a_j^+ |\Phi_j^{(\mathrm{i})}\rangle$, where a_j^+ is the creation operator for the j-th maser. Hence the inner product is

$$\langle \Phi_j^{(\mathrm{i})} | \Phi_j^{(\mathrm{f})} \rangle \rightarrow \langle \Phi_j^{(\mathrm{i})} | a_j^+ | \Phi_j^{(\mathrm{i})} \rangle,$$

and in terms of a more general density matrix formalism we have

$$\langle \Phi^{(\mathrm{i})} | \Phi^{(\mathrm{f})} \rangle \rightarrow \sum_n \sqrt{n+1}\, \rho^{(\mathrm{i})}_{n,\, n+1} .$$

Thus we see that an off-diagonal density matrix is needed for the production of interferences. For example, a thermal field having indeterminate photon number would not lead to interferences since the photon number distribution is diagonal in this case.

The atomic-interference experiment in connection with one-atom maser cavities is a rather complicated scheme for a «Welcher Weg» detector. There is a much simpler possibility which we will discuss briefly in the following. This is based on the logic of the famous «Ramsey fringe» experiment. In this experi-

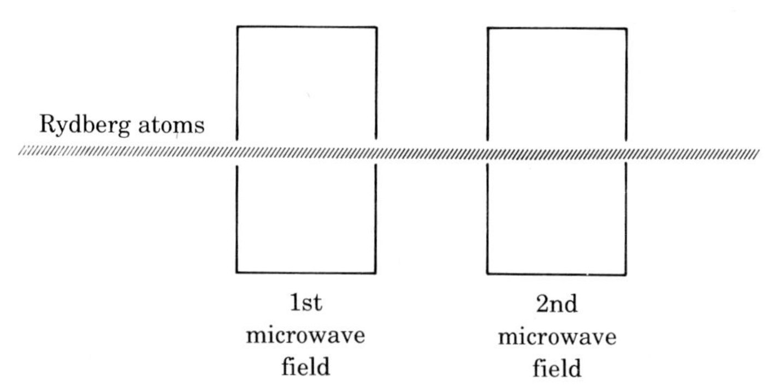

Fig. 6. – Setup for the Ramsey experiment.

ment two microwave fields are applied to the atoms one after the other. The interference occurs since the transition from an upper state to a lower state may either occur in the first or in the second interaction region. In order to calculate the transition probability, we must sum the two amplitudes and then square, thus leading to an interference term. We will show here only the principle of this experiment; a more detailed discussion is the subject of another paper [30]. In the setup we discuss here, the two Ramsey fields are two one-atom maser cavities (see fig. 6). The atoms enter the first cavity in the upper state and are weakly driven into a lower state $|b\rangle$. That is, each microwave cavity induces a small transition amplitude governed by $m\tau$, where m is the atom-field coupling constant and τ is the time of flight across the cavity.

Now, if the quantum state of the initial (final) field in the j-th cavity is given by $\Phi_j^{(\mathrm{i})}$ ($\Phi_j^{(\mathrm{f})}$), then the state of the atom + maser 1 + maser 2 at the various relevant times is given in terms of the coupling constant m_j and interaction time τ_j, and initial $|\Phi_j^{(\mathrm{i})}\rangle$ and final states of the j-th maser by

$$|\psi(0)\rangle = |a, \Phi_1^{(\mathrm{i})}, \Phi_2^{(\mathrm{i})}\rangle,$$

$$|\psi(\tau_1)\rangle \approx |a, \Phi_1^{(\mathrm{i})}, \Phi_2^{(\mathrm{i})}\rangle - im_1\tau_1|b, \Phi_1^{(\mathrm{f})}, \Phi_2^{(\mathrm{i})}\rangle,$$

$$|\psi(\tau_1+T)\rangle \simeq |a, \Phi_1^{(\mathrm{i})}, \Phi_2^{(\mathrm{i})}\rangle - im_1\tau_1|b, \Phi_1^{(\mathrm{f})}, \Phi_2^{(\mathrm{i})}\rangle \exp[-i\Delta\omega T],$$

$$|\psi(\tau_1+T+\tau_2)\rangle \approx |a, \Phi_1^{(\mathrm{i})}, \Phi_2^{(\mathrm{i})}\rangle - im_1\tau_1|b, \Phi_1^{(\mathrm{f})}, \Phi_2^{(\mathrm{i})}\rangle \exp[-i\Delta\omega T] -$$

$$-im_2\tau_2|b, \Phi_1^{(\mathrm{i})}, \Phi_2^{(\mathrm{f})}\rangle,$$

where $\Delta\omega$ is the atom-cavity detuning and $T \gg \tau_j$ the time of flight between the two cavities and a and b denote the lower and upper states of the atoms, respectively. If we ask for P_b, the probability that the atom exits cavity 2 in the lower state $|b\rangle$, this is given by

$$P_b = [\langle \Phi_1^{(f)}, \Phi_2^{(i)} | m_1^* \tau_1 \exp[i \Delta\omega T] + \langle \Phi_1^{(i)}, \Phi_2^{(f)} | m_2^* \tau_2] \cdot$$

$$\cdot [| \Phi_1^{(f)}, \Phi_2^{(i)} \rangle m_1 \tau_1 \exp[- i \Delta\omega T] + | \Phi_1^{(i)}, \Phi_2^{(f)} \rangle m_2 \tau_2] =$$

$$= m_1^* m_1 \tau_1^2 + m_2^* m_2 \tau_2^2 + (m_1^* m_2 \tau_1 \tau_2 \exp[i \Delta\omega T] \langle \Phi_1^{(f)}, \Phi_2^{(i)} | \Phi_1^{(i)}, \Phi_2^{(f)} \rangle + \text{c.c.}).$$

Now in the usual Ramsey experiment $| \Phi_j^{(i)} \rangle = | \Phi_j^{(f)} \rangle = | \alpha_j \rangle$, where $| \alpha_j \rangle$ is the *coherent state* in the j-th maser, which is not changed by the addition of a single photon. Thus the «fringes» appear going as $\exp[i\Delta\omega T]$. However, consider the situation in which $| \Phi_j^{(i)} \rangle$ is a *number state, e.g.* the state $| 0_j^{(i)} \rangle$ having no photons in the j-th cavity initially; now we have

$$P = m_1^* m_1 \tau_1^2 + m_2^* m_2 \tau_2^2 + (m_1^* m_2 \tau_1 \tau_2 \exp[i\Delta\omega T] \langle 1_1, 0_2 | 0_1, 1_2 \rangle + \text{c.c.}).$$

In this case, the one-atom masers are now acting as «Welcher Weg» detectors, and the interference term vanishes due to the atom-maser quantum correlation.

We note that the more usual Ramsey fringe experiment involves a strong field «$\pi/2$-pulse» interaction in the two regions. This treatment is more involved than necessary for the present purposes. A more detailed analysis of the one-atom maser Ramsey problem is given elsewhere [31].

We conclude this section by emphasizing again that this new and potentially experimental example of wave-particle duality and observation in quantum mechanics displays a feature which makes it distinctly different from the Bohr-Einstein recoiling-slit experiment. In the latter the coherence, that is the interference, is lost due to a phase disturbance of the light beams. In the present case, however, the loss of coherence is due to the correlation established between the system and the one-atom maser. Random-phase arguments never enter the discussion. We emphasize that the argument of the number state not having a well-defined phase is not relevant here; the important dynamics are due to the atomic transition. It is the fact that «which-path» information is made available which washes out the interference cross-terms [29].

6. – Experiments with trapped ions—the Paul trap.

In contrast to neutral atoms, ions can easily be influenced by electromagnetic fields because of their charge. In most of the experiments the Paul trap is used. It consists of a ring electrode and two end caps as shown in fig. 7. Trapping can be achieved if time-varying electric fields [32] are applied between the ring and the caps (the two caps are electrically connected). A d.c. voltage in addition changes the relation of the potential depth along the symmetry axis (vertical direction in fig. 7) to that in a perpendicular direction. The equation of mo-

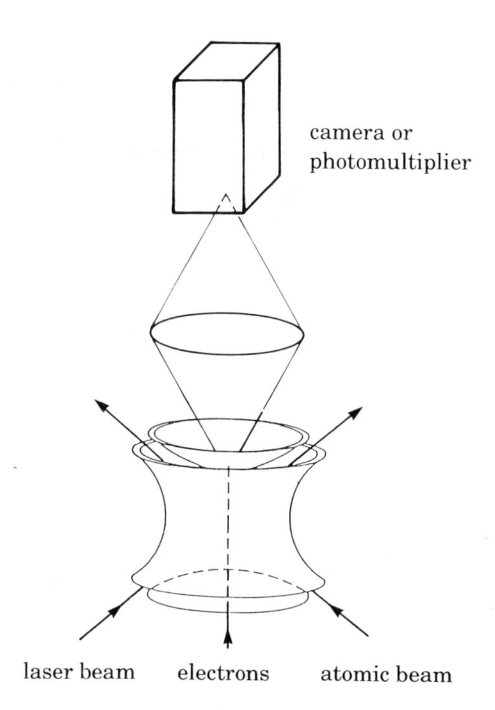

Fig. 7. – Sketch of the Paul trap. The fluorescence light is observed through a hole in the upper electrode.

tion of an ion in such a situation is the Mathieu differential equation, well known in classical mechanics, which—depending on the voltages applied to the trap (d.c. and radiofrequency voltages)—allows stable and unstable solutions. Another way to achieve trapping is the use of a constant magnetic field aligned along the symmetry axis leading to the Penning trap [1, 33]. In this case only a d.c. voltage has to be applied between ring and cap electrodes.

In order to produce the ions in the Paul trap, a neutral atomic beam is directed through the trap centre and ionized by electrons. Unfortunately, the resulting trapped ions have a lot of kinetic energy rendering them useless for most applications, such as spectroscopy; therefore, the ions have to be cooled. This is done by laser light. For this purpose the laser frequency ν is tuned below the resonance frequency, so that the energy of the photon is not sufficient to excite the atom. Crudely, the ion can extract the missing energy from its motion and thus reduce its kinetic energy. In other words, the atomic velocity Doppler-shifts the atom into resonance to bridge the detuning gap Δ between laser and resonance frequency and the atom absorbs the photon of momentum $\hbar k = h\nu/c$. After the absorption process the momentum of the atom is reduced, lowering its kinetic energy. This leads to a net cooling effect since the re-emission due to atomic fluorescence is isotropic in space. The lowest temperature achievable is determined by the Doppler limit [34] which is in the millikelvin region. The low temperatures can be obtained within a fraction of a second.

The results discussed in this lecture were obtained using a Paul trap with a ring diameter of 5 mm and an end-cap separation of 3.54 mm [14, 35]. This trap is larger than most of the other ion traps used in laser experiments [1, 36]. The radiofrequency of the field used for dynamic trapping is 11 MHz. The trap is mounted inside a stainless-steel, ultrahigh-vacuum chamber. We can obtain storage times of hours using a background gas at a pressure of 10^{-10} mbar. The ions are loaded into the trap by means of a thermal beam of neutral atoms (magnesium atoms in our case) which are then ionized close to the centre of the trap by an electron beam entering the trap through a small hole in the lower end cap (see fig. 7). In order not to distort the trap potential, the hole is covered by a fine molybdenum mesh. The neutral Mg beam and the laser beam pass through the gaps between the end cap and the ring electrodes. The laser frequency is shifted by an amount Δ below the $3S_{1/2} \rightarrow 3P_{3/2}$ resonance transition of $^{24}Mg^+$ at 280 nm to extract kinetic energy from the ions by radiation pressure as discussed above. In this way, a single ion can be cooled to a temperature below 10 mK. The fluorescence from the ions is observed through a hole in the upper end cap (see fig. 7), again covered by a molybdenum mesh. The large size of the trap allows a large solid angle for detecting the fluorescence radiation, either with a photomultiplier, or by means of a photon-counting imaging system. To observe the ions, the cathode of the imaging system is placed in the image plane of the microscope objective attached to the trap; in this way images of the ions could be obtained [37, 38].

7. – Order *vs.* chaos: crystal *vs.* cloud.

The existence of phase transitions in a Paul trap manifests itself by significant jumps in the fluorescence intensity of the ions as a function of the detuning Δ between the laser frequency and the atomic-transition frequency. These discontinuities are indicated in fig. 8 by vertical arrows, and occur between two types of spectra: a broad and a narrow one, analogous to the fluorescence spectrum of a single, cooled ion. We interpret the broad spectrum as a fingerprint of an ion cloud and the narrow spectrum as being characteristic for an ordered many-ion situation with a «single-ion signature». Thus the jumps clearly indicate a transition from a state of erratic motion within a cloud to a situation where the ions arrange themselves in regular structures. In such a crystalline structure the mutual ion Coulomb repulsion is compensated by the external, dynamic trap potential. The regime of detunings in which such crystals exist is depicted in fig. 8 by the horizontal arrow. The existence of the two phases—crystal and cloud—can be verified experimentally by direct observation with the help of a highly sensitive

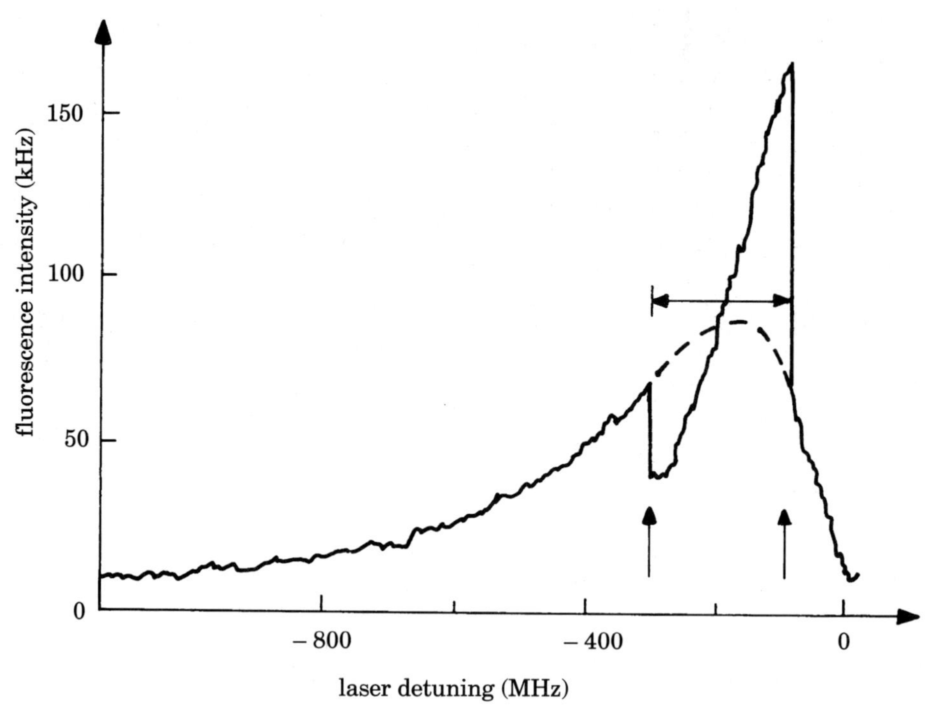

Fig. 8. – Fluorescence intensity, that is, photon counts per second, from five ions as a function of the laser detuning Δ. The vertical arrows indicate the detunings where phase transitions occur. The horizontal arrow shows the range of detunings in which a stable five-ion crystal is observed. The spectrum was scanned from left to right.

imaging system, and theoretically by analysing ion trajectories via Monte Carlo computer simulations [37, 38].

The excitation spectrum of fig. 8 and the jumps in it were recorded by altering the detuning Δ from large negative values to zero. When we scan the spectrum in the opposite direction, the jumps occur at different values of Δ, which means there is hysteresis associated with these phase transitions [37]. Such a hysteresis behaviour can be expected with laser-cooled ions because the cooling power of the laser strongly depends upon the details of the velocity distribution of the ions. The behaviour of the ions in the trap is governed by the trap voltage, the laser detuning and the laser power. Hysteresis loops appear whenever one of these parameters is changed up or down whilst the others are kept constant.

Figure 9 shows the ion structures as measured with the imaging system. For the measurements only a radiofrequency voltage was applied to the trap electrodes: in this case the potential is a factor of two deeper along the symmetry axis than perpendicular to it, and, therefore, plane ion structures are observed being perpendicular to the symmetry axis (for details see [37, 38]).

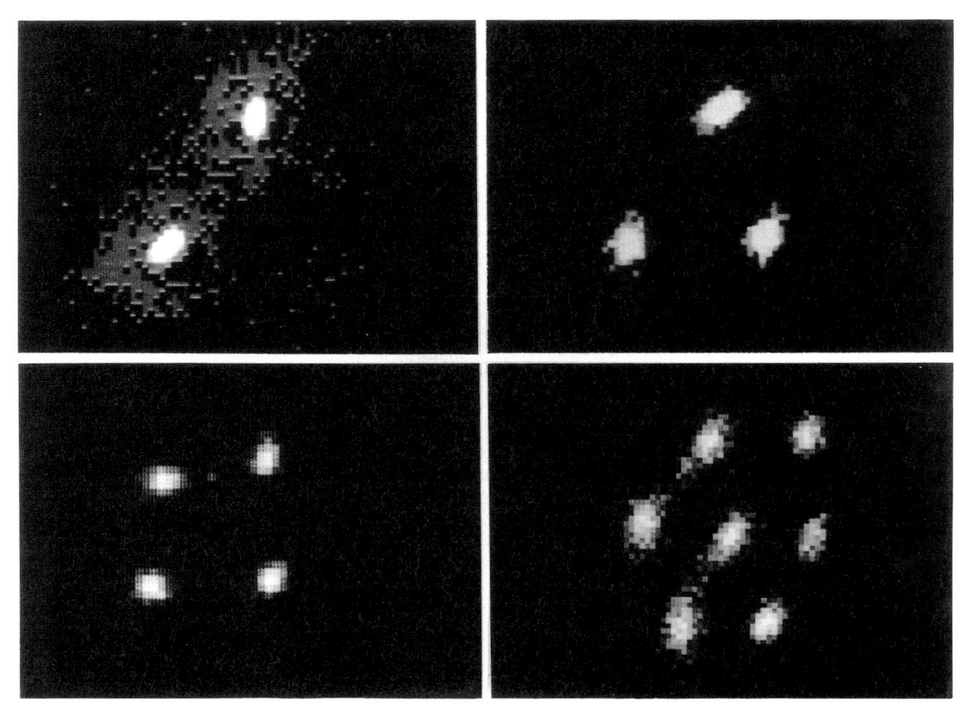

Fig. 9. – Two, three, four and seven ions confined by the dynamical potential of a Paul trap and crystallized into an ordered structure in a plane perpendicular to the symmetry axis of the trap. The average separation of the ions is 20 μm.

After we have accepted the existence of the ion crystals and the corresponding phase transitions, how do they actually occur? Would the cooling laser not force any cloud immediately to crystallize? A heating mechanism balancing the cooling effect of the laser must be the answer to this puzzle, but what heating mechanism? Since the early days of Paul traps this so-called radiofrequency heating has repeatedly been cited [1]. A deeper understanding, however, was missing and was provided only recently by a detailed study of the dependence of the cloud → crystal and crystal → cloud phase transitions on the relevant parameters [37-39].

The ions are subjected to essentially four different forces: the first one arising from the trapping field, then the Coulomb interaction between the ions, the laser cooling force, and finally a random force, arising from the spontaneously emitted photons. Using these forces, computer simulations of the motion of the ions can be performed [37]. Depending on the external parameters such as the laser power, the laser detuning and the radiofrequency voltage, the experimentally observed phenomena could be reproduced. Some of the results of the simulations are summarized in fig. 10. Plotted is the radiofrequency heating parameter κ of five ions *vs.* their mean separation [39]. For zero laser power and

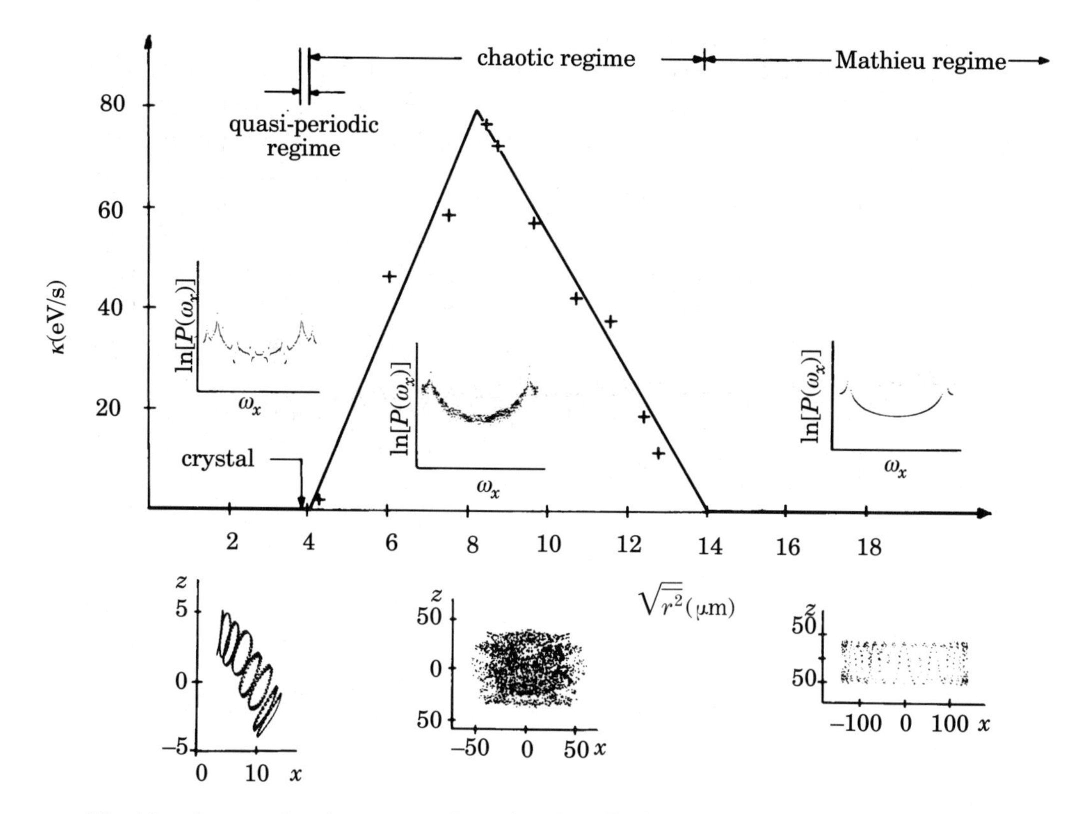

Fig. 10. – Average heating rate κ of two ions in a Paul trap *vs.* mean ion separation. The insets show the power spectrum and the corresponding stroboscopic Poincaré sections (plane perpendicular to the symmetry axis) of relative separation of the two ions in three characteristic domains: the crystal state, the chaotic regime and the Mathieu regime. The units on the axes are in μm. In order to calculate the power spectrum of the «crystal» shown on the left-hand side, the distance of the two ions was displaced by 1 μm from the equilibrium position. The Mathieu regime shown on the right is dominated by the secular motion.

large r, we did not observe any net heating of the ions. This is confirmed by our experiments, in which, even in the absence of a cooling laser, large clouds of ions can be stored in a Paul trap over several hours without being heated out of the trap. When the ions are far apart, the Coulomb force is small, and on short time scales the ions behave essentially like independent single stored ions. For this reason, we call this part of the heating diagram the Mathieu regime [38, 39]. Turning on a small laser, the r.m.s. radius r reduces drastically, but comes to a halt at about 14 μm. At this distance the nonlinear Coulomb force between the ions plays an important role and the motion of the ions gets chaotic. In this situation the power spectrum of the ions becomes a continuum leading to a radiofrequency heating process.

Increasing the laser power further results in an even smaller cloud. The smaller cloud produces more chaotic radiofrequency heating, as seen clearly by the negative slope of the heating curve in the range $8\,\mu\text{m} < r < 14\,\mu\text{m}$. Finally, in the range $4\,\mu\text{m} < r < 8\,\mu\text{m}$ there is still chaotic heating, but the slope of the heating curve is positive. As a consequence of the resulting triangular shape of the heating curve at a laser power of about $P = 150\,\mu\text{W}$, corresponding to $r \approx \approx 8\,\mu\text{m}$, the chaotic-heating power can no longer balance the cooling power of the laser light and the cloud collapses into the crystalline state located at $r \approx 3.8\,\mu\text{m}$. At this point the amplitude of the ion motion is so small that the nonlinear part of the repulsive Coulomb force is negligible again, so that chaotic heating disappears and the phase transition occurs.

Due to this collapse of the cloud state, the behaviour of the heating rate in the range $3.8\,\mu\text{m} < r < 8\,\mu\text{m}$ cannot be studied by balancing laser cooling and radiofrequency heating. In this case we start out from the crystal state and slightly displace the ions to explore the vicinity of the crystal. We observe no heating for $3.8\,\mu\text{m} < r < 4\,\mu\text{m}$, but quasi-periodic motion, and thus dub this regime the «quasi-periodic» regime. We call the upper edge of the quasi-periodic regime ($r \approx 4\,\mu\text{m}$) the «chaos threshold». An initial condition beyond the chaos threshold, *i.e.* satisfying $r \approx 4\,\mu\text{m}$, leads to heating, and expansion of the ion configuration and numerical data relevant for the shape of the heating curve can be taken during this expansive phase. The laser power P is set to zero for this type of experiment. We conjecture that—apart from the trivial case of a single stored ion—the heating curve is universal, *i.e.* its qualitative shape, including the existence of the chaotic regime, does not depend on the number of simultaneously trapped ions, and even applies to systems as remotely connected as, *e.g.*, Rydberg atoms in strong electromagnetic fields.

For the quasi-periodic, the chaotic and the Mathieu regimes, respectively, we display the corresponding type of power spectrum as the insets above the abscissa of fig. 10. The data were actually taken for the case of two ions, but would not look much different in the five-ion case. We obtain a discrete spectrum in the quasi-periodic regime and a complicated noisy spectrum in the chaotic regime. The spectrum in the Mathieu regime is again quite simple and dominated by the secular-motion frequency. We also show stroboscopic pictures of the locations of the ions in the plane perpendicular to the symmetry axis of the trap characterizing the three regions (insets below the abscissa in fig. 10).

As a function of increasing r.f. voltage, the chaos threshold moves towards the radius of the crystalline configuration. However, before the chaos threshold reaches the crystal radius, the equations of motion of the ions become unstable in the direction of the symmetry axis, indicating that in this particular situation the particles would fall out of the trap. In order to achieve proper melting without losing particles, the crystal radius has to be enlarged artificially by noise so that the size of the distorted crystal overlaps the region of chaotic heating.

Proper melting without the assistance of noise should be possible if we start in a quasi-periodic state typified in the bottom left inset of fig. 10. Such configurations have a larger radius in the beginning, and the chaos threshold could be reached before the onset of the single-particle Mathieu instability, not otherwise [39].

8. – The ion storage ring.

A completely new era of accelerator physics could begin if it were possible to produce, store and accelerate Coulomb crystals in particle accelerators and storage rings. To work with crystals instead of the usual dilute, weakly coupled particle clouds has at least one advantage: the luminosity of accelerators (storage rings) could be greatly enhanced, and (nuclear) reactions whose cross-sections are too small to be investigated in currently existing accelerators would become amenable to experimentation.

In the following, we would like to discuss very briefly our recent experiments using a miniature quadrupole storage ring. The storage ring is similar to the one described by DREES and PAUL [40] or by CHURCH [41]. We can observe phase transitions of the stored ions and the observed ordered ion structures are quite similar to the ones expected in relativistic storage rings, however, much easier to achieve. The motivation for building this small storage ring came from

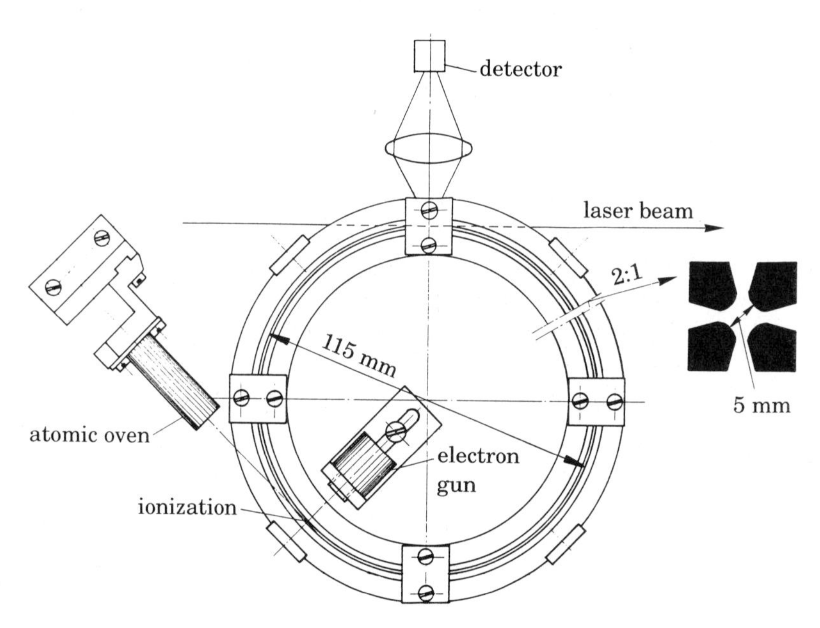

Fig. 11. – Quadrupole storage ring. The cross-section of the electrode configuration is shown in the insert on the right-hand side of the figure. The diameter of the ring is 113 mm and the distance between the electrodes is 5 mm.

the fact that micromotion perturbs the ion structures in a Paul trap and only a single trapped ion is free of micromotion [1]. The ring trap used consists of a quadrupole field, leading to harmonic binding of the ions in a plane transversal to the electrodes of the quadrupole and no confinement along the axis (see fig. 11). Confinement along the axis is achieved, however, by the Coulomb interaction between the ions when the ring is filled; then the total number of ions in the ring determines the average distance between them.

A scheme of the ring trap used for our experiment is shown in fig. 11 [42, 43]. It consists of four electrode rings shown in the insert on the right. The hyperbolic cross-section of the electrodes required for an ideal quadrupole field was approximated by a circular one. The experiments were also performed with Mg^+ ions. The ions were produced between the ring electrodes by ionizing the atoms of a weak atomic beam produced in an oven which injected the atoms tangentially into the trap region. The electrons used for the ionization came from an electron gun the electron beam of which was perpendicular to the direction of

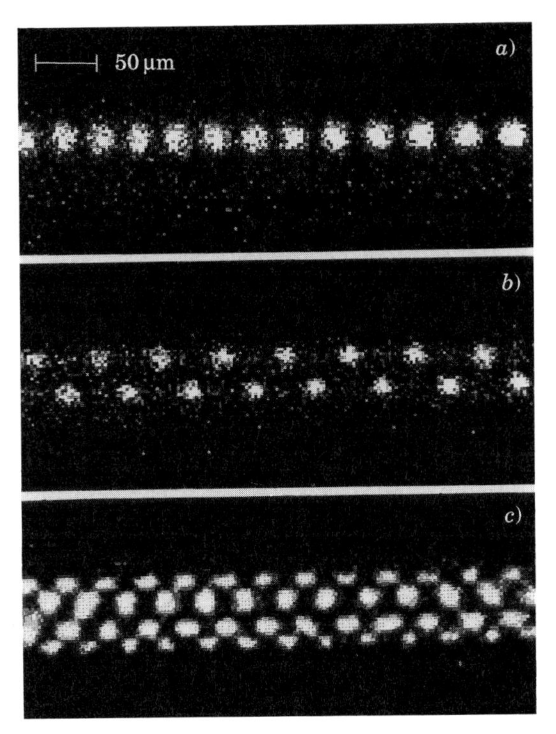

Fig. 12. – Crystalline structures of laser-cooled $^{24}Mg^+$ ions in the quadrupole storage ring. At a low ion density ($\lambda = 0.29$) the ions form a string along the field axis (*a*)). Increasing the ion density transforms the configuration to a zigzag (*b*), $\lambda = 0.92$). At still higher ion densities, the ions form ordered helical structures on the surface of a cylinder, *e.g.*, three interwoven helices at $\lambda = 2.6$ (*c*)). As the fluorescent light is projected onto the plane of observation, in this case the inner spots are each created by two ions seated on opposite sides of the cylindrical surface, resulting in a single, bright spot.

the atomic beam. A shutter in front of the atomic-beam oven allowed the interruption of the atomic flux. The ultrahigh-vacuum chamber was pumped by an ion getter pump. After baking the chamber a vacuum of 10^{-10} mbar could be reached. Under these conditions the number of ions stored in the trap stayed practically constant for several hours.

When laser cooling of the ions is started, a sudden change in the fluorescence intensity is observed, resembling very much that seen with stored ions in a Paul trap (fig. 8) which indicates a phase transition and the formation of an ordered structure of ions. The ion structure can also be observed using an ultrasensitive imaging system. Pictures of typical ion structures are shown in fig. 12. The ions are excited by a frequency-tunable laser beam which enters the storage ring tangentially. In the linear configuration the ions are all sitting in the centre of the quadrupole field; therefore, they do not show micromotion and it is possible to cool them further to temperatures in the microkelvin region. The new cooling methods proposed by DALIBARD *et al.* [34] can be applied to the Mg^+ ions so that the single-photon recoil limit can be achieved for the cooling process. At this limit the kinetic energy of the ions is smaller than the energy resulting from the «zero-energy» motion of the harmonically bound ions; the ion structure then reaches its vibrational ground state, *i.e.* a Mössbauer situation is generated. The observed ion configurations in the quadrupole ring trap are described in a recent paper by BIRKL *et al.* [44]. We will review the major results reported in this paper in the next section and compare the observed ordered structure to the results of molecular-dynamics calculations [45].

9. – Ordered structures in the storage ring.

In the molecular-dynamics calculations [45], a cylindrically symmetric, static harmonic potential is assumed to describe the confining field. Each particle is subjected to the Coulomb forces of all other particles and to the confining field. The classical equations of motion are integrated for a system of several thousand particles, starting with random positions and velocities, to give the time evolution of the system. Cooling of the stored particles is simulated by scaling down the resulting velocities of the stored particles at defined instants of the integration process. After sufficient cooling, ordered structures such as strings, zigzags, shells and multiple shells should arise owing to the confining field's harmonic potential. Our experiments are well suited to checking these predictions. To compare the experimental results with theory, we can use the normalized «linear particle density» λ which is given by the ion density multiplied by the ratio of Coloumb repulsion and confining force of the trap [45]. Low λ values correspond to a deep potential well or a small number of ions, resulting in an equilibrium structure closely confined to the field axis comprising a string of ions (fig. 12*a*)). This is the micromotion-free configuration discussed above, and

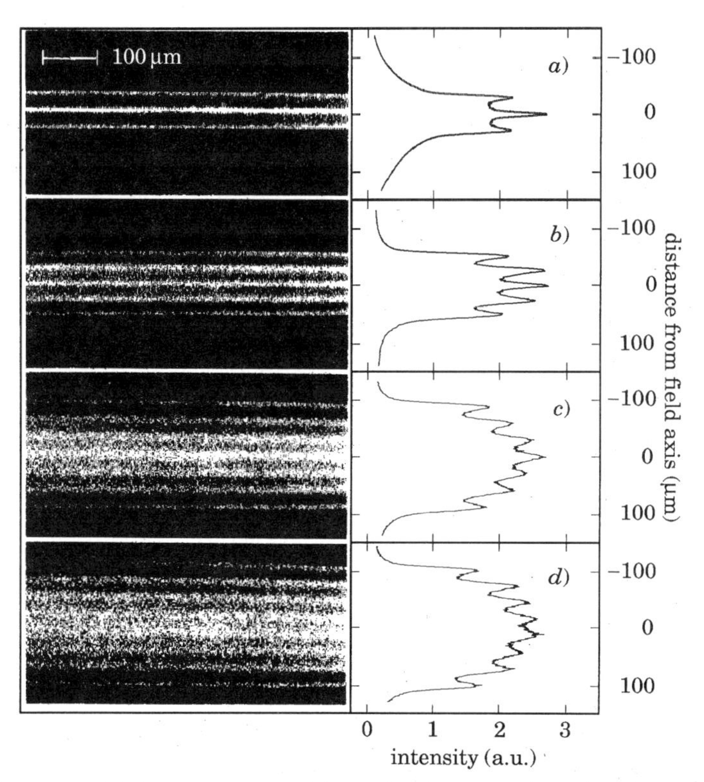

Fig. 13. – Images and intensity profiles of (from the top): one shell plus string (a), $\lambda \approx 4.3$), two shells plus string (b), $\lambda \approx 12.2$), three shells plus string (c), $\lambda \approx 26$) and four shells (d), $\lambda \approx 34$). There are up to $\approx 8 \cdot 10^5$ ions stored in the ring for the four-shell structure.

the analogue of the single stored ion in a Paul trap, as in both cases the ions sit in the potential minimum and show no micromotion. For higher values of λ, the structures extend more and more into the off-axis region, giving rise to (in the order of increasing λ) a plane zigzag structure (fig. 12*b*)) and cylindrical structures with the ions forming helices on cylindrical surfaces. The structure in fig. 12*c*) consists of three interwoven helices with six ions per pitch. The string and the zigzag have also been observed with laser-cooled Hg^+ ions in a linear trap [46].

Increasing further the number of ions leads to structures with smaller spacings between the ions where we cannot optically resolve individual ions any more. Images of these structures are presented in fig. 13. The radial intensity profiles displayed on the right-hand side of the figure provide information about the structures as they give a measure of the radial distribution of the ions. For increasing λ it becomes energetically more favourable to create a string inside the first ion shell (fig. 13*a*)) to provide space for more particles. This results in a structure which is a three-dimensional analogue of the plane seven-ion crystal for a Paul trap (fig. 9). The inner string turns into a second shell at higher densities: a string then develops inside this second shell (fig. 13*b*)), and so on. Figures 13*c*) and *d*) show structures consisting of three shells plus string and four shells, respectively. We have been able to observe all possible structures, from the string up to four shells plus string. The formation of multiple-shell crystalline structures in the quadrupole storage ring contrasts with the observation of shell structures in Penning traps, where the ions do not occupy fixed positions inside the shells [47].

10. – Comparison with theory.

Figure 14*a*) gives a summary of experimental data for all recorded images in which the ions were individually resolved. The depth of ψ_0 of the potential well and the ion density per unit length are the experimental parameters. The theoretical boundaries between the different shell structures, predicted in ref. [45] in terms of the functional dependence of λ on ψ_0 and the ion density, are given by the straight lines with constant λ. String structures are expected for $\lambda < 0.709$, zigzag structures in the range $0.709 < \lambda < 0.964$ and single shells in the range $0.964 < \lambda < 3.10$. Many different structures which are degenerate in energy are expected within the single-shell regime. We obtained stable configurations near $\lambda = 1.3$ and $\lambda = 2.0$ (two interwoven helices) and near $\lambda = 3.0$ (three interwoven helices, fig. 12*c*)). The observed structures agree with the predicted scheme for a large range of experimental parameters, thus confirming the theoretical results.

A summary of the experimental observations for ordered shell structures with up to four shells plus string and without resolution of individual ions is presented in fig. 14*b*). The depth ψ_0 of the confining potential well is again one of the experimental parameters. As the ion density cannot be determined directly from the images, the radius ρ of the structures is used instead as the second parameter. The theoretically predicted boundaries between the different shell structures are again given as straight lines of constant λ following [45] where the dependence of ρ on λ and ψ_0 was established. The observed ion configurations are seen once more to be determined by λ for a wide range of potential depths and ion densities.

Our results have important implications for two very different fields. Consider first the physics of low-energy particles: an ion string in a quadrupole ring, being free of micromotion, can be cooled to its vibrational ground state in the confining potential using recently proposed laser-cooling techniques [34]. This would place the string in the Lamb-Dicke regime with a vanishing first-order Doppler effect because the spatial amplitude of the motion is smaller than the wavelength of the atomic transition. Furthermore, the second-order Doppler effect, which can only be reduced by further cooling, also disappears

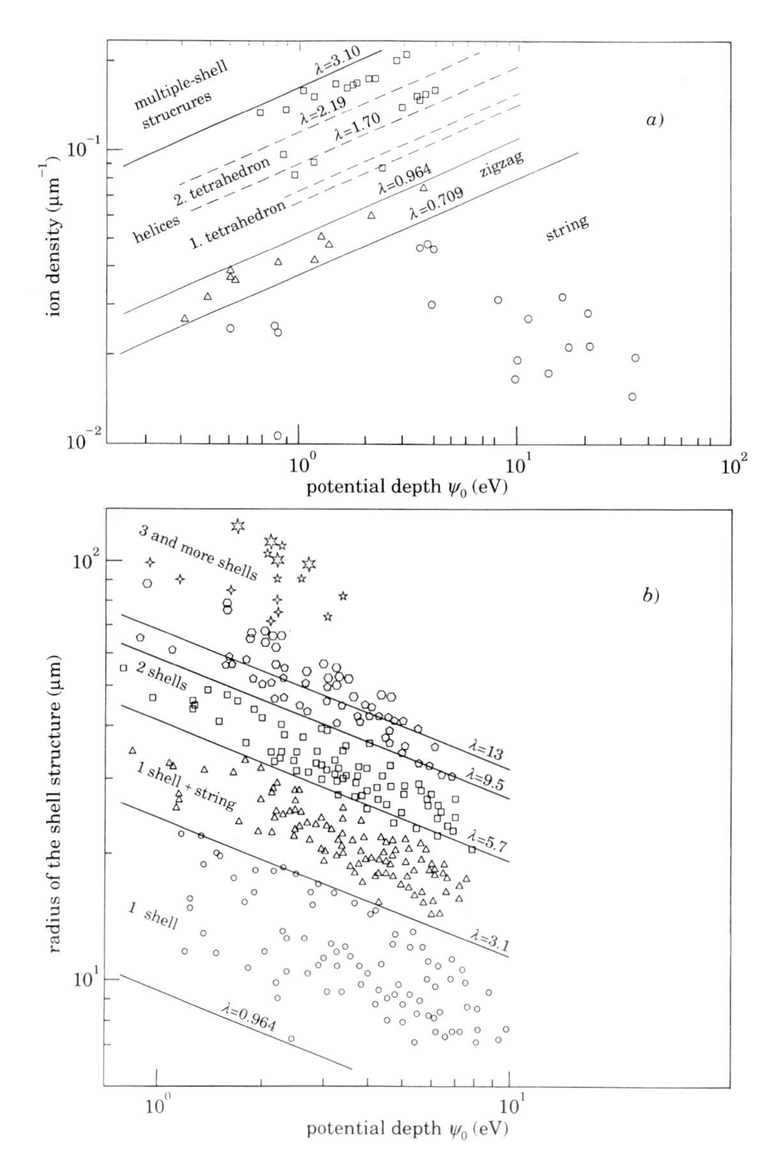

Fig. 14. – Summary of the experimental results. *a*) Individual ions resolved, where the observed structures are characterized by the ion density per unit length and the depth of the potential well ψ_0. These two parameters can be combined to give the normalized linear particle density λ which fully determines the ion configuration. The straight lines show critical λ values separating the regions corresponding to the various theoretically expected structures. The observed configurations are labelled with different symbols for each structure: ○ strings, △ zigzags, □ helices. *b*) Individual ions unresolved, where the observed shell structures with up to four shells plus string are characterized by their radius ρ and the potential depth ψ_0. The various structures are again separated by lines of theoretically determined critical λ; ○ 1 shell, △ 1 shell + string, □ 2 shells, ⬠ 2 shells + string, ⬡ 3 shells, ✧ 3 shells + string, ☆ 4 shells, ✡ 4 shells + string.

making the stored ions very interesting for frequency standards. The large number of ultracold ions available in the ring will lead to a high signal-to-noise ratio. Finally, cooled ions in the ring represent a quantum object of macroscopic dimensions (a Wigner crystal).

Second, with the experimental confirmation that the ordered structures expected in high-energy ion storage rings can indeed be formed, a completely new era of accelerator physics will emerge if it is possible to reproduce such Coulomb crystals in these rings. The enhanced luminosity of the corresponding beams would allow studies of ionic reactions whose cross-sections are too small for investigations in existing accelerators.

REFERENCES

[1] D. J. Wineland, W. M. Itano, J. C. Bergquist, J. J. Bollinger and J. D. Prestage: in *Atomic Physics 9,* edited by R. S. van Dyck jr. and E. N. Fortson (World Scientific Publishing, Singapore, 1984).

[2] S. Chu, J. E. Bjorkholm, A. Ashkin and A. Cable: in *Atomic Physics 10,* edited by H. Narumi and S. Shimamura (North-Holland, Amsterdam, 1986); D. E. Pritchard, K. Helmerson and A. G. Martin: in *Atomic Physics 11,* edited by S. Haroche, J. C. Gay and G. Grynberg (World Scientific Publishing, Singapore, 1988).

[3] D. Meschede, H. Walther and G. Müller: *Phys. Rev. Lett.*, **54**, 551 (1985).

[4] For a review, see the articles by S. Haroche and J. M. Raimond: in *Advances in Atomic and Molecular Physics,* Vol. **20** (Academic Press, New York, N.Y., 1985), p. 350; J. A. Gallas, G. Leuchs, H. Walther and H. Figger: in *Advances in Atomic and Molecular Physics,* Vol. **20** (Academic Press, New York, N.Y., 1985), p. 413.

[5] E. T. Jaynes and F. W. Cummings: *Proc. IEEE*, **51**, 89 (1963).

[6] See, for example, J. H. Eberly, N. B. Narozhny and J. J. Sanchez-Mondragon: *Phys. Rev. Lett.*, **44**, 1323 (1980), and references therein.

[7] G. Rempe, H. Walther and N. Klein: *Phys. Rev. Lett.*, **58**, 353 (1987).

[8] D. F. Walls: *Nature (London)*, **280**, 451 (1979); see also articles in *Photons and Quantum Fluctuations*, edited by E. R. Pike and H. Walther (Hilger, Bristol, 1988).

[9] D. F. Walls: *Nature (London)*, **306**, 141 (1983); **324**, 210 (1986); see also the various articles in *Squeezed and Nonclassical Light,* edited by P. Tombesi and E. R. Pike (Plenum Press, New York, N.Y., 1988).

[10] First demonstration of photon antibunching: H. J. Kimble, M. Dagenais and L. Mandel: *Phys. Rev. Lett.*, **39**, 691 (1977); *Phys. Rev. A,* **18**, 201 (1978); see also J. D. Cresser, J. Häger, G. Leuchs, M. Rateike and H. Walther: in *Dissipative Systems in Quantum Optics* (Springer, Berlin, 1982), p. 21.

[11] First demonstration of sub-Poissonian photon statistics: R. Short and L. Mandel: *Phys. Rev. Lett.*, **51**, 384 (1983).

[12] First demonstration of squeezing: R. E. Slusher, L. W. Hollberg, B. Yurke, J. C. Mertz and J. F. Valley: *Phys. Rev. Lett.*, **55**, 2409 (1985); for reviews, see, *e.g.*, H. J. Kimble and D. Walls: *J. Opt. Soc. Am. B*, **4**, 1449 (1987); R. Loudon and P. L. Knight: *J. Mod. Opt.*, **34**, 707 (1987).

[13] H. J. Carmichael and D. F. Walls: *J. Phys. B*, **9**, 1199 (1976).

[14] F. Diedrich and H. Walther: *Phys. Rev. Lett.*, **58**, 203 (1987).

[15] P. FILIPOWICZ, J. JAVANAINEN and P. MEYSTRE: *Opt. Commun.*, **58**, 327 (1986); *Phys. Rev. A,* **34**, 3077 (1986); *J. Opt. Soc. Am. B,* **3**, 906 (1986).
[16] L. LUGIATO, M. O. SCULLY and H. WALTHER: *Phys. Rev. A*, **36**, 740 (1987).
[17] J. KRAUSE, M. O. SCULLY and H. WALTHER: *Phys. Rev. A*, **36**, 4547 (1987); J. KRAUSE, M. O. SCULLY, T. WALTHER and H. WALTHER: *Phys. Rev. A,* **39**, 1915 (1989).
[18] P. MEYSTRE: *Opt. Lett.*, **12**, 669 (1987); in *Squeezed and Nonclassical Light,* edited by P. TOMBESI and E. R. PIKE (Plenum, New York, N.Y., 1988), p. 115.
[19] P. MEYSTRE, G. REMPE and H. WALTHER: *Opt. Lett.*, **13**, 1078 (1988).
[20] J. J. SLOSSER, P. MEYSTRE and E. M. WRIGHT: *Opt. Lett.*, **15**, 233 (1990).
[21] G. REMPE and H. WALTHER: *Phys. Rev. A*, **42**, 1650 (1990).
[22] G. REMPE, F. SCHMIDT-KALER and H. WALTHER: *Phys. Rev. Lett.*, **64**, 2783 (1990).
[23] L. MANDEL: *Opt. Lett.*, **4**, 205 (1979).
[24] See, for example, D. BOHM: *Quantum Theory* (Prentice Hall, Englewood Cliffs, N.J., 1951) or M. JAMMER: *The Philosophy of Quantum Mechanics* (Wiley, New York, N.Y., 1974).
[25] A detailed analysis of Einstein's version of the double-slit experiment is given by W. WOOTTERS and W. ZUREK: *Phys. Rev. D,* **19**, 473 (1979); see also ref. [26].
[26] For an excellent presentation of the Bohr-Einstein dialogue see Chapt. 1 in J. A. WHEELER and W. H. ZUREK: *Quantum Theory and Measurement* (Princeton University Press, Princeton, N.J., 1983) and, in particular, the article by N. BOHR: *Discussion with Einstein on epistemological problems in atomic physics.*
[27] B.-G. ENGLERT, J. SCHWINGER and M. O. SCULLY: *Found. Phys.*, **18**, 1045 (1988); J. SCHWINGER, M. O. SCULLY and B.-G. ENGLERT: *Z. Phys. D*, **10**, 135 (1988); M. O. SCULLY, B.-G. ENGLERT and J. SCHWINGER: *Phys. Rev. A,* **40**, 1775 (1989).
[28] M. O. SCULLY and H. WALTHER: *Phys. Rev. A*, **39**, 5229 (1989).
[29] M. O. SCULLY, B.-G. ENGLERT and H. WALTHER: *Nature (London)*, **351**, III (1991).
[30] B.-G. ENGLERT, H. WALTHER and M. O. SCULLY: *Appl. Phys. B*, **54**, 366 (1992).
[31] J. KRAUSE, M. O. SCULLY and H. WALTHER: *Phys. Rev. A*, **34**, 2032 (1986).
[32] W. PAUL, O. OSBERGHAUS and E. FISCHER: *Ein Ionenkäfig, Forschungsber. des Wirtschafts- und Verkehrsministeriums Nordrhein-Westfalen,* 415 (1958); E. FISCHER: *Z. Phys.,* **156**, 1 (1959).
[33] H. G. DEHMELT: in *Adv. At. Mol. Phys.*, **3**, edited by D. R. BATES and I. ESTERMANN (Academic Press, New York, N.Y., 1967), p. 53; F. M. PENNING: *Physica,* **3**, 873 (1936).
[34] J. DALIBARD, C. SALOMON, A. ASPECT, E. ARIMONDO, R. KAISER, N. VANSTEENKISTE and C. COHEN-TANNOUDJI: in *Atomic Physics,* **11**, edited by S. HAROCHE, J. C. GAY and G. GRYNBERG (World Scientific Publishing, Singapore, 1988), p. 199, and other contributions to this volume.
[35] F. DIEDRICH, E. PEIK, J. M. CHEN, W. QUINT and H. WALTHER: *Phys. Rev. Lett.*, **59**, 2931 (1987).
[36] W. NEUHAUSER, M. HOHENSTATT, P. TOSCHEK and H. DEHMELT: *Phys. Rev. Lett.*, **41**, 233 (1978); *Phys. Rev. A,* **22**, 1137 (1980).
[37] R. BLÜMEL, J. M. CHEN, W. QUINT, W. SCHLEICH, Y. R. SHEN and H. WALTHER: *Nature (London)*, **334**, 309 (1988).
[38] R. BLÜMEL, J. M. CHEN, F. DIEDRICH, E. PEIK, W. QUINT, W. SCHLEICH, Y. R. SHEN and H. WALTHER: in *Atomic Physics,* **11**, edited by S. HAROCHE, J. C. GAY and G. GRYNBERG (World Scientific Publishing, Singapore, 1988), p. 243.
[39] R. BLÜMEL, C. KAPPLER, W. QUINT and H. WALTHER: *Phys. Rev. A*, **40**, 808 (1989).

[40] J. DREES and W. PAUL: *Z. Phys.*, **180**, 340 (1964).
[41] D. A. CHURCH: *J. Appl. Phys.*, **40**, 3127 (1969).
[42] H. WALTHER: in *Proceedings of the Workshop on Light Induced Kinetic Effects on Atoms, Ions and Molecules,* edited by L. MOI, S. GOZZINI, C. GABBANINI, E. ARIMONDO and F. STRUMIA (ETS Editrice, Pisa, 1991), p. 261.
[43] I. WAKI, S. KASSNER, G. BIRKL and H. WALTHER: *Phys. Rev. Lett.*, **68**, 2007 (1992).
[44] G. BIRKL, S. KASSNER and H. WALTHER: *Nature (London)*, **357**, 310 (1992).
[45] R. W. HASSE and J. P. SCHIFFER: *Ann. Phys. (N.Y.)*, **203**, 419 (1990); A. RAHMAN and J. P. SCHIFFER: *Phys. Rev. Lett.*, **57**, 1133 (1986).
[46] M. G. RAIZEN, J. M. GILLIGAN, J. C. BERGQUIST, W. M. ITANO and D. J. WINELAND: *J. Mod. Opt.*, **39**, 233 (1992).
[47] S. L. GILBERT, J. J. BOLLINGER and D. J. WINELAND: *Phys. Rev. Lett.*, **60**, 2022 (1988).

Single Ions for Metrology and Quantum Optics.

P. E. TOSCHEK
Institut für Laser-Physik, Universität Hamburg - D-20355 Hamburg, B.R.D.

1. – Introduction.

The spatial confinement of atomic particles in electromagnetic fields, or «trapping», is an old concept. Various techniques have been designed for the storage of ensembles of charged particles in a small volume for an extended time in order to facilitate precise electromagnetic measurements among which mass spectrometry stands out [1]. However, ion trapping has drawn particular attention for spectroscopic studies during the past decade [2, 3], since light sources have become available whose spectral purity and brightness allow the spectroscopist to fully utilize the remarkable features of the localized particles. They include i) long, sometimes indefinite times of interaction with the impinging radiation, ii) the absence of wall effects, iii) the potential of substantially reducing the particles' speed and speed-related effects, and iv) the opportunity to single out individual atomic particles. The ensuing benefits of these features are precise measurements on a well-defined material probe which may even be a *single* ion [4, 5]. Thus repeated measurements by a macroscopic device on a microphysical system have become feasible. Consequently, principal motivations of those measurement are i) the demonstration of frequency standards free of systematics, a crucial problem of metrology, and ii) the study of the interaction of microphysical systems with light, *i.e.* the challenges of quantum optics.

A substantial number of microwave-optical double-resonance experiments on trapped ion clouds have been reported [6]. These experiments allow the precise determination of the hyperfine-resonance frequencies of ionic species with many significant figures. When one of these high-Q resonances is used for the frequency control of the microwave oscillator, the locked system makes a clock. Such a clock, working on the ground-state hyperfine transition of Be^+ at 303 MHz, has been demonstrated on some 10^4 ions in a Penning trap [7]. In a recent version, the ion sample was collisionally (or «sympathetically») cooled by a simultaneously trapped cloud of optically cooled Mg^+. The detection of the hyperfine resonance was by two r.f. pulses of 100 s separation. The signal frequency when detuned from resonance gives rise to a phase shift, that is accumulated during the pulse delay time τ. This phase shift appears between the freely precessing hyperfine coherence, generated by the first pulse, and the second signal pulse, and it causes a related periodicity both in the population of the upper hyperfine level of the ground state, and in the laser-excited fluorescence, which is known as temporal «Ramsey fringes». The width of the central fringe was 0.9 mHz, and the frequency stability, for τ up to 10^4 s, was found to be $3 \cdot 10^{-12} \tau^{-1/2}$ by comparison with a hydrogen maser.

Another recent example of an ion clock is based on the ground-state hyperfine transition of $^{171}Yb^+$ at 12.6 GHz [8]. A r.f. (or Paul) trap contained some 10^6 Yb^+ ions which were exposed to the microwave radiation, and to coherent light at 369.4 nm wavelength, generated by frequency-doubling red dye-laser light. This light excites the $^2S_{1/2}$-$^2P_{1/2}$ resonance line and the pertaining fluorescence and pumps the ions into the $F = 1$ level of the ground state, such that the fluorescence disappears in the steady state. On resonance, the microwave induces transitions to the $F = 0$ level and makes the fluorescence reappear. The spectrum of the resonance, *i.e.* the rate of fluorescence *vs.* microwave tuning, shows a 400 mHz wide line whose Q is $3 \cdot 10^{10}$ with $S/N \simeq 50$. Accordingly, the potential minimum instability of an oscillator that is locked to this line is $\sigma_y = (Q \times \times S/N)^{-1} \simeq 6 \cdot 10^{-13} \tau^{-1/2}$. Such a frequency-controlled oscillator was set to work by a feedback loop in the form of a computer code; its observed frequency instability was on the order of $10^{-11} \tau^{-1/2}$. Although Q is open to improvement, this figure still exceeds the performance of commercial cesium clocks. Moreover, a twenty-time narrower resonance has been observed recently, when magnetic stray fields that cause inhomogeneous line broadening were screened with extreme care [9].

These examples show that the fluorescence of the resonance signal of an ionic cloud may be recorded with such a satisfactory signal-to-noise ratio that oscillators controlled by it make clocks that are competitive with conventional ones. However, such a cloud is prone to the widely varying conditions to which individual ions are subject at different locations inside the cloud. These conditions include the strength of the trapping field which increases with the distance from the centre, the field of space charge and the speed of the ions—both of their driven or «micro» motion, and of their secular motion in the pseudopotential of the trap. Therefore, spectra of the ion cloud are complicated by line shifts and inhomogeneous broadenings of various kinds. Moreover, it has turned out that laser-cooling the ions in a cloud makes their minimum temperature severely limited by the pickup of energy during ion collisions, and Doppler shifts keep contributing appreciably to the width of optical lines. In addition, ion-ion collisions may perturb the response of the ensemble and give rise to more spectral broadening.

This is why, after all, a *single* trapped ion seems a natural sample for spectroscopic measurements of ultra-high resolution. The photon count rate of an ion's laser-excited fluorescence runs as high as 10^5/s. If limited by quantum noise only, the detection warrants a favourable signal-to-noise ratio. The individual trapped ion then represents a microphysical object under continued scrutiny. It is the prime object for testing quantum modelling of the measuring processes and its epistemological corollaries: An individual trajectory of the object in phase space becomes observable, as contrasted with the expectation values of conventional quantum measurement on large ensembles. Thus what follows is restricted to observations on a single ion in a r.f. trap.

2. – Line shapes of a single ion.

All species of single atomic particles so far observed have been trapped ions—Ba^+ [4], Mg^+ [10], Hg^+ [11], Sr^+ [12] and recently Yb^+, see below—, and they have been detected—with one exception [13]—by their laser-excited fluorescence. The required high rate of excitation and fluorescence seems to conflict with high spectral resolution, and in particular with the observation of those narrow lines in the excitation spectrum which are dipole-forbidden. This dilemma is overcome by pursuing one out of two strategies: to detect either resonance fluorescence that is quantized by the events of weak-line absorption in a single-ion version of double resonance, or the excitation of a dark line in the excitation spectrum which is the signature of a «trapped state» of the ion (see fig. 1).

The excitation of a *single* atomic particle on a narrow and weak dipole-forbidden line makes quench the excitation and fluorescence on a resonance line that shares a level with the weak signal line for as long as the atom stays in the metastable state that pertains to this signal line [14]. The excitation to that metastable level leaves the atom unavailable for the excitation of fluorescence

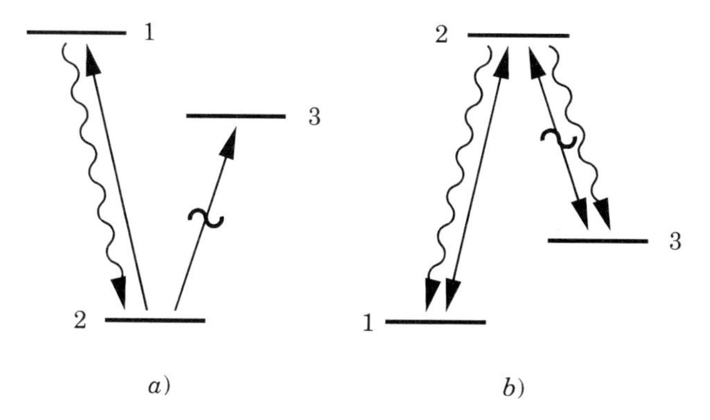

Fig. 1. – Fluorescence on $1 \to 2$ shows dark periods when signal light is tuned to transition 1-3 (*a*)), on $2 \to 1$ a dark line when detunings of the light fields from their respetive resonances are equal (*b*)).

TABLE I. – *Quantized double resonance*: Resonance fluorescence $^2P_{1/2}$-$^2S_{1/2}$ observed *vs.* tuning of laser across signal transition $^2S_{1/2}$-$^2D_{5/2}$ of width $\Delta\nu_{\text{res}}$.

	λ_0	$\Delta\nu_0$	λ_{laser}	$\Delta\nu_{\text{laser}}$	$\Delta\nu_{\text{res}}$	References
$^{199}Hg^+ (F=0\rightarrow 2)$	282 nm	1.7 Hz	563 nm	few Hz	172 Hz	[11]
Ba^+	1.76 μm	4 mHz	1.76 μm	15 kHz	35 kHz	[18]
Sr^+	674 nm	500 mHz	674 nm	1 MHz	few MHz	[12]

on the strong resonance line, as verified by observation [15-17]. In this concept the signal resonance is defined by the maximum rate of dark intervals, *i.e.* of those pauses when the fluorescence is switched off. Examples for the detection of ultra-narrow resonances by quantized fluorescence are given in table I. These observations involve, with the indicated species, the respective $E2$ transition $^2S_{1/2}$-$^2D_{5/2}$ as the signal line, and detection of the fluorescence on the $E1$ resonance line $^2S_{1/2}$-$^2P_{1/2}$. The fluorescence from the metastable $^2D_{5/2}$ level is far too weak to be detected, but the stochastic modulation of the resonance fluorescence yields a signal whose probability of detection is 100%. Note the 172 Hz wide resonance of $^{199}Hg^+$, the smallest width of an optical line observed so far, which is still dominated by power broadening and residual frequency drifts of the light sources.

The recording of an ultra-narrow line requires the ion to be almost at rest in the trap. This state is accomplished by laser cooling [19, 20]. In fact, the Hg^+ ion was laser-cooled down to 50 μK, corresponding to the ion being 95% of the time in the lowest vibrational level of the trap potential, by an arrangement of two alternate cooling processes. This scheme includes Doppler pre-cooling [21] by down-tuned laser light on the resonance line, and additional cooling on the weak and narrow signal line. Since this line is even narrower than the frequency spacing of the vibrational sidebands which result from the residual motion of the ion in the trapping potential, only the first lower sideband may be excited by the light, whereas spontaneous re-emission is centred about the carrier frequency, and the energy defect cools the ion. This cyclic process is efficient and makes a very low final temperature achievable («sideband cooling» [22]).

The second strategy for recording narrow lines of a single ion makes use of two-photon-resonant electronic Raman-type excitation of an $E2$ signal line to a metastable state by bichromatic laser light. Its two frequencies are close to resonance with an intermediate level, usually $^2P_{1/2}$, whereas the metastable level is $^2D_{3/2}$ (see fig. 1*b*)). When the difference frequency of the light components meets the Raman resonance $^2S_{1/2}$-$^2D_{3/2}$, the laser-excited fluorescence vanishes: this is what has been called «dark line» of a Λ-shaped level configuration [23-25]. Semi-classical theory allows one to model this phenomenon.

The rate of fluorescence measures the population of the upper resonance level, which is calculated from a steady-state solution of the three-level rate equations

$$
(1)\quad \begin{cases}
\dot\rho_{11} = -\frac{i}{2}\Omega_g(\rho_{12}-\rho_{21}) + \Gamma_g\rho_{22}\,,\\
\dot\rho_{33} = -\frac{i}{2}\Omega_r(\rho_{32}-\rho_{23}) + \Gamma_r\rho_{22}\,,\\
\dot\rho_{22} = -\dot\rho_{11} - \dot\rho_{33}\,,\\
\dot\rho_{12} = -i\Delta_g\rho_{12} - \frac{1}{2}(\Gamma_r+\Gamma_g)\rho_{12} - \frac{i}{2}\Omega_r\rho_{13} + \frac{i}{2}\Omega_g(\rho_{22}-\rho_{11})\,,\\
\dot\rho_{32} = -i\Delta_r\rho_{32} - \frac{1}{2}(\Gamma_r+\Gamma_g)\rho_{23} - \frac{i}{2}\Omega_g\rho_{31} + \frac{i}{2}\Omega_r(\rho_{22}-\rho_{33})\,,\\
\dot\rho_{13} = -i(\Delta_g-\Delta_r)\rho_{13} + \frac{i}{2}\Omega_g\rho_{23} - \frac{i}{2}\Omega_r\rho_{12} - \gamma\rho_{13}\,.
\end{cases}
$$

Here $\rho_{ij} = \langle i|\psi\rangle\langle\psi|j\rangle$ is a population (for $i=j$) or a coherent superposition ($i\neq j$) of states, $|\psi\rangle = \sum_k c_k \exp[-i\omega_k t]\,|k\rangle$, where c_k is complex, $i, j, k = \{1, 2, 3\}$, and γ is the decay rate of the «Raman» coherence 1-3. Ω, Δ and Γ stand for the Rabi frequencies, light detunings and decay rates, respectively, on the two transitions which are labelled «green» and «red» in anticipation of the experiment. The equations show that the ion dynamics is symmetric in the levels 1 and 3 as well as in the transitions 1-2 and 2-3. Whereas the optical coherences ρ_{12}, ρ_{23} are excited by the light fields E_g, E_r via $\Omega_g = |d_{12}|E_g/\hbar$, $\Omega_r = |d_{23}|E_r/\hbar$, respectively ($d_{ij}$ are dipole matrix moments), the Raman coherence requires simultaneous excitation by both light fields. A typical solution for the population of the upper level, *vs.* the detuning of one of the two light frequencies, is shown in fig. 2*a*) for equal Rabi frequencies, $\Omega_g = \Omega_r$ and in fig. 2*b*) for unequal ones. Note the narrow dark line at equal detunings of the two frequency components of the light which corresponds to a null of the fluorescence. The reason why the fluorescence vanishes is uncovered when we transform the Bloch equations with the help of a new basis set of eigenfunctions

$$
(2)\qquad |2\rangle,\qquad |c\rangle = \alpha|1\rangle + \beta|3\rangle,\qquad |n\rangle = \beta|1\rangle - \alpha|3\rangle,
$$

with $\alpha = \Omega_g/\sqrt{\Omega_g^2+\Omega_r^2}$, $\beta = \Omega_r/\sqrt{\Omega_g^2+\Omega_r^2}$. We find, for the special case of equal Rabi frequencies, $\Omega_r = \Omega_g = \Omega$, and $\gamma_{13} = 0$,

$$
(3)\quad \begin{cases}
\dot\rho_{cc} = (i/2)\,\Omega(\rho_{2c}-\rho_{c2}) + (1/2)(\Gamma_g+\Gamma_r)\rho_{22} + (i/2)(\Delta_g-\Delta_r)(\rho_{cn}-\rho_{nc})\,,\\
\dot\rho_{nn} = (1/2)(\Gamma_g+\Gamma_r)\rho_{22} - (i/2)(\Delta_g-\Delta_r)(\rho_{cn}-\rho_{nc}).
\end{cases}
$$

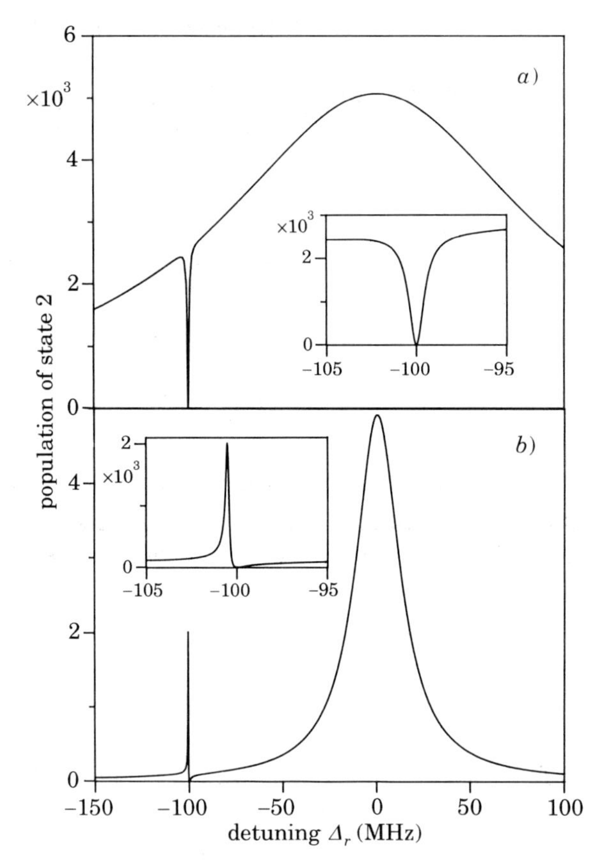

Fig. 2. – Population of intermediate level 2, and rate of fluorescence, *vs.* detuning Δ_r of one of the light frequencies from resonance, for three-level atom (Λ-shaped energy levels). $\Gamma_g = \Gamma_r = \Gamma/2 = 2\pi \times 10.2$ MHz, $\Delta_g = -100$ MHz. *a*) $\Omega_g = \Omega_r = \Gamma/2$. *b*) $\Omega_g = 10\,\Omega_r = 1.4\Gamma_g$.

These equations of motion reveal a specific asymmetry of the states $|n\rangle$ and $|c\rangle$: The coupling state c interacts with the light which transfers population, via the excited coherence ρ_{2c}, to the intermediate state 2. In contrast, the noncoupling state n does not interact with the light at all. Thus optical pumping fills the latter state up whose population remains «trapped».

For very unequal Rabi frequencies, $\Omega_g \gtrless \Omega_r$, the shape of the dark line becomes dispersive (see fig. 2*b*)). Although the fluorescence null remains at $\Delta_g = \Delta_r$, the side peak represents the resonance of state c which is light-shifted by the unbalanced part of the light intensity. This particular feature will be detailed in the next section.

In a spectroscopic study on a single trapped Ba^+ ion, the dark lines corresponding to «trapped states» of the ion have been observed, and their shapes have been compared with the solutions of Bloch equations [26]. The ion was pre-

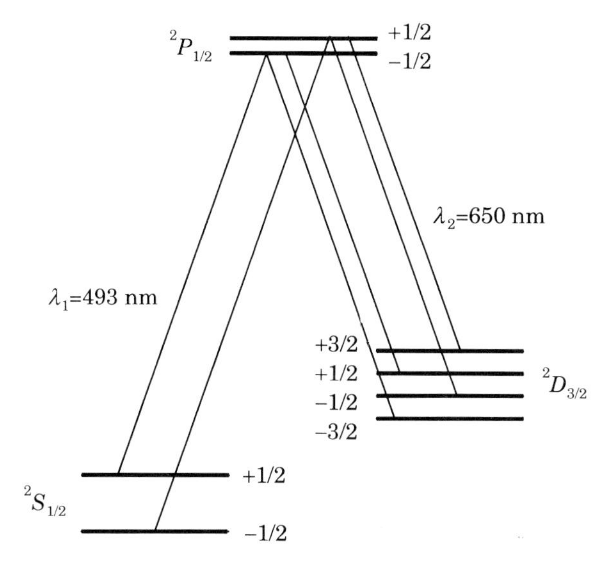

Fig. 3. – Three-level system including Zeeman sublevels. Indicated transitions are those allowed for $\Delta m_J = \pm 1$.

pared, by electron impact on evaporated Ba atoms, inside a 1 mm wide r.f. trap that consists of a ring connected to the r.f. circuit and two grounded tips opposing each other on the symmetry axis; both are made of 0.2 mm thick wire. Of utmost importance is the precise positioning of the ion in the electric centre of the trap, *i.e.* at the saddle point of the time-varying potential. To this end, the photon-counting signal of the fluorescence light is made to provide the stop pulse of a time-to-digital converter whose start pulse is derived from a selected phase of the r.f. trapping voltage, which also represents a particular phase of the ion's voltage-driven micromotion. The distribution of photon counts over the delay times is modulated, since the periodic Doppler shift of the ion's resonance frequencies shows up as varying probability of excitation of the fluorescence with the given tuning of the lasers. Four pin-shaped electrodes around the trap, loaded by adjustable voltage, allow one to spatially shift the ion. Vanishing modulation is the signature of the ion being placed in the electric centre of the trap. The residual imprecision of this method of ion positioning is as small as 10 nm.

For the recording of excitation spectra of the ionic fluorescence, a nonzero magnetic field is indispensable since otherwise optical pumping carries the ion into those Zeeman substates of the $^2D_{3/2}$ level which are unaffected by the laser light—*e.g.*, the $m = \pm 3/2$ states, with linear polarization of the light. The actual magnetic field at the ion's position—a few gauss—is the vectorial sum of the ambient field—which is made up of the Earth magnetic field and small residual magnetization of the electrodes—and the applied field generated by

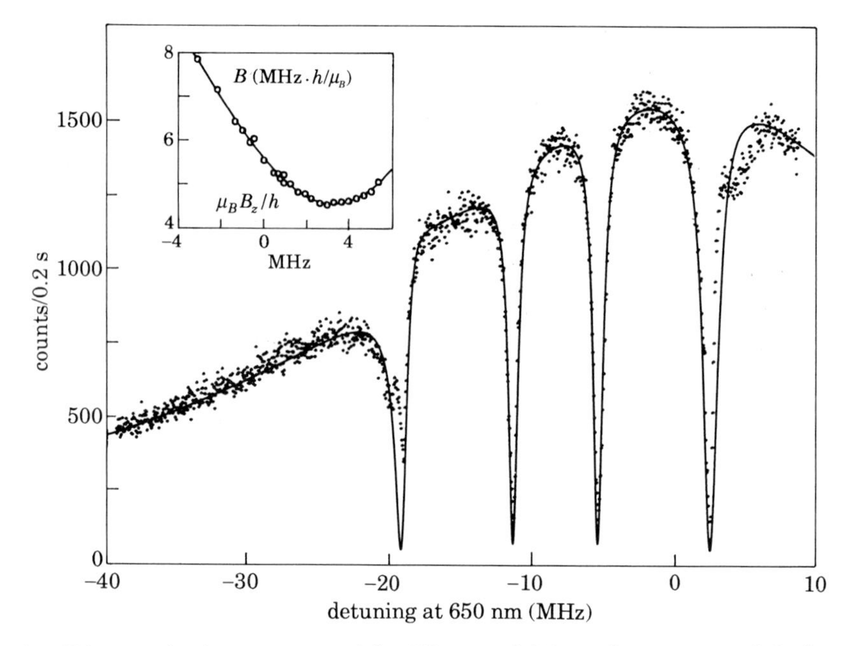

Fig. 4. – 650 nm excitation spectrum of the 493 nm and 650 nm fluorescence of single cooled Ba^+ ion in a r.f. trap showing a Zeeman-split dark line. Components at $\pm 11u/5$, $\pm 3u/5$ off the green-light tuning, $-$ 8.5 MHz off the resonance line. Parameters: $u = \mu_B B/h =$ $= 4.7$ MHz ($B = 0.35$ mT), $I_g = 10\,\text{mW/cm}^2$, $I_r = 25\,\text{mW/cm}^2$. $\gamma_{\text{sd}} = 12\,\text{kHz} \times 2\pi$, $\beta = 90°$. Insert: determined local field B (from splitting) *vs.* applied magnetic field B_z (from ref. [26]).

coils. It is due to this superposition of fields that the modulus of the total magnetic field at the ion site varies as the square of the applied field as long as that field is small.

The magnetic field splits the three electronic Ba^+ levels $^2S_{1/2}$,$^2P_{1/2}$ and $^3D_{3/2}$ in eight sublevels (fig. 3) which give rise to four dark resonances when the linear polarizations of the two coaxial laser beams are identical and perpendicular to the direction of the magnetic field, and $\Delta m = \pm 1$ transitions become excited. (Light polarization parallel with the field, which supports only two dark resonances, would again give rise to optical pumping into the disconnected Zeeman levels of maximum $|m_J|$ and make the fluorescence quench.) An observed excitation spectrum is shown in fig. 4. The green light at 493 nm was downtuned from the $^2S_{1/2}$-$^2P_{1/2}$ resonance by 8.5 MHz, whereas the red light (650 nm) was scanned across its resonance $^2P_{1/2}$-$^2D_{3/2}$. The total magnetic field determined from the observed Zeeman splitting *vs.* the applied field is shown in the insert.

The Zeeman effect complicates the comparison with theory. The Bloch equations of an eight-level atom have been solved for the populations of the two P sublevels whose sum varies as the detected fluorescence signal. The Rabi fre-

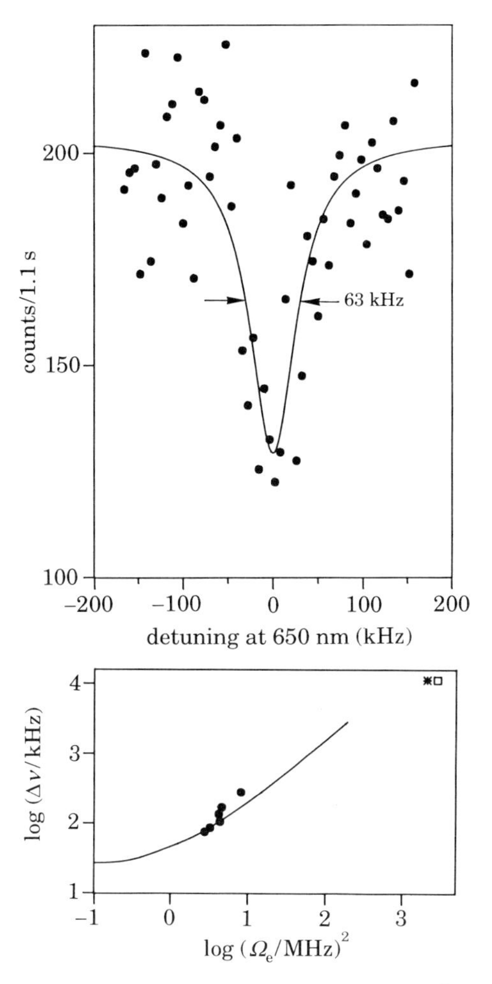

Fig. 5. – Dark line at low light intensity. Line $Q \simeq 6 \cdot 10^9$. $I_g = 0.8\,\mathrm{mW/cm^2}$, $I_r = 0.5\,\mathrm{mW/cm^2}$. Insert: full width $\Delta\nu$ of dark line *vs.* $\Omega_e^2 = \Omega_g^2 + \Omega_r^2$, calculated (—, *) and experimental data with Ω_e evaluated from fit of recorded dark line (●) and from measured I_e (□) (from ref. [26]).

quencies corresponding to the green and red light, the magnetic field and an effective width γ of the trapped state n are taken as free parameters. Calculated spectra have been made fit the recorded ones (full line in fig. 4), and values for these parameters have been determined. The result for $\gamma = 12\,\mathrm{kHz} \times 2\pi$ makes up for the upper limit of the absolute emission bandwidths of the two lasers that are about 6 kHz, from acoustical noise of the control cavities and from fluctuations of the ambient magnetic field. With arbitrary directions of polarization of the light beams and/or of the magnetic field, up to eight dark resonances appear, whose maximum quenching of fluorescence, in the steady state, is less complete as a consequence of complex optical pumping. All these recordings fit in with calculated spectra.

The decay rate of the trapped states n, *i.e.* the natural linewidth, contributes to γ a few mHz only, since the lifetime of the $^2D_{3/2}$ level is at least 17 s [27]. The width of the dark lines remains dominated by power broadening even at intensity levels on the order of a mW/cm^2. To make such a line control an oscillator—which in fact would be the beat note of the green and red lasers corresponding to 2.05 μm wavelength—one needs both the line $Q = \omega/\Delta\omega$ and the signal-to-noise ratio of the detected signal to be high. The narrowest dark line recorded so far is 75 kHz wide at S/N $\simeq 3.5$ and corresponds to $Q \simeq 6 \cdot 10^9$ (fig. 5). Note that the calculated values of the power-broadened width agree with the observed data derived from fitting the line spectra *including* the absorption *depth* of the dark lines, as well as with a value derived from a calibration of the light intensity at the ion's location. The i.r. difference frequency corresponding to $\lambda = 2.05$ μm when controlled by this line would show the figure of instability

$$\sigma_y = (Q_{\text{i.r.}} \times \text{S/N})^{-1} \simeq 1.3 \cdot 10^{-10}\, \tau^{-1/2}, \tag{4}$$

where σ_y is the frequency standard deviation, and τ the probing time. Improvement by factor ten seems feasible and would provide frequency *stability* comparable with current commercial standards.

3. – Line shifts.

With the representation of a standard frequency, its *absolute* precision, or «accuracy», is particularly significant and determines the results of long-term measurements. Accuracy is characterized by residual systematic shifts of the

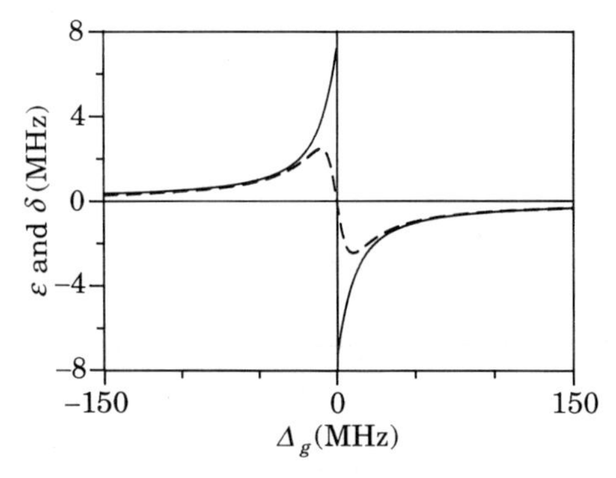

Fig. 6. – Shift of «bright line» ε, and Raman light shift δ, *vs.* detuning Δ_g, for $\Omega_g = 10\,\Omega_r = 1.4\,\Gamma_g$.

line centre, *e.g.* by the applied light. Although the fluorescence null is not affected by light shifts, the associated «bright» line is, and it may affect the symmetry of the neighbouring dark line. The $E2$ resonance of a three-level atom in Λ configuration, excited by two light fields, detuned from their one-photon resonances by Δ_g, Δ_r, undergoes a two-photon or «Raman» light shift

$$\delta = \frac{\Delta_g \Omega_r^2}{4\Delta_g^2 + \Gamma^2} - \frac{\Delta_r \Omega_g^2}{4\Delta_r^2 + \Gamma^2}, \tag{5}$$

where Γ is the decay rate of the intermediate resonance level (fig. 6) [28]. This expression becomes $\Delta(4\Delta^2 + \Gamma^2)^{-1}(\Omega_r^2 - \Omega_g^2)$ for $\Delta_g \simeq \Delta_r = \Delta \gg \delta$. With unbalanced light intensities, *i.e.* $\Omega_g \lessgtr \Omega_r$, and for large detuning $\Delta \gg \Omega_g, \Omega_r, \Gamma$, the shift of the bright line $\varepsilon = \Delta_r^{\max} - \Delta_g$ coincides with the Raman light shift δ. On the other hand, at central tuning of light g, the shift

$$\varepsilon(\Delta_g = 0) = \pm \frac{1}{2} (\Omega_g^2 + \Omega_r^2)^{3/4} / \sqrt{\Omega_g} \tag{6}$$

characterizes the dynamic Stark splitting of the resonance line g and its upper level probed by light r (see fig. 6).

Excitation spectra of the fluorescence recorded at unbalanced light intensities show one bright line associated with and shifted from each dark line

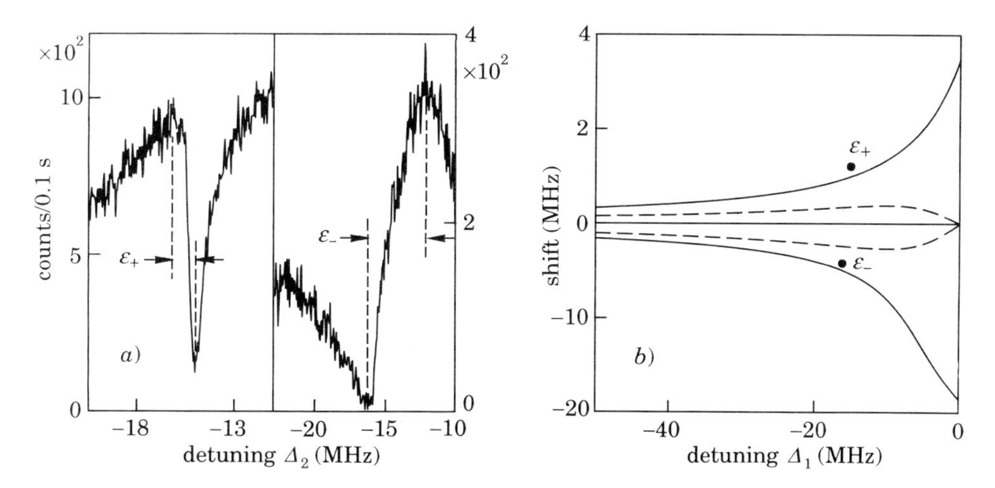

Fig. 7. – *a*) Excitation spectra with unbalanced light intensities (Ω_1, Ω_2 are Rabi frequencies Ω_g, Ω_r, respectively, for the particular dark lines): $\Omega_1 = 6.0\,\text{MHz} \times 2\pi$, $\Omega_2 = 2.2\,\text{MHz} \times 2\pi$ (left); $\Omega_1 = 5.3\,\text{MHz} \times 2\pi$, $\Omega_2 = 15.6\,\text{MHz} \times 2\pi$ (right). *b*) Shift ε of «bright line» (——) and Raman light shift δ (– – –) *vs.* fixed-frequency detuning Δ off the Zeeman-shifted 1-photon resonance, calculated with the parameter values of the above spectra. Observed ε are indicated. (From ref. [26].)

(fig. 7*a*)). They allow, from the measurement of the shift ε, the evaluation of the Raman light shift δ (fig. 7*b*)). The asymmetry of the dark line caused by the shift ε may be reduced, as far as permitted by the available S/N ratio, by using only the deepest part of the line, close to resonance, for the locking of the oscillator. Alternatively, the Rabi frequencies of the two light beams may be balanced, for the same purpose, by suitably adjusting the local intensities of the two light components.

It is instructive to review other possible systematic perturbations. The Zeeman shifts of the dark lines cancel when the locking cycle simultaneously uses corresponding symmetrically shifted dark-line components, such that fluctuations of the magnetic field do not result in a net shift.

Stark splitting by the trapping fields is negligible since the ion is made to reside in the field node.

The first-order Doppler effect of the cold ion is absent, and the second-order Doppler effect contributes a few parts in 10^{-19} only.

The light-induced electric quadrupole of the ionic $S_{1/2}$-$D_{3/2}$ coherence in fact interacts with a field gradient. In the particular case of a d.c. quadrupole field, this interaction shifts the resonances on the order of 1 Hz per 1 V applied to the ring electrode of the trap. The largest remaining systematic effect, as it seems, is the light shift induced by the black-body radiation, a shift of a few times 10^{-16} which may be taken into account as a correction.

4. – Detection of photon correlation as dynamic spectroscopy.

So far, the recorded spectra of the laser-excited fluorescence of a single ion document its steady state. The *dynamics* of the ion's interaction with the light is probed, on the other hand, by recording the autocorrelation functions of the fluorescent light. The second-order correlation, or photon correlation function, is equivalent to the conditional probability of counting a photon in channel 2 of detection at time $t+\tau$ after having counted a photon in the other channel 1 at time t [29]:

$$g^{(2)}(\tau) = \frac{\langle n_1(t)\, n_2(t+\tau)\rangle}{\langle n_1(t)\rangle\langle n_2(t+\tau)\rangle}\,, \tag{7}$$

where $n(t)$ is the number of photon counts during the time interval between t and $t+\Delta t$. It is preferable for the photon counter to have such a temporal resolution that $n=0$ in most of the time increments Δt equal to the resolution time, and two (or more) unresolved events in the same time increment are rare.

The photon correlation function depends in a characteristic manner on the statistical nature of the light. The intensity correlation of thermal light, whose photon distribution is $P_n^{\text{th}} = \overline{n}^n(1+\overline{n})^{-(n+1)}$, is two for $\tau = 0$ and approaches

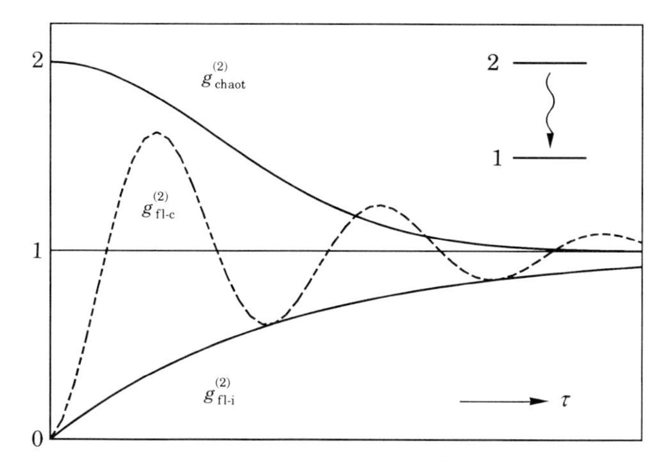

Fig. 8. – Photon correlation function for thermal, coherent and sub-Poissonian light.

unity for large τ; coherent light with a Poissonian distribution, $P_n^{\text{cl}} = (\bar{n}^n/n!)\cdot$ $\cdot \exp[-\bar{n}]$, has $g^{(2)} \equiv 1$. The correlation function of the fluorescence of a single two-level atom increases with τ from zero at small delay τ («antibunching» [30, 31]) and the light shows a sub-Poissonian distribution [32] (fig. 8). The photon correlation function reflects the time-varying interaction of the light with this atom following a preparing act of emission when the atom was put into the ground state. Wiggles at small τ indicate Rabi oscillations that are damped when the system approaches the stationary state. When the number of levels that take part in the interaction with the light exceeds two, as in Ba^+, the photon correlation function is expected to display qualitatively different features in addition. They include very large peak values of $g^{(2)}$ and protracted sub-Poissonian correlation with $g^{(2)} < 1$ for long delay times. These features arise from optical pumping which appears in a system with *more* than two levels, and from building up two-photon coherence (ρ_{13}) in the atom, respectively.

The experimental device for the detection of $g^{(2)}$ on a single Ba^+ ion [33] is shown in fig. 9. One of the channels of detection admits green fluorescence whereas the other channel admits both red and green fluorescence. This arrangement allows one to observe both the photon correlation function, when the green start photon has prepared the ion in the S ground state, and the different function that emerges after having prepared the ion either in the S or D state. The normalized photon correlation function does not depend on the colour of the *stop* photon: This is so since the emission rates of photons in the two colours both vary in proportion with the upper-state population ρ_{22}, *i.e.* they keep probing the same population in the $P_{1/2}$ level.

An example of observed photon correlation data is shown in fig. 10. The full line gives a correlation function calculated with the parameters of the corresponding spectrum and with no extra fit. The intensity of the green light is al-

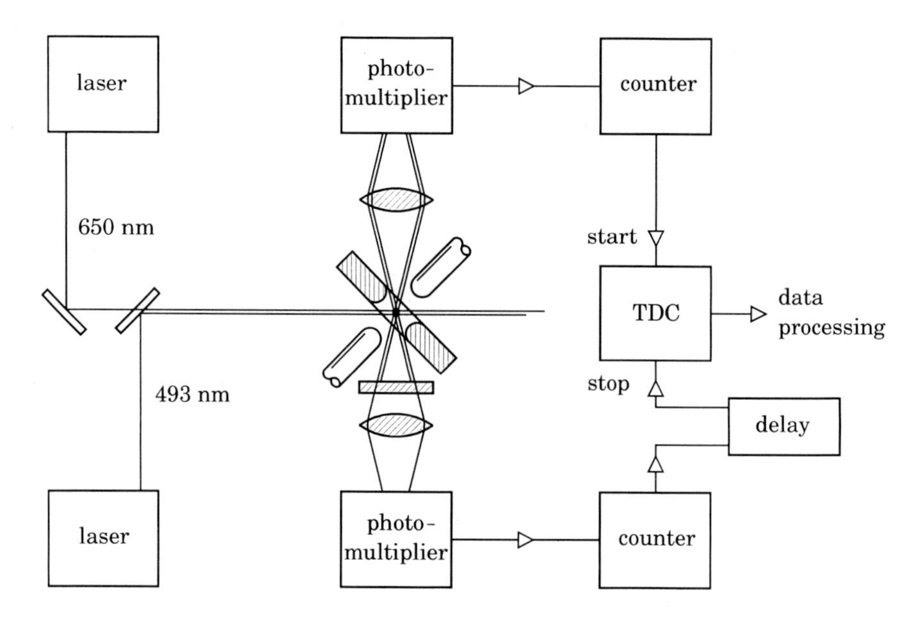

Fig. 9. – Experiment for the recording of photon correlation functions.

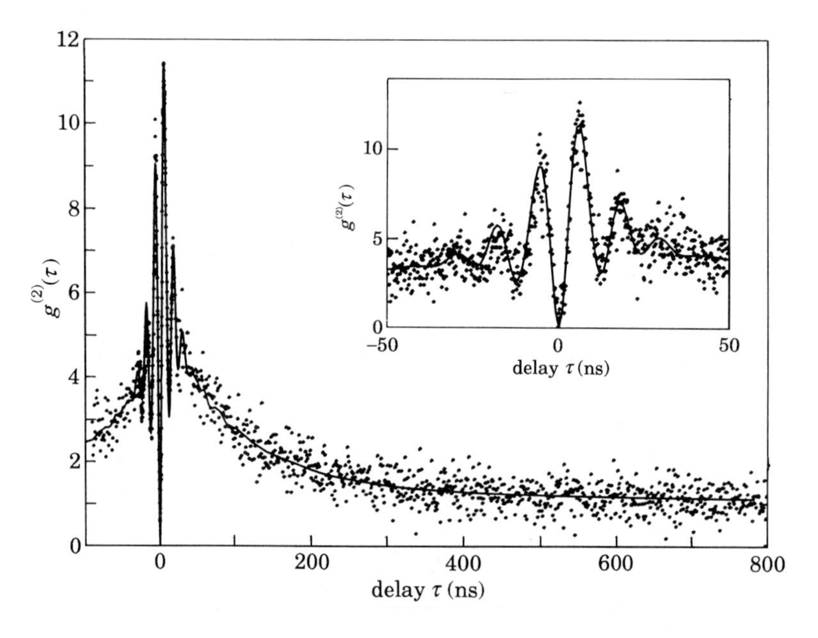

Fig. 10. – Photon correlation function $g^{(2)}(\tau)$. Red light detuned far off resonance. Parameters: $I_g = 2.2\ \mathrm{W/cm^2}$, $I_r = 660\ \mathrm{mW/cm^2}$, $\Delta_g = -40\ \mathrm{MHz}$, $\Delta_r = -110\ \mathrm{MHz}$. Optical pumping into the $D_{3/2}$ level leaves $g^{(2)}(\tau) = 1 \ll g^{(2)}_{\max}$ for large τ. Inset: Detail near $\tau = 0$. (From ref. [33].)

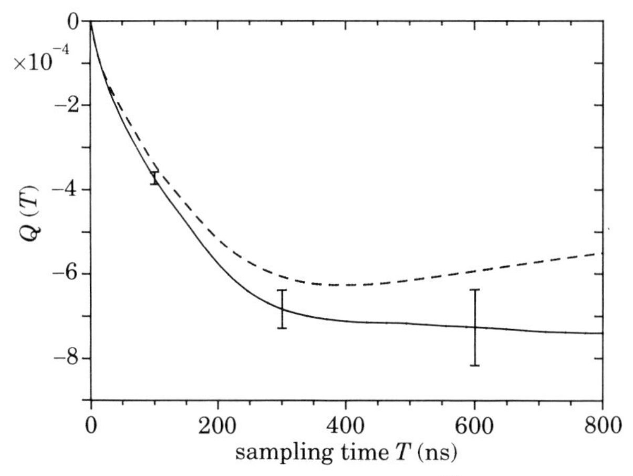

Fig. 11. – Mandel's Q function derived from recorded $g^{(2)}$ data (solid line) and calculated with the parameter values from corresponding spectral fit for strong red and weak green light (dashed line) (from ref. [33]).

most four times the intensity of the red one, and the «start» photon is green, for $\tau > 0$. In this situation, optical pumping into the metastable $^2D_{3/2}$ level dominates the internal dynamics of the ion. There is initial antibunching, in agreement with observations on two-level particles. The maximum value of $g^{(2)}(\tau)$ largely exceeds two, and, with other combinations of the parameters, values as great as 60 have been observed. In addition, the time constant for the approach of the steady state is much longer than the decay time of the resonance level $^2P_{1/2}$ and indicates the pumping time. In contrast, with strong *red* light the ion, after reduction to the ground state by emission of a green photon, forms an S-D coherence by a stimulated Raman-Stokes process immediately after re-excitation of the resonance dipole, which circumvents the population of the $P_{1/2}$ level and the concomitant fluorescence for a time that also largely exceeds the $P_{1/2}$ lifetime. In other words, this type of dynamics gives rise to $g^{(2)}(\tau) < 1$ for a substantial range of delay times τ, or to protracted sub-Poissonian probability for the second fluorescence photon. The temporal distribution of the fluorescence light is characterized by the variance $\langle(\Delta n(T))^2\rangle$ of photon counting within counting intervals of length T, or by Mandel's Q-function [32] $Q(T) = \langle(\Delta n(T))^2\rangle/\langle n(T)\rangle - 1$. Negative values of Q indicate sub-Poissonian statistics. The sub-Poissonian correlations of all delay times contribute to Q according to the relationship

$$Q(T) = 2\langle n(T)\rangle T^{-2}\int_0^T (T-\tau)(g^{(2)}(\tau)-1)\,\mathrm{d}\tau\,, \tag{8}$$

such that Q becomes negative. Values of the Q function *vs.* the length of the counting intervals are shown in fig. 11 both from direct observation of the counting variance and calculated, according to eq. (8), from a photon correlation function recorded at weak intensity of the green and strong intensity of the red light. Over the inspected range of counting intervals, the photon statistics remains sub-Poissonian, and, therefore, nonclassical.

5. – A trapped ion as a Schrödinger cat.

The above observations make by no means the only example where features of the fluorescence light of a real Ba^+ ion deviate from those of the emission of a two-level model atom. It is the existence of Zeeman sublevels that allows the observation of more unusual if not unexpected effects.

Let us consider a gas of atoms with an upper level 0 ($J_0 = 0$) and a lower level 1 ($J_1 = 1$). Spontaneous emission at frequency ν_1 from the atoms in state 0 is composed of $\sigma^{\pm}$ and π light. If we choose to detect the light past a circular polarizer, the *observed* gas atoms emerge, due to the selective detection, in state $m_1 = +1$ *or* -1, *i.e.* with a magnetic moment. If instead we place a linear polarizer in the detection channel whose orientation is perpendicular to the axis of quantization marked by a weak magnetic field, we observe atoms in superpositions of the states $m_1 = \pm 1$. The entire gas sample is left in a *mixture* of these states after the detection of the decay, since the phases of the corresponding individual atomic quadrupoles that evolve at twice the Larmor frequency are distributed over the full angle. The detection of light from an individual particle, however, eliminates the averaging, and the evolving quadrupole is ready for probing by a second interaction with light. This probing may result from the observation of spontaneous decay of the lower level 1 to a nondegenerate ground state 2 at frequency ν_2. When phase matching is provided by the detection of ν_1 and ν_2 in opposite directions, the dispersion of the phases is eliminated, and even gas atoms display a superposition state.

Coincidence measurements of photons at ν_1 and ν_2 have revealed distributions of photon counts *vs.* delay, which show a beat note with light beams of parallel polarizations that are perpendicular with the axis of quantization [34]. This beat note may be described as interference between the cascade decay paths via the $m_1 = +1$ and -1 sublevels. The Ba^+ ion, on the other hand, decays from the $m_p = +1/2$ sublevel of the $P_{1/2}$ state to the $m_d = +3/2$ and $-1/2$ sublevels of the metastable $D_{3/2}$ state, and from $m_p = -1/2$ to $m_d = +1/2$, $-3/2$, when red σ^+ and σ^- polarized fluorescence is emitted (fig. 12). The metastable D levels cannot be probed by immediate spontaneous emission. Instead, the red laser light re-excites the ion and allows observation of the subsequent fluorescence, which provides the second photon count.

Observed photon correlation functions, with and without a linear polarizer in the «start» channel, are shown in fig. 12. The beat note in the recording with

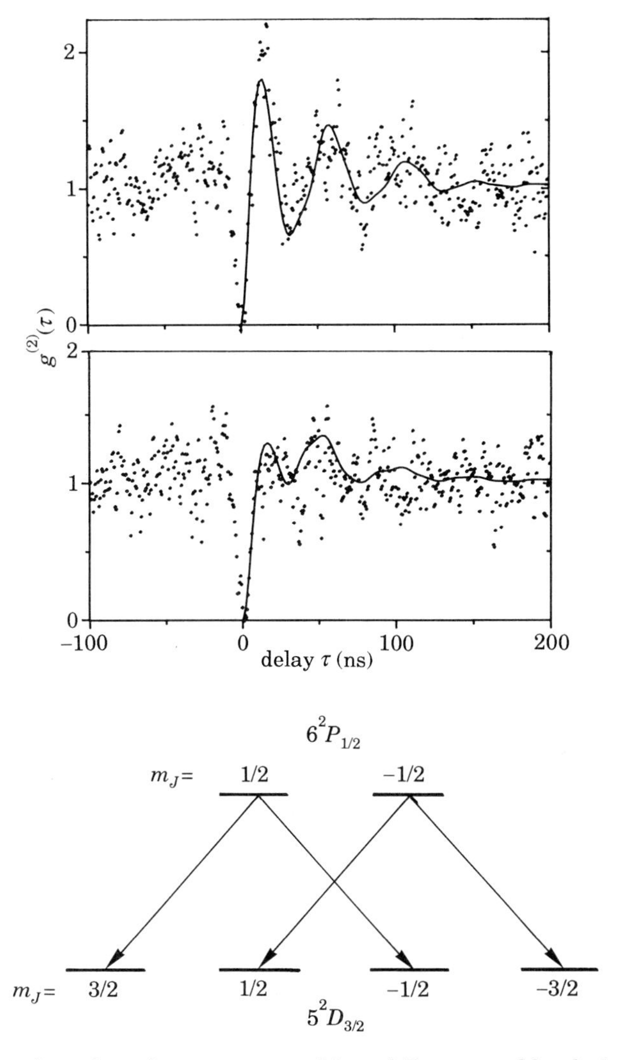

Fig. 12. – Generation of a coherent superposition of Zeeman sublevels in state $D_{3/2}$ by spontaneous decay: With linear polarizer (top), no polarizer (centre). Formation of superposition states (bottom).

the linear polarizer in place marks the interference, on the photocathode, of the amplitudes of Raman-anti-Stokes scattering from the sublevels $m_d = +1/2$ and $-3/2$ (or from $m_d = +3/2$ and $-1/2$) into the ground state after the ion has been prepared as a superposition of these Zeeman states by the emission of the first detected photon. The Zeeman splitting of the sublevels is about 16 MHz, corresponding to the observed 60 ns period length of the beat note. With the linear polarizer removed, mixtures of states are generated whose phases are distributed over 2π. This spread of phases amounts to the preparation of mere population in the D sublevels, but no Zeeman coherence, *i.e.* a mixture of states.

The generation of a coherent superposition of the D sublevels requires, as a necessary condition, that the routes for the generation of the individual states are indistinguishable. Indeed, this condition holds for the trapped Ba^+ ion: The σ^+ and σ^- polarizations cannot be distinguished with the linear polarizer. The 20 MHz Zeeman splitting is not resolved since the photon counting is a time-resolved observation (628 ps resolution), and a Fourier transformation requires spectral resolution to be not better than 250 MHz. The difference in the ion recoil for the emission events that prepare the ion in the superimposed state is $\Delta p = \hbar\Delta\omega/c$ which, according to the uncertainty relation, can only be resolved when the ion is not spatially localized. In fact, for the two contributing Zeeman-split resonances $P(1/2)$-$D(3/2)$, $D(-1/2)$ to be distinguished, spatial uncertainty Δx must exceed $\hbar/\Delta p = c/\Delta\omega \simeq 5$ m. In contrast, the perfect localization of the ion allows the preparation of the ion in the corresponding D sublevels indistinguishable even via detection of the photon recoil.

In connection with the foundations of quantum mechanics, possible realizations of a physical object, which is both a closed, macroscopic system—or «classical» object—*and* a quantum superposition, have been discussed. Certainly, the notion of such an object, a «Schrödinger cat», seems meaningful only if it survives decoherence [35] for an appreciable lapse of time that is needed for probing. As a rule, multiparticle systems dephase quickly. An exception may be spin resonance where macroscopic magnetization has been observed precessing in the magnetic field for hours [36]. This phenomenon results, however, from negligible interaction of the spins with each other.

It seems that the qualification of an object as «macroscopic» should be redefined as «repeatedly preparable for the direct interaction with a laboratory detector». In this sense, the individual Ba^+ ion is a macroscopic system, and when it is in the superposition state of the D sublevels, it is a Schrödinger cat (precisely: a compound of two cats (see fig. 12, bottom). The 100 ns decay time of the superposition is due to the coupling with the laser fields. With no red laser light, the cat's lifetime should be much longer and could be measured by a modified recording of the photon correlation function. This measurement includes switching on the red light with a variable delay, after the detection of a red fluorescence photon (which prepares the Zeeman coherence), and finally detecting a green photon for the read-out of the surviving coherence. Very little dephasing is anticipated, and the beat note should be visible even after delay times of many seconds. The ultimate dephasing time should equal twice the lifetime of the $D_{3/2}$ level, 34 s [27]. Unfortunately, the counting rate of such a recording is small.

It is worth noting that the superpositions of Zeeman substates generated in the intermediate level of a well-detectable fluorescence cascade [34] only lives on the order of, say, 100 ns. Moreover, longer decay times would give rise to inevitable quick dephasing of the intermediate superposition.

6. – Observations on single trapped ytterbium ion.

Recently, in the Hamburg group, a single Yb^+ ion has been prepared in a radiofrequency trap, and its fluorescence has been detected. A simplified level scheme of Yb^+ is shown in fig. 13. The excitation of fluorescence is by c.w. UV light at 369.4 nm wavelength which is generated by frequency-doubling, in an external cavity, the output of a red-dye ring laser. With the branching ratio 1:300, the resonance level $^2P_{1/2}$ decays into the metastable level $^2D_{3/2}$. In order to prevent optical pumping into the latter state and to keep up the UV fluorescence, the ion is additionally illuminated by light from an F_A colour centre laser at 2.44 μm, based on Li-doped KCl. As in the analogous level scheme of Ba^+, this excitation with bichromatic light generates a dark line which has been observed, although its Zeeman splitting is not yet resolved.

Collisions by background gas, and also off-resonant Raman pumping by the i.r. light, can excite transitions on the fine-structure line $^2D_{3/2}$-$^2D_{5/2}$. Such a transition event would make a dark interval appear in the fluorescence of a Ba^+ ion whose $^2D_{5/2}$ level is metastable. In contrast, the corresponding level of Yb^+ decays in 30 ms into the ground state, or via a dipole-allowed transition into the almost stable state $6s^2\,^2F_{7/2}$, whose lifetime is more than a week. Consequently, we expect two very different kinds of dark intervals to appear in the laser-ex-

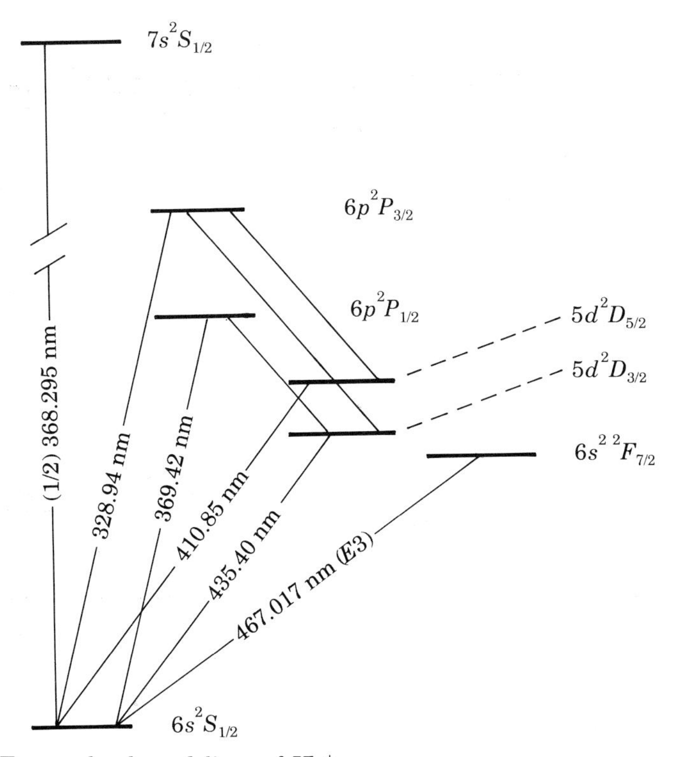

Fig. 13. – Energy levels and lines of Yb^+.

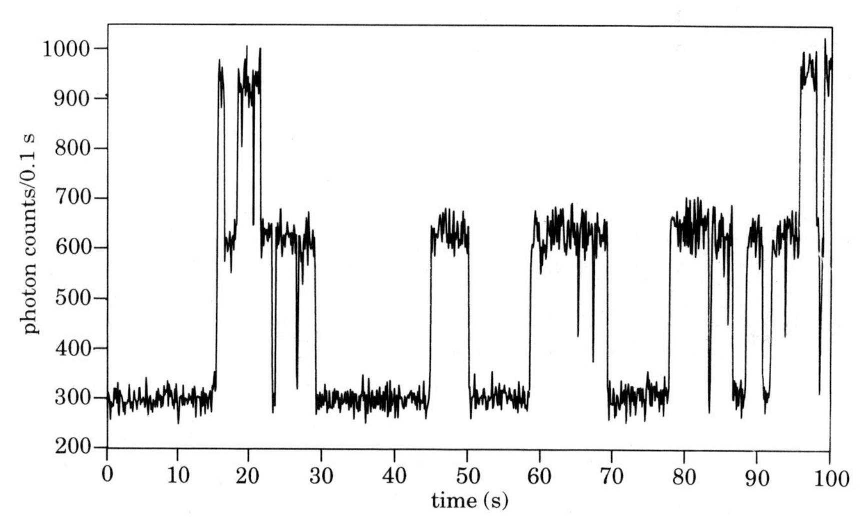

Fig. 14. – Resonance fluorescence of single Yb^+ or two Yb^+ *vs.* time. Dark intervals start upon collisionally induced transition on the fine-structure line $D_{3/2} \rightarrow D_{5/2}$ followed by decay to the ground state (short pauses) or to the $F_{7/2}$ state (long pauses).

cited UV fluorescence: Short ones with subsecond mean duration appear when the ion decays into the ground state and is immediately re-excited, and very long ones that show up upon the ion's stay in the $F_{7/2}$ level, terminated by an inelastic collision with an atom of the background gas. Also, several resonances with high-lying states exist whose off-resonant excitation by the UV light allows the ion to leave the $F_{7/2}$, at a small rate, for the ground state again. A recording of the fluorescence of an Yb^+ ion is shown in fig. 14. For part of the recording time, the trap housed two ions. Note the well-distinguished two classes of dark intervals which characterize the alternative decay routes.

So far the interruption of fluorescence and the corresponding «quantum jumps» have been observed with a Ba^+, Sr^+, or Hg^+ ion [12, 15-17]. The Yb^+ ion is an interesting addition to this small class of ions since it has various narrow optical resonances some of which may be excited even by diode laser light, and the scheme of detection via quantized fluorescence is crucial for their application in precise frequency control.

Another subject of considerable interest is the ion's kinetics in the trap and its dynamics under the action of light forces. Such experiments do not critically depend on the particular ion species. The external degrees of freedom of the ion in the trap, namely its vibrations in the potential well, are coupled to the light via vibrational Raman transitions or, in Mössbauer terminology, n-phonon lines. The excitation of these transitions offers a handle for the preparation and manipulation of the ion's vibrational state in the pseudopotential of the trap. This kind of studies, on a single particle, is remarkable in its own right [37]. As it seems, single trapped ions make well-suited microphysical test objects in the laboratory for an ever growing variety of motivations.

7. – Summary.

Single trapped ions are microphysical objects of quantitative spectroscopy whose internal and external degrees of freedom can be probed, at the present state of the art, with high spectral and temporal resolution. When such an ion is used for the control of an oscillator frequency, systematic fractional shifts larger than 10^{-16} can be eliminated, and those larger than 10^{-18} can be controlled. Therefore, a frequency standard based on a single trapped ion has a remarkable potential for the absolute representation of a frequency.

Excitation spectra of the fluorescence of an ion by bichromatic light have revealed the formation of superpositions of (meta)stable energy eigenstates with same parity (Raman coherences, «trapped» states), when dark lines show up. These lines are known from spectral excitation of gases by bichromatic light, however, only an individual isolated atomic particle renders these lines unshifted and least broadened.

The internal dynamics of a single ion is unravelled by the photon correlation function of its fluorescence. With a two-level ion, antibunching and Rabi flopping show up in the photon correlation function. A three-level ion displays a much richer assortment of novel nonclassical features which are the effects of optical pumping and the excitation of Raman coherences.

Optical pumping in a three-level ion (Λ scheme) affects also the Zeeman sublevels. Zeeman *coherence* is generated by spontaneous decay following the pump excitation. It is probed upon re-excitation by the other light, and it shows up as a beat note when recording the photon correlation function under conditions that forbid identification of the ion in one particular of the Zeeman states which form the superposition. This repeatedly preparable single-ion superposition state that is read out by the detection of the green light spontaneously emitted by anti-Stokes scattering comes close to what is considered a Schrödinger cat.

Ions that feature even more complex level schemes, like Yb^+, contribute other phenomena, among which is the appearance of two different types of quantum jumps that interrupt the laser-induced fluorescence.

* * *

The unpublished data of fig. 12 have been recorded by I. SIEMERS and M. SCHUBERT, those of fig. 14 by V. ENDERS. Figures 1, 8 and 9 have been prepared by J. SIERKS, figures 13 and 14 by V. ENDERS, all other figures by I. SIEMERS. I thank I. SIEMERS for critically reading the manuscript.

REFERENCES

[1] R. E. March and R. J. Hughes: *Quadrupole Storage Mass Spectroscopy* (Wiley, New York, N.Y., 1989).

[2] P. E. Toschek: in *Les Houches Session XXXVIII, 1982 - Tendances actuelles en physique atomique,* edited by G. Grynberg and R. Stora (Elsevier, Amsterdam, 1984), p. 381.

[3] See *Workshop and Symposium on the Physics of Low-Energy Stored and Trapped Particles, Proceedings, Phys. Scr.,* **T22** (1988).

[4] W. Neuhauser, M. Hohenstatt, P. E. Toschek and H. Dehmelt: *Phys. Rev. A*, **22**, 1137 (1980).

[5] P. E. Toschek and W. Neuhauser: in *Atomic Physics 7,* edited by D. Kleppner and F. M. Pipkin (Plenum, New York, N.Y., 1981), p. 529.

[6] See, *e.g.*, G. Werth: *Atomic Physics XIII,* to be published.

[7] J. J. Bollinger, J. D. Prestage, W. M. Itano and D. J. Wineland: *Phys. Rev. Lett.*, **54**, 1000 (1985).

[8] R. Casdorff, V. Enders, R. Blatt, W. Neuhauser and P. E. Toschek: *Ann. Phys. (Leipzig)*, **48**, 41 (1991).

[9] D. Schnier, A. Bauch, R. Schröder and Chr. Tamm: in *Proceedings of the 6th European Frequency Time Forum, Noordwijk, 1992,* p. 415.

[10] D. J. Wineland and W. M. Itano: *Phys. Lett. A*, **82**, 75 (1981).

[11] F. Diedrich, J. C. Bergquist, W. M. Itano and D. J. Wineland: *Phys. Rev. Lett.*, **62**, 403 (1989); J. C. Bergquist, F. Diedrich, W. M. Itano and D. J. Wineland: in *Laser Spectroscopy IX,* edited by M. S. Feld, J. E. Thomas and A. Mooradian (Academic Press, Boston, Mass., 1989), p. 274.

[12] A. A. Madej and J. D. Sankey: *Opt. Lett.*, **15**, 634 (1990).

[13] D. J. Wineland, W. M. Itano and J. C. Bergquist: *Opt. Lett.*, **12**, 389 (1987).

[14] H. Dehmelt: *Bull. Am. Phys. Soc., Ser. 2*, **20**, 60 (1975).

[15] W. Nagourney, J. Sandberg and H. Dehmelt: *Phys. Rev. Lett.*, **56**, 2797 (1986).

[16] Th. Sauter, W. Neuhauser, R. Blatt and P. E. Toschek: *Phys. Rev. Lett.*, **57**, 1696 (1986).

[17] J. C. Bergquist, R. C. Hulet, W. M. Itano and D. J. Wineland: *Phys. Rev. Lett.*, **57**, 1699 (1986).

[18] W. Nagourney, Nan Yu and H. Dehmelt: *Opt. Commun.*, **79**, 176 (1990).

[19] W. Neuhauser, M. Hohenstatt, P. E. Toschek and H. Dehmelt: *Appl. Phys.*, **17**, 125 (1978); *Phys. Rev. Lett.*, **41**, 233 (1978).

[20] D. J. Wineland, R. E. Drullinger and F. L. Walls: *Phys. Rev. Lett.*, **40**, 1639 (1978).

[21] T. W. Hänsch and A. L. Schawlow: *Opt. Commun.*, **13**, 68 (1975).

[22] D. J. Wineland and H. Dehmelt: *Bull. Am. Phys. Soc.*, **20**, 637 (1975).

[23] Th. Hänsch, R. Keil, A. Schabert, Ch. Schmelzer and P. E. Toschek: *Z. Phys.*, **226**, 293 (1969).

[24] Th. Hänsch and P. E. Toschek: *Z. Phys.*, **236**, 213 (1970).

[25] G. Alzetta, A. Gozzini, L. Moi and G. Orriols: *Nuovo Cimento B*, **36**, 5 (1976).

[26] I. Siemers, M. Schubert, R. Blatt, W. Neuhauser and P. E. Toschek: *Europhys. Lett.*, **18**, 139 (1992).

[27] R. Schneider and G. Werth: *Z. Phys. A*, **293**, 103 (1979).

[28] G. Orriols: *Nuovo Cimento B*, **53**, 1 (1979).

[29] R. Loudon: *The Quantum Theory of Light* (Clarendon, Oxford, 1981).

[30] H. J. Kimble, M. Dagenais and L. Mandel: *Phys. Rev. Lett.*, **39**, 691 (1977).

[31] F. DIEDRICH and H. WALTHER: *Phys. Rev. Lett.*, **58**, 203 (1987).
[32] L. MANDEL: *Opt. Lett.*, **4**, 204 (1979).
[33] M. SCHUBERT, I. SIEMERS, R. BLATT, W. NEUHAUSER and P. E. TOSCHEK: *Phys. Rev. Lett.*, **68**, 3016 (1992).
[34] A. ASPECT, J. DALIBARD, P. GRANGIER and G. ROGER: *Opt. Commun.*, **49**, 429 (1984).
[35] R. OMNÈS: *Rev. Mod. Phys.*, **64**, 339 (1992).
[36] T. E. CHUPP and R. J. HOARE: *Phys. Rev. Lett.*, **64**, 2261 (1990).
[37] R. GLAUBER: in *Proc. S.I.F.*, Course CXVIII (North-Holland, Amsterdam, 1992), p. 643.

Precision Hyperfine Spectroscopy in Ion Traps.

G. WERTH

Institut für Physik - Staudinger Weg 7, 6500 Mainz 1, B.R.D.

1. – Introduction.

The confinement of ions by electromagnetic fields to a small volume in space has a number of significant advantages compared to other spectroscopic techniques. Most noteworthy is the fact that the coherence time in interaction with radiation fields can be made almost arbitrarily long, which results in extremely narrow linewidth, if transitions are induced between stable or long-living states. Moreover, if the resonance wavelength between these states is larger than the ion motional amplitude (Dicke criterion), no first-order Doppler effect shows up apart from sidebands at the ion oscillation frequency symmetrically around an unshifted and unbroadened carrier. These two properties make the ion trapping technique particularly well suited for ground-state hyperfine spectroscopy: The energy levels are stable and the transition wavelength in many ions is of the order of several cm—larger than a typical ion cloud diameter of a few mm in a trap. This holds for static (Penning) as well as for dynamic (Paul) traps, and the particular properties of the ion motion in these traps are of little influence in hyperfine spectroscopy. The traps serve as a mere container for charged particles. For their detailed discussion the reader is referred to the literature [1,2]. We just note that the storage time easily exceeds several hours. The traps may be operated under UHV conditions or at moderate ($\sim 10^{-6}$ mbar) buffer gas pressures. While in Paul traps collisions with light buffer gases tend to cool the ions and thus increase their storage time, Penning traps are very sensitive to poor background gas conditions and in general a rapid ion loss is observed above 10^{-8} mbar. The average kinetic energy of an uncooled ion cloud depends on the trapping conditions. As a rule of thumb it is 1/10 of the trap potential well depth [3,4]. The trapped-ion number is limited by space charge. In a typical trap potential well depth of a few 10 eV it does not exceed 10^6 ions/ /cm^3. This low density leads to very small absorption of light from a laser, but this is easily overcome by the repetitive use of the same ions. The fluorescence after laser excitation on a strong resonance transition at saturation intensity is as strong as about 10^8 photons per second from a single particle. Even at moderate detection efficiencies of $10^{-4} \div 10^{-5}$, limited mainly by solid angle and detector quantum efficiency, quite strong signals can be obtained. It makes even possible the observation of a single trapped ion by the naked eye [4].

The standard method in hyperfine spectroscopy in ion traps is the optical-microwave double resonance. In many cases the energy spacing between different ionic ground-state hyperfine levels is larger than the optical Doppler width of $(1 \div 3)$ GHz for uncooled ions in a trap. Then one of the hyperfine levels may selectively be excited by the laser. By repeated excitation it will be completely depleted after a short time (optical pumping), unless relaxation by collisions with buffer gas atoms leads to repopulation. Fortunately such relaxation cross-sections in particular with noble-gas atoms are generally small ($\leqslant 10^{-18}$ cm^2) and even at 10^{-6} mbar buffer gas pressures relaxation time constants of many seconds can be obtained. The experimental indication for optical pumping is the decrease of fluorescence intensity with decreasing ground-state population. A microwave-induced transition to a different ground-state hyperfine level is indicated by the increase in fluorescence intensity. This technique works particularly well for ions of the earth-alkali elements. They have an alkalilike level structure with strong resonance transitions, generally in the near ultraviolet, large hyperfine splittings of their $^2S_{1/2}$ ground state, which in addition is particularly insensitive to relaxation collisions. A slight drawback is the fact that in most of the cases (except for Mg^+ and Be^+) long-living metastable D-levels are between the $S_{1/2}$ ground and the P-excited states, in which the excited ion may

TABLE I. – *Ground-state hyperfine separations of alkalilike ions measured by microwave-optical double resonance in ion traps.*

Ion	Hyperfine separation (Hz)	
$^9Be^+$	1250017674.096	(0.008)
$^{25}Mg^+$	1788763128	(162)
$^{87}Sr^+$	5002368354	(24)
$^{131}Ba^+$	9107913698.97	(0.50)
$^{133}Ba^+$	9925453554.59	(0.10)
$^{135}Ba^+$	7183340234.90	(0.57)
$^{137}Ba^+$	8037741667.69	(0.36)
$^{171}Yb^+$	12642812124.3	(1.4)
$^{173}Yb^+$	10491720239.55	(0.09)
$^{199}Hg^+$	40507347997.8	(1.0)

decay. It then would be lost from the pumping cycle. Repumping of the ions by additional lasers at the *P-D* wavelength or quenching of the *D*-states by buffer gas collision are the ways to bring the ions back into the pumping cycle.

In recent years many isotopes of the alkalilike ions have been investigated. The results are listed in table I. As is evident, the experimentally obtained hyperfine splittings are extremely precise, in most cases limited only by the second-order Doppler effect or by the available reference frequencies. Because of the long coherence time the linewidth may be extremely narrow. As a recent example we may quote a hyperfine transition in $^{171}Yb^+$ [5], where the full linewidth is as low as 16 mHz at a transition frequency of 13 GHz (fig. 1). A similar impressive result is reported on $^{199}Hg^+$, where at 40.5 GHz transition frequency a linewidth of 40 mHz [6] has been obtained. The value of these results is not the numbers in themselves, since our understanding of hyperfine-splitting constants is limited to about $10^{-3} \div 10^{-4}$ by nuclear-structure effects. More important are differential effects. One of them is the relative insensitivity of the hyperfine-transition frequencies to outer perturbations. It leads to the development of frequency standards, which have already surpassed the stability of hydrogen masers [5-7]. Other differential effects are related to fundamental physical test. The dependence of the hyperfine-transition frequency on the rotation of the Earth has been investigated by a group at NIST, Boulder, to set limits on the spatial anisotropy of space [8]. The same group looked also for a frequency dependence on the state population, which would be caused by nonlinear terms in the Hamiltonian [9], and for anomalous spin-dependent forces [10]. Finally the ratio of magnetic hyperfine-interaction constants for two different isotopes in general differs from the ratio of the corresponding nuclear moments by the Bohr-Weisskopf effect as the result of the distribution of nuclear magnetization over the extended nuclear volume. The possibility of determining this quantity over chains of isotopes using ion traps has recently been discussed [11].

It is quite natural from the experimental point of view that the experiments

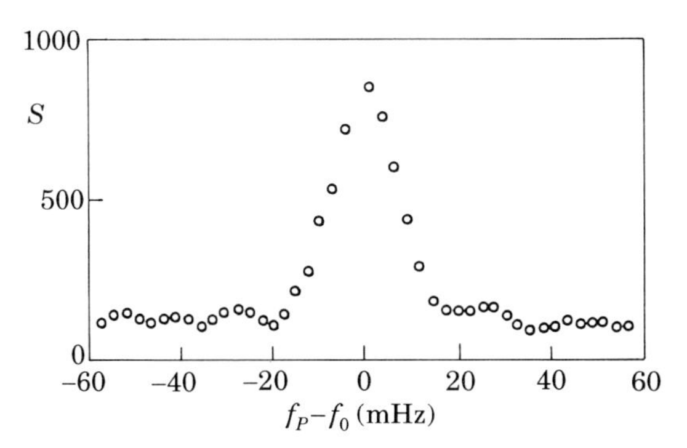

Fig. 1. – $F = 0$, $m_F = 0$-$F = 1$, $m_F = 0$ ground-state hyperfine transition in $^{171}Yb^+$. The transition frequency in 12.6 GHz and the full linewidth is 16 mHz (courtesy D. SCHNIER).

on the alkalilike ions represent the first generation of precision hyperfine-structure measurements in ion traps, since these systems have a number of very favourable properties. The experimental technique has been refined in recent years to a state where «second generation» experiments may be performed on systems which offer much less favourable conditions. The first two of such experiments will be described below, which may illustrate the present state of the art of experiments on large trapped ion clouds.

2. – Hyperfine structure of $^{207}Pb^+$.

The lead ion has an extremely simple level structure as is evident from the insert of fig. 3. In contrast to the alkalilike ions its ground level is a $6\,^2P_{1/2}$ state, split by hyperfine interaction with its spin-1/2 nucleus into two hyperfine lev-

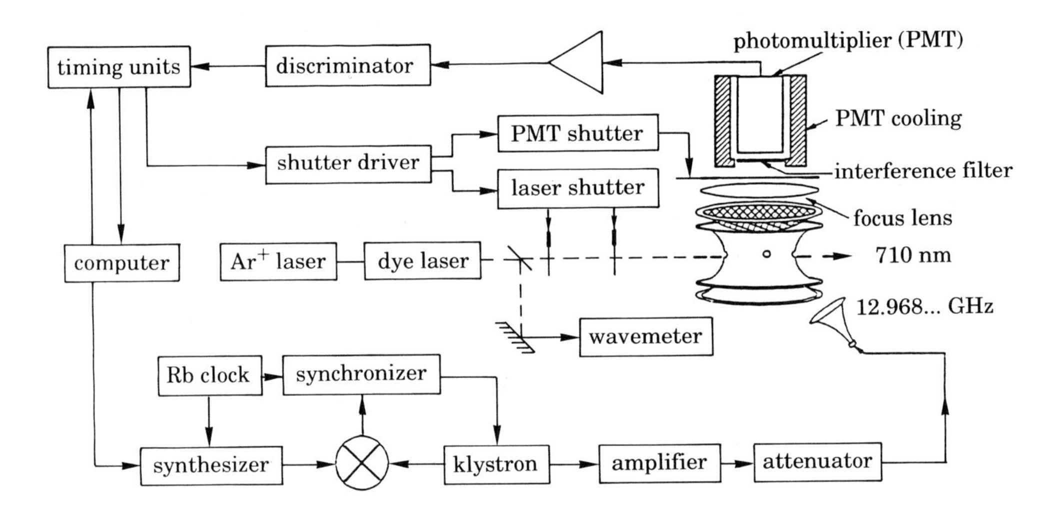

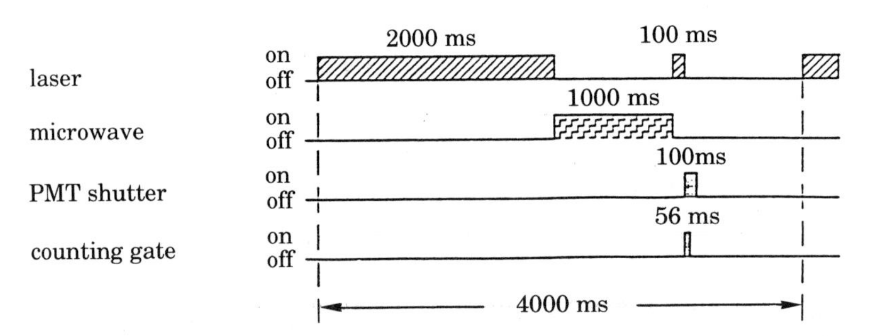

Fig. 2. – Experimental setup and timing sequence for a optical-microwave double resonance on $^{207}Pb^+$.

els $F = 0$ and 1. The experimental difficulty arises from the fact that the first excited state and the only one accessible by available laser radiation is a $6\,^2P_{3/2}$ state. The corresponding $M1$ fine-structure transition probability at 710 nm is very low. The lifetime of the excited $^2P_{3/2}$ state has been determined to 47 ms [12]. Thus the fluorescence count rate from this state, if excited at saturation intensity, is about 7 orders of magnitude smaller than that for alkalilike systems. On the other hand, the long lifetime of the excited state allows time-separated fluorescence detection essentially free of any background, when the exciting laser is pulsed. The low fluorescence count rate can be in part compensated by long average times, since the storage of Pb^+ is extremely simple due to its large mass. In our experiment the storage time usually exceeded 1 week. Figure 2 shows our experimental setup including the timing and fig. 3 a fluorescence spectrum, obtained after about 1 h of averaging. The spectrum shows the hyperfine structure of the ground and excited states as well as an unshifted component from an impurity of $^{208}Pb^+$ in our sample from which the isotope shift can be derived. The linewidth of 2 GHz corresponds to an ion temperature of 1500 K. The count rate at peak intensity is about 20 per cycle (= 2 s). The sensitivity may be best illustrated by fig. 4, where the $F = 0$-$F' = 1$ part of the $P_{1/2}$-$P_{3/2}$ transition is shown with higher resolution. On the right-hand side of this line a small component shows up from the $F = 0$-$F' = 2$ transition, which originates from a small electric-quadrupole component in the $M1$ transition am-

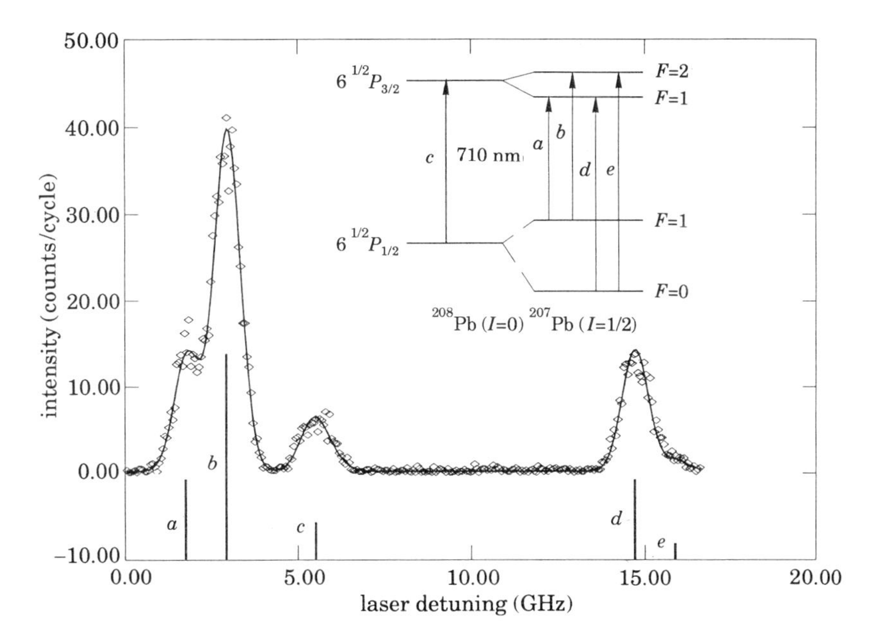

Fig. 3. – Fluorescence spectrum of stored $^{207}Pb^+$ after excitation of the $^2P_{1/2}$-$^2P_{3/2}$ resonance transition at 710 nm.

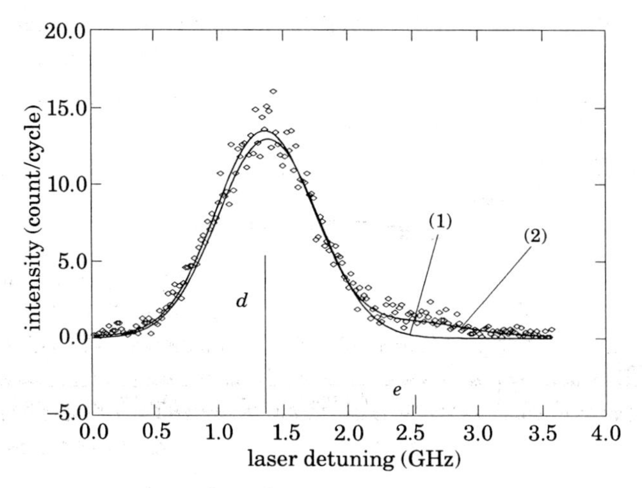

Fig. 4. – $F = 0$-$F' = 1$ part of the ${}^2P_{1/2}$-${}^2P_{3/2}$ $M1$ resonance transition at 710 nm of ${}^{207}Pb^+$, showing a small $F = 0$-$F' = 2$ component due to an admixture of $E2$ radiation.

plitude. The maximum count rate in this component is below 1 Hz. The $E2$ amplitude could be determined to 5% of the total transition amplitude [13].

The precise measurement of the ground-state hyperfine splitting was performed by microwave-optical double resonance in the same manner as described above [14]. In spite of the very low count rate a precision could be obtained

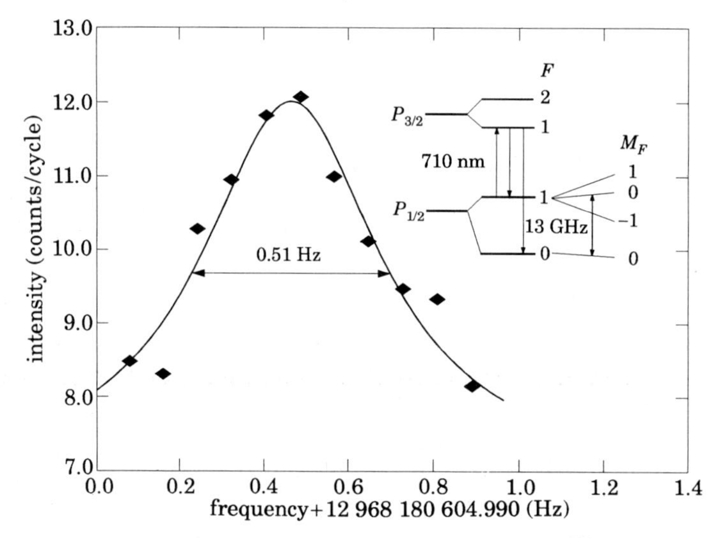

Fig. 5. – $F = 0$, $m_F = 0$-$F = 1$, $m_F = 0$ hyperfine transition in ${}^{207}Pb^+$. The linewidth of 0.51 Hz in the 12.9 GHz transition frequency is limited by phase noise of the microwave oscillator. A Lorentzian lineshape is fitted through the experimental points.

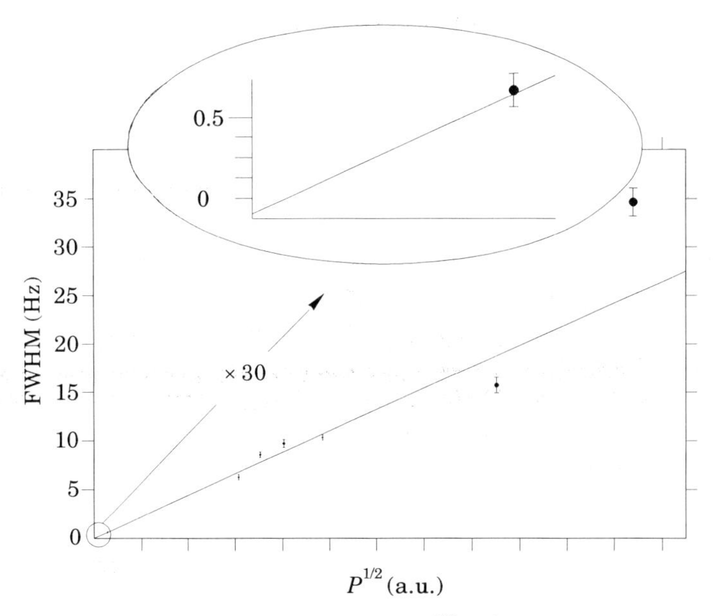

Fig. 6. – Linewidth of the hyperfine transition of $^{207}Pb^+$ *vs.* the square root of the microwave power P. The intersection at $P = 0$ indicates a linewidth of 13 mHz due to relaxation.

which matches that of the alkalilike system:

$$\Delta W_{\mathrm{HFS}} = 12\,968\,180\,601.61\,(0.22)\,\mathrm{Hz}\,.$$

The error mainly comes from the second-order Doppler shift of the buffer gas cooled ion cloud. The minimum linewidth obtained in this experiment was about 0.5 Hz (fig. 5), limited by the spectral bandwidth of the microwave generator. Measurements of the linewidth as a function of microwave power (fig. 6) indicate that the relaxation-limited linewidth in a He buffer gas atmosphere at 10^{-6} mbar would be lower than 13 mHz.

The motivation to investigate $^{207}Pb^+$ was that its simple level structure and its easily available excitation wavelength of 710 nm could make it an interesting candidate for a frequency standard. The problem of cooling this ion could be overcome by sympathetic cooling, if a different ion with suitable $E1$ cooling transitions would be stored simultaneously in the same trap as demonstrated in other systems [15]. Furthermore the experiment on Pb^+ can be regarded as a feasibility test for future experiments on trapped highly charged ions [15]. For hydrogenlike systems of $Z = 60 \div 70$ the ground-state hyperfine-splitting energy is in the optical spectral region. The corresponding $M1$ transitions are of the same type and intensity as those investigated in singly charged Pb^+.

3. – Hyperfine structure of Eu^+ isotopes.

In contrast to Pb^+ the stable odd Eu^+ isotopes have a very complex level structure: The $^9S_4$ ground state couples the nuclear spin of $I = 5/2$ to 6 hyperfine levels $F = 13/2$ to $F = 3/2$. Experimental difficulties arise from the fact that the only excited ^9P states accessible by available lasers partially decay into several long-living metastable ^9D states, from which they have to be removed by quenching collision with buffer gas atoms. Collisions, on the other hand, may disturb the ground-state optical-pumping process by relaxation. Fortunately there is a small pressure gap around 10^{-6} mbar of N_2 buffer gas, where quenching from the metastable states is sufficiently fast ($\tau \leqslant 0.1$ s) and ground-state hyperfine relaxation is negligible. Under these conditions the optical-excitation spectrum of a natural mixture of Eu^+ isotopes is just sufficiently well resolved to identify the single hyperfine components (fig. 7). Microwave-induced transitions between different hyperfine levels, detected by the increase of fluorescence quanta, are split by the residual magnetic field in our trap ($B = 270$ mG) into many Zeeman components. Figure 8 shows an example. The centre of the whole Zeeman pattern gives the desired hyperfine-transition frequency, which can be determined to about 50 Hz. The precision here does not match that of the previous example of $^{207}Pb^+$ and of those listed in table I, because due to the even total electronic angular momentum $J = 4$ and the odd nuclear moment $I = 5/2$ the total quantum number $F = I + J$ has to be odd. Then no Zeeman substates $m_F = 0$ exist, which would be independent to

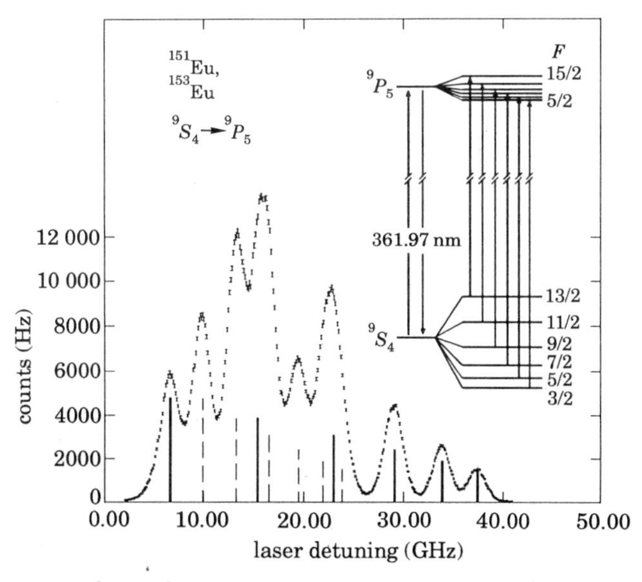

Fig. 7. – Fluorescence from a laser-excited natural mixture of Eu^+ isotopes in a Paul trap. The observed transition is $^9S_4$-$^9P_5$ at 382 nm.

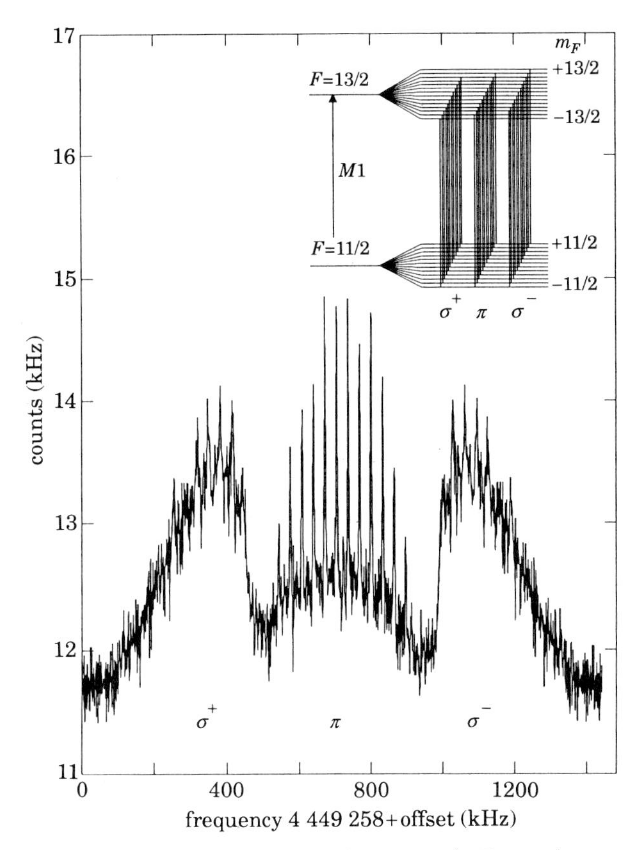

Fig. 8. – $\Delta m_F = 0$, ± 1 Zeeman components in the $F = 13/2$-$F = 11/2$ ground-state hyperfine transition of trapped $^{151}Eu^+$. The centre of the pattern at a transition frequency of 4.4 GHz can be determined to 50 Hz.

first order of the magnetic-field strength and in particular not broadened by any B field inhomogeneity. In the case of Eu^+ we observe a linewidth proportional to the g-factor of the Zeeman transition, which clearly indicates that the residual magnetic-field inhomogeneity limits our accuracy. The precision, however, is sufficient to derive the hyperfine-coupling constants. If we write the hyperfine energy W_{HFS} in the standard way as

$$W_{HFS} = A(K/2) + B\,\frac{3K(K+1) - 4I(I+1)J(J+1)}{8I(2I-1)J(2J-1)} + Cf(I, J, F),$$

$$K = F(F+1) - I(I+1) - J(J+1),$$

where the constants A, B and C denote the strength of the magnetic-dipole, electric-quadrupole and magnetic-octupole interactions, respectively, we obtain after correction of the measured frequencies by second-order hyperfine interaction the values [16]

$$^{151}\mathrm{Eu}^+: \quad A = 1\ 540\ 297\ 394\ (13)\ \mathrm{Hz}\,,$$

$$B = -\ 660\ 862\ (231)\ \mathrm{Hz}\,,$$

$$C = 26\ (23)\ \mathrm{Hz}\,;$$

$$^{153}\mathrm{Eu}^+: \quad A = 684\ 565\ 993\ (9)\ \mathrm{Hz}\,,$$

$$B = -\ 1\ 752\ 868\ (84)\ \mathrm{Hz}\,,$$

$$C = 3\ (7)\ \mathrm{Hz}\,.$$

The values are somewhat preliminary due to the complexity of the second-order hyperfine correction.

The specific interest in Eu^+ arises from the fact that Eu^+ offers the possibility of systematically investigating the change of A factors over a long series of isotopes. Nine isotopes with nonzero nuclear spin exist which live longer than 1 week. Those isotopes can be investigated off line from the production area by standard techniques as was demonstrated previously on then 10 day isotope $^{131}Ba^+$. The ratio of A factors for two isotopes differs from that of the corresponding nuclear magnetic moments by the change in distribution of the magnetization over the nuclear volume [17] (Bohr-Weisskopf effect). The order of magnitude for that change is $10^{-2} \div 10^{-4}$. It is expressed by the differential hyperfine anomaly $^1\Delta^2 = (A_1/A_2)(\mu_1/\mu_2) - 1$, where the indices 1 and 2 refer to 2 different isotopes.

The precision obtained in the stable Eu^+ isotopes is by far enough to determine $^1\Delta^2$ to better than 1% even in cases where it is small, if the corresponding nuclear magnetic moments are known well enough. These come from nuclear-resonance experiments or may be the subject of measurements in ion traps with strong superimposed magnetic fields [18]. Our theoretical understanding of the hyperfine anomaly is poor to date, but systematic investigation in chains of isotopes may shed new light on this problem.

4. – Conclusion.

The status of microwave and radiofrequency spectroscopy in ion traps using optical methods for state preparation and detection has reached a level where complex and experimentally difficult systems can be successfully investigated. $^{207}Pb^+$ and Eu^+ are first examples, where ground-state hyperfine splittings in nonalkalilike ions have been measured. In spite of the experimental difficulty rather high precision has been obtained in both cases.

* * *

The measurements on Pb^+ ions in our laboratory were performed by F. XIN and G. Z. LI, those on Eu^+ by O. BECKER and K. ENDERS. The second-order hyperfine correction is based on calculations by J. DEMBCZYNSKI, Poznan, Poland. We gratefully acknowledge the support of the Deutsche Forschungsgemeinschaft for our experiments.

REFERENCES

[1] P. H. DAWSON: *Quadrupole Mass Spectrometry and its Application* (Elsevier, Amsterdam, 1976).
[2] L. S. BROWN and G. GABRIELSE: *Rev. Mod. Phys.*, **58**, 233 (1986).
[3] H. W. SCHAAF, U. SCHMELING and G. WERTH: *Appl. Phys.*, **25**, 249 (1981).
[4] W. NEUHAUSER, M. HOHENSTATT, P. E. TOSCHEK and H. G. DEHMELT: *Phys. Rev. Lett.*, **41**, 233 (1978).
[5] D. SCHNIER: Thesis, Physikalisch-Technische Bundesanstalt, Braunschweig, PTB-Bericht, Opt-37 (1992).
[6] J. D. PRESTAGE, R. C. TJOELKER, G. J. DICK and L. MALECKI: *J. Mod. Opt.*, **39**, 221 (1992).
[7] L. S. CULTER, R. P. GIFFARD, P. J. WHEELER and G. WINKLER: in *Proceedings of the 41st Annual Frequency Control Symposium, Philadelphia, Penn., 1987*, p. 12.
[8] J. D. PRESTAGE, J. J. BOLLINGER, W. M. ITANO and D. J. WINELAND: *Phys. Rev. Lett.*, **54**, 2387 (1985).
[9] J. J. BOLLINGER, D. J. HEINZEN, W. M. ITANO, S. L. GILBERT and D. J. WINELAND: *Phys. Rev. Lett.*, **63**, 1031 (1991).
[10] D. J. WINELAND, J. J. BOLLINGER, D. J. HEINZEN, W. M. ITANO and M. G. RAIZEN: *Phys. Rev. Lett.*, **67**, 1735 (1991).
[11] F. ARBES, O. BECKER, H. KNAB, K. H. KNÖLL and G. WERTH: *Nucl. Instrum. Methods B*, **70**, 494 (1992).
[12] A. ROTH, GH. GERZ, D. WILSDORF and G. WERTH: *Z. Phys. D*, **8**, 235 (1988).
[13] X. FENG, G. Z. LI, R. ALHEIT and G. WERTH: *Phys. Rev. A*, **46**, 327 (1992).
[14] X. FENG, G. Z. LI and G. WERTH: *Phys. Rev. A*, **46**, 2959 (1992).
[15] D. J. LARSON, J. C. BERGQUIST, J. J. BOLLINGER, W. M. ITANO and D. J. WINELAND: *Phys. Rev. Lett.*, **57**, 70 (1986).
[16] O. BECKER, K. ENDERS, J. DEMBCZYNSKI and G. WERTH: *Phys. Rev. A*, **48**, 3546 (1993).
[17] A. BOHR and V. WEISSKOPF: *Phys. Rev.*, **77**, 94 (1950).
[18] H. KNAB, K. H. KNÖLL, F. SCHEERER and G. WERTH: *Z. Phys. D*, **25**, 205 (1993).

COOLING

Laser Cooling from the Semi-Classical to the Quantum Regime.

J. DALIBARD and Y. CASTIN

Laboratoire de Spectroscopie Hertzienne de l'ENS (*)
24 rue Lhomond, F-75231 Paris Cedex 05, France

1. – Introduction.

The control of atomic motion by laser light is a field which has expanded very rapidly over the last few years. One of the most spectacular achievements in this domain is the possibility of reaching extremely low atomic kinetic temperatures, in the microkelvin range, by irradiating an atomic vapour with multiple quasi-resonant laser beams [1]. The limits of laser cooling in these so-called *optical molasses* correspond to r.m.s. velocities $\bar{v}$ of the order of only a few photon recoil velocities [2, 3]:

$$\bar{v} \simeq \text{a few } \frac{\hbar k}{M}, \tag{1}$$

where $\hbar k$ is the momentum of a photon involved in the cooling process and M is the atomic mass. One can even pass beyond this *recoil limit* using some improved cooling schemes [4, 5].

The combination of these low temperatures with the possibility of trapping the atoms around a given point in space offers a new unique tool for atomic spectroscopy and quantum optics, and many fields of atomic physics can benefit from these new techniques: metrology using atomic fountains [6, 7], collision physics [8-11], nonlinear optics [12, 13], etc.

These ultra-low temperatures also allow one to reach situations where the quantum nature of the atomic motion plays an important role. The atomic de

(*) Unité de recherche de l'Ecole Normale Supérieure et de l'Université Paris 6, associée au CNRS.

Broglie wavelength

$$\Lambda_{\text{dB}} = \frac{h}{Mv} \tag{2}$$

is indeed quite large, of the order of a fraction of optical wavelength. This may be of great help for atomic interferometry experiments [14]. For sufficiently high atomic densities, this might also offer a way of observing collective quantum effects in a sample of cold neutral atoms.

This course is devoted to a description of various approaches to laser cooling of neutral atoms. We restrict here to the case of a closed atomic transition between a stable ground state g and an excited state e with a lifetime Γ^{-1} (see fig. 1). These two energy levels are separated by an energy $\hbar\omega_A$ and they may both have a Zeeman degeneracy. The cooling laser field is supposed to be monochromatic, with an angular frequency ω_L.

The laser-cooling problem constitutes a model case for dissipation in quantum mechanics (see fig. 2): The *atom + laser field* system evolves coherently due to absorption and stimulated-emission processes, and it is dissipatively coupled to a *reservoir* formed by the quantized field in its ground state. This dissipative coupling corresponds to spontaneous-emission processes, and plays an essential role for interaction times longer than the excited lifetime Γ^{-1}.

We start in sect. 2 with a brief survey of the semi-classical description of

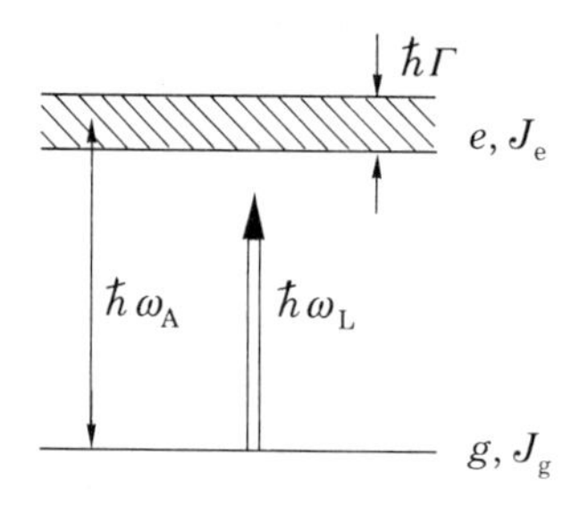

Fig. 1. – Closed atomic transition with a stable ground state g (angular momentum J_g) and an excited state e (angular momentum J_e) with a lifetime Γ^{-1}.

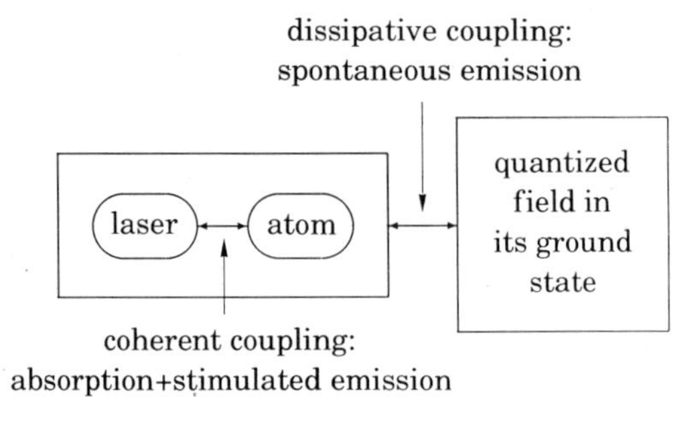

Fig. 2. – The interacting systems in a laser-cooling problem.

laser cooling, in which the atomic centre of mass is treated as a classical, point-like particle. We present two cooling mechanisms, the Doppler-cooling scheme and the Sisyphus-cooling scheme. We then show that the limit of this second cooling mechanism requires a quantum treatment, which is presented in sect. 3. Finally sect. 4 is devoted to the description of a new, general approach to dissipative processes in quantum optics based on Monte Carlo wave functions and to a discussion of its application to laser-cooling problems.

2. – Semi-classical description of laser cooling.

2·1. *Validity of the semi-classical treatment.* – In order to treat the atom as a moving classical pointlike particle, two assumptions are necessary. The first condition ensures that the atomic position is well defined on the shortest spatial scale of variation of the laser field parameters (phase, polarization or intensity), *i.e.* the optical wavelength λ. This can be written

$$\Delta x \ll \lambda \,, \tag{3}$$

where Δx denotes the size of the atomic wave packet or, more precisely, the coherence length of the atomic-density operator (*e.g.*, the thermal de Broglie wavelength for a Boltzmann distribution).

The second condition ensures that the atomic velocity is well defined with respect to the velocity width of the atomic resonance. More precisely we require an uncertainty $k\Delta v$ on the Doppler shift much smaller than the width Γ of the atomic resonance:

$$k\Delta v \ll \Gamma \,. \tag{4}$$

Condition (4) can also be seen as the requirement that the spatial spreading of the atomic wave packet between two successive spontaneous emissions remains small as compared to λ. We assume here that spontaneous-emission processes occur with a rate Γ (saturated transition) and we use the fact that each of these processes decreases the coherence length of the atomic-density matrix to less than λ [15].

Clearly the two assumptions (3) and (4) are compatible with the Heisenberg inequality

$$M\Delta x \Delta v \geqslant \frac{\hbar}{2} \tag{5}$$

only if the following relation holds:

$$E_{\rm R} = \frac{\hbar^2 k^2}{2M} \ll \hbar\Gamma \,. \tag{6}$$

This is the so-called *broad-line condition* which requires the recoil energy $E_{\rm R}$ to be much smaller than the energy width $\hbar\Gamma$ of the excited atomic level. We have indicated in table I the value of the ratio $\hbar\Gamma/E_{\rm R}$ for the resonance line of three

TABLE I. – *Value of the «broad-line parameter» $\hbar\Gamma/E_R$ for various atoms used in laser-cooling experiments.*

Atom	$\hbar\Gamma/E_R$
H*	40
Na	400
Cs	2600

«typical» atoms, the metastable helium atom in the $2\,^3S_1$ state, and the sodium and cesium atoms. We see that the broad-line condition is well satisfied for Na and Cs, but is only marginal for He*.

2·2. *The average radiative forces.* – Once the two conditions (3) and (4) hold, one can derive an expression for the average radiative force $\boldsymbol{f}$ acting on an atom located in a given position $\boldsymbol{r}$ and with a given velocity $\boldsymbol{v}$. Using Ehrenfest theorem, one gets [16, 17]

$$\boldsymbol{f} = \sum_{i=x,y,z} d_i \boldsymbol{\nabla} E_i(\boldsymbol{r}). \tag{7}$$

The force is proportional to the average atomic dipole $\boldsymbol{d}$ and to the gradient of the laser electric field $\boldsymbol{E}$ at the atom location. We now discuss these two contributions.

1) The laser electric field is assumed here to be monochromatic, and can, therefore, be written as

$$\boldsymbol{E}(\boldsymbol{r}) = \mathcal{E}(\boldsymbol{r})\,\boldsymbol{\varepsilon}(\boldsymbol{r})\exp[-i(\omega t - \phi(\boldsymbol{r}))] + \text{c.c.} \tag{8}$$

The force related to the gradient of the phase $\phi(\boldsymbol{r})$ is called the radiation pressure force or scattering force, the force related to the gradient of the real amplitude $\mathcal{E}(\boldsymbol{r})$ is the dipole or gradient force, and finally the force related to the gradient of the complex unit vector $\boldsymbol{\varepsilon}(\boldsymbol{r})$ is simply named polarization gradient force.

2) The mean atomic dipole $\boldsymbol{d}$ is obtained from the average value of the atomic-dipole operator:

$$\boldsymbol{d} = \langle \boldsymbol{D} \rangle = \mathrm{Tr}(\rho_{\mathrm{at}} \boldsymbol{D}). \tag{9}$$

For a two-level atom without Zeeman degeneracy (*), the dipole operator $\boldsymbol{D}$ is

(*) This situation can be achieved in practice using a σ_+ polarized laser beam acting on a $J_g \leftrightarrow J_e = J_g + 1$ transition, in which case the atom is optically pumped on the transition $|g\rangle = |g, m_g = J_g\rangle \leftrightarrow |e\rangle = |e, m_e = J_e\rangle$. Note that there is no polarization gradient force in this case.

given by

$$D = d_0(|e\rangle\langle g| + |g\rangle\langle e|), \tag{10}$$

where d_0 is the reduced dipole moment of the transition. For a more complex atomic transition in which the angular momenta of the ground and excited levels are taken into account, this atomic-dipole operator involves Clebsch-Gordan coefficients between ground and excited Zeeman sublevels [18]. The average value in (9) is taken over the steady-state atomic-density operator, calculated from the optical Bloch equations. This steady-state density operator can be calculated either for an atom at rest in r or for an atom dragged with a velocity v and passing in r at a given time.

We give here the expression of the radiative forces acting on a two-level atom at rest without Zeeman degeneracy. The radiation pressure force f_{RP} and the dipole force f_{dip} are given by [17]

$$f_{RP} = \frac{\hbar\Gamma}{2}\frac{s}{1+s}\nabla\phi, \tag{11}$$

$$f_{dip} = -\frac{\hbar\delta}{2}\frac{\nabla s}{1+s}, \tag{12}$$

where we have introduced the detuning $\delta = \omega_L - \omega_A$ between the laser and atom frequencies, the Rabi frequency $\Omega = 2\mathcal{E} d_0\, \varepsilon/\hbar$, and the saturation parameter

$$s = \frac{\Omega^2/2}{\delta^2 + \Gamma^2/4}. \tag{13}$$

2·3. *Fokker-Planck equation for the atomic phase space distribution.* – For slowly moving atoms, we obtain from (7) the force at first order in velocity

$$f(r, v) = f(r, 0) - [\alpha(r)] \cdot v, \tag{14}$$

where $[\alpha(r)]$ is the friction tensor, describing the damping of atomic motion in the optical molasses. In order to study the limit of cooling in these molasses, we need also to take into account the counterpart of cooling, which is the heating due to the randomness of spontaneous-emission processes.

In the semi-classical approach, this can be done by writing a Fokker-Planck equation for the evolution of the atomic centre-of-mass phase space distribution. Such an equation can be obtained in the limit

$$T_{int} \ll T_{ext}, \tag{15}$$

which corresponds to a situation where the internal atomic variables have a time response T_{int} much smaller than the time of evolution of external variables T_{ext}. It is valid for slow atoms ($k|v|T_{int} \ll 1$) and it is obtained by eliminating the internal dynamics adiabatically to get an equation for the external atomic phase space distribution, or, more precisely, the atomic-density matrix in the Wigner representation $w(\boldsymbol{r}, \boldsymbol{p}, t)$ [19,20]:

$$(16)\qquad \frac{\partial w}{\partial t} = -\boldsymbol{v}\cdot\boldsymbol{\nabla}_{\boldsymbol{r}}\, w - \sum_{i=x,y,z} \frac{\partial}{\partial p_i}\left(\left(f_i(\boldsymbol{r}, 0) - \alpha_{ij}(\boldsymbol{r})\, v_j\right) w\right) + \sum_{i=x,y,z} D_{ij}(\boldsymbol{r})\, \frac{\partial^2 w}{\partial p_i\, \partial p_j}\,.$$

This equation contains a free-flight term, a force term (see (14)), and finally a diffusion term describing the heating due to the randomness of spontaneous-emission processes. $D_{ij}(\boldsymbol{r})$ is the tensorial momentum diffusion coefficient.

The steady state of (16) gives the position and momentum equilibrium distributions for the atom and allows one in particular to derive a temperature for laser-cooled atoms. A simple approximate solution of this equation is obtained by assuming a spatially uniform distribution w. Due to the spatial periodicity of optical molasses, $\boldsymbol{f}(\boldsymbol{r}, 0)$ averages to 0, and one is left in the isotropic case with the simpler equation

$$(17)\qquad \frac{\partial w}{\partial t} = \boldsymbol{\nabla}_{\boldsymbol{p}}\,(\bar{\alpha}\boldsymbol{v}\, w) + \overline{D}\, \boldsymbol{\nabla}_{\boldsymbol{p}}^2\, w\,,$$

whose solution is Gaussian, with an effective temperature which can be written in terms of the spatially averaged friction and diffusion coefficients:

$$(18)\qquad k_{\mathrm{B}}\, T = \frac{\overline{D}}{\bar{\alpha}}\,.$$

Let us briefly conclude with some comments on the validity of (18). First, we have to check in each particular situation that this kinetic energy $k_{\mathrm{B}}\, T$ is larger than the depth of the wells which may be created by $\boldsymbol{f}(\boldsymbol{r}, 0)$. If it is not the case, the assumption of a spatially uniform steady-state distribution is clearly not valid. Second, we have to give some estimate for the validity condition (15) of the general Fokker-Planck equation (16). A typical internal time is $T_{\mathrm{int}} \sim \Gamma^{-1}$ for a two-level atom without Zeeman degeneracy, while external time will be found in the following to be of the order of $M/\hbar k^2$ or longer. The broad-line condition then ensures that (15) is verified. For an atom with Zeeman degeneracy in the ground state, this is not true any more because there may appear very long pumping times which reverse the inequality of (15) and make this Fokker-Planck approach unapplicable.

2·4. *Doppler cooling.* – We consider here the one-dimensional situation represented in fig. 3. An atom is moving in the field created by two counterpropagating plane running waves, detuned red from resonance ($\omega_{\mathrm{L}} < \omega_{\mathrm{A}}$) [21,22]. We assume that the two running waves are weak and do not saturate the atomic transition. Then we can obtain the total force acting on the atom by adding in-

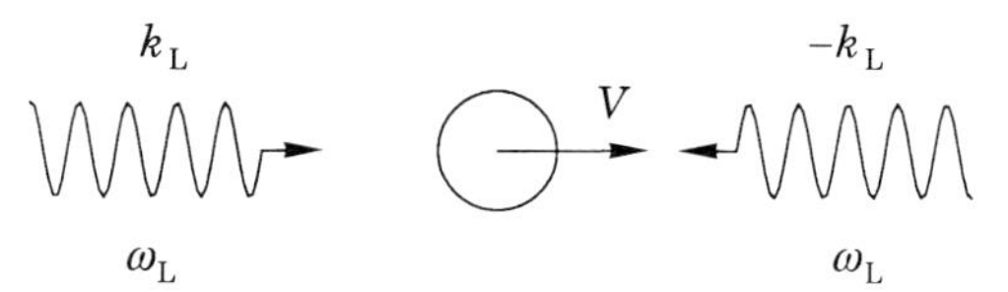

Fig. 3. – Doppler cooling in 1D optical molasses.

dependently the radiation pressure forces created by each running wave [23]. Due to the choice $\omega_L < \omega_A$ and because of Doppler effect, a moving atom sees the counterpropagating wave closer to resonance than the copropagating one. Consequently, for an atom moving, for instance, to the right as in fig. 3, the radiation pressure force created by the counterpropagating wave coming from the right is larger than the radiation pressure force exerted by the copropagating wave coming from the left. Therefore, a moving atom feels a net force opposed to its velocity. This friction force damps the atomic motion and the atom is cooled.

Using the expression (11) for the radiation pressure force created by each travelling wave and replacing δ by $\delta \pm kv$ to take into account Doppler effect, we obtain for the net force acting on a moving atom in the limit $k|v| \ll |\delta|$

$$f = -\bar{\alpha} v, \tag{19}$$

where the friction coefficient $\bar{\alpha}$ is given by

$$\bar{\alpha} = \hbar k^2 s_0 \frac{2|\delta|\Gamma}{\delta^2 + \Gamma^2/4}. \tag{20}$$

We note, as announced above, that the external time $M/\bar{\alpha}$ associated with this damping is larger than $M/\hbar k^2$, since $s_0 \ll 1$.

The momentum diffusion coefficient can be estimated simply by noting that, due to the randomness of the number of absorbed photons per unit time and the randomness of the momentum carried away by the spontaneously emitted photon, the atomic momentum performs a random walk with a step $\hbar k$ and a rate $\sim 2\Gamma s_0$. This gives

$$\bar{D} = \hbar^2 k^2 \Gamma s_0 . \tag{21}$$

Using (18), we now get the temperature of Doppler-cooled atoms. The temperature is minimal for a detuning $\delta = \omega_L - \omega_A$ equal to $-\Gamma/2$, and it is given by [24]

$$k_B T = \frac{\hbar \Gamma}{2}. \tag{22}$$

This gives temperatures in the range of $100\,\mu K$. More precisely, for the 3 atoms considered above, one obtains $36\,\mu K$ for He*, $240\,\mu K$ for Na and $120\,\mu K$ for Cs. We note that, for atoms cooled at the Doppler limit (22), the r.m.s. velocity $\bar{v}$ and the de Broglie wavelength (*i.e.* coherence length Δx) are given by

$$\bar{v} = \sqrt{\frac{\hbar\Gamma}{2M}}, \tag{23}$$

$$\Lambda_{\text{dB}} = \Delta x = 2\pi\sqrt{\frac{2\hbar}{M\Gamma}}, \tag{24}$$

so that both conditions (3) and (4) are simultaneously fulfilled in the broad-line limit:

$$\frac{\Delta x}{\lambda} = \sqrt{\frac{2\hbar k^2}{M\Gamma}} \ll 1, \tag{25}$$

$$\frac{k\bar{v}}{\Gamma} = \sqrt{\frac{\hbar k^2}{2M\Gamma}} \ll 1. \tag{26}$$

This shows that, in the broad-line limit, it is legitimate to use a semi-classical treatment to derive the Doppler-cooling limit.

2·5. *Sisyphus cooling.* – Sisyphus cooling is the simplest example of laser cooling with polarization gradients. It uses the Zeeman structure of the ground atomic state to provide a cooling which is much more efficient than Doppler cooling. The search for cooling mechanisms other than Doppler has been initiated by the discovery of anomalously low temperature in optical molasses, well below the Doppler-cooling limit [2], and the role of polarization gradients has been emphasized shortly after [25, 26].

In one dimension, two types of polarization gradient, corresponding to two different cooling mechanisms, have been identified [27-29]. The first one is obtained with two plane waves with orthogonal linear polarizations (fig. 4*a*)). The axes of the resulting polarization are constant, oriented at 45° with respect to

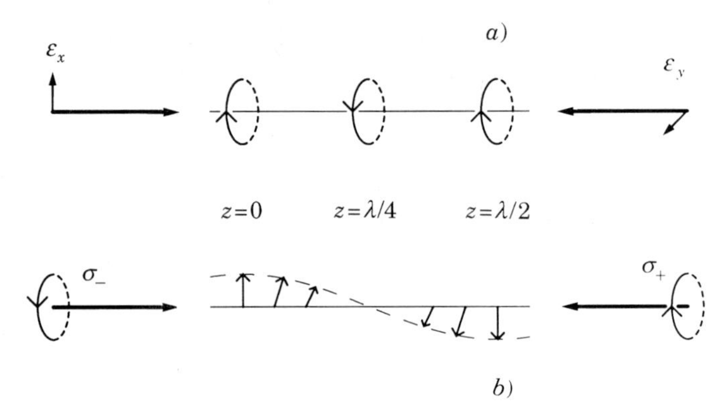

Fig. 4. – The two limiting cases of polarization gradient in one dimension. *a*) The lin ⊥ lin configuration, with a gradient of ellipticity and constant axis of polarization. *b*) The σ_+-σ_- configuration, with a resulting rotating linear polarization.

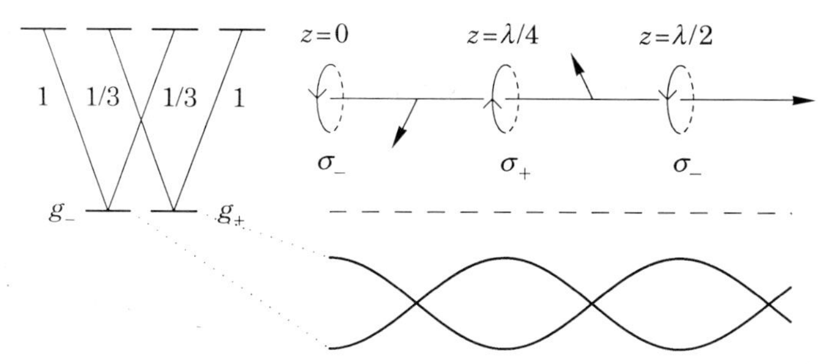

Fig. 5. – Light-shifted ground-state energy levels of a $J_g = 1/2 \leftrightarrow J_e = 3/2$ atom in a lin ⊥ lin laser field. Due to the gradient of ellipticity of the light, the two Zeeman sublevels oscillate in phase opposition with a period $\lambda/2$.

the polarization axis of the incoming beams. The ellipticity of the resulting polarization varies in space, going from circular to linear over a distance of $\lambda/8$. This configuration leads to Sisyphus cooling, as we show below. The second configuration is obtained with two incoming waves with orthogonal circular polarizations (fig. 4*b*)). The resulting polarization in this σ_+-σ_- configuration is linear everywhere (no gradient of ellipticity), and its direction rotates with a period $\lambda/2$. This configuration leads to orientational cooling.

We now focus on the situation of fig. 4*a*) and we consider the motion of an atom with an angular momentum $J_g = 1/2$ in the ground state and $J_e = 3/2$ in the excited state (see fig. 5) (*). We restrict ourselves to the low-saturation domain:

$$s_0 = \frac{\Omega^2/2}{\delta^2 + \Gamma^2/4} \ll 1 , \tag{27}$$

which is known experimentally to lead to the lowest temperatures. In (27), $\Omega = 2 d_0 \mathcal{E}_0 / \hbar$, where d_0 is the reduced dipole moment of the transition, and $\mathcal{E}_0$ the field amplitude in each travelling wave. When (27) is fulfilled, the atoms remain mostly in their internal ground-state sublevels. In the following, we also restrict to situations where the Doppler shifts can be neglected compared to Γ. Therefore, we ignore here the Doppler-cooling mechanism presented in subsect. 2˙4. We decompose the effect of the laser light on the atoms into two parts. We first consider the reactive part, *i.e.* the shifts of the levels caused by the light. We then study the dissipative part of this coupling, corresponding to the real transitions between the ground-state sublevels associated with spontaneous-emission processes.

We consider as for Doppler cooling a negative detuning, so that the two Zee-

(*) The simpler atomic transition $J_g = 0 \leftrightarrow J_e = 1$ would not lead to any additional cooling with respect to Doppler cooling.

man substates are shifted downwards. The key point is that the size of the shift of each substate depends on the location of the atom. If the atom is located at a place where the light is σ_- polarized ($z = 0$ in fig. 5), the shift of level $|g_-\rangle = |g, m = -1/2\rangle$ is three times bigger than the shift of level $|g_+\rangle = |g, m = 1/2\rangle$, because of the intensity factors (squares of Clebsch-Gordan coefficients) of the $m_e - m_g = -1$ transitions, as indicated in fig. 5. At a place where the light is σ_+ polarized ($z = \lambda/4$ in fig. 5), the conclusion is reversed and the level $|g_+\rangle$ is shifted three times more than the level $|g_-\rangle$. In a place where the light is linear, one finds by symmetry that the two shifts are equal. Going through a small algebra, we obtain that the reactive part of the atom-laser coupling consists in a periodic potential $U_\pm(z)$, depending on the atomic ground-state sublevel $g_\pm$:

$$U_\pm(z) = \frac{U_0}{2}(-2 \pm \cos 2kz) \qquad \text{with } U_0 = -\frac{2}{3}\hbar\delta s_0 . \tag{28}$$

We now consider the transitions between g_+ and g_- caused by spontaneous emissions; these optical-pumping processes also depend on the location of the atom. Suppose that the atom is moving towards the right and starts in $z = 0$ in level g_- (fig. 6). At this place since the light is σ_-, the atom cycles on the transition $|g, m = -1/2\rangle \leftrightarrow |e, m = -3/2\rangle$ and can never jump to the level g_+. Therefore, it has to climb uphill until it reaches a place where there is a sufficient amount of σ_+ light so that the atom can be optically pumped to g_+ by a sequence $|g, m = -1/2\rangle \to |e, m = 1/2\rangle \to |g, m = 1/2\rangle$. This occurs preferentially around $z = \lambda/4$, at the top of $U_-(z)$, where the light is purely σ_+, and the atom is then put in a valley for $U_+(z)$. Once in level g_+ in z close to $\lambda/4$, the atom has little chance to come back to g_-, because the light is essentially σ_+ polarized

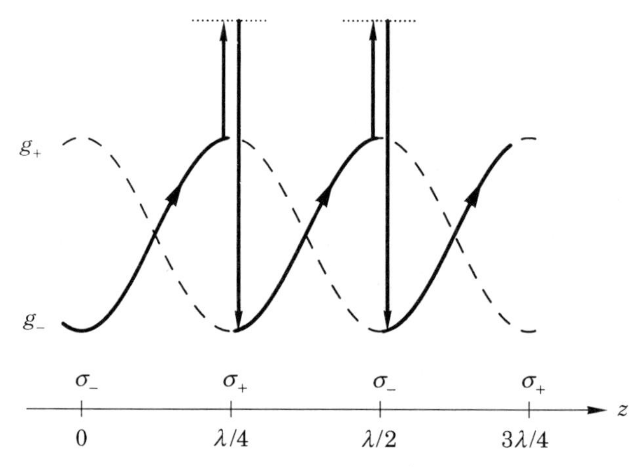

Fig. 6. – Sisyphus effect: due to the spatial variation of the optical-pumping rates, a moving atom climbs more than it goes down in its energy diagram. This causes a damping of its velocity in a much more efficient way than Doppler cooling.

at this place. The atom has, therefore, to climb again in $U_+(z)$ until it reaches a place where there is a noticeable fraction of σ_- light, which can pump it back to g_-.

It is clear that the atom loses energy in this process since it climbs more than it goes down. This is in close analogy with the Sisyphus myth in the Greek mythology where Sisyphus was sentenced by the Gods to push a rock forever to the top of a mountain, the rock rolling back to the valley each time it had reached the top. For this atomic Sisyphus process, it is instructive to make a balance of momentum and energy exchange. As the atom climbs uphill, it converts kinetic energy into potential energy, its total energy remaining constant. The decrease of momentum of the atom is due to a redistribution of photons between the two travelling waves forming the polarization gradient. Then, when the atom jumps from the top of a hill to a valley, it spontaneously emits a photon whose energy is higher than the laser photon energy by the height of the hill. This transition decreases the potential energy of the atom, while leaving its kinetic energy unchanged if one neglects the recoil associated with the spontaneous emission of the photon.

If we now take into account the recoil in spontaneous-emission processes, we have to add to the previous reasoning the corresponding heating, which puts a threshold for the potential depths U_0 [15, 18]. Each fluorescence cycle, *i.e.* absorption of a laser photon spontaneous emission of a fluorescence photon, can be shown to lead to an increase of kinetic energy $((41/30)E_R)$. Among those cycles, only a fraction of $1/6$ contributes to cooling, by changing the internal atomic state: $g_+ \to e \to g_-$ or $g_- \to e \to g_+$. The average loss of energy for these cooling cycles is $U_0/2$, so that we find that there is a net cooling effect only if

$$U_0 > (82/5)E_R \,. \tag{29}$$

The intuitive limit of Sisyphus cooling corresponds to a situation where the energy of the atom is of the order of or smaller than the potential modulation depth U_0. In this case, the atom does not have a sufficient kinetic energy to climb the potential hills, and it gets trapped in the potential valleys of $U_\pm(z)$. This intuitive reasoning is confirmed by a semi-classical treatment where one calculates a friction coefficient and a momentum diffusion coefficient, averaged over a wavelength [28]. The friction coefficient is found to be independent of the light intensity:

$$\bar{\alpha} = 3\hbar k^2 \, \frac{-\delta}{\Gamma} \tag{30}$$

and the equilibrium temperature is given by

$$k_B T = \frac{3}{8} U_0 \,. \tag{31}$$

This indicates that small temperatures should be obtained with small laser intensities and large detunings.

Let us discuss briefly the validity of this semi-classical result. First we note that the result (31) can only be qualitatively correct since we obtained it assuming a uniform spatial atomic distribution, and we predict at the same time a temperature of the order of the potential-well depth U_0 (subsect. 2·3). We now compare the optical-pumping relaxation time for internal variables, $T_{\text{int}} \simeq \simeq 1/\Gamma s_0$, with the characteristic time T_{ext} for external variables. We take here

$$T_{\text{ext}} \simeq M/\bar{\alpha} \text{ or } 1/\Omega_{\text{osc}}\,, \tag{32}$$

which corresponds to either the velocity damping time or the oscillation period in the bottom of the potential well (28). Both choices lead to the same expression for the validity condition (15):

$$T_{\text{int}} \ll T_{\text{ext}} \Leftrightarrow U_0 \gg \frac{\delta^2}{\Gamma^2} E_{\text{R}}\,. \tag{33}$$

One can check that this validity condition is also equivalent to the requirement $k\bar{v}\, T_{\text{int}} \ll 1$, where $\bar{v}$ is the r.m.s. velocity deduced from (31). This last condition ensures that it is legitimate to keep only the first-order velocity components in the expression of the force acting on the atom, for all the velocity classes populated in steady state (see (14)). For a given detuning, the validity condition (33) puts a lower bound on the intensity and, therefore, on the temperature that one can predict in this model:

$$k_{\text{B}} T \gg k_{\text{B}} T_{\text{min}} \sim E_{\text{R}} \frac{\delta^2}{\Gamma^2}\,. \tag{34}$$

We finally note that, between the threshold (29) and the lower bound (33), there is for $|\delta| \gg \Gamma$ a large range of values of U_0 for which Sisyphus cooling is expected to work, but cannot be described by the previous semi-classical treatment. We show in the next section how to fill this gap using a quantum approach.

3. – The quantum regime of Sisyphus cooling.

3·1. *Experimental and numerical results.* – Let us now compare the predictions (31), (34) of the simple semi-classical model of Sisyphus cooling presented above with results obtained either experimentally or numerically. We have plotted in fig. 7 temperature measurements which have been obtained with Cs atoms [3]. These data show a very good qualitative agreement with (31). The temperature varies linearly with $U_0 \sim \Omega^2/|\delta|$ over a wide range of laser intensities and detunings, and the proportionality coefficient differs by less than a

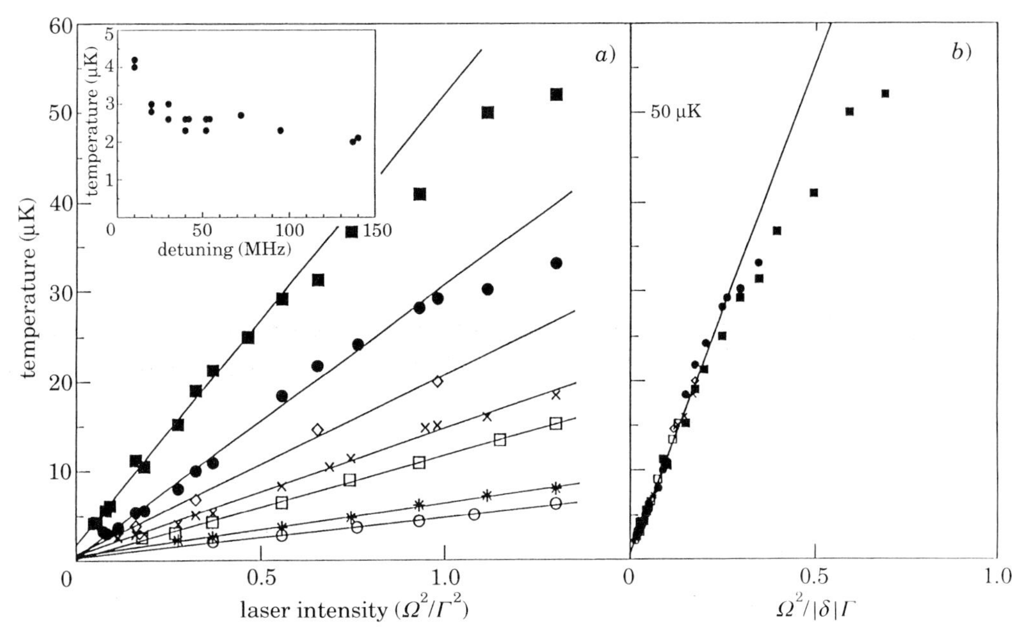

Fig. 7. – Temperature measurements of Cs atoms in optical molasses, as a function of laser intensity and detuning. *a*) Temperature *vs.* intensity at fixed detuning. Insert: lowest temperature achieved as a function of detuning. *b*) Temperatures of *a*) plotted against $\Omega^2/|\delta|$. The straight line is a fit to the points with small $\Omega^2/|\delta|$. The «universal law» $k_B T \sim \Omega^2/|\delta|$ is valid until $k_B T \sim 10 E_R$; $|\delta|/2\pi$: ■ 10 MHz, ● 20 MHz, ◇ 30 MHz, × 40 MHz, □ 54 MHz, * 95 MHz, ○ 140 MHz.

factor 3 from the one appearing in (31). This agreement is very remarkable if one notes that temperature measurements were done in 3D, while (31) is a 1D prediction. Also the Cs transition is $J_g = 4 \leftrightarrow J_e = 5$, quite far from our $J_g = 1/2 \leftrightarrow J_e = 3/2$ model.

However, an important difference appears between the semi-classical model and those experimental results, that we already noticed at the end of the previous section and which indicates the need for a more elaborate treatment. The lowest temperatures are obtained for detunings large compared to Γ. For such a detuning, as one decreases the laser intensity, the molasses works well until one reaches a temperature of the order of $10E_R$. This limit, which is in good qualitative agreement with (29), is independent of detuning and is well below the bound (34). This indicates that many points of fig. 7, and in particular the coldest ones, are not within the validity region of the semi-classical treatment given by (33).

This difference also appears in a complete numerical treatment of 1D Sisyphus cooling for a $J_g = 1/2 \leftrightarrow J_e = 3/2$ transition in which one looks for the steady state of the total (internal + external) atomic-density matrix [15, 30]. The steady-state r.m.s. velocity is plotted in fig. 8 in units of the recoil velocity,

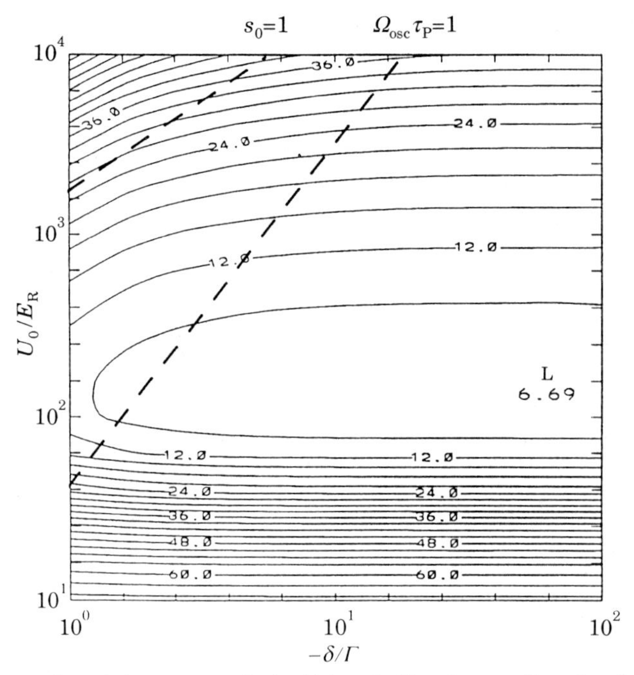

Fig. 8. – Contour plot of the r.m.s. velocity $\bar{v}$ in 1D Sisyphus cooling for the Cs atom parameters. The lowest velocities are of the order of 6.7 recoil velocities. They are obtained for a set of parameters δ, U_0 well outside the validity region of the semi-classical treatment. This validity region is below the high-saturation domain indicated by the line $s_0 = 1$, and above the dotted line $\Omega_{osc}\tau_P = 1$, where $\tau_P = 9/2\Gamma s_0$ is the optical-pumping time.

as a function of the detuning $|\delta|/\Gamma$ and the optical-well depth U_0/E_R. We have also shown in this figure the validity domain of the semi-classical approach (33). One clearly sees that the smallest r.m.s. velocities are obtained for a set of parameters outside the semi-classical validity region (*).

It is, therefore, necessary to elaborate another theoretical treatment for describing laser cooling in the regime $T_{int} > T_{ext}$. One could first think of an improved semi-classical treatment, by taking into account the velocity dependence of friction and diffusion, and also the spatial variation of the atomic distribution. However, it is actually easier to go directly to a quantum treatment of atomic motion, which will in addition lead to the predictions of new quantum effects related to the trapping of the atoms in the optical wells.

3·2. *Principle of a quantum treatment.* – We have seen both experimentally and numerically that the parameters δ and Ω, or equivalently δ and U_0, leading

(*) For numerical reasons, this plot has been calculated with a simplified dipole radiation pattern, which slightly overestimates the heating due to spontaneous photon recoil.

to the lowest temperatures were such that

$$\Omega_{\text{osc}} T_{\text{int}} \gg 1 . \tag{35}$$

In a quantum description of atomic motion in the molasses, this condition has a clear interpretation. The quantity $\hbar\Omega_{\text{osc}}$ is the energy difference between two neighbouring levels of the Hamiltonian which describes the atomic motion in the optical wells (28):

$$H_\varepsilon = \frac{P^2}{2M} + U_\varepsilon(Z), \qquad \varepsilon = \pm , \tag{36}$$

where Z and P are the atomic position and momentum operators. The quantity $\hbar/T_{\text{int}}$ is the energy width of these quantized levels due to optical-pumping processes. The quantum interpretation of (35) is that the splitting between the levels is much larger than their width. In this case, it is well known that the steady-state density matrix can be described only using the populations of those levels: this is the so-called secular approximation. This situation is quite similar to the one of ion cooling theory, except that the confining potential is created in our case by the light shift itself.

We now proceed in two steps [15, 31]. First we derive the expression for the eigenstates of the Hamiltonian H_ε. Then we calculate the transition rates between those eigenstates due to optical-pumping processes, and we determine the population of each eigenstate in steady state.

3·3. *Eigenstates of the Hamiltonian* H_ε. – The Hamiltonian H_ε describes the motion of a particle in a periodic potential $U_\varepsilon(z)$. Its energy spectrum consists of allowed energy bands separated by forbidden gaps. The eigenstates can be labelled as $|n, q, \varepsilon\rangle$, where n is an integer $\geqslant 0$ labelling the band and q is the Bloch index, chosen in the first Brillouin zone $(-k < q \leqslant k)$.

Due to the condition

$$U_+(z) = U_-\left(z + \frac{\lambda}{4}\right) \tag{37}$$

the two spectra of H_+ and H_- are identical, and the eigenstates of H_- can be deduced from the ones of H_+ by

$$E_{n,q,+} = E_{n,q,-} = E_{n,q} , \tag{38}$$

$$\langle z|n, q, +\rangle = |g_+\rangle \psi_{n,q}(z) , \tag{39}$$

$$\langle z|n, q, -\rangle = |g_-\rangle \psi_{n,q}\left(z - \frac{\lambda}{4}\right). \tag{40}$$

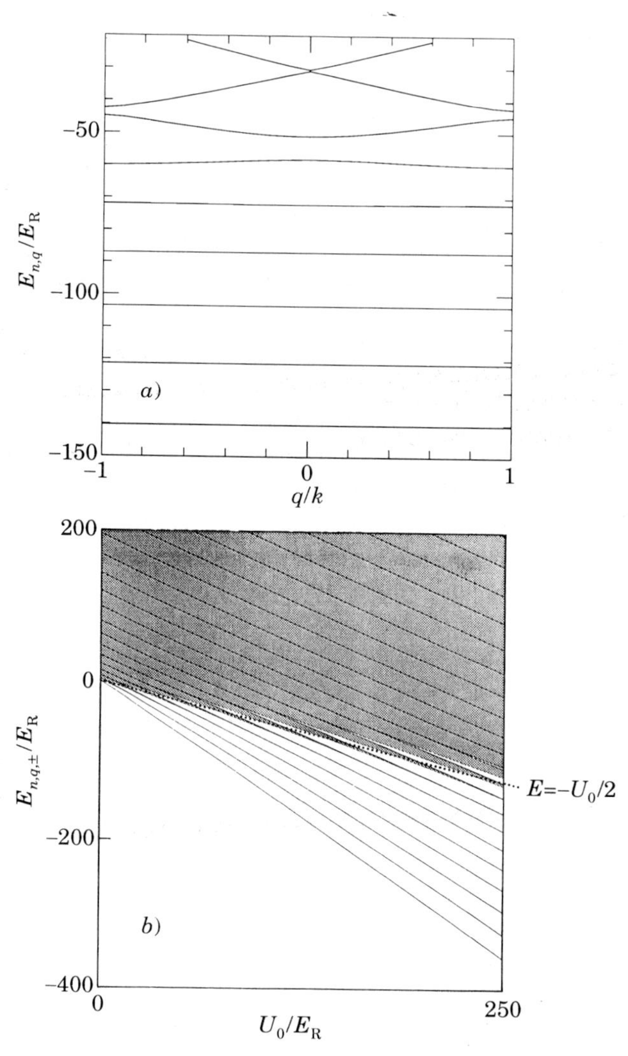

Fig. 9. – Band structure of the energy spectrum of H_ε for $U_0/E_R = 100$ (a)) and as a function of U_0/E_R (b)). The shaded areas correspond to allowed energies. For a given U_0, the energies above $-U_0/2$ corresponding to an above-barrier motion are mostly allowed (quasi-free motion). On the opposite, the energy bands corresponding to a bound classical motion ($-3U_0/2 < E < -U_0/2$) are very narrow, except in the immediate vicinity of $U_0/2$.

The eigenvalue problem

$$-\frac{\hbar^2}{2M}\frac{d^2\psi_{n,q}(z)}{dz^2} + U_+(z)\psi_{n,q}(z) = E_{n,q}\psi_{n,q}(z) \tag{41}$$

can be cast into a universal one using the reduced units $\zeta = kz$, U_0/E_R, E_{nq}/E_R:

$$-\frac{d^2\psi_{n,q}(\zeta)}{d\zeta^2} + \frac{U_0}{2E_R}(-2+\cos 2\zeta)\psi_{n,q}(\zeta) = \frac{E_{n,q}}{E_R}\psi_{n,q}(\zeta). \tag{42}$$

The spectrum of H_ε is indicated in fig. 9a) for $U_0/E_R = 100$, which is close to the optimum situation appearing in fig. 8. We find 6 bands corresponding to bound states ($E_{n,q} < -U_0/2$); the width of the lowest band $n = 0$ is extremely small ($10^{-6} E_R$). A plot of the energy spectrum as a function of U_0/E_R is indicated in fig. 9b). It shows that the number of «bound bands» increases as $\sqrt{U_0/E_R}$, as does the splitting $\hbar\Omega_{osc}$ between two adjacent bands.

3·4. *Steady-state populations.* – We now take into account the transitions induced by optical-pumping processes between the various $|n, q, \varepsilon\rangle$. As we have emphasized above, the density matrix in the secular approximation can be described only in terms of the populations $\pi_{n,q,\varepsilon}$ of those eigenstates.

In steady state, we have

$$0 = \dot{\pi}_{n,q,\varepsilon} = -\pi_{n,q,\varepsilon}\sum_{n',q',\varepsilon'}\gamma(n,q,\varepsilon\to n',q',\varepsilon') + \sum_{n',q',\varepsilon'}\gamma(n',q',\varepsilon'\to n,q,\varepsilon)\pi_{n',q',\varepsilon'}. \tag{43}$$

The rates $\gamma(n, q, \varepsilon \to n', q', \varepsilon')$ are derived from the master equation describing optical-pumping processes. We will not perform a detailed calculation for

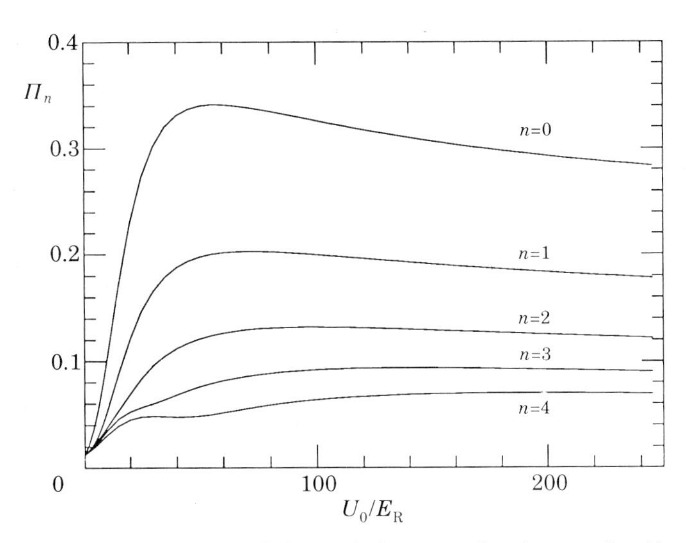

Fig. 10. – Steady-state populations of the various energy bands, as a function of U_0/E_R. This calculation has been done taking into account the first 80 bands, with 6 values for q in each band.

these rates here (see [15,31]). Let us just indicate that each rate appears as a product of Γs_0 by functions of matrix elements between $|n, q, \varepsilon\rangle$ and $|n', q', \varepsilon'\rangle$ of operators such as $\exp[ikZ]$, which depend only on U_0/E_R. Consequently, Γs_0 can be factorized out of (43), so that the steady-state populations $\pi_{n,q,\varepsilon}$ depend only on U_0.

Two types of transition are involved in the rate equations (43). The first ones correspond to transitions $\varepsilon \to -\varepsilon$, which are at the origin of the Sisyphus cooling (cf. subsect. 2·5). If these transitions were the only ones, one would find a strong accumulation of the atoms in the lowest band $n = 0$. In fact, this accumulation is counterbalanced by transitions $\varepsilon \to \varepsilon$ with different values of q and n. These transitions correspond to a heating, due, for instance, to the randomness of the momenta of the emitted fluorescence photons.

We note that the symmetry of the rates in the exchange $\varepsilon \leftrightarrow \varepsilon'$ implies in steady state

$$(44)\quad \gamma(n, q, \varepsilon \to n', q', \varepsilon') = \gamma(n, q, \varepsilon' \to n', q', \varepsilon) \Rightarrow \pi_{n,q,+} = \pi_{n,q,-}\,.$$

We have plotted in fig. 10 the total steady-state population π_n of each of the first five bands. We see that, for increasing U_0, each population goes through a maximum and then decreases. For instance, π_0 is maximal around $U_0 = 60\,E_R$ and reaches 0.34: more than 1/3 of the atoms are then localized in the $n = 0$ band. Using these populations and the momentum or position representations of the eigenstates $|n, q, \varepsilon\rangle$, we can now reconstruct the steady-state momentum and position distributions. We have shown in fig. 11 the results obtained for $U_0/E_R = 100$. We have also plotted in fig. 12

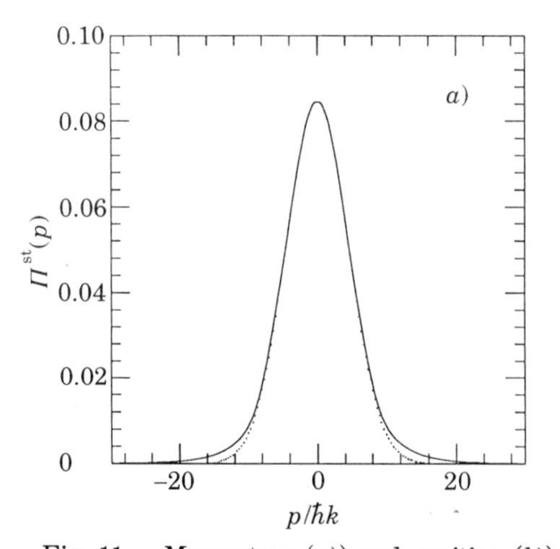

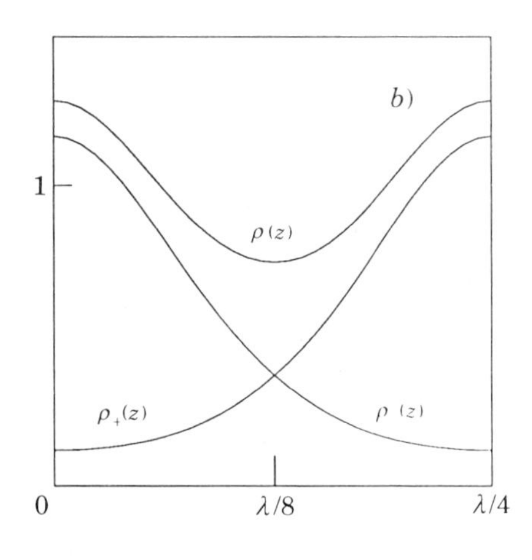

Fig. 11. – Momentum (a)) and position (b)) steady-state distributions for $U_0/E_R = 100$. The dotted line in a) is the Gaussian curve with the same width at $\exp[-1/2]$. In b), we have indicated the position distribution for each sublevel $\rho_\pm(z)$, as well as the total distribution $\rho(z) = \rho_+(z) + \rho_-(z)$.

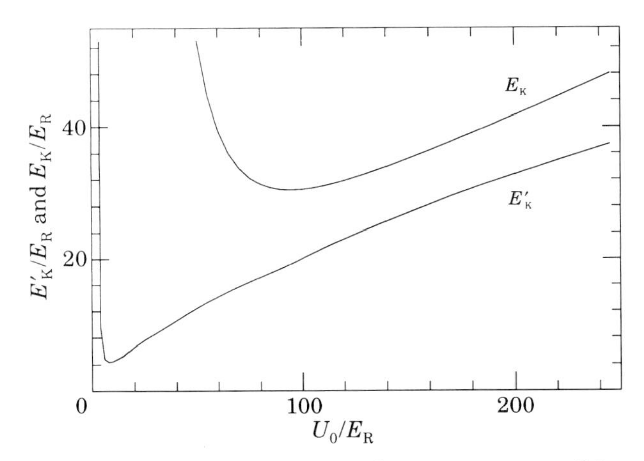

Fig. 12. – Steady-state kinetic energies $E_K = M\bar{v}^2/2$ and $E'_K = M\,\delta v^2/2$ (where δv is the half-width at $\exp[-1/2]$ of the velocity distribution) as a function of U_0. These two quantities would be equal for Gaussian velocity distributions.

the average kinetic energy $E_K = M\bar{v}^2/2$ as a function of U_0. It is minimal for $U_0/E_R = 95$, with $\bar{v} \simeq 5.5\,\hbar k/M$.

3·5. *Conclusions for this approach.* – The general features of this quantum treatment are in good qualitative agreement with experimental and numerical results of subsect. 3·1. The most striking result is the existence of a universal parameter U_0/E_R. This universality appears clearly in fig. 8 where the contour lines are parallel with the δ axis in the region $\Omega_{osc}T_{int} \ll 1$, indicating that here also the results depend only on U_0/E_R. Also both the variations of the steady-state atomic kinetic energy with U_0 and the order of magnitude of the minimum $\bar{v}$ are in agreement with the numerical results shown in fig. 8.

This approach also gives access to other observable quantities of laser-cooled atomic samples, more deeply connected to the quantization of atomic motion, such as the discrete structure of the energy spectrum. These quantum features have recently been observed in two spectroscopy experiments, one dealing with the absorption spectrum of the cold atomic sample [32], the other one with the fluorescence spectrum [33]. Also an experiment in Stony Brook has shown the existence of strong magnetic r.f. resonances in laser-cooled samples produced by M.I.L.C. (magnetically induced laser cooling), which may be related to transitions between the quantized levels [34].

This treatment can clearly be generalized to other transitions and other cooling schemes. We have extended it to the 1D Sisyphus cooling of a $J_g = 1 \leftrightarrow J_e = 2$ transition [15], and also to the study of Sisyphus cooling in 2D [35]. BERGEMAN has used a similar approach to study the M.I.L.C. situation [36]. Also COURTOIS and GRYNBERG have used such an approach to give a quantitative analysis of the spectroscopy experiments showing the atomic quantization mentioned above [37].

4. – A Monte Carlo wave function approach.

We now turn to the last part of this lecture which is devoted to the description of a new general method which can be used for the theoretical study of dissipative processes in quantum optics and atomic physics, such as laser cooling. Usually the dissipative coupling between a small system and a large reservoir can be treated by a master-equation approach [38-41]; one writes a linear equation for the time evolution of the reduced-system density matrix $\rho_S = \mathrm{Tr}_{\mathrm{res}}(\rho)$, trace over the reservoir variables of the total density matrix. If we denote H_S the Hamiltonian for the system, this equation can be written

$$\dot{\rho}_S = \frac{i}{\hbar}[\rho_S, H_S] + \mathscr{L}_{\mathrm{relax}}(\rho_S). \tag{45}$$

In (45), $\mathscr{L}_{\mathrm{relax}}$ is the relaxation superoperator, acting on the density operator ρ_S. It is assumed here to be local in time, which means that $\dot{\rho}_S(t)$ depends only on ρ_S at the same time (Markov approximation). All the system dynamics can be deduced from (45). One can calculate one-time average values of a system operator A: $a(t) = \langle A\rangle(t) = \mathrm{Tr}(\rho_S(t)A)$, and also, using the quantum regression theorem [42], multitime correlation functions such as $\langle A(t+\tau)B(t)\rangle$.

We present here an alternative treatment based on a Monte Carlo evolution of wave functions of the small system (MCWF) [43-46]. This evolution consists of two elements: evolution with a non-Hermitian Hamiltonian, and randomly decided «quantum jumps», followed by wave function normalization. This approach, which is equivalent to the master-equation treatment, has two main interests. First, new physical insight may be gained, in particular in the studies of the behaviour of a single quantum system. Second, if the relevant Hilbert space of the quantum system has a dimension N large compared to 1, the number of variables involved in a wave function treatment ($\sim N$) is much smaller than the one required for calculations with density matrices ($\sim N^2$). For the problem of laser cooling in the quantum regime which is of interest here, N stands for the number of internal + external atomic states, and is indeed $\gg 1$. The MCWF approach may, therefore, bring an important gain in computing time compared with the density matrix treatment used, for instance, for getting the results plotted in fig. 8.

4.1. *The* MCWF *procedure.* – The class of relaxation operators that we consider here is the following:

$$\mathscr{L}_{\mathrm{relax}}(\rho_S) = -\frac{1}{2}\sum_m (C_m^\dagger C_m \rho_S + \rho_S C_m^\dagger C_m) + \sum_m C_m \rho_S C_m^\dagger. \tag{46}$$

This type of relaxation operators is very general and is found in most of the quantum optics problems involving dissipation. In (46), the C_m's are operators acting in the space of the small system. Depending on the nature of the problem there can be one, a few or an infinity of these operators.

For the particular case of spontaneous emission by a two-level system without Zeeman degeneracy, there is just a single operator $C_1 = \sqrt{\Gamma}\sigma^-$ in the relaxation operator (46):

$$\mathcal{L}_{\text{relax}}(\rho_S) = -\frac{\Gamma}{2}(\sigma^+\sigma^-\rho_S + \rho_S\sigma^+\sigma^-) + \Gamma\sigma^-\rho_S\sigma^+ \tag{47}$$

with

$$\sigma^+ = |e\rangle\langle g|\,, \qquad \sigma^- = |g\rangle\langle e|\,. \tag{48}$$

One can check that this form of $\mathcal{L}_{\text{relax}}$ indeed leads to the well-known relaxation part of the optical Bloch equations:

$$\begin{cases}(\dot{\rho}_S)_{ee} = -\Gamma(\rho_S)_{ee}\,, \\ (\dot{\rho}_S)_{gg} = \Gamma(\rho_S)_{ee}\,,\end{cases} \qquad \begin{cases}(\dot{\rho}_S)_{eg} = -(\Gamma/2)(\rho_S)_{eg}\,, \\ (\dot{\rho}_S)_{ge} = -(\Gamma/2)(\rho_S)_{ge}\,.\end{cases} \tag{49}$$

We now present the procedure for evolving wave functions of the small system. Consider at time t that the system is in a state with the normalized wave function $|\phi(t)\rangle$. In order to get the wave function at time $t + \delta t$, we proceed in two steps:

1) First we calculate the wave function $|\phi^{(1)}(t + \delta t)\rangle$ obtained by evolving $|\phi(t)\rangle$ with the non-Hermitian Hamiltonian

$$H = H_S - \frac{i\hbar}{2}\sum_m C_m^\dagger C_m\,. \tag{50}$$

This gives for sufficiently small δt

$$|\phi^{(1)}(t + \delta t)\rangle = \left(1 - \frac{iH\delta t}{\hbar}\right)|\phi(t)\rangle. \tag{51}$$

Since H is not Hermitian, this new wave function is clearly not normalized. The square of its norm is

$$\langle\phi^{(1)}(t+\delta t)|\phi^{(1)}(t+\delta t)\rangle = \langle\phi(t)|\left(1 + \frac{iH^\dagger\delta t}{\hbar}\right)\left(1 - \frac{iH\delta t}{\hbar}\right)|\phi(t)\rangle = 1 - \delta p\,, \tag{52}$$

where δp reads

$$\delta p = \delta t\,\frac{i}{\hbar}\langle\phi(t)|\,H - H^\dagger\,|\phi(t)\rangle = \sum_m \delta p_m\,, \tag{53}$$

$$\delta p_m = \delta t\langle\phi(t)|\,C_m^\dagger C_m\,|\phi(t)\rangle \geqslant 0\,. \tag{54}$$

The magnitude of the step δt is adjusted so that this calculation at first order is valid; in particular it requires $\delta p \ll 1$.

For the particular case of the two-level atom problem (47), the non-Hermitian Hamiltonian is

$$H = H_S - \frac{i\hbar\Gamma}{2}\,|e\rangle\langle e|\,. \tag{55}$$

This amounts to adding the imaginary term $-i\hbar\Gamma/2$ to the energy of the unstable excited state, as usual in scattering theory.

2) The second step of the evolution of $|\phi\rangle$ between t and $t+\delta t$ consists in a possible «quantum jump» (fig. 13). The various possible «directions» for those jumps are given by the C_m operators, and the probability for making a jump in the «direction» of a particular C_m is δp_m given in (54). The new normalized wave function after such a jump is given by

$$|\phi(t+\delta t)\rangle = \frac{C_m\,|\phi(t)\rangle}{\|C_m\,|\phi(t)\rangle\|} \qquad \text{with probability } \delta p_m\,. \tag{56}$$

Using (53), we find that the total probability for making a jump is δp. In the no-jump case, which occurs then with a probability $1-\delta p$, we take as new normalized wave function at time $t+\delta t$:

$$|\phi(t+\delta t)\rangle = \frac{|\phi^{(1)}(t+\delta t)\rangle}{\|\,|\phi^{(1)}(t+\delta t)\rangle\|} \qquad \text{with probability } 1-\delta p = 1-\sum_m \delta p_m\,. \tag{57}$$

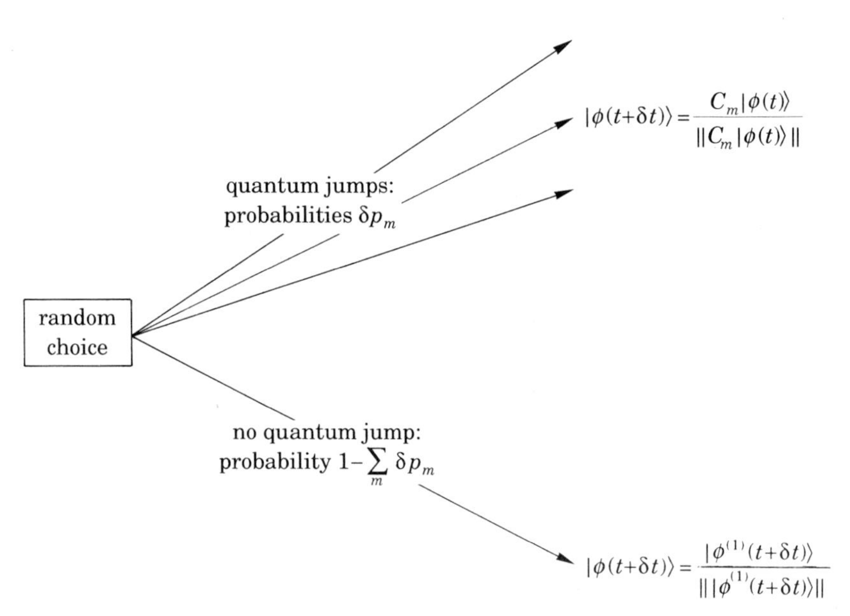

Fig. 13. – The possible quantum jumps in the Monte Carlo evolution.

Consider again as an example the particular case of the spontaneous emission of a two-level atom. The wave function at time t can be written as

$$|\phi(t)\rangle = \alpha(t)|e\rangle + \beta(t)|g\rangle. \tag{58}$$

Since there is a single C_m operator in this case, there is only one possible type of quantum jump. The probability for this quantum jump is

$$\delta p = \Gamma|\alpha|^2\,\delta t, \tag{59}$$

and the wave function after the jump, deduced from (48) and (56), is simply $|\phi(t+\delta t)\rangle = |g\rangle$. If no jumps occur, the wave function at time $t+\delta t$ is similar to (58), with the coefficients $\alpha(t+\delta t)$ and $\beta(t+\delta t)$ deduced from $\alpha(t)$ and $\beta(t)$ using the evolution with the non-Hermitian Hamiltonian (55). Therefore, we see for this particular case that the Monte Carlo evolution can be understood as the stochastic evolution of the atomic wave function if a continuous detection of the emitted photons is performed. The probability of detecting a photon during a particular time step δt is indeed equal to δp given in (59), and the new wave function after the detection, according to the standard quantum measurement theory, corresponds to the atom in its ground state g.

It is actually quite a general result that the Monte Carlo evolution outlined above represents a possible history of the system wave function with a suitable continuous-detection process taking place [43, 45]. Consequently, although this procedure does not make any reference to measurements on the system, it is often useful, in order to get some physical understanding for the result of the simulation, to refer to such a continuous-detection process as if it was really performed.

4·2. *Equivalence with the master equation.* – With this set of rules we can propagate a wave function $|\phi(t)\rangle$ in time, and we now show that this procedure is equivalent to the master equation (45). More precisely we consider the quantity $\overline{\sigma}(t)$ obtained by averaging $\sigma(t) = |\phi(t)\rangle\langle\phi(t)|$ over the various possible outcomes at time t of the MCWF evolutions all starting in $|\phi(0)\rangle$, and we prove that $\overline{\sigma}(t)$ coincides with $\rho_S(t)$ at all times t, provided they coincide at $t=0$.

Consider a MCWF $|\phi(t)\rangle$ at time t. At time $t+\delta t$, the average value of $\sigma(t+\delta t)$ is

$$\overline{\sigma(t+\delta t)} = (1-\delta p)\,\frac{|\phi^{(1)}(t+\delta t)\rangle}{\||\phi^{(1)}(t+\delta t)\rangle\|}\,\frac{\langle\phi^{(1)}(t+\delta t)|}{\||\phi^{(1)}(t+\delta t)\rangle\|} + \\ + \sum_m \delta p_m\,\frac{C_m|\phi(t)\rangle}{\|C_m|\phi(t)\rangle\|}\,\frac{\langle\phi(t)|C_m^\dagger}{\|C_m|\phi(t)\rangle\|} \tag{60}$$

which gives, using (51), (52) and (56),

$$\overline{\sigma(t+\delta t)} = \sigma(t) + \frac{i\,\delta t}{\hbar}\,[\sigma(t), H_{\rm S}] + \delta t\, \mathcal{L}_{\rm relax}(\sigma(t))\,. \tag{61}$$

We now average this equation over the possible values of $\sigma(t)$ and we obtain

$$\frac{\mathrm{d}\bar{\sigma}}{\mathrm{d}t} = \frac{i}{\hbar}\,[\bar{\sigma}, H_{\rm S}] + \mathcal{L}_{\rm relax}(\bar{\sigma})\,. \tag{62}$$

This equation is identical to the master equation (45). If we assume that $\rho_{\rm S}(0) = |\phi(0)\rangle\langle\phi(0)|$, $\bar{\sigma}(t)$ and $\rho_{\rm S}(t)$ coincide at any time, which demonstrates the equivalence between the two points of view. In the case where $\rho_{\rm S}(0)$ does not correspond to a pure state, one has first to decompose it as a statistical mixture of pure states, $\rho(0) = \sum p_i\,|\chi_i\rangle\langle\chi_i|$, and then randomly choose the initial MCWFs among the $|\chi_i\rangle$ with the probability law p_i.

As mentioned in the introduction, the master-equation approach and the reduced density matrix give access to one-time average values $a(t) = \langle A\rangle(t) = \mathrm{Tr}(\rho_{\rm S}(t)A)$, which can now also be obtained with the MCWF method. One calculates, for several outcomes $|\phi^{(i)}(t)\rangle$ of the MCWF evolution, the quantum average $\langle\phi^{(i)}(t)|A|\phi^{(i)}(t)\rangle$, and one takes the mean value of this quantity over the various outcomes $|\phi^{(i)}(t)\rangle$:

$$\langle A\rangle_{(n)}(t) = \frac{1}{n}\sum_{i=1}^{n}\langle\phi^{(i)}(t)|A|\phi^{(i)}(t)\rangle\,. \tag{63}$$

For n sufficiently large, (62) implies that $\langle A\rangle_{(n)}(t) \simeq \langle A\rangle(t)$.

As an example of the agreement between the master-equation approach and the MCWF approach, we have calculated by those two methods the excited-state population of a two-level atom coupled to a coherent laser field. The parameters for this Rabi nutation are a zero detuning δ between the laser and atomic frequencies, and a Rabi frequency $\Omega = 3\Gamma$. In fig. 14*a*), we show this excited-state population for a single «history» for $|\phi(t)\rangle$. One finds, as expected, a continuous evolution for this population oscillating between 0 and 1, with random quantum jumps projecting the atomic wave function into the ground state. In fig. 14*b*), we indicate the MCWF result obtained with the average of 100 wave functions. It shows a damped oscillation as a result of the dephasing of the individual oscillations due to the randomness of the various quantum jumps. This MCWF result is in very good agreement with the one derived from the master equation (optical Bloch equations).

As appears clearly in the proof, the equivalence of the master-equation and MCWF approaches does not depend on the particular value of the time step δt. From a practical point of view, the largest possible δt is preferable, and one might benefit from using a generalization of (51) to a higher order in δt, as, for example, a 4th-order Runge-Kutta-type calculation. The only requirement on δt

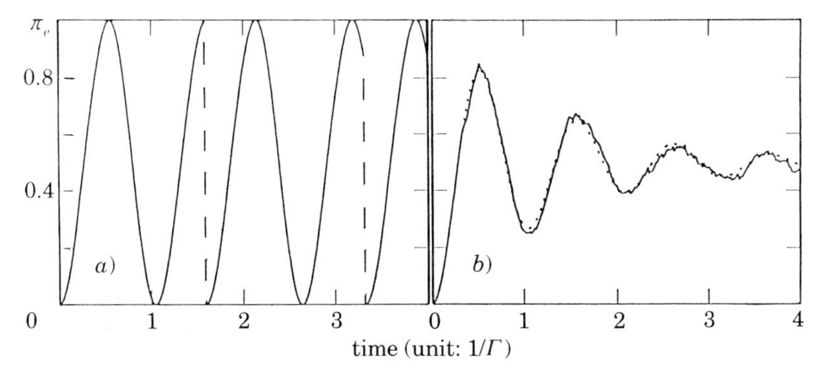

Fig. 14. – *a*) Time evolution of the excited-state population of a two-level atom in the MCWF approach. The dashed lines indicate the projection of the atomic wave function onto the ground state (quantum jump). *b*) Excited-state population averaged over 100 MCWF starting all in ground state at time 0. The dotted line represents the master-equation result.

is that the various $\eta_i \delta t$, where the $\hbar\eta_i$ are the eigenvalues of H, should be small compared to 1. Of course, we assume here that those eigenvalues have been simplified as much as possible in order to eliminate the bare energies of the eigenstates of H_S. For instance, for a two-level atom with a transition frequency ω_A coupled to a laser field with frequency ω_L, one makes the rotating-wave approximation in the rotating frame so that the $|\eta_i|$'s are of the order of the natural width Γ, the Rabi frequency Ω or the detuning $\delta = \omega_L - \omega_A$; they are consequently much smaller than ω_A.

One might wonder whether there is a minimal size for the time step δt. In the derivation presented above, it can be chosen arbitrarily small. However, one should remember that the derivation of (45) involves a coarse-grain average of the real density operator evolution. The time step of this coarse-grain average has to be much larger than the correlation time τ_c of the reservoir, which is typically an optical period for the problem of spontaneous emission. Therefore, one should be cautious when considering any result derived from this MCWF approach involving details with a time scale of the order of or shorter than τ_c, and only δt larger than τ_c should be applied. This appears clearly if one starts directly from the interaction Hamiltonian between the system and the reservoir in order to generate the stochastic evolution for the system wave function [43]. The condition $\delta t \gg \tau_c$ is then required to prevent quantum Zeno-type effects [47]. This restriction is discussed in detail in [48] in connection with quantum measurement theory.

4·3. *Connection with previous works.* – The problem of stochastic wave function evolution in connection with the treatment of dissipative systems in quantum optics has recently received a lot of attention. In the context of nonclassical field generation, CARMICHAEL [45] has proposed an approach named «quantum trajectories», inspired by the theory of photoelectron counting sequences [49] and quite similar to the spirit of the present work.

For simple atomic systems (2 or 3 levels) coupled to the electromagnetic field, the dynamics can be interpreted in terms of one or a few *delay functions*, which give the probability distribution of the time intervals between the emission of two successive photons [50-52]. When this function is known analytically, it can generate a very efficient Monte Carlo analysis of the process: just after the emission of the n-th fluorescence photon at time t_n, the atom is in its ground state and the choice of a single random number is sufficient to determine the time t_{n+1} of emission of the $(n+1)$-th photon. This type of Monte Carlo analysis has been used in [53] to simulate an atomic-beam cooling experiment, and in [51] to prove numerically the existence of dark periods in the fluorescence of a 3-level atom (quantum jumps). Very recently, laser cooling of atoms using velocity-selective coherent population trapping [54] and lasing without inversion [55] have been analysed by this type of Monte Carlo method.

Unfortunately, the delay function cannot be calculated analytically for complex systems involving a large number of levels. Nevertheless, it is possible to generate a Monte Carlo solution for this problem in which a single random number determines the time of emission of each fluorescence photon [46]. The evolution of the system between two quantum jumps has to be integrated step by step numerically, so that the amount of calculation involved is similar to the one required by the method presented in this lecture.

Another class of stochastic equations for system wave functions, which is also equivalent to the master equation (45), has been introduced by GISIN and PERCIVAL [56] (see also the work by DIÓSI [57]). In this approach, only continuous stochastic equations are considered. CARMICHAEL has shown that, for the particular case of the homodyne detection of the fluorescence light, the quantum jump formalism can be transformed into such a continuous stochastic equation [45]. Actually this proof can be extended to the most general case [58].

4·4. *Application to laser cooling.* – We now focus on the case of 1D Doppler cooling of a two-level atom, for which we present some numerical results. This will give an illustration of the effectiveness of the MCWF method as compared with the master-equation approach for studying laser cooling in the quantum regime.

4·4.1. The model. We consider here Doppler cooling of a two-level atom, as has already been described in subsect. 2·3: Doppler cooling originates from the fact that an atom moving in the field of a standing wave is closer to resonance with the counterpropagating component of the wave than with the copropagating one; the atom, therefore, feels a net radiation pressure force opposed to its velocity. This picture works well at nonsaturating laser intensities, where one can add the effect of the two waves independently. At higher intensities this semi-classical analysis becomes more complicated [59, 60] and a quantum treatment of the atomic external motion is a good alternative. We present here the result of such a treatment using both a master-equation and a MCWF approach.

The Hamiltonian H_S reads here, using the rotating-wave approximation,

$$H_S = \frac{P^2}{2M} + \hbar\Omega\cos(kZ)(\sigma^+ + \sigma^-) - \hbar\delta P_e\,, \tag{64}$$

where Z and P are the atomic position and momentum operators and Ω is the Rabi frequency of each travelling wave forming the standing wave. We choose the initial wave function $|\phi(0)\rangle$ equal to $|g, p=0\rangle$. At a time t, $|\phi(t)\rangle$ can be written

$$|\phi(t)\rangle = \sum_n \alpha_n(t)|g, p = p_0 + 2n\hbar k\rangle + \beta_n(t)|e, p = p_0 + (2n+1)\hbar k\rangle, \tag{65}$$

where the momentum p_0 depends on the random recoils which have occurred between 0 and t, and remains constant between two quantum jumps. According to subsect. 2·2, the evolution of α_n and β_n consists of sequences of two steps. First the wave function evolves linearly with the non-Hermitian Hamiltonian $H = H_S - i\hbar\Gamma|e\rangle\langle e|/2$:

$$i\dot{\alpha}_n = \frac{(p_0 + 2n\hbar k)^2}{2M\hbar}\alpha_n + \frac{\Omega}{2}(\beta_n + \beta_{n-1}), \tag{66}$$

$$i\dot{\beta}_n = \left(\frac{(p_0 + (2n+1)\hbar k)^2}{2M\hbar} - \delta - \frac{i\Gamma}{2}\right)\beta_n + \frac{\Omega}{2}(\alpha_n + \alpha_{n+1}). \tag{67}$$

Then we randomly decide whether a quantum jump occurs. The probability δp for a jump is proportional to the total excited-state population:

$$\delta p = \Gamma\sum_n |\beta_n|^2\,\delta t\,. \tag{68}$$

If no quantum jump occurs, we simply normalize the wave function. If a quantum jump occurs, the momentum $\hbar k'$ along the z-axis of the fluorescence photon is chosen randomly with a probability law deduced from the dipole radiation pattern, which leads to [44]

$$\alpha_n(t + \delta t) = \mu\beta_n(t), \tag{69}$$

$$\beta_n(t + \delta t) = 0\,, \tag{70}$$

$$p_0 \to p_0 - \hbar k'\,, \tag{71}$$

where μ is a normalization coefficient. We note that in this way the recoil due to spontaneous emission is treated in an exact manner. In the master-equation approach, an exact treatment of the spontaneous recoil requires a discretization of atomic momenta on a grid with a step size smaller than $\hbar k$. This increases the amount of calculation of the master equation with respect to the MCWF one, in addition to the N *vs.* N^2 argument mentioned in the introduction (*).

In order to make a fair comparison between the two approaches, we have chosen a coarse discretization for the atomic momentum, with a step size $\hbar k$, *i.e.* $k' = -k$, 0 or k and a probability law $1/5:3/5:1/5$, which gives an optimum representation of the diffusion rate due to the directional distribution of spontaneously emitted photons.

4.4.2. Numerical results. We have considered the case of sodium atoms for which the Doppler-cooling limit (22) corresponds to $p_{\text{r.m.s.}} \simeq 8.4\,\hbar k$. We have discretized the momentum between $-50\,\hbar k$ and $+50\,\hbar k$ which corresponds to a basis with 202 eigenstates in total, with at any time 101 nonzero coefficients α_n and β_n (see (65), where p_0 is either an odd or even multiple of $\hbar k$).

The results for the evolution of the sample mean $\langle P^2 \rangle_{(n)}$, defined as

$$\langle P^2 \rangle_{(n)}(t) = \frac{1}{n} \sum_{i=1}^{n} \langle \phi^{(i)}(t) | P^2 | \phi^{(i)}(t) \rangle , \tag{72}$$

are given in fig. 15 together with the results for $\langle P^2 \rangle(t)$ obtained using the master-equation treatment. These results correspond to the parameters $\Omega = -\delta = \Gamma/2$. The MCWF results have been obtained with the average of $n = 500$ evolutions.

We have indicated in fig. 15 the statistical error $\delta P^2_{(n)}$ on the determination of $\langle P^2 \rangle_{(n)}$ (see [44] for details). This quantity $\delta P^2_{(n)}$ gives an estimate of the quality of the result, and, with $n = 500$ wave functions, the *signal-to-noise* ratio in the range of 20 is quite satisfactory.

With a scalar machine, we have found that the time required for the calculation with 500 wave functions is equal to the time required for the master-equation evolution. With a vectorial compiler, we have found that there is an additional gain of a factor 15 in the benefit of the MCWF procedure. Therefore, even for this relatively simple 1D problem with «only» 200 levels, the MCWF method is at least as efficient as the master-equation approach for determining cooling limits with a good precision.

We clearly see in fig. 15 the existence of two regimes in the evolution of $\langle P^2 \rangle(t)$. At short time ($t < 200\,\Gamma^{-1}$, see insert of fig. 15), the number of spontaneous emissions is small, and the physics involved is essentially the diffraction of the plane atomic de Broglie wave by the grating formed by the laser standing wave [61]. For longer interaction times, dissipation comes into play [62, 63] and $\langle P^2 \rangle(t)$ tends to a steady-state value, of the order of $(11\,\hbar k)^2$. This value for

---

(*) This is the reason why fig. 8 has been obtained with a simplified dipole radiation diagram.

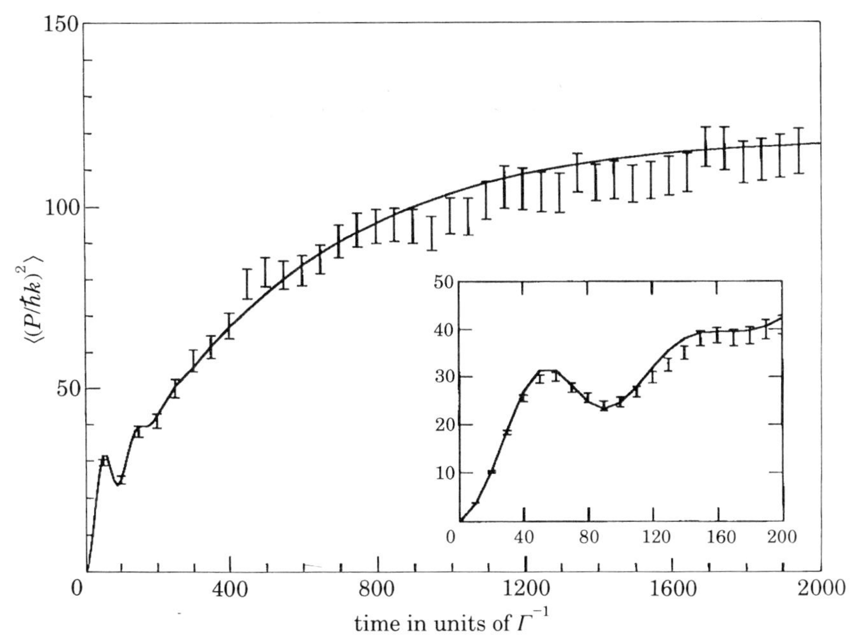

Fig. 15. – Time evolution of $\langle P^2 \rangle$ in Doppler cooling of Na atoms, with $\Omega = -\delta = \Gamma/2$. The points represent the Monte Carlo results, obtained by averaging $n = 500$ MCWF evolutions. The solid curve is the result of the density matrix approach. The insert details the short-time regime corresponding to the diffraction of the atomic de Broglie wave by the laser standing wave.

$p_{\text{r.m.s.}}$ is larger than the Doppler-cooling limit ($8.4\hbar k$) because of saturation effects.

4'5. *Conclusion for the* MCWF *approach.* – We have presented a stochastic evolution for the wave function of a system coupled to a reservoir in the Markovian regime. Each time step in this stochastic evolution consists of two parts: a non-Hermitian evolution and a possible quantum jump. We have proved the equivalence of this Monte Carlo wave function approach with the master-equation treatment.

This approach provides a computational tool which is often more efficient than the standard master-equation treatment for systems with a number of states $N \gg 1$ (for a detailed discussion see [44]). Indeed a wave function involves only N components while a density matrix is described by N^2 terms. It is, therefore, particularly well adapted to the study of laser cooling in the quantum regime. We have presented here the 1D example of Doppler cooling. This method has also been applied successfully to determine the cooling limits of the Sisyphus mechanism in 2 dimensions [35], and also to calculate the spectrum of the light emitted by an assembly of cold atoms [64]. Problems such as the study of collisions between cold atoms, or nonlinear mixing of quantum fields may also benefit from such an approach.

We have emphasized that this simulation is in many practical cases directly connected to a measurement sequence performed on the system. Each Monte Carlo trajectory is a possible history for the individual quantum system. In this respect, the noise appearing when one simulates with this method the measurement of a given observable A is also interesting. The fluctuations in the number of occurrences of a given eigenvalue a_i of A correspond to the quantum noise that one would get in a real experiment, performing the relevant detection scheme on an individual quantum system. Since more and more quantum optics and atomic-physics experiments are now performed with a single system (single ion or atom, single mode of a cavity), Monte Carlo wave function methods should, therefore, have many applications, since they lead to predictions closer to actual experimental signals than the master equation, which rather deals with ensemble averages.

* * *

The authors are very grateful to C. COHEN-TANNOUDJI and K. MØLMER for their participation to various parts of this lecture.

REFERENCES

[1] S. CHU, L. HOLLBERG, J. BJORKHOLM, A. CABLE and A. ASHKIN: *Phys. Rev. Lett.*, **55**, 48 (1985).
[2] P. LETT, R. WATTS, C. WESTBROOK, W. PHILLIPS, P. GOULD and H. METCALF: *Phys. Rev. Lett.*, **61**, 169 (1988).
[3] C. SALOMON, J. DALIBARD, W. D. PHILLIPS, A. CLAIRON and S. GUELLATI: *Europhys. Lett.*, **12**, 683 (1990).
[4] A. ASPECT, E. ARIMONDO, R. KAISER, N. VANSTEENKISTE and C. COHEN-TANNOUDJI: *Phys. Rev. Lett.*, **61**, 826 (1988).
[5] M. KASEVICH and S. CHU: *Phys. Rev. Lett.*, **69**, 1741 (1992).
[6] M. KASEVICH, E. RIIS, S. CHU and R. DE VOE: *Phys. Rev. Lett.*, **63**, 612 (1989).
[7] A. CLAIRON, C. SALOMON, S. GUELLATI and W. D. PHILLIPS: *Europhys. Lett.*, **16**, 165 (1991).
[8] P. L. GOULD, P. D. LETT, P. S. JULIENNE, W. D. PHILLIPS, H. R. THORSHEIM and J. WEINER: *Phys. Rev. Lett.*, **60**, 788 (1988).
[9] M. PRENTISS, A. CABLE, J. E. BJORKHOLM, S. CHU, E. RAAB and D. PRITCHARD: *Opt. Lett.*, **13**, 452 (1988).
[10] A. GALLAGHER and D. E. PRITCHARD: *Phys. Rev. Lett.*, **63**, 957 (1989).
[11] P. D. LETT, P. S. JESSEN, W. D. PHILLIPS, S. L. ROLSTON, C. I. WESTBROOK and P. L. GOULD: *Phys. Rev. Lett.*, **67**, 2139 (1991).
[12] D. GRISON, B. LOUNIS, C. SALOMON, J.-Y. COURTOIS and G. GRYNBERG: *Europhys. Lett.*, **15**, 149 (1991).
[13] J. TABOSA, G. CHEN, Z. HU, R. LEE and H. J. KIMBLE: *Phys. Rev. Lett.*, **66**, 3245 (1991).

[14] See, *e.g.*, the special issue of *Applied Physics B* of May 1992 on *Optics and Interferometry with Atoms*, edited by J. MLYNECK, V. BALYKIN and P. MEYSTRE.
[15] Y. CASTIN: PhD Thesis, Université Paris 6 (February 1992).
[16] R. J. COOK: *Phys. Rev. A*, **20**, 224 (1979).
[17] J. P. GORDON and A. ASHKIN: *Phys. Rev. A*, **21**, 1606 (1980).
[18] C. COHEN-TANNOUDJI: in *Fundamental Systems in Quantum Optics, Les Houches Summer School 1990*, edited by J. DALIBARD, J.-M. RAIMOND and J. ZINN-JUSTIN (North-Holland, Amsterdam, 1992), p. 1.
[19] J. DALIBARD and C. COHEN-TANNOUDJI: *J. Phys. B*, **18**, 1661 (1985).
[20] S. STENHOLM: *Rev. Mod. Phys.*, **58**, 699 (1986).
[21] T. HÄNSCH and A. SCHAWLOW: *Opt. Commun.*, **13**, 68 (1975).
[22] D. WINELAND and H. DEHMELT: *Bull. Am. Phys. Soc.*, **20**, 637 (1975).
[23] V. G. MINOGIN and O. T. SERIMAA: *Opt. Commun.*, **30**, 373 (1979).
[24] D. J. WINELAND and W. M. ITANO: *Phys. Rev. A*, **20**, 1521 (1979).
[25] S. CHU, D. WEISS, Y. SHEVY and P. UNGAR: in *Atomic Physics 11*, edited by S. HAROCHE, J. C. GAY and G. GRYNBERG (World Scientific, Singapore, 1989), p. 636.
[26] J. DALIBARD, C. SALOMON, A. ASPECT, E. ARIMONDO, R. KAISER, N. VANSTEENKISTE and C. COHEN-TANNOUDJI: in *Atomic Physics 11*, edited by S. HAROCHE, J. C. GAY and G. GRYNBERG (World Scientific, Singapore, 1989), p. 199.
[27] P. J. UNGAR, D. S. WEISS, E. RIIS and S. CHU: *J. Opt. Soc. Am. B*, **6**, 2058 (1989).
[28] J. DALIBARD and C. COHEN-TANNOUDJI: *J. Opt. Soc. Am. B*, **6**, 2023 (1989).
[29] More complex polarization configurations leading to mixed types of cooling can also be considered: V. FINKELSTEIN, P. R. BERMAN and J. GUO: *Phys. Rev. A*, **45**, 1829 (1992).
[30] Y. CASTIN, J. DALIBARD and C. COHEN-TANNOUDJI: *Proceedings of the LIKE Workshop 1990*, in *Light Induced Kinetic Effects on Atoms, Ions and Molecules*, edited by L. MOI, S. GOZZINI, C. GABBANINI, E. ARIMONDO and F. STRUMIA (ETS Editrice, Pisa, 1991), p. 5.
[31] Y. CASTIN and J. DALIBARD: *Europhys. Lett.*, **14**, 761 (1991).
[32] P. VERKERK, B. LOUNIS, C. SALOMON, C. COHEN-TANNOUDJI, J.-Y. COURTOIS and G. GRYNBERG: *Phys. Rev. Lett.*, **68**, 3861 (1992).
[33] P. S. JESSEN, C. GERZ, P. D. LETT, W. D. PHILLIPS, S. L. POLSTON, R. J. C. SPREEUW and C. I. WESTBROOK: *Phys. Rev. Lett.*, **69**, 49 (1992).
[34] R. GUPTA, S. PADUA, T. BERGEMAN and H. METCALF: in *Proc. S.I.F., Course CXVIII*, edited by E. ARIMONDO, W. D. PHILLIPS and F. STRUMIA (North-Holland, Amsterdam, 1992), p. 345; R. GUPTA, S. PADUA, C. XIE, H. BATELAAN, T. BERGEMAN and H. METCALF: poster at ICAP XIII (August 1992).
[35] C. BERG-SØRENSEN, K. MØLMER, Y. CASTIN and J. DALIBARD: poster at ICAP XIII (August 1992).
[36] T. BERGEMAN: poster at ICAP XIII (August 1992).
[37] J.-Y. COURTOIS and G. GRYNBERG: *Phys. Rev. A*, **46**, 7060 (1992).
[38] W. H. LOUISELL: *Quantum Statistical Properties of Radiation* (Wiley, New York, N.Y., 1973).
[39] F. HAAKE: *Statistical Treatment of Open Systems by Generalized Master Equations, Springer Tracts in Modern Physics*, Vol. **66**, edited by G. HOHLER (Springer, Berlin, 1973), p. 98.
[40] C. COHEN-TANNOUDJI: in *Les Houches 1975, Frontiers in Laser Spectroscopy*, edited by R. BALIAN, S. HAROCHE and S. LIBERMAN (North Holland, Amsterdam, 1977), p. 3.
[41] C. W. GARDINER: *Handbook of Stochastic Methods* (Springer, Berlin, 1983).

[42] M. Lax: *Phys. Rev.*, **172**, 350 (1968).
[43] J. Dalibard, Y. Castin and K. Mølmer: *Phys. Rev. Lett.*, **68**, 580 (1992).
[44] K. Mølmer, Y. Castin and J. Dalibard: *J. Opt. Soc. Am. B*, **10**, 524 (1993).
[45] H. J. Carmichael: lecture notes at U.L.B., Fall 1991 (unpublished); see also H. J. Carmichael and L. Tian: in *OSA Annual Meeting Technical Digest* 1990, p. 3.
[46] R. Dum, P. Zoller and H. Ritsch: *Phys. Rev. A*, **45**, 4879 (1992); R. Dum, A. S. Parkins, P. Zoller and C. W. Gardiner: *Phys. Rev. A*, **46**, 4382 (1992).
[47] B. Misra and E. C. G. Sudarshan: *J. Mat. Phys. (N.Y.)*, **18**, 756 (1977).
[48] G. C. Hegerfeldt and T. S. Wilser: in *II International Wigner Symposium, July 1991, Goslar*, proceedings to be published by World Scientific.
[49] P. L. Kelley and W. H. Kleiner: *Phys. Rev.*, **136**, A316 (1964).
[50] C. Cohen-Tannoudji and J. Dalibard: *Europhys. Lett.*, **1**, 441 (1986).
[51] P. Zoller, M. Marte and D. F. Walls: *Phys. Rev. A*, **35**, 198 (1987).
[52] H. J. Carmichael, S. Singh, R. Vyas and P. R. Rice: *Phys. Rev. A*, **39**, 1200 (1989).
[53] R. Blatt, W. Ertmer, P. Zoller and J. L. Hall: *Phys. Rev. A*, **34**, 3022 (1986).
[54] C. Cohen-Tannoudji, F. Bardou and A. Aspect: in *Laser Spectroscopy X, 1991*, edited by M. Ducloy, E. Giacobino and G. Camy (World Scientific, Singapore, 1992), p. 3.
[55] C. Cohen-Tannoudji, B. Zambon and E. Arimondo: *C. R. Acad. Sci.*, **314**, 1139, 1293 (1992).
[56] N. Gisin: *Phys. Rev. Lett.*, **52**, 1657 (1984); *Helv. Phys. Acta*, **62**, 363 (1989); N. Gisin and I. Percival: *Phys. Lett. A*, **167**, 315 (1992).
[57] L. Diósi: *J. Phys. A*, **21**, 2885 (1988).
[58] Y. Castin, J. Dalibard and K. Mølmer: *Proceedings of the I.C.A.P. XIII*, edited by T. W. Hänsch and H. Walther (August 1992).
[59] V. G. Minogin: *Sov. Phys. JETP*, **53**, 1164 (1981).
[60] K. Berg-Sørensen, E. Bonderup, K. Mølmer and Y. Castin: *J. Phys. B*, **25**, 4195 (1992).
[61] C. Tanguy, S. Reynaud and C. Cohen-Tannoudji: *J. Phys. B*, **17**, 4623 (1984).
[62] M. Wilkens, E. Schumacher and P. Meystre: *Opt. Commun.*, **86**, 34 (1991).
[63] S. Stenholm: in *Proc. S.I.F., Course CXVIII*, edited by E. Arimondo, W. D. Phillips and F. Strumia (North-Holland, Amsterdam, 1992), p. 29.
[64] P. Marte, R. Dum, R. Taieb, P. Lett and P. Zoller: *Phys. Rev. Lett.*, **71**, 1335 (1993).

ATOM INTERFEROMETERS

Atom Optics.

M. SIGEL †, C. S. ADAMS and J. MLYNEK

Universität Konstanz - 7750 Konstanz, B.R.D.

1. – Introduction.

1·1. *Motivation for atom optical experiments.* – Atom optics, in analogy to electron or neutron optics, is concerned with the manipulation of atomic-matter waves. Such experiments on atomic beams have been performed since the nineteen-twenties, but recently the improvement of tools, like free-standing microstructures and tunable laser light, has led to a rapid development of the field. Beams of atoms were split coherently, focussed and reflected from mirrors. In the last years various schemes for atom interferometers were demonstrated. This lecture is an attempt to provide an introductory overview of a number of demonstration experiments in the field of atom optics. Short overviews can also be found in ref. [1-4].

Particle optics is different from classical optics in that it uses massive particles that do not necessarily have to be treated relativistically. However, particle beams exhibit many phenomena well known from classical optics like diffraction and refraction. The centre-of-mass motion of a particle is described by classical mechanics (which is analogous to classical ray optics) or by the propagation of de Broglie waves (which is analogous to classical wave optics). Particle optics based on *electrons* and *neutrons* are well-established fields. Experiments with these particles have played an important role in exploring the puzzling wave-particle duality. Both neutron and electron interferometers have been used in many beautiful experiments which have improved our understanding of quantum mechanics. The interactions of neutron and electron beams with matter have become indispensable tools in the analysis of structures and surfaces, consider, *e.g.*, neutron scattering and the electron microscope. For an overview of neutron or electron optics see, *e.g.*, ref. [5-7].

Atoms are in many ways different from electrons and neutrons and particle optics with atoms promises a wealth of new effects and applications. Like neu-

† Deceased on July 4, 1993.

trons, atoms are neutral and, therefore, cannot be manipulated using static electromagnetic fields as easily as electrons or ions. In return beams of atoms and neutrons are less susceptible to electromagnetic stray fields. Atoms are heavier, which makes them more suitable as gravitational and inertial sensors and which also means that very short de Broglie wavelengths can be obtained for lower particle energies. This may be of interest to achieve high resolution with a low-damage surface probe. Atoms come in many different species, which can be either bosons or fermions, and may have a total angular momentum or magnetic moment much larger than neutrons or electrons. Beams of atoms are easier and cheaper to produce than neutron beams as there is no need for a nuclear reactor. And maybe most importantly, atoms have a complex internal structure that can be probed and modified using resonant laser light or static electromagnetic fields.

In recent years the availability of intense tunable laser radiation has focussed interest on experiments in atom optics based on the interaction of atomic transitions with resonant light fields. In addition, the spontaneous decay of atoms offers a possibility not available to electron or neutron optics. Spontaneous emission is a dissipative process and can be utilized for the preparation of atomic beams by laser-cooling techniques. Laser cooling and trapping of atoms has developed into an exciting field of research in itself; for an introductory overview see, *e.g.*, ref. [8-10]. In the context of atom optics cooling and imaging techniques are important for the preparation of well-collimated, dense atomic beams with a well-defined velocity. Spontaneous emission can on the other hand be unwelcome when it introduces unwanted random momentum kicks that give rise to diffusive aberration. Considerable theoretical interest in the mechanical effects of light, that goes well beyond merely using laser radiation as a tool in atom optics, has been stimulated by the experimental feasibility of such experiments.

1·2. *Optical elements for atoms.* – As with any optical experiment, a typical atom optical experiment consists of a source, optical elements and a detector. The source is supposed to provide an intense, well-collimated, monochromatic atomic beam. Sources can be categorized into either fast or slow. Among sources of fast atoms are thermal expansions that produce atomic beams exhibiting basically a Boltzmann distribution of velocities and supersonic expansions that produce a Gaussian velocity distribution with a narrow width. The main advantages of these traditional sources are high brightness (up to 10^{22} atoms/sr cm^2 s) and simplicity. Slow beams are produced by launching laser-cooled atoms from a trap. It is possible to launch as many as $5 \cdot 10^7$ atoms with temperatures as low as 30 μK [11]. The advantages of this technique are large de Broglie wavelengths and small velocity spread.

A large number of different schemes have been used for the detection of atomic beams. The preferred technique depends on the atomic species and the desired spatial or velocity resolution. Ground-state atoms can be detected with surface ionization devices (*e.g.*, hot-wire detectors) or by electron impact ionization (*e.g.*, Bayard-Alpert ionization gauge). Fluorescence detection or resonant ionization are used to detect atoms in a particular electronic state. Metastable states can be detected by ionization followed by Auger neutralization at a surface (*e.g.*, secondary electron multiplier) and Rydberg states by ionization in static electric fields. An overview of atomic- and molecular-beam methods including information on fast sources and detectors can be found in ref. [12].

In recent years there has been considerable activity in developing optical elements for atoms such as beam splitters, lenses and mirrors. The focus of this contribution will be to report on experimental demonstrations of optical elements that do not rely on spontaneous emission. Until recently few techniques had been developed for manipulating atoms. In traditional optics photons are deflected by transmission through an interface between materials with different refractive indices. Electrons, due to their charge, are easily deflected by static electric or magnetic fields. Neutrons penetrate through solids and can be diffracted from crystalline structures. However, as atoms are neutral and do not penetrate through matter, none of these approaches are suitable for atom optics.

The approaches that have been pursued so far to provide nondissipative optical elements for atoms may be categorized as diffractive, refractive or «single-photon recoil». *Diffractive optical elements* achieve the desired distortion of an incoming wave front by interaction with small structures, thus exploiting the wave nature of the atomic centre-of-mass motion. In the limit of very large «obstacles» the outgoing waves interfere constructively only in one direction—this is the regime of geometrical optics. *Refractive optics* is based on introducing phase shifts by a spatial modulation of the index of refraction. In analogy to standard optics the index of refraction is defined as the ratio of a particle's group velocity and the vacuum group velocity. The effective index of refraction seen by an atom can be changed by shifting the potential energy of the atom. Intuitively this means: If the internal energy of an atom is changed, energy conservation demands an acceleration or deceleration of the centre-of-mass motion. Using static electromagnetic fields the internal energy can be changed by a Zeeman or Stark shift and using resonant light by the a.c. Stark shift (*i.e.* light shift). As an example a plot of a potential acting as a lens for particles is displayed in fig. 1. «*Single-photon recoil*» refers to the change in atomic momentum caused by the absorption or emission of a single photon.

The diffraction of atoms from the surface of single crystals was studied by STERN *et al.* as early as 1929 [13]. In 1969 LEAVITT and BILLS [14] observed a single-slit diffraction pattern using a thermal potassium beam. The progress in microfabrication technology now permits the production of structures sufficiently fine to diffract thermal atoms, whose de Broglie wavelengths are below 1 Å.

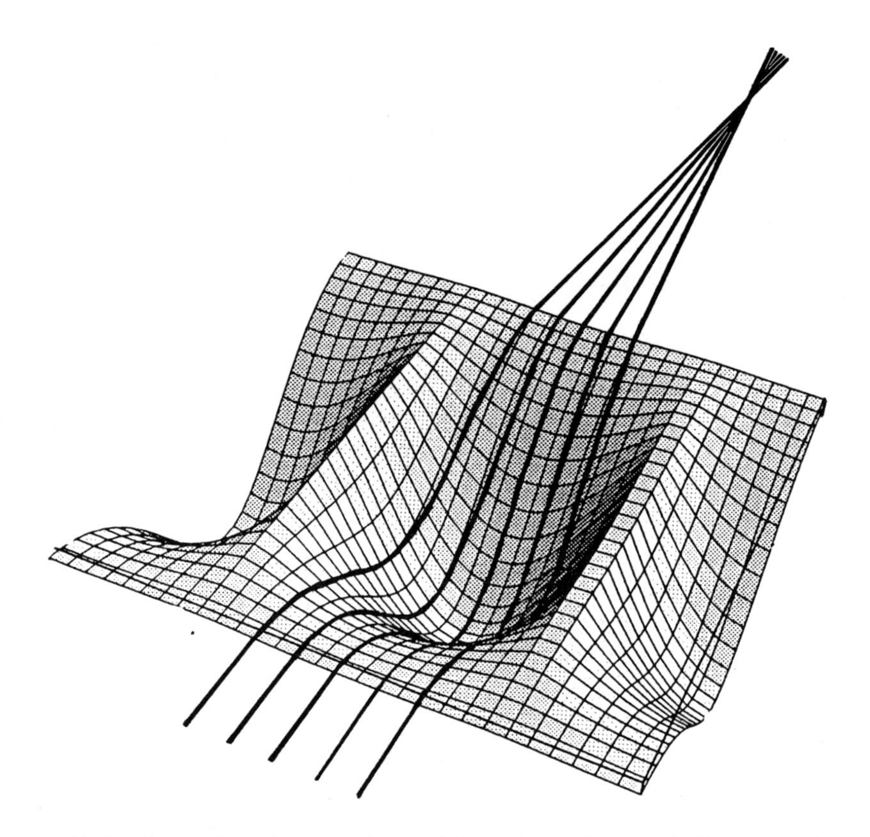

Fig. 1. – Refractive optical elements for particles rely on changes in the external motion induced by shifting the potential energy of the particle. The potential displayed acts as a lens for particles moving along the depicted trajectories.

The diffraction of thermal atoms from free-standing microfabricated gratings has been studied by various groups [15, 16]. The first deflection of an atomic beam by the light pressure force was demonstrated by FRISCH in 1933 [17]. While it was very difficult to examine the mechanical effects of atom-light interactions with thermal light sources, such experiments became easier to perform with the introduction of laser light.

1·3. *Outline.* – This lecture is organized as follows: In sect. **2** some prerequisites for the discussion of atom optical elements are given. First the evolution of the atomic centre-of-mass wave function in the presence of diffracting screens and potentials will be outlined. A rudimentary description of the dipole force (arising when an atom interacts with an inhomogeneous light field) in terms of the «dressed-atom picture» is given. Within this discussion the concept of optical potentials is introduced, which provides a clear intuitive picture of atom optical elements relying on the dipole force. Section **3** deals with beam splitters for atoms. A number of different schemes based on diffraction from microfabricated or light gratings and absorption or emission of single photons is discussed. Beam splitters have been applied in interferometers for atoms. A number of interferometer experiments are described in sect. **4**. Another optical element which is essential in optical systems is the lens. Lenses for atoms are discussed in sect. **5**. The reflection of atoms from polished surfaces and reflection using the dipole force are discussed in sect. **6**. The lecture concludes with a short summary and outlook.

2. – General principles.

2·1. *Refraction and reflection of matter waves.* – In the following two subsections a brief introduction to the propagation of matter waves is given. The treatment will remain rather rudimentary, the interested reader may consult more detailed introductory articles [5, 6].

The external motion of a nonrelativistic particle of mass m is described by the time-dependent one-body Schrödinger equation

$$\left\{-\frac{\hbar^2}{2m}\nabla^2 + V(\boldsymbol{r}, t)\right\}\Psi(\boldsymbol{r}, t) = i\hbar\frac{\partial}{\partial t}\Psi(\boldsymbol{r}, t), \tag{1}$$

where $\Psi(\boldsymbol{r}, t)$ is the scalar wave function in position representation and $V(\boldsymbol{r}, t)$ is the potential energy of the particle. For particles with an internal structure the Schrödinger equation is a vectorial equation; in the adiabatic approximation the effects of external fields can be incorporated in the scalar potential $V(\boldsymbol{r}, t)$ such that the Schrödinger equation decouples in scalar equations. For a time-independent potential, the time dependence can be separated by substituting $\Psi(\boldsymbol{r}, t) = \psi(\boldsymbol{r})\exp[-iEt/h]$. The time-independent Schrödinger equation then reads

$$\nabla^2\psi(\boldsymbol{r}) + \frac{2m}{\hbar^2}[E - V(\boldsymbol{r})]\psi(\boldsymbol{r}) = 0. \tag{2}$$

This equation has the form of a Helmholtz equation; the similarity with the well-known equation for electromagnetic fields explains the close analogy between optics with massive particles and optics with photons. It is this analogy that justifies the term «particle optics».

If a wave vector or k-vector is introduced by

$$|\boldsymbol{k}(\boldsymbol{r})| = \sqrt{\frac{2m}{\hbar^2}(E - V(\boldsymbol{r}))}, \tag{3}$$

eq. (2) reduces to the standard Helmholtz form

$$\{\nabla^2 + \boldsymbol{k}^2(\boldsymbol{r})\}\psi(\boldsymbol{r}) = 0. \tag{4}$$

For a free particle ($V(\boldsymbol{r}) < E$) the solution of eq. (4) can be written in terms of plane waves

$$\psi(\boldsymbol{r}) = A \exp[i\boldsymbol{k} \cdot \boldsymbol{r}]. \tag{5}$$

The *de Broglie wavelength* of the plane wave is $\lambda_{\mathrm{dB}} = 2\pi/k = h/p$. For $V(\boldsymbol{r}) = 0$ the energy is $E = \hbar\omega = (\hbar k)^2/2m$, where ω is the de Broglie angular frequency. For nonrelativistic particles the group velocity

$$v = \frac{\partial \omega}{\partial k} = \frac{\hbar k}{m} \tag{6}$$

can be identified with the velocity of the classical particle.

The index of refraction for a plane wave with a group velocity $v(\boldsymbol{r})$, defined as the ratio of $v(\boldsymbol{r})$ and the vacuum group velocity v_0, is given by

$$n(\boldsymbol{r}) := \frac{v(\boldsymbol{r})}{v_0} = \sqrt{1 - V(\boldsymbol{r})/E}. \tag{7}$$

This index of refraction describes the *effective* interaction of an atom with the field in analogy to the index of refraction which describes the interaction of photons with atoms, *e.g.*, in a glass prism or lens.

If the absolute value of the wave function varies slowly on the scale of the wavelength, the «short-wavelength approximation» or «ray approximation» can be introduced and the eikonal equation (which is the central equation in geometrical optics [18]) can be obtained from eq. (4). In quantum mechanics this approximation is referred to as the «WKB approximation». In the limit of zero de Broglie wavelength particle trajectories obey Newtonian mechanics. The rays of geometrical particle optics thus correspond to the classical trajectories of particles. Spatial variations of the potential $V(\boldsymbol{r})$ (*i.e.* of the index of refraction) lead to a deflection of particles. At an interface between regions with a different potential ($V(\boldsymbol{r}) = 0$ for $x < 0$ and $V(\boldsymbol{r}) = V$ for $x > 0$) this deflection is called reflection for $V > E$ and total internal reflection or refraction (depending on the angle of incidence) for $V < E$. Partial reflection requires a potential step with a width comparable to the de Broglie wavelength.

2·2. *Diffraction of matter waves.* – In the regime where the short-wavelength approximation is no longer valid, diffraction, the hallmark of wavelike behaviour, is obtained. The first experimental evidence that massive particles can behave like waves was obtained in 1927 by DAVISSON and GERMER in the US [19] and THOMSON in England [20]. They observed diffraction of electrons by reflection from a metallic crystal and by transmission through thin membranes, respectively. Diffraction of atoms and molecules was first observed by STERN in 1929 [13].

The diffraction of electromagnetic radiation is a well-defined boundary-value problem involving the solution to Maxwell's equations with the appropriate boundary conditions at the diffracting obstacle. However, analytical solutions have been found only for idealized cases. For most practical circumstances approximation methods have to be used, of which Kirchhoff's formulation of the Huygens-Fresnel principle is the most important [18]. *Kirchhoff's theory* is based on the assumption that the amplitude of the wave can be described by a scalar function ψ, with the simple boundary conditions that ψ and grad ψ vanish at the surface of an obstacle and are undisturbed everywhere else. The diffraction of *matter waves* is described completely by the solution of the Schrödinger equation with the appropriate boundary conditions. Kirchhoff's theory is applicable to the diffraction of matter waves because the propagation of electromagnetic and matter waves is described by the same Helmholtz equation and consequently by the same Green's functions $\exp[i\boldsymbol{kr}]/|\boldsymbol{r}|$, which can be identified with Huygens' secondary wavelets. Are the conditions of Kirchhoff's theory valid for the diffraction of atoms? Firstly, the diffracting apertures must have dimensions large compared to the wavelength. This appears to be appropriate for most practical cases in atom optics. Secondly, for particles with total angular momentum, the description using a scalar amplitude is not more unjustified than it is for photons. And thirdly, according to Kirchhoff's boundary conditions radiation which is not transmitted is supposed to be absorbed by the obstacle. This condition is incompatible with electromagnetic theory but appears appropriate for atoms because they are either adsorbed or diffusely scattered by most surfaces.

A typical diffraction problem is displayed in fig. 2. Waves originating from a

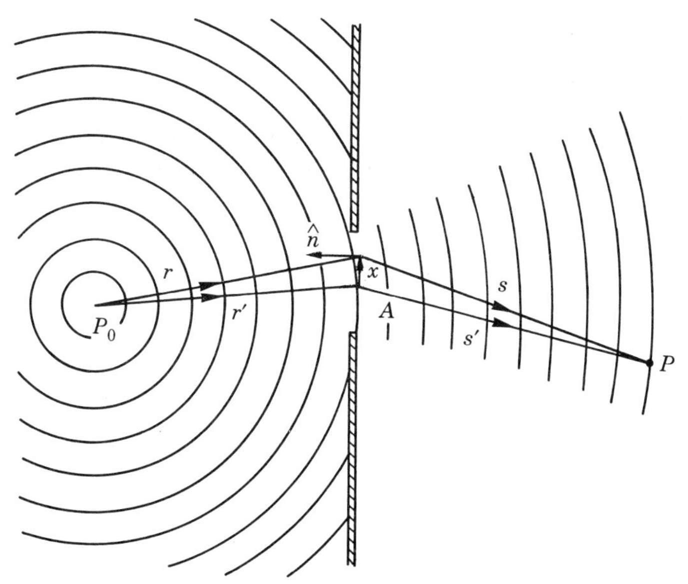

Fig. 2. – Schematic of a typical diffraction problem: Waves originating from a monochromatic point source P_0 are diffracted by the aperture A. The amplitude at the observation point P is given by the coherent sum over all secondary wavelets originating from the aperture.

monochromatic point source P_0 are diffracted by an aperture A. The amplitude of the wave at an observation point P is the sum of all secondary wavelets originating from the aperture:

$$\psi(P) = -\frac{i\psi_0}{2\lambda}\int_A \frac{\exp[ik(r+s)]}{r\cdot s}\{\cos(\boldsymbol{n},\boldsymbol{r}) - \cos(\boldsymbol{n},\boldsymbol{s})\}\,dA\,. \tag{8}$$

Equation (8) is known as the *Fresnel-Kirchhoff diffraction formula.* In the far field the denominator of the integrand does not change appreciably over the aperture, while the numerator oscillates rapidly with the phase difference term $\chi(\boldsymbol{x}) = k\cdot(r+s) - k\cdot(r'+s')$, where $\boldsymbol{x}$ is the coordinate in the aperture plane. r' and s' equal r and s for $\boldsymbol{x} = 0$. For small angles the difference of the cosines in the integral differs only slightly from 1. Equation (8) can, therefore, be simplified to

$$\psi(P) = -\frac{i\psi_0}{\lambda}\frac{\exp[ikr']}{r'}\frac{\exp[iks']}{s'}\int_A \exp[i\chi(\boldsymbol{x})]\,dA\,. \tag{9}$$

The more general case of diffraction from a plane with a spatially dependent partial transmission as well as a spatially dependent phase shift can be expressed by introducing a complex amplitude transmission function $t(\boldsymbol{x})$ and integrating over the entire diffracting plane:

$$\psi(P) = -\frac{i\psi_0}{\lambda}\frac{\exp[ikr']}{r'}\frac{\exp[iks']}{s'}\int_\infty t(\boldsymbol{x})\exp[i\chi(\boldsymbol{x})]\,d\boldsymbol{x}\,. \tag{10}$$

Amplitude structures, which are described by the modulus of $t(\boldsymbol{x})$, can be realized by absorbing microstructures. Phase structures, which are described by the phase of $t(\boldsymbol{x})$, are obtained by a spatially dependent shift of the potential energy of the particle.

In diffraction phenomena two regimes are distinguished: *Fraunhofer* and *Fresnel diffraction.* The phase difference $\chi(\boldsymbol{x})$ can be expanded in the aperture variable $\boldsymbol{x}$. In the case where $\chi(\boldsymbol{x})$ can be approximated as linear in $\boldsymbol{x}$

$$\chi(\boldsymbol{x}) = k\cdot\left(\frac{\boldsymbol{r}'}{r'} - \frac{\boldsymbol{s}'}{s'}\right)\cdot\boldsymbol{x} = \boldsymbol{q}\boldsymbol{x} \tag{11}$$

(*i.e.* if the incoming and the outgoing wave are superpositions of plane waves) one talks of Fraunhofer diffraction. An approximate criterion is

$$\chi(\boldsymbol{x})_{\max} = (k\boldsymbol{x}^2_{\max}/2)(1/r' + 1/s') \ll 2\pi\,. \tag{12}$$

The diffraction formula in the Fraunhofer approximation is

$$\psi(P) = -\frac{i\psi_0}{\lambda}\frac{\exp[ikr']}{r'}\frac{\exp[iks']}{s'}\int_\infty t(\boldsymbol{x})\exp[i\boldsymbol{q}\boldsymbol{x}]\,dA\,. \tag{13}$$

Thus for Fraunhofer diffraction $\psi(P)$ is given by the Fourier transform of the amplitude transmission function $t(\boldsymbol{x})$.

The Fraunhofer approximation becomes inadequate whenever the curvature of the wave fronts contributes nonnegligible phase differences in the Fresnel-Kirchhoff diffraction formula. If only the quadratic phase dependence of $\chi(\boldsymbol{x})$ is retained, one talks of Fresnel diffraction. In that case the amplitude distribution arising from an aperture in Cartesian coordinates will be represented by the well-known Fresnel integrals.

As for most experiments with light, the quantity observed in particle optics is the intensity I, *i.e.* the modulus square of the amplitude. The intensity at the observation point P for an incoherent extended source is given by the incoherent sum over the source points $n(\boldsymbol{x}_S)$ and the frequency spectrum $n(\nu)$:

$$I(P) \propto \int \mathrm{d}\nu\, n(\nu) \int \mathrm{d}A_S\, n(\boldsymbol{x}_S) |\psi(P)|^2 . \tag{14}$$

2·3. *Interaction of atoms with light fields.* – In the last two decades the mechanical effects of near-resonant light on atoms have attracted considerable experimental and theoretical attention [21]. As photons carry momentum $p = h/\lambda$, absorption and emission processes cause variations in the atomic centre-of-mass momentum. Averaged over many radiative lifetimes, these variations can be discussed in terms of radiative forces fluctuating around a mean value. The mean radiative force acting on a two-level atom can be discussed as consisting of two components: the *radiation pressure force*, which is associated with the phase gradient of the light field, and the *dipole force*, which is associated with the intensity gradient of the light field.

In the «photon picture» the radiation pressure force arises from absorption and spontaneous-emission cycles, while the *dipole force* arises from absorption and stimulated-emission cycles, *i.e.* the redistribution of photons in the light field. It is interesting to note that for atom optical elements based on the dipole force the roles of light and matter are interchanged with respect to classical optics. For both the refraction of photons from glass and the refraction of atoms from light the refraction effect stems from stimulated-scattering processes.

The radiation pressure force arises from the incoherent scattering of photons. In scattering one photon an atom first suffers a recoil $\hbar k$ in the direction of the incident photon and then $-\hbar k$ in the direction of the scattered photon. Since the spontaneous emission of photons is isotropic, the momentum change of the atom averaged over n scattering events is $n\hbar k$. The momentum change of an atom induced by spontaneous scattering was first observed by FRISCH in 1933 [17]. Even though the photon momentum $\hbar k$ is small, the radiation pressure force can be quite substantial (thousands of times the Earth's gravitational force), because photons may be scattered at a rate equal to the natural linewidth of the transition (typically around $10^7\ \mathrm{s}^{-1}$). Making use of its velocity dependence the radiation pressure force can be used to cool atoms by two counterpropagating laser beams detuned slightly below the atomic-transition frequency, as first suggested by HÄNSCH and SCHAWLOW in 1975 [22] and demonstrated for the three-dimensional case by CHU *et al.* at AT&T Bell Labs in 1985 [23]. More recently, new cooling methods like, *e.g.*, polarization gradient cooling have been demonstrated that allow the cooling of atoms to much lower temperatures [9]. Laser-cooling techniques play an important role for the preparation of atomic beams, however, these «dissipative optical elements» will not be discussed in this lecture.

The dipole force arises from the coherent redistribution of photons within the light field. In the absence of spontaneous emission and nonadiabatic evolution, the dipole force is conservative and can be understood as the gradient of an *optical potential*; it is often referred to as the *gradient force*. The dipole force is well suited to design nondissipative optical elements for atoms. A «*dressed-atom* picture» of the dipole force has been given by DALIBARD and COHEN-TANNOUDJI [24]. As this approach is both intuitive and provides a quantitative understanding, it will be used in this lecture to discuss applications of the dipole force in atom optics. A short account of the dressed-atom approach to the dipole force following the description of Dalibard and Cohen-Tannoudji is given below.

In the dressed-atom picture the interaction of a two-level atom with a quantized radiation field is described by considering the energy levels of the combined interacting system of atom and field. The effect of spontaneous emission is included as population transfer between these dressed levels; however, for our discussion spontaneous emission will be neglected. In this case the total Hamiltonian is the sum of the atomic Hamiltonian H_{A}, the Hamiltonian of the radiation field H_{R} and the atom-field coupling H_{I}:

$$H = H_{\mathrm{A}} + H_{\mathrm{R}} + H_{\mathrm{I}} \,. \tag{15}$$

The Hamiltonian for a two-level atom is

$$H_{\mathrm{A}} = \frac{\boldsymbol{p}^2}{2m} + \hbar\omega_0 b^+ b \,, \tag{16}$$

where b and b^+ are the atomic lowering and raising operators $b = |g\rangle\langle e|$ and $b^+ = |e\rangle\langle g|$, g and e label ground and excited state. The Hamiltonian of the radiation field is the sum of the contributions for all modes λ:

$$H_{\mathrm{R}} = \sum_{\lambda} \hbar\omega_{\lambda} a_{\lambda}^+ a_{\lambda} \,, \tag{17}$$

where a_{λ} and a_{λ}^+ are the destruction and creation operators of a photon in mode λ. In the electric-dipole and rotating-wave approximations the coupling between the atomic electric-dipole operator $\boldsymbol{d}$ and the positive $(\mathcal{E}_{\lambda}(\boldsymbol{R})\, a_{\lambda})$ and negative $(\mathcal{E}_{\lambda}^*(\boldsymbol{R})\, a_{\lambda}^+)$ frequency components of the electric field of mode λ can be written as

$$H_I(\boldsymbol{R}) = -\boldsymbol{d}\cdot[b^+ \mathcal{E}_\lambda(\boldsymbol{R})a_\lambda + b\mathcal{E}_\lambda^*(\boldsymbol{R})a_\lambda^+]. \tag{18}$$

The dressed-atom Hamiltonian for an atom with resonance frequency ω_0 in a laser field with the angular frequency ω_L is

$$H_{DA}(\boldsymbol{r}) = \hbar(\omega_L - \Delta)b^+b + \hbar\omega_L a_L^+ a_L - \boldsymbol{d}\cdot[b^+ \mathcal{E}_L(\boldsymbol{r})a_L + b\mathcal{E}_L^*(\boldsymbol{r})a_L^+], \tag{19}$$

where Δ is the atom-light detuning. It is assumed that $\Delta = \omega_L - \omega_0 \ll \omega_L, \omega_0$. The atomic motion is treated semi-classically such that the atomic-position operator $\boldsymbol{R}$ is replaced by its expectation value $\boldsymbol{r}$. The atom-laser coupling connects the states $|e, n\rangle$ and $|g, n+1\rangle$, where n refers to the number of photons in the laser mode. Lasers emit a coherent state of the light field, *i.e.* the photon number is given by a Poissonian distribution of photon number states. As in most experimental situations the photon number is extremely large, the fluctuations around the mean can be neglected.

The coupling can be characterized by the Rabi frequency $\omega_R(\boldsymbol{r})$ defined by

$$\omega_R(\boldsymbol{r})\exp[i\varphi(\boldsymbol{r})] := \frac{2}{\hbar}\langle e, n|H_I(\boldsymbol{r})|g, n+1\rangle = -2\sqrt{n+1}\,\frac{\boldsymbol{d}\cdot\mathcal{E}_I(\boldsymbol{r})}{\hbar}. \tag{20}$$

$\varphi(\boldsymbol{r})$ is the phase of the field. The eigenenergies of the dressed atom are

$$E_{\pm,n}(\boldsymbol{r}) = (n+1)\hbar\omega_L - \frac{\hbar}{2}\Delta \pm \frac{\hbar}{2}\Omega(\boldsymbol{r}), \tag{21}$$

where

$$\Omega(\boldsymbol{r}) = \sqrt{\omega_R^2(\boldsymbol{r}) + \Delta^2} \tag{22}$$

is called the *effective Rabi frequency*. In fig. 3 the eigenenergies as a function of position in a Gaussian laser beam are plotted for positive detuning. The eigenvalues $E_{\pm,n}(\boldsymbol{r})$ can be interpreted as *optical potentials*, a concept central to the discussion of optical elements based on the dipole force.

For constant field phase $\varphi(\boldsymbol{r})$ the corresponding (dressed) eigenstates are

$$|+, n; \boldsymbol{r}\rangle = \cos\frac{\Theta(\boldsymbol{r})}{2}|e, n\rangle + \sin\frac{\Theta(\boldsymbol{r})}{2}|g, n+1\rangle, \tag{23}$$

$$|-, n; \boldsymbol{r}\rangle = -\sin\frac{\Theta(\boldsymbol{r})}{2}|e, n\rangle + \cos\frac{\Theta(\boldsymbol{r})}{2}|g, n+1\rangle, \tag{24}$$

where the angle $\Theta(\boldsymbol{r})$ is defined as

$$\cos\Theta(\boldsymbol{r}) = -\frac{\Delta}{\Omega(\boldsymbol{r})}, \tag{25}$$

$$\sin\Theta(\boldsymbol{r}) = \frac{\omega_R(\boldsymbol{r})}{\Omega(\boldsymbol{r})}. \tag{26}$$

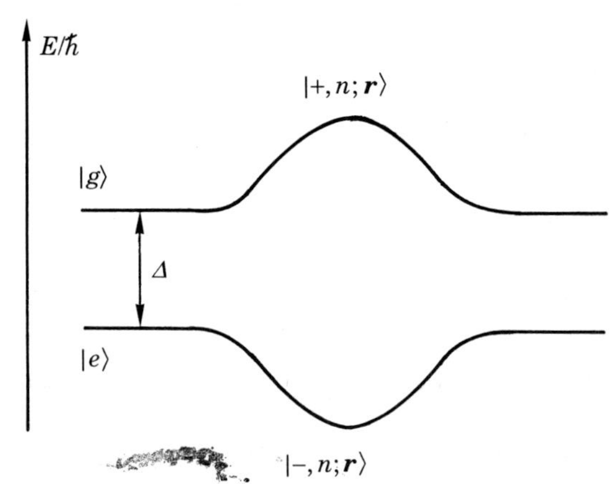

Fig. 3. – Variation of the dressed-atom energy eigenvalues in a Gaussian laser beam (n photons). Both the eigenenergies and the eigenstates are a function of position. Outside of the laser beam the eigenstates are identified with the uncoupled eigenstates of the atom and the light field.

The mean force on an atom is given by Ehrenfest's equation

$$f(r) = \left\langle \frac{i}{h}[H_{\mathrm{I}}(r), P] \right\rangle = \langle -\nabla H_{\mathrm{I}}(r) \rangle . \tag{27}$$

In the absence of nonadiabatic evolution and spontaneous emission the force f on the dressed states $|+\rangle$ and $|-\rangle$ is given by

$$f_{\pm}(r) = \mp\nabla \frac{\hbar\Omega(r)}{2} . \tag{28}$$

This means that the $|+\rangle$ dressed state is a low-field seeker while $|-\rangle$ is a high-field seeker. For adiabatic evolution and positive («blue») detuning, the population of $|g\rangle$ is transferred into $|+\rangle$, while for negative («red») detuning it is transferred into $|-\rangle$. Consequently the dipole force attracts a ground-state atom towards regions of high intensity for $\Delta < 0$ and expels it for $\Delta > 0$. For large detuning Δ, the effective Rabi frequency $\Omega(r)$ is approximately proportional to ω_{R}^2 (see eq. (22)), *i.e.* proportional to the light intensity; for smaller Δ it is approximately proportional to ω_{R}, *i.e.* proportional to the root of the intensity. For sufficiently small Δ nonadiabatic transitions into the other dressed state may occur and the force is no longer necessarily conservative.

3. – Beam splitters for atoms.

3·1. *Motivation.* – Beam splitters for atoms transform an incoming momentum state into a coherent linear superposition of different momentum states. Despite their name, beam splitters combine beams as much as they split them.

In atom optics various schemes for beam splitters are already being used to split and recombine atomic beams in interferometers. This important application is discussed in sect. **4**. They could furthermore be used for correlation measurements (*e.g.*, an atomic Hanbury-Brown and Twiss experiment) or as in//outcoupler for atomic cavities [2] (see subsect. **6**˙1). In analogy to classical optics a beam splitter whose outgoing beams are in orthogonal internal states may be called a «polarizing beam splitter». For polarizing beam splitters the coupling into the output states depends on the polarization of the incoming state, as, for example, in the magnetic Stern-Gerlach effect. In atom optics, unlike in classical optics, spontaneous emission in the beam-splitting process may introduce random phases in the outgoing states and, therefore, destroy their coherence. This must be avoided in most applications, one then talks of *coherent beam splitters*. Beam splitters for atoms have been produced using reflection from crystalline surfaces, diffraction from microfabricated transmission gratings (amplitude gratings), diffraction from standing-wave light fields (phase gratings), using the optical Stern-Gerlach effect, by «single-photon recoil», and using the magnetic Stern-Gerlach effect. A short review of beam splitters based on the dipole force can be found in ref. [25].

3˙2. *Reflection from a crystal surface.* – ESTERMANN and STERN obtained a beam splitter for helium and molecular hydrogen by diffraction from cleaved ionic crystals as early as 1929 [13, 26]. Their experiments were originally aimed at furnishing proof for the wave nature of the centre-of-mass motion of massive particles. In order to work with long de Broglie wavelengths, the experiments were carried out with He and H_2. As attempts to observe diffraction by reflection from a ruled grating remained unsuccessful, STERN, KNAUER and later ESTERMANN tried the diffraction from cleaved surfaces of LiF and NaCl. The equipotential surfaces of these crystals, as probed by the atoms, are corrugated and provide a two-dimensional grating. In their experiments the de Broglie wavelength was adjusted by cooling or heating the source and diffraction angles up to 20° were obtained. Today diffraction of helium beams from surfaces is a useful tool in surface analysis [27].

3˙3. *Diffraction from a transmission grating.* – Diffraction of atoms from a transmission grating was first observed by KEITH *et al.* in 1988 [15]. The grating used in their experiment was made from a 0.5 μm thick gold foil and had a grating period of 0.2 μm. The splitting angle between the zeroth- and first-order diffraction peaks was around 100 μrad. Our group has demonstrated the diffraction of metastable helium by a transmission grating [16]. The splitting angle was of the same order as in Keith's experiment. A plot of the atomic count rate against the transverse position of the detector slit is displayed in fig. 4. No appreciable de-excitation of the metastable atoms by transmission through the grating was observed.

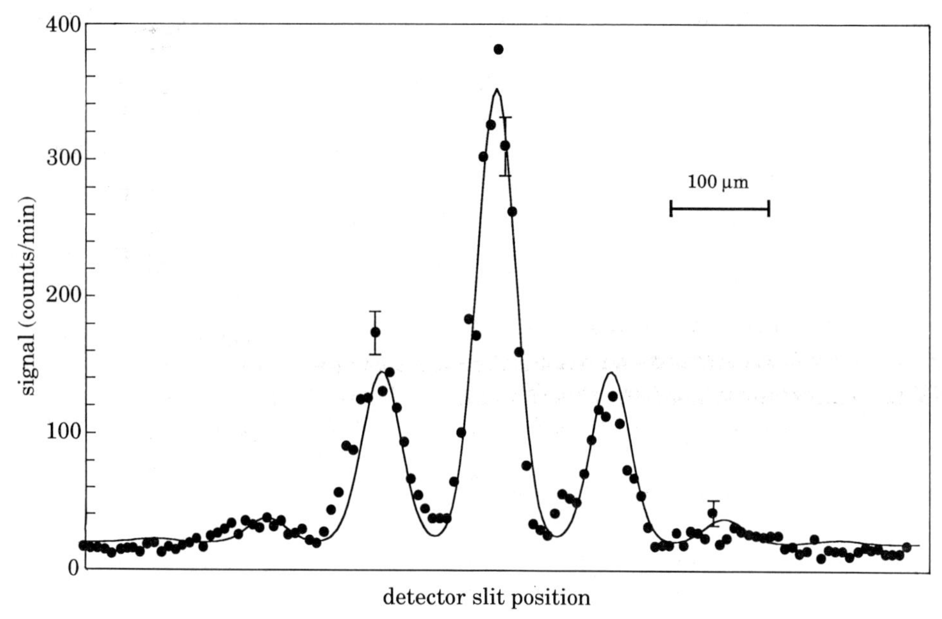

Fig. 4. – Atomic-intensity distribution obtained by diffraction of metastable helium (de Broglie wavelength $\lambda_{\mathrm{dB}} = 1.03$ Å) from a 0.5 μm period grating. The solid line is a semi-empirical fit.

Transmission gratings are easy to handle and can, as they only rely on the external motion, be used for any atomic species. The splitting angles obtained are larger than in diffraction from an optical standing wave (see subect. 3˙4) because the period of a microfabricated transmission grating can be made smaller than optical wavelengths. Free-standing gratings with periods as small as 100 nm have been produced [28]. The disadvantages are the existence of various diffraction orders, the absorption of atoms by the grating, and the fact that diffraction is highly dispersive. A grating relies on amplitude splitting. In atom optics as in classical optics, coherent beam splitters may also be produced by wave front splitting, *e.g.* by the combination of a single and a double slit (see subsect. 4˙2).

3˙4. *Diffraction from an optical standing wave.* – In this lecture the diffraction of an atomic beam from optical standing waves is discussed as the diffraction from a phase grating formed by the optical potential (see fig. 5), other treatments can be found in ref. [29, 30]. A significant simplification of the problem is possible when the transverse kinetic-energy term in the atomic Hamiltonian can be neglected. This is known as the *Raman-Nath approximation* [18, 31]. This approximation is valid for short interaction times, *i.e.* sharply focussed laser beams. To illustrate the regime of validity of the Raman-Nath approximation an intuitive picture of the stimulated-scattering process in terms

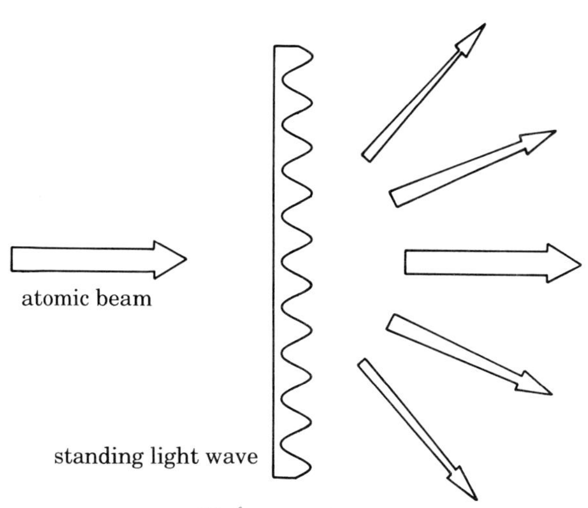

Fig. 5. – Schematic of the diffraction of atoms from a near-resonant standing-wave light field. The light field acts as a phase grating for atoms and splits an incoming plane wave into a superposition of plane waves separated by integer multiples of the running-wave photon momentum $\hbar k$.

of discrete momentum transfer of multiples of $\hbar k$ caused by the absorption and stimulated emission of photons can be used [32, 33]: For a tightly focussed laser beam there is a broad distribution of k-vectors such that the atom can easily choose the «right» component to simultaneously fulfil energy and momentum conservation, as depicted in fig. 6*a*) (this situation is equivalent to the interaction with a pulsed standing wave, where for short interaction times the energy conservation constraint is relaxed through the uncertainty relationship). In contrast, for a broad laser beam, *i.e.* a narrow distribution of k-vectors, energy conservation can only be fulfilled for specific angles of the incoming beam in which case Bragg scattering may be observed (see fig. 6*b*)). The Bragg regime is considered in subsect. 3·5.

The energy eigenvalues of the dressed-state Hamiltonian of a two-level atom are given in eq. (21). In a standing wave these eigenvalues display a modulation with a periodicity of $\lambda/2$. For an atom passing through the standing wave at position x, the phase shift is given by the time-integral over the dressed-state energy eigenvalue $E(x, t)$, where x is the transverse position in the standing wave and t models the passage through the standing wave,

$$\Delta\Phi(x) = \frac{1}{\hbar}\int E(x, t)\,\mathrm{d}t\,. \tag{29}$$

The transmission function of a phase grating is $t(x) = \exp[i\,\Delta\Phi(x)]$. The momentum distribution (and the far-field diffraction pattern) of a plane wave scattered from the phase grating is thus given by the Fourier transform of the exponen-

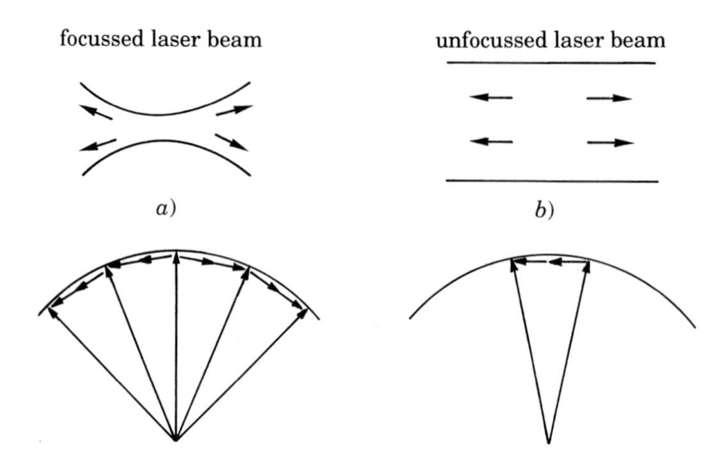

Fig. 6. – Schematic of energy and momentum conservation for diffraction of atoms from an optical standing wave in momentum space. The outgoing momentum states lie on a circle. The angular distribution of k-vectors in a focussed laser beam is described by a Gaussian. a) A tightly focussed laser beam has a large angular uncertainty in the k-vector direction. Energy conservation can, therefore, be fulfilled for a large number of diffraction orders independent of the angle of the incoming beam. b) In a broad laser beam the probability of finding photons with a large-angle k-vector is very small. Thus diffraction is only possible if the incoming momentum fulfils the Bragg condition. First-order Bragg scattering is displayed.

tial of the phase shift (eq. (13))

$$\mathrm{FT}\{\exp[i\,\Delta\Phi(x)]\} = \int \exp\left[i\int \Omega(x,\,t)\,\mathrm{d}t\right]\exp[ikx]\,\mathrm{d}x\,. \tag{30}$$

For large detuning Δ all incoming ground states are adiabatically transferred into one dressed state, whose energy eigenvalue displays a sinusoidal behaviour with a periodicity of $\lambda/2$. Consequently the phase shift $\Delta\Phi(x)$ displays the same sinusoidal behaviour such that the interaction can then be viewed as diffraction from a sinusoidal phase grating. Due to the periodicity of the grating the transverse-momentum distribution is a comb of δ-functions spaced by $2\hbar k$. For a sinusoidal phase grating the intensity of the n-th momentum component is given by the square of the n-th-order Bessel function J_n [34]:

$$I_n \propto \left| J_n\left(\int \Omega_{\max}(t)/2\,\mathrm{d}t\right)\right|^2, \tag{31}$$

where $\Omega_{\max}(t)$ is the time-dependent effective Rabi frequency experienced by an atom passing through an antinode of the standing wave. A plot of such a momentum distribution is displayed in fig. 7 for $\int\Omega_{\max}(t)/2\,\mathrm{d}t = 10$. The total momentum width of the diffracted beam increases linearly with the number of Rabi periods. For many periods the population is distributed among a large num-

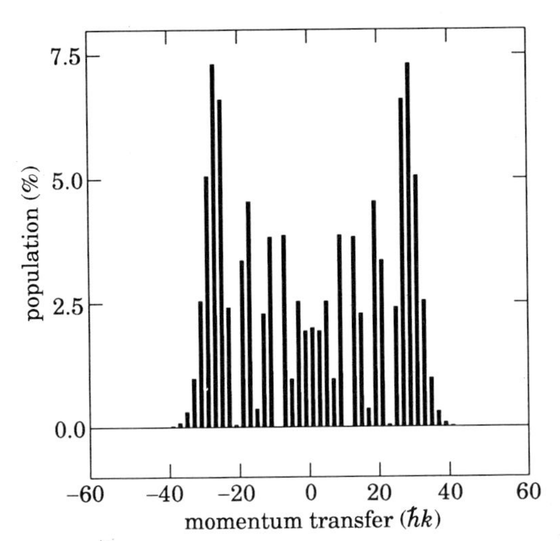

Fig. 7. – Momentum distribution for the diffraction of a plane wave by an optical standing wave obtained from eq. (31) for large detuning. The diffraction orders (represented by bars) have a spacing of $2\hbar k$; the population of the diffraction order n is given by the square of a Bessel function with index n. The average width of the momentum distribution is proportional to the number of Rabi periods ($\int \omega_{\max}(t)\,\mathrm{d}t = 2\pi \cdot 10$ in this plot).

ber of diffraction orders. Diffraction from a standing wave can, therefore, be looked upon as an effective beam splitter only for momentum transfers of a few $\hbar k$.

The first convincing experimental demonstration of diffraction of atoms from a standing light wave was achieved by MOSKOWITZ *et al.* in 1983 using a beam of sodium atoms [35]. In 1986 GOULD *et al.* reported on improved experimental results [36]. In these experiments the detuning Δ was chosen such that all incoming atoms evolved adiabatically into one dressed state. The spatial modulation of the corresponding eigenvalue was almost sinusoidal and the number of spontaneous emissions was less than unity. The transition from coherent to incoherent scattering, *i.e.* the transition from diffraction to diffusion, has been treated theoretically by TANGUY *et al.* [37] and SCHUMACHER *et al.* [38] and was demonstrated experimentally by GOULD *et al.* [39].

3·5. *Bragg scattering.* – If the atoms gain too much transverse momentum in the interaction with an optical standing wave, the transverse kinetic energy can no longer be neglected and the Raman-Nath approximation breaks down. In this case energy conservation places a severe constraint on the possible momentum transfer: The absolute values of the transverse momentum of incoming and outgoing atom must be equal. As illustrated in fig. 6*b*), efficient scattering is only possible if the transverse-momentum difference between the two allowed outgoing beams is an integer multiple of $2\hbar k$ [32, 40]. This corresponds to the Bragg condition of X-ray or neutron scattering. In the phase grating picture efficient scattering can only occur if both the in- and outgoing transverse-momentum states correspond to reciprocal-lattice vectors. There is a time-dependent phase factor, which brings about a periodic oscillation of the population between the two beams. The oscillation frequency for m-th-order Bragg scattering scales as $(\omega_R^2/\Delta)^m$. In neutron scattering this is referred to as «Pendellösung» oscillations.

MARTIN *et al.* reported the first experimental observation of Bragg scattering of atoms in 1988 [33]. They observed first- and second-order Bragg diffraction of sodium atoms and have also been able to demonstrate the variation of the splitting ratio as a function of the laser power. Bragg diffraction has been discussed as a beam splitter for atomic interferometry; it has two input and two output ports—unfortunately the momentum splitting is comparably small.

It has been proposed to use *Doppleron resonances* or *velocity-tuned resonances* [41, 42] as beam splitters for atoms because the rate of momentum transfer is considerably higher than in Bragg scattering [43, 44]. Doppleron resonances are multiphoton transitions in an optical standing wave that transfer an atom from the ground state to the excited state or *vice versa.*

3·6. *Magneto-optical beam splitter.* – Very recently our group has proposed [45] and demonstrated [46] a new beam-splitting scheme that relies on the interaction of three-level atoms with counterpropagating light beams of orthogonal linear polarization and a magnetic field applied parallel to the light beams. Diffraction of atoms from a standing light wave fails to produce a clearly two-peaked splitting of an incoming plane wave because the atom cannot distinguish between the two running waves and, therefore, has equal probability of «absorbing or emitting photons» from either direction. A more effective beam splitter could be achieved by allowing the atom to distinguish between the two running waves, such that, once it has absorbed a photon from one direction, this component continues to absorb from that direction and to emit into the other. The two running waves become distinguishable if they have different polarizations. An atom can only appreciate the concept of polarization if it has polarization-selective transitions (the simplest case being a three-level atom). In addition a magnetic field applied parallel to the running waves is required to select the preferred momentum transfer process.

The energy eigenvalues for a three-level atom («V-system»), subject to a homogeneous magnetic field and counterpropagating orthogonally polarized travelling waves, are shown in fig. 8 as a function of position in the light field. Three different ratios of the Rabi and Larmor frequency ω_R and ω_L are displayed (ω_R is defined as in [45]; in terms of eqs. (20) ω_R is the one-beam Rabi frequency). For ω_R much smaller or much larger than ω_L the influence of the light field or the magnetic field, respectively, can be understood as perturbations and all three levels display an approximately sinusoidal modulation with the period of the light field (see fig. 8*a*), *c*)). If the laser field is not turned on too quickly, in-

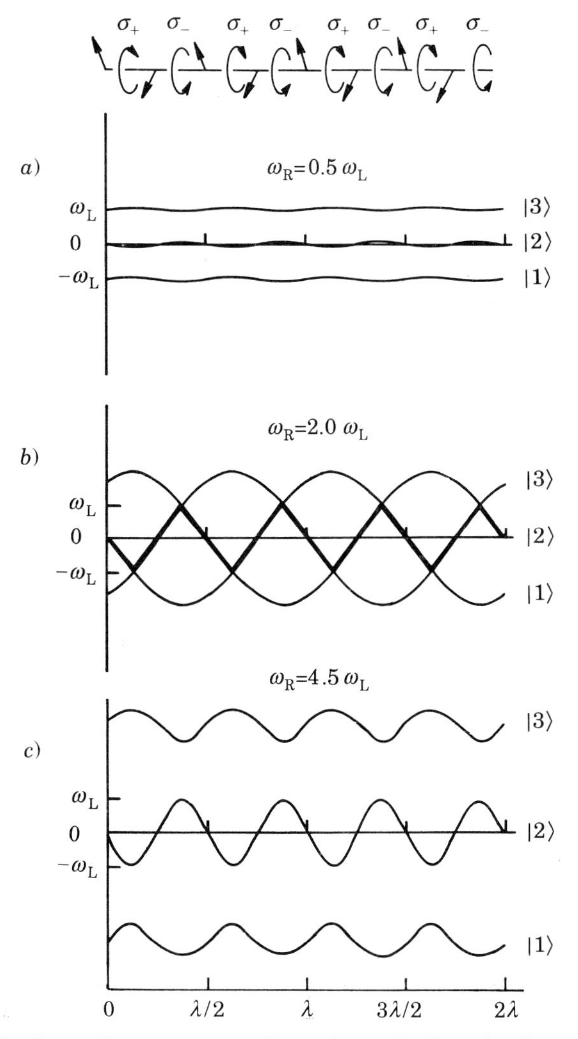

Fig. 8. – Plot of the dressed-state energy eigenvalues of a three-level atom as a function of position in two counterpropagating light beams of orthogonal linear polarization for three different values of the Rabi and Larmor frequency ω_R and ω_L. For $\omega_R = 2\omega_L$ the central eigenstate $|2\rangle$ exhibits a triangular shape (thicker line). Diffraction of atoms from this triangular phase grating produces a clearly two-peaked splitting.

coming ground-state atoms are transferred into the dressed state corresponding to the central eigenvalue. For $\omega_R = 2\omega_L$ the atoms experience an approximately triangle-shaped optical potential (see fig. 8*b*)). In the Raman-Nath regime the linear regions of this triangular phase grating will split an incoming beam into two beams with a splitting proportional to the Rabi frequency (see fig. 9*a*)). Even for Gaussian light beams and a homogeneous magnetic field, where the condition $\omega_R = 2\omega_L$ is met at no more than two points in the Gaussian

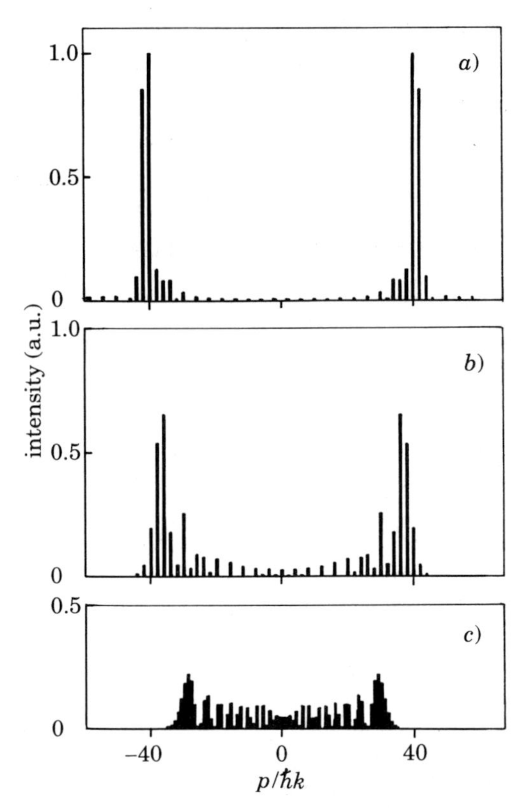

Fig. 9. – Histogram of the calculated momentum distribution for the magneto-optical beam splitter for $\omega_R = 2\omega_L$. The field amplitude and interaction time were chosen such that $\int \omega_R \, dt = 2\pi \cdot 10$. *a*) Calculation for a rectangular light beam profile. *b*) Gaussian profile: Lower-order momentum states are somewhat populated, but there is still a clearly two-peaked splitting. *c*) For comparison: diffraction from an optical standing wave.

light field, the diffraction pattern still displays a clearly two-peaked splitting. The momentum distribution for a Gaussian standing-wave profile is displayed in fig. 9*b*). The interaction time must be shorter than the excited-state lifetime for spontaneous emission to be avoided. For helium in the triplet ground state, a momentum splitting of $80\hbar k$ has been predicted for reasonable experimental conditions [45]. In a recent preliminary experiment we have used the magneto-optical interaction to demonstrate a momentum splitting of $40\hbar k$ using the $2\,^3S$ to $2\,^3P$ transition in helium [46].

3·7. *Optical Stern-Gerlach effect.* – The optical Stern-Gerlach effect is analogous to the familiar transverse magnetic Stern-Gerlach effect except that the magnetic-field gradient is replaced by a gradient in the intensity of a resonant light field. An incoming polarized beam is transferred into a superposition of two orthogonal eigenstates which subsequently become spatially separated in the field gradient. In the magnetic Stern-Gerlach effect a beam of polarized magnetic moments enters a transverse-magnetic-field gradient such that they do not follow the field adiabatically but evolve into a coherent superposition of the energy eigenstates. The different forces on the eigenstates result in a spatial separation of the components. In the optical Stern-Gerlach effect a two-level atom enters a near-resonant light field with a transverse gradient. If the resonance detuning Δ is smaller than a «critical detuning», transitions will populate both the $|+\rangle$ and $|-\rangle$ eigenstates. In the purely nonadiabatic regime the initial eigenstates are projected onto the dressed states $|+\rangle$ and $|-\rangle$. In this case the relative populations are given by the coefficients in eqs. (23) and (24), which are a function of the detuning. For $\Delta = 0$ the dressed states $|+\rangle$ and $|-\rangle$ are equal superpositions of ground and excited state such that a 50:50 beam splitter is obtained. In a transverse gradient of the optical potential $\pm \hbar\Omega/2$, the two eigenstate components experience dipole forces in opposite directions and, therefore, become spatially separated. It is important that no spontaneous emission occurs during the process as this would randomize the phase of the atomic wave functions. The split beams are orthogonal—thus a Stern-Gerlach beam splitter is a polarizing beam splitter. However, after the interaction the excited-state component of both beams decays spontaneously; this means that the excited-state component is incoherently transferred into the ground state. Both momentum states then consist of sums of a coherent and an incoherent ground-state compo-

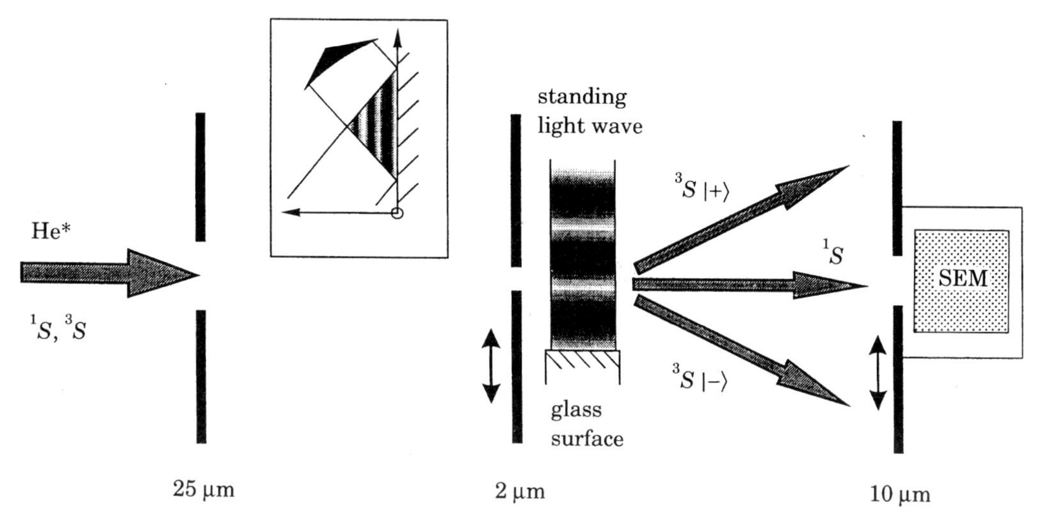

Fig. 10. – Schematic diagram of the experimental setup for the demonstration of the optical Stern-Gerlach effect. Metastable helium atoms (He*) in the triplet state ^3S evolved nonadiabatically in a superposition of eigenstates $|+\rangle$ and $|-\rangle$. The linear parts of an optical standing wave subsequently separated the eigenstates, while the undeflected ^1S singlet states served as a marker. A large-period standing wave was produced by glancing reflection of a laser beam (see inset). The He* atoms were detected with a secondary electron multiplier (SEM).

nent with coefficients depending on the detuning Δ. Therefore, for an interferometer using an optical Stern-Gerlach beam splitter, interference fringes with reduced visibility are expected.

The first experimental demonstration of the optical Stern-Gerlach effect was made by our group in 1991 [47]. In the experiment a beam of metastable helium atoms entered a light field with a Gaussian profile in the longitudinal direction (see fig. 10). The transverse gradient in ω_R was obtained in the linear part of a large-period standing wave. Three scans of the atomic count rate over the transverse position of a detector slit are displayed in fig. 11. The central peak corresponds to singlet-state atoms, which were also present in the atomic beam (as these states do not interact with the laser light they served as a marker for the position of the undeflected beam). The symmetrical splitting of the atomic beam can be seen clearly. We verified in a sequence of scans for various Δ that

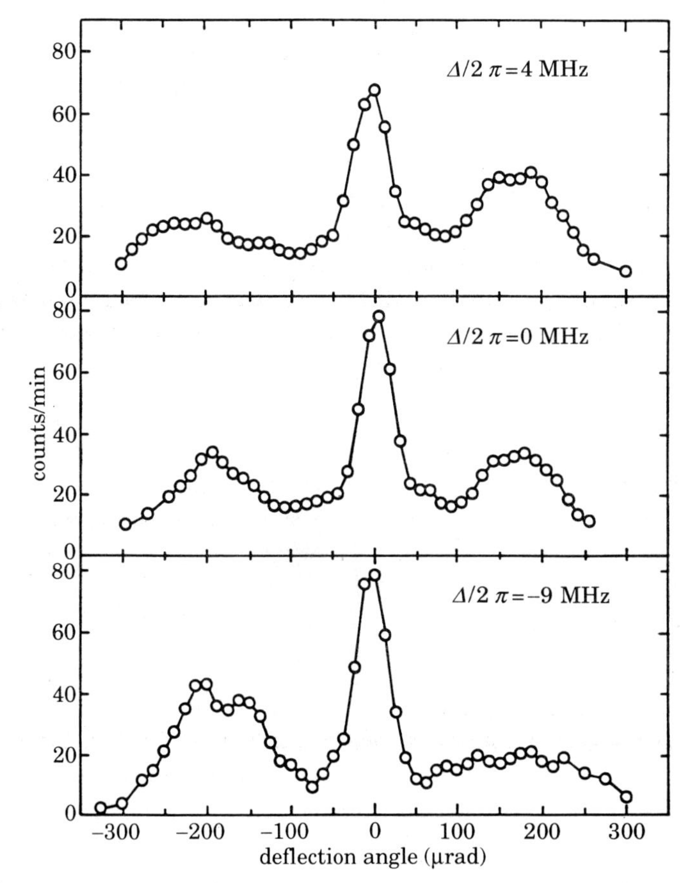

Fig. 11. – Atomic-intensity profiles for the optical Stern-Gerlach effect for small detuning Δ. It can be seen that the splitting ratio depends on Δ. The central peak is due to the undeflected singlet-state helium atoms.

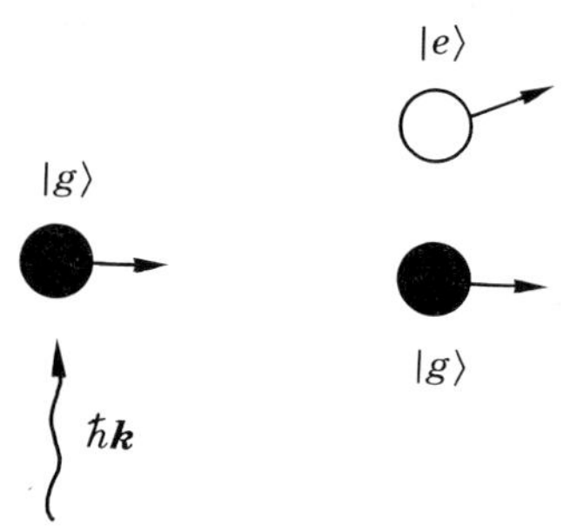

Fig. 12. – A ground-state atom is transformed into a superposition of momentum states split by the photon momentum $\hbar k$ by exciting it into a superposition of the ground ($|g\rangle$) and the excited ($|e\rangle$) state.

the splitting ratio depends on Δ in the way expected from theory. The splitting between the two emerging beams in the displayed scans was around $8\hbar k$. It should be noted that this splitting is only limited by the available laser power, which gives rise to the hope that very large splitting angles can be obtained in the future.

3·8. *Beam splitters based on single-photon recoil.* – A conceptionally simple beam splitter for two-level atoms involves exciting an atom in a coherent superposition of ground and excited state by applying a resonant radiation pulse. As the excited-state component carries the photon momentum, there is a momentum splitting between the two components of one-photon momentum (see fig. 12). This beam splitter is analogous to the splitting of a light beam by a travelling-wave acousto-optic modulator. The length of the interaction time determines the splitting ratio: A $\pi/2$-pulse excitation [48] represents a 50:50 beam splitter. Usually transverse running waves are used, in which case the interaction time is defined by the transit time of the atom through the beam. Running waves have the advantage that the direction of momentum transfer is well defined. It is necessary to work with long-lived excited states because spontaneous decay of the excited-state component has to be avoided. The momentum splitting in a photon recoil beam splitter is limited to only $1\hbar k$. A larger splitting can be achieved for multilevel atoms, *e.g.* a beam splitter based on Raman transitions produces a splitting of $2\hbar k$ (see subsect. 4·5). Photon recoil beam splitters have successfully been used in atom interferometers as discussed in subsect. 4·4 and 4·5.

4. – Interferometers for atoms.

4·1. *Motivation.* – As wave functions are complex valued, they are characterized by two real numbers: amplitude and phase. In most experiments amplitudes are measured—interferometry gives access to the phase. Interferometry with light has been an important tool for more than a century. Matter wave interferometry with electrons and neutrons [49] has allowed many important demonstrations of quantum-mechanical effects, like the Aharonov-Bohm effect or the sign reversal of a fermion's wave function after a 2π rotation, to give just two examples. Atom interferometry was proposed some time ago, but it is experimentally challenging—mainly due to the short de Broglie wavelengths. However, in recent years a number of atom interferometers were reported within a short period of time. For a short overview see ref. [50, 51].

Possible applications of interferometers for atoms fall into two categories: applications in which atoms are used as test particles for general effects and the examination of interactions specific to atoms. Atoms may allow measurements with vastly increased sensitivity over other test particles. As a consequence of their large mass and potentially low velocity, atoms make ideal inertial sensors (for example, to measure gravity or rotations). For inertial sensors the phase shift is proportional to the area enclosed by the two paths in an interferometer. A portable device to map the Earth's gravitational acceleration g with high precision would be useful in prospecting for oil or minerals. Using atom interferometers, phase shifts quadratic in g or the search for a «fifth force» may become accessible. Atom interferometers have the potential to provide a much higher accuracy in measuring rotational phase shifts (Sagnac effect) than laser gyros. Rotational sensors are not only interesting in navigation but could also be used to study general-relativistic effects, *e.g.* the search for a cosmologically preferred frame or for yet unobserved effects, like the geodetic and Lense-Thirring rotations to test metric gravitation theories [52, 53]. Another interesting rotational effect accessible to atom interferometry is the phase shift due to spin-rotation coupling, where the low velocity and large total angular momentum of the atom could provide a high sensitivity.

Atom interferometers can be applied to test a whole range of fundamental concepts in quantum mechanics previously discussed in the context of neutron interferometry [5, 6]. For example, atom interferometers may be used for the examination of topological effects (*e.g.*, Berry's phase [54]), which have stimulated considerable interest [55]. Further examples are tests of quantum mechanics like the search for nonlinear terms in the Schrödinger equation. A subject of continuing interest are «which path» and «delayed choice» experiments [56]. Atoms allow the investigation of «entangled states» produced, *e.g.*, by the interaction between an atom and a light field. Entangled states that persist after the atom and the field become spatially separated provide a means to study the effects of nonlocality [57].

Finally, atom interferometry allows precise measurements of atomic-structure properties, for example the d.c. or a.c. polarizability, the charge neutrality of the atom, collisional phase shifts or phase shifts as a consequence of spontaneous emission [58]. Atom interferometers may be a useful tool for investigating the interaction of atoms with surfaces. Also questions of coherence of the wave function could be examined.

As beam splitters are the essential component of interferometers, atom interferometers can be categorized into groups based on the beam-splitting technique used. Up to now diffraction from transmission structures, single-photon recoil and the magnetic Stern-Gerlach effect have been used to realize interferometers. Atom interferometers have been demonstrated using thermal beams as well as slow atoms launched from a trap. There is also interest in interferometry within a trap.

4'2. *Double-slit interferometer.* – In 1990 our group realized an atom interferometer in a simple Young's double-slit configuration [59]: A double slit was used to split the wave front of the incoming atomic beam. Interference fringes were recorded by transversely scanning a detection slit through the interference region. A schematic of the setup is displayed in fig. 13. The count rate was significantly increased using a detection grating of the same periodicity as the interference pattern (see experimental result in fig. 14). The spacing of the double slit was 8 μm, the width of the individual slits 1 μm. A 2 μm entrance slit was used to illuminate the double slit coherently. The microstructures used were produced by Heidenhain GmbH [60]. We used a supersonic nozzle beam of metastable helium with a relative full width of the velocity distribution of about 5%. The de Broglie wavelength could be tuned by adjusting the temperature of the expansion nozzle. For mean wavelengths of 0.56 and 1.03 Å we observed a periodicity in the interference pattern of 8.5 and 4.5 μm. The fringe visibilities were up to 60%. A major inconvenience with the setup was the low count rate.

As the period of the interference pattern is inversely proportional to the atomic wavelength, it can be increased by using slow atoms. SHIMIZU *et al.* have performed an experiment in which they released laser-cooled metastable neon atoms from a magneto-optical trap placed vertically above a double slit [61]. The

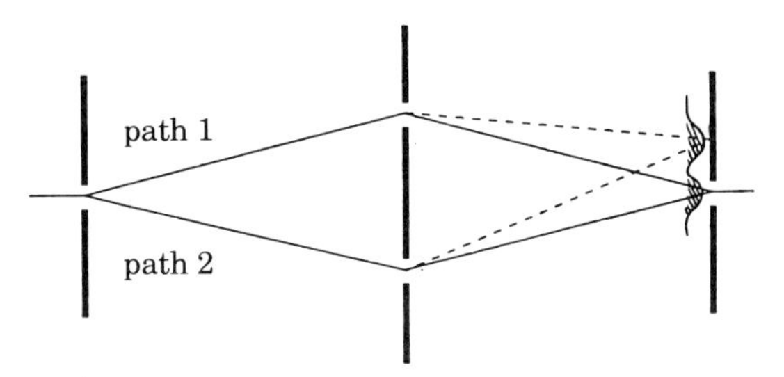

Fig. 13. – Schematic of the double-slit interferometer consisting of an entrance slit, a double slit and a detector slit. An atom moves along two spatially separated paths from the source to the detector. Coherent superposition of the cylindrical waves originating from the double slits produces interference fringes. The width of the entrance slit is 2 μm, the double slits are 1 μm in width and their spacing is 8 μm.

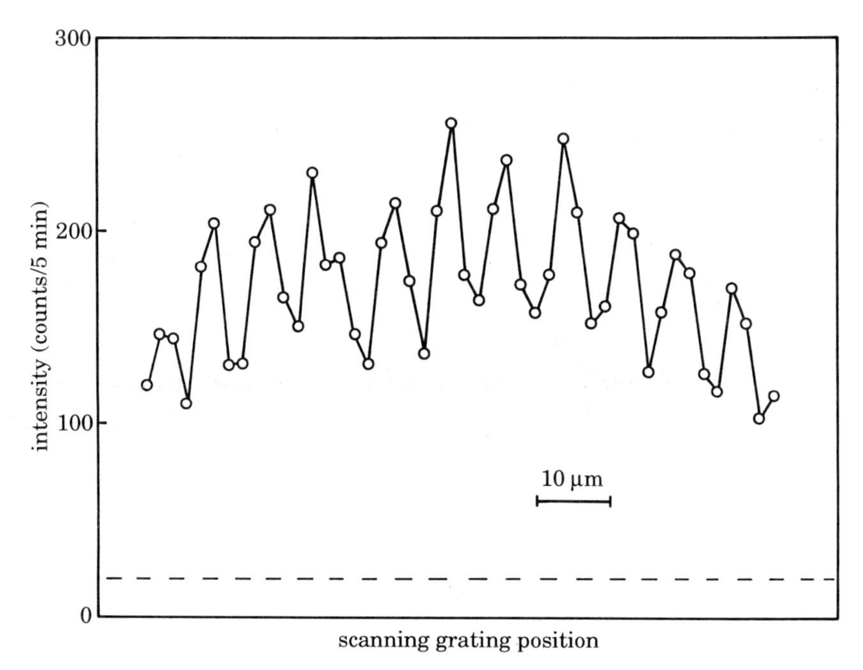

Fig. 14. – Atomic-intensity profile in the double-slit interferometer monitored with a grating (period 8 μm, slit width 4 μm) for a de Broglie wavelength of 1.03 Å. The line connecting the experimental points is a guide for the eye. The dashed line gives the detector background.

slit spacing and width were 6 μm and 2 μm, respectively. The interference pattern was observed using a microchannelplate detector placed below the double slit. As the atoms were accelerated by gravity, the period of the interference fringes depended on the vertical distance to the detection plane. Periods in the 100 μm range were observed. SHIMIZU *et al.* also demonstrated the deflection of the interference pattern by the cylindrical electric field produced by a charged wire placed next to the double slit [62].

Two other wave front splitting interferometers that may be of interest are the atomic biprism and Lloyd's mirror [58]. The interference fringes obtained with a biprism may be looked upon as the hologram of a phase object placed in one of the paths. A Lloyd's mirror arrangement would be useful in studying phase shifts induced by reflection of atoms from an atomic mirror.

4·3. *Three-grating interferometer.* – An alternative scheme for an interferometer based on microstructures is the three-grating arrangement demonstrated by KEITH *et al.* [63]. A schematic of their setup is given in fig. 15. The configuration is comparable to the Mach-Zehnder interferometer of classical optics. A three-grating interferometer is achromatic, which is particularly useful in atom optics due to the usually broad velocity distribution of atomic beams.

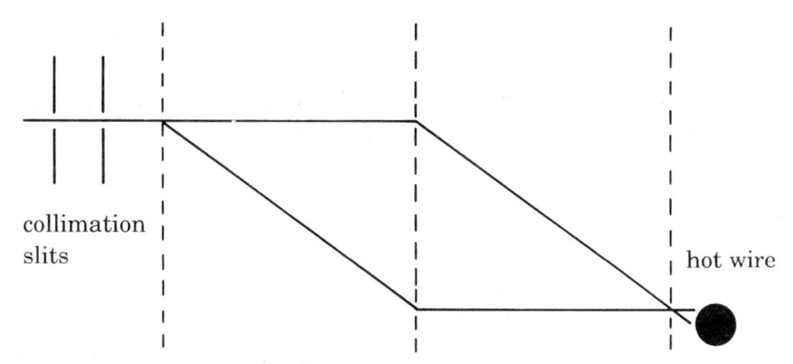

Fig. 15. – Schematic of a three-grating interferometer. The first grating is used to split a collimated atomic beam, the second grating diffracts both components, such that a third grating can be used to monitor the interference formed by overlapping diffraction orders. Only the paths leading to interference fringes are depicted.

They used sodium in a supersonic argon jet ($\lambda = 0.16$ Å). The gratings had a period of 400 nm (later 200 nm) and a relative spacing of 0.66 m. Interference fringes were recorded as a function of the transverse position of the second grating by measuring the count rate on a 25 μm hot wire placed behind the third grating. The fringe visibility obtained was 13% (later 25%). The transverse splitting of the atoms at the central grating was 27 μm (later 55 μm), the width of the beam was 30 μm. The gratings were produced from a silicon nitride membrane by electron beam lithography and reactive ion etching. Extremely good gratings are required in a three-grating interferometer; the distortions must be small compared to the grating period over the whole area of the grating. In their experiment KEITH *et al.* used an active stabilization scheme to compensate for relative vibrations and drifts of the gratings: An *optical* three-grating interferometer was set up on the same translation stages as the «atomic» gratings.

4·4. *Ramsey interferometer.* – The types of interferometers described in subsect. 4·2 and 4·3 rely only on the centre-of-mass motion and were demonstrated for neutrons in the 80's. As discussed in subsect. 3·8, beam splitters can also be produced by exciting atoms in a superposition of two electronic eigenstates using light pulses. In this case components in different momentum states are also in different internal states and, therefore, the two paths of the interferometer are labelled by the internal state emerging from the beam splitter. This allows resonant interactions with only one path even though the beams may overlap in space. Also the interference pattern can be measured by counting only atoms in one internal state. Experimentally this is often easier than spatially resolving narrowly spaced interference fringes.

The interaction of an atomic beam with two pairs of travelling waves in a Ramsey excitation geometry has been studied in the context of ultra-high-resolution spectroscopy for some time [64]; it was pointed out by BORDÉ in 1989 that the Ramsey fringes can be interpreted as atomic interference [65]. In order to avoid spontaneous emission and to increase the sensitivity, long-lived atomic transitions are required. Ramsey-type interferometers were reported in 1991 by RIEHLE *et al.* for calcium at the Physikalisch-Technische Bundesanstalt in Braunschweig [66] and in 1992 by STERR *et al.* for magnesium at the Universität Bonn [67]. A schematic setup is shown in fig. 16. There are two possibilities for an incoming ground state $|g, 0\rangle$ to be split and recombined at the fourth interaction zone; only these paths are depicted in fig. 16. Each trapezoid can be interpreted as a distinct Mach-Zehnder interferometer. In each interferometer the probability for an atom to be detected oscillates between the two output ports as a function of the phase shift between the two paths. The excited-state component can be detected by fluorescence from the states $|e, -1\rangle$ and $|e, 1\rangle$ after the fourth interaction zone. Interference fringes in the fluorescence are recorded as a function of the laser detuning. As the transverse-momentum states differ by $2\hbar k$, the fringes are separated by twice the recoil shift. For the resolution of small phase shifts it may be disturbing that the resulting pattern is the sum of two fringe systems. However, as the two interferometers are in different in-

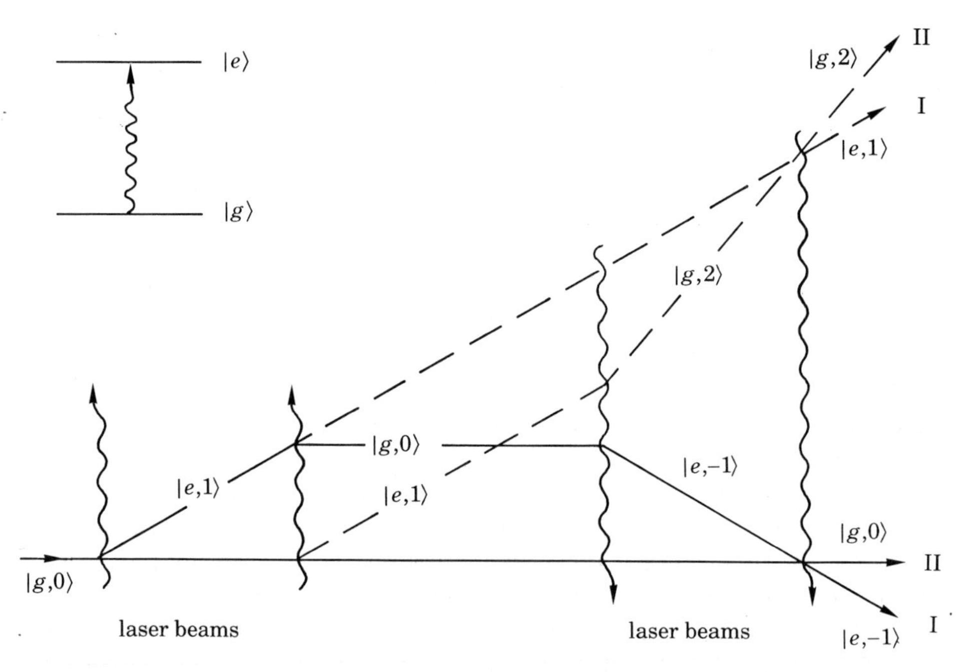

Fig. 16. – Interferometric interpretation of optical Ramsey excitation of an atomic beam by four travelling laser fields. An incoming ground-state atom with transverse momentum $0\,\hbar k\,|g, 0\rangle$ is excited into a sequence of superpositions of internal and momentum states. The solid and dashed trapezoids correspond to distinct Mach-Zehnder-type interferometers.

ternal states between the second and third laser pulse, either of the recoil components can be suppressed by depopulation or deflection of one state using a second laser or due to spontaneous emission of the excited-state component, as has been demonstrated by both groups [67, 68]. In the experiments so performed so far the transverse splittings have been much smaller than the width of the atomic beams.

In the Braunschweig experiment the phase shift resulting from a rotation of the entire experimental apparatus, *i.e.* the Sagnac effect, was measured [66]. The fact that the two paths are in different internal states in parts of the interferometer allows the measurement of phase shifts introduced by interactions that discriminate between the internal states. Both the Braunschweig and the Bonn group investigated phase shifts as a consequence of the dipole force. They applied a laser beam resonant with the strong ${}^1S_0(|g\rangle)$-${}^1P_1$ transition such that the ground state $|g\rangle$ experienced an optical potential $\hbar\Omega$. Depending on the detuning Δ, $|g\rangle$ moved over a hill or through a valley in the optical potential and, therefore, experienced an advance or retardation of its phase. The Bonn group also examined the phase shift as a consequence of the Stark shifts of the ground and excited states in a d.c. electric field, hence measuring the difference of the d.c. polarizabilities of the two states [69].

4·5. *Interferometer using stimulated Raman transitions.* – In an interferometer using light pulses as beam splitters, a long-lived and hence ultra-narrow transition is necessary. In the optical regime it is not easy to provide the required ultra-stable lasers. In the r.f. regime the frequency stabilization is easier, but the momentum of an r.f. photon is extremely small. KASEVICH and CHU have developed an atom interferometer based on stimulated Raman transitions, which combines the advantages of the long lifetime associated with an r.f. transition with the large photon recoil of an optical transition [70, 71]. Their interferometer is based on stimulated Raman transitions between the two hyperfine components of the sodium ground state. The transitions are driven by two counterpropagating laser beams with frequencies ω_1 and ω_2. Both lasers are detuned by the same amount from the optical transition frequency (see fig. 17*a*)). In this case it is not the absolute laser frequency but the frequency difference (1.7 GHz for sodium) which must be stable. Experimentally the difference frequency can be set very accurately by generating the two frequencies using a stable r.f. oscillator to phase-modulate the laser beam. An atom transferred from one hyperfine level to the other undergoes a momentum change of $2\hbar k$, because it absorbs a photon from one travelling-wave laser field and emits into the other. The two-photon transition provides phase compensation, such that a $\pi/2$-π-$\pi/2$ pulse sequence can be used corresponding to a beam splitter, two mirrors and another beam splitter as in an optical Mach-Zehnder interferometer (see fig. 17*b*)). As in the «Ramsey interferometer», the momentum states are labelled by the internal states. In the experiment the sodium atoms were injected

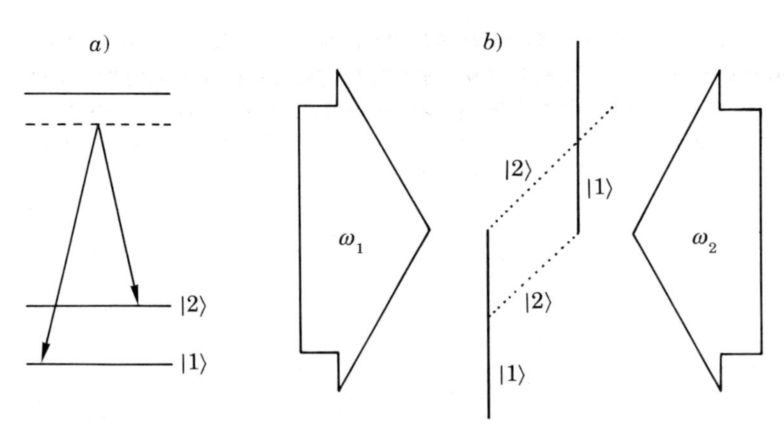

Fig. 17. – Basic principle of the interferometer using stimulated Raman transitions between hyperfine components $|1\rangle$ and $|2\rangle$ of the atomic ground state: An atom in the state $|1\rangle$ is excited in a coherent superposition of $|1\rangle$ and $|2\rangle$ by a Raman pulse consisting of two antiparallel running waves with frequencies ω_1 and ω_2. A second laser pulse reverses the internal and momentum states of the two components. The third Raman pulse mixes the populations such that interference fringes in the population of $|1\rangle$ or $|2\rangle$ as a function of relative phase or frequency of the Raman pulses can be observed.

into the interferometer in the $|1\rangle$ $(F = 1)$ state and are detected behind the third Raman pulse in the $|2\rangle$ $(F = 2)$ state by resonant photoionization. In contrast to the 4-zone Ramsey interferometer no atoms are lost into other paths. Interference fringes can be obtained, for example, as a function of the phase of the final $\pi/2$ pulse.

As a source KASEVICH and CHU used laser-cooled atoms in an «atomic fountain». Atoms were stored in a magneto-optical trap, then cooled in a polarization gradient optical molasses, and finally launched vertically. This geometry allows measurement times in the 0.1 s range and consequently provides very good resolution. Recently KASEVICH and CHU demonstrated that stimulated Raman pulses can also be used to cool sodium atoms below the photon recoil [72]. Both a longitudinal and a transverse geometry for the interferometer are possible, *i.e.* the Raman beams can be arranged perpendicular or parallel to the atomic trajectories. The transverse geometry shown in fig. 17*b*) may be used to realize an atomic gyroscope. A transverse geometry with four $\pi/2$ interaction zones was used by the Stanford group to measure the ratio of $\hbar$ to the mass of cesium [73]. KASEVICH and CHU operated the three-zone interferometer in a longitudinal geometry with the atoms launched vertically out of the trap. In this case there is a gravitationally induced phase shift $\Delta\Phi = -\boldsymbol{kg}\,\Delta t^2$, where $\boldsymbol{k}$ is the difference between the wave vectors of the laser fields and Δt the time between subsequent Raman pulses. For long interaction times it was necessary to shift $\boldsymbol{k}$ between the pulses to compensate for the changing Doppler shift. For $\Delta t = 0.05$ s (maximum wave packet separation about 3 mm) a phase uncertainty of $3 \cdot 10^{-3}$ cycles and consequently a sensitivity to changes in g of $\Delta g/g = 3 \cdot 10^{-8}$ was obtained for a total scan time of 2000 s. This accuracy level approaches the resolution of state-of-the-art gravimeters using the «falling corner cube» technique [74]. Thus atom interferometry starts to become interesting for gravity tests and geophysical applications.

4·6. *Longitudinal Stern-Gerlach interferometer.* – In the magnetic Stern-Gerlach effect an incoming atomic state is split in different Zeeman sublevels. An atom interferometer based on a Stern-Gerlach beam splitter can be divided into four zones: Firstly the incoming beam must be polarized. It is then prepared in a coherent superposition of magnetic sublevels by nonadiabatic passage into a magnetic field orthogonal to the field in the polarizer. The sublevels are split by a longitudinal field gradient. As the sublevels are orthogonal, they have to be re-mixed in order to interfere. Finally an analyser is used to select a particular polarization. Interference fringes in the population of the selected substate are obtained as a function of the phase difference arising from the different potentials experienced by the sublevels in the magnetic field between the mixing regions.

ROBERT *et al.* used a longitudinal magnetic-field gradient to split and recombine a partially polarized metastable $2s_{1/2}$ hydrogen beam ($\lambda_{dB} = 0.4$ Å) [75, 76]. The quenching of two hyperfine states of $2s_{1/2}$ hydrogen atoms due to the motional Stark shift in a transverse magnetic field was used as a partial analyser and polarizer. Interference fringes were observed by measuring the intensity of one polarization component as a function of the magnetic field (the interaction length was 85 mm). The longitudinal Stern-Gerlach interferometer has been applied to demonstrate the manifestation of a topological phase for a nonadiabatic cyclic evolution [77]. The longitudinal Stern-Gerlach interferometer resembles the experiment by SOKOLOV, who used rapid passage of 20 keV hydrogen atoms through electric fields to induce nonadiabatic transitions between the $2s_{1/2}$ and $2p_{1/2}$ states in a measurement of the Lamb shift [78].

5. – Lenses for atoms.

5·1. *Motivation.* – Lenses are essential elements for many optical systems. Possible applications of lenses for atoms are microscopy and lithography [79]. The resolution of a diffraction-limited microscope is determined by the wavelength. The success of the electron microscope is that wavelengths much smaller than optical wavelengths can be achieved for electrons; atomic resolution is possible with energies in the keV range. However, such energies can produce significant damage to the sample. A particle with kinetic energy E has a de Broglie wavelength $\lambda_{dB} = h/\sqrt{2mE}$. Thus increasing the mass by, for example, using atoms offers the possibility of achieving high resolution with only moderate energies and consequently less damage to the sample. In addition, the internal structure of atoms may be used to probe new aspects of atom-surface interactions like magnetization or chemical reactivity. Metastable rare-gas atoms are already used to perform spectroscopy on surfaces [80, 81]. Atom optics can add spatial resolution to such experiments. Unlike many other probes atoms do not penetrate the surface, and, therefore, the interaction provides detail of the uppermost layer. A second important application of lenses is atom lithography, where atoms are deposited onto a surface with a high spatial resolution. Both applications can be envisaged using either imaging or scanning techniques. Sharply focussed atomic beams would also be useful in classical atomic-beam experiments, *e.g.* studies of atomic scattering.

A sharply focussed beam can be produced by de-magnifying a source object as in classical or electron optics. In atom optics, in addition to lenses, laser-cooling techniques can be used for beam compression [82]. However, it appears that the highest resolution will be achieved using nondissipative lenses because of the problem of dissipative aberration. Dissipative techniques may turn out to be useful for the preparation of the beam prior to the interaction with a lens.

The fundamental principle of any lens can be illuminated by a discussion in terms of Fermat's principle, which states that the optical path length is the same for all rays from a source point to its image point (modulo 2π in case of a zone plate). Putting it another way, a lens introduces larger phase shifts for rays close to the optical axis than for rays further away. For a thin lens the phase shift is proportional to the square of the distance r from the optical axis. In atom optics such a phase shift can be achieved by shifting the internal energy of the atoms.

The resolution of an ideal optical system is limited by diffraction from the aperture. According to the widely used Rayleigh criterion, the resolution Δd of a circular aperture of diameter D is

$$\Delta d \approx \frac{1.22\lambda f}{D} = 1.22\lambda F\,, \tag{32}$$

where F is called the F-number of the lens. A small F-number gives high resolution and a large collection angle.

Lenses for atoms have been demonstrated using microstructures (Fresnel zone plate), static fields and the dipole force. For a lens based on refraction from a potential, the potential should have a parabolic shape, *i.e.* $V(r) \propto r^2$. Higher-order deviations can be treated as spherical aberration. The maxima of $\hbar\Omega(r)$ in both a focussed laser beam and in the antinode of a standing wave have been applied to focus atoms. Also it has been proposed to use the local minimum of $\hbar\Omega(r)$ inside a focussed TEM_{01}^* laser beam [83, 84]. As in classical optics, curved mirrors can also be used as imaging elements.

In general, atom lenses suffer from chromatic aberration. For all «refractive» lenses the transverse-momentum transfer on the respective component depends on the interaction time, which is inversely proportional to the group velocity of the atom v. In addition, the deflection angle depends on the ratio of the transverse-momentum change to the longitudinal momentum. The focal length, which is determined by the deflection angle for a given displacement of an atom from the optical axis, is, therefore, proportional to v^2. It follows from eq. (32) that the resolution is proportional to v, such that it is advantageous to work with slow atoms. For a Fresnel zone plate the focal length depends linearly on v because the transverse-momentum change is independent of v. Thus by combining a focussing zone plate with a defocussing refractive lens an achromatic lens can be constructed. The focal length of a curved mirror is independent of the atomic velocity—the angle of incidence equals the angle of reflection for all velocities. The effect of chromatic aberration can be reduced by compressing the width of the velocity distribution of the atomic beam using laser-cooling techniques [85].

5˙2. *Focussing using a Fresnel zone plate.* – It is well known from classical optics that a Fresnel zone plate acts as a diffraction-limited lens for waves. The principle is to admit only contributions from the source plane which interfere constructively in the image plane. This is achieved by blocking the paths that would interfere destructively with a sequence of concentric opaque rings. It follows that the radius R_n of the n-th zone is given by

$$R_n = \sqrt{n}\, R_1 \,. \tag{33}$$

The focal length f_m (which obeys the lens equation) for the diffraction order m is given by

$$f_m = \frac{R_1^2}{m\lambda} \,. \tag{34}$$

The resolution of a zone plate for plane waves can be approximately expressed as the width of the finest (outermost) ring divided by the diffraction order m. In contrast to refractive lenses, the resolution achieved using a zone plate is independent of the atomic velocity.

In our group we have used a zone plate with $R_1 = 9.4\ \mu\text{m}$ and a diameter of $220\ \mu\text{m}$ [60] to demonstrate the focussing of $\lambda_{\text{dB}} = 1.96\ \text{Å}$ helium atoms [86]. A SEM picture of the zone plate is shown in fig. 18. The focal length f_1 for this wavelength was 0.45 m and the F-number approximately 2000. We imaged both a single and a double slit (see fig. 19). The resolution was limited by chromatic aberrations; we used a supersonic expansion where the relative width of the velocity distribution was approximately 7%.

This focussing technique exploits the wave nature of the atomic centre-of-mass motion and does not depend on the internal state. Therefore, a zone plate can be readily used for any atom. A zone plate is diffraction limited, *i.e.* lens errors other than chromatic aberration do not play any significant role. The disad-

Fig. 18. – Scanning-electron-microscope picture of the zone plate used to focus an atomic beam. Only the innermost zones are visible. The diameter of the central zone is 19 μm.

vantage is that only 10% of the incoming atoms are diffracted into the first order, resulting in a loss of intensity along with an undesired halo. Finally the minimum spot size that can be obtained with such a lens is roughly equal to the width of the smallest outer ring, independent of the wavelength, and thus is limited by the state of the art in microfabrication technology. Currently this is around 100 nm for free-standing zone plates. In any case atomic resolution will not be achievable using this technique.

5·3. *Focussing in static electric or magnetic fields.* – The permanent magnetic-dipole moment or induced electric-dipole moment of atoms can be used to focus atomic beams. As there are no local maxima of static electric or magnetic fields in free space (Earnshaw's theorem), it is impossible to focus high-field-seeking states (which includes the ground state or energetically lower-lying magnetic substates) using this method [87]. However,

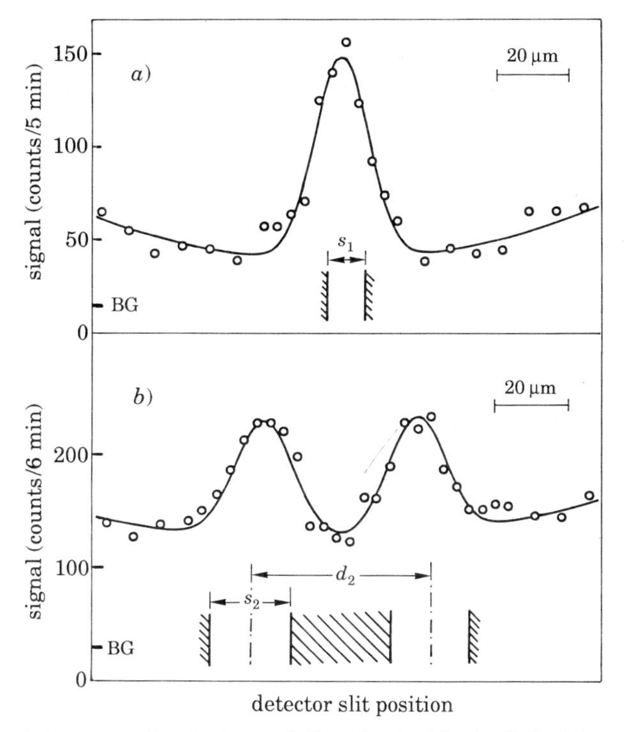

Fig. 19. – «Atomic images» of a single slit (a)) and a double slit (b)) obtained using a Fresnel zone plate. The imaged structures are shown to scale in the insets: $s_1 = 10\,\mu\text{m}$, $s_2 = 22\,\mu\text{m}$ and $d_2 = 49\,\mu\text{m}$. The ratio of object and image distance was 8/7. BG denotes the detector background. The solid lines are numerical fits.

low-field seekers (*e.g.*, higher-lying substates) may be focussed using a field configuration with a local minimum on the optical axis.

A magnetic-hexapole arrangement produces a quadratic field $B \propto r^2$ and consequently a quadratic Zeeman shift $\Delta E \propto r^2$. FRIEDBURG and PAUL demonstrated an atomic lens based on a magnetic-hexapole arrangement in 1951 [88, 89]. They achieved an atomic spot size of around 0.2 mm. For (moderate) electric fields the Stark shift is quadratic; a parabolic potential is thus obtained for a linear electric field $E \propto r$. This can be produced by an electric-quadrupole configuration as used by GORDON *et al.* to focus excited ammonia molecules in the first maser [90].

Strong-field seekers can be focussed using oscillating fields, in analogy to the trapping of charged particles in a Paul trap. The focussing of neutral atoms by an oscillating electric field has recently been demonstrated by SHIMIZU *et al.* [91].

5·4. *Focussing in a co-propagating laser beam.* – An atomic beam propagating along a near-resonant laser beam is focussed for «red» detuning ($\Delta < 0$) and

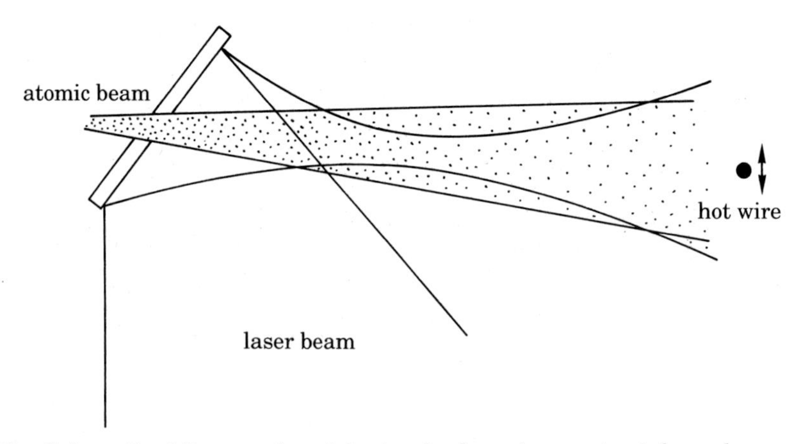

Fig. 20. – Schematic of the experimental setup for focussing an atomic beam by a co-propagating laser beam. The laser and the atomic beam are overlapped using a mirror with a central aperture. For red detuning the atoms are attracted towards higher light intensity and thus focussed.

de-focussed for «blue» detuning ($\Delta > 0$). The focussing and de-focussing of sodium atoms in a laser beam was first reported by BJORKHOLM *et al.* in 1978 [92, 93]. This experiment was the first demonstration of the manipulation of atoms by the dipole force. A schematic of the experimental setup is shown in fig. 20.

For long interaction times spontaneous emission established a steady-state population where the high-field-seeking dressed eigenstate $|+\rangle$ is more populated than the low-field-seeking state $|-\rangle$, such that the net dipole force acts towards the beam axis. In this case the force is a population-weighted average of the potential gradient experienced by each dressed state. For an explicit expression see, *e.g.*, [24, 92]. For $\Delta > 0$ the populations of the dressed states are reversed and the atomic beam is defocussed. In the final version of their experiment BJORKHOLM *et al.* achieved an atomic spot size of 28 μm using a laser spot size of 100 μm [94]. As the interaction zone is very long, their setup must be treated as a thick lens. Spontaneous emission gives rise to considerable diffusive aberration. For spontaneous emission to be avoided the laser beam would have to be focussed very strongly.

5'5. *Cylindrical lens based on an optical standing wave.* – Our group has demonstrated a cylindrical lens for metastable helium atoms based on the interaction with the antinode of a «red» ($\Delta < 0$) detuned transverse optical standing wave [95]. The spherical aberrations for this configuration are smaller than in case of a focussed Gaussian beam or a TEM_{01}^* mode. In our experiment we produced a standing wave with a period of 45 μm by glancing-angle reflection of a laser beam. The absolute value of the detuning was chosen smaller than ω_R such that the nodes of the effective Rabi frequency Ω became broader than the antinodes. The optical potential of this light field configuration is displayed in fig. 1 as an example for a refractive optical element for atoms. A 25 μm slit centred at an antinode defined the aperture of the lens (see fig. 21 for a schematic of the setup). The laser beam waist in the direction of the atomic beam was 40 μm, the interaction time consequently was sufficiently short to avoid spontaneous emission. We imaged a 2 μm single slit and a 8 μm grating with a focal length $f = 0.285$ m. In fig. 22 it can be seen how the atomic image of the 2 μm slit broadened as the focal length of the lens was tuned by varying the laser power. After deconvolving the contribution of the detection slit the width of the image was around 4 μm. This scheme can be readily extended to two dimensions by using two orthogonally crossed cylindrical lenses. Also the setup can be scaled down to produce a spot size of approximately 10 nm using a 0.5 μm standing wave.

An array of cylindrical lenses has been used by TIMP *et al.* to focus a sodium beam to a series of lines on a Si substrate [96]. Through investigating the diffraction of light from the surface of the substrate they were able to confirm that the reflectivity displayed the expected periodicity of $\lambda/2 = 294$ nm.

5·6. *Focussing using a curved mirror.* – As in classical optics, atoms can be focussed by reflection from a curved mirror. BERKHOUT *et al.* have performed an experiment where a highly divergent beam of cold ($T < 0.5$ K) hydrogen atoms was reflected from an optical-quality hemispherical concave substrate coated

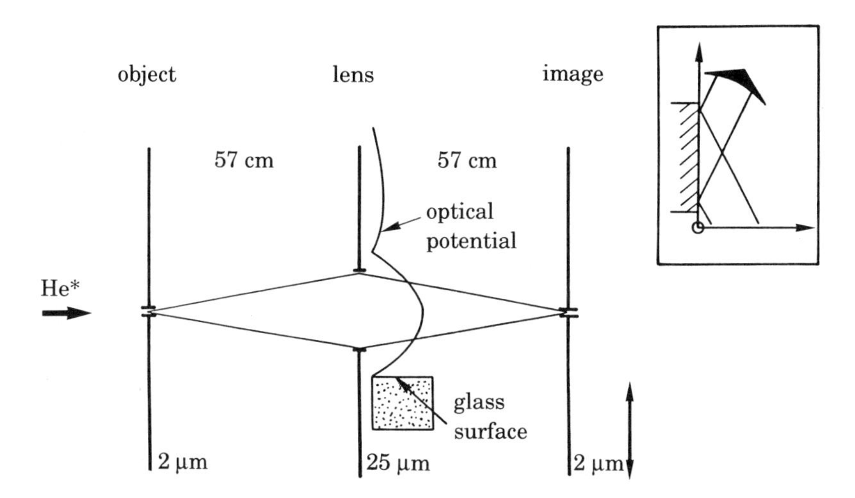

Fig. 21. – Experimental setup of a cylindrical lens for atoms. In an antinode of an optical standing wave the optical potential is parabolic, such that for red detuning atoms that pass through the antinode are focussed. A large-period standing wave was produced by reflection of a laser beam under a glancing angle (see inset). The atomic species used was helium in the metastable triplet state.

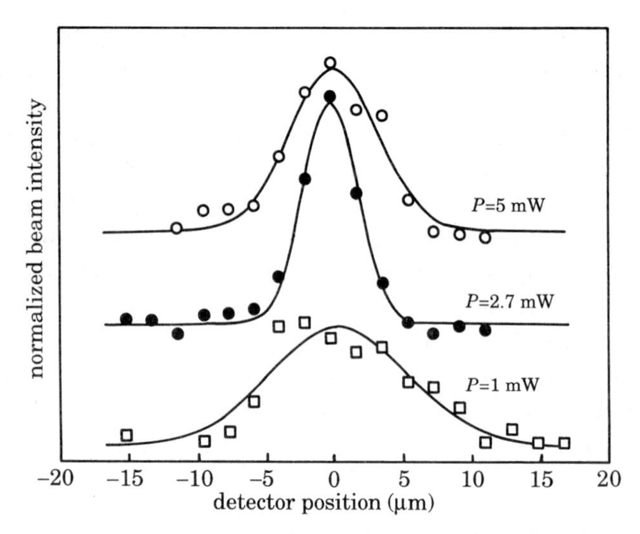

Fig. 22. – «Atomic» image of a 2 μm slit formed by a cylindrical lens based on the dipole force. It can be seen that the image broadened as the laser power and thus the focal length of the lens was varied.

with a film of liquid helium [97]. If the source was placed at the centre of curvature of the mirror, at least 80% of the effusing atoms were reflected back into the source.

6. – Mirrors for atoms.

6·1. *Motivation.* – Another optical element, which is important for atom optics, is a mirror. Mirrors would be useful in atom interferometers or in cavities for atoms [98]. In analogy to an optical cavity, an atomic cavity could be either a standing atomic wave between two mirrors or a running atomic wave using three or more mirrors. Partially reflecting mirrors could serve as input and output couplers. Unlike photons, very slow atoms are deflected by gravity. For this reason it should also be possible to trap atoms in a gravitational field by bouncing them off one parabolic mirror [99]. Cavities for atoms would be of interest for storing cold atoms. A gas of very cold atoms at high densities should exhibit quantum-statistical effects: If the atomic wave packets overlap, their indistinguishability will modify their collective behaviour. The interaction of this state with radiation may be interesting. It may be possible for such a cavity to emit a coherent beam of atoms [2]. A cavity for atoms may furthermore be useful as an interferometric device; due to the cavity enhancement the sensitivity to phase shifts could be very high.

6·2. *Reflection from surfaces.* – For the «direct» specular reflection of atoms from a surface two conditions have to be met: Firstly, like in usual optics, the projection of the surface roughness onto the direction of the incoming beam must be smaller than the wavelength. This condition can be met for very plane surfaces using atoms with very low normal velocities. Secondly, the average time an atom spends on the reflecting surface must be sufficiently small. If the atom sticks on the surface for too long, the «information» about its incoming momentum will be lost and it will be re-emitted diffusely with an intensity proportional to the sine of the angle θ between the direction of the outgoing atoms and the surface. For this reason atoms with larger sticking coefficients are harder to reflect.

As early as 1929 KNAUER and STERN reported on a 5% reflection of thermal beams of H_2 and He from a metal surface for 1 mrad incidence angle [100]. In 1930 ESTERMANN and STERN reflected helium atoms from the surface of cleaved ionic crystals for much larger angles [26]. As discussed in subsect. 5·6, BERKHOUT *et al.* used a concave mirror coated with superfluid helium to focus a divergent hydrogen beam [97]. In 1986 ANDERSON *et al.* reported the reflection of thermal cesium atoms from a polished glass surface [101]. For a glass surface with a flatness of better than 25 Å over an area of a mm^2, they measured a reflectivity of more than 50% for $\theta = 40$ mrad. However, the atoms had an increased probability of emerging from the scattering process under an angle smaller than the incidence angle.

6·3. *Reflection from an evanescent wave.* – Reflection of atoms has also been demonstrated using light forces. For the applications mentioned in subsect. 6·1 a coherent mirror is required, which means spontaneous emission must be avoided. Using the dipole force it is clear that any barrier in the optical poten-

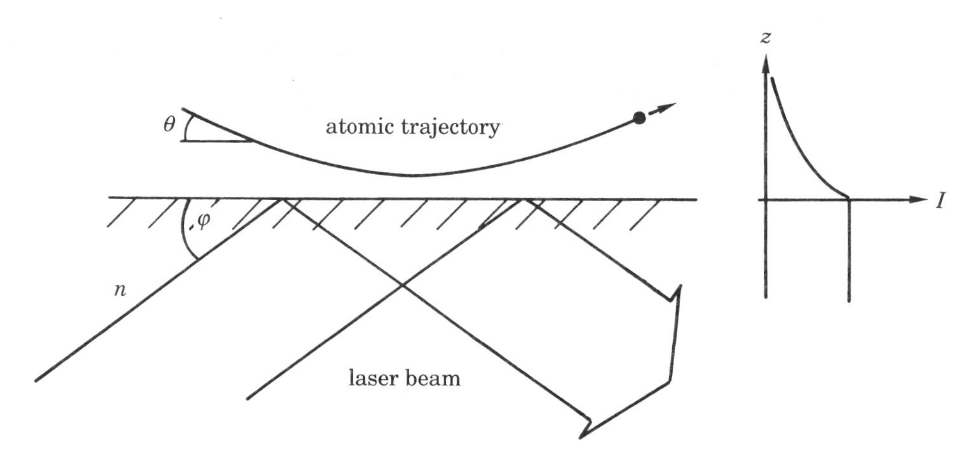

Fig. 23. – Basic principle of the reflection of atoms from an optical evanescent wave: The evanescent wave is produced by total internal reflection of a laser beam from the interface between a dielectric medium with index of refraction n and vacuum. For blue detuning the evanescent wave acts as a potential barrier. If the barrier is sufficiently high the atom is reflected.

tial $\hbar\Omega$ larger than the transverse kinetic energy of the atom will act as a mirror. In 1982 COOK and HILL suggested the use of a blue detuned evanescent light wave as a mirror for atoms [102]. It is interesting to note that, while the index of refraction for light is larger in glass than in vacuum, the atoms experience a

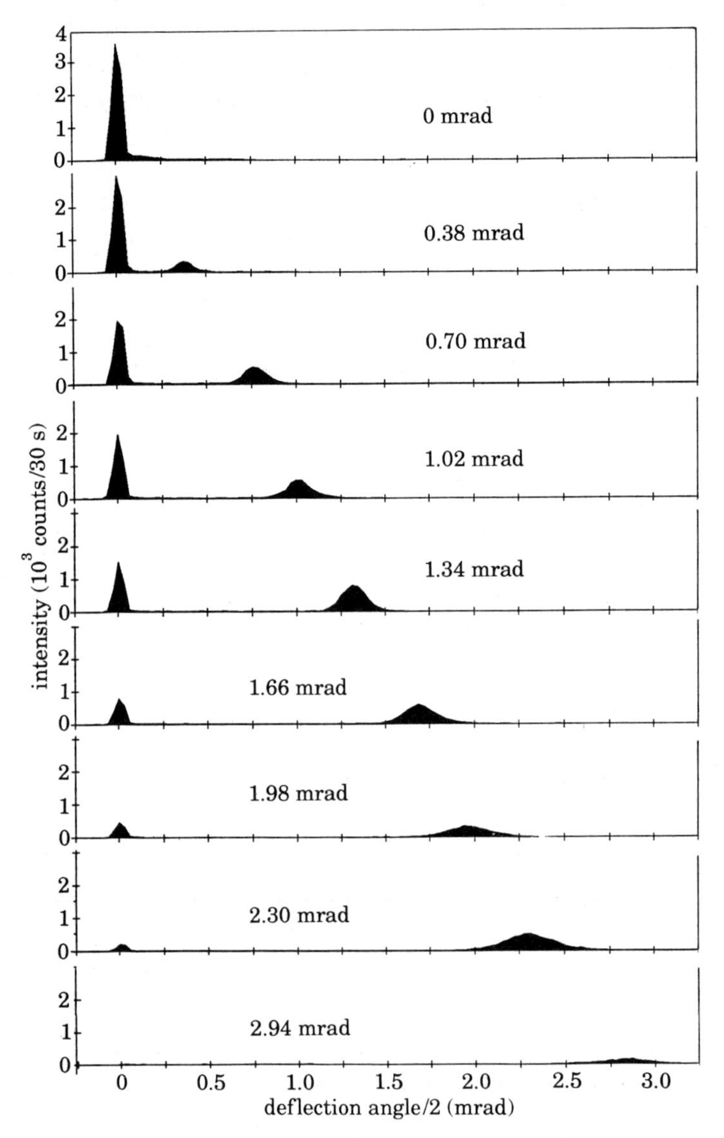

Fig. 24. – Experimental results of the reflection of metastable argon atoms from an evanescent wave: Intensity profiles of reflected and unreflected atoms. The angle of incidence θ was varied between 0 and 3 mrad in multiples of 0.3(2) mrad. It can be seen that the reflection law is obeyed. The change in the relative intensity of the two peaks as a function of θ is due to geometrical factors. The undeflected atoms do not intersect with the mirror.

decrease of «their» index of refraction in the evanescent wave. Consequently both the laser beam and the atomic beam experience total internal reflection from an «optically thick» medium. An evanescent wave produced by total internal reflection of a laser beam at a dielectric interface (see fig. 23) provides a large intensity gradient such that atoms can be reflected within a short distance. This is important in order to avoid spontaneous emission. For an evanescent wave the equipotential surfaces can be shaped conveniently by shaping the dielectric surface. The electric-field amplitude E in an evanescent wave produced by a plane wave of angular frequency ω_{L} for incidence angle φ is given by

$$E(z) = E(0)\cdot \exp[-\alpha z], \qquad \text{where } \alpha = \frac{\omega_{\mathrm{L}}}{c}\sqrt{n^2\cos^2\varphi - 1}\,. \tag{35}$$

The maximum transverse-momentum transfer scales with the fourth root of the intensity of the standing wave. A number of techniques such as build-up cavities, surface plasmons, planar waveguide structures and stratified media may be used to enhance evanescent waves.

In 1987 BALYKIN *et al.* reported on the deflection of thermal sodium atoms from an evanescent wave [103, 104]. The reflectivity was nearly 100% for small angles of incidence θ and decreased to around 10% for $\theta = 6$ mrad. KASEVICH *et al.* have demonstrated an «atomic trampoline» by bouncing slow sodium atoms off an evanescent wave under normal incidence [105]. The atoms were released from a magneto-optical trap centred 2 cm above the evanescent wave. Only a small fraction of the atoms are reflected back towards the trap region, however, after two bounces they were able to recapture about 0.03%.

In our group we have recently investigated the reflection of a well-collimated beam of metastable argon atoms from an evanescent wave for angles of incidence up to 5.5 mrad [106]. Typical results are displayed in fig. 24. By comparing the reflection efficiency for «open» and «closed» electronic transitions, the average number of spontaneous-emission processes occurring during the reflection process could be estimated to around 0.25. This result demonstrates that «coherent» reflection is possible and indicates that evanescent-wave mirrors could become standard elements in atom optical systems.

7. – Conclusion.

In summary there has been tremendous progress in the field of atom optics since the development of tunable laser radiation and free-standing microstructures. A number of optical elements for atoms have been demonstrated, but there remains considerable scope for their improvement. Atom interferometers have been realized and applied to the measurement of phase shifts. As the sensitivity of atom interferometers continues to improve, they are bound to find widespread application in pure and applied fields. The element of spatial resolution that atom optics adds to the investigation of the interactions of atoms with surfaces could prove to be useful in both microscopy and lithography. If the current degree of interest and activity is sustained, one can expect the field to continue to advance rapidly.

One of the great attractions of atom optics is its conceptual simplicity. Atom optics deals with the motion of atoms in real space. Although atoms are composite particles, they are relatively well understood and their interaction with the electromagnetic field provides very «clean» experimental conditions. The outcome of atom optical experiments can be predicted very accurately by quantum mechanics. So what is the motivation for atom optics in theoretical physics? At the current level of accuracy, atom optics cannot compete with other techniques (for example, ultra-high-resolution spectroscopy) as a test of quantum theory. However, the side-by-side progress of experiment and theory in atom optics provides a good example of how progress in science has been obtained through mutual stimulation. For example, in studies of the mechanical effects of light considerable theoretical interest and subsequently enhanced theoretical models were generated by «unexpected» experimental results. Finally, even though we have come to live with quantum mechanics, live very well in fact, quantum mechanics produces a lot of «strange» effects that we still feel somewhat uneasy about. Atom optics is ideally placed to do «quantum-mechanical experiments» to illustrate quantum phenomena and to further improve our understanding.

* * *

This work was supported by the Deutsche Forschungsgemeinschaft. One of us (CSA) is grateful for financial support provided by the Royal Society (London).

REFERENCES

[1] V. I. BALYKIN and V. S. LETOKHOV: *Laser Optics of Neutral Atomic Beams*, *Phys. Today*, 23 (April 1989).

[2] D. E. PRITCHARD: *Atom optics*, in R. R. LEWIS and J. C. ZORN, Editors: *Proceedings of the Twelfth International Conference on Atomic Physics 1990*, *A.I.P. Series*, Vol. **239** (American Institute of Physics, New York, N.Y., 1990), p. 165.

[3] O. CARNAL and J. MLYNEK: *Europhys. News*, **23**, 149 (1992).

[4] Special issue of *Appl. Phys. B* (April 1992), *Optics and Interferometry with Atoms*, edited by J. MLYNEK, V. BALYKIN and P. MEYSTRE.

[5] V. SEARS: *Neutron Optics* (Oxford University Press, New York, N.Y., Oxford, 1989).

[6] A. G. KLEIN and S. A. WERNER: *Rep. Prog. Phys.*, **46**, 259 (1983).

[7] P. W. HAWKES and E. KASPER: *Principles of Electron Optics* (Academic Press Limited, London, 1989).

[8] E. ARIMONDO, W. D. PHILLIPS and F. STRUMIA, Editors: *Proc. S.I.F.*, *Course CXVIII* (North-Holland, Amsterdam, 1991).

[9] C. N. COHEN-TANNOUDJI and W. D. PHILLIPS: *Phys. Today*, 33 (October 1990).
[10] S. CHU: *Science*, **253**, 861 (1991).
[11] M. KASEVICH and S. CHU: *Appl. Phys. B*, **54**, 321 (1992).
[12] G. SCOLES, Editor: *Atomic and Molecular Beam Methods*, Vol. I (Oxford University Press, New York, N.Y., 1988).
[13] O. STERN: *Naturwissenschaften*, **17**, 391 (1929).
[14] J. A. LEAVITT and F. A. BILLS: *Am. J. Phys.*, **37**, 905 (1969).
[15] D. W. KEITH, M. L. SCHATTENBURG, H. I. SMITH and D. E. PRITCHARD: *Phys. Rev. Lett.*, **61**, 1580 (1988).
[16] O. CARNAL, A. FAULSTICH and J. MLYNEK: *Appl. Phys. B*, **53**, 88 (1991).
[17] O. R. FRISCH: *Z. Phys.*, **86**, 42 (1933).
[18] M. BORN and E. WOLF: *Principles of Optics*, 6th edition (Pergamon Press, Oxford, 1980).
[19] C. DAVISSON and L. H. GERMER: *Phys. Rev.*, **30**, 705 (1927).
[20] G. P. THOMSON: *Proc. R. Soc. London, Ser. A*, **117**, 600 (1928).
[21] Special issue of the *J. Opt. Soc. Am. B*, **2**, *Mechanical Effects of Light* (November 1985).
[22] T. W. HÄNSCH and A. L. SCHAWLOW: *Opt. Commun.*, **13**, 68 (1975).
[23] S. CHU, L. HOLLBERG, J. E. BJORKHOLM, A. CABLE and A. ASHKIN: *Phys. Rev. Lett.*, **55**, 48 (1985).
[24] J. DALIBARD and C. COHEN-TANNOUDJI: *J. Opt. Soc. Am. B*, **2**, 1707 (1985).
[25] C. S. ADAMS, T. PFAU and J. MLYNEK: in H. WALTHER, T. W. HÄNSCH and B. NEÍZERT, Editors: *Proceedings of the Thirteenth International Conference on Atomic Physics 1992* (American Institute of Physics, New York, N.Y., 1993), p. 200.
[26] I. ESTERMANN und O. STERN: *Z. Phys.*, **61**, 95 (1930).
[27] *E.g.*, B. POELSEMA and G. COMSA: *Scattering of Thermal Energy Atoms from Disordered Surfaces*, Springer Tracts Mod. Phys., Vol. **115** (Springer, Berlin, 1989).
[28] *E.g.*, CH. R. EKSTROM, D. W. KEITH and D. E. PRITCHARD: *Appl. Phys. B*, **54**, 369 (1992).
[29] *E.g.*, R. J. COOK and A. F. BERNHARDT: *Phys. Rev. A*, **18**, 2533 (1978).
[30] S. STENHOLM: in E. ARIMONDO, W. D. PHILLIPS and F. STRUMIA, Editors: *Proc. S.I.F., Course CXVIII* (North-Holland, Amsterdam, 1992), p. 29.
[31] E.g., M. V. BERRY: *The Diffraction of Light by Ultrasound* (Academic, London, 1966).
[32] P. J. MARTIN, B. G. OLDAKER, A. H. MIKLICH and D. E. PRITCHARD: *Phys. Rev. Lett.*, **60**, 515 (1988).
[33] P. J. MARTIN, P. L. GOULD, B. G. OLDAKER, A. H. MIKLICH and D. E. PRITCHARD: *Physica B*, **151**, 255 (1988).
[34] *E.g.*, M. ABRAMOWITZ and A. STEGUM: *Handbook of Mathematical Functions* (Dover Inc., New York, N.Y., 1965), 9.1.21.
[35] P. E. MOSKOWITZ, P. L. GOULD, S. R. ATLAS and D. E. PRITCHARD: *Phys. Rev. Lett.*, **51**, 370 (1983).
[36] P. L. GOULD, G. A. RUFF and D. E. PRITCHARD: *Phys. Rev. Lett.*, **56**, 827 (1986).
[37] C. TANGUY, S. REYNAUD, M. MATSUOKA and C. COHEN-TANNOUDJI: *Opt. Commun.*, **44**, 249 (1983).
[38] E. SCHUMACHER, M. WILKENS, P. MEYSTRE and S. GLASGOW: *Appl. Phys. B*, **54**, 451 (1992).
[39] P. L. GOULD, P. J. MARTIN, G. A. RUFF, R. E. STONER, J.-L. PICQUE and D. E. PRITCHARD: *Phys. Rev. A*, **43**, 585 (1991).
[40] A. F. BERNHARDT and B. W. SHORE: *Phys. Rev. A*, **23**, 1290 (1981).

[41] E. Kyrölä and S. Stenholm: *Opt. Commun.*, **22**, 123 (1977).
[42] D. E. Pritchard and P. L. Gould: *J. Opt. Soc. Am. B*, **2**, 1799 (1985).
[43] E. Schumacher, M. Wilkens, P. Meystre and S. Glasgow: *Appl. Phys. B*, **54**, 451 (1992).
[44] S. Glasgow, P. Meystre, M. Wilkens and E. M. Wright: *Phys. Rev. A*, **43**, 2455 (1991).
[45] T. Pfau, C. S. Adams and J. Mlynek: *Europhys. Lett.*, **21**, 439 (1993).
[46] T. Pfau, C. Kurtsiefer, C. S. Adams, M. Sigel and J. Mlynek: *Phys. Rev. Lett.*, **71**, 3427 (1993).
[47] T. Sleator, T. Pfau, V. Balykin, O. Carnal and J. Mlynek: *Phys. Rev. Lett.*, **68**, 1996 (1992).
[48] *E.g.*, L. Allen and J. H. Eberly: *Optical Resonance and Two-Level Atoms* (Wiley Interscience, New York, N.Y., 1975).
[49] G. Badurek, H. Rauch and A. Zeilinger, Editors: *Matter wave interferometry*, in *Proceedings of the International Workshop on Matter Wave Interferometry, Physica B*, **151** (1988).
[50] B. Levy: *Phys. Today*, 17 (July 1991).
[51] J. Helmcke, F. Riehle, A. Witte and Th. Kisters: *Interferometry with atoms*, in *Proceedings of the 23rd EGAS Conference, Torun, Poland, 9-12 July, 1991, Phys. Scr.*, T **40**, 32 (1992).
[52] M. O. Scully, M. S. Zubairy and M. P. Haugan: *Phys. Rev. A*, **24**, 2009 (1981).
[53] C. M. Will: *Phys. Rep.*, **113**, 345 (1984).
[54] M. V. Berry: *Proc. R. Soc. London, Ser. A*, **392**, 45 (1984).
[55] A. Shapere and F. Wilczek, Editors: *Geometrical Phases in Physics, Advanced Series in Mathematical Physics*, Vol. **5** (World Scientific, Singapore, 1989).
[56] *E.g.*, J. A. Wheeler: in G. Toraldo di Francia, Editor: *Proc. S.I.F., Course LXXII* (North-Holland, Amsterdam 1979), p. 395.
[57] *E.g.*, P. Storey, M. Collett and D. Walls: *Phys. Rev. Lett.*, **68**, 472 (1992).
[58] T. Sleator, O. Carnal, T. Pfau, A. Faulstich, H. Takuma and J. Mlynek: in *Proceedings of the Tenth International Conference on Laser Spectroscopy, Font-Romeu, France*, edited by M. Ducloy, E. Giacobino and G. Camy (World Scientific, Singapore, 1991), p. 264.
[59] O. Carnal and J. Mlynek: *Phys. Rev. Lett.*, **66**, 2689 (1991).
[60] Heidenhain GmbH, Traunreut, Germany.
[61] F. Shimizu, K. Shimizu and H. Takuma: *Phys. Rev. A*, **46**, R17 (1992).
[62] F. Shimizu, K. Shimizu and H. Takuma: *Jpn. J. Appl. Phys.*, **31**, 46 (1992).
[63] D. W. Keith, C. R. Ekstrom, Q. A. Turchette and D. E. Pritchard: *Phys. Rev. Lett.*, **66**, 2693 (1991).
[64] J. Helmcke, D. Zevgolis and B. Ü. Yen: *Appl. Phys. B*, **28**, 83 (1982); Ch. J. Bordé, Ch. Salomon, S. Avrillier, A. van Lerberghe, Ch. Bréant, D. Bassi and G. Scoles: *Phys. Rev. A*, **30**, 1836 (1984).
[65] Ch. J. Bordé: *Phys. Lett. A*, **140**, 10 (1989).
[66] F. Riehle, Th. Kister, A. Witte, J. Helmcke and Ch. J. Bordé: *Phys. Rev. Lett.*, **67**, 177 (1991).
[67] U. Sterr, K. Sengstock, J. H. Müller, D. Bettermann and W. Ertmer: *Appl. Phys. B*, **54**, 341 (1992).
[68] F. Riehle, A. Witte, Th. Kisters and J. Helmcke: *Appl. Phys. B*, **54**, 333 (1992).
[69] V. Rieger, K. Sengstock, U. Sterr, J. H. Müller and W. Ertmer: *Opt. Commun.*, **99**, 172 (1993).

[70] M. KASEVICH and S. CHU: *Phys. Rev. Lett.*, **67**, 181 (1991).
[71] M. KASEVICH and S. CHU: *Appl. Phys. B*, **54**, 321 (1992).
[72] M. KASEVICH and S. CHU: *Phys. Rev. Lett.*, **69**, 1741 (1992).
[73] S. CHU: personal communication.
[74] J. E. FALLER and I. MARSON: *Metrologia*, **25**, 49 (1988).
[75] J. ROBERT, CH. MINIATURA, S. LE BOITEUX, J. REINHARDT, V. BOCVARSKI and J. BAUDON: *Europhys. Lett.*, **16**, 29 (1991).
[76] CH. MINIATURA, J. ROBERT, S. LE BOITEUX, J. REINHARDT and J. BAUDON: *Appl. Phys. B*, **54**, 347 (1992).
[77] CH. MINIATURA, J. ROBERT, O. GORCEIX, V. LORENT, S. LE BOITEUX, J. REINHARDT and J. BAUDON: *Phys. Rev. Lett.*, **69**, 261 (1992).
[78] YU. L. SOKOLOV and V. P. YAKOVLOV: *Sov. Phys. JETP*, **56**, 7 (1982).
[79] *E.g.*, R. POOL: *Science*, **255**, 1513 (1992).
[80] W. SESSELMANN, H. CONRAD, G. ERTL, J. KÜPPERS, B. WORATCHEK and H. HABERLAND: *Phys. Rev. Lett.*, **50**, 446 (1983).
[81] M. W. HART, M. S. HAMMOND, F. B. DUNNING and G. K. WALTERS: *Phys. Rev. B*, **39**, 5488 (1989).
[82] J. NELLESSEN, J. WERNER and W. ERTMER: *Opt. Commun.*, **78**, 300 (1990).
[83] V. I. BALYKIN and V. S. LETOKHOV: *Opt. Commun.*, **64**, 151 (1987).
[84] J. J. MCCLELLAND and M. R. SCHEINFEIN: *J. Opt. Soc. Am. B*, 8, 1975 (1991).
[85] A. FAULSTICH, A. SCHNETZ, M. SIGEL, T. SLEATOR, O. CARNAL, V. BALYKIN, H. TAKUMA and J. MLYNEK: *Europhys. Lett.*, **17**, 393 (1992).
[86] O. CARNAL, M. SIGEL, T. SLEATOR, H. TAKUMA and J. MLYNEK: *Phys. Rev. Lett.*, **67**, 3231 (1991).
[87] W. KETTERLE and D. E. PRITCHARD: *Appl. Phys. B*, **54**, 403 (1992).
[88] H. FRIEDBURG und W. PAUL: *Naturwissenschaften*, **38**, 159 (1951).
[89] H. FRIEDBURG: *Z. Phys.*, **130**, 493 (1951).
[90] J. P. GORDON: *Phys. Rev.*, **99**, 1253 (1955).
[91] F. SHIMIZU: personal communication.
[92] J. E. BJORKHOLM, R. R. FREEMAN, A. ASHKIN and D. B. PEARSON: *Phys. Rev. Lett.*, **41**, 1361 (1978).
[93] D. B. PEARSON, R. R. FREEMAN, J. E. BJORKHOLM and A. ASHKIN: *Appl. Phys. Lett.*, **36**, 99 (1980).
[94] J. E. BJORKHOLM, R. R. FREEMAN, A. ASHKIN and D. B. PEARSON: *Opt. Lett.*, **5**, 111 (1980).
[95] T. SLEATOR, T. PFAU, V. BALYKIN and J. MLYNEK: *Appl. Phys. B*, **54**, 375 (1992).
[96] G. TIMP, R. E. BEHRINGER, D. M. TENNANT and J. E. CUNNINGHAM: *Phys. Rev. Lett.*, **69**, 1636 (1992).
[97] J. J. BERKHOUT, O. J. LUITEN, I. D. SETIJA, T. W. HIJMANS, T. MIZUSAKI and J. T. M. WALRAVEN: *Phys. Rev. Lett.*, **63**, 1689 (1989).
[98] V. I. BALYKIN and V. S. LETOKHOV: *Appl. Phys. B*, 48, 517 (1989).
[99] H. WALLIS, J. DALIBARD and C. COHEN-TANNOUDJI: *Appl. Phys. B*, **54**, 407 (1992).
[100] F. KNAUER und O. STERN: *Z. Phys.*, **53**, 779 (1929).
[101] A. ANDERSON, S. HAROCHE, E. A. HINDS, W. JHE, D. MESCHEDE and L. MOI: *Phys. Rev. A*, **34**, 3513 (1986).
[102] R. J. COOK and R. K. HILL: *Opt. Commun.*, **43**, 258 (1982).
[103] V. I. BALYKIN, V. S. LETOKHOV, YU. B. OVCHINNIKOV and A. I. SIDOROV: *JETP Lett.*, **45**, 282 (1987).

[104] V. I. BALYKIN, V. S. LETOKHOV, YU. B. OVCHINNIKOV and A. I. SIDOROV: *Phys. Rev. Lett.*, **60**, 2137 (1988).
[105] M. A. KASEVICH, D. S. WEISS and S. CHU: *Opt. Lett.*, **15**, 607 (1990).
[106] W. SEIFERT, C. S. ADAMS, V. BALYKIN, C. HEINE, YU. OVCHINNIKOV and J. MLYNEK: *Phys. Rev. A*, to be published.

QUANTUM OPTICS

Cavity Quantum Electrodynamics.

S. HAROCHE

Laboratoire de Spectroscopie Hertzienne de l'École Normale Supérieure
24 rue Lhomond, 75231 Paris, Cedex 05, France

1. – Introduction. Aim of this lecture.

«Cavity quantum electrodynamics» studies the radiative properties of atomic systems confined by boundaries to a limited region of space. It deals with small numbers of atoms and photons interacting together in a box. The mere fact that the electromagnetic field is confined, instead of being free to escape from the source to infinity as is usually assumed to be the case in quantum optics, leads to interesting, unusual and possibly useful physical effects. This is a fundamental field of physics, of course, as the name QED tends to suggest, but it is also applied physics, since many ideas and concepts of cavity QED are now being increasingly used to develop new laser systems with interesting and unusual features.

In this lecture, I review various aspects of cavity QED, recalling the basic concepts and describing some recently performed experiments in the field. Clearly, I have been obliged to make choices and I will have to be very brief on each experiment I have chosen to describe. The details I cannot give here may be found in the references. Obviously too, my presentation is biased by my own perceptions of what I find interesting, and I may give undue coverage to the work I have been directly involved with. I have also adopted as a rule in these notes to avoid equations and detailed calculations and to restrict the analysis to physical pictures and back of the envelope estimates of the various effects. The reader interested in more detailed theoretical developments will also be given references to satisfy his curiosity. A general review work covering the field of cavity QED up to 1990 may be found in my Les Houches lecture notes (session LIII) [1]. Other tutorial references can also be found in the literature [2-7].

When an atom is placed in a small volume limited by metallic boundaries, its radiative properties are expected to be affected in two ways. First, the rate of photon emission or absorption by this atom is changed with respect to its value in free space, with a possibility of strong spontaneous-emission enhancement or inhibition. The former effect was predicted long ago by PURCELL [8] and the latter computed in details in 1981 by KLEPPNER [9]. Second, the atom energies, and hence its radiation frequencies, are altered and also become different from their «free space» unperturbed values. Calculations of these shifts can be found in numerous references, some of which have been published long ago [10, 11]. These effects may seem at first sight puzzling if one is used to think about atomic radiation and spontaneous emission as an intrinsic property of the atomic electrons alone. They become less strange if one realizes that the radiative properties of electric charges also reflect the structure of the field surrounding them, which in turn depends upon the field boundaries. Even if the radiation field is initially in its ground vacuum state as is the case in spontaneous-emission processes, it exhibits physically important fluctuations, whose properties depend upon the boundary geometry. These fluctuations perturb an atom immersed in them and this leads to important alterations of the atomic radiative properties. These alterations depend upon the atom electric-dipole polarization, the cavity geometry and the respective values of the atom and cavity resonance frequencies. By playing on the value of these various parameters, one can literally «tailor» spontaneous emission and more generally manipulate the atom radiative properties in ways leading to interesting new applications.

In fact the change in the emission rates and frequencies is only part of the story. Very important is also the fact that the photon, which in free space would escape to infinity at the speed of light, is, in a closed space, confined within a small volume. As a result, a single emitted photon corresponds to the existence of a large electric field at the position of the emitting atom. Moreover, if the cavity has a large Q-factor, this photon is stored around the atom for a relatively long time, allowing for a coherent atom-field coupling at the single-atom/single-photon level. The situation we are thus naturally led to is one where nonlinear phenomena become manifest at the quantum level, which is quite unusual in quantum optics, where, generally, nonlinear phenomena require intense fields made up of a very large number of photons.

As already noticed above, some of these ideas are not really new. An atom in front of a mirror, for example, is a simple enough system, which has been considered by theorists long ago. LENNARD-JONES, back in the thirties, has shown that the atomic energy levels are shifted by the van der Waals interaction between the atom and its instantaneous image in the mirror [10]. This shift is predicted to be inversely proportional to the cube of the atom-mirror distance and proportional to a proper combination of the mean-square dipole moment components of the atomic electron. At the same time, the radiative damping of the atom is changed, in a way depending upon the orientation of the dipole relative to the surface. The field of a dipole parallel to the mirror and close enough to it interferes destructively with the field of its image, and as a result the system does not radiate at all. The same atom, oriented normal to the metal, emits a field which interferes constructively with its image and thus radiates twice as fast as in free space. These effects are not intrinsically quantum in nature. They are mere consequences of classical Maxwell electromagnetism, taking field boundary conditions into account. Radio-engineers, for example, have known, at least since the beginning of this century, that the radiation pattern of an antenna is affected by the reflection of the Earth surface on which it stands. The new aspects I discuss in this lecture are the manifestations of these classical «mundane» effects, and the new refinements they lead to, at the single-photon/ /single-atom level.

2. – Control of spontaneous emission in a cavity. Two complementary interpretations.

The main difficulty one encounters when trying to study these effects on atomic systems is the precise control of the system geometry. How to define and keep constant the atom-to-mirror separation? A clever solution to this problem has been found in the seventies by DREXHAGE, who realized a pioneering study in this field [12]. His idea was to calibrate the atom-mirror distance by using monolayers of fatty acids as «spacers». Using the well-known techniques of Blodgett and Langmuir monolayer deposition, he made films with well-defined thicknesses, deposited fluorescing dye molecules on top and studied the radiation pattern and lifetime of these molecules as a function of the number of underlying layers. In spite of their simplicity and cleverness, these experiments remained essentially qualitative for a while, one reason being that the spacer layers complicated the physics. They have an index of refraction of their own, which is moreover anisotropic, giving rise to birefringent effects which have to be taken into account in the theory of these effects. Moreover, the method applies only to deposited molecules and not to free atoms far from a physical surface. Studies of atomic emission in a cavity vacuum had to wait a few more years for the progresses in atomic-beam laser spectroscopy methods, especially those involving large Rydberg atoms.

Before describing some of these experiments, it is at this stage important to stress another important idea of cavity QED, illustrated in fig. 1. I have chosen to discuss it on the prototype example of an atom in a cavity made of two plane parallel mirrors. In fact, cavity QED effects can be interpreted in two complementary ways [1]. The first (illustrated in fig. 1*a*)) is the one we have emphasized so far: the fluctuating electron in the atom «feels» its fluctuating images in the mirrors, which here are produced by multiple reflections. The radiation then results from an interference effect between the atom and its images. A complete calculation predicts in this case a fully destructive interference for a mirror separation L smaller than half the atom radiation wavelength. But there is another interpretation (fig. 1*b*)). The atom also «feels» the vacuum field fluctuations between the mirrors, which have a spectral distribution different from what they are in free space. In fact, the low-frequency modes of the vacuum

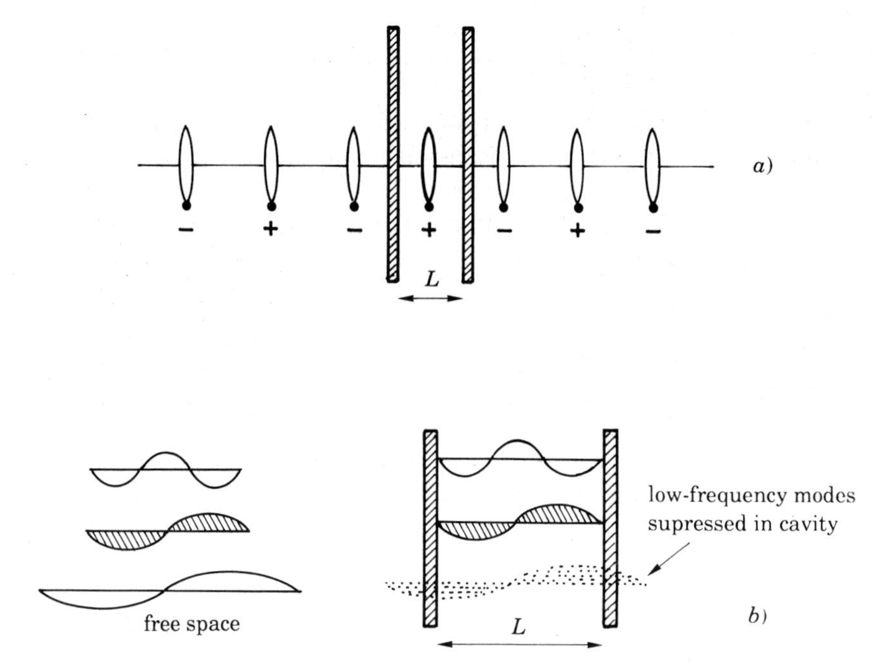

Fig. 1. – Dual aspect of cavity QED between parallel mirrors: *a*) atomic dipole at equal distances of mirrors interacts with its images resulting from multiple reflections; *b*) the vacuum field mode distribution (sketched in free space at left) is changed in mirror gap (sketched at right), with a cut-off of low-frequency components. The vacuum field perturbing effect on the atom is accordingly modified.

field, which have an electric field parallel to the mirrors and a wavelength larger than twice the mirror separation, are cut off by the boundary conditions and atoms which «would like» to radiate at such wavelengths cannot do so and must retain their excitation. We thus again understand, from this point of view, the inhibition of spontaneous emission. In fact this vacuum mode analysis is very close to the one made in 1948 by CASIMIR in his pioneering paper on the force between two metallic surfaces [13]. It is important to realize that both pictures are equally valid, in fact they correspond to the two sides of the same coin. In a consistent quantum-mechanical description of electrodynamics, both charges and fields have quantum fluctuations, intimately related to one another, and the same phenomenon may be understood with reference to either of these two complementary fluctuations. In the following, I use one or the other of these points of view, whichever happens to lead to the simplest picture.

Several experiments have, after Drexhage's work, demonstrated spontaneous-emission alteration effects on simple atomic or molecular systems. A first group of experiments has been perfomed on atoms crossing Fabry-Perot resonators, either in the microwave or in the optical domain [14, 15] (I come back below to the interest of using large Rydberg atoms for these experiments). Another set of experiments has been performed on the cyclotron resonance of free electrons in a trap, in the setup used for the beautiful $g-2$ experiments of Dehmelt's group in Seattle [16]. In these experiments, the electrodes of the trap itself (a ring and two hyperboloidal caps) made up the cavity which altered the field mode distribution around the electron and changed its cyclotron damping rate. A last set of experiments [17-19], the simplest in their geometry, involved atoms or molecules between two closely spaced metallic mirrors, the structure I have briefly discussed above.

I have chosen to describe briefly one experiment of this last kind, performed at Yale [18]. Its principle, sketched in fig. 2, is very simple. A beam of cesium atoms is sent through a gap made of two gold-coated mirrors at a distance of about 1 μm from each other. The atoms travel about 8 mm down the gap before emerging on the other side, which corresponds to an extremely high beam collimation, whose realization constitutes the main difficulty of the experiment. Before entering the gap, the atoms are excited by laser absorption followed by a radiative cascade into the $5D$ level (called e in fig. 2). This level decays spontaneously, with a branching ratio equal to one, on a 3.5 μm wavelength transition towards the final $6P$ state (called f in fig. 2). The atomic dipole is prepared parallel to the mirrors by a proper choice of laser polarization and frequency (so as to excite the hyperfine components of the $5D$ level which can radiate essentially on transitions polarized parallel to the metal). According to the above discussion, the $5D$ state radiation is inhibited in the gap and the atoms emerge in the excited state at the other end, after a delay of the order of thirteen spontaneous-emission lifetimes. The excited atoms are detected by another laser which selectively excites level $5D$ to a Rydberg state, which is subsequently ionized and counted. This experiment thus achieves—at least for excited atoms—the Holy Grail of immortality!

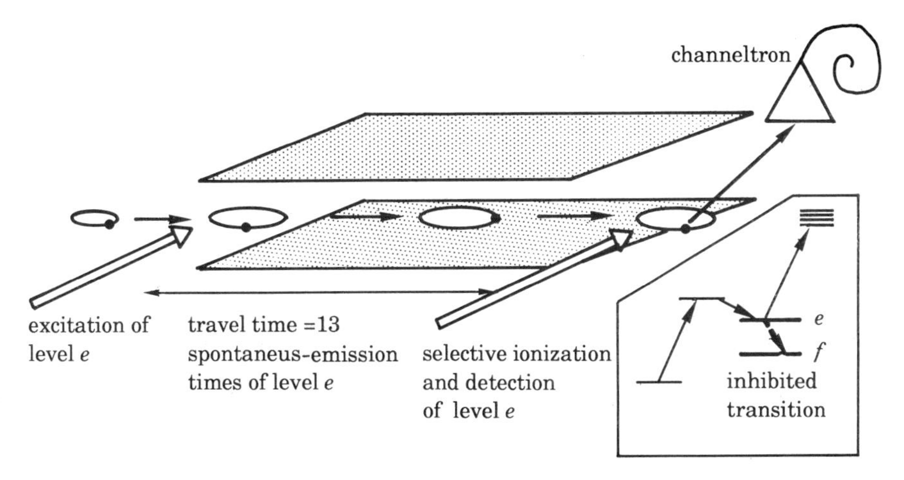

Fig. 2. – Sketch of experimental setup demonstrating suppression of spontaneous emission at optical frequencies (adapted from ref. [18]).

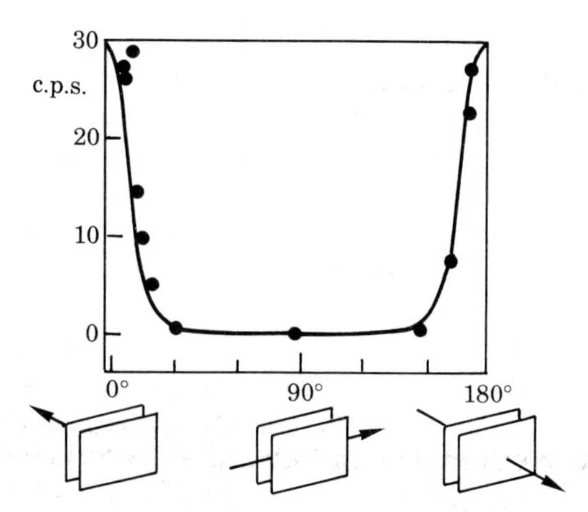

Fig. 3. – Transmission of atomic excited state through gap between mirrors as a function of magnetic-field orientation with respect to the normal to the mirrors. The magnetic field rotates the atomic dipole in the gap and spontaneous emission is restored as soon as the dipole possesses a component normal to the metal surfaces. Points are experimental and curve is theoretical (from ref. [18]).

This kind of experiment also demonstrates the possibility of tailoring at will spontaneous emission by playing on atom and cavity parameters. Such a possibility is illustrated in fig. 3. We have applied a magnetic field to the atoms crossing the gap. If the field is normal to the mirrors, the Larmor precession of the dipoles leaves them in a plane parallel to the mirrors and the spontaneous emission remains inhibited for all magnetic-field strengths. If, on the other hand, the field is oriented along the surface, the Larmor precession brings the dipole, after a quarter of a turn, in the direction normal to the mirrors, a configuration in which it can radiate, even faster than in free space. The excited atom is then lost between the mirrors and does not reach the detector. We see in fig. 3 the excited-state gap transmission *vs.* the angle between the magnetic field and the normal to the mirrors. The points are experimental and the curve theoretical. This experiment illustrates dramatically the fact that an atom spontaneous-emission rate depends on the atom environment. Here we have shown that the vacuum fluctuations between the mirrors are anisotropic, very large for electric fields normal to the surface and vanishingly small for electric fieds along the mirrors.

3. – Energy shifts in cavity QED: van der Waals forces and other effects.

We have considered so far dissipative effects in cavity QED. At the same time that emission rates are changed, the atomic-emission frequencies must also be altered. In other words, the real as well as the imaginary part of the atomic energy is changed by the presence of the boundaries around the atom. We have also studied at Yale this effect [20] in the mirror configuration of the previous experiment. Figure 1*a*) shows an atomic dipole between two mirrors with its successive images. The frequency shift appears, for small mirror separation at least, as resulting from the near-field «instantaneous» perturbation of the images on the dipole. The Lennard-Jones theory gives the following shift at mid gap for an S state [1, 3, 20]:

$$(\Delta E)_{ns} = -\frac{4e^2}{3L^3}\left[1+\frac{1}{3^3}+\frac{1}{5^3}+\ldots\right]\langle nS\,|\,r_e^2\,|\,nS\rangle . \tag{1}$$

L is the gap width, r_e the electron separation from the nucleus and $e^2 = q^2/4\pi\varepsilon_0$, where q is the electron charge. The series in brackets, which sums up to about 1.05, accounts for the cumulative effect of successive images. The shift is in fact minimal at mid gap where the atom is at the largest possible distance from either mirror and diverges when the atom approaches the mirrors. The experiment to measure these shifts is sketched in fig. 4. The atomic beam—here sodium—crosses as in the previous experiment the 8 mm long mirror gap, but the atoms are now in their ground state ($3S$). Just before exiting the gap, the atoms are excited to a Rydberg nS state by a stepwise process (two lasers excite successively the $3S \rightarrow 3P$ and $3P \rightarrow nS$ resonances). The frequency of the second laser is scanned through the Rydberg resonance and an excitation spec-

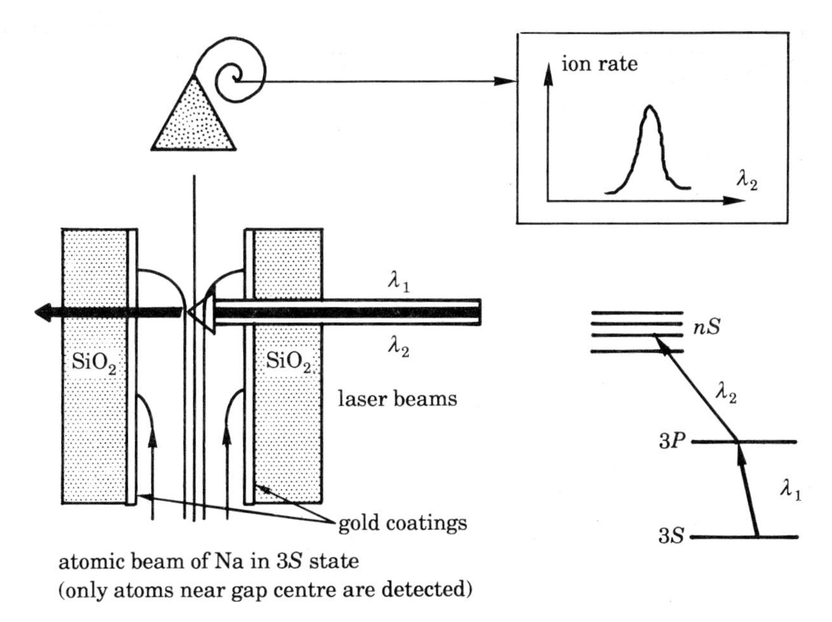

Fig. 4. – Sketch of experimental setup for measuring the van der Waals interaction between atom and metal surfaces in a mirror gap. The relevant atomic energy levels of Na are shown in the inset (adapted from ref. [20]).

trum is recorded by field-ionizing these excited atoms and collecting the ions in a channeltron. The position of the spectral line reflects the shifted energy of the nS state and thus allows one to measure the van der Waals interaction. In fact, it is remarkable that one measures in this way the interaction precisely at mid gap. If an atom strays away from the mid-gap position, it is strongly attracted to the closest mirror by the van der Waals force and does not reach the detector. Only those atoms which cross the gap very near the potential maximum escape attraction to the mirrors and are detected. The results of this experiment are shown in fig. 5. The left part (fig. 5*a*)) shows the excitation spectra of the 13S state for decreasing L values (from top to bottom). The unshifted line at the right of each recording is an unperturbed reference obtained in free space in an auxiliary atomic beam. The van der Waals shift is clearly observed. The right part (fig. 5*b*)) shows the shifts *vs.* mirror spacings for four nS states ($n = 10$ to 13). The circles are experimental and the lines correspond to the Lennard-Jones theory without any adjustment. Here for the first time the various important features of the van der Waals interaction between a free atom and a metal are directly tested. In the perspective of cavity QED, the lifetime and frequency shift experiments illustrate the important fact that an atom in a cavity is a «dressed» system whose real and imaginary parts of the energy are both altered by the proximity of the metallic surfaces. By controlling the experimental parameters, it is possible to manipulate these complex energies and to perform interesting studies in this way.

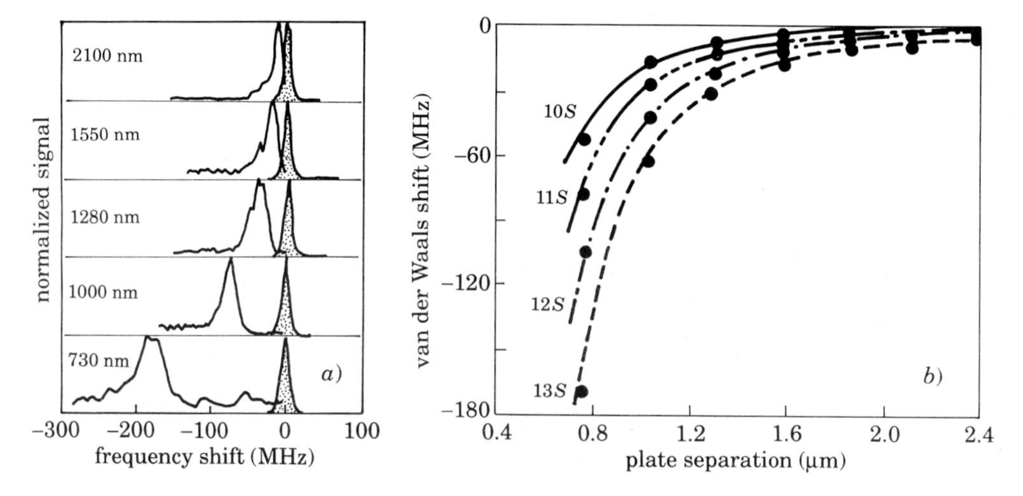

Fig. 5. – Measurement of van der Waals interaction between Na atom and metal surfaces in mirror gap. *a*) Excitation spectra of 13S state for various L values. The shifted line corresponds to the van der Waals interaction at mid gap. The unshifted line is a reference obtained outside the gap in an auxiliary beam. *b*) Shift *vs.* plate separation for nS states with n ranging from 10 to 13. Points are experimental and lines correspond to Lennard-Jones theory (adapted from ref. [20]).

The van der Waals perturbation is only one of the possible frequency shift effects encountered in cavity QED. Van der Waals shifts are due to «near-field» instantaneous image perturbations varying as $1/z^3$, where z is the atom to metal surface distance. In a microcavity, mostly the first images contribute to the shift. What happens if the cavity size increases and becomes larger than a fraction of the characteristic atomic emission/absorption wavelengths? One way of understanding the physics is to say that the interaction between the atom and its image becomes «retarded», the reflected field experiencing a dephasing during its round trip from the atom to the mirrors. This retardation effect leads, for ground atomic states, to the replacement of the van der Waals interaction by the famous «Casimir-Polder» interaction [21] varying as $1/z^4$ instead of $1/z^3$. I will not discuss it any further in this lecture. For the excited atomic states which I am mostly interested in here, retardation means that the reflected field is no longer the instantaneous electrostatic field varying as $1/z^3$, but the «far-field» radiation, decreasing as $1/z$ and oscillating as $\cos(2\pi z/\lambda)$ [22]. (In fact, the far-field shift is a sum of $\cos(2\pi z/\lambda_i)$ functions where the λ_i's are the wavelengths of the transitions connecting the level of interest to more bound states; we assume here for the sake of simplicity that one transition has a dominant strength and we call λ its wavelength.) A plot of the frequency shift *vs.* separation (fig. 6) thus presents two quite different parts. For small separation, it varies as $1/z^3$ (dashed-line part) and, for large z's (solid line), it becomes an oscillating function with an amplitude varying as $1/z$. Although the shift becomes very small at a large distance from a single mirror, it can be strongly enhanced by resonance effects, which is not the case of the van der Waals «instantaneous term». If the mirror distance is set so that one cavity mode is resonant with the atomic transition, the «far-field» contributions of all the images add up constructively, giving rise to a $1/m$ series (where m are integers), which diverges logarithmically if the cavity finesse (or Q-factor) increases to infinity [1, 22]. The atom is then so strongly affected by its coupling to the cavity that a pertur-

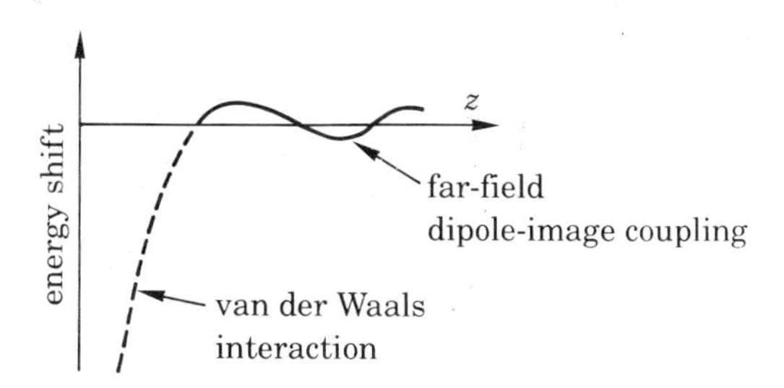

Fig. 6. – Typical variation of energy shift experienced by an atomic excited state, as a function of the separation z from a plane mirror. At short range (dashed line), the van der Waals interaction varies as $1/z^3$. At large distances, the interaction varies as $\cos(2\pi z/\lambda)/z$, where λ is the atomic wavelength associated to the main transition linking the excited state to more bound levels (solid line). The transition between the two regimes occurs for $z = \lambda/2\pi$.

bative approach such as the one we have followed so far is no longer applicable.

4. – Resonant coupling of a two-level atom and a cavity: normal-mode splitting effect.

It is then more appropriate to abandon the charge image fluctuation picture and to adopt the atom-vacuum fluctuation one. The atom and field subsystems we are considering are represented in fig. 7*a*), with their energy levels indicated in fig. 7*b*). We are now considering a very-high-Q cavity which subtends one mode of the field, resonant or nearly resonant with the atomic transition between two atomic levels e and g. As is well known, a mode of the field is a harmonic oscillator, whose degree of excitation are photon numbers. We are thus dealing with a two-level system resonantly or nearly resonantly coupled to a harmonic oscillator, the prototype model of quantum optics known as the Jaynes-Cummings one [23]. If there is initially no photon in the field, the oscillator is in its ground state and the two-level atomic system is merely coupled to the single-mode vacuum field. Such a model is quite reminiscent of the simple problem of two degenerate coupled oscillators or pendulums, represented in fig. 7*c*). When sweeping the frequency of one oscillator (say the cavity mode) around the value of the other supposed to be kept fixed (say the atom), one obtains for the coupled system the typical anticrossing frequency diagram shown in fig. 7*d*). Due to the coupling, the combined system eigenfrequencies avoid each other and are split, at resonance, by an amount 2Ω. The frequency 2Ω also represents the rate at which the two subsystems exchange their excitation energy, supposed to be initially stored in one of them. If ω_0 is the common atom and field mode frequency, the frequencies of the combined atom-cavity system (the ones which would be observed in a spectroscopy experiment on this system) have become $\omega_0 \pm \Omega$. Of course, this corresponds to physically important effects only if Ω is large enough. This normal-mode splitting effect has been predicted and analysed theoretically in several articles [24, 25].

A discussion about orders of magnitude is at this stage useful. The atom-cavity coupling parameter appears in fact as the product of the atomic dipole D—which measures, so to speak, the size of the atomic electron fluctuation—by the r.m.s. vacuum field in the cavity E_{vac}—which measures the field fluctuations [1, 2, 26]. The dipole size is proportional to the square of the atom principal quantum number—which explains the interest in doing cavity QED experiments with large Rydberg atoms. E_{vac}, on the other hand, is proportional to $(\omega/V)^{1/2}$, where V is the cavity volume. The highest the frequency and the strongest the confinement of the field, the larger the vacuum fluctuations will be. This leads to two possible strategies. Cavity QED experiments can be performed with Rydberg atoms resonant with superconducting millimetre wave

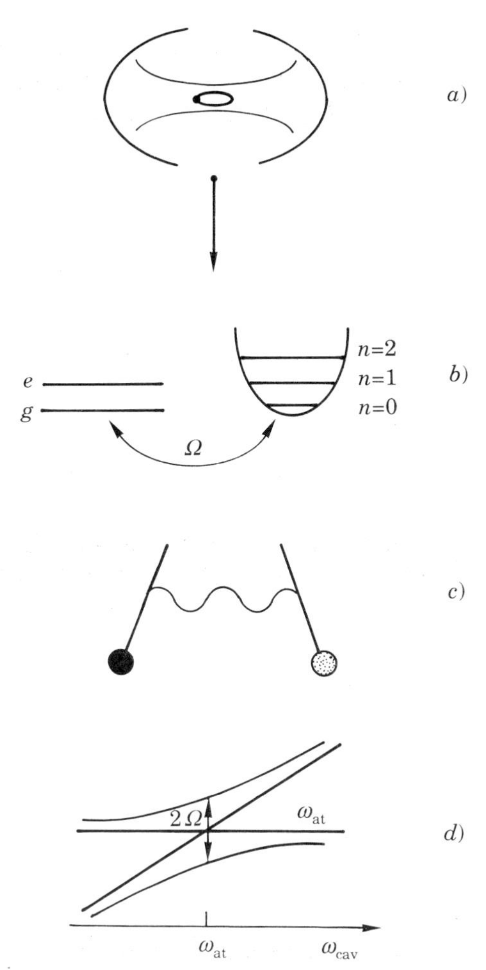

Fig. 7. – An atom in a resonant or nearly resonant high-Q cavity: *a*) Sketch of the system. *b*) Energy levels of the atom and field parts: the atom is a two-level spinlike system (e, g) and the field a harmonic oscillator. The field ground state $n = 0$ is the vacuum. *c*) Mechanical analogy: the atom and the field are two resonant or nearly resonant pendulums coupled together. *d*) Frequency diagram of coupled systems as a function of detuning, exhibiting the «normal-mode splitting effect». The strength of the «anticrossing» at resonance measures the rate Ω at which the two subsystems can exchange energy.

cavities (the Rydberg frequencies fall indeed in the millimetre wave part of the spectrum). The atom is then about 1000 Å big and the cavity of centimetre size. D is then very large (1000 qa_0, where q is the electron charge and a_0 the Bohr radius, for a Rydberg atom with principal quantum number n around 30), but E_{vac} is rather small (10^{-2} V/m typically). Or, one can study an optical transition corresponding to a small dipole (of the order qa_0) in a very small Fabry-Perot optical resonator for which E_{vac} is relatively big (several V/m). It turns out that

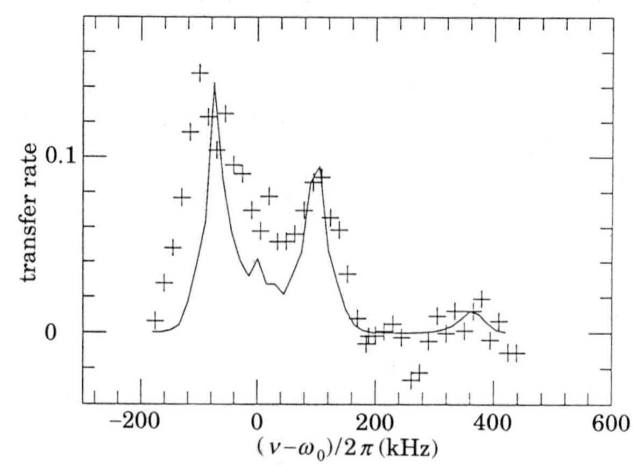

Fig. 8. – Normal-mode splitting effect observed with three atoms in a resonant microwave cavity. The relative transfer rate $39S \rightarrow 39P_{3/2}$ is plotted *vs.* $\nu - \omega_0$ (where ν is the probe frequency and ω_0 is the common atom and cavity frequency). The points are experimental and the curve is the result of a numerical simulation (the asymmetry of the doublet and the existence of a third component are slight complications due to the existence of a third hyperfine Rydberg level interacting with the cavity field). (From ref. [27].)

in both cases Ω values in the range 10^5 to $10^6\ \text{s}^{-1}$ can be achieved, which means that a single atom and the cavity couple to each other within a characteristic time of about $1\ \mu\text{s}$. This coupling leads to observable spectacular effects if it exceeds the field and cavity damping rates.

Two direct and very recent illustrations of the mode splitting effect in cavity QED are shown in fig. 8 and 9. Figure 8 corresponds to a Rydberg atom-superconducting cavity experiment performed in Paris in 1991 [27]. The excitation spectrum of Rydberg atoms between the $39S$ and $39P$ levels of rubidium is recorded *vs.* microwave frequency, for atoms placed inside a microwave cavity exactly resonant with the atomic transition. The probing microwave is fed into the cavity through a connecting waveguide. We clearly see that the atomic absorption line is split into two main components about 100 kHz apart. The number of atoms present at the same time in the cavity is 3 in this experiment. Even more spectacular is the recording of fig. 9, which represents the transmission spectrum of a Fabry-Perot optical resonator containing on average only one sodium atom inside (experiment by the group of Kimble at Caltech) [28]. Here again, the atom-cavity mode splitting is clearly resolved. It is interesting to interpret these experiments in two complementary ways. One can say that, in the Caltech experiment, the cavity transmission frequency is pulled by the index-of-refraction effect of a single atom in the resonator. Alternatively, one can also say that the atomic-line frequencies of the Paris experiment are split apart by the dynamical Stark effect produced by the cavity vacuum! These experiments, which would have been considered as unachievable goals only a few years ago,

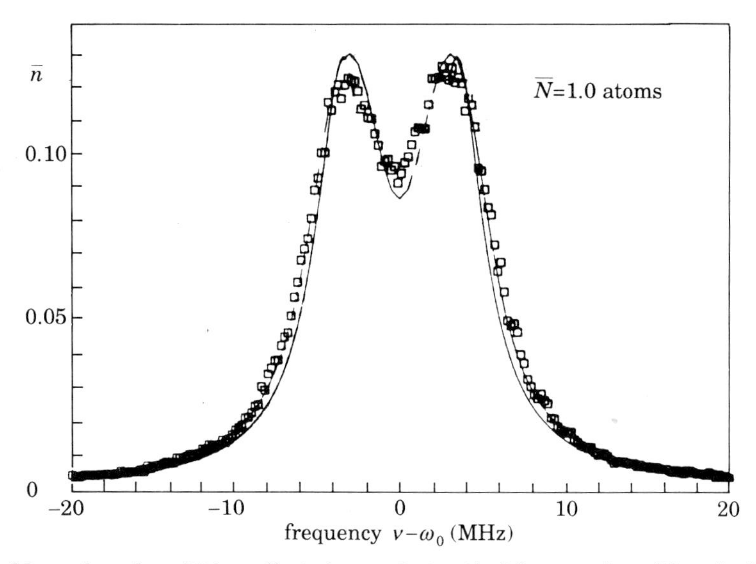

Fig. 9. – Normal-mode splitting effect observed at optical frequencies with a single sodium atom coupled to a small, very-high-Q Fabry-Perot resonator. The signal is the transmission of a laser beam crossing the cavity mirrors plotted as a function of $\nu - \omega_0$ ($\nu =$ laser frequency, $\omega_0 =$ common cavity mode and atom transition frequency). The transmission is expressed in terms of the average number n of intracavity photons. Curve is theoretical and points experimental (from ref. [28]).

tell us a lot about the recent progresses of lasers, atomic-beam and cavity technology. The Caltech experiment has required the use of super-high-quality mirrors, with finesses in the range 10^5 to 10^6. Similar experiments, with larger atomic samples (few tens to few hundreds atoms at same time in the cavity), have also been performed [29, 30]. The atoms in these samples behave as a single collective system [26] and the normal-mode splitting in this case is proportional to $\sqrt{N}$, where N is the atom number (note that, in the microwave experiment [27] where $N = 3$, there is also a small splitting enhancement due to collective effects).

5. – «Rabi oscillation» and the micromaser.

I have considered so far the simple situation where the cavity is initially in its vacuum state. What happens if the cavity field is already excited and contains n photons prior to the introduction of an atom in the mode? The resonant atom-field coupling is then increased by $\sqrt{n+1}$, which expresses the fact that the rate of atom-field energy exchange is proportional to the electric-field amplitude. As a result, the atom-cavity anticrossings become wider by the same factor [1, 2, 26]. At resonance, the normal-mode splitting is $2\Omega\sqrt{n+1}$. As a consequence, an atom introduced at resonance in a cavity containing n photons exchanges its energy with the field at a rate $2\Omega\sqrt{n+1}$. The probability of finding the atom in level g a time t after initial excitation in level e is [1,2,26]

$$P_{e\to g}(t) = \sin^2(\Omega\sqrt{n+1}\,t). \tag{2}$$

This is called the «Rabi oscillation» of the atomic population. Such oscillations have been observed in many cavity QED experiments, involving fields with very small photon numbers down to a few units [31].

The Rabi oscillation is the basic phenomenon at work in the micromaser, a very interesting device in cavity QED physics first operated in the Munich group of Walther and colleagues [32], which is sketched in fig. 10. A beam of atoms is sent through a high-Q superconducting cavity. The atoms are prepared by laser excitation in the upper level of a Rydberg transition resonant with the cavity mode. The atoms interact one by one with the cavity field, which builds up as the result of successive atom-field interactions. After the cavity, the atoms are field-ionized and detected in two different zones, which separately count them in levels e and g. In fact, the field properties are deduced from the atom-counting statistics and the atoms play a double role in the device. They are the source of the field as in any laser/maser system, and they also are the detectors of the field. All the basic ingredients of a laser/maser system are gathered, but this is a very special laser/maser in which the cavity is most of the time empty. Three characteristic times are important in this device. The average time between atoms, t_{at}, is controlled by merely adjusting the Rydberg atom laser excitation rate. The atom-cavity crossing time t_{int} can also be controlled by selecting the atom velocity with a rotating-wheel selector. The cavity damping time t_{cav} is the inverse of the rate at which the photons produced in the cavity are lost. It can also be controlled, to a point, by varying the superconduc-

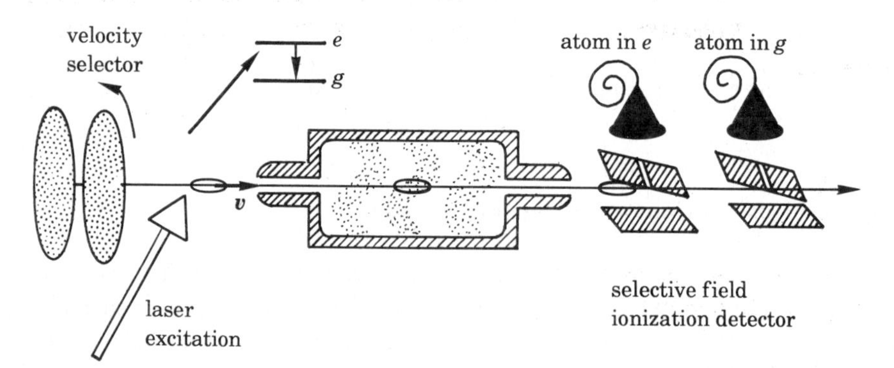

Fig. 10. – Sketch of a Rydberg atom micromaser. Atoms are prepared in the upper Rydberg state e before crossing one at a time the high-Q superconducting cavity. Velocity selection can be achieved with a set of spinning wheels with slots intercepting the atom beam. Selective state atomic detection is performed by field ionization after the atoms have left the cavity.

tor temperature. True micromaser operation requires the following conditions:

(3a) $$t_{\text{at}} > t_{\text{int}} ,$$

(3b) $$t_{\text{cav}} > t_{\text{at}} .$$

Condition (3a) means that there is at most one atom at a time in the cavity and (3b) implies that photons emitted by successive atoms accumulate in the mode, giving rise to a measurable maser action. An order of magnitude of the number of photons in the cavity is obviously, in steady-state operation, equal to $t_{\text{cav}}/t_{\text{at}}$. The probability for each atom crossing the cavity to release a photon depends on the actual photon number distribution. A reasonable order of magnitude for this probability is obtained by computing the probability $P_{e \to g}(t)$ of eq. (2) with n being replaced by $t_{\text{cav}}/t_{\text{at}}$. This probability is appreciable only if the following condition is satisfied:

(3c) $$\theta_{\text{int}} = \Omega t_{\text{int}} \sqrt{\frac{t_{\text{cav}}}{t_{\text{at}}}} \gtrsim 1 .$$

θ_{int} is (in order of magnitude) the typical Rabi precession angle undergone by the atom-cavity field system when an atom crosses the cavity. The compatibility between conditions (3a), (3b) and (3c) clearly comes from the fact that t_{cav} is very long in superconducting cavities (up to a fraction of a second) whereas the single atom-cavity coupling time $1/\Omega$ can be as short as 10^{-5} s. A more quantitative theory of the micromaser can be found in ref. [33, 34].

What makes the micromaser special, beyond the spectacular fact that the cavity is mostly void of atoms and contains only from a few ten to a few hundred photons under normal operation, is that one can precisely control in this device the various sources of noise and randomness of a quantum oscillator. In ordinary maser/laser systems, the atomic excitation can be dissipated in side modes. This contributes to destroy at random times the coherent coupling between each atom and the lasing mode. No such effect exists in the micromaser since the small cavity is closed and subtends only the masing mode. Second, there is also a complete control over the atom-field interaction time via velocity selection, so that all atoms undergo exactly the same Rabi oscillation in the cavity, in sharp contrast with ordinary laser/masers in which there is usually a wide distribution of interaction times. In fact, the micromaser operation can be described by a very simple semi-classical model, as shown in fig. 11. The solid curve represents the probability that one atom leaves a photon in the cavity, an oscillating function of $\sqrt{n+1}$, which represents the «gain per atom» of the device. The average cavity loss during time t_{at}, on the other hand, is a linear function of the field intensity represented by the dashed line. The stable operating points of the maser are the intersections of these two curves marked by open circles (gain = loss and loss slope larger than gain slope). We see that the micro-

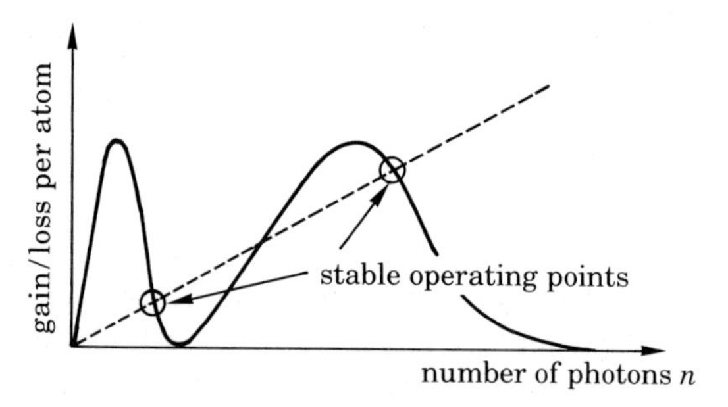

Fig. 11. – Semi-classical model of micromaser operation. The photon gain per atom is represented by a solid line, as a function of the number of photons in the cavity. The cavity losses are represented by a dashed line. The intersections between these two functions corresponding to stable operating points are shown by open circles. Multistable regime is clearly exhibited.

maser is usually a multistable device which can present a hysteretic behaviour (the actual operating regime depends upon the system previous history). We also remark that, in a wide range of operating conditions, the maser oscillates at points for which the differential gain is negative (gain per atom decreases when the field intensity increases). This is a consequence of the fact that the atoms which cross the cavity undergo a well-defined Rabi oscillation.

A very interesting consequence of these properties is that the micromaser field is usually sub-Poissonian [33, 34]. In usual laser systems, the photon number fluctuation in the cavity reflects in fact the shot noise of the atomic excitation. In the micromaser, the atomic excitation has also a Poissonian noise, due to the intrinsic randomness of the atomic-beam flux. But these pump fluctuations are now squeezed by the mere fact that the maser «likes» to operate with a negative differential gain. Assume, for example, that a positive fluctuation increases the pumping rate at a given time. The number of photons in the cavity will at first have a tendency to increase too, since more atoms emit photons in it. However, each atom will have a smaller probability of emitting a photon if the differential gain is negative. As a result, the relative photon number fluctuation will be smaller than the pump one and the field will have a sub-Poisson character. In fact, the field is not directly accessible in the experiment. What is measured is the rate of atoms exiting the cavity in level e or g. One can show [35] that the sub-Poisson character of n results in a sub-Poisson statistics for N_g, the number of atoms detected during a given time interval in the lower level g. So, indirectly the photon statistics in the cavity can be inferred from this measurement.

Figures 12a) and b) present more quantitative results. The left-hand curve shows the photon variance in the cavity as a function of the dimentionless parameter θ_{int}, defined in eq. (3c), and measuring basically the number of Rabi cycles undergone by the atom in the cavity. This is a theoretical result [33] which

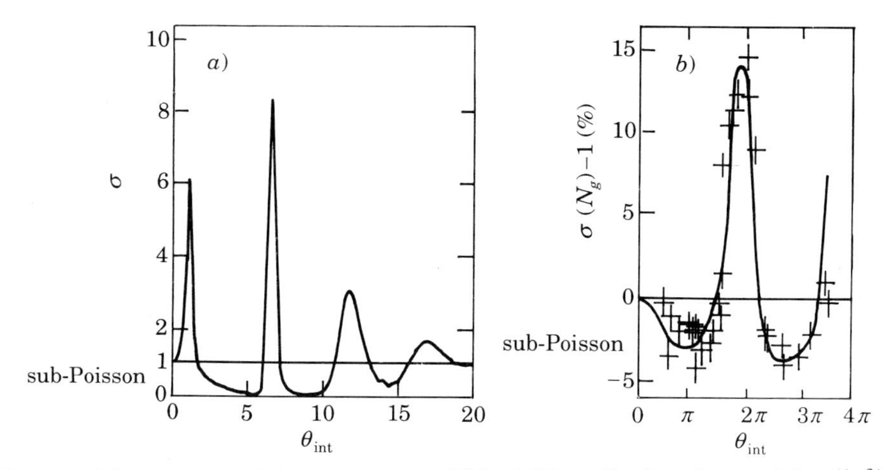

Fig. 12. – The variance of the micromaser field. *a*) Normalized variance $\sigma(n) = (\langle n^2 \rangle - \langle n \rangle^2)/\langle n \rangle$ of photon number in the cavity as a function of the dimensionless maser parameter $\theta_{\text{int}} = \Omega t_{\text{int}} \sqrt{t_{\text{cav}}/t_{\text{at}}}$. Regimes of sub-Poisson operation ($\sigma < 1$) are predicted (from ref. [33]). *b*) Normalized variance $\sigma(N_g) - 1$ of atoms detected in the lower maser level g as a function of the maser parameter θ_{int} around the 2π value. Points are experimental and curve is theoretical (adapted from ref. [35]).

confirms that, in a wide range of operating parameters, the field photon number variance is sub-Poissonian. Then the right-hand figure shows the measured atom number variance in level g as a function of the same parameter. In the solid line, the theoretical curve shows that the atomic variance reflects the field one. The experimental crosses show a remarkable agreement between experiment and theory. These results have been obtained recently by the Munich group [36].

Many other experiments have been performed on the micromaser, which illustrate the new possibilities offered by the cavity QED concepts. One of these experiments, made in Paris a few years ago, has exploited the possibility of manipulating spontaneous emission in a confined space to achieve for the first time maser action on a two-photon transition [37]. The search for a two-photon maser or laser has been going on, in fact, since the early sixties, when SOROKIN and PROKHOROV independently described such a device and noted that it had interesting properties [38, 39]. The typical level scheme for a two-photon maser/laser is shown in fig. 13*a*). The amplification must occur between two levels e and f of the same parity, with a third level i of opposite parity somewhere in between, the two-photon transition amplitude being inversely proportional to the energy mismatch Δ between the field frequency and the one-photon $e \rightarrow i$ transition. In order to achieve two-photon oscillation, one tunes the cavity at exactly half the energy of the e-to-f transition and one tries to take advantage of the cavity Q-factor at this frequency. Unfortunately, the one-photon $e \rightarrow i$ transition is so much more probable that, in a conventional cavity, it usually depletes the upper level e before the two-photon gain can take over. In a closed-cavity

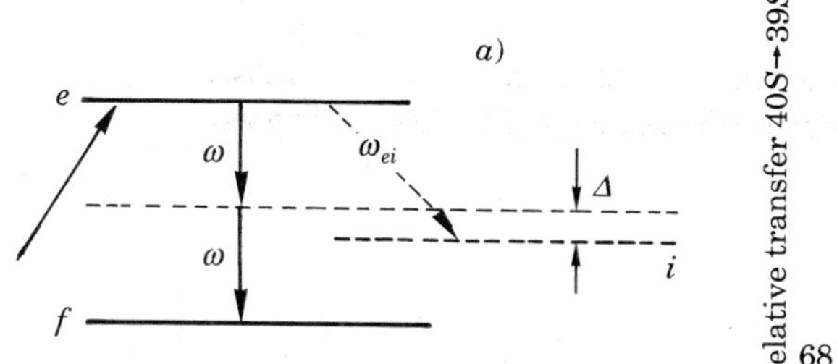

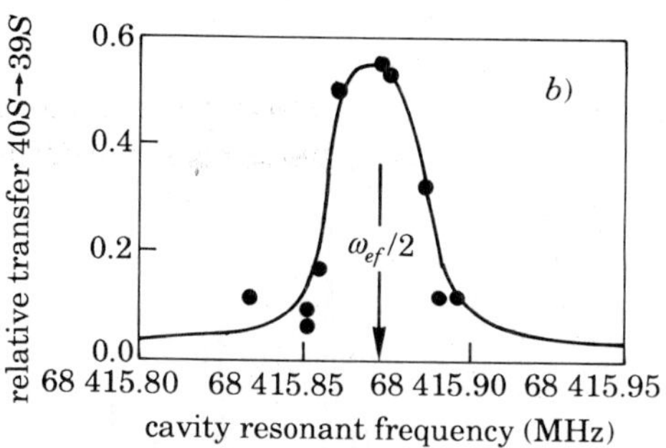

Fig. 13. – The two-photon Rydberg atom micromaser. *a*) Scheme of relevant energy levels. In the experiment, e is the $40S$ state in rubidium-85, f is the $39S$ state and i the $39P_{3/2}$ state ($\Delta/2\pi$ is only 39 MHz in this favourable case). *b*) Relative population transfer $40S \to \to 39S$ as a function of cavity tuning, revealing the two-photon resonance at frequency $\omega_{ef}/2$ (adapted from ref. [37]).

micromaser, though, this cannot happen. If the high-Q cavity is tuned at exactly $\omega_{ef}/2$, the two-photon amplitude is strongly enhanced and, at the same time, the competing one-photon emission is totally inhibited since there is no mode in the cavity at the corresponding frequency. We have operated in this way the first true two-photon oscillator in 1987. The transfer from $40S$ to $39S$ level in Rb is shown to be resonant when the cavity, tuned by mechanical deformation of its walls, has a frequency exactly equal to $\omega_{ef}/2$ (fig. 13*b*)). There is a lot of interesting features in this device [40] which emits photons in pairs, but their discussion is beyond the scope of this lecture. Let me just add that a c.w. two-photon laser, operating under somewhat different conditions, has been recently realized at Oregon state university by Mossberg group [41].

6. – Dispersive effects and manipulation of photons in cavity QED [42-45].

Let us now turn to nonresonant atom-cavity coupling effects. Consider a two-level atom e-g coupled to a slightly off-resonant cavity mode (with a frequency mismatch δ between the field and the atomic frequency). The energy shifts of the levels e and g, when the field contains n photons, are [45]

$$\Delta E_e(n) = -\frac{\hbar\Omega^2}{\delta}(n+1), \tag{4a}$$

$$\Delta E_g(n) = \frac{\hbar\Omega^2}{\delta}\,n\,. \tag{4b}$$

These formulae, resulting from simple second-order perturbation theory, express that the off-resonant shift is merely proportional to the square of the coupling divided by the frequency mismatch δ. Equations (4) are valid if δ is larger than $\Omega\sqrt{n+1}$. The shift then exhibits a dispersive character (changes sign with detuning). The «1 term» contribution in the excited-state shift expression ($4a$) represents the vacuum effect, a kind of Lamb shift induced by the cavity modified vacuum, whereas the «n term» is a «light shift» term, proportional to the field intensity. When an atom crosses slowly enough a cavity containing a nonresonant field (this field being produced by an external classical microwave generator or merely by a black-body source), it will not undergo any e-to-g transition, but simply have its energies shifted, in a transient stage, by an amount proportional to n or $n+1$ (with a different sign for the upper and lower states). Note that the atom-cavity interaction time should be long enough so that the adiabatic approximation could be invoked [1] (the atom-cavity levels must continuously branch into the perturbed states as the atom gets into the cavity and evolve back into the unperturbed atom-field states as the atom leaves the cavity). If one could measure the transient energy shift experienced by the atom, one would realize a measurement of n. This measurement would be of a very interesting kind indeed. Whereas usually measuring a photon means destroying it, here I am talking about a nondestructive way of measuring. The photon is still there after the measurement since the atom cannot have absorbed it. This is what is called a quantum nondemolition measurement. An experiment to demonstrate this effect is in progress in Paris. I present simply here the general principle of this method, which is, I believe, a nice illustration of the quantum theory of measurement.

Assume that a field containing an unknown number of photon is stored in a high-Q cavity and send through this cavity an atom whose de Broglie wave packet is represented in fig. 14a) before the atom enters the cavity. Clearly, the nonresonant energy shift of the atom has the effect of «retarding» or «advancing» the matter wave propagation (depending on the state in which the atom is prepared). After cavity exit, the de Broglie wave is dephased with respect to the position it would have if the cavity had been removed (fig. 14b)). In other words, the cavity field acts as an index plate for the matter wave which has a «thickness» proportional to the photon number in the field (or the photon number plus one). Measuring this dephasing is our goal. As we will see, this mea-

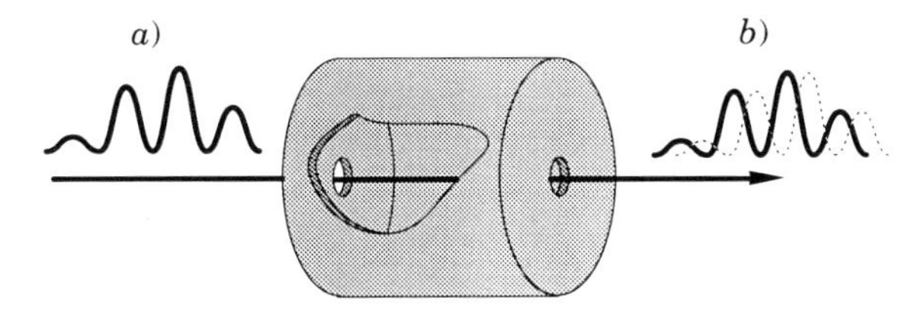

Fig. 14. – A high-Q cavity with a slightly nonresonant field is equivalent to a «retarding» (or «advancing») plate for the atom de Broglie wave; a) an atom wave packet impinges on the cavity, b) after crossing the cavity, the wave packet has been dephased with respect to its state if the cavity had been removed (the «crests» of the atom wave are slightly displaced, by an amount proportional to $n+1$ (if the atom is in the upper level of the atomic transition close to resonance with the cavity mode).

surement will require to pass many atoms through the cavity and, during this time, we have to assume that the field relaxation remains negligible.

One conceptually simple way of measuring the matter wave dephasing is shown in fig. 15*a*). Of course, the obvious method to measure a dephasing is interferometry. Here we are talking about atomic interferometry, a very hot topic these days and so we are in fact dreaming of marrying atomic interferometry with cavity QED [42]. We perform a Young double-slit experiment with our matter wave and we design the setup so that the atoms are sensitive to the dephasing cavity field in only one of the two channels (for example, we place our cavity so that the field presents an antinode in front of one of the slits and a node in front of the other). We record the matter wave fringes on an array of detectors and measure the fringe position. Obviously the fringe pattern will be translated by an amount proportional to n (or $n + 1$) and thus, by recording the positions of the fringes, one can determine n. I will not analyse here the process in details, since such an analysis can be found in other references [42, 43]. I will just stress the fact that it becomes really interesting if the field does not have initially a well-defined photon number (this is the case, for example, of a classical coherent field). In this situation, the phase of the fringe pattern is not fixed at the beginning of the experiment, but reveals itself progressively, as the number of photons gets «pinned down» to a single value. We have simulated the whole process with a computer. We start with a Poissonian photon number distribution (fig. 15*b*), top frame). The first atom is detected at a random position on the screen (fig. 15*c*), second frame from top), but, once it has been detected, we have acquired information on n. We know for sure that some n values are excluded, namely those which would give a dark fringe at the position where the atom has been observed. As a result, the photon number distribution has been changed (fig. 15*b*), second frame). The process goes on as more and more atoms are collected, until only one photon number is left (fig. 15*b*), bottom frame). The field has then collapsed into a Fock state, which is a natural result when one performs a quantum measurement of the photon number. At the same time, the atomic fringe pattern has acquired a well-defined phase (fig. 15*c*), bottom frame), which precisely corresponds to the found n value. The whole process appears as an ideal model of a quantum measurement.

What would be the interest in performing such a QND measurement of n? One possibility would be to observe quantum jumps of the field. Of course, the photon number cannot remain absolutely fixed in a realistic experiment. One good reason is that the cavity Q, however large, is finite. After a while, n will thus change by discrete increments. Also the temperature of the cavity walls, however small, is not absolutely zero and so thermal photons may be created. When such a process occurs, the phase of the fringe pattern is suddenly translated by a finite amount, revealing a quantum jump of n. We have also simulated such processes with a computer. The field damping appears in this case as a staircase evolution, each step occurring at a random time.

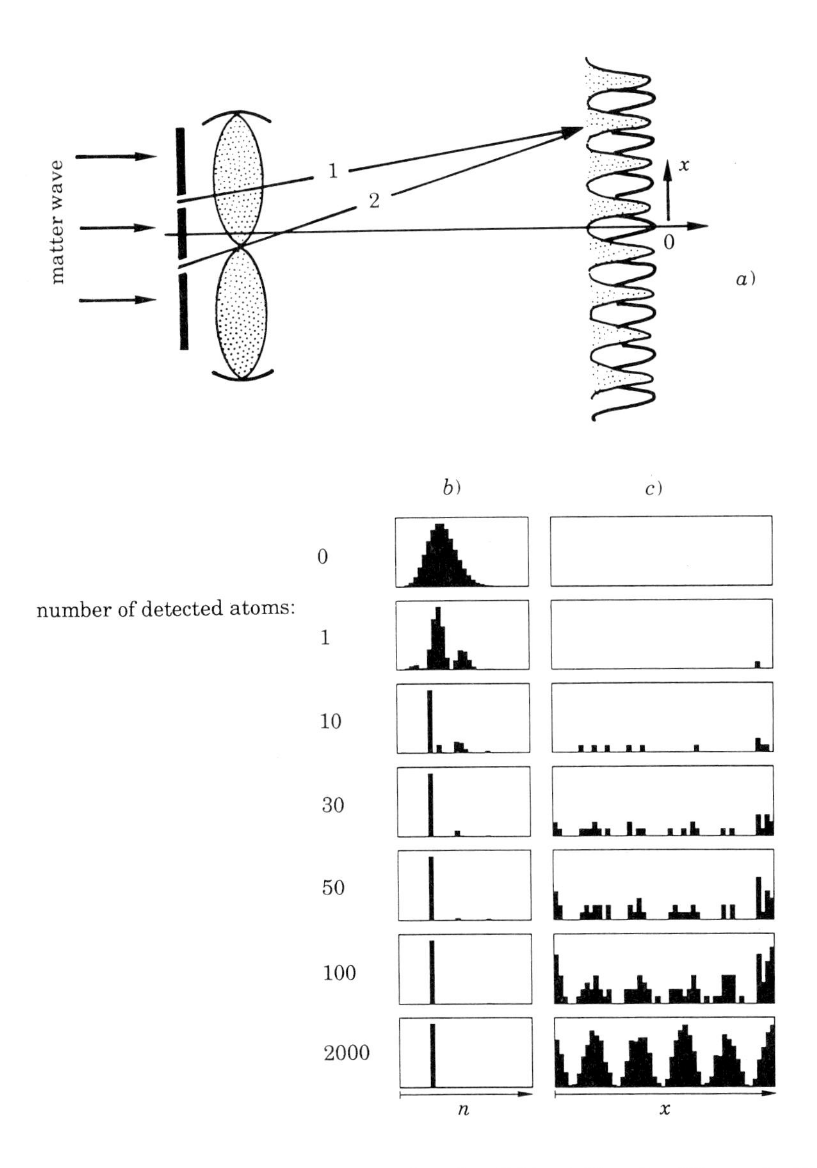

Fig. 15. – Marrying atomic interferometry and cavity QED. *a*) Sketch of Young double-slit interferometer which could be used to measure the atom de Broglie wave dephasing, and hence the photon number in the cavity. *b*) Evolution of the photon number distribution in a simulated measuring sequence, as a function of the number of detected atoms. The initial field in the cavity is a coherent state with an average number of 10 photons (the top frame shows the initial Poisson distribution of the field). *c*) Distribution of detected atoms on the array of detectors, for the same measuring sequence as the one shown in *b*). The fringes are «revealed» progressively as the photon number gets «pinned down» to a single value (from ref. [42]).

We are now trying to perform such quantum nondemolition measurements at ENS in Paris. The fringes we are looking for are not Young fringes but Ramsey fringes detected on Rydberg atoms [44, 45]. The mathematics is quite the same in both cases but the experiment, still difficult, is much simpler with the Ramsey than with the Young design.

Another fascinating prospect of these cavity QED atomic interferometric experiments is to make possible the preparation and study of so-called «Schrödinger cat states» of the field [46, 47]. We have so far focussed on the atom wave function dephasing. Obviously, the field itself also experiences a dephasing when a nonresonant atom crosses the cavity. This «back action» effect can be seen as the «index»-of-refraction effect of the atom on the field. Assume now that the initial field in the cavity is classical (Glauber state) with a relatively well-defined phase (this means, of course, that its photon number is not perfectly defined). We will describe it by a c-number α, whose amplitude and phase represent the average amplitude and phase of the field. Send one atom through the interferometer. If the atom follows path 1 (in fig. 15a)), the cavity field experiences a phase shift which we will call ϕ. If it follows path 2, on the other hand, there is no coupling with the field and hence no phase shift. As a result, the total wave function of the system when the atom reaches the detection plane is given by the following expression:

$$|\psi\rangle = \int \mathrm{d}x\,[C_1(x)\,|\,\alpha \exp[i\phi];\, x\rangle + C_2(x)\,|\,\alpha;\, x\rangle]\,. \tag{5}$$

It is a quantum-mechanical superposition of contributions, one for each possible x position of the atom and each of these terms is in turn a sum of two interfering terms, one associated to a dephased field ($\alpha \exp[i\phi]$) and one to an unperturbed field (α). The amplitudes $C_1(x)$ and $C_2(x)$ describe the matter waves diffracted by slits 1 and 2, respectively. Let us now detect the atom. It will turn out to be somewhere, say at point x_0. Immediately afterwards, quantum mechanics tells us that the field has collapsed into the following expression:

$$|\psi(\text{field})\rangle = C_1(x_0)\,|\,\alpha \exp[i\phi]\rangle + C_2(x_0)\,|\,\alpha\rangle\,. \tag{6}$$

It is a superposition of the initial field, with amplitude $C_2(x_0)$, and of the phase-shifted field, with amplitude $C_1(x_0)$. Such quantum superpositions of classical fields with different phases or amplitudes are known as «Schrödinger cats», from a clear analogy with the famous cat paradox developed by Schrödinger in the thirties [46]. Instead of asking the question «is the cat dead or alive?», we now have the less dramatic but still interesting question «is the phase of the field shifted or not?». The quantum coherences between the two parts of this wave function relax in fact with a speed proportional to the average photon number [45]. The quantum coherence is thus lost very fast if the field is large. Hence, such quantum states could be practically observable only with relatively small n values. The study of these nonclassical superpositions as they relax in the cavity would be a very interesting test of quantum mechanics at the frontier of the quantum and the classical worlds [48].

7. – Extension to optical effects in microcavities.

The effects discussed up to now are rather fundamental physics. They point, however, directly to interesting applications in the field of more applied quantum optics. Let us simply mention several directions. It is now possible to prepare by molecular-epitaxy methods quantum well structures embedded in a cavity which has a size of the order of one optical wavelength. The cavity mirrors are obtained by piling up dielectric layers of different indices, realizing in fact Bragg mirrors [49, 50]. It has been shown that the spontaneous-emission pattern of excitons is modified by the field confinement in these structures [51]. Vertical microlasers, in which the lasing action takes place normal to the substrate wafer, have also been recently realized [50]. In these lasers again, the field is confined in a small volume and the losses in transverse nonlasing modes are minimized, giving such lasers a very low current threshold.

A usual optical Fabry-Perot cavity with multidielectric mirrors or a microlaser semiconductor cavity is made, as we have seen, of periodic layers of dielectric material making up Bragg-type mirrors. These are one-dimensional periodic structures. It has been noted a couple of years ago that this kind of periodic medium can be generalized to three dimensions in order to perform a new kind of light guiding and confinement. This is the idea of photonic band gaps [52]. It has been shown that 3D periodic structures made by drilling holes in a high-index medium, or by piling together high-index cylinders or spheres, may present forbidden frequency gaps for the propagation of light in the same way as a 3D semiconducting medium presents forbidden bands for the propagation of electrons. Such media have been realized at a macroscopic scale by mechanical fabrication methods and photonic band gaps for microwaves have been demonstrated [53]. If one then realizes a defect in the structure, for example by cutting a small piece of the otherwise periodic material, this has the effect of forming a local electromagnetic mode in the vicinity of the defect [54]. Strong photon confinement can then be obtained and low-threshold emitting laser diodes may be built by applying these ideas. For microwaves, the sizes of the holes are in the millimetre domain and can be done mechanically. In the optical domain, we are talking of micrometre-sized channels which can be made by ion etching techniques. Such channels have recently been realized by Dr. A. SCHERER and co-workers at Bellcore. Application of these structures to the design of single-mode light-emitting diodes and microlasers is now being seriously considered.

Another direction of cavity QED consists in the study of surface modes of small dielectric spheres [55]. The idea here is to confine the electromagnetic field at the surface of dielectric materials, by studying Mie resonances, also called whispering-gallery modes which propagate around the surface when the circumference matches resonant conditions. These surface modes may have very high Q-factors and correspond to photon confinement in very small volumes. One can thus perform nonlinear optics and cavity QED type of experiments on the surface or in the evanescent wave around the sphere. The sphere can be either a liquid droplet [56] or a glass ball [55, 57]. Several studies of spontaneous-emission rate alterations in such structures have been recently performed [58, 59] and lasing action of such modes reported [60]. A promising direction is the achievement of extremely high Q-factors, in order to realize photon trap experiments with such systems. The setup designed to study these modes is quite simple in its principle. The sphere is coupled to a light beam via an evanescent wave obtained by total internal reflection in a high-index prism. The resonances of the sphere result in a decrease of the light reflected by the prism. Narrow resonances corresponding to Q-factors in the range $10^9 \div 10^{10}$ have been obtained recently in Paris [57], following earlier studies by BRAGINSKI *et al.* [55]. With such high Q's, associated to a high field confined in a small volume, nonlinear phenomena occur at low laser fluxes (bistability, optical hysteresis become observable very easily). We do not believe that we have reached the highest possible Q-factors yet and we find these whispering-gallery-mode resonances very promising for future cavity QED work. Related whispering-gallery-mode resonances, obtained on semiconductor microstructures of much smaller size, have been reported recently [61].

8. – Concluding remarks.

It is always a dangerous game to try to predict the directions in which a field of research is going. Let us try nevertheless to list in conclusion what we believe to be the most promising fields in cavity QED. One important area is, of course, the manipulation of spontaneous-emission rates and patterns for specific applications [62]. Another area is the realization of nonlinear-optics experiments at the photon level, including bistability, hysteretic behaviour, and more generally the study of atom field dynamics at the quantum level in a dissipative setting [63]. Another interesting area is the development of experiments aiming at storing photons in a cavity and manipulating these photons to generate nonclassical fields, Schrödinger cats and the like [42, 44]. Connected to these ideas is the possibility of coupling cavity QED concepts with particle traps and atomic interferometry. What happens in particular if the atom-cavity coupling, which depends upon the position of the atom in the cavity, becomes sensitive to quantum uncertainties in the atom position? Field and atom quantum fluctuations become then entangled in a quite interesting new way. At last let us list the new techniques more or less directly related to cavity QED ideas: photonic band gaps, microlasers, microspheres.

REFERENCES

[1] S. HAROCHE: in *Fundamental Systems in Quantum Optics* (Les Houches Session LIII), edited by J. DALIBARD, J. M. RAIMOND and J. ZINN-JUSTIN (Elsevier Science Publishers, Amsterdam, 1992), p. 767.

[2] S. HAROCHE and J. M. RAIMOND: in *Advances in Atomic and Molecular Physics*, Vol. **20**, edited by D. BATES and B. BEDERSON (Academic Press, New York, N.Y., 1985), p. 347.

[3] E. A. HINDS: in *Advances in Atomic and Molecular Physics*, Vol. **28**, edited by D. BATES and B. BEDERSON (Academic Press, New York, N.Y., 1990), p. 237.

[4] S. HAROCHE and D. KLEPPNER: *Phys. Today*, **42**, 24 (1989).

[5] S. HAROCHE: *Phys. World*, **4**, 33 (1991).

[6] H. WALTHER: *Phys. Scr.*, **T23**, 165 (1988).

[7] P. MEYSTRE: in *Progress in Optics*, Vol. **30**, edited by E. WOLF (Elsevier Science Publishers B.V., Amsterdam, 1992), p. 261.

[8] E. M. PURCELL: *Phys. Rev.*, **69**, 681 (1946).

[9] D. KLEPPNER: *Phys. Rev. Lett.*, **47**, 233 (1981).

[10] J. E. LENNARD-JONES: *Trans. Faraday Soc.*, **28**, 334 (1932).

[11] H. MORAWITZ: *Phys. Rev.*, **187**, 1792 (1969); *Phys. Rev. A*, **7**, 1148 (1973); H. MORAWITZ and M. R. PHILPOTT: *Phys. Rev. A*, **10**, 4863 (1974); G. BARTON: *Proc. R. Soc. London, Ser. A*, **320**, 251 (1970); M. BABIKER and G. BARTON: *Proc. R. Soc. London, Ser. A*, **326**, 255 (1972); *J. Phys A*, **9**, 129 (1976); M. R. PHILPOTT: *Chem. Phys. Lett.*, **19**, 435 (1973); P. MILONNI, J. R. ACKERHALT and W. A. SMITH: *Phys. Rev. Lett.*, **31**, 958 (1973); P. MILONNI and P. KNIGHT: *Opt. Commun.*, **9**, 119 (1973); G. BARTON: *J. Phys B*, **7**, 2134 (1974); E. A. POWER and T. THIRUNAMACHANDRAN: *Phys. Rev. A*, **25**, 2473 (1982); C. LUTKEN and F. RAVNDAL: *Phys. Scr.*, **28**, 209 (1983); *Phys. Rev. A*, **31**, 2082 (1985); J. M. WYLIE and J. E. SIPE: *Phys. Rev. A*, **30**, 1185 (1984); **32**, 2030 (1985).

[12] K. H. DREXHAGE: in *Progress in Optics XII*, edited by E. WOLF (North-Holland, New York, N.Y., 1974), p. 163.

[13] H. B. G. CASIMIR: *Proc. K. Ned. Akad. Wet.*, **51**, 793 (1948).

[14] P. GOY, J. M. RAIMOND, M. GROSS and S. HAROCHE: *Phys. Rev. Lett.*, **50**, 1903 (1983).

[15] D. J. HEINZEN, J. J. CHILDS, J. E. THOMAS and M. S. FELD: *Phys. Rev. Lett.*, **58**, 1320 (1987).

[16] G. GABRIELSE and H. DEHMELT: *Phys. Rev. Lett.*, **55**, 67 (1985).

[17] R. G. HULET, E. S. HILFER and D. KLEPPNER: *Phys. Rev. Lett.*, **55**, 2137 (1985).

[18] W. JHE, A. ANDERSON, E. HINDS, D. MESCHEDE, L. MOI and S. HAROCHE: *Phys. Rev. Lett.*, **58**, 666 (1987).

[19] F. DE MARTINI, G. INNOCENTI, G. R. JACOBOVITZ and P. MATALONI: *Phys. Rev. Lett.*, **59**, 2955 (1987).

[20] V. SANDOGHDAR, C. SUKENIK, E. HINDS and S. HAROCHE: *Phys. Rev. Lett.*, **68**, 3432 (1992).

[21] H. B. G. CASIMIR and D. POLDER: *Phys. Rev.*, **73**, 360 (1948).

[22] G. BARTON: *Proc. R. Soc. London, Ser. A*, **410**, 175 (1987).

[23] E. T. JAYNES and F. W. CUMMINGS: *Proc. IEEE*, **51**, 89 (1963).
[24] J. J. SANCHEZ-MONDRAGON, N. B. NAROZHNY and J. H. EBERLY: *Phys. Rev. Lett.*, **51**, 550 (1983).
[25] G. S. AGARWAL: *Phys. Rev. Lett.*, **53**, 1732 (1984).
[26] S. HAROCHE: in *New Trends in Atomic Physics* (Les Houches Session XXXVIII), edited by G. GRYNBERG and R. STORA (North-Holland, Amsterdam, 1984), p. 193.
[27] F. BERNARDOT, P. NUSSENZVEIG, M. BRUNE, J. M. RAIMOND and S. HAROCHE: *Europhys. Lett.*, **17**, 33 (1992).
[28] R. J. THOMPSON, G. REMPE and H. G. KIMBLE: *Phys. Rev. Lett.*, **68**, 1132 (1992).
[29] M. G. RAIZEN, R. J. THOMPSON, R. J. BRECHA, H. J. KIMBLE and H. J. CARMICHAEL: *Phys. Rev. Lett.*, **63**, 240 (1989).
[30] Y. ZHU, D. J. GAUTHIER, S. E. MORIN, QILIN WU, H. J. CARMICHAEL and T. W. MOSSBERG: *Phys. Rev. Lett.*, **64**, 2499 (1990).
[31] G. REMPE, H. WALTHER and N. KLEIN: *Phys. Rev. Lett.*, **58**, 353 (1987).
[32] D. MESCHEDE, H. WALTHER and G. MÜLLER: *Phys. Rev. Lett.*, **54**, 551 (1985).
[33] P. FILIPOWICZ, J. JAVANAINEN and P. MEYSTRE: *Phys. Rev. A*, **34**, 3077 (1986).
[34] L. A. LUGIATO, M. O. SCULLY and H. WALTHER: *Phys. Rev. A*, **36**, 740 (1987).
[35] G. REMPE and H. WALTHER: *Phys. Rev. A*, **42**, 1650 (1990).
[36] G. REMPE, F. SCHMIDT-KALER and H. WALTHER: *Phys. Rev. Lett.*, **64**, 2783 (1990).
[37] M. BRUNE, J. M. RAIMOND, P. GOY, L. DAVIDOVICH and S. HAROCHE: *Phys. Rev. Lett.*, **59**, 1899 (1987).
[38] P. P. SOROKIN and N. BRASLAU: *IBM J. Res. Dev.*, 8, 177 (1964).
[39] A. M. PROKHOROV: *Science*, **149**, 828 (1965).
[40] L. DAVIDOVICH, J. M. RAIMOND, M. BRUNE and S. HAROCHE: *Phys. Rev. A*, **36**, 3771 (1987).
[41] D. GAUTHIER, QILIN WU, S. MORIN and T. MOSSBERG: *Phys. Rev. Lett.*, **68**, 464 (1992).
[42] S. HAROCHE, M. BRUNE and J. M. RAIMOND: *Appl. Phys. B*, **54**, 355 (1992).
[43] S. HAROCHE, M. BRUNE and J. M. RAIMOND: *J. Phys. (Paris)*, **2**, 659 (1992).
[44] M. BRUNE, S. HAROCHE, V. LEFEVRE, J. M. RAIMOND and N. ZAGURY: *Phys. Rev. Lett.*, **65**, 976 (1990).
[45] M. BRUNE, S. HAROCHE, J. M. RAIMOND, L. DAVIDOVICH and N. ZAGURY: *Phys. Rev. A*, **45**, 5193 (1992).
[46] E. SCHRÖDINGER: *Naturwissenschaften*, **23**, 807, 823, 844 (1935) (English translation by J. D. TRIMMER: *Proc. Am. Phys. Soc.*, **124**, 3235 (1980)).
[47] B. YURKE and D. STOLER: *Phys. Rev. Lett.*, **57**, 13 (1986).
[48] W. H. ZUREK: *Phys. Today*, **44**, No. 10, 36 (1991).
[49] Y. YAMAMOTO, S. MACHIDA, K. IGETA and Y. HORIKOSHI: in *Coherence and Quantum Optics VI*, edited by L. MANDEL, E. WOLF and J. H. EBERLY (Plenum, New York, N.Y., 1990), p. 1249.
[50] J. L. JEWELL, J. P. HARBISON and A. SCHERER: *Sci. Am.*, **265**, No. 5, 56 (1991).
[51] Y. YAMAMOTO, S. MACHIDA, Y. HORIKOSHI, K. IGETA and G. BJÖRK: *Opt. Commun.*, **80**, 337 (1991).
[52] E. YABLANOVITCH: *Phys. Rev. Lett.*, 58, 2059 (1987).
[53] E. YABLANOVITCH and T. G. GMITTER: *Phys. Rev. Lett.*, **63**, 1950 (1989).
[54] E. YABLANOVITCH, T. J. GMITTER, R. D. MEADE, A. M. RAPPE, K. D. VROMMER and J. D. JOANNOPOULOS: *Phys. Rev. Lett.*, **67**, 3380 (1991).
[55] V. B. BRAGINSKY, M. L. GORODETSKY and V. S. ILCHENKO: *Phys. Lett. A*, **137**, 393 (1989); S. SCHILLER and P. L. BYER: *Opt. Lett.*, **16**, 1138 (1991).
[56] S. X. QIAN and R. K. CHANG: *Phys. Rev. Lett.*, **56**, 926 (1986).

[57] L. COLLOT, V. LEFEVRE, M. BRUNE, J. M. RAIMOND and S. HAROCHE: *Europhys. Lett.*, **23**, 327 (1993).
[58] Y. Z. WANG: in *Laser Spectroscopy X, Proceedings of the Tenth International Conference on Laser Spectroscopy*, edited by M. DUCLOY, E. GIACOBINO and G. CAMY (World Scientific, Singapore, 1992), p. 205.
[59] A. J. CAMPILLO, J. D. EVERSOLE and H. B. LIN: *Phys. Rev. Lett.*, **67**, 437 (1991).
[60] H. M. TZENG, K. F. WALL, M. B. LONG and R. K. CHANG: *Opt. Lett.*, **9**, 499 (1984).
[61] S. L. MCCALL, A. F. J. LEVI, R. E. SLUSHER, S. J. PEARTON and R. A. LOGAN: *Appl. Phys. Lett.*, **60**, 289 (1992).
[62] Y. ZHU, A. LEZAMA, T. W. MOSSBERG and M. LEWENSTEIN: *Phys. Rev. Lett.*, **61**, 1946 (1988).
[63] J. H. KIMBLE: in *Fundamental Systems in Quantum Optics* (Les Houches Session LIII), edited by J. DALIBARD, J. M. RAIMOND and J. ZINN-JUSTIN (Elsevier Science Publishers, Amsterdam, 1992), p. 545.

Squeezed States of Light.

E. GIACOBINO

Laboratoire de Spectroscopie Hertzienne (*), *Université Pierre et Marie Curie F-75252 Paris Cedex 05, France*

1. – Introduction.

Contrary to many other domains in physics, optics is a field where quantum fluctuations are easily observed. In many experimental situations, they give rise to fluctuations in the detected quantity, usually a current in a photodetector, which are known as the «shot noise». In the past few years, the improvement in the laser sources and in the detectors has been such that the shot noise is encountered more and more often in precision measurements in spectroscopy and interferometry, thus limiting the sensitivity of the detection.

These quantum fluctuations have long been considered as an insuperable limit. But, starting in the mid-eighties, experiments have shown that, if the quantum noise could not be suppressed, it could be circumvented. The best quantum noise reduction attainable at present ranges from 50% to 70%, depending on the type of experimental situation considered.

In this lecture, we are going to show how the quantum fluctuations can be manipulated by various techniques related to nonlinear optics so as to produce «squeezed light», that is light in which the quantum fluctuations are decreased on one of the components of the field (phase or amplitude, for example).

2. – Quantum fluctuations and squeezed states.

A plane wave of frequency ω_L, and of given direction and polarization, corresponding to a particular mode of the electromagnetic field, can be described

(*) The Laboratoire de Spectroscopie Hertzienne de l'Ecole Normale Supérieure et de l'Université Pierre et Marie Curie is associated with the CNRS (URA 0018).

by its electric field [1]:

$$E(t) = E_1 \cos \omega_L t + E_2 \sin \omega_L t , \tag{1}$$

where E_1 and E_2 are the two quadrature components of the field (defined with respect to some phase reference). When the field is quantized, the two quadrature components are operators that can be expressed as functions of the creation and annihilation operators $a^\dagger$ and a of the mode (these operators add or remove one photon of energy $\hbar\omega_L$):

$$E_1 = e_0 (a + a^\dagger), \tag{2a}$$

$$E_2 = i e_0 (a - a^\dagger), \tag{2b}$$

where e_0 is the «field of one photon»

$$e_0 = \sqrt{\frac{\hbar\omega_L}{2\varepsilon_0 V}} . \tag{3}$$

As usual in quantum optics, the field is quantized in a box of volume V. We will also use the positive and negative frequency components of the field, $\mathcal{E}^+(t)$ and $\mathcal{E}^-(t)$ defined by

$$\mathcal{E}^+(t) = e_0 a \exp[-i\omega_L t] \quad \text{and} \quad \mathcal{E}^-(t) = e_0 a^\dagger \exp[i\omega_L t]. \tag{4}$$

Since the commutator of a and $a^\dagger$ is

$$[a, a^\dagger] = 1 , \tag{5}$$

E_1 and E_2 do not commute. They are conjugate quantities, as the position q and the momentum p of a particle. As a result, they cannot be measured at the same time with infinite precision, and their mean-square dispersions ΔE_1 and ΔE_2 in any quantum state obey a Heisenberg inequality:

$$\Delta E_1 \, \Delta E_2 \geqslant e_0^2 . \tag{6}$$

We can represent the field in a phase space diagram in which the coordinates are the two quadrature components in $\sin \omega_L t$ and in $\cos \omega_L t$. In such a diagram, a classical field is represented by a vector, whose length and argument are the amplitude and phase of the field. A quantum field cannot be so well determined. The results of measurements of E_1 or E_2 fluctuate by amounts which are of the order of ΔE_1 and ΔE_2. The end of the vector representing the field belongs to an area in phase space the surface of which is necessarily larger than the product of the dispersions given by eq. (6) (fig. 1). In the same way, the measurements of the phase or of the amplitude fluctuate by quantities which are of the order of the dimension of the dispersion area respectively in the direction of the mean field and in the direction perpendicular to the mean field.

In the case of a particle, the Heisenberg inequality concerns two variables,

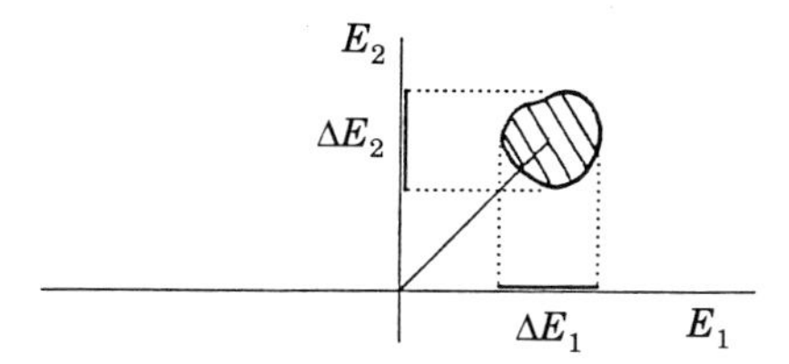

Fig. 1. – Phase space representation of a quantum field. The coordinates are the two quadrature components of the field. The whole diagram rotates at the optical frequency.

the position q and the momentum p which are of different physical nature. In contrast, the two conjugate variables of the electromagnetic field, E_1 and E_2, are of the same nature, and most of the usual light sources have no preferred quadrature component. In other words, the fluctuations of the electromagnetic field usually have no natural phase reference. As a result, in this case, the dispersions ΔE_1 and ΔE_2 are equal and independently fulfil an inequality:

$$\Delta E_1 = \Delta E_2 \geqslant e_0 \,. \tag{7}$$

The quantum fluctuations of the electromagnetic field are of order e_0.

2·1. *Coherent state and vacuum state.* – This state that is the closest to the classical field (shown in fig. 2*a*)) is the one where the fluctuations in the two

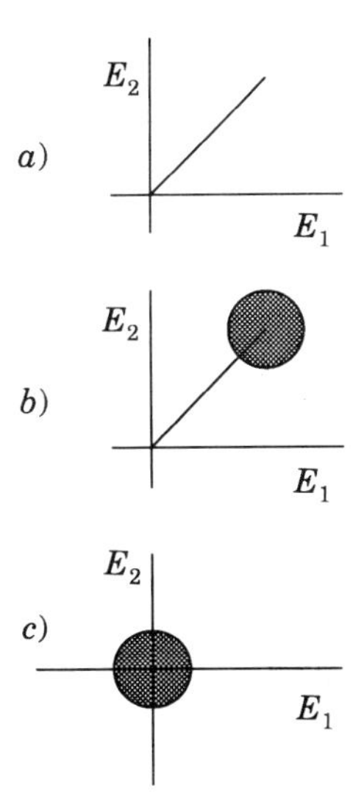

Fig. 2. – Phase space representations of a classical field (*a*)), of a coherent field (*b*)) and of the vacuum field (*c*)) (the fluctuations shown in *b*) in the two representations are not to scale). The time-dependent representation gives the E_1 component as a function of time, taking into account the fact that the phase space diagram rotates at the optical frequency of the considered mode.

quadratures are equal and equal to the minimum allowed by eq. (7). Then the dispersion area has the shape of a disk of diameter e_0 in phase space (fig. 2*b*)). Such a minimum-uncertainty field is the so-called *coherent state*. A perfectly stabilized laser would emit a coherent field. In the representation of the E_1 quadrature component of the electric field as a function of time, this corresponds to a sine curve having some finite width.

The field corresponding to the state with lowest energy of a mode has a zero mean value, but it cannot have zero fluctuations, because of the Heisenberg relation eq. (6). It can be shown that this *vacuum state* is a particular coherent state, with zero mean field. Its representation in phase space is given by a disk of diameter e_0 centred on the origin (fig. 2*c*)).

2·2. *Squeezed states.* – To surpass the limit set by eq. (7), it is necessary to break the symmetry between the two quadratures. The Heisenberg inequality does not forbid to decrease ΔE_1 below e_0, provided that ΔE_2 is increased. States having that type of property are called *squeezed states*. For example, they may fulfil

$$\Delta E_1 < e_0 \,, \qquad \Delta E_2 \geqslant e_0^2 / \Delta E_1 \,. \tag{8}$$

In phase space, a squeezed state has an elongated shape, for example elliptical, with a smaller dispersion in one of its components. The fluctuations may be

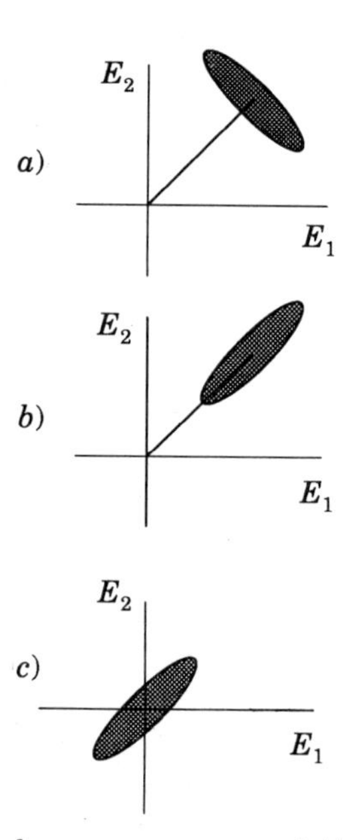

Fig. 3. – Squeezed states in the phase space representations. The dispersion area has an elliptical shape with its smaller axis in the squeezing direction. In *a*), the amplitude quadrature is squeezed, in *b*), the phase quadrature is squeezed. In *c*), squeezed vacuum is shown.

squeezed in one of the quadratures E_1 or E_2, or in the amplitude quadrature (parallel to the mean field) (fig. 3*a*)) or in the phase quadrature (perpendicular to the mean field) (fig. 3*b*)) or in any other component of the field. The vacuum field can also be squeezed, as shown in fig. 3*c*).

2·3. *Intensity measurements.* – The well-known photon noise or shot noise is associated with the quantum fluctuations detected in intensity measurements. If the mean field is much larger than the field fluctuations, one can calculate the intensity fluctuations in a linear approximation:

$$\Delta I^2 = 4E_1^2\, \Delta E_1^2 \,, \tag{9}$$

where we have supposed that the mean field was aligned along the E_1 axis. The standard photon noise corresponds to the case in which the amplitude fluctuations are equal to the coherent-field fluctuations $\Delta E_1^2 = e_0^2$. The photon noise is characterized by a variance proportional to the mean intensity:

$$(\Delta I^2)_{\mathrm{st}} = E_1^2 4e_0^2 \,. \tag{10}$$

This noise can be reduced by squeezing the fluctuations of the E_1 component of the field. More generally, in every optical measurement, it can be shown that the fluctuations correspond to some component of the field, which can be, for example, a quadrature component of a field. To reduce this noise, one must squeeze the fluctuations of the measured component.

2·4. *Model for a laser beam.* – Many measurements in optics deal with quasi-monomode light beams rather than standing waves in closed cavities. The central frequency of the mode is ω_{L}, which corresponds to the only filled mode, the neighbouring modes containing no photons, or almost no photons; they will introduce fluctuations at nonzero frequencies around the central frequency. In such a case, it is convenient to use the Fourier transforms of the fields for «noise» frequencies ω around ω_{L}

$$\mathcal{E}^{+}(t) = e_{\omega_{\mathrm{L}}} \int \frac{\mathrm{d}\omega}{2\pi}\, a_{\omega_{\mathrm{L}}+\omega} \exp[-i\omega t] \exp[-i\omega_{\mathrm{L}} t] \,, \tag{11}$$

$$\mathcal{E}^{-}(t) = e_{\omega_{\mathrm{L}}} \int \frac{\mathrm{d}\omega}{2\pi}\, a^{\dagger}_{\omega_{\mathrm{L}}-\omega} \exp[-i\omega t] \exp[i\omega_{\mathrm{L}} t] \,. \tag{12}$$

The commutation relation of the Fourier transform operators $a_{\omega_{\mathrm{L}}+\omega}$ and $a^{\dagger}_{\omega_{\mathrm{L}}-\omega}$ is

$$[a_{\omega_{\mathrm{L}}+\omega},\, a^{\dagger}_{\omega_{\mathrm{L}}-\omega'}] = 2\pi\delta(\omega+\omega')$$

and e_{ω_L} is defined by

$$e_{\omega_L} = \sqrt{\frac{\hbar\omega_L}{2\varepsilon_0 Sc}} \tag{13}$$

where S is the cross-section area of the beam. We see that the mean value of the product $\langle a^{\dagger}_{\omega_L - \omega} a_{\omega_L + \omega} \rangle$ is a number of photons per unit time in the beam. The Heisenberg relation for the mean-square dispersions of the quadrature components averaged over some finite time T can now be shown to be

$$\Delta E_1 \Delta E_2 \geqslant e^2_{\omega_L}/T\,, \tag{14}$$

which is the same as eq. (6) for a mode of volume ScT. Applying this to the calculation of the photon noise, or shot noise, (eq. (10)), we find the well-known result that the shot noise is reduced by using longer measurement times. In this lecture, we will mainly be interested in the Fourier components of the field, and more precisely of its fluctuations, and in the dependence of these fluctuations on the frequency, *i.e.* the noise spectra of the field.

To show the effect of the quantum fluctuations in optical measurements, we are first going to consider a very simple system, the beamsplitter.

3. – Quantum noise at the output of a beamsplitter.

When a beam of light passes through a 50/50 beamsplitter, entering through port A (fig. 4), quantum noise is added to the two outputs. This can be viewed as due to the particle nature of light. Each photon has a probability 1/2 of being transmitted or reflected if nothing enters in the other input channel (channel B). Because of the random character of the transmission or reflection processes, there is a partition noise which is independent of the photon statistics of the input beam. Then, it can be shown that the noise in the difference between the intensities in the two output channels is the standard quantum noise, independently of the photon statistics of the input beam. This property can also be derived by calculating the fluctuations of the fields at the output ports of the beamsplitter, as shown below.

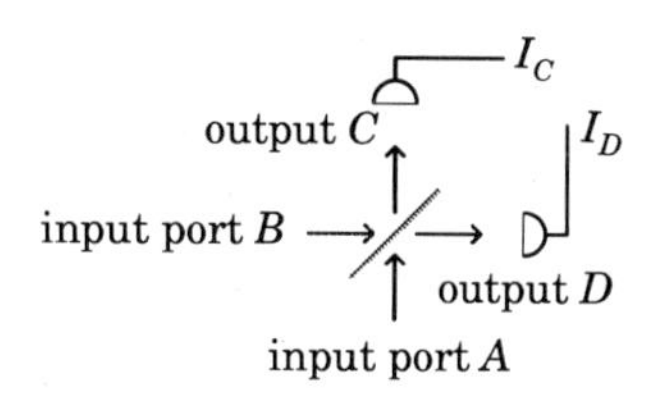

Fig. 4. – The fields in the input ports A and B are partly transmitted and partly reflected and are recombined in the output ports C and D.

3'1. *Homodyne measurements.* – Let us now consider a homodyne measurement performed with a beamsplitter: the incident beam entering in channel A is mixed on the beamsplitter with the field entering in channel B (fig. 4). The beamsplitter has amplitude transmission and reflection coefficients t and r (with $r^2 = 1 - t^2$), and we assume that the relations between the fields going out of the system in channels C and D is given by classical optics:

$$E_C = tE_A + rE_B \, , \tag{15a}$$

$$E_D = -rE_A + tE_B \, , \tag{15b}$$

where the minus sign in eq. (15*b*) comes from the π phase shift between the reflections from air off glass and from glass off air undergone by beams A and B.

Let us first consider a 50/50 beamsplitter. Then

$$r = t = 1/\sqrt{2} \, . \tag{16}$$

From now on, we treat the fields as classical fields. Assuming that field E_A is much more intense than field E_B, the intensities in the output channels C and D are

$$I_C = (1/2)(|E_A|^2 + 2E_A E_{B1}) \, , \tag{17a}$$

$$I_D = (1/2)(|E_A|^2 - 2E_A E_{B1}) \, , \tag{17b}$$

where E_{B1} is the quadrature component of E_B in phase with E_A. The measurement of the intensity difference $I = I_C - I_D$ between the two channels gives

$$I = I_C - I_D = 2E_A E_{B1} \, . \tag{18}$$

The intensity difference I is a signal resulting from the homodyning of the quadrature component E_{B1} by the local oscillator E_A. If one now calculates the fluctuations, those of the local oscillator field E_A disappear in the difference between the output beams:

$$\Delta I^2 = 4E_A^2 \, \Delta E_{B1}^2 \, . \tag{19}$$

A rigorous demonstration using quantum fields gives the same result.

Let us assume now that no field is sent into the input channel B. In quantum optics, this means that the vacuum fluctuations having $\Delta E_{B1}^2 = e_0^2$ enter into channel B and are homodyned by the local oscillator. It can be shown that the vacuum fluctuations can be emulated by a classical random field having the same mean-square fluctuations. Then, the noise in the intensity difference I is given by

$$\Delta I^2 = 4E_A^2 e_0^2 \, . \tag{20}$$

Comparing this result with eq. (10), we see that this noise is the same as the standard quantum noise of the input beam A, the local oscillator. It is independent of the actual fluctuations of the local oscillator, and is the same whether the latter has an intensity noise which is larger than the standard photon noise, or has an intensity squeezed noise. This provides a way of measuring directly the standard quantum noise associated with any field.

It is also possible to get a noise in the intensity difference that is smaller than the standard quantum noise of the local oscillator. If squeezed vacuum is entered into input B in place of regular vacuum, we see that the noise in I is changed. If the squeezed quadrature is the one in phase with the local oscillator, the fluctuations in E_B are smaller than the vacuum fluctuations and the noise in I is smaller than the standard quantum noise. If the squeezed quadrature is not along the local oscillator, the fluctuations in I are larger than the standard fluctuations.

This property was used in the experiments described below to demonstrate the generation of squeezed vacuum. This set-up has another very interesting property. It can be noticed that, since the noise in the intensity difference after the beamsplitter has been reduced, which means that the two output beams C and D now have intensity correlations at the quantum level. By entering a squeezed field in channel B, the partition noise of the photons on the beamsplitter can be decreased to a level which depends on the degree of squeezing of the squeezed vacuum. Such beams having high intensity correlations are called twin photon beams. We will see below that twin beams can also be generated directly in experiments of nonlinear optics.

3·2. *Effects of linear losses on squeezed light.* – Losses have a drastic effect on squeezed light. We can model the losses that reduce the intensity of a light beam by a factor T with a beamsplitter having an amplitude transmission coefficient t, such that $t^2 = T$.

Let us consider the beamsplitter of fig. 4, and suppose that a field with nonzero mean value enters in channel A. The vacuum field enters in channel B, and eq. (15a) implies that

$$\Delta E_{C1}^2 = t^2 \Delta E_{A1}^2 + r^2 e_0^2 \,. \tag{21}$$

The beamsplitter mixes the fluctuations of the two incident fields. If the fluctuations of the input field are at shot noise ($\Delta E_{A1}^2 = e_0^2$), we find that the fluctuations of the output field E_C are at shot noise as well, since $t^2 + r^2 = 1$. But, if the input field has fluctuations lower than shot noise, the lossy medium tends to bring the fluctuations back to the vacuum fluctuations. Note that this is true also for a field having excess noise, that is a noise larger than shot noise.

In a detection system, a quantum efficiency smaller than 1 has the same effect, since a detector with a quantum efficiency T can be considered as a perfect detector preceded by a beamsplitter with transmission coefficient T.

4. – Generation of squeezed states.

Squeezed states can be generated by several nonlinear optical systems. In most cases the involved phenomena are related to parametric processes.

The parametric effect can come from a $\chi^{(2)}$-type or a $\chi^{(3)}$-type nonlinearity. The former case corresponds to «three-wave mixing»: when irradiated by a pump wave at frequency ω_0, a nonlinear $\chi^{(2)}$ material emits two signal waves which have frequencies ω_1 and ω_2, fulfilling the condition $\omega_0 = \omega_1 + \omega_2$ for energy conservation. One photon of the pump disappears, while two photons are created in the signal waves.

The latter case corresponds to «four-wave mixing»: in a $\chi^{(3)}$ material, two photons of the pump wave at frequency ω_0 disappear, while two photons are created in the signal waves at frequencies ω_1 and ω_2. Energy conservation requires that $2\omega_0 = \omega_1 + \omega_2$.

We will see in the following that the systems may be operated either in the degenerate regime, in which the emitted photons are identical, or in the nondegenerate regime, in which the two signal modes have different frequencies or different polarizations.

A different technique has also been used to squeeze the intensity of a light field. It does not use a parametric generator, but a laser the output noise of which is reduced by regulating the noise of the pumping system. Practically, this idea has been implemented with a laser diode having a regulated pumping current. If the quantum efficiency of the electron-to-photon conversion is large enough, the laser emits a regulated flow of photons, resulting in an intensity squeezed noise. Here, we will concentrate on parametric systems and we will not give any further detail on this technique, which has led to a noise reduction of about 85% in the intensity of the field emitted by a laser diode [2].

4'1. *Degenerate parametric generation.* – The first observation of squeezing has been made in 1985 in an experiment of parametric generation involving four-wave mixing in sodium vapour [3]. The nonlinearity was enhanced by placing the nonlinear medium in an optical cavity. In 1986, squeezed light was generated in a similar system, but with three-wave mixing in a nonlinear $\chi^{(2)}$ crystal [4]. In both cases, the systems were operated in the *degenerate* regime.

4'1.1. Theory. Let us first briefly discuss the physical process that leads to squeezing in parametric amplification. We are going to deal with this system in a way similar to the one we used for the beamsplitter: *we calculate the fields with the classical equations of electromagnetism*, taking into account the fact that the *input fields have stochastic fluctuations which emulate the quantum fluctuations*. In particular, when there is no input classically, we have to consider that the vacuum fluctuations are entering the system.

For $\chi^{(2)}$ and $\chi^{(3)}$ parametric interactions, this procedure has been shown to be completely equivalent to the standard quantum treatment when the fluctuations can be treated in the linear approximation. The effect can then be pictured in a simple way (at least in the limit of low pumping rates): the system processes the vacuum noise input and squeezes it. The output, having zero mean field, is expected to be squeezed vacuum [5].

To enhance the nonlinearity, the parametric medium, pumped with a laser, can be placed in an optical cavity. Such a system can oscillate like a laser if the pump power is larger than some threshold value [6]. In this section, we suppose that the pump power is low enough for the cavity to stay below the oscillation threshold, and we neglect the pump depletion. Then the change of the signal field α_1 in one round trip in the cavity can be written as a function of the cavity losses, the parametric gain and the input field α_1^{in}. Here, since there is no input field with a finite value in the signal mode, the input field is the vacuum field coming into the cavity through the coupling mirror [7]:

$$\tau \partial \alpha_1 / \partial t = -\gamma \alpha_1 + \eta \alpha_1^* + t \alpha_1^{in} , \tag{22}$$

where the α's are the complex classical components (slowly varying amplitudes) of the fields, τ is the cavity round-trip time, γ is the dimensionless damping coefficient of the field in the cavity, which is related to the amplitude transmission coefficient t (assumed to be small) of the output mirror by

$$\gamma = 1 - r = t^2/2 , \tag{23}$$

η is the parametric gain, which is proportional to the pump field, to the $\chi^{(2)}$ coefficient and to the length of the crystal. The parametric gain term in α_1 involves α_1^*, which is characteristic of parametric amplification, and which is the reason for the phase dependence of the process.

We express the solution in terms of the quadrature components:

$$q_1 = (\alpha_1 + \alpha_1^*)/\sqrt{2} , \qquad p_1 = (\alpha_1 - \alpha_1^*)/\sqrt{2}\, i , \tag{24}$$

with similar notations for q_1^{in} and p_1^{in}. Taking the Fourier transform of eq. (22) and of its complex conjugate, we obtain

$$\widetilde{q}_1(\omega) = \frac{t\widetilde{q}_1^{in}(\omega)}{\gamma - \eta + i\omega\tau} , \qquad \widetilde{p}_1(\omega) = \frac{t\widetilde{p}_1^{in}(\omega)}{\gamma + \eta + i\omega\tau} , \tag{25}$$

where ω is the noise frequency about the optical frequency. On the other hand, the outgoing field α_1^{out} is related to the field inside the cavity and to the incoming vacuum field by a reflection-transmission relation similar to eq. (15*b*):

$$\alpha_1^{out} = t\alpha_1 - \alpha_1^{in} , \tag{26}$$

where the reflection coefficient r has been approximated by one. The output field is the superposition of the cavity field transmitted through the coupling mirror and of the vacuum fluctuations reflected by the mirror. Using eqs. (25)

and (26), one gets

$$\tilde{q}_1^{\text{out}}(\omega) = \frac{\gamma + \eta - i\omega\tau}{\gamma - \eta + i\omega\tau}\,\tilde{q}_1^{\text{in}}(\omega), \tag{27a}$$

$$\tilde{p}_1^{\text{out}}(\omega) = \frac{\gamma - \eta - i\omega\tau}{\gamma + \eta + i\omega\tau}\,\tilde{p}_1^{\text{in}}(\omega). \tag{27b}$$

Equations (27) clearly show that the input vacuum field has been considerably modified: quadrature q_1 is amplified, while quadrature p_1 is squeezed and even goes to zero for $\omega = 0$ and $\gamma = \eta$, that is for zero noise frequency and when the oscillation threshold is approached. The vacuum field is indeed squeezed by the system.

4·1.2. Experiments. In the experiment (fig. 5) performed in ref. [4], a nonlinear crystal of $MgO:LiNbO_3$ is pumped with a single-mode doubled YAG laser at 0.532 μm. The crystal is placed in an optical cavity which is resonant for the infrared subharmonic field at 1.06 μm. The field going out of the cavity is analysed in a homodyne detector in which it is mixed with a local oscillator as described in subsect. 3·2. The local oscillator is a part of the fundamental mode of the pump YAG laser at 1.06 μm. The photodetection signal is fed into a spectrum analyser.

The signal observed at a fixed noise frequency depends on the relative phase of the local oscillator and of the light emitted by the cavity. It oscillates around the shot noise level as a function of the phase of the local oscillator. The noise reduction has been verified to be improved when the incident pump power tends to its threshold value. The optimum noise reduction obtained with this technique is 63% [4]. The best figure obtained in a similar experiment using the $\chi^{(3)}$ coefficient of sodium atoms near resonance in an atomic beam is 25% [3].

Squeezing of about 20% has also been observed using forward four-wave mixing in a long silica optical fibre [8]. However, the experiments have been

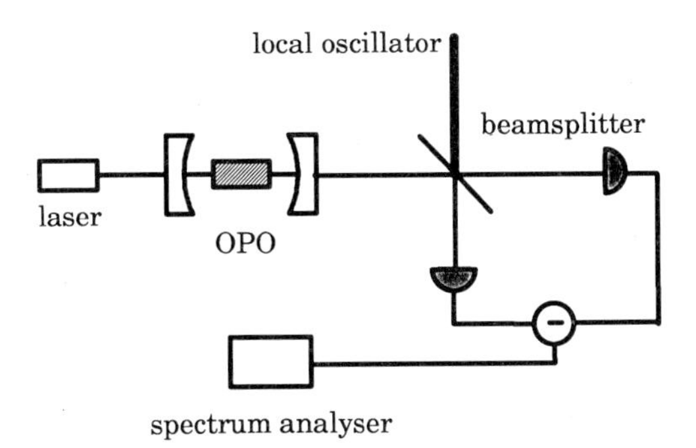

Fig. 5. – Experimental scheme using parametric generation in a cavity to produce squeezed vacuum. The squeezed vacuum is analysed by combining it with a local oscillator in a homodyne detector made of a 50/50 beamsplitter and two photodetectors.

hampered by index fluctuations due to Brillouin scattering. Short pulses can be squeezed more efficiently with this method because their interaction time is too fast to be sensitive to Brillouin scattering. Short-pulse [9] and soliton [10] squeezing have been observed recently in fibres. Other experiments without an optical cavity use pulsed pumping in a nonlinear crystal [11, 12].

Up to now, the systems having $\chi^{(2)}$ nonlinearities seem to yield better squeezing than the $\chi^{(3)}$ systems. This comes mainly from two reasons: the pump frequency is far from the signal frequencies, which allows easy elimination of spurious fields due to pump scattering; in the doubling crystals, the nonlinearity is not resonant, which means that there are no fluctuations due to the nonlinear medium, except for those associated with linear losses.

4.2. *Bistability.* – One of the earliest proposals for squeezing generation was to use the properties of a nonlinear cavity in the vicinity of the bistability threshold [13]. In the limit where the $\chi^{(3)}$ nonlinearity is purely parametric, the effect has been treated in ref. [14]. It is well known that a critical divergence of the fluctuations occurs near the bistability turning point. This divergence takes place in one quadrature of the field. For the quantum fluctuations, the total area of the fluctuation distribution in phase space is conserved, which implies that the other quadrature of the field is squeezed. This can be seen from the form of the semi-classical input-output transformation for the field in a one-ended cavity [15]. The input and output fields α^{in} and α^{out} and the field inside the cavity α verify the equations

$$t\alpha^{\text{in}} = (1 - r\exp[-i\Phi])\,\alpha\,, \tag{28a}$$

$$t\alpha^{\text{out}} = (\exp[-i\Phi] - r)\,\alpha\,, \tag{28b}$$

where r and t are the amplitude reflection and transmission coefficients of the coupling mirror and Φ is the total phase shift around the cavity, including the linear dephasing Φ_0 due to the round trip in the cavity, and the phase shift due to the nonlinear medium of Kerr coefficient K:

$$\Phi = \Phi_0 + K|\alpha|^2\,. \tag{29}$$

The transformation $\alpha^{\text{in}} \rightarrow \alpha^{\text{out}}$ is a mere rotation about the origin by an intensity-dependent angle. Its result on the probability distribution is a distortion from a disk shape to an elongated shape having the same area (fig. 6). As a result, one quadrature of the output field is squeezed. This squeezing increases in the vicinity of the bistability turning point.

To calculate the squeezing spectrum, we use a semi-classical method similar to the one presented in the previous section, in which the fields are treated as classical fields and their fluctuations are considered as classical stochastic vari-

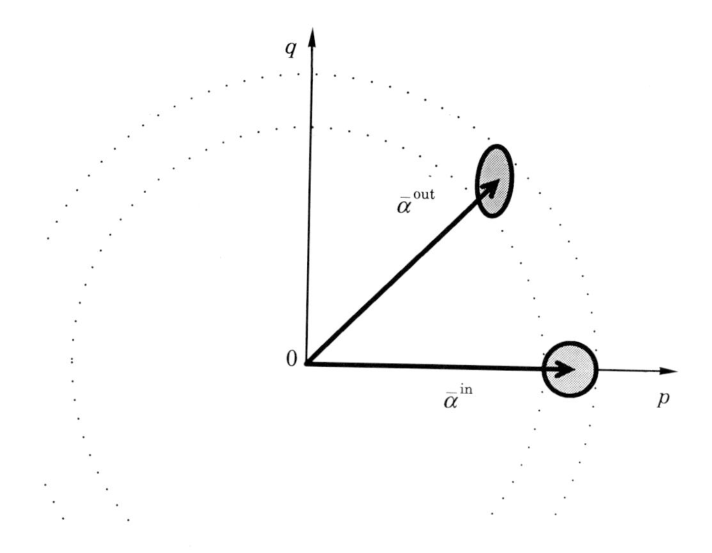

Fig. 6. – Phase space representation of the change of a field after passing through a cavity containing a nonlinear medium. The incoming coherent field undergoes an intensity-dependent phase shift that leads to a distortion of the probability distribution.

ables, which are driven by the quantum fluctuations of the incoming fields. But in contrast with subsect. **4**'1 where the equations for the fluctuations were linear, here we have to deal with nonlinear equations. We derive the linear input-output transformation that gives the fluctuations in the output field as a function of the fluctuations of the input fields by differentiating the equations for the field around the mean values. The model we use in this section was developed in ref. [15].

For the sake of simplicity, we will assume that the cavity has a good finesse. The amplitude reflection coefficient of the coupling mirror is $1-\gamma$, where γ is small compared to 1. Subsequently, the transmission coefficient of the coupling mirror can be expressed as $\sqrt{2\gamma}$. The round-trip time of the light in the cavity is τ, the decay rate of the field in the cavity is γ/τ. For an ideal Kerr medium (without losses), the classical equations relating the input and output fields α^{in} and α^{out} to the cavity field α are

$$\tau\partial\alpha(t)/\partial t = -(\gamma + i\Phi_0 + iK|\alpha(t)|^2)\,\alpha(t) + \sqrt{2\gamma}\,\alpha^{\text{in}}(t)\,, \tag{30}$$

$$\alpha^{\text{out}}(t) = \sqrt{2\gamma}\,\alpha(t) - \alpha^{\text{in}}(t)\,. \tag{31}$$

Differentiating these equations around the chosen working point, characterized by the mean cavity field α_0, one gets equations for the small fluctuating part $\delta\alpha(t)$ of the fields, as a function of the fluctuations of the input field. Then, tak-

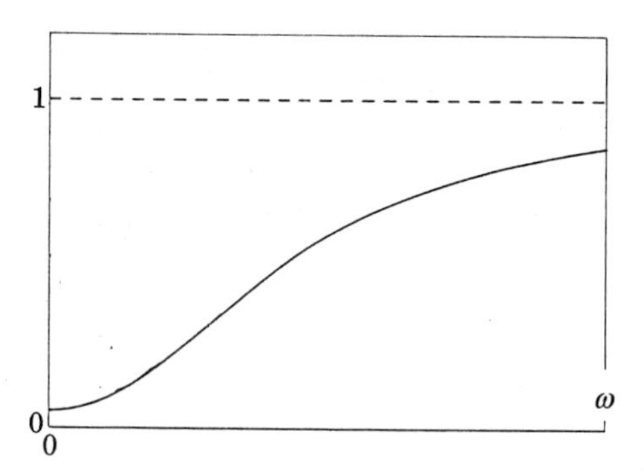

Fig. 7. – «Optimum» squeezing of a field going out of a cavity containing a Kerr medium. Optimum squeezing means that for each value of the noise frequency ω the quadrature exhibiting the largest squeezing has been plotted.

ing the Fourier transform yields

$$(32a)\quad (\gamma + i\Phi_0 + i2K|\alpha_0|^2 - i\omega\tau)\,\delta\tilde{\alpha}(\omega) + iK\alpha_0^2\,\delta\tilde{\alpha}^*(\omega) = \sqrt{2\gamma}\,\delta\tilde{\alpha}^{\text{in}}(\omega),$$

$$(32b)\quad (\gamma - i\Phi_0 - i2K|\alpha_0|^2 - i\omega\tau)\,\delta\tilde{\alpha}^*(\omega) - iK\alpha_0^{*2}\,\delta\tilde{\alpha}(\omega) = \sqrt{2\gamma}\,\delta\tilde{\alpha}^{\text{in}*}(\omega).$$

The input-output equations relating the Fourier component $\delta\tilde{\alpha}^{\text{out}}(\omega)$ of the output field fluctuations at frequency ω (relative to the driving laser frequency) to $\delta\tilde{\alpha}^{\text{in}}(\omega)$ are then derived using eq. (31). It must be noticed that the nonlinear Kerr effect couples the two quadrature components of the fluctuations, which is at the origin of squeezing. It can be shown that the optimum squeezing in one quadrature of the output field tends to zero at zero frequency when one approaches the bistability turning points (fig. 7).

We now consider that the medium introduces linear losses. The losses are modelled by a second output port that introduces an additional decay rate γ' and that has a small transmission $\sqrt{2\gamma'}$ for another input field $\delta\tilde{\alpha}'^{\text{in}}(\omega)$ with zero mean value and with fluctuations which are the vacuum fluctuations. With these assumptions, eq. (32*a*) is transformed into

$$(33)\quad (\gamma + \gamma' + i\Phi_0 + i2K|\alpha_0|^2 - i\omega\tau)\,\delta\tilde{\alpha}(\omega) + iK\alpha_0^2\,\delta\tilde{\alpha}^*(\omega) = \\ = \sqrt{2\gamma}\,\delta\tilde{\alpha}^{\text{in}}(\omega) + \sqrt{2\gamma'}\,\delta\tilde{\alpha}'^{\text{in}}(\omega)$$

with a similar change for eq. (32*b*). The presence of the last term on the right-hand side of this equation corresponds to additional noise and degrades the amount of squeezing.

Squeezed-state generation has been observed in a related system, made of two-level atoms placed in a high-finesse cavity. The nonlinearity cannot be considered as purely parametric, since the atomic medium is driven close to atomic resonance, and additional effects come into play. A noise reduction of 30% has been observed experimentally [16].

4.3. *Nondegenerate parametric generation.* – As mentioned earlier, in parametric generation, two signal photons are emitted at the same time, while one or two pump photons are annihilated. The signal photons are created in pairs, which results in photon correlation and subsequently in intensity correlations between the two signal beams [17]. In the nondegenerate operation, the two signal beams, which are called «twin beams», differ by either their frequencies or their polarizations and can be separated from each other by means of a polarizer or of a prism.

Experiments on twin beams have been performed in various conditions: with low-intensity c.w. pump lasers, or with high-power pump lasers. In the former case, the crystal emits parametric fluorescence. A maximum correlation of 67% has been obtained [18]. In the latter case [19], intense twin beams are emitted by the crystal yielding up to 75% noise reduction [20].

To increase the effective parametric interaction length, the nonlinear crystal can be placed in an optical resonant cavity. When pumped with a high enough intensity, the system oscillates like a laser. The optical parametric oscillator (OPO) emits twin laserlike beams, having a high degree of quantum correlation due to the photon pair emission. However, the cavity introduces some decorrelation between the twin photons, since they do not necessarily go out of the cavity after the same number of round trips. Then, if the time duration of a measurement is short compared to the cavity storage time, one photon of a pair may be detected in one beam, whereas the other photon of the pair arrives later and is not detected in the second beam. On the contrary, if the measurement time is long compared to the storage time, all the pairs are detected and the correlation is recovered. The experiments do not rely on photon counting, since the beams going out of the OPO are laserlike beams and photon counting is impossible. In such a case, it is better to analyse the frequency spectrum of the intensity fluctuations. In the high-frequency range, that is for frequencies higher than the cavity bandwidth (inverse of the cavity storage time), the intensity noises of the two beams are not expected to be correlated, since high frequencies correspond to short measurement times. The noise in the difference between the intensities of the two signal beams is then expected to be equal to the standard quantum limit (shot noise) of a beam with an intensity equal to the sum of the intensities of the two beams. But for frequencies within the bandwidth of the cavity, corresponding to long measurement times, the noise in the difference between the intensities of the two signal beams is reduced below the standard quantum limit.

4.3.1. Theory. To theoretically determine the quantum fluctuations in this case, the semi-classical method used for the OPO below threshold can be extended to the derivation of the above-threshold case, but the treatment must take into account the depletion of the pump field α_0 [21]. As in eq. (22), the equations for the signal fields α_1 and α_2 are driven by the vacuum fluctuations

α_1^{in} and α_2^{in} entering the cavity through the coupling mirror. On the other hand, the equation for the pump field is driven by the external input field α_0^{in}, which has a nonzero mean value:

$$\tau\partial\alpha_1/\partial t = -\gamma\alpha_1 - \chi\alpha_2^*\alpha_0 + t\alpha_1^{in}, \tag{34a}$$

$$\tau\partial\alpha_2/\partial t = -\gamma\alpha_2 - \chi\alpha_1^*\alpha_0 + t\alpha_2^{in}, \tag{34b}$$

$$\tau\partial\alpha_0/\partial t = -\gamma_0\alpha_0 + \chi\alpha_1\alpha_2 + t_0\alpha_0^{in}, \tag{34c}$$

where χ is the parametric coupling coefficient, γ and γ_0 are the cavity damping coefficients for the signal beams and for the pump related to the transmission coefficients of the coupling mirror t and t_0 by eq. (23). The cavity has been assumed to be resonant with all three fields.

First, the mean value of the steady-state solution can be found by solving eqs. (34) neglecting the fluctuations α_1^{in} and α_2^{in}. Below a threshold incident pump power $|\alpha_0^{in}|^2 = |\alpha_0^{thr}|^2$ equal to

$$|\alpha_0^{thr}|^2 = \gamma_0\gamma^2/2\chi^2 \tag{35}$$

one finds for α_1 and α_2 a solution which is identically zero. Above the threshold, the solution is given by

$$\alpha_0 = -\gamma/\chi, \tag{36}$$

$$|\alpha_1|^2 = |\alpha_2|^2 = (\gamma_0\gamma/\chi^2)(\sigma - 1), \tag{37}$$

where the parameter σ is given by

$$\sigma = \sqrt{2\chi^2/\gamma_0\gamma^2}\,\alpha_0^{in}. \tag{38}$$

The fluctuations are found by linearizing eqs. (34) around the steady state. We define the quadrature components q_i and p_i by

$$q_1 = (\alpha_i + \alpha_i^*)/\sqrt{2}, \qquad p_i = (\alpha_i - \alpha_i^*)/\sqrt{2}\,i \tag{39}$$

for $i = 0, 1, 2$, and we introduce the quantities

$$q_- = (q_1 - q_2)/\sqrt{2}, \qquad p_- = (p_1 - p_2)/\sqrt{2}. \tag{40}$$

We assume the mean fields α_i to be real, and the fluctuations δq_i and δp_i in q_i and p_i are the fluctuations in the amplitude and phase quadratures. The equation for the fluctuations δq_- in the amplitude difference q_- reads

$$\tau\partial\delta q_-/\partial t = -\gamma\delta q_- + \chi q_0/\sqrt{2}\,\delta q_- + t\delta q_-^{in}. \tag{41}$$

We see that the fluctuations in the amplitude difference are not coupled to the other quantities, and we can expect them to have a particularly simple behaviour. Taking the Fourier transform and using the input-output relations on the coupling mirror of the type of eq. (26), one gets

$$\delta\tilde{q}_{-}^{\text{out}}(\omega) = \frac{\gamma + \chi q_0/\sqrt{2} - i\omega\tau}{\gamma - \chi q_0/\sqrt{2} + i\omega\tau}\,\delta\tilde{q}_{-}^{\text{in}}(\omega)\,. \tag{42}$$

Since $\gamma = -\chi q_0/\sqrt{2}$ (eq. (36)) when the OPO operates above threshold, it can be seen that the fluctuations in the difference between the amplitudes of the signal fields go to zero for zero noise frequency. The same is true for the noise in the intensity difference, which is proportional to the noise in the amplitude difference. The noise spectrum in the intensity difference is obtained from eq. (42):

$$S_{I_1 - I_2}(\omega) = \frac{\omega^2\tau^2}{\omega^2\tau^2 + 4\gamma^2}\,, \tag{43}$$

where the shot noise has been normalized to 1. The noise spectrum in the difference between the intensities of the twin beams has a Lorentzian line shape, starting from zero at zero frequency and going back up to shot noise at frequencies much larger than the cavity bandwidth. This property is a very robust one: it does not depend on the noise of the pump beam, whose contribution cancels out in the difference, nor on the pump power: in contrast to quadrature squeezing below threshold, noise reduction in the twin beams takes place even far from threshold.

The only processes which may degrade the squeezing are the losses, which are not included in the preceding formula. Their effect can be understood by considering that, each time a photon is lost inside the cavity, it degrades the correlation between the twin beams. Taking into account the cavity losses as well as the losses outside the cavity in the photodetection system, one finds

$$S_{I_1 - I_2}(\omega) = 1 - \frac{\xi\eta}{1 + \omega^2/4\gamma'^2}\,, \tag{44}$$

where η is the photodetection quantum efficiency, and ξ is the OPO output coupling efficiency given by

$$\xi = \frac{\gamma}{\gamma'}\,, \tag{45}$$

where γ' is the total cavity damping coefficient, including the coupling mirror and the internal losses. The predicted squeezing at zero noise frequency is not perfect any more. The remaining noise at zero frequency is equal to the ratio of the internal losses in the cavity to the total cavity damping coefficient.

4.3.2. Experiment. The experimental set-up used in ref. [22, 23] comprises a nonlinear KTP crystal pumped by an Ar^+ laser or a doubled Nd:YAG

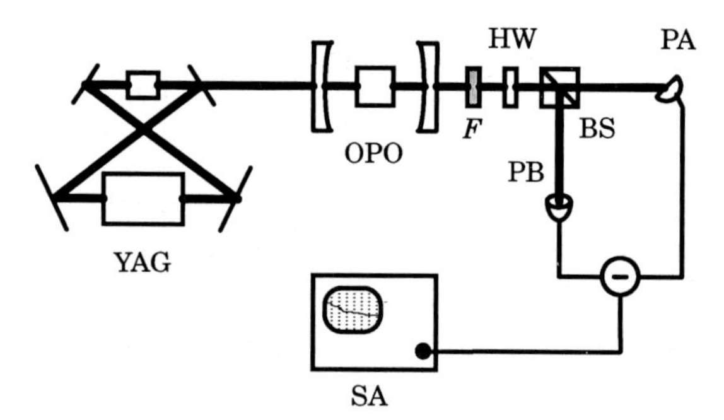

Fig. 8. – Experimental scheme for the generation of twin beams (YAG: intracavity doubled Nd:YAG laser; F: green filter; HW: half-wave plate; BS: beamsplitter; PA, PB: photodiodes; SA: spectrum analyser).

laser and placed in an optical cavity (fig. 8). The cavity has a higher finesse for the signal and idler fields than for the pump.

Due to type-II phase matching in the crystal, the twin beams emitted by the OPO have orthogonal polarizations and are separated at the output of the optical cavity with a polarizing beamsplitter. The intensities of the two beams are detected with photodiodes and the photocurrents are subtracted. This signal is fed into a spectrum analyser.

The shot noise reference is obtained by rotating the polarizations of the two beams by 45° before the polarizing beamsplitter. The polarizing beamsplitter is now a 50/50 beamsplitter for each of the beams. If there was only one beam, the

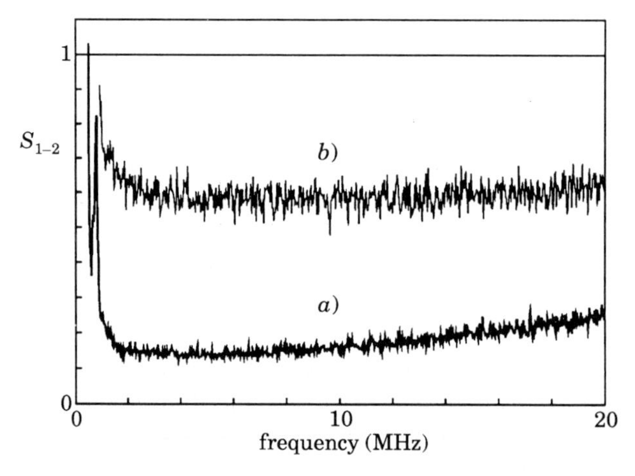

Fig. 9. – Noise spectrum of the difference between the intensities of the twin beams emitted by the nondegenerate OPO (a)). The shot noise has been normalized to 1. In b), the detection efficiency has been reduced by a factor of 2, resulting in a degradation of the squeezing by the same factor.

noise detected in the intensity difference between the two output channels would be the shot noise of that beam. With two beams having very different frequencies, there is no contribution of the interference term in the studied frequency range, and the shot noise powers add independently to yield the required shot noise reference.

The noise in the intensity difference is found to be squeezed on a frequency range of the order of the inverse of the cavity storage time (fig. 9). Noise reductions up to 86% have been observed in such conditions [23] with a ξ parameter (eq. (45)) of 0.91 (curve *a*)). The detection efficiency of the photodiodes is 94%. If the detection efficiency is reduced by inserting a 50% attenuator in the beams, the squeezing is degraded by a factor of 2, as can be seen in curve *b*) of fig. 9.

5. – Applications of squeezing to optical measurements.

Applications of squeezed light to interferometry and to ultra-high-sensitivity intensity measurements have been proposed and demonstrated.

5˙1. *Interferometric measurements.* – The principle which underlies the use of squeezed light in interferometers [24] can be understood from the properties of the beamsplitter exposed in subsect. 3˙2. The interferometer as a whole behaves like a beamsplitter with reflection and transmission coefficients which depend on the operating point.

Let us consider the case of a setting at half-maximum of a fringe; there, any change in the phase difference between the two arms of the interferometer is detected as a change in the difference of the intensities in the two output channels. The sensitivity of the phase difference measurement is limited by the standard quantum noise in the output. Reduction of this quantum noise is obtained by injecting squeezed vacuum in the second input port of the interferometer [25].

5˙2. *Intensity measurements.* – Twin photons can be used to increase the sensitivity in intensity measurements. This effect [26] has been demonstrated by placing an amplitude modulator in one of the twin beams emitted by an optical parametric oscillator. Then the intensity difference between the twin beams is measured. Because of the high quantum correlation of the beams this technique allows the detection of subshot noise modulation signals. A similar experiment has been performed [18] to measure turbidity in a liquid-crystal cell.

6. – Conclusion.

In recent years, there has been significant progress in squeezing. The quantum noise is now reduced by one order of magnitude in several systems. Most of the experiments have been performed using bulk nonlinear systems. But squeezing potentialities have been shown in fibres, especially for short pulses and soliton pulses. It should also be possible to use low-loss integrated systems to generate squeezed light.

Applications of quantum noise reduction are foreseen. They concern, on the one hand, ultrasensitive measurements in physics (detection of gravitational waves, new spectroscopic techniques) and intrinsically new phenomena (quantum nondemolition measurements, two-photon interferences, inhibited relaxation). Squeezed light should also be useful for more technical purposes in communication systems.

* * *

This work has been supported in part by the EEC contracts ESPRIT BRA 3186 and 6934.

REFERENCES

[1] C. Cohen-Tannoudji, J. Dupont-Roc et G. Grynberg: *Processus d'interaction entre photons et atomes* (Interéditions/Editions du CNRS, Paris, 1988) (*Atom-Photon Interactions* (Wiley, New York, N.Y., 1991)).
[2] S. Machida and Y. Yamamoto: *Phys. Rev. Lett.*, **60**, 792 (1988); W. H. Richardson and Y. Yamamoto: *Phys. Rev. Lett.*, **66**, 1963 (1991); W. H. Richardson, S. Machida and Y. Yamamoto: *Phys. Rev. Lett.*, **66**, 2867 (1991).
[3] R. E. Slusher, L. W. Hollberg, B. Yurke, J. C. Mertz and J. F. Valley: *Phys. Rev. Lett.*, **55**, 2409 (1985); R. E. Slusher, B. Yurke, P. Grangier, A. LaPorta, D. F. Walls and M. Reid: *J. Opt. Soc. Am.*, **4**, 1453 (1987).
[4] L. A. Wu, H. J. Kimble, J. L. Hall and H. Wu: *Phys. Rev. Lett.*, **57**, 2520 (1986); L. A. Wu, Min Xiao and H. J. Kimble: *J. Opt. Soc. Am. B*, **4**, 1465 (1988).
[5] H. P. Yuen and J. H. Shapiro: *Opt. Lett.*, **4**, 334 (1979).
[6] Y. R. Shen: *The Principles of Nonlinear Optics* (Wiley, New York, N.Y., 1984).
[7] S. Reynaud and A. Heidmann: *Opt. Commun.*, **71**, 209 (1989).
[8] R. M. Shelby, M. D. Levenson, S. H. Perlmutter, R. G. DeVoe and D. F. Walls: *Phys. Rev. Lett.*, **57**, 691 (1986); B. L. Schumaker, S. H. Perlmutter, R. M. Shelby and M. D. Levenson: *Phys. Rev. Lett.*, **58**, 357 (1987).
[9] K. Bergman and H. A. Haus: *Opt. Lett.*, **16**, 663 (1991).
[10] M. Rosenbluh and R. M. Shelby: *Phys. Rev. Lett.*, **66**, 153 (1991).
[11] R. E. Slusher, P. Grangier, A. LaPorta, B. Yurke and M. J. Potasek: *Phys. Rev. Lett.*, **59**, 2566 (1987).
[12] T. Hirano and M. Matsuoka: *Opt. Lett.*, **15**, 1153 (1990).
[13] L. Lugiato and G. Strini: *Opt. Commun.*, **41**, 67, 374 (1982).
[14] M. J. Collett and D. F. Walls: *Phys. Rev. A*, **32**, 2887 (1985).
[15] S. Reynaud, C. Fabre, E. Giacobino and A. Heidmann: *Phys. Rev. A*, **40**, 1440 (1989).
[16] M. G. Raizen, L. A. Orozco, Min Xiao, T. L. Boyd and H. J. Kimble: *Phys. Rev. Lett.*, **59**, 198 (1987).

[17] S. REYNAUD, C. FABRE and E. GIACOBINO: *J. Opt. Soc. Am. B*, **4**, 1520 (1987).
[18] J. G. RARITY and P. R. TAPSTER: in *International Conference on Quantum Electronics, Technical Digest Series*, Vol. 8 (Optical Society of America, Washington, D.C., 1990), p. 8.
[19] I. ABRAM, R. K. RAJ, J. L. OUDAR and G. DOLIQUE: *Phys. Rev. Lett.*, **57**, 2516 (1987).
[20] O. AYTUR and P. KUMAR: *Phys. Rev. Lett.*, **65**, 1551 (1990).
[21] C. FABRE, E. GIACOBINO, A. HEIDMANN and S. REYNAUD: *J. Phys. (Paris)*, **50**, 1209 (1989).
[22] A. HEIDMANN, R. J. HOROWICZ, S. REYNAUD, E. GIACOBINO, C. FABRE and G. CAMY: *Phys. Rev. Lett.*, **59**, 2555 (1987); T. DEBUISSCHERT, S. REYNAUD, A. HEIDMANN, E. GIACOBINO and C. FABRE: *Quantum Opt.*, **1**, 3 (1989).
[23] J. MERTZ, T. DEBUISSCHERT, A. HEIDMANN, C. FABRE and E. GIACOBINO: *Opt. Lett.*, **16**, 1234 (1991).
[24] C. M. CAVES: *Phys. Rev. D*, **21**, 1963 (1981).
[25] MIN XIAO, LING-AN WU and H. J. KIMBLE: *Phys. Rev. Lett.*, **59**, 278 (1987).
[26] C. D. NABORS and R. M. SHELBY: *Phys. Rev. A*, **42**, 556 (1990).

CHAOS

Quantum Chaos and Laser Spectroscopy.

D. KLEPPNER

Department of Physics and Research Laboratory of Electronics
Massachusetts Institute of Technology - Cambridge, Mass.

1. – Introduction.

The goal of «quantum chaos» is to understand the connections between quantum mechanics and nonlinear dynamics in the regimes of nonintegrable, disorderly (*i.e.* chaotic) motion. A primary goal of laser spectroscopy is to study the structure of quantum systems in fine detail, preferably with totally resolved spectra. There is an inherent conflict between these goals: detailed behavior is unimportant in disorderly motion, while classical behavior has no obvious connection to the structure of eigenvalues. The goals are in some sense complementary. Both are essential for a full understanding of the system but, at least for the present, pursuing one goal requires neglecting the other.

In spite of the notable advances in the study of quantum chaos, the subject remains controversial. The controversy ultimately focuses on a value judgement as to whether or not the subject is fundamentally important. The arguments *pro* and *con* can be summarized briefly as follows.

PRO: Recognition of the roles that nonlinear dynamics and chaos play in broad areas of the physical and biological worlds constitutes one of the most important intellectual advances in the last half century. Understanding its implications for quantum mechanics is, therefore, essential.

CON: Chaos is a classical phenomenon but Nature obeys the laws of quantum mechanics. Reconciling classical and quantum-mechanical descriptions may be a satisfying exercise, but it is fundamentally an academic exercise. Who cares whether classical physics works when we have quantum mechanics to fall back upon? In a nutshell, quantum chaos does not exist!

This summary exaggerates the extremes of the debate, but the fact remains that there is a conflict between the subject of quantum chaos and the goals of laser spectroscopy. Here the term «quantum chaos» is intended to be taken in its commonly used sense: the study of those particular aspects of the quantum behavior that are associated with the fact that a system's classical behavior is chaotic (or irregular, to use a less colorful phrase). «Laser spectroscopy» is intended not only literally to denote an experimental technique capable of enormous precision, but also figuratively as an expression of the drive in atomic physics to understand systems in all the detail that quantum mechanics permits.

The contrast between chaotic and quantum-mechanical behavior can be visualized by considering a particle moving freely on a plane but confined by a perfectly rigid wall that forms a circular enclosure—«circular billiards», as it is sometimes known. The particle's trajectory, a precessing polygon, is predictable. Furthermore, the motion is stable: a small change in initial conditions produces a trajectory that departs only trivially from the initial trajectory. However, the behavior changes drastically if a small semicircular bump is affixed to the wall. The particle's motion is unaffected until it happens to hit the bump, at which point it careens off in some totally new direction. A small change in the initial direction will lead to a completely new final path. The motion is so sensitive that it cannot be predicted for long. The system is chaotic.

Now consider the quantum properties of such a system. The lowest mode of the unperturbed system has a circular wave function with an antinode at the center and a node at the circumference, the next higher state has one circular node, and so on. Introducing a small protrusion on the wall slightly perturbs the energies of the low-lying states, but otherwise the states are essentially unaffected. There is nothing comparable with the qualitative change in behavior of the classical system.

The circular billiards illustrates two features of quantum chaos. First, chaos arises because a classical system can be sensitive to arbitrarily small features in space. Quantum-mechanical systems, however, are insensitive to features in phase space that occupy volumes smaller than h^3. Quantum mechanics tends to «smooth» out the small irregularities that typically give rise to chaotic phenomena. Second, in order to observe a quantum signature for chaos in the circular billiards, one must look at states so highly excited that their wavelengths are comparable to the size of the irregularity. Mimicking the motion of a particle would require creating a wave packet composed of many of these high-lying eigenstates. For these reasons, regimes of classical chaos invariably correspond to quantum systems with high quantum numbers.

In systems with high quantum numbers one might reasonably expect the correspondence principle, or some similar approach, to provide a natural bridge between quantum and classical mechanics. So far, however, the bridge is lacking. Although quantum mechanics is commonly believed to predict classical behavior when applied to classical problems, the assumption is not justified. Bridges between quantum and classical worlds are missing except for some special cases. The correspondence principle is often assumed to constitute such a bridge, but in fact it is really little more than an assertion that such a bridge must exist. The problem is that the correspondence principle offers no specific procedure for assuring that the quantum and classical descriptions are somehow consistent. In fact, except for a few special cases such as the hydrogen atom, the rigid rotor and some simple radiative phenomena, the correspondence principle is useless.

Ehrenfest's theorem is a second possible bridge, but it, too, is less general than commonly assumed. Ehrenfest's theorem asserts that the expectation values of quantum operators obey classical equations of motion. Recall that its starting point is the Heisenberg equation of motion for an operator $\widehat{O}$:

$$\frac{\mathrm{d}\widehat{O}}{\mathrm{d}t} = \frac{-i}{\hbar}[H, \widehat{O}] + \frac{\partial \widehat{O}}{\partial t}. \tag{1}$$

Taking expectation values, we obtain

$$\frac{\mathrm{d}\langle\widehat{O}\rangle}{\mathrm{d}t} = \frac{-i}{\hbar}\langle[H, \widehat{O}]\rangle + \left\langle\frac{\partial \widehat{O}}{\partial t}\right\rangle. \tag{2}$$

Taking the simple example $H = \widehat{p}^2/2m + V(\boldsymbol{r})$, we obtain

$$\frac{\mathrm{d}\langle\boldsymbol{p}\rangle}{\mathrm{d}t} = -\langle\nabla V\rangle = \langle\boldsymbol{F}(\boldsymbol{r})\rangle. \tag{3}$$

However, the classical equation of motion requires that

$$\frac{\mathrm{d}\langle\boldsymbol{p}\rangle}{\mathrm{d}t} = \boldsymbol{F}(\langle\boldsymbol{r}\rangle). \tag{4}$$

This is true if $\boldsymbol{F}(\langle\boldsymbol{r}\rangle) = \langle\boldsymbol{F}(\boldsymbol{r})\rangle$, which happens to be true in some special cases such as the harmonic oscillator, but it is not generally true. Consequently, Ehrenfest's theorem does not provide a satisfying connection between classical and quantum mechanics.

Finally, one sometimes hears the comment that quantum mechanics evolves into classical mechanics as $\hbar \to 0$. This is not a meaningful statement. Consider, for instance, what would happen to the fine-structure constant $\alpha = e^2/\hbar c$! Changing a single fundamental constant essentially changes the nature of the Universe.

2. – Energy levels and energy level statistics.

Among the most impressive discoveries of quantum chaos is the universal behavior of energy level fluctuations of disorderly systems. The underlying ideas rest on the following considerations.

If energy levels «cross» as some parameter such as an applied field or an internuclear separation is slowly varied, then, unless the crossing is purely accidental, there must be a quantum number that distinguishes the two states at the point of degeneracy. This quantum number is the eigenvalue of an operator whose conservation law reflects some symmetry. Consequently, the system possesses an additional constant of motion. In simple systems, such a constant is adequate to assure that the system is integrable.

A notable example of such behavior is the hydrogen atom in an electric field [1, 2]. The Hamiltonian is separable in parabolic co-ordinates and the solution introduces a new quantum number. Physically, the quantum number is proportional to the permanent dipole moment which the hydrogen atom can possess. The behavior is shown in fig. 1*a*), which presents a «Stark map» for the $m = 1$ states of lithium. (The azimuthal quantum number m is a «good» quantum number because of the rotational invariance of the Hamiltonian about the axis of the field.) These states are essentially hydrogenic. The energy levels evolve in a simple fashion «passing through» each other without interacting. Figure 1*b*) shows the Stark structure of the $m = 0$ states of lithium. The $l = 0$ components of the wave function are significantly displaced due to the effect of the inner-shell electrons (*i.e.* by «penetration of the ionic core» by the valence electron). The symmetry is destroyed, and the energy levels strongly repel. Such repulsions are characteristic of disorderly classical motion. The classical motion under the potential used in fig. 1*b*) is, in fact, disorderly.

The distinction between an orderly and a disorderly spectrum can be quantified through study of the fluctuations in the intervals between adjacent levels. To study the distribution of these intervals, the spectrum is first «linearized» to remove global variations of the density of states. (Such variations are usually small and can be eliminated by a linear or quadratic adjustment of the energy scale.) Denoting the interval between adjacent levels as s, where the mean value of s is taken to be unity, one then plots a histogram of the values of s.

If the spectrum is regular and the density of states is sufficiently high that the levels are not simply periodic, then the probability of finding a level in some interval is proportional to the length of the interval. Consequently, the distribution of intervals is given by a simple Poisson function:

$$P(s) = \exp[-s]. \tag{5}$$

However, for the disorderly spectrum of a Hamiltonian that obeys time-reversal symmetry, the form of the distribution is given by the GOE (Gaussian orthogonal ensemble) distribution which, to excellent approximation, is given by the Wigner distribution

$$P(s) = \frac{\pi}{2}\, s \exp[-(\pi s^2/4)]. \tag{6}$$

Because this behavior has been observed in molecular, atomic and nuclear spectra, as well as numerous model systems, the presence of GOE energy level statistics is often taken as the signature of quantum chaos [3]. The mathemat-

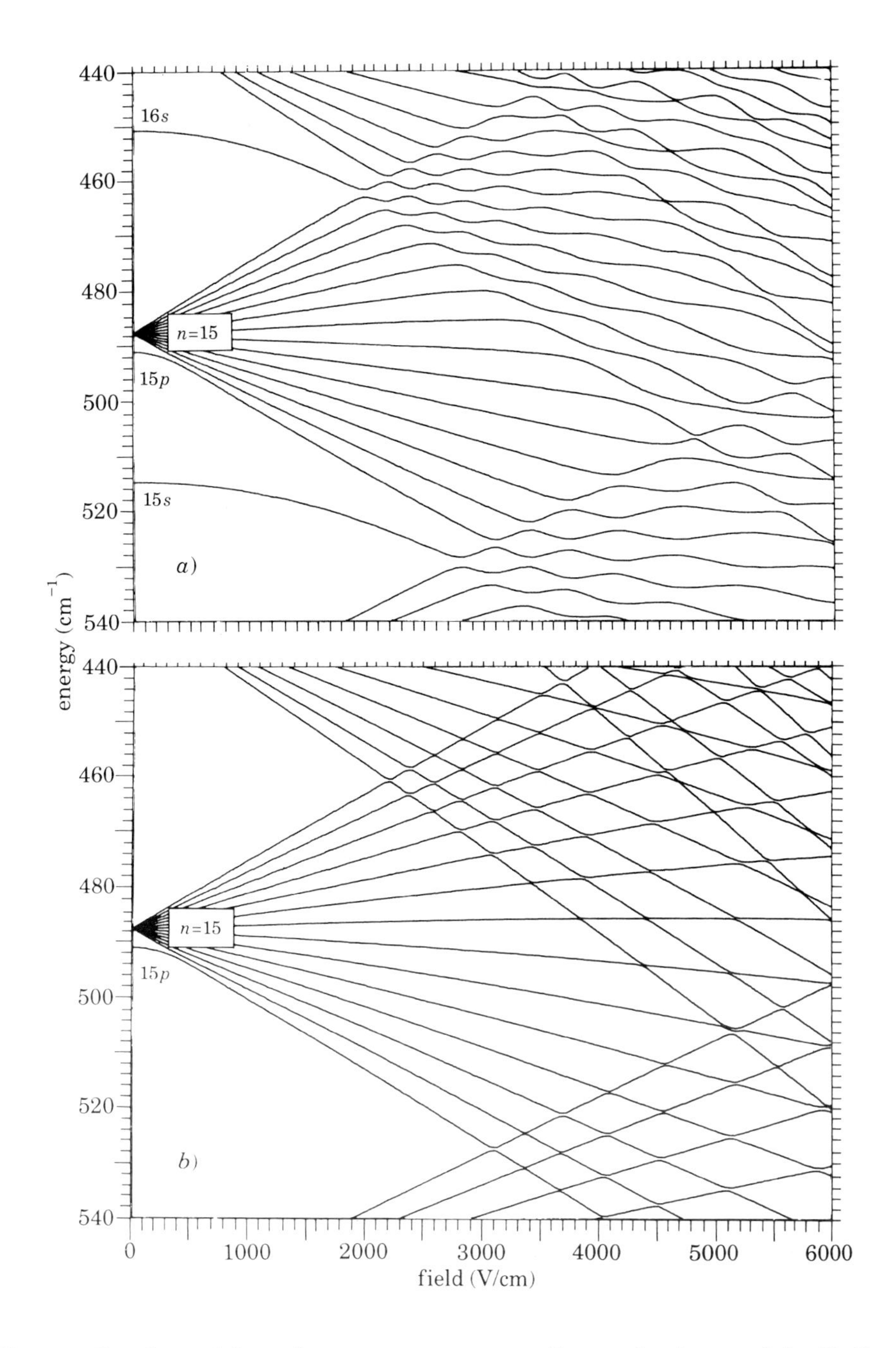

Fig. 1. – Regular and irregular quantum structure. Energy level map of the Rydberg states of lithium in the vicinity of $n = 15$, in an electric field. *a*) Stark structure for $|m| = 1$. *b*) Stark structure for $m = 0$. (From ref. [2].)

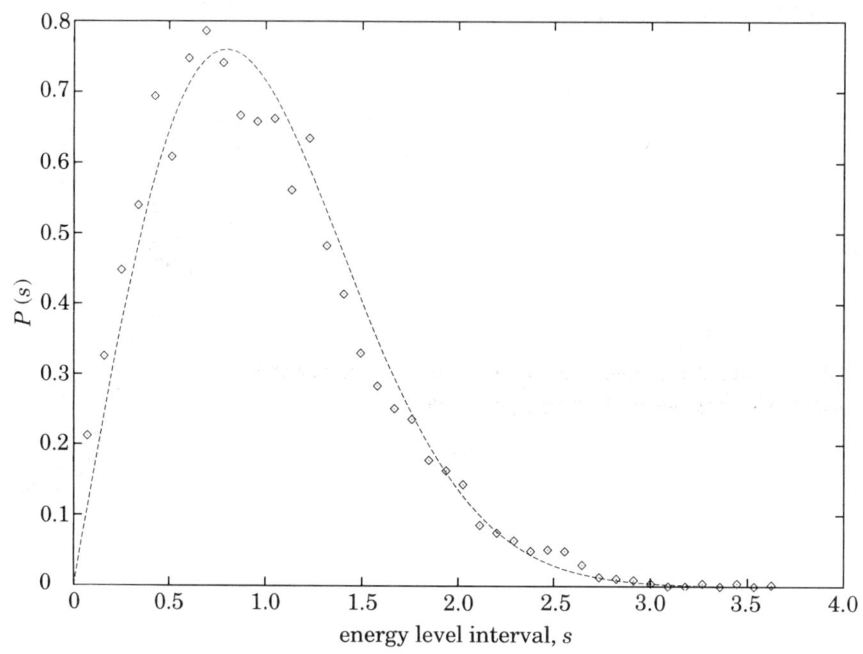

Fig. 2. – The distribution of energy level separations for a disorderly system. The dashed line is the Wigner distribution. The data were calculated for the diamagnetic hydrogen atom for energies between $-22\ \text{cm}^{-1}$ and $-12\ \text{cm}^{-1}$, comprising 159 levels. The energies were computed for fields in the range of 4.5 T to 5 T, at intervals of 0.005 T. (Courtesy of M. Courtney and A. Hashiomoto.)

ical basis for this behavior is described by Haake [4]. An example of this behavior is shown in fig. 2, which presents the distribution of energy level intervals for the diamagnetic hydrogen atom, a disorderly system which will be described in the following sections.

The Wigner distribution, eq. (6), vanishes at the origin reflecting the fact that in disorderly regions energy levels obey the «no crossing» theorem. This behavior is evident in fig. 1*b*).

The distribution of distance between adjacent energy levels in only one of many statistical properties that can be studied. Tests also exist for higher-order correlations between energy levels. In principle, these tests can reveal system-specific features of behavior.

Although universal behavior is always impressive, the fact is that, from the experimenter's point of view, rather little information comes from a great deal of work. Huge numbers of lines need to be measured in order to carry out a satisfactory statistical analysis, and the spectrum must be uncontaminated by signals from states of a different symmetry class. Most serious, however, is the fact that the energy level statistics may inadvertently conceal important features of the quantum-mechanical behavior.

3. – The diamagnetic hydrogen atom.

To turn to a concrete problem, consider the diamagnetic hydrogen atom. It is among the simplest of systems that can display chaos. For our purposes the system consists of an electron, a proton and a static magnetic field. Spin, relativity and other such troublesome effects can be safely neglected. Taking $\boldsymbol{B}$ to point along the z-axis, the Hamiltonian, in atomic units, can be taken as

$$H = \frac{p^2}{2} - \frac{1}{r} + L_z B + \frac{1}{8} B^2 \rho^2 . \tag{7}$$

The atomic unit of magnetic field is $2.35 \cdot 10^5$ T. The only good quantum numbers are $m = L_z$ and parity. For the special case $m = 0$, the Hamiltonian simplifies to

$$H = \frac{p^2}{2} - \frac{1}{r} + \frac{1}{8} B^2 \rho^2 . \tag{8}$$

The goals for studying this problem are twofold: first, to obtain quantum-mechanical solutions for one of the simplest nonseparable problems in quantum mechanics, and, second, to investigate its connections with quantum chaos. An extensive theoretical and experimental literature now exists on this innocently simple looking behavior [5-8], for both its classical and quantum-mechanical properties. We shall concentrate chiefly on some recent developments, but to introduce them we summarize some of the salient properties.

3.1. *The classical diamagnetic hydrogen atom.* – The description of systems governed by eq. (8) is greatly simplified by introducing the reduced, or scaled, variables: $\tilde{\boldsymbol{r}} = B^{2/3}\boldsymbol{r}$, $\tilde{\boldsymbol{p}} = B^{-1/3}\boldsymbol{p}$. The Hamiltonian becomes

$$\widetilde{H} = \frac{\tilde{p}^2}{2} - \frac{1}{\tilde{r}} + \frac{1}{8} \tilde{\rho}^2 , \tag{9}$$

where $\tilde{\rho}^2 = \tilde{x}^2 + \tilde{y}^2$. The magnetic field has disappeared from the Hamiltonian, so that the full panoply of classical behavior is exhibited as the scaled energy $\varepsilon = B^{-2/3} E_0$ varies over its allowed range. Physically, $\varepsilon^{3/2}$ is proportional to the ratio of the period for Kepler motion to the period for cyclotron motion.

Low magnetic field corresponds to the limit $\varepsilon \to -\infty$. At zero field the motion is periodic. At slightly higher fields, regions *a*) and *b*) of fig. 3, the motion is no longer periodic but it remains regular. For instance, trajectories plotted on a Poincaré surface of section form regular closed paths, indicating that the full trajectories in phase space lie on a torus. As ε increases, the trajectories start to become irregular in region *c*). Here the spectral features are reminiscent of those in fig. 1*b*). One would expect that the energy level statistics would display GOE behavior in this regime, and fig. 2 reveals that this is the case.

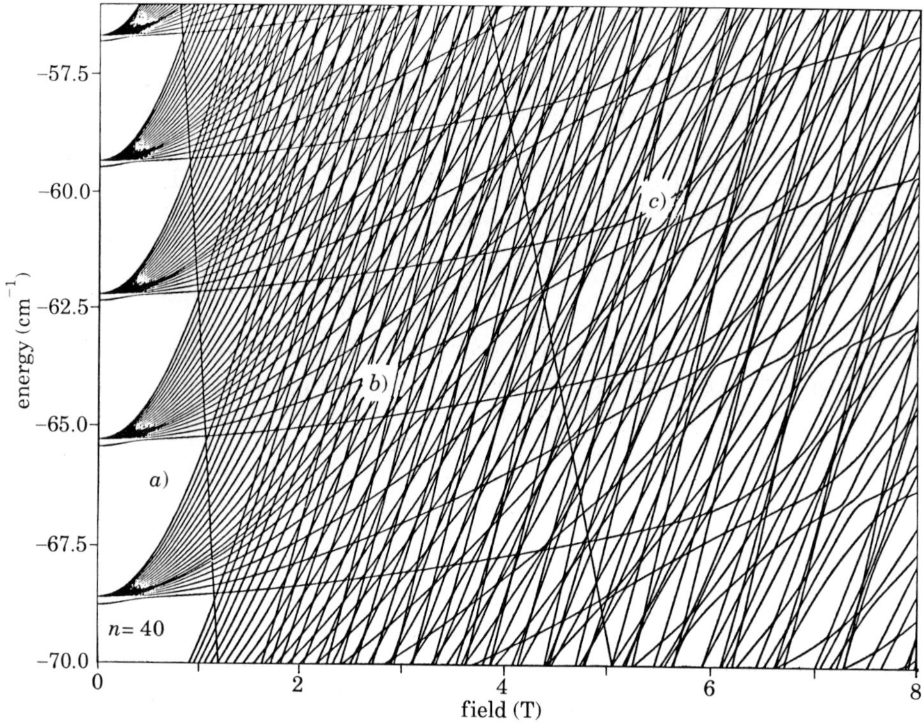

Fig. 3. – A panoramic display of the energy level structure of the diamagnetic lithium atom, $m = 0$, odd-parity states. In region *a*) the spectrum is periodic, or quasi-periodic. In region *b*) the overlapping periodic spectra show no visible interactions, due to an approximate symmetry in the system. In this region, the distribution of energy level intervals is approximately Poissonian. In region *c*) the spectrum becomes disorderly, indicating the onset of irregular motion in the classical system. (Courtesy of C. Iu and M. Courtney.)

The problem assumes a transparent form in the so-called oscillator representation. This employs the semi-parabolic coordinates

$$\mu = (-2E)^{1/4}(r+z)^{1/2}\,, \qquad \nu = (-2E)^{1/4}(r-z)^{1/2}\,,$$

and the parameter

$$\lambda = B^2/(-2E)^2\,. \tag{10}$$

The Hamiltonian eventually becomes

$$H = \left(\frac{p_\mu^2}{2} + \frac{\mu^2}{2}\right) + \left(\frac{p_\nu^2}{2} + \frac{\nu^2}{2}\right) + \frac{1}{8}\,\lambda\mu^2\nu^2(\nu^2+\nu^2) = 2\varepsilon\,, \tag{11}$$

where $\varepsilon = 1/(-2E)^{1/2}$. In this representation the system appears as two harmonic oscillators that are coupled by a nonlinear interaction. The strength of the coupling is proportional to the magnetic field. The motion at sufficiently low field is evidently regular, but it is hardly surprising that the system becomes irregular at high field. Essentially, the magnetic field provides a «knob» for chaos. The ability to vary the strength of the nonlinear coupling experimentally (and, of course, theoretically) makes this problem particularly attractive.

3'2. *The quantum diamagnetic hydrogen atom.* – The simplicity of eq. (8) is deceptive. The Hamiltonian is nonseparable and general methods for dealing with nonseparable problems are still lacking. For relatively weak fields, *i.e.* for $\varepsilon < -0.5$, solutions can be obtained by methods that are essentially perturbative. Historically, the problem first attracted interest in this regime because of interest in atomic structure in astrophysical fields [9]. Solutions for laboratory fields have been made possible by the recognition of an approximate symmetry that exists at low fields [8]. However, in the regime of classical chaos, $\varepsilon > -0.1$, and particularly at positive energy, perturbative methods are useless. As will be described below, recently there has been substantial progress on this problem.

4. – High-resolution spectroscopy of the diamagnetic Rydberg atom.

Achieving highly resolved spectra for Rydberg states of hydrogen is difficult due to the limitations of laser sources. The task is much simpler for the alkali metal atoms. The odd-parity states of Li are a natural choice for study, for the largest quantum defect involved, for the p state, is only $4 \cdot 10^{-2}$. The experimental approach, which is described elsewhere [10], is straightforward: An atomic beam is employed, directed along the axis of a superconducting solenoid. The beam reduces motional Stark effects and Doppler broadening. The atoms are excited by a series of laser-induced transitions. The magnetic field is calibrated from the spectrum of low-lying states, and the energy is measured by reference to a molecular «atlas», which is itself calibrated against the low-field spectrum.

A spectrum covering the range of $-30\,\text{cm}^{-1}$ to $+30\,\text{cm}^{-1}$ at a field of 6 T is shown in fig. 4. The energy was determined to a relative accuracy of $1 \cdot 10^{-3}\,\text{cm}^{-1}$ and an absolute accuracy of $5 \cdot 10^{-3}\,\text{cm}^{-1}$. The magnetic field was determined to an accuracy of $1 \cdot 10^{-3}\,\text{T}$. There are two possible uses for such a spectrum. The first is to check the energy level statistics. The region is disorderly, and one expects GOE statistics. The second is to confirm the validity of quantum-mechanical calculations. In reality, the scientific path is somewhat more roundabout than this. In the absence of data, developing techniques for calculating the spectrum is not necessarily a rewarding task. Once experimental spectra are available, the task becomes more urgent. A number of theoretical and computational approaches have now been pursued successfully. Furthermore, once the validity of the calculations is established, it turns out to be

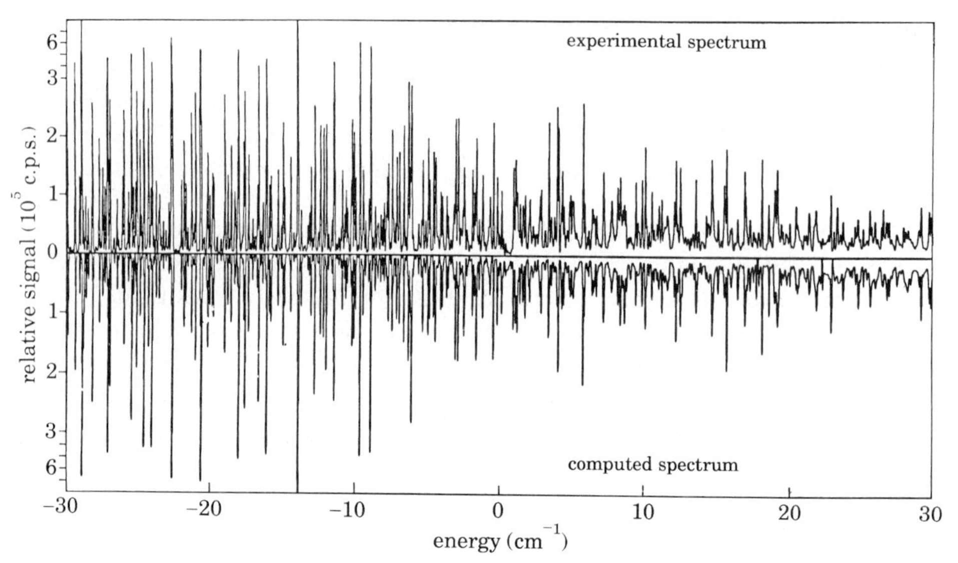

Fig. 4. – The measured spectrum of lithium, odd parity, $m = 0$. Reflected below a computed spectrum. (From ref. [11].)

much easier to check the energy level statistics from the computed spectrum than from the experimental data. This is the approach that has been followed so far. The data presented in fig. 2 were generated in this fashion.

The calculated spectrum in fig. 4 was obtained by DELANDE, BOMMIER and GAY [12], who developed a computational method that employed the oscillator representation of hydrogen and the complex rotation method. Figure 5 shows a blow-up of a portion of the spectrum which reveals that the agreement is generally excellent, particularly considering that the calculation is for hydrogen while the experiment employed lithium. Discrepancies between theory and experiment are difficult to discern, though more refined calculations reveal that there are some small departures near the limit of resolution, 10^{-3} cm^{-1}.

A number of other calculational advances have occurred based on approaches that naturally introduce the final Landau states of the free electron [13-15]. These are applicable to the alkali metal atoms as well as to hydrogen. They reveal some small departures from hydrogenic behavior that would be barely discernable in fig. 5. The method of Watanabe and Komine [14] actually reveals that the broad feature in fig. 5 is actually an unresolved high-Rydberg progression.

The ability to account quantitatively for the spectrum represents progress in quantum calculations, and there is reason to believe that the methods can be applied to related but more complicated problems in atomic and molecular theory. However, aside from validating that the level fluctuations display the expected GOE behavior, as described above, it provides relatively little insight into possible connections with classical dynamics.

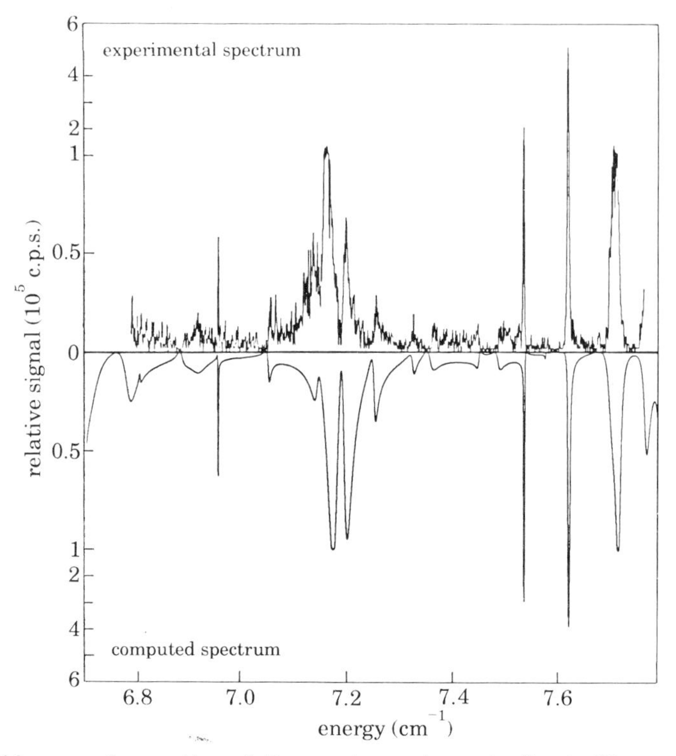

Fig. 5. – A blow-up of a portion of the spectrum shown in fig. 4. (From ref. [10].)

5. – Orderly structures.

A much fuller experimental picture of the problem than that provided by a single spectrum can be obtained by creating energy level «maps»—experimental pictures of the energy level structure formed by taking spectra at successively higher fields and then plotting them together. This approach has the advantage of calling attention to generic features that are invisible in individual spectra. Such a map is shown in fig. 6. The regime corresponds to hard chaos, and the overall impression is consistent with this: a disorderly spectrum with rapid variations in oscillator strength and strong level repulsions. However, within the map, in the regions labeled by *a*), *b*) and *c*) are a series of well-defined and orderly Rydberg progressions.

The Rydberg progressions have a straightforward interpretation. At very high magnetic fields, the problem can be viewed as a combination of rapid motion around the z-axis due to the magnetic field superimposed on slow motion along the z-axis due to the Coulomb potential. The result is a series of free electron states—Landau levels—each of which supports a Rydberg progression. In

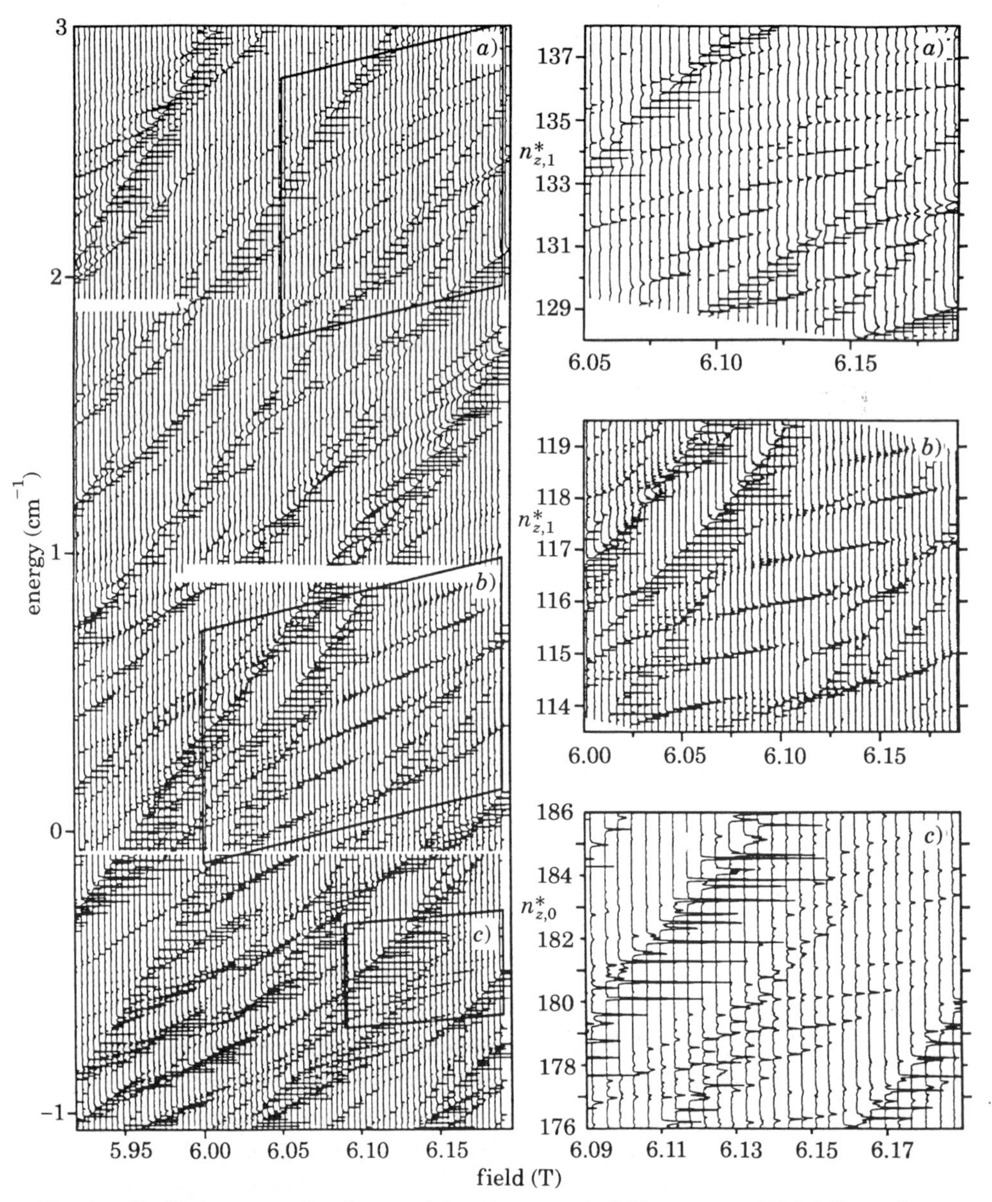

Fig. 6. – (Left) An energy level map of the diamagnetic lithium atom. (Right) Several portions of the map, plotted in terms of the reduced quantum number described in the text, which display Rydberg progressions. (From ref. [16].)

this regime the energy of $m = 0$ state is given by

$$E(n_\rho, n_z^*) = \left(n_\rho + \frac{1}{2}\right)B - \frac{1}{2(n_z^*)^2}, \tag{12}$$

where n_ρ is a nonnegative integer. The effective quantum number n_z^* can be written in terms of a quantum defect as $n_z^* = n_z - \mu$, where n_z is a positive integer. Although it might not be expected that such behavior would be observed at low magnetic fields, in fact it is. To enhance the features of the Rydberg states one can eliminate the global effect of the magnetic field by plotting the data in terms of the effective quantum number, n_z^*. Using the notation $n^{*z,j}$, where $j = n_\rho$, we have

$$n^{*z,j} \equiv ((2j+1)B - 2E)^{-1/2} . \tag{13}$$

The data on the right in fig. 6 are plotted in this representation. The unit interval between the lines is the signature of a Rydberg progression: their small slope indicates that the quantum defect varies slowly with field. Some of the Rydberg progressions reveal a regular fine structure.

The orderly progressions shown in fig. 6 would not be revealed by a traditional analysis of energy level fluctuations: Their number is too small to affect

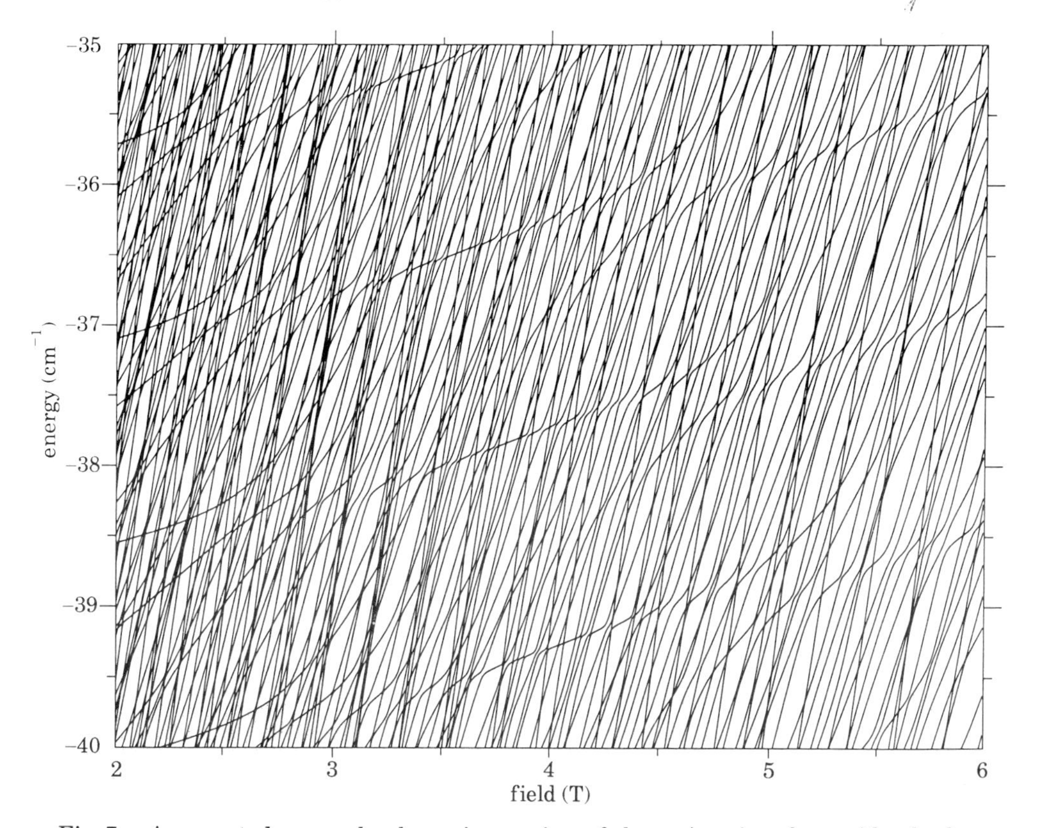

Fig. 7. – A computed energy level map in a regime of chaos. A series of «quasi-levels» become visible when the illustration is regarded from close to the page, from the left. These are believed to be related to the Rydberg progressions shown in fig. 6. (Courtesy of M. COURTNEY.)

the overall distribution. Nevertheless, the structures are immediately apparent to the eye.

A natural question arises as to how the one-dimensional Rydberg progressions evolve from low to high field. Apparently, the states are periodically demolished and reconstructed as the field is increased [17]. The states are evident in the computed energy level structure shown in fig. 7. They appear as periodic quasi-structures formed by a series of narrow anticrossings. These structures appear to be closely related to the classical periodic orbits directed along the z-axis. The periodic orbits pass through regimes of stability and instability that mirror the quantum structure, though the detailed correspondence is not yet understood.

6. – Periodic-orbit spectroscopy.

Experimental interest in the diamagnetic-hydrogen problem antedates the development of modern laser spectroscopy. The subject was launched in 1969 by the discovery by GARTON and TOMKINS [18] that the absorption spectrum of barium in a magnetic field revealed periodic oscillations near and above the ionization threshold. These are clearly visible in fig. 8. At the ionization limit ($E = 0$) the oscillation frequency is $1.5\,\omega_c$, where ω_c is the cyclotron frequency. As was initially pointed out by EDMONDS [19], this is identical to the period of an electron moving in a trajectory lying in the (x, y)-plane under the combined Coulomb and Lorentz forces. Numerous investigations based on semi-classical approximations refined this view, but its full significance was not appreciated until a series of studies on the spectroscopy of hydrogen in a magnetic field by WELGE and his colleagues in the mid '80s [20-22].

It is fortunate that the original spectra of the diamagnetic Rydberg atom had poor resolution, for under conditions of high resolution the periodic oscillations in fig. 8 would not be apparent. For instance, the spectrum for hydrogen with a resolution of $30 \cdot 10^{-3}\ \text{cm}^{-1}$, shown in fig. 9*a*), displays no obvious regularities. However, if one takes the Fourier transform of the spectrum, so as to observe the system's response in time, a series of well-defined peaks emerges. The unit of time is the cyclotron period, $1/\omega_c$. The first peak occurs at a time of $2/3$, the same period as observed by GARTON and TOMKINS. In addition, a series of longer-period oscillations is evident. Each of these can be associated with the period of the orbit of an electron moving in a classical trajectory that begins and ends at the nucleus, as sketched in fig. 9*b*).

The discovery of the periodicities in the spectrum of the diamagnetic hydrogen atom has generated a flowering of interest in what has come to be called «periodic-orbit spectroscopy». The theoretical foundations for this approach were created by GUTZWILLER [23], and also by BALIAN and BLOCH [24]. They have been refined and applied to this problem in a series of papers by DELOS and his colleagues [25-27].

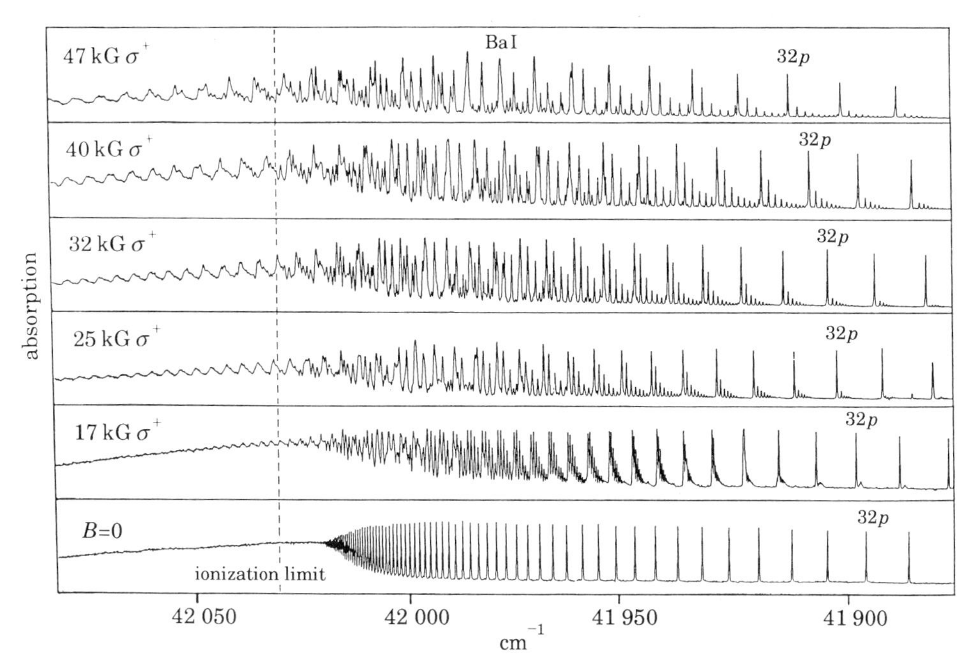

Fig. 8. – The photoabsorption spectrum of barium in increasing magnetic fields. The modulations in the absorption rate at positive energy occur with an interval of $1.5\,\omega_c$, where ω_c is the cyclotron frequency. Their period corresponds to the shortest-period classical orbit of an electron in a plane perpendicular to the magnetic field, passing through the origin. (Courtesy of F. TOMKINS.)

The goal of periodic-orbit spectroscopy is to relate the global features of the absorption spectrum to periodic classical motions of an electron. Near the atom, the electron is described by a wave that emerges from a small spherical surface, but, once it has left the vicinity of the atom, it is described by a particle that follows a classical path. Certain paths return to the spherical region, where they are again described by waves whose phase depends upon the action along the classical path. The interference between the incoming and outgoing waves gives rise to the photoabsorption spectrum. Considered in the temporal domain, as in fig. 9*b*), the returning electrons converge at a series of times that grow successively longer with the length of the trajectory, and whose intensity diminishes due to the spreading of the wave.

Periodic-orbit spectroscopy is fundamentally semi-classical but it is applied in regimes where one expects the semi-classical approximation to be extremely accurate. In this approximation, the energy level density at energy E is given by [24, 25]

$$\rho(E) = \overline{\rho(E)} + \sum_i \sum_{k=1}^{x} a_{ik}(E) \cos[k\{S_i(E)/\hbar - \mu_i\}]. \tag{14}$$

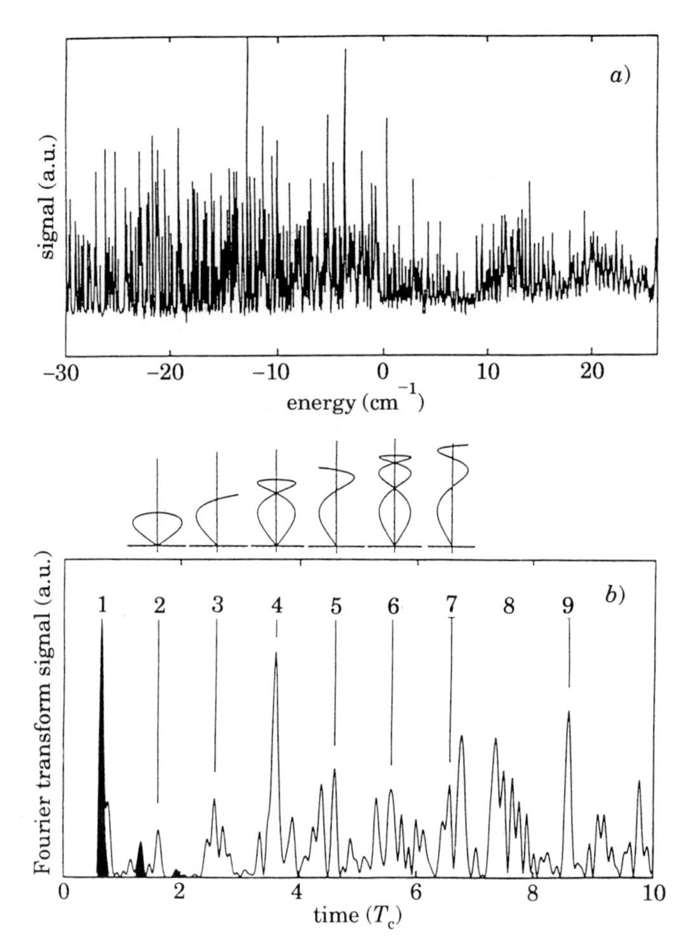

Fig. 9. – *a*) The photoabsorption spectrum of hydrogen in a magnetic field of 6 T. *b*). The Fourier transform of the spectrum in *a*). The unit of time is the period for a cyclotron orbit. The time associated with each peak corresponds to the period of an allowed classical orbit, as sketched in ρ-z coordinates above the peak. The first peak, with a period of $2/3$, corresponds to a trajectory in the (x, y)-plane. (Courtesy of K. H. WELGE.)

The index i denotes a particular periodic trajectory, and the index k signifies the number of recurrences—*i.e.* the number of times the electron returns to the origin. The factor a_{ik} is a modulation amplitude that reflects the stability of the orbit. For unstable orbits, a_{ik} decays exponentially, and in many cases only one recurrence is important. $S_i(E)$ is the classical action along orbit i, and μ_i is a phase related to the boundary conditions of the wave and to geometrical factors such as the phase shift introduced at a caustic surface.

The absorption spectrum depends not only on the density of states but also on the density of oscillator strength for transitions to each state. DELOS and DU[25] have shown that the oscillator strength density $Df(E)$ can be expressed as

$$Df(E) = \overline{Df(E)} + \sum_i A_i \sin(T_i E/\hbar + \Delta_i), \tag{15}$$

where the first term on the right-hand side is a smoothly varying function, and the second carries information about the orbits. An orbit with period T_i gives rise to a modulation in the spectrum with period $\Delta E_i = h/T_i$.

To implement this approach, it is essential to employ scaled variables in which the spectrum at a reduced energy is plotted as a function of the action. Experimentally, this requires simultaneously scanning the energy and the magnetic field. The success of this method is illustrated in fig. 10, in which the spectrum computed from 70 periodic orbits is compared with the experimental spectrum. The agreement is impressive, but the results also point to an underlying problem. The experimental data have been smoothed to reduce the resolution to match the theory, so that the full spectrum is not represented. The resolution is set by the period of the longest orbit considered, so that increasing the resolution requires introducing longer-period orbits. Whether or not this can be done is an open question. The problem is that the number of orbits proliferates exponentially with the period. Furthermore, they become increasingly difficult to locate due to their inherent instability.

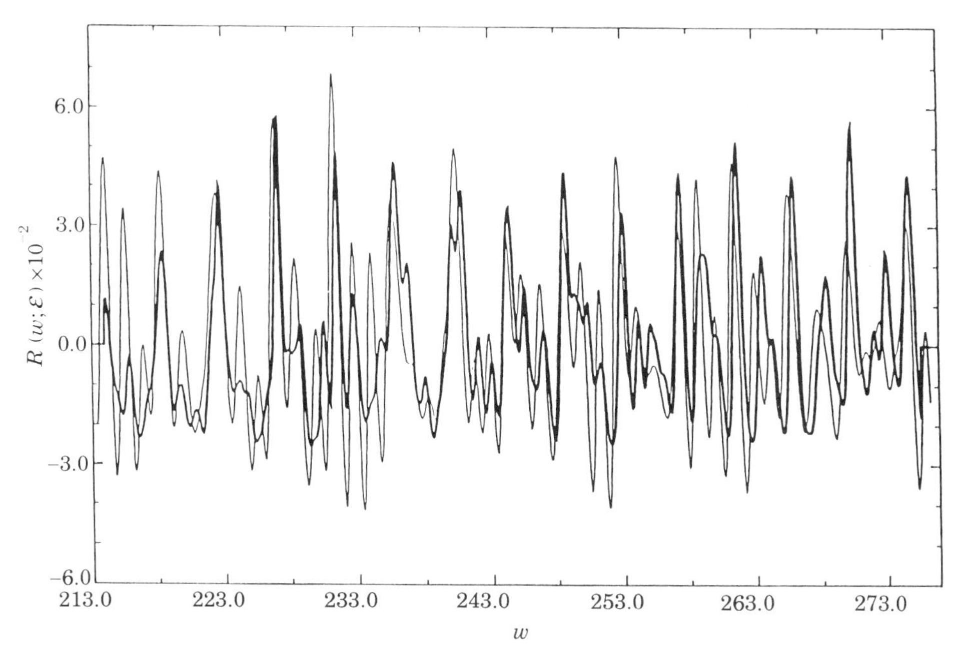

Fig. 10. – Hydrogen absorption spectrum in a field of 6 T plotted in terms of the reduced action. Light line: experimental data (smoothed); heavy line: spectrum calculated from periodic-orbit theory, using 70 orbits. (From ref. [27].)

The ultimate question is whether one can predict the complete quantum spectrum starting with the trace formula. If so, one would have a bridge between classical and quantum mechanics for a nonlinear system that would constitute a plausible solution to the problem of quantum chaos. At present, however, it is not known whether the trace formula actually converges as the number of orbits is increased. Ultimately, one might speculate that the required number of periodic orbits is equal to the number of eigenstates in the system, a number in the hundreds or thousands. Fundamentally, however, the answer depends on the validity of the semi-classical approximation. If one looks for the complete spectrum, it seems unlikely that the semi-classical methods will suffice. However, as discussed below, this may be a false goal.

7. – Conclusion.

Laser spectroscopy of the diamagnetic hydrogen atom illustrates both the strengths and shortcomings of contemporary studies of quantum chaos. The spectroscopy has led to quantum calculations of tremendous detail, providing an extremely rich picture of the structure of a nonlinear system. The spectroscopy has revealed regularities in the structure in a regime of hard chaos whose significance is not yet fully understood. Periodic-orbit spectroscopy provides a complementary approach to the problem, for it provides a direct link between the absorption spectrum and the classical dynamics of the system. The subject is moving forward rapidly, particularly in the understanding of how new orbits abruptly come into existence as the period is lengthened. The orbits suddenly bifurcate, and the process of bifurcation, particularly in its relation to quantum behavior, is currently being investigated [27].

Underlying these advances, however, is the fact that the ultimate goal of quantum chaos is not yet clear. The term «bridging the gap» between quantum and classical behavior has a pleasing metaphorical ring, but no obvious meaning. Presumably, as is often said for chaos itself, you will know it when you see it, but such an assertion really begs the issue.

Quantum chaos is usually regarded as the study of the special quantum features that are associated with irregular classical motion. However, one might argue that this is putting things backward, and that a proper goal for quantum chaos would be to predict classical behavior from quantum principles. For chaotic motion, it would be natural for the predictions to be statistical. However, such an approach has not yet been pursued.

Underlying all these considerations is the observations that, in the field of quantum chaos, theory outweighs experiment by approximately one hundred to one. Physics is only healthy when there is a reasonable balance between experiment and theory. There is an obvious need for new experiments.

* * *

The research at MIT is due to the efforts of M. M. KASH, G. R. WELCH, C. H. IU and M. COURTNEY. This work has been sponsored by the National Science Foundation and the Office of Naval Research.

REFERENCES

[1] H. A. BETHE and E. E. SALPETER: *Quantum Mechanics of One- and Two-Electron Atoms* (Academic Press Inc., New York, N.Y., 1957); R. F. STEBBINGS and F. B. DUNING, Editors: *Rydberg States of Atoms and Molecules* (Cambridge University Press, Cambridge, 1983).
[2] M. L. ZIMMERMAN, M. G. LITTMAN, M. M. KASH and D. KLEPPNER: *Phys. Rev. A*, **20**, 2251 (1979).
[3] Cf. M. J. GIANNONI, A. VOROS and J. SINN-JUSTIN: *Chaos in Quantum Physics* (Elsevier Science Publishers, Amsterdam, 1991).
[4] F. HAAKE: *Quantum Signatures of Chaos* (Springer-Verlag, Berlin, 1991).
[5] H. FRIDRICH and D. WINTGEN: *Phys. Rep.*, **183**, 37 (1989).
[6] H. HASEGAWA, M. ROBNIK and G. WUNNER: *Prog. Theor. Phys.*, **98**, 198 (1989).
[7] *Irregular Atomic Systems and Quantum Chaos*, edited by J.-C. GAY (Gordon and Breach, New York, N.Y., 1992).
[8] D. DELANDE: in *Fundamental Systems in Quantum Optics*, edited by J. DALIBARD, J.-M. RAIMOND and J. ZINN-JUSTIN (North-Holland, Amsterdam, 1992), p. 381.
[9] R. H. GARSTANG: *Rep. Prog. Phys.*, **40**, 105 (1977).
[10] D. KLEPPNER: in *Fundamental Systems in Quantum Optics*, edited by J. DALIBARD, J.-M. RAIMOND and J. ZINN-JUSTIN (North-Holland, Amsterdam, 1992), p. 417.
[11] C. IU, G. R. WELCH, M. M. KASH, D. KLEPPNER, D. DELANDE and J.-C. GAY: *Phys. Rev. Lett.*, **66**, 145 (1991).
[12] D. DELANDE, A. BOMMIER and J.-C. GAY: *Phys. Rev. Lett.*, **66**, 141 (1991).
[13] P. F. O'MAHONY and F. MOTA-FURTADO: *Phys. Rev. Lett.*, **67**, 2283 (1991); *Atomic Physics 13*, edited by H. WALTHER, T. W. HÄNSCH and B. NEIZERT (American Institute of Physics, New York, N.Y., 1993), p. 449.
[14] S. WATANABE and H. KOMINE: *Phys. Rev. Lett.*, **67**, 3227 (1991).
[15] M. H. HALLEY, D. DELANDE and K. T. TAYLOR: *J. Phys. B*, **25**, L525 (1992).
[16] C. IU, G. R. WELCH, M. M. KASH, L. HSU and D. KLEPPNER: *Phys. Rev. Lett.*, **63**, 113 (1989).
[17] J. DELOS: private communication.
[18] W. R. S. GARTON and F. S. TOMKINS: *Astrophys. J.*, **158**, 839 (1969).
[19] A. R. EDMONDS: *J. Phys. (Paris) Colloq.*, **31**, C4 71 (1970).
[20] A. HOLLE, G. WIEBUSCH, J. MAIN, B. HAGER, H. ROTTKE and K. H. WELGE: *Phys. Rev. Lett.*, **56**, 2594 (1986).
[21] A. HOLLE, J. MAIN, G. WIEBUSCH, H. ROTTKE and K. H. WELGE: *Phys. Rev. Lett.*, **61**, 161 (1988).
[22] J. MAIN, G. WIEBUSCH and K. H. WELGE: in *Irregular Atomic Systems and Quantum Chaos*, edited by J.-C. GAY (Gordon and Breach, New York, N.Y., 1992), p. 241.

[23] M. C. GUTZWILLER: *Chaos in Classical and Quantum Mechanics* (Springer-Verlag, New York, N.Y., 1990).
[24] R. BALIAN and C. BLOCH: *Ann. Phys.* (*N.Y.*), **69**, 76 (1972).
[25] M. L. DU and J. B. DELOS: *Phys. Rev. A*, **38**, 1896, 1913 (1988).
[26] J.-M. MAO and J. B. DELOS: *Phys. Rev. A*, **45**, 1746 (1992).
[27] J. MAIN, G. WIEBUSCH, K. WELGE, J. SHAW and B. DELOS: *Phys. Rev. A* (in press).

PROCEEDINGS OF THE INTERNATIONAL SCHOOL OF PHYSICS «ENRICO FERMI»

Course I
Questioni relative alla rivelazione delle particelle elementari, con particolare riguardo alla radiazione cosmica
edited by G. PUPPI

Course II
Questioni relative alla rivelazione delle particelle elementari, e alle loro interazioni con particolare riguardo alle particelle artificialmente prodotte ed accelerate
edited by G. PUPPI

Course III
Questioni di struttura nucleare e dei processi nucleari alle basse energie
edited by C. SALVETTI

Course IV
Proprietà magnetiche della materia
edited by L. GIULOTTO

Course V
Fisica dello stato solido
edited by F. FUMI

Course VI
Fisica del plasma e relative applicazioni astrofisiche
edited by G. RIGHINI

Course VII
Teoria della informazione
edited by E. R. CAIANIELLO

Course VIII
Problemi matematici della teoria quantistica delle particelle e dei campi
edited by A. BORSELLINO

Course IX
Fisica dei pioni
edited by B. TOUSCHEK

Course X
Thermodynamics of Irreversible Processes
edited by S. R. DE GROOT

Course XI
Weak Interactions
edited by L. A. RADICATI

Course XII
Solar Radioastronomy
edited by G. RIGHINI

Course XIII
Physics of Plasma: Experiments and Techniques
edited by H. ALFVÉN

Course XIV
Ergodic Theories
edited by P. CALDIROLA

Course XV
Nuclear Spectroscopy
edited by G. RACAH

Course XVI
Physicomathematical Aspects of Biology
edited by N. RASHEVSKY

Course XVII
Topics of Radiofrequency Spectroscopy
edited by A. GOZZINI

Course XVIII
Physics of Solids (Radiation Damage in Solids)
edited by D. S. BILLINGTON

Course XIX
Cosmic Rays, Solar Particles and Space Research
edited by B. PETERS

Course XX
Evidence for Gravitational Theories
edited by C. MØLLER

Course XXI
Liquid Helium
edited by G. CARERI

Course XXII
Semiconductors
edited by R. A. SMITH

Course XXIII
Nuclear Physics
edited by V. F. WEISSKOPF

Course XXIV
Space Exploration and the Solar System
edited by B. ROSSI

Course XXV
Advanced Plasma Theory
edited by M. N. ROSENBLUTH

Course LVIII
Dynamics Aspects of Surface Physics
edited by F. O. GOODMAN

Course LIX
Local Properties at Phase Transitions
edited by K. A. MÜLLER and A. RIGAMONTI

Course LX
C*-Algebras and their Applications to Statistical Mechanics and Quantum Field Theory
edited by D. KASTLER

Course LXI
Atomic Structure and Mechanical Properties of Metals
edited by G. CAGLIOTI

Course LXII
Nuclear Spectroscopy and Nuclear Reactions with Heavy Ions
edited by H. FARAGGI and R. A. RICCI

Course LXIII
New Directions in Physical Acoustics
edited by D. SETTE

Course LXIV
Nonlinear Spectroscopy
edited by N. BLOEMBERGEN

Course LXV
Physics and Astrophysics of Neutron Stars and Black Holes
edited by R. GIACCONI and R. RUFFINI

Course LXVI
Health and Medical Physics
edited by J. BAARLI

Course LXVII
Isolated Gravitating Systems in General Relativity
edited by J. EHLERS

Course LXVIII
Metrology and Fundamental Constants
edited by A. FERRO MILONE, P. GIACOMO and S. LESCHIUTTA

Course LXIX
Elementary Modes of Excitation in Nuclei
edited by A. BOHR and R. A. BROGLIA

Course LXX
Physics of Magnetic Garnets
edited by A. PAOLETTI

Course LXXI
Weak Interactions
edited by M. BALDO CEOLIN

Course LXXII
Problems in the Foundations of Physics
edited by G. TORALDO DI FRANCIA

Course LXXIII
Early Solar System Processes and the Present Solar System
edited by D. LAL

Course LXXIV
Development of High-Pozer Lasers and their Applications
edited by C. PELLEGRINI

Course LXXV
Intermolecular Spectroscopy and Dynamical Properties of Dense Systems
edited by J. VAN KRANENDONK

Course LXXVI
Medical Physics
edited by J. R. GREENING

Course LXXVII
Nuclear Structure and Heavy-Ion Collisions
edited by R. A. BROGLIA, R. A. RICCI and C. H. DASSO

Course LXXVIII
Physics of the Earth's Interior
edited by A. M. DZIEWONSKI and E. BOSCHI

Course LXXIX
From Nuclei to Particles
edited by A. MOLINARI

Course LXXX
Topics in Ocean Physics
edited by A. R. OSBORNE and P. MALANOTTE RIZZOLI

Course LXXXI
Theory of Fundamental Interactions
edited by G. COSTA and R. R. GATTO

Course LXXXII
Mechanical and Thermal Behaviour of Metallic Materials
edited by G. CAGLIOTI and A. FERRO MILONE

Course LXXXIII
Positrons in Solids
edited by W. BRANDT and A. DUPASQUIER

Course LXXXIV
Data Acquisition in High-Energy Physics
edited by G. BOLOGNA and M. VINCELLI

Course LXXXV
Earhquakes: Observation, Theory and Interpretation
edited by H. KANAMORI and E. BOSCHI

Course LXXXVI
Gamow Cosmology
edited by F. MELCHIORRI and R. RUFFINI

Course LXXXVII
Nuclear Structure and Heavy-Ion Dynamics
edited by L. MORETTO and R. A. RICCI

Course LXXXVIII
Turbulence and Predictability in Geophysical Fluid Dynamics and Climate Dynamics
edited by M. Ghil, R. Benzi and G. Parisi

Course LXXXIX
Highlights of Condensed-Matter Theory
edited by F. Bassani, F. Fumi and M. P. Tosi

Course XC
Physics of Amphiphiles: Micelles, Vesicles and Microemulsions
edited by V. Degiorgio and M. Corti

Course XCI
From Nuclei to Stars
edited by A. Molinari and R. A. Ricci

Course XCII
Elementary Particles
edited by N. Cabibbo

Course XCIII
Frontiers in Physical Acoustics
edited by D. Sette

Course XCIV
Theory of Reliability
edited by A. Serra and R. E. Barlow

Course XCV
Solar-Terrestrial Relationships and the Earth Environment in the Last Millennia
edited by G. Cini Castagnoli

Course XCVI
Excited-State Spectroscopy in Solids
edited by U. M. Grassano and N. Terzi

Course XCVII
Molecular-Dynamics Simulations of Statistical-Mechanical Systems
edited by G. Ciccotti and W. G. Hoover

Course XCVIII
The Evolution of Small Bodies in the Solar System
edited by M. Fulchignoni and Ľ. Kresák

Course XCIX
Synergetics and Dynamic Instabilities
edited by G. Caglioti and H. Haken

Course C
The Physics of* NMR *Spectroscopy in Biology and Medicine
edited by B. Maraviglia

Course CI
Evolution of Interstellar Dust and Related Topics
edited by A. Bonetti and J. M. Greenberg

Course CII
Accelerated Life Testing and Experts Opinions in Reliability
edited by C. A. Clarotti

Course CIII
Trends in Nuclear Physics
edited by P. Kienle, R. A. Ricci and A. Rubbino

Course CIV
Frontiers and Borderlines in Many-Particle Physics
edited by R. A. Broglia and J. R. Schrieffer

Course CV
Confrontation between Theories and Observations in Cosmology: Present Status and Future Programmes
edited by J. Audouze and F. Melchiorri

Course CVI
Current Trends in the Physics of Materials
edited by G. F. Chiarotti, F. Fumi and M. Tosi

Course CVII
The Chemical Physics of Atomic and Molecular Clusters
edited by G. Scoles

Course CVIII
Photoemission and Absorption Spectroscopy of Solids and Interfaces with Synchrotron Radiation
edited by M. Campagna and R. Rosei

Course CIX
Nonlinear Topics in Ocean Physics
edited by A. R. Osborne

Course CX
Metrology at the Frontiers of Physics and Technology
edited by L. Crovini and T. J. Quinn

Course CXI
Solid-State Astrophysics
edited by E. Bussoletti and G. Strazzulla

Course CXII
Nuclear Collisions from the Mean-Field into the Fragmentation Regime
edited by C. Detraz and P. Kienle

Course CXIII
High-pressure Equation of State: Theory and Applications
edited by S. Eliezer and R. A. Ricci

Course XXVI
Selected Topics on Elementary Particle Physics
edited by M. CONVERSI

Course XXVII
Dispersion and Absorption of Sound by Molecular Processes
edited by D. SETTE

Course XXVIII
Star Evolution
edited by L. GRATTON

Course XXIX
Dispersion Relations and their Connection with Causality
edited by E. P. WIGNER

Course XXX
Radiation Dosimetry
edited by F. W. SPIERS and G. W. REED

Course XXXI
Quantum Electronics and Coherent Light
edited by C. H. TOWNES and P. A. MILES

Course XXXII
Weak Interactions and High-Energy Neutrino Physics
edited by T. D. LEE

Course XXXIII
Strong Interactions
edited by L. W. ALVAREZ

Course XXXIV
The Optical Properties of Solids
edited by J. TAUC

Course XXXV
High-Energy Astrophysics
edited by L. GRATTON

Course XXXVI
Many-Body Description of Nuclear Structure and Reactions
edited by C. BLOCH

Course XXXVII
Theory of Magnetism in Transition Metals
edited by W. MARSHALL

Course XXXVIII
Interaction of High-Energy Particles with Nuclei
edited by T. E. O. ERICSON

Course XXXIX
Plasma Astrophysics
edited by P. A. STURROCK

Course XL
Nuclear Structure and Nuclear Reactions
edited by M. JEAN and R. A. RICCI

Course XLI
Selected Topics in Particle Physics
edited by J. STEINBERGER

Course XLII
Quantum Optics
edited by R. J. GLAUBER

Course XLIII
Processing of Optical Data by Organisms and by Machines
edited by W. REICHARDT

Course XLIV
Molecular Beams and Reaction Kinetics
edited by CH. SCHLIER

Course XLV
Local Quantum Theory
edited by R. JOST

Course XLVI
Physics with Intersecting Storage Rings
edited by B. TOUSCHEK

Course XLVII
General Relativity and Cosmology
edited by R. K. SACHS

Course XLVIII
Physics of High Energy Density
edited by P. CALDIROLA and H. KNOEPFEL

Course IL
Foundations of Quantum Mechanics
edited by B. D'ESPAGNAT

Course L
Mantle and Core in Planetary Physics
edited by J. COULOMB and M. CAPUTO

Course LI
Critical Phenomena
edited by M. S. GREEN

Course LII
Atomic Structure and Properties of Solids
edited by E. BURSTEIN

Course LIII
Developments and Bonderlines of Nuclear Physics
edited by H. MORINAGA

Course LIV
Developments in High-Energy Physics
edited by R. R. GATTO

Course LV
Lattice Dynamics and Intermolecular Forces
edited by S. CALIFANO

Course LVI
Experimental Gravitation
edited by B. BERTOTTI

Course LVII
History of 20th Century Physics
edited by C. WEINER

Course CXIV
Industrial and Technological Applications of Neutrons
edited by M. FONTANA and F. RUSTICHELLI

Course CXV
The Use of EOS for Studies of Atmospheric Physics
edited by J. C. GILLE and G. VISCONTI

Course CXVI
Status and Perspectives of Nuclear Energy: Fission and Fusion
edited by R. A. RICCI, C. SALVETTI and E. SINDONI

Course CXVII
Semiconductor Superlattices and Interfaces
edited by A. STELLA

Course CXVIII
Laser Manipolation of Atoms and Ions
edited by E. ARIMONDO, W. D. PHILLIPS and F. STRUMIA

Course CXIX
Quantum Chaos
edited by G. CASATI, I. GUARNERI and U. SMILANSKY